Solutions Manual for

Advanced Mathematics

An Incremental Development

Second Edition

John H. Saxon, Jr.

SAXON PUBLISHERS, INC.

Printed in the United States of America

ISBN: 1-56577-042-0

29 0928 22

4500844444

Preface

This manual contains solutions to every problem in the second edition of John Saxon's *Advanced Mathematics* textbook. The solutions are designed to be representative of students' work, but please keep in mind that many problems will have more than one correct solution. We have attempted to stay as close as possible to the methods and procedures outlined in the textbook. Early solutions of problems of a particular type contain every step. Later solutions omit simpler steps. The final answers are set in boldface for ease of grading.

Below each problem number is at least one number in parentheses, called a *Lesson Reference Number*. Each number refers to a lesson in which the concept for a particular problem type is taught. Some *Lesson Reference Numbers* are denoted as (*R*). This means "review problem"; students should know the concepts involved in solving such problems before taking *Advanced Mathematics*.

The following Saxon employees were instrumental in the development of this solutions manual, and we gratefully acknowledge their contributions: Adriana Maxwell, Linda Etheridge, Adrian Goodey, Darrin McDaniel, and Dan O'Connor for working the solutions and proofreading the various revisions; Angela Johnson, David LeBlanc, and Heather Shaver for typesetting the manual; Aaron Lauve, Michael Lott, and Travis Southern for creating the graphics; and John Chitwood for producing the cover art.

We also thank teacher-consultant Diana Stolfus for providing us with the initial draft of the solutions on which this manual is based.

Problem Set 1

1. (1)
$90 - A = \text{complement of angle } A$
$180 - A = \text{supplement of angle } A$

$$\begin{aligned} 5(90 - A) &= 150 + (180 - A) \\ 450 - 5A &= 330 - A \\ 4A &= 120 \\ A &= \mathbf{30°} \end{aligned}$$

2. (1)
$90 - B = \text{complement of angle } B$
$180 - B = \text{supplement of angle } B$

$$\begin{aligned} 7(90 - B) &= 2(180 - B) + 220 \\ 630 - 7B &= 360 - 2B + 220 \\ 5B &= 50 \\ B &= \mathbf{10°} \end{aligned}$$

3. (1)
$90 - A = \text{complement of angle } A$

$$\begin{aligned} 4(90 - A) &= 200 \\ 360 - 4A &= 200 \\ 4A &= 160 \\ A &= \mathbf{40°} \end{aligned}$$

4. (R)
If 20% fused, then 80% did not fuse.

$$\begin{aligned} 0.80T &= 1420 \\ T &= \mathbf{1775\ grams} \end{aligned}$$

5. (R)
Pusillanimous brave men $= p$
Oxymorons $= x$

$$\begin{aligned} \frac{x}{x + p} &= \frac{2}{2 + 17} \\ \frac{x}{342} &= \frac{2}{19} \\ 19x &= 684 \\ x &= \mathbf{36} \end{aligned}$$

6. (R)
$$\frac{2^{-3}x^0(x^2)}{x^{-3}xy^{-3}y} = \frac{(1)x^2x^3y^3}{2^3xy} = \frac{x^5y^3}{8xy} = \mathbf{\frac{x^4y^2}{8}}$$

7. (R)
$$-(-3 - 2) + 4(-2) + \frac{1}{-2^{-3}} - (-2)^{-3}$$
$$= -(-5) + (-8) + \left(-2^3\right) - \frac{1}{(-2)^3}$$
$$= 5 - 8 + (-8) - \frac{1}{-8}$$
$$= -11 + \frac{1}{8} = -\frac{88}{8} + \frac{1}{8} = \mathbf{-\frac{87}{8}}$$

8. (R)
$$\frac{xy}{y^{-2}} - \frac{3x^4y^4}{x^3y} + \frac{7xy^{-2}}{xy^{-3}} = xyy^2 - 3xy^3 + \frac{7y^3}{y^2}$$
$$= xy^3 - 3xy^3 + 7y = \mathbf{7y - 2xy^3}$$

9. (R)
$$\frac{x^0y^{-2}x}{x^3y}\left(\frac{x^2y}{m} - \frac{3x^4y^2}{m^{-2}}\right)$$
$$= \frac{1}{x^2y^3}\left(\frac{x^2y}{m} - 3x^4m^2y^2\right)$$
$$= \mathbf{\frac{1}{y^2m} - \frac{3x^2m^2}{y}}$$

10. (R)
$$\begin{aligned} 3^0(2x - 5) + (-x - 5) &= -3\left(x^0 - 2\right) \\ 1(2x - 5) + (-x - 5) &= -3(1 - 2) \\ 2x - 5 - x - 5 &= 3 \\ x &= \mathbf{13} \end{aligned}$$

11. (R)
$$xy - x\left(x - y^0\right) = 2\left(-\frac{1}{2}\right) - 2(2 - 1) = \mathbf{-3}$$

12. (R)
$$\frac{2}{x} + \frac{x}{x + 1} = \frac{2(x + 1) + x(x)}{x(x + 1)}$$
$$= \mathbf{\frac{x^2 + 2x + 2}{x^2 + x}}$$

13. (R)
$$(2x + 3)\left(2x^2 + 2x + 2\right)$$
$$= 2x\left(2x^2 + 2x + 2\right) + 3\left(2x^2 + 2x + 2\right)$$
$$= 4x^3 + 4x^2 + 4x + 6x^2 + 6x + 6$$
$$= \mathbf{4x^3 + 10x^2 + 10x + 6}$$

14. (R)
Multiply the first equation by –5 and add to the second equation. Then solve for N_D.

$$\begin{aligned} -5N_N - 5N_D &= -250 \\ 5N_N + 10N_D &= 450 \\ \hline 5N_D &= 200 \\ N_D &= \mathbf{40} \end{aligned}$$

Now substitute this back into the second equation.

$$\begin{aligned} 5N_N + 10N_D &= 450 \\ 5N_N + 10(40) &= 450 \\ 5N_N &= 50 \\ N_N &= \mathbf{10} \end{aligned}$$

15. (R) Solve the first equation for N_P.

$$N_P + N_D = 50$$
$$N_P = 50 - N_D$$

Now substitute into the second equation.

$$N_P + 10N_D = 140$$
$$(50 - N_D) + 10N_D = 140$$
$$9N_D = 90$$
$$\mathbf{N_D = 10}$$

Substitute this value back into the first equation.

$$N_P + N_D = 50$$
$$N_P + (10) = 50$$
$$\mathbf{N_P = 40}$$

16. (R) First we write the two equations.

$$\begin{cases} \text{(a)}\ N_N = N_D + 7 \\ \text{(b)}\ 5N_N + 10N_D = 155 \end{cases}$$

Now we can use substitution to solve.

(b) $$5N_N + 10N_D = 155$$
$$5(N_D + 7) + 10N_D = 155$$
$$5N_D + 35 + 10N_D = 155$$
$$15N_D = 120$$
$$\mathbf{N_D = 8}$$

(a) $$N_N = N_D + 7$$
$$N_N = (8) + 7$$
$$\mathbf{N_N = 15}$$

17. (1) **Two line segments are congruent if the segments have equal length.**

18. (1) **An acute angle is an angle whose measure is greater than 0° and less than 90°.**

19. (1) If the lengths of two sides of a triangle are equal, then the angles opposite those sides are congruent. Thus, $\mathbf{x = 36}$. Since the sum of the interior angles of a triangle equals 180°, we can solve for y.

$$36 + 36 + y = 180$$
$$\mathbf{y = 108}$$

20. (1) $$A + 60 = 180$$
$$\mathbf{A = 120}$$

$$2B = 60$$
$$\mathbf{B = 30}$$

$$3C = A$$
$$3C = 120$$
$$\mathbf{C = 40}$$

21. (1) $$4A = 80$$
$$\mathbf{A = 20}$$

$$2B + 80 = 180$$
$$2B = 100$$
$$\mathbf{B = 50}$$

$$2C = 80$$
$$\mathbf{C = 40}$$

22. (1) $$\frac{3}{4} = \frac{x}{8}$$
$$4x = 24$$
$$\mathbf{x = 6}$$

23. (1) $$A_{\text{circle}} = A_{\text{triangle}}$$
$$\pi r^2 = \frac{1}{2}BH$$
$$\pi(\pi)^2 = \frac{1}{2}(\pi)H$$
$$H = \mathbf{2\pi^2\ cm = 19.74\ cm}$$

24. (1) Length of a side (base) $= 12 \div 3 = 4$

$$A = \frac{1}{2}BH = \frac{1}{2}(4)(2\sqrt{3}) = \mathbf{4\sqrt{3}\ in.^2}$$

25. (1) $$A = \frac{1}{2}BH$$
$$B = \frac{2A}{H} = \frac{2(18\sqrt{3})}{3\sqrt{6}} \cdot \frac{\sqrt{6}}{\sqrt{6}} = 6\sqrt{2}$$

Perimeter $= 3B = 3(6\sqrt{2}) = \mathbf{18\sqrt{2}\ cm}$

26. (1) $$A_{\text{shaded sector}} = \frac{40}{360}\pi(3\sqrt{3})^2 = \frac{1}{9}\pi(27)$$
$$= \mathbf{3\pi\ m^2 = 9.42\ m^2}$$

27. (1) $$A_{\text{shaded region}} = \pi(18 + 8)^2 - \pi(8)^2 - \pi(18)^2$$
$$= 676\pi - 64\pi - 324\pi$$
$$= \mathbf{288\pi\ cm^2 = 904.78\ cm^2}$$

28. (1) First, find the radius of one circle to use in finding the length of a side of the square.

$$A_{\text{circles}} = 52\pi$$
$$9\pi r^2 = 52\pi$$
$$r^2 = \frac{52}{9}$$
$$r = \frac{\sqrt{52}}{3} = \frac{2\sqrt{13}}{3}$$

$$L_{\text{side}} = 6r = 6\left(\frac{2\sqrt{13}}{3}\right) = 4\sqrt{13}$$

Now subtract the area of the circles from the area of the square.

$$A_{\text{shaded portion}} = A_{\text{square}} - A_{\text{circles}}$$
$$= (4\sqrt{13})^2 - 52\pi$$
$$= \mathbf{(208 - 52\pi)\ m^2 = 44.64\ m^2}$$

29. (1) A: $\sqrt{\frac{1}{4}} + \sqrt{\frac{1}{25}} = \frac{1}{2} + \frac{1}{5} = \frac{7}{10} = 0.7$

B: $\sqrt{\frac{1}{4} + \frac{1}{25}} = \sqrt{\frac{25}{100} + \frac{4}{100}} = \sqrt{\frac{29}{100}} = 0.54$

$0.7 > 0.54$

Therefore the answer is **A.**

30. (1) First we subtract B from A.

$$A - B = y - x - (y + x) = -2x$$

Since x must be positive, $-2x$ is negative. Therefore the answer is **B.**

Problem Set 2

1. (1) $90 - A$ = complement of angle A

$180 - A$ = supplement of angle A

$$90 - A = \frac{1}{2}(180 - A) - 20$$
$$90 - A = 90 - \frac{1}{2}A - 20$$
$$\frac{1}{2}A = 20$$
$$A = \mathbf{40°}$$

2. (1) $180 - A$ = supplement of angle A

$$3(180 - A) = 450$$
$$540 - 3A = 450$$
$$3A = 90$$
$$A = \mathbf{30°}$$

3. (R) First we write the two equations.

$$\begin{cases} \text{(a)}\ 3N_R = 2N_B - 1 \\ \text{(b)}\ N_R + N_B = 13 \end{cases}$$

(b) $N_R + N_B = 13$

$$N_B = 13 - N_R$$

Now we can use substitution to solve.

(a) $3N_R = 2N_B - 1$

$$3N_R = 2(13 - N_R) - 1$$
$$3N_R = 26 - 2N_R - 1$$
$$5N_R = 25$$
$$\mathbf{N_R = 5}$$

(b) $N_R + N_B = 13$

$$(5) + N_B = 13$$
$$\mathbf{N_B = 8}$$

4. (R) Tulips = t

Roses = r

$$\frac{t}{t + r} = \frac{17}{17 + 11}$$
$$\frac{t}{3444} = \frac{17}{28}$$
$$28t = 58{,}548$$
$$t = \mathbf{2091}$$

5. (R) If 40% was consumed, then 60% was not consumed.

$$0.60T = 1215$$
$$T = \mathbf{2025\ gallons}$$

6. (R) $\frac{5^2a^3b^{-2}a}{5^1a^{-3}b^2b^0} = \frac{5^2a^3a^3a}{5b^2b^2(1)} = \mathbf{\frac{5a^7}{b^4}}$

7. (R) $-(-3 - 2) + 4(-5) - \frac{1}{4^{-2}} + (-4)^2$

$$= -(-5) + (-20) - 4^2 + 16$$
$$= 5 - 20 - 16 + 16 = \mathbf{-15}$$

8. (R) $\frac{4r^3s}{r} - \frac{2s^2}{r^{-1}} - \frac{3r^3s}{r^2s^{-1}} + \frac{2r^4}{r^2s^{-1}}$

$$= 4r^2s - 2rs^2 - 3rs^2 + 2r^2s$$
$$= \mathbf{6r^2s - 5rs^2}$$

9. (R) $\frac{c^1d^2c}{c^{-1}d}\left(\frac{cd^3f}{d^2c^2} - \frac{c^{-1}dcd}{c^2d}\right)$

$$= c^3d\left(\frac{df}{c} - \frac{d}{c^2}\right) = \mathbf{c^2d^2f - cd^2}$$

10. (R) $2^2(x - 7) + (3 - 2x) = -5(2x + 1)$
$$4x - 28 + 3 - 2x = -10x - 5$$
$$12x = 20$$
$$x = \frac{20}{12}$$
$$x = \mathbf{\frac{5}{3}}$$

11. (R) $x^2 + y^2 - xy(x + y)$
$$= 1^2 + 2^2 - (1)(2)(1 + 2)$$
$$= 1 + 4 - 2(3) = \mathbf{-1}$$

12. (R) $(3x - 2y)(x^2 + xy - y^2)$
$$= 3x(x^2 + xy - y^2) - 2y(x^2 + xy - y^2)$$
$$= 3x^3 + 3x^2y - 3xy^2 - 2x^2y - 2xy^2 + 2y^3$$
$$= \mathbf{3x^3 + x^2y - 5xy^2 + 2y^3}$$

13. (R) Multiply the first equation by –2 and add to the second equation.
$$-2N_R - 2N_B = -140$$
$$2N_R + 3N_B = 190$$
$$\mathbf{N_B = 50}$$

Now substitute this back into the first equation.
$$N_R + N_B = 70$$
$$N_R + (50) = 70$$
$$\mathbf{N_R = 20}$$

14. (R) Solve the first equation for N_W.
$$N_W + 2N_G = 7$$
$$N_W = 7 - 2N_G$$

Now substitute into the second equation.
$$3N_W - 5N_G = 10$$
$$3(7 - 2N_G) - 5N_G = 10$$
$$21 - 6N_G - 5N_G = 10$$
$$11N_G = 11$$
$$\mathbf{N_G = 1}$$

Substitute this value back into the first equation.
$$N_W + 2N_G = 7$$
$$N_W + 2(1) = 7$$
$$\mathbf{N_W = 5}$$

15. (R) $\frac{20 + 17 + 24 + 18 + 29 + x}{6} = 22$
$$108 + x = 132$$
$$x = \mathbf{24}$$

16. (1) A scalene triangle is **a triangle which has sides all of different lengths.**

17. (1) $y + 4y + 35 = 180$
$$5y = 145$$
$$\mathbf{y = 29}$$
$$35 + x = 180$$
$$\mathbf{x = 145}$$

18. (1) First we write three equations.
$$\begin{cases} (a)\ y = z \\ (b)\ x + y = 180 \\ (c)\ y + (y + z) = 180 \end{cases}$$

Now we can use substitution to solve.

(c) $y + (y + z) = 180$
$$2y + z = 180$$
$$2(z) + z = 180$$
$$3z = 180$$
$$\mathbf{z = 60}$$

(a) $y = z$
$$\mathbf{y = 60}$$

(b) $x + y = 180$
$$x + (60) = 180$$
$$\mathbf{x = 120}$$

19. (1) $2A + 70 = 180$
$$\mathbf{A = 55}$$
$$5B = 70$$
$$\mathbf{B = 14}$$
$$C = 2A$$
$$C = 2(55)$$
$$\mathbf{C = 110}$$

20. (1) $\frac{5}{15} = \frac{7}{x}$
$$5x = 105$$
$$x = \mathbf{21}$$

21. (1) $A = \frac{1}{2}BH$
$$16\sqrt{3} = \frac{1}{2}(8)H$$
$$H = \mathbf{4\sqrt{3}\ m}$$

22. (1) $A_{\text{shaded sectors}} = 2\left[\frac{70}{360}\pi(5)^2\right]$
$$= \mathbf{\frac{175\pi}{18}\ m^2 = 30.54\ m^2}$$

23. (1) $A_{\text{shaded region}} = \pi(15 + 6)^2 - \pi(15)^2 - \pi(6)^2$
$= 441\pi - 225\pi - 36\pi = \mathbf{180\pi\ m^2} = \mathbf{565.49\ m^2}$

24. (2) $V_{\text{cylinder}} = \pi r^2 h$
$= \pi(4)^2(8) = \mathbf{128\pi\ m^3} = \mathbf{402.12\ m^3}$

$V_{\text{sphere}} = \frac{2}{3}V_{\text{cylinder}}$
$= \frac{2}{3}(128\pi) = \mathbf{\frac{256\pi}{3}\ m^3} = \mathbf{268.08\ m^3}$

25. (2) $V_{\text{cylinder}} = \pi r^2 H$
$588\pi = \pi(7)^2 H$
$H = \mathbf{12\ cm}$

26. (2) $V_{\text{cylinder}} = (A_{\text{base}})(h)$
$h = \frac{V_{\text{cylinder}}}{A_{\text{base}}}$
$h = \frac{9\pi}{\frac{1}{2}(2\pi)(3) + \frac{60}{360}\pi(3)^2}$
$h = \frac{9\pi}{3\pi + \frac{3}{2}\pi} = \frac{9\pi}{\frac{9}{2}\pi} = \mathbf{2\ cm}$

27. (2) $V_{\text{cone}} = \frac{1}{3}(A_{\text{base}})(h)$
$= \frac{1}{3}\left[\frac{1}{2}\pi(3)^2 + 5(6)\right](8)$
$= \frac{1}{3}\left(\frac{9}{2}\pi + 30\right)(8)$
$= \mathbf{(80 + 12\pi)\ m^3} = \mathbf{117.70\ m^3}$

28. (2) $V_{\text{cylinder}} = (A_{\text{base}})(h)$
$= \left[\frac{\pi(10)^2}{2} + 11(10) + \frac{\pi(5)^2}{2}\right](9)$
$= \left(110 + \frac{125\pi}{2}\right)(9)$
$= \mathbf{\left(\frac{1980 + 1125\pi}{2}\right)\ cm^3} = \mathbf{2757.15\ cm^3}$

$A_{\text{surface}} = 2(A_{\text{base}}) + \text{(perimeter)(height)}$
$= 2\left(110 + \frac{125\pi}{2}\right) + (10\pi + 5\pi + 32)(9)$
$= 220 + 125\pi + 135\pi + 288$
$= \mathbf{(508 + 260\pi)\ cm^2} = \mathbf{1324.81\ cm^2}$

$V_{\text{cone}} = \frac{1}{3}(A_{\text{base}})(h)$
$= \frac{1}{3}\left(110 + \frac{125\pi}{2}\right)(9)$
$= \mathbf{\left(\frac{660 + 375\pi}{2}\right)\ cm^3} = \mathbf{919.05\ cm^3}$

29. (1) A: $\sqrt{\frac{4}{9}} + \sqrt{\frac{9}{16}} = \frac{2}{3} + \frac{3}{4} = \frac{17}{12} = 1.42$

B: $\sqrt{\frac{4}{9} + \frac{9}{16}} = \sqrt{\frac{64 + 81}{144}} = \sqrt{\frac{145}{144}} = 1.00$

$1.42 > 1.00$

Therefore the answer is **A.**

30. (1) The base and the height of the two triangles will always be equal. Since the area of any triangle is one half the base times the height, the areas will also be equal. Therefore the answer is **C.**

Problem Set 3

1. (1) $90 - A = \text{complement of } A$
$180 - A = \text{supplement of } A$
$180 - A = 20 + 2(90 - A)$
$180 - A = 200 - 2A$
$A = \mathbf{20°}$

2. (R) $u = \text{ubiquitous}$
$s = \text{seldom}$
$\frac{u}{s} = \frac{7}{1}$
$\frac{u}{s + u} = \frac{7}{1 + 7}$
$\frac{u}{296} = \frac{7}{8}$
$8u = 2072$
$u = \mathbf{259}$

3. (R) $T = \text{total}$
$F \times \text{(of)} = \text{(is)}$
$0.86(T) = 4300$
$T = 5000$

$5000 - 4300 = \mathbf{700}$

4. First we write the two equations.
(R)

$$\begin{cases} \text{(a)} & N_N = N_D + 13 \\ \text{(b)} & 5N_N + 10N_D = 185 \end{cases}$$

Now we can use substitution to solve.

(b)
$$\begin{aligned} 5N_N + 10N_D &= 185 \\ 5(N_D + 13) + 10N_D &= 185 \\ 65 + 15N_D &= 185 \\ 15N_D &= 120 \\ \mathbf{N_D} &= \mathbf{8} \end{aligned}$$

(a)
$$\begin{aligned} N_N &= N_D + 13 \\ N_N &= (8) + 13 \\ \mathbf{N_N} &= \mathbf{21} \end{aligned}$$

5. (3)

$c^2 \bigcirc a^2 + b^2$

$12^2 \bigcirc 7^2 + 6^2$

$144 \bigcirc 49 + 36$

$144 > 85$

Since the square of the longest side is greater than the sum of the squares of the other two sides, the triangle is an **obtuse triangle.**

6. First we find the scale factor in both directions.
(3)

$3 \cdot \overleftarrow{SF} = 6$

$\overleftarrow{SF} = 2 \rightarrow \overrightarrow{SF} = \frac{1}{2}$

Now we solve for x and y.

$x = 4 \cdot \overleftarrow{SF}$ $\qquad y = 7 \cdot \overrightarrow{SF}$

$x = 4 \cdot 2$ $\qquad y = 7 \cdot \frac{1}{2}$

$\mathbf{x = 8}$ $\qquad \mathbf{y = \frac{7}{2}}$

7. First we find the scale factor in both directions.
(3)

$7 \cdot \overrightarrow{SF} = 3$

$\overrightarrow{SF} = \frac{3}{7} \rightarrow \overleftarrow{SF} = \frac{7}{3}$

Now we solve for x and y.

$x = 15 \cdot \overrightarrow{SF}$ $\qquad y = 5 \cdot \overleftarrow{SF}$

$x = 15 \cdot \frac{3}{7}$ $\qquad y = 5 \cdot \frac{7}{3}$

$\mathbf{x = \frac{45}{7}}$ $\qquad \mathbf{y = \frac{35}{3}}$

8. First we find the scale factor in both directions.
(3)

$15 \cdot \overrightarrow{SF} = 21$

$\overrightarrow{SF} = \frac{7}{5} \rightarrow \overleftarrow{SF} = \frac{5}{7}$

Solve for x using the Pythagorean Theorem.

$x = \sqrt{15^2 - 12^2}$

$x = \sqrt{225 - 144}$

$\mathbf{x = 9}$

Now solve for y and z using scale factors.

$y = 9 \cdot \overrightarrow{SF}$ $\qquad z = 12 \cdot \overrightarrow{SF}$

$y = 9 \cdot \frac{7}{5}$ $\qquad z = 12 \cdot \frac{7}{5}$

$\mathbf{y = \frac{63}{5}}$ $\qquad \mathbf{z = \frac{84}{5}}$

9. First, separate the triangles.
(3)

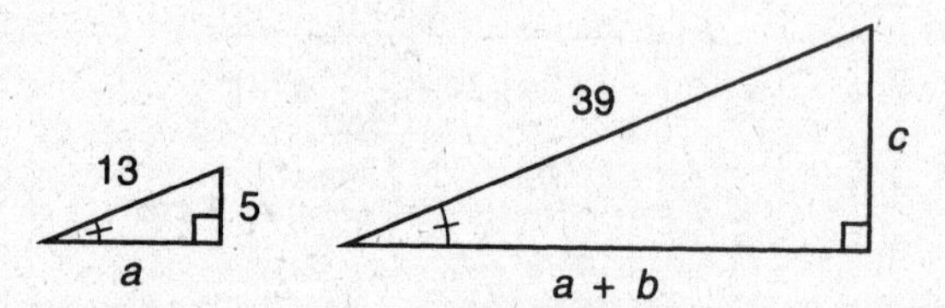

Solve for a using the Pythagorean Theorem.

$a^2 = 13^2 - 5^2$

$a = \sqrt{169 - 25}$

$\mathbf{a = 12}$

Use the ratios for similar triangles to solve for b and c.

$\frac{5}{13} = \frac{c}{39}$ $\qquad \frac{5}{12} = \frac{15}{12 + b}$

$13c = 195$ $\qquad 60 + 5b = 180$

$\mathbf{c = 15}$ $\qquad 5b = 120$

$\mathbf{b = 24}$

10. First, separate the triangles.
(3)

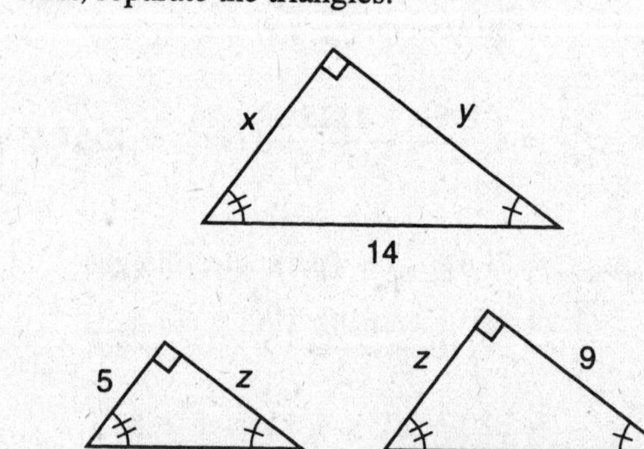

Next, write the ratios using the largest and smallest right triangles and solve for x.

$$\frac{x}{5} = \frac{14}{x}$$
$$x^2 = 70$$
$$\mathbf{x = \sqrt{70}}$$

Now use the Pythagorean Theorem to solve for the other sides.

$$y^2 = 14^2 - (\sqrt{70})^2 \qquad z^2 = (\sqrt{70})^2 - 5^2$$
$$y^2 = 196 - 70 \qquad z^2 = 70 - 25$$
$$y = \sqrt{126} \qquad z = \sqrt{45}$$
$$\mathbf{y = 3\sqrt{14}} \qquad \mathbf{z = 3\sqrt{5}}$$

11. (R) $\dfrac{7^4 p^4 q^{-1} p q^{-2}}{7^3 p^{-2} q^4} = \dfrac{7^4 p^4 p p^2}{7^3 q^4 q q^2} = \mathbf{\dfrac{7p^7}{q^7}}$

12. (R) $\dfrac{6^3 a^0 b^{-3} a^2}{6^2 a^{-4} b^2 b^{-4}} = \dfrac{6^3 (1) a^2 a^4 b^4}{6^2 b^2 b^3} = \mathbf{\dfrac{6a^6}{b}}$

13. (R) Solve the first equation for x.

$$x - 3y = -6$$
$$x = 3y - 6$$

Now substitute into the second equation.

$$2x + 5y = 21$$
$$2(3y - 6) + 5y = 21$$
$$6y - 12 + 5y = 21$$
$$11y = 33$$
$$\mathbf{y = 3}$$

Substitute this value back into the first equation.

$$x - 3y = -6$$
$$x - 3(3) = -6$$
$$\mathbf{x = 3}$$

14. (R) $2\frac{1}{4}x - 3\frac{1}{2} = -\frac{1}{16}$

$$16\left(\frac{9}{4}x - \frac{7}{2}\right) = 16\left(-\frac{1}{16}\right)$$
$$36x - 56 = -1$$
$$36x = 55$$
$$x = \mathbf{\frac{55}{36}}$$

15. (R) $-2(-x^0 - 4^0) - 3x(2 - 6^0)$

$$= x(-2 - 3^2 - 2) - x(-2 - 2^0)$$
$$-2(-1 - 1) - 3x(2 - 1) = x(-2 - 9 - 2) - x(-2 - 1)$$
$$4 - 3x = -13x + 3x$$
$$7x = -4$$
$$x = \mathbf{-\frac{4}{7}}$$

16. (R) $\dfrac{3}{x-2} + \dfrac{4}{x-1} - \dfrac{1}{x}$

$$= \frac{3x(x-1) + 4x(x-2) - (x-2)(x-1)}{x(x-1)(x-2)}$$
$$= \frac{3x^2 - 3x + 4x^2 - 8x - x^2 + 3x - 2}{x(x-1)(x-2)}$$
$$= \mathbf{\frac{6x^2 - 8x - 2}{x(x-1)(x-2)}}$$

17. (R) $(x^2 - 1)(4x^3 - 3x^2 + 5)$

$$= x^2(4x^3 - 3x^2 + 5) - 1(4x^3 - 3x^2 + 5)$$
$$= 4x^5 - 3x^4 + 5x^2 - 4x^3 + 3x^2 - 5$$
$$= \mathbf{4x^5 - 3x^4 - 4x^3 + 8x^2 - 5}$$

18. (R) $\dfrac{4x^{-2}y^{-2}}{z^2}\left(\dfrac{3x^2y^2z^2}{4} + \dfrac{2x^0y^{-2}}{z^2y^2}\right)$

$$= \frac{4}{x^2y^2z^2}\left[\frac{3x^2y^2z^2}{4} + \frac{2(1)}{y^4z^2}\right]$$
$$= \mathbf{3 + \frac{8}{x^2y^6z^4}}$$

19. (R) $-2^0 - 3^0(-2 - 5^0) - \dfrac{1}{-2^{-2}} + x^2y - xy$

$$= -1 - 1(-2 - 1) - (-2^2) + (-2)^2(3) - (-2)(3)$$
$$= -1 - (-3) - (-4) + 12 - (-6) = \mathbf{24}$$

20. (1) $\dfrac{4}{5} = \dfrac{6}{x}$

$$4x = 30$$
$$x = \mathbf{\frac{15}{2}}$$

21. (1) $A_{\text{shaded sectors}} = 2\left[\dfrac{140}{360}\pi(\sqrt{3})^2\right]$

$$= \mathbf{\frac{7\pi}{3}\ cm^2 = 7.33\ cm^2}$$

22. $A_{\text{shaded region}}$
(1)

$= \pi(\sqrt{12} + \sqrt{3})^2 - \pi(\sqrt{3})^2 - \pi(\sqrt{12})^2$

$= \pi(2\sqrt{3} + \sqrt{3})^2 - 3\pi - 12\pi$

$= 27\pi - 15\pi$

$= \mathbf{12\pi\ cm^2} = \mathbf{37.70\ cm^2}$

23. $A_{\text{shaded region}} = 2^2 - 4\left[\pi\left(\frac{1}{2}\right)^2\right]$
(1)

$= 4 - 4\left(\frac{1}{4}\pi\right)$

$= \mathbf{(4 - \pi)\ in.^2} = \mathbf{0.86\ in.^2}$

24. We must find the radius of a small semicircle, and subtract the area of a small semicircle from half of the large circle's area.
(1)

$A_{\text{large circle}} = \pi(r_{\text{large circle}})^2$

$(r_{\text{large circle}})^2 = \frac{1}{\pi}A_{\text{large circle}}$

$(r_{\text{large circle}})^2 = \frac{1}{\pi}(9\pi)$

$(r_{\text{large circle}})^2 = 9$

$r_{\text{large circle}} = 3$

$r_{\text{semicircle}} = \frac{1}{3}r_{\text{large circle}} = \frac{1}{3}(3) = 1$

$A_{\text{shaded region}} = \frac{1}{2}A_{\text{large circle}} - A_{\text{semicircle}}$

$= \frac{1}{2}(9\pi) - \frac{1}{2}(1\pi)$

$= \frac{9}{2}\pi - \frac{1}{2}\pi$

$= \mathbf{4\pi\ in.^2} = \mathbf{12.57\ in.^2}$

25. First, draw the picture and label the missing sides.
(2)

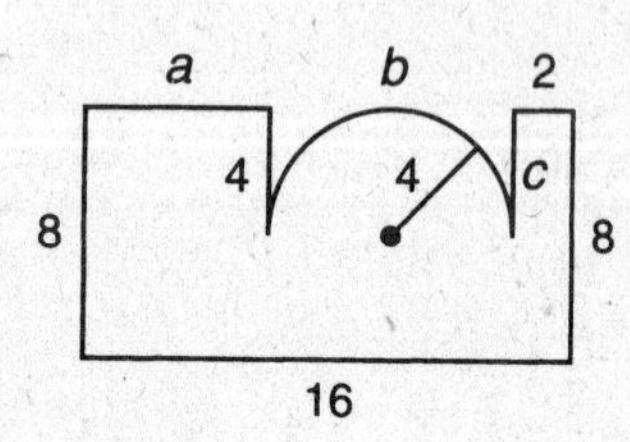

Sum the known sides.

$16 + 8 + 4 + 2 + 8 = 38$

Now find the length of a.

$a = 16 - 2(4) - 2 = 6$

We see that c is equal to 4 and we can find b by calculating the length of a semicircle of radius 4.

$b = \frac{1}{2}(2\pi r) = \frac{1}{2}(2\pi 4) = 4\pi$

Now find the total.

$38 + 6 + 4 + 4\pi = \mathbf{(48 + 4\pi)\ ft} = \mathbf{60.57\ ft}$

26.
(2)

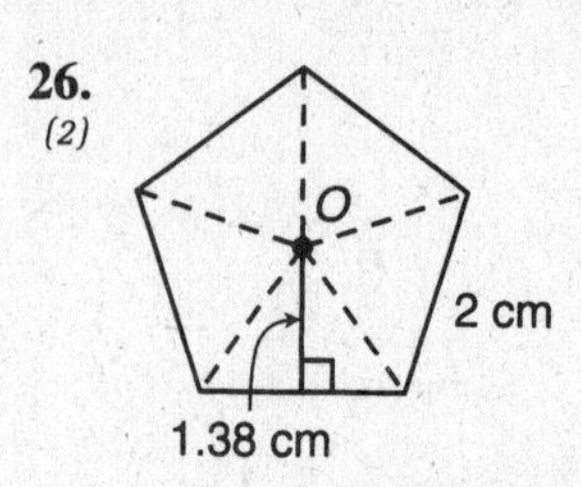

$A_{\text{figure}} = 5\left[\frac{1}{2}(2)(1.38)\right]$

$= 5(1.38)$

$= \mathbf{6.9\ cm^2}$

27. $V_{\text{difference}} = V_{\text{cylinder}} - V_{\text{sphere}}$
(2)

$= \pi r^2 h - \frac{2}{3}\pi r^2 h$

$= \frac{1}{3}\pi(18)^2(36)$

$= \mathbf{3888\pi\ cm^3} = \mathbf{12{,}214.51\ cm^3}$

28. $V_{\text{cone}} = \frac{1}{3}(A_{\text{base}})(h)$
(2)

$= \frac{1}{3}\left[\frac{1}{2}(6)(7) + 3(7)\right](5)$

$= \frac{1}{3}(42)(5) = \mathbf{70\ m^3}$

29. A: $7^{26} - 7^{25} = 7^{25}(7 - 1) = 7^{25}(6)$
(1)

B: $7^{25}(6)$

$7^{25}(6) = 7^{25}(6)$

Therefore the answer is **C**.

30. A: $\pi(8)^2 = 64\pi$
(1)

B: $2[\pi(4)^2] = 32\pi$

$64\pi > 32\pi$

Therefore the answer is **A**.

Problem Set 4

1. (1)

$90 - A = \text{complement of angle } A$

$180 - A = \text{supplement of angle } A$

$$3(90 - A) = 40 + (180 - A)$$
$$270 - 3A = 220 - A$$
$$2A = 50$$
$$A = \mathbf{25°}$$

2. (R)

$f = \text{flatterers}$

$s = \text{sycophants}$

$$\frac{s}{f} = \frac{10}{7}$$
$$\frac{s}{1106} = \frac{10}{7}$$
$$7s = 11{,}060$$
$$s = \mathbf{1580}$$

3. (R)

$R = \text{total respondents}$

$$F \times (\text{of}) = (\text{is})$$
$$0.72(R) = 594$$
$$R = \mathbf{825}$$

4. (R) First we write the two equations.

$$\begin{cases} \text{(a)} & N_D + N_Q = 20 \\ \text{(b)} & 10N_D + 25N_Q = 335 \end{cases}$$

(a) $N_D + N_Q = 20$

$$N_D = 20 - N_Q$$

Now we can use substitution to solve.

(b)
$$10N_D + 25N_Q = 335$$
$$10(20 - N_Q) + 25N_Q = 335$$
$$200 + 15N_Q = 335$$
$$15N_Q = 135$$
$$N_Q = \mathbf{9}$$

(a)
$$N_D + N_Q = 20$$
$$N_D + (9) = 20$$
$$N_D = \mathbf{11}$$

5. (R) First we write the two equations.

$$\begin{cases} \text{(a)} & N_O + N_B = 44 \\ \text{(b)} & 2N_B = N_O + 4 \end{cases}$$

(a) $N_O + N_B = 44$

$$N_O = 44 - N_B$$

Now we can use substitution to solve.

(b)
$$2N_B = N_O + 4$$
$$2N_B = (44 - N_B) + 4$$
$$3N_B = 48$$
$$N_B = \mathbf{16}$$

(a)
$$N_O + N_B = 44$$
$$N_O + (16) = 44$$
$$N_O = \mathbf{28}$$

6. (3)

$$c^2 \bigcirc a^2 + b^2$$
$$10^2 \bigcirc 9^2 + 3^2$$
$$100 \bigcirc 81 + 9$$
$$100 > 90$$

Since the square of the longest side is greater than the sum of the squares of the other two sides, the triangle is an **obtuse triangle.**

7. (4)

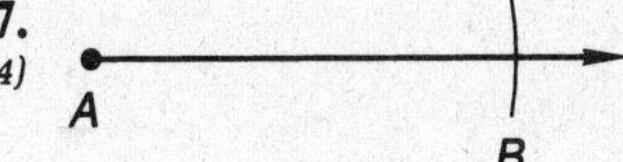

8. (4)

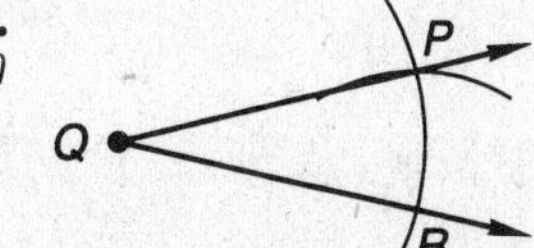

9. (4)

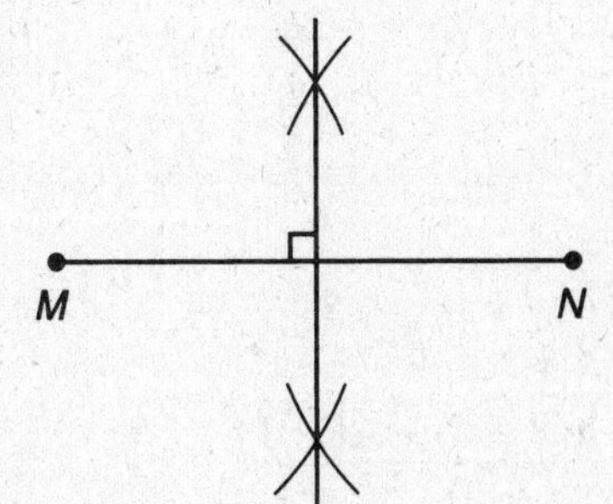

10. (4)

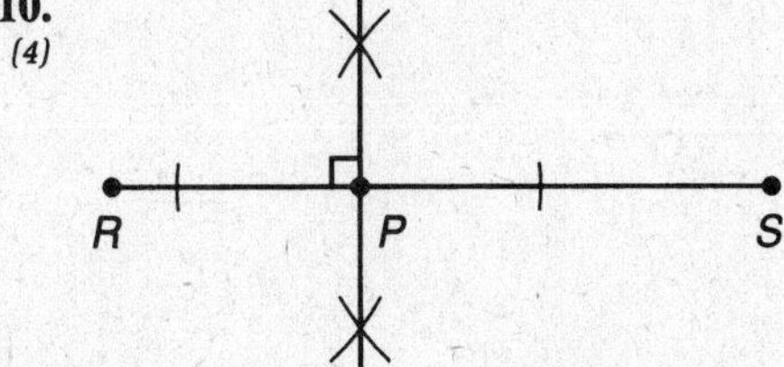

11. (4)

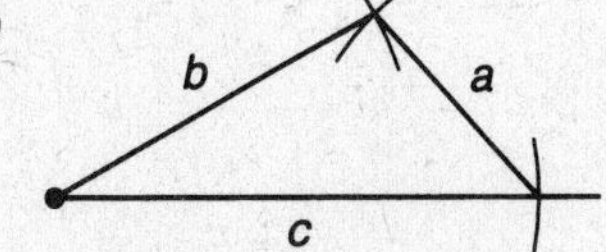

12. (R) Solve the second equation for y.

$$3x + y = 35$$
$$y = 35 - 3x$$

Now substitute into the first equation.

$$2x - 3y = 5$$
$$2x - 3(35 - 3x) = 5$$
$$2x - 105 + 9x = 5$$
$$11x = 110$$
$$\boldsymbol{x = 10}$$

Substitute this value back into the second equation.

$$3x + y = 35$$
$$3(10) + y = 35$$
$$\boldsymbol{y = 5}$$

13. (R)
$$\frac{3}{2}x + \frac{1}{5} = \frac{3}{10}$$
$$10\left(\frac{3}{2}x + \frac{1}{5}\right) = 10\left(\frac{3}{10}\right)$$
$$15x + 2 = 3$$
$$15x = 1$$
$$x = \boldsymbol{\frac{1}{15}}$$

14. (R)
$$-3\left(-x^0 - 4\right) + 2(-x)^0 = 7\left(x - 3^2 + 4^2\right)$$
$$-3(-1 - 4) + 2(1) = 7(x - 9 + 16)$$
$$15 + 2 = 7x + 7(7)$$
$$7x = -32$$
$$x = \boldsymbol{-\frac{32}{7}}$$

15. (R)
$$\frac{3}{x^2} + \frac{1}{x} - \frac{2}{x + 1}$$
$$= \frac{3(x + 1) + x(x + 1) - 2\left(x^2\right)}{x^2(x + 1)}$$
$$= \frac{3x + 3 + x^2 + x - 2x^2}{x^2(x + 1)}$$
$$= \boldsymbol{\frac{-x^2 + 4x + 3}{x^2(x + 1)}}$$

16. (R)
$$\frac{9s^2t^{-3}}{s^{-2}t}\left(\frac{3^{-1}s^{-1}t}{s^2} - \frac{s^3t^4}{t^{-1}}\right)$$
$$= \frac{9s^4}{t^4}\left(\frac{t}{3s^3} - s^3t^5\right)$$
$$= \boldsymbol{\frac{3s}{t^3} - 9s^7t}$$

17. (R)
$$\frac{3^0 4^{-1}}{2^{-4}}x^{-1}y + 2^{-1}xy^0 - 3x + y^2$$
$$= \frac{(1)2^4 y}{4x} + \frac{x(1)}{2} - 3x + y^2$$
$$= \frac{4(2)}{(1)} + \frac{(1)}{2} - 3(1) + (2)^2$$
$$= 8 + \frac{1}{2} - 3 + 4 = \boldsymbol{\frac{19}{2}}$$

18. (3) First we find the scale factor in both directions.

$$4 \cdot \overleftarrow{SF} = 6$$
$$\overleftarrow{SF} = \frac{6}{4} = \frac{3}{2} \rightarrow \overrightarrow{SF} = \frac{2}{3}$$

Now we solve for x and y.

$$x = 6 \cdot \overleftarrow{SF} \qquad y = 10 \cdot \overrightarrow{SF}$$
$$x = 6 \cdot \frac{3}{2} \qquad y = 10 \cdot \frac{2}{3}$$
$$\boldsymbol{x = 9} \qquad \boldsymbol{y = \frac{20}{3}}$$

19. (3) First we find the scale factor in both directions.

$$3 \cdot \overrightarrow{SF} = 6$$
$$\overrightarrow{SF} = \frac{6}{3} = 2 \rightarrow \overleftarrow{SF} = \frac{1}{2}$$

Solve for a using the Pythagorean Theorem.

$$a = \sqrt{5^2 - 3^2}$$
$$a = \sqrt{16}$$
$$\boldsymbol{a = 4}$$

Now solve for b and c using scale factors.

$$b = 5 \cdot \overrightarrow{SF} \qquad c = a \cdot \overrightarrow{SF}$$
$$b = 5 \cdot 2 \qquad c = 4 \cdot 2$$
$$\boldsymbol{b = 10} \qquad \boldsymbol{c = 8}$$

20. (3) First, separate the triangles.

Solve for c using the Pythagorean Theorem.

$$c = \sqrt{3^2 + 4^2}$$
$$c = \sqrt{25}$$
$$\boldsymbol{c = 5}$$

Now use the ratios for similar triangles to solve for a and b.

$$\frac{3}{4} = \frac{a}{6} \qquad \frac{4}{5} = \frac{6}{b+5}$$

$$4a = 18 \qquad 4b + 20 = 30$$

$$\mathbf{a = \frac{9}{2}} \qquad 4b = 10$$

$$\mathbf{b = \frac{5}{2}}$$

21. (3) First, separate the triangles.

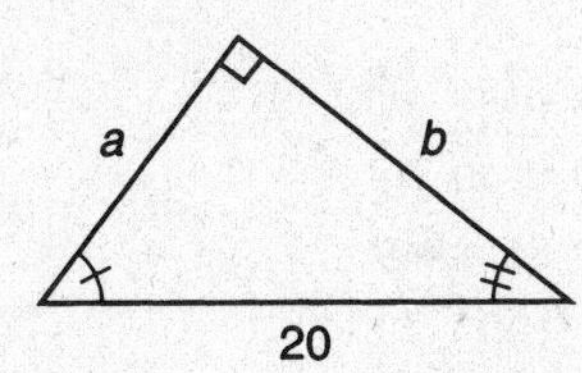

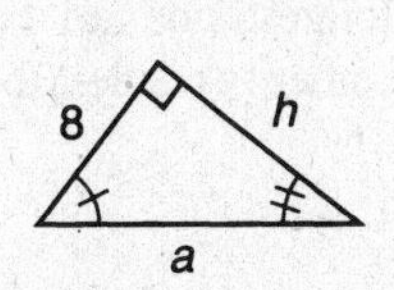

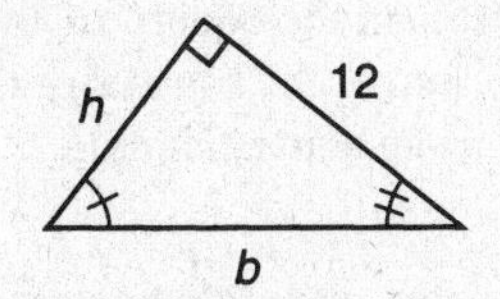

All three triangles are similar. The ratios of sides of similar triangles are equal. We use this fact to solve for a, b, and h.

$$\frac{20}{a} = \frac{a}{8} \qquad \frac{b}{20} = \frac{12}{b} \qquad \frac{h}{8} = \frac{12}{h}$$

$$a^2 = 160 \qquad b^2 = 240 \qquad h^2 = 96$$

$$\mathbf{a = 4\sqrt{10}} \qquad \mathbf{b = 4\sqrt{15}} \qquad \mathbf{h = 4\sqrt{6}}$$

22. (1)

$$\frac{6}{7} = \frac{5}{d}$$

$$6d = 35$$

$$\mathbf{d = \frac{35}{6}}$$

23. (2) First draw the figure and label the missing sides.

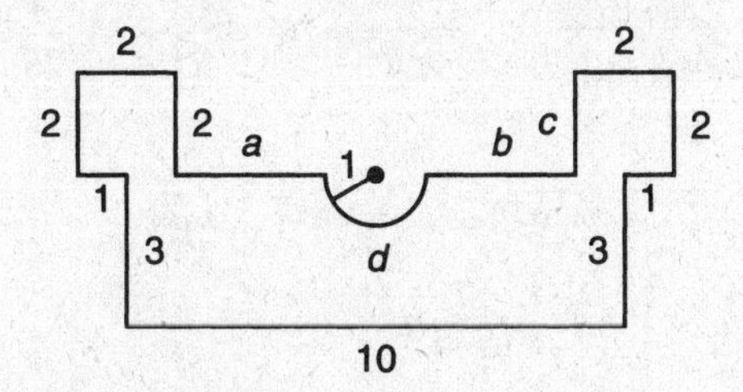

Sum the known sides.

$$10 + 2(3) + 2(1) + 5(2) = 28$$

Now find the length of $a + b$.

$$a + b = 10 - (2 - 1) - (2 - 1) - 2 = 6$$

We see that c is equal to 2 and we can find d by calculating the length of the semicircle of radius 1.

$$d = \frac{1}{2}(2\pi r) = \frac{1}{2}[2\pi(1)] = \pi$$

Now find the total.

$$28 + 6 + 2 + \pi = \mathbf{(36 + \pi)\ m = 39.14\ m}$$

24. (2)

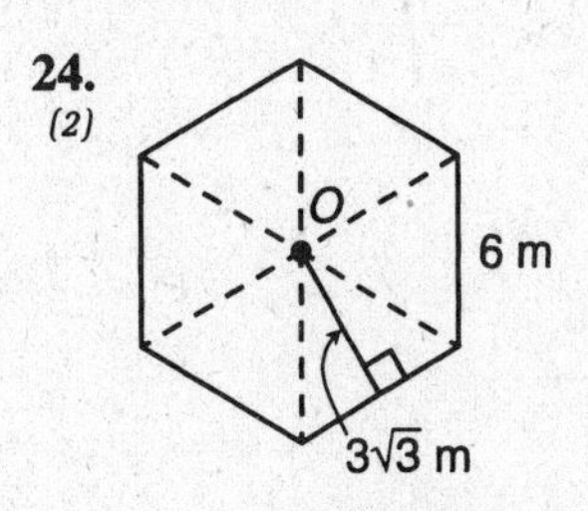

$$A_{\text{figure}} = 6\left[\frac{1}{2}(6)(3\sqrt{3})\right] = 6(9\sqrt{3}) = \mathbf{54\sqrt{3}\ m^2}$$

25. (2)

$$V_{\text{slice}} = (A_{\text{sector}})(h)$$

$$= \left[\frac{60}{360}\pi(3)^2\right](12)$$

$$= \left[\frac{1}{6}\pi(9)\right](12)$$

$$= \mathbf{18\pi\ m^3 = 56.55\ m^3}$$

26. (2)

$$V_{\text{cone}} = \frac{1}{3}(A_{\text{base}})(h)$$

$$= \frac{1}{3}[\pi(10)^2](10)$$

$$= \mathbf{\frac{1000\pi}{3}\ cm^3 = 1047.20\ cm^3}$$

$$l = \sqrt{10^2 + 10^2} = 10\sqrt{2}$$

$$A_{\text{surface}} = A_{\text{base}} + \pi rl$$

$$= \pi(10)^2 + \pi(10)(10\sqrt{2})$$

$$= \mathbf{(100\pi + 100\sqrt{2}\pi)\ cm^2 = 758.45\ cm^2}$$

27. (2)

$$V_{\text{sphere}} = \frac{4}{3}\pi r^3$$

$$= \frac{4}{3}\pi(10)^3$$

$$= \mathbf{\frac{4000\pi}{3}\ m^3 = 4188.80\ m^3}$$

$$A_{\text{surface}} = 4\pi r^2$$

$$= 4\pi(10)^2$$

$$= \mathbf{400\pi\ m^2 = 1256.64\ m^2}$$

28. (2) $V_{\text{pyramid}} = \frac{1}{3}(A_{\text{base}})(h) = \frac{1}{3}\left(\frac{20}{4}\right)^2(5) = \mathbf{\frac{125}{3}\ cm^3}$

29. (1) $m\angle A < m\angle C$

The greater the angle, the greater the length of the segment opposite it. Since $m\angle C$ is greater than $m\angle A$, the segment AB is greater than segment BC. Therefore the answer is **A.**

30. (1)

$$\text{A} - \text{B} = \frac{x+1}{y+1} - \left(\frac{x}{y} + 1\right)$$
$$= \frac{x+1}{y+1} - \frac{x+y}{y}$$
$$= \frac{(x+1)(y) - (x+y)(y+1)}{y(y+1)}$$
$$= \frac{xy + y - (xy + y^2 + x + y)}{y(y+1)}$$
$$= \frac{-y^2 - x}{y^2 + y}$$

Since $y > 1$, the value of the denominator must be positive. However, since the value of the numerator depends on the value of x, which can be positive or negative, we cannot determine the sign. Therefore, the answer is **D.**

Problem Set 5

1. (1) $90 - \theta$ = complement of angle θ

$180 - \theta$ = supplement of angle θ

$$2(180 - \theta) = 104 + 4(90 - \theta)$$
$$360 - 2\theta = 104 + 360 - 4\theta$$
$$2\theta = 104$$
$$\theta = \mathbf{52°}$$

2. (R) A = alloy

$$0.65A = 1508$$
$$A = 2320$$

Alloy − Other = Titanium

2320 − 1508 = **812 grams**

3. (R) First we write the two equations.

$$\begin{cases} \text{(a)}\ N_D + N_Q = 24 \\ \text{(b)}\ 10N_D + 25N_Q = 390 \end{cases}$$

(a) $N_D + N_Q = 24$

$N_D = 24 - N_Q$

Now we can use substitution to solve.

(b)
$$10N_D + 25N_Q = 390$$
$$10(24 - N_Q) + 25N_Q = 390$$
$$240 - 10N_Q + 25N_Q = 390$$
$$15N_Q = 150$$
$$N_Q = \mathbf{10}$$

(a) $N_D + N_Q = 24$

$N_D + (10) = 24$

$N_D = \mathbf{14}$

4. (3)

$c^2 \bigcirc a^2 + b^2$

$11^2 \bigcirc 9^2 + 7^2$

$121 \bigcirc 81 + 49$

$121 < 130$

Since the square of the longest side is less than the sum of the squares of the other two sides, the triangle is an **acute triangle.**

5. (R) First we write the two equations.

$$\begin{cases} \text{(a)}\ \frac{N_B}{N_W} = \frac{7}{5} \\ \text{(b)}\ N_B = N_W + 12 \end{cases}$$

Now we can use substitution to solve.

(a)
$$\frac{N_B}{N_W} = \frac{7}{5}$$
$$5N_B = 7N_W$$
$$5(N_W + 12) = 7N_W$$
$$2N_W = 60$$
$$N_W = \mathbf{30}$$

(b) $N_B = N_W + 12$

$N_B = (30) + 12$

$N_B = \mathbf{42}$

6. (5)
$$\frac{A_{\text{little circle}}}{A_{\text{big circle}}} = \frac{\pi\left(\frac{2}{5}R\right)^2}{\pi R^2} = \frac{\frac{4}{25}R^2}{R^2} = \mathbf{\frac{4}{25}}$$

7. (5)
$$\sqrt{x^3y^2}\sqrt[4]{xy^3} = (x^3y^2)^{\frac{1}{2}}(xy^3)^{\frac{1}{4}}$$
$$= x^{\frac{3}{2}}yx^{\frac{1}{4}}y^{\frac{3}{4}} = \mathbf{x^{\frac{7}{4}}y^{\frac{7}{4}}}$$

8. (5)
$$x^{\frac{1}{2}}x^{\frac{3}{4}}\sqrt{xy}\sqrt[3]{x^4} = x^{\frac{5}{4}}(xy)^{\frac{1}{2}}x^{\frac{4}{3}}$$
$$= x^{\frac{15}{12}}x^{\frac{6}{12}}y^{\frac{1}{2}}x^{\frac{16}{12}} = \mathbf{x^{\frac{37}{12}}y^{\frac{1}{2}}}$$

9. (5) $\dfrac{a^{\frac{x}{2}}(y^{2-x})^{\frac{1}{2}}}{a^{4x}y^{-2x}} = a^{\frac{x}{2}}y^{\frac{2-x}{2}}a^{-4x}y^{2x}$

$= a^{\frac{x}{2}}a^{\frac{-8x}{2}}y^{\frac{2-x}{2}}y^{\frac{4x}{2}} = \mathbf{a^{\frac{-7x}{2}}y^{\frac{2+3x}{2}}}$

10. (5) $\dfrac{y^{x+3}y^{\left(\frac{x}{2}-1\right)}z^a}{y^{\left(\frac{x-a}{2}\right)}z^{\left(\frac{x-a}{3}\right)}} = y^{x+3}y^{\left(\frac{x}{2}-1\right)}z^a y^{-\left(\frac{x-a}{2}\right)}z^{-\left(\frac{x-a}{3}\right)}$

$= y^{\frac{(2x+6)+(x-2)+(a-x)}{2}}z^{\frac{3a+(a-x)}{3}} = \mathbf{y^{\frac{a+2x+4}{2}}z^{\frac{4a-x}{3}}}$

11. (5) $2\sqrt{\dfrac{3}{2}} - 3\sqrt{\dfrac{2}{3}} + 2\sqrt{24}$

$= 2\sqrt{\dfrac{3}{2}}\cdot\sqrt{\dfrac{2}{2}} - 3\sqrt{\dfrac{2}{3}}\cdot\sqrt{\dfrac{3}{3}} + 2\sqrt{4\cdot 6}$

$= \dfrac{2}{2}\sqrt{6} - \dfrac{3}{3}\sqrt{6} + 4\sqrt{6} = \mathbf{4\sqrt{6}}$

12. (5)

$$\begin{array}{r} \sqrt{3} + \sqrt{x} \\ \sqrt{3} + \sqrt{x} \\ \hline 3 + \sqrt{3x} \\ \sqrt{3x} + x \\ \hline \mathbf{3 + 2\sqrt{3}\sqrt{x} + x} \end{array}$$

13. (5)

$$\begin{array}{l} \sqrt{2} - \sqrt{x} \\ 1 - \sqrt{x} \\ \hline \sqrt{2} - \sqrt{x} \\ \quad - \sqrt{2x} + x \\ \hline \sqrt{2} - \sqrt{x} - \sqrt{2}\sqrt{x} + x = \mathbf{\sqrt{2} - (1 + \sqrt{2})\sqrt{x} + x} \end{array}$$

14. (5) $\dfrac{x^{-2} + y^{-2}}{(xy)^{-1}} = \dfrac{\left(\dfrac{1}{x^2} + \dfrac{1}{y^2}\right)}{\dfrac{1}{xy}} = \left(\dfrac{x^2 + y^2}{x^2y^2}\right)xy$

$= \mathbf{\dfrac{x^2 + y^2}{xy}}$

15. (5) $3i^4 + 2i^5 + 3i^3 + 2i^2$

$= 3(ii)(ii) + 2(ii)(ii)i + 3(ii)i + 2(ii)$

$= 3(-1)(-1) + 2(-1)(-1)i + 3(-1)i + 2(-1)$

$= 3 + 2i - 3i - 2 = \mathbf{1 - i}$

16. (5) $\sqrt{-3}\sqrt{4} + 3\sqrt{-2}\sqrt{-9} = (\sqrt{3}i)(2) + (3\sqrt{2}i)(3i)$

$= \mathbf{-9\sqrt{2} + 2\sqrt{3}i}$

17. (5) $2\sqrt{-2}\sqrt{2} + 3i\sqrt{2} = 2(\sqrt{2}i)\sqrt{2} + 3\sqrt{2}i$

$= 4i + 3\sqrt{2}i = \mathbf{(4 + 3\sqrt{2})i}$

18. (1) First we write two equations.

$$\begin{cases} \text{(a)}\ 20x + 5y = 110 \\ \text{(b)}\ 2x + 5y = 180 - 110 - 50 = 20 \end{cases}$$

Now we can use elimination to solve for x.

$$\begin{array}{lrcl} \text{(b)} & 2x + 5y & = & 20 \\ -1\text{(a)} & -20x - 5y & = & -110 \\ \hline & -18x & = & -90 \\ & \mathbf{x} & = & \mathbf{5} \end{array}$$

Use this value of x to solve for y.

$$\begin{array}{lrcl} \text{(b)} & 2x + 5y & = & 20 \\ & 2(5) + 5y & = & 20 \\ & \mathbf{y} & = & \mathbf{2} \end{array}$$

19. (5) First find BC.

$(BC)^2 = 3^2 + 2^2$

$(BC)^2 = 13$

$BC = \sqrt{13}$

Now find AC.

$(AC)^2 = (\sqrt{13})^2 + n^2$

$(AC)^2 = 13 + n^2$

$AC = \mathbf{\sqrt{13 + n^2}}$

20. (1) $A_{\text{shaded}} = A_{\text{semicircle}} - A_{\text{triangle}} + A_{\text{sector}}$

$= \dfrac{1}{2}\pi(4)^2 - \dfrac{1}{2}(8)(4) + \dfrac{60}{360}\pi(4)^2$

$= 8\pi - 16 + \dfrac{8}{3}\pi$

$= \mathbf{\left(\dfrac{32\pi}{3} - 16\right)\text{cm}^2 = 17.51\ \text{cm}^2}$

21. (1) $A_{\text{shaded}} = A_{\text{triangle}} - A_{\text{sector}}$

$= \dfrac{1}{2}(5)\pi - \dfrac{120}{360}\pi(2)^2$

$= \dfrac{5}{2}\pi - \dfrac{4}{3}\pi = \mathbf{\dfrac{7\pi}{6}\ \text{m}^2 = 3.67\ \text{m}^2}$

22. (1) First find the radius of one circle to use in finding the length of a side of the square.

$4\pi r^2 = 36\pi$

$r^2 = 9$

$r = 3$

$L_{\text{side}} = 4r = 12$

$A_{\text{square}} = 12^2 = \mathbf{144\ m^2}$

23. (3) First, separate the triangles.

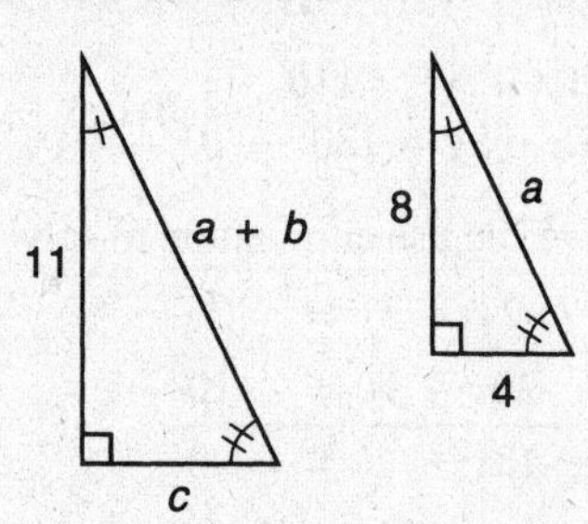

Solve for a using the Pythagorean Theorem.

$$a = \sqrt{8^2 + 4^2}$$
$$a = \sqrt{80}$$
$$\mathbf{a = 4\sqrt{5}}$$

Now use the ratios for similar triangles to solve for c and b.

$$\frac{8}{4} = \frac{11}{c}$$
$$8c = 44$$
$$\mathbf{c = \frac{11}{2}}$$

$$\frac{8}{11} = \frac{4\sqrt{5}}{4\sqrt{5} + b}$$
$$32\sqrt{5} + 8b = 44\sqrt{5}$$
$$8b = 12\sqrt{5}$$
$$\mathbf{b = \frac{3\sqrt{5}}{2}}$$

24. (4)

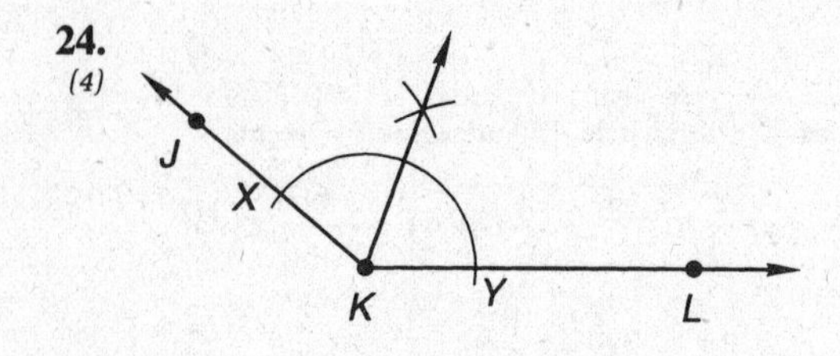

25. (4)

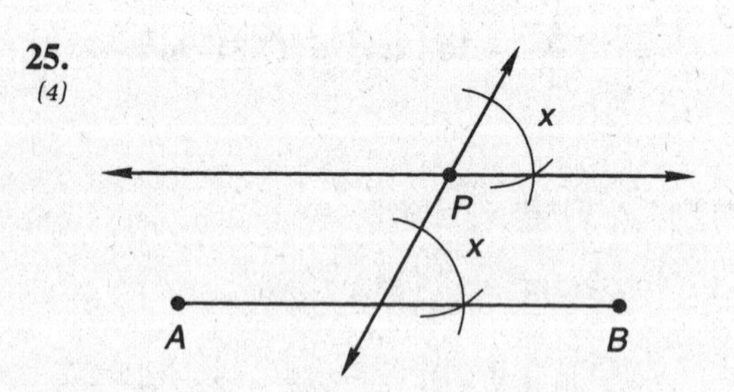

26. (4)

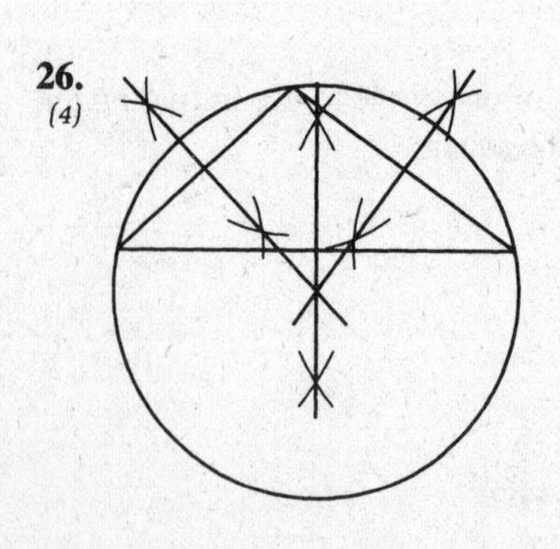

27. (2)
$$V_{\text{cone}} = \frac{1}{3}(A_{\text{base}})(h)$$
$$= \frac{1}{3}\left[\frac{1}{2}\left(\sqrt{8^2 - 6^2}\right)(6) + \frac{30}{360}\pi(6)^2\right](12)$$
$$= \frac{1}{3}\left[3\sqrt{28} + \frac{1}{12}(36)\pi\right](12)$$
$$= \frac{1}{3}(6\sqrt{7} + 3\pi)(12)$$
$$= \mathbf{(24\sqrt{7} + 12\pi)\ m^3 = 101.20\ m^3}$$

28. (2)
$$V_{\text{cone}} = \frac{1}{3}(A_{\text{base}})(h)$$
$$= \frac{1}{3}\left[6\left(\frac{1}{2}\right)(8)\left(4\sqrt{3}\right)\right](14)$$
$$= \frac{1}{3}(96\sqrt{3})(14)$$
$$= \mathbf{448\sqrt{3}\ cm^3 = 775.96\ cm^3}$$

$$l = \sqrt{(4\sqrt{3})^2 + (14)^2} = \sqrt{244} = 2\sqrt{61}$$

$$A_{\text{surface}} = A_{\text{base}} + 6A_{\text{side}}$$
$$= 6\left[\left(\frac{1}{2}\right)(8)\left(4\sqrt{3}\right)\right] + 6\left[\left(\frac{1}{2}\right)(8)\left(2\sqrt{61}\right)\right]$$
$$= \mathbf{(96\sqrt{3} + 48\sqrt{61})\ cm^2 = 541.17\ cm^2}$$

29. (1) A: $2y(x + z) = 2y(0 + z) = 2yz$

B: $x(y + z) = 0(y + z) = 0$

Both y and z are greater than 1, so $2yz$ must be postitive. Therefore the answer is **A.**

30. (1) A: $2(3) - 18\left(\frac{1}{6}\right) = 6 - 3 = 3$

B: $3(3) - 36\left(\frac{1}{6}\right) = 9 - 6 = 3$

A and B are equal. Therefore the answer is **C.**

Problem Set 6

1. (R) First we write the two equations.

$$\begin{cases} \text{(a)} & 0.1N_R + 0.2N_B = 24 \\ \text{(b)} & 3N_R = N_B + 20 \end{cases}$$

(b) $3N_R = N_B + 20$
$$N_B = 3N_R - 20$$

Now we can use substitution to solve.

(a) $$0.1N_R + 0.2N_B = 24$$

$$0.1N_R + 0.2(3N_R - 20) = 24$$

$$N_R + 6N_R - 40 = 240$$

$$7N_R = 280$$

$$\mathbf{N_R = 40}$$

$$N_B = 3N_R - 20$$

$$N_B = 3(40) - 20$$

$$\mathbf{N_B = 100}$$

2. (1)

$$4(180 - A) = 8(90 - A) + 28$$

$$720 - 4A = 720 - 8A + 28$$

$$4A = 28$$

$$A = \mathbf{7°}$$

3. (R) First we write the two equations.

$$\begin{cases} \text{(a)} & \dfrac{N_R}{N_G} = \dfrac{5}{16} \\ \text{(b)} & 6N_R = N_G + 112 \end{cases}$$

(b) $$6N_R = N_G + 112$$

$$N_G = 6N_R - 112$$

Now we can use substitution to solve.

(a) $$\frac{N_R}{N_G} = \frac{5}{16}$$

$$5N_G = 16N_R$$

$$5(6N_R - 112) = 16N_R$$

$$30N_R - 560 = 16N_R$$

$$14N_R = 560$$

$$\mathbf{N_R = 40}$$

$$N_G = 6N_R - 112$$

$$N_G = 6(40) - 112$$

$$\mathbf{N_G = 128}$$

4. (3)

$$c^2 \bigcirc a^2 + b^2$$

$$13^2 \bigcirc 5^2 + 12^2$$

$$169 \bigcirc 25 + 144$$

$$169 = 169$$

Since the square of the longest side is equal to the sum of the squares of the other two sides, the triangle is a **right triangle.**

5. (6)

$$\frac{4}{7} + \frac{3}{x + 3} = \frac{5}{3}$$

$$21(x + 3)\left(\frac{4}{7} + \frac{3}{x + 3}\right) = \left(\frac{5}{3}\right)(x + 3)21$$

$$3(4)(x + 3) + 21(3) = 7(5)(x + 3)$$

$$12x + 36 + 63 = 35x + 105$$

$$-23x = 6$$

$$x = \mathbf{-\frac{6}{23}}$$

6. (6)

$$\frac{5}{3} - \frac{2}{x - 4} = \frac{1}{2}$$

$$6(x - 4)\left(\frac{5}{3} - \frac{2}{x - 4}\right) = \left(\frac{1}{2}\right)(x - 4)6$$

$$2(5)(x - 4) - 6(2) = 3(x - 4)$$

$$10x - 40 - 12 = 3x - 12$$

$$7x = 40$$

$$x = \mathbf{\frac{40}{7}}$$

7. (6)

$$\frac{1}{x - 7} + \frac{1}{4} = \frac{1}{3}$$

$$12(x - 7)\left(\frac{1}{x - 7} + \frac{1}{4}\right) = \left(\frac{1}{3}\right)(x - 7)12$$

$$12 + 3(x - 7) = 4(x - 7)$$

$$x - 7 = 12$$

$$x = \mathbf{19}$$

8. (6)

$$\sqrt{s - 7} + \sqrt{s} = 7$$

$$\sqrt{s - 7} = 7 - \sqrt{s}$$

$$s - 7 = 49 - 14\sqrt{s} + s$$

$$14\sqrt{s} = 56$$

$$\sqrt{s} = 4$$

$$s = \mathbf{16}$$

9. (6)

$$\sqrt{s - 27} + \sqrt{s} = 9$$

$$\sqrt{s - 27} = 9 - \sqrt{s}$$

$$s - 27 = 81 - 18\sqrt{s} + s$$

$$18\sqrt{s} = 108$$

$$\sqrt{s} = 6$$

$$s = \mathbf{36}$$

10. (6) First, we write the two equations.

$$\begin{cases} \frac{3}{7}x + \frac{2}{5}y = 11 \\ 0.03x - 0.2y = -0.37 \end{cases}$$

Then we simplify the equations.

$$\begin{array}{l} \text{(a)} \\ \text{(b)} \end{array}\begin{cases} 15x + 14y = 385 \\ 3x - 20y = -37 \end{cases}$$

Now we can use elimination to solve.

$$\begin{array}{rrcl} \text{(a)} & 15x + 14y & = & 385 \\ -5\text{(b)} & -15x + 100y & = & 185 \\ \hline & 114y & = & 570 \\ & \mathbf{y} & = & \mathbf{5} \end{array}$$

Use this value for y in equation (a).

$$\begin{array}{rrcl} \text{(a)} & 15x + 14y & = & 385 \\ & 15x + 14(5) & = & 385 \\ & 15x + 70 & = & 385 \\ & 15x & = & 315 \\ & \mathbf{x} & = & \mathbf{21} \end{array}$$

11. (6) First, label the three equations.

$$\begin{array}{l} \text{(a)} \\ \text{(b)} \\ \text{(c)} \end{array}\begin{cases} 2x + 3y = -1 \\ x - 2z = -3 \\ 2y - z = -4 \end{cases}$$

Now we can use elimination to solve.

$$\begin{array}{rrcl} \text{(a)} & 2x + 3y & = & -1 \\ -2\text{(b)} & -2x \quad + 4z & = & 6 \\ \hline \text{(d)} & 3y + 4z & = & 5 \\ 4\text{(c)} & 8y - 4z & = & -16 \\ \hline & 11y & = & -11 \\ & \mathbf{y} & = & \mathbf{-1} \end{array}$$

Use this value for y in equations (a) and (c).

$$\begin{array}{rrcl} \text{(a)} & 2x + 3y & = & -1 \\ & 2x + 3(-1) & = & -1 \\ & 2x & = & 2 \\ & \mathbf{x} & = & \mathbf{1} \end{array}$$

$$\begin{array}{rrcl} \text{(c)} & 2y - z & = & -4 \\ & 2(-1) - z & = & -4 \\ & \mathbf{z} & = & \mathbf{2} \end{array}$$

12. (6) First, label the three equations.

$$\begin{array}{l} \text{(a)} \\ \text{(b)} \\ \text{(c)} \end{array}\begin{cases} x - 2y + z = -2 \\ 2x - 2y - z = -3 \\ x + y - 2z = 1 \end{cases}$$

Now we can use elimination to solve.

$$\begin{array}{rrcl} \text{(a)} & x - 2y + z & = & -2 \\ -1\text{(c)} & -x - y + 2z & = & -1 \\ \hline \text{(d)} & -3y + 3z & = & -3 \end{array}$$

$$\begin{array}{rrcl} 2\text{(a)} & 2x - 4y + 2z & = & -4 \\ -1\text{(b)} & -2x + 2y + z & = & 3 \\ \hline \text{(e)} & -2y + 3z & = & -1 \\ -1\text{(d)} & 3y - 3z & = & 3 \\ \hline & \mathbf{y} & = & \mathbf{2} \end{array}$$

Use this value for y in equation (d) to solve for z.

$$\begin{array}{rrcl} \text{(d)} & -3y + 3z & = & -3 \\ & -3(2) + 3z & = & -3 \\ & 3z & = & 3 \\ & \mathbf{z} & = & \mathbf{1} \end{array}$$

Use the values for y and z in equation (a) to solve for x.

$$\begin{array}{rrcl} \text{(a)} & x - 2y + z & = & -2 \\ & x - 2(2) + (1) & = & -2 \\ & \mathbf{x} & = & \mathbf{1} \end{array}$$

13. (5) $\sqrt[3]{ab^4}(ab^3)^{\frac{1}{5}}b^2 = a^{\frac{1}{3}}b^{\frac{4}{3}}a^{\frac{1}{5}}b^{\frac{3}{5}}b^2$

$= a^{\frac{1}{3}+\frac{1}{5}}b^{\frac{4}{3}+\frac{3}{5}+2} = \mathbf{a^{\frac{8}{15}}b^{\frac{59}{15}}}$

14. (5) $\dfrac{x^{\left(\frac{a}{3}-2\right)}y^{\left(\frac{b-2}{3}\right)}}{x^{2a}\left(y^{\frac{1}{3}}\right)^{2a}} = x^{\frac{a-6}{3}}y^{\frac{b-2}{3}}x^{-2a}y^{-\frac{2a}{3}}$

$= x^{\frac{a-6}{3}-2a}y^{\frac{b-2}{3}-\frac{2a}{3}} = \mathbf{x^{\frac{-5a-6}{3}}y^{\frac{b-2a-2}{3}}}$

15. (5) $2\sqrt{\frac{7}{3}} - \sqrt{\frac{3}{7}} - 2\sqrt{84}$

$= 2\sqrt{\frac{7}{3}\cdot\frac{3}{3}} - \sqrt{\frac{3}{7}\cdot\frac{7}{7}} - 2\cdot 2\sqrt{21}$

$= \left(\frac{2}{3}\sqrt{21} - \frac{1}{7}\sqrt{21} - 4\sqrt{21}\right)\frac{21}{21}$

$= \frac{\sqrt{21}}{21}(14 - 3 - 84) = \mathbf{\frac{-73\sqrt{21}}{21}}$

16. (5) $\dfrac{x^{-1} + y^{-1}}{x^{-1}y} = \dfrac{\frac{1}{x} + \frac{1}{y}}{\frac{y}{x}} = \left(\dfrac{x + y}{xy}\right)\left(\dfrac{x}{y}\right) = \mathbf{\dfrac{x + y}{y^2}}$

17. (5)

$$\begin{array}{r} \sqrt{2} - x \\ 2 - \sqrt{x} \\ \hline 2\sqrt{2} - 2x \\ -\sqrt{2x} + x\sqrt{x} \\ \hline \mathbf{2\sqrt{2} - 2x - \sqrt{2}\sqrt{x} + x\sqrt{x}} \end{array}$$

18. (5) $\sqrt{-3}\sqrt{3} - \sqrt{2}i - \sqrt{-3}\sqrt{-2} - 2$

$= (\sqrt{3}i)\sqrt{3} - \sqrt{2}i - (\sqrt{3}i)\sqrt{2}i - 2$

$= 3i - \sqrt{2}i - \sqrt{6}(ii) - 2$

$= 3i - \sqrt{2}i - (-1)\sqrt{6} - 2$

$= \mathbf{(\sqrt{6} - 2) + (3 - \sqrt{2})i}$

19. (5) $2i^3 + 3i^2 + 2i - 2\sqrt{2}i$

$= 2(ii)i + 3(ii) + 2i - 2\sqrt{2}i$

$= 2(-1)i + 3(-1) + 2i - 2\sqrt{2}i$

$= \mathbf{-3 - 2\sqrt{2}i}$

20. (5) First find BC.

$(BC)^2 = 4^2 + 5^2$

$(BC)^2 = 41$

$BC = \sqrt{41}$

Now find AC.

$(AC)^2 = (\sqrt{41})^2 + 3^2$

$(AC)^2 = 50$

$AC = \sqrt{50} = \mathbf{5\sqrt{2}}$

21. (4)

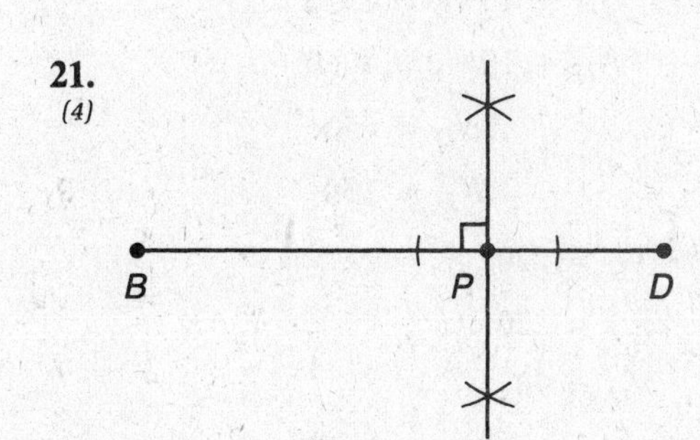

22. (4)

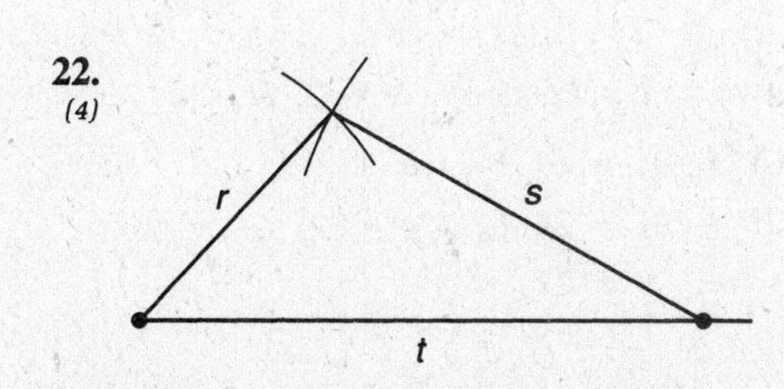

23. (4)

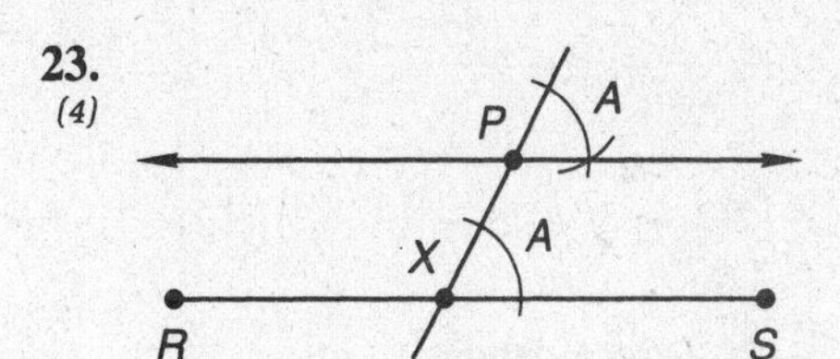

24. (3) First we find the scale factor in both directions.

$\overleftarrow{SF} \cdot 5 = 3$

$\overleftarrow{SF} = \frac{3}{5} \rightarrow \overrightarrow{SF} = \frac{5}{3}$

Now we solve for a and b.

$a = 6 \cdot \overleftarrow{SF}$ $\quad$ $b = 4 \cdot \overrightarrow{SF}$

$a = 6 \cdot \frac{3}{5}$ $\quad$ $b = 4 \cdot \frac{5}{3}$

$\mathbf{a = \frac{18}{5}}$ $\quad$ $\mathbf{b = \frac{20}{3}}$

25. (3) First, separate the triangles.

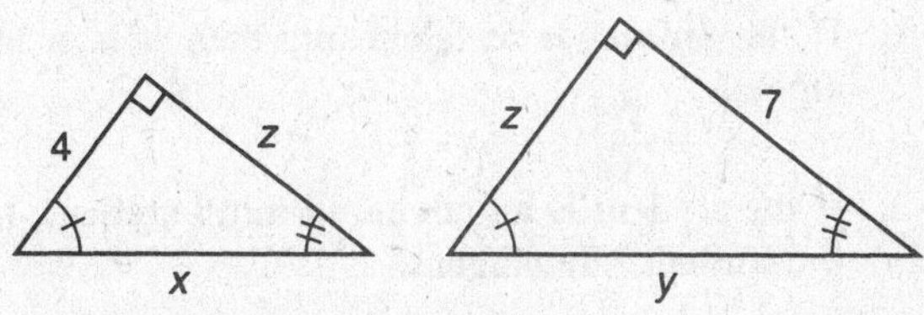

All three triangles are similar. The ratios of sides of similar triangles are equal. We use this fact to solve for x, y, and z.

$\frac{x}{11} = \frac{4}{x}$ $\quad$ $\frac{y}{11} = \frac{7}{y}$ $\quad$ $\frac{z}{4} = \frac{7}{z}$

$x^2 = 44$ $\quad$ $y^2 = 77$ $\quad$ $z^2 = 28$

$\mathbf{x = 2\sqrt{11}}$ $\quad$ $\mathbf{y = \sqrt{77}}$ $\quad$ $\mathbf{z = 2\sqrt{7}}$

26. (1) $A_{\text{shaded region}} = A_{\text{total}} - A_{\text{unshaded}}$

$= \pi(6)^2 - \frac{90}{360}\pi(6)^2 - 2\left(\frac{1}{2}\right)\pi(3)^2$

$= \pi(6)^2 - \frac{1}{4}\pi(6)^2 - \pi(3)^2 = \frac{3}{4}\pi(36) - \pi(9)$

$= \mathbf{18\pi\ cm^2 = 56.55\ cm^2}$

27. (2) $V_{\text{pyramid}} = \frac{1}{3}(A_{\text{base}})(h)$

$= \frac{1}{3}[2(2 + 3 + 2) + (3 \cdot 2)](12)$

$= \frac{1}{3}(20)(12) = \mathbf{80\ cm^3}$

28. (2) $A_{\text{surface}} = A_{\text{base}} + 3(A_{\text{side}})$

$= \frac{1}{2}(2)(\sqrt{3}) + 3\left[\frac{1}{2}(2)(5)\right]$

$= \sqrt{3} + 15 = \mathbf{16.73\ m^2}$

29. (1) A: $\dfrac{\frac{24 + 3}{4}}{\frac{8 - 5}{4}} = \frac{27}{4} \cdot \frac{4}{3} = 9$

B: $3^2 = 9$

A and B are equal. Therefore the answer is **C.**

30. (1) A: $\dfrac{\sqrt{0.49} + \frac{3}{4} + 0.8}{3} = \dfrac{0.7 + 0.75 + 0.8}{3} = 0.75$

B: $75\% = 0.75$

A and B are equal. Therefore the answer is **C.**

Problem Set 7

1. (7) **If the animal is an elephant, then it is a large animal.**

2. (7) **If the student is an advanced math student, then the student is intelligent.**

3. (7) **If an animal is a seal, then it is not green.**

4. (7) **If the coach is not happy, then the team did not win.**

5. (7) The argument is **invalid.** The minor premise identifies Lori as having a general property which, according to the major premise, she shares with a particular set of members. This does not constitute Lori as being a member of this set.

6. (7) **If the motor is on, then the car is moving.** The contrapositive of the major premise identifies a property about the set of motors that are on, and the minor premise identifies this motor as a member of the set mentioned in the contrapositive. That makes the property associated with the set neccessary. The conclusion explicitly states that this same property is true. Therefore, the argument is **valid.**

7. (R) First we write the two equations.

$$\begin{cases} \text{(a)} & \dfrac{N_R}{N_G} = \dfrac{4}{11} \\ \text{(b)} & 0.6N_G = N_R + 52 \end{cases}$$

(a) $\dfrac{N_R}{N_G} = \dfrac{4}{11}$

$11N_R = 4N_G$

(b) $0.6N_G = N_R + 52$

$N_R = 0.6N_G - 52$

Now we can use substitution to solve.

(a) $11N_R = 4N_G$

$11(0.6N_G - 52) = 4N_G$

$6.6N_G - 572 = 4N_G$

$2.6N_G = 572$

$\mathbf{N_G = 220}$

$11N_R = 4N_G$

$11N_R = 4(220)$

$\mathbf{N_R = 80}$

8. (R) First we write the two equations.

$$\begin{cases} \text{(a)} & N_N + N_D = 220 \\ \text{(b)} & 5N_N + 10N_D = 1750 \end{cases}$$

(a) $N_N + N_D = 220$

$N_N = 220 - N_D$

Now we can use substitution to solve.

(b) $5N_N + 10N_D = 1750$

$5(220 - N_D) + 10N_D = 1750$

$1100 - 5N_D + 10N_D = 1750$

$5N_D = 650$

$\mathbf{N_D = 130}$

(a) $N_N + N_D = 220$

$N_N + (130) = 220$

$\mathbf{N_N = 90}$

9. (R) Consecutive even integers: N, $N + 2$, $N + 4$

$6[N + (N + 4)] = 24 + 11(N + 2)$

$12N + 24 = 24 + 11N + 22$

$N = 22$

22, 24, 26

10. (6)
$$\frac{2}{5} - \frac{3}{x-4} = \frac{5}{6}$$
$$30(x-4)\left(\frac{2}{5} - \frac{3}{x-4}\right) = \left(\frac{5}{6}\right)(x-4)30$$
$$12(x-4) - 90 = 25(x-4)$$
$$-90 = 13(x-4)$$
$$-90 = 13x - 52$$
$$13x = -38$$
$$\mathbf{x = -\frac{38}{13}}$$

11. (6) First we write the two equations.
$$\begin{cases} \frac{4}{5}x - \frac{2}{3}y = \frac{7}{30} \\ 0.01x + 0.1y = 0.03 \end{cases}$$

Then we simplify the equations.
$$\begin{matrix}(a)\\(b)\end{matrix}\begin{cases} 24x - 20y = 7 \\ x + 10y = 3 \end{cases}$$

Now we can use elimination to solve for x.
$$\begin{array}{lrcl} (a) & 24x - 20y & = & 7 \\ 2(b) & 2x + 20y & = & 6 \\ \hline & 26x & = & 13 \end{array}$$
$$\mathbf{x = \frac{1}{2}}$$

Use this value for x in equation (b) to solve for y.
$$(b) \quad x + 10y = 3$$
$$\left(\frac{1}{2}\right) + 10y = 3$$
$$10y = \frac{5}{2}$$
$$\mathbf{y = \frac{1}{4}}$$

12. (6)
$$\sqrt{2x-7} + \sqrt{25} = 7$$
$$\sqrt{2x-7} + 5 = 7$$
$$\sqrt{2x-7} = 2$$
$$2x - 7 = 4$$
$$2x = 11$$
$$\mathbf{x = \frac{11}{2}}$$

13. (6) First label the three equations.
$$\begin{matrix}(a)\\(b)\\(c)\end{matrix}\begin{cases} 2x + 3y - z = 6 \\ 3x - y + z = 1 \\ x + y + z = 1 \end{cases}$$

Now we can use elimination to solve.
$$\begin{array}{lrcl} (a) & 2x + 3y - z & = & 6 \\ (b) & 3x - y + z & = & 1 \\ \hline (d) & 5x + 2y & = & 7 \end{array}$$
$$\begin{array}{lrcl} (a) & 2x + 3y - z & = & 6 \\ (c) & x + y + z & = & 1 \\ \hline (e) & 3x + 4y & = & 7 \\ -2(d) & -10x - 4y & = & -14 \\ \hline & -7x & = & -7 \end{array}$$
$$\mathbf{x = 1}$$

Use this value for x in equation (e) to solve for y.
$$(e) \quad 3x + 4y = 7$$
$$3(1) + 4y = 7$$
$$\mathbf{y = 1}$$

Use the values for x and y in equation (c) to solve for z.
$$(c) \quad x + y + z = 1$$
$$(1) + (1) + z = 1$$
$$\mathbf{z = -1}$$

14. (5)
$$yx^{\frac{2}{3}}\sqrt[3]{x^2}\left(x^2y\right)^{\frac{1}{5}} = y^1x^{\frac{2}{3}}x^{\frac{2}{3}}x^{\frac{2}{5}}y^{\frac{1}{5}}$$
$$= x^{\frac{2}{3}+\frac{2}{3}+\frac{2}{5}}y^{\frac{1}{5}+1} = \mathbf{x^{\frac{26}{15}}y^{\frac{6}{5}}}$$

15. (5)
$$\begin{array}{r} \sqrt{2} + \sqrt{x} \\ \sqrt{2} + \sqrt{x} \\ \hline 2 + \sqrt{2x} \quad\quad \\ + \sqrt{2x} + x \\ \hline \mathbf{2 + 2\sqrt{2}\sqrt{x} + x} \end{array}$$

16. (5)
$$\frac{a^{-2} + b^{-1}}{a^{-1}b} = \frac{\frac{1}{a^2} + \frac{1}{b}}{\frac{b}{a}} = \left(\frac{b + a^2}{a^2b}\right)\left(\frac{a}{b}\right)$$
$$= \mathbf{\frac{b + a^2}{ab^2}}$$

17. (5)
$$\left(x^{\frac{1}{2}} + y^{\frac{1}{2}}\right)^2 = x^{\frac{1}{2}}\left(x^{\frac{1}{2}} + y^{\frac{1}{2}}\right) + y^{\frac{1}{2}}\left(x^{\frac{1}{2}} + y^{\frac{1}{2}}\right)$$
$$= \left(x^{\frac{1}{2}}\right)^2 + 2\left(x^{\frac{1}{2}}\right)\left(y^{\frac{1}{2}}\right) + \left(y^{\frac{1}{2}}\right)^2$$
$$= \mathbf{x + 2x^{\frac{1}{2}}y^{\frac{1}{2}} + y}$$

18. (5) $\sqrt{2}\sqrt{-2} + \sqrt{-3}\sqrt{-3} + \sqrt{-4} - i$

$= \sqrt{2}\sqrt{2}i + \sqrt{3}\sqrt{3}(ii) + \sqrt{4}i - i$

$= 2i + 3(-1) + 2i - i$

$= \mathbf{-3 + 3i}$

19. (5) $(3 - i + \sqrt{-4})(2 - i - \sqrt{-9})$

$= (3 - i + 2i)(2 - i - 3i)$

$= (3 + i)(2 - 4i)$

$= 3(2 - 4i) + i(2 - 4i)$

$= 6 - 12i + 2i - 4(ii)$

$= 6 - 12i + 2i - 4(-1)$

$= \mathbf{10 - 10i}$

20. (5) $3i^4 - 2i^3 + 4i - 5\sqrt{-16}$

$= 3(ii)(ii) - 2(ii)i + 4i - 5\sqrt{16}\,i$

$= 3(-1)(-1) - 2(-1)i + 4i - 5(4i)$

$= 3 + 2i + 4i - 20i$

$= \mathbf{3 - 14i}$

21. (5) $\dfrac{\text{Side}_L}{\text{Side}_S} = \dfrac{3}{1}$

$\dfrac{\text{Area}_L}{\text{Area}_S} = \left(\dfrac{\text{Side}_L}{\text{Side}_S}\right)^2 = \left(\dfrac{3}{1}\right)^2 = \mathbf{\dfrac{9}{1}}$

22. (5) $(AB)^2 = 4^2 + m^2$

$(AB)^2 = 16 + m^2$

$AB = \sqrt{16 + m^2}$

$(AC)^2 = \left(\sqrt{16 + m^2}\right)^2 + 2^2$

$(AC)^2 = 16 + m^2 + 4$

$AC = \mathbf{\sqrt{20 + m^2}}$

23. (4)

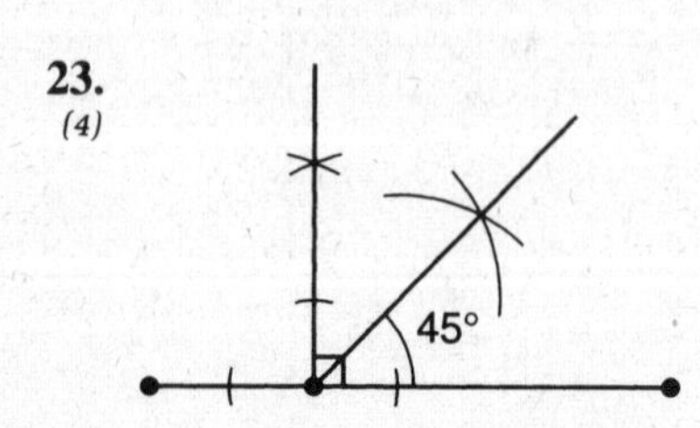

24. (3) First, separate the triangles.

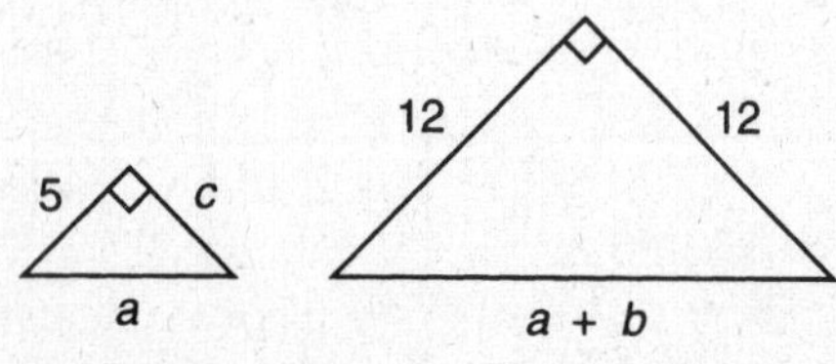

Use the ratios for similar triangles to solve for c.

$\dfrac{5}{12} = \dfrac{c}{12}$

$\mathbf{c = 5}$

Solve for a using the Pythagorean theorem.

$a = \sqrt{5^2 + c^2}$

$a = \sqrt{5^2 + 5^2}$

$\mathbf{a = 5\sqrt{2}}$

Now use ratios for similar triangles to solve for b.

$\dfrac{a}{a + b} = \dfrac{5}{12}$

$\dfrac{5\sqrt{2}}{5\sqrt{2} + b} = \dfrac{5}{12}$

$60\sqrt{2} = 25\sqrt{2} + 5b$

$5b = 35\sqrt{2}$

$\mathbf{b = 7\sqrt{2}}$

25. (4)

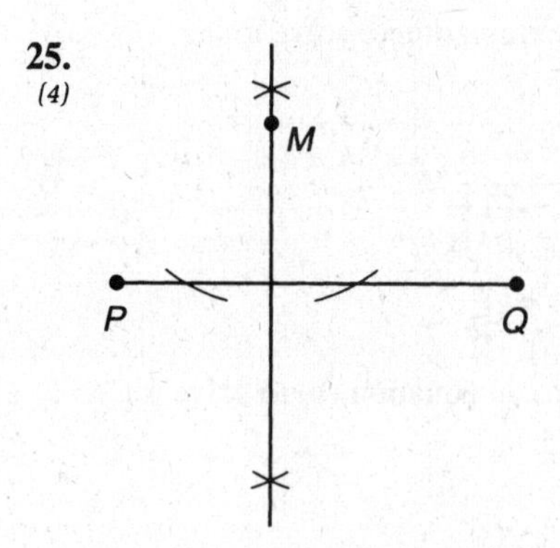

26. (1) $A_{\text{shaded}} = \dfrac{360 - 162}{360}\pi(2\sqrt{15})^2$

$= \dfrac{198}{360}\pi(60)$

$= \mathbf{33\pi\ m^2 = 103.67\ m^2}$

27. (1)

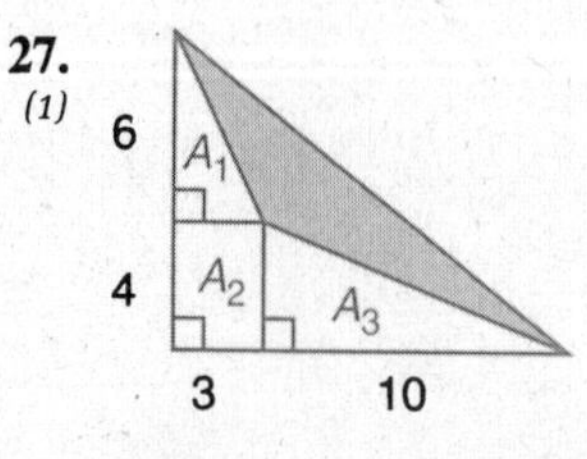

$A_{\text{shaded}} = A_{\text{big}\,\Delta} - A_1 - A_2 - A_3$

$= \dfrac{1}{2}(4 + 6)(3 + 10) - \dfrac{1}{2}(3)(6) - (3)(4) - \dfrac{1}{2}(4)(10)$

$= 65 - 9 - 12 - 20 = \mathbf{24\ m^2}$

28. (2)
$$V_{\text{slice}} = (A_{\text{base}})(h) = \left[\frac{45}{360}\pi(4)^2\right](16) = \mathbf{32\pi\ cm^3} = \mathbf{100.53\ cm^3}$$

29. (1)
$$a - b = a + b$$
$$-b = b$$

A: $b = 0$

B: 1

We eliminate a to solve for b. The only case where $-b = b$ is when b is equal to zero. Zero is less than 1, therefore the answer is **B.**

30. (1) A: $4\pi(1)^2\text{ cm}^2 = 4\pi\text{ cm}^2$

B: $\pi(2)^2\text{ cm}^2 = 4\pi\text{ cm}^2$

A and B are equal. Therefore the answer is **C.**

Problem Set 8

1. (R) First we write the two equations.

$$\begin{cases} \text{(a)}\ 0.30N_G + 0.15N_P = 21 \\ \text{(b)}\ 4N_P = 10 + 3N_G \end{cases}$$

(a) $0.30N_G + 0.15N_P = 21$
$$2N_G + N_P = 140$$
$$N_P = 140 - 2N_G$$

Now we can use substitution to solve.

(b) $4N_P = 10 + 3N_G$
$$4(140 - 2N_G) = 10 + 3N_G$$
$$11N_G = 550$$
$$\mathbf{N_G = 50}$$

(b) $4N_P = 10 + 3N_G$
$$4N_P = 10 + 3(50)$$
$$\mathbf{N_P = 40}$$

2. (R) First we write the two equations.

$$\begin{cases} \text{(a)}\ \dfrac{N_W}{N_H} = \dfrac{7}{3} \\ \text{(b)}\ 3N_H = 2N_W - 40 \end{cases}$$

(a) $\dfrac{N_W}{N_H} = \dfrac{7}{3}$
$$N_H = \frac{3}{7}N_W$$

Now we can use substitution to solve.

(b) $3N_H = 2N_W - 40$
$$3\left(\frac{3}{7}N_W\right) = 2N_W - 40$$
$$\frac{5}{7}N_W = 40$$
$$N_W = \mathbf{56}$$

3. (R) First we write the two equations.

$$\begin{cases} \text{(a)}\ N_N + N_Q = 173 \\ \text{(b)}\ 5N_N + 25N_Q = 1125 \end{cases}$$

(a) $N_N + N_Q = 173$
$$N_N = 173 - N_Q$$

Now we can use substitution to solve.

(b) $5N_N + 25N_Q = 1125$
$$5(173 - N_Q) + 25N_Q = 1125$$
$$20N_Q = 260$$
$$\mathbf{N_Q = 13}$$

(a) $N_N + N_Q = 173$
$$N_N + (13) = 173$$
$$\mathbf{N_N = 160}$$

4. (1) First we write the two equations.

$$\begin{cases} \text{(a)}\ 2(180 - A) + 3(180 - B) = 500 \\ \text{(b)}\ B = 10 + A \end{cases}$$

Now we can use substitution to solve.

(a) $2(180 - A) + 3(180 - B) = 500$
$$360 - 2A + 540 - 3B = 500$$
$$-2A - 3B = -400$$
$$2A + 3B = 400$$
$$2A + 3(10 + A) = 400$$
$$5A = 370$$
$$\mathbf{A = 74°}$$

(b) $B = 10 + A$
$$B = 10 + (74)$$
$$\mathbf{B = 84°}$$

5. (7) **Invalid.** The major premise identifies a property of the set of all authors but the minor premise does not identify John as a member of that set.

6. The major premise identifies a property of the set of all reds, and the minor premise identifies Joe Bob as a member of that set. That means that he has the specific property that the major premise indicates. The conclusion explicitly claims Joe Bob has this same property. Therefore the argument is **valid.**
(7)

7. **If it is blue, then it is a dog.** The contrapositive of the major premise identifies a property about the set of things that are blue, and the minor premise identifies Jim as a member of the set mentioned in the contrapositive. That means that he has the specific property that the major premise indicates. The conclusion claims that Jim has this exact property. Therefore, the argument is **valid.**
(7)

8. Consecutive odd integers: N, $N + 2$, $N + 4$
(7)

$$3[N + (N + 2)] = 4(N + 4)$$
$$6N + 6 = 4N + 16$$
$$2N = 10$$
$$N = 5$$

5, 7, 9

9. First, find similar triangles.
(8)

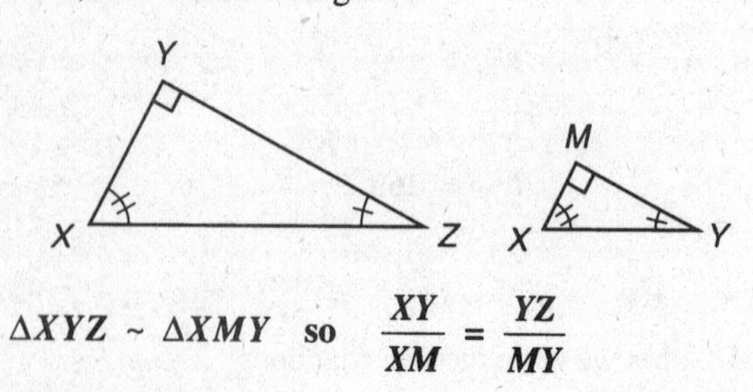

$$\Delta XYZ \sim \Delta XMY \quad \text{so} \quad \frac{XY}{XM} = \frac{YZ}{MY}$$

10. (8)

$$\frac{3x - 1}{12} = \frac{x + 11}{21}$$
$$21(3x - 1) = 12(x + 11)$$
$$63x - 21 = 12x + 132$$
$$51x = 153$$
$$x = \mathbf{3}$$

11. (8)

$$\frac{8}{x} = \frac{12}{8}$$
$$12x = 64$$
$$x = \frac{64}{12}$$
$$x = \frac{\mathbf{16}}{\mathbf{3}}$$

12. First, label the three equations.
(6)

$$\begin{aligned} &(a)\\ &(b)\\ &(c) \end{aligned} \begin{cases} x - y = 1 \\ y - 2z = 1 \\ 3x - 4z = 7 \end{cases}$$

Now we can use elimination to solve.

$$\begin{array}{lrcl} \text{(a)} & x - y & = & 1 \\ \text{(b)} & y - 2z & = & 1 \\ \hline \text{(d)} & x - 2z & = & 2 \end{array}$$

$$\begin{array}{lrcl} \text{(c)} & 3x - 4z & = & 7 \\ -3\text{(d)} & -3x + 6z & = & -6 \\ \hline & 2z & = & 1 \\ & z & = & \mathbf{\frac{1}{2}} \end{array}$$

Use this value for z in equations (b) and (d).

(b) $y - 2z = 1$ (d) $x - 2z = 2$

$$y - 2\left(\frac{1}{2}\right) = 1 \qquad x - 2\left(\frac{1}{2}\right) = 2$$

$$\mathbf{y = 2} \qquad \mathbf{x = 3}$$

13. First, label the three equations.
(6)

$$\begin{aligned} &(a)\\ &(b)\\ &(c) \end{aligned} \begin{cases} 2x + 2y - z = 9 \\ 3x + 3y + z = 16 \\ x - 2y = -1 \end{cases}$$

Now we can use elimination to solve.

$$\begin{array}{lrcl} \text{(a)} & 2x + 2y - z & = & 9 \\ \text{(b)} & 3x + 3y + z & = & 16 \\ \hline \text{(d)} & 5x + 5y & = & 25 \end{array}$$

$$\begin{array}{lrcl} \text{(c)} & x - 2y & = & -1 \\ -\frac{1}{5}\text{(d)} & -x - y & = & -5 \\ \hline & -3y & = & -6 \\ & \mathbf{y} & = & \mathbf{2} \end{array}$$

Use this value for y in equation (d) to solve for x.

(d) $$5x + 5y = 25$$
$$5x + 5(2) = 25$$
$$\mathbf{x = 3}$$

Use the values for x and y in equation (b) to solve for z.

(b) $$3x + 3y + z = 16$$
$$3(3) + 3(2) + z = 16$$
$$\mathbf{z = 1}$$

14. First we write the two equations.
(6)

$$\begin{cases} \frac{2}{3}x + \frac{1}{4}y = \frac{49}{12} \\ 0.05x + 0.2y = 0.85 \end{cases}$$

Then we simplify the equations.

$$\begin{aligned} &(a)\\ &(b) \end{aligned} \begin{cases} 8x + 3y = 49 \\ 5x + 20y = 85 \end{cases}$$

Now we can use elimination to solve.

$$\begin{aligned} -5(a) \quad -40x - 15y &= -245 \\ 8(b) \quad 40x + 160y &= 680 \\ \hline 145y &= 435 \\ \mathbf{y} &= \mathbf{3} \end{aligned}$$

Use this value for y in equation (a).

$$\begin{aligned} (a) \quad 8x + 3y &= 49 \\ 8x + 3(3) &= 49 \\ 8x &= 40 \\ \mathbf{x} &= \mathbf{5} \end{aligned}$$

15. (6)

$$\begin{aligned} \frac{3}{2} + \frac{4}{x+1} &= \frac{1}{3} \\ 2(3)(x+1)\left(\frac{3}{2} + \frac{4}{x+1}\right) &= \left(\frac{1}{3}\right)(2)(3)(x+1) \\ 3(3)(x+1) + 4(6) &= 2(x+1) \\ 9x + 9 + 24 &= 2x + 2 \\ 7x &= -31 \\ x &= -\mathbf{\frac{31}{7}} \end{aligned}$$

16. (6)

$$\begin{aligned} \sqrt{3x+1} + \sqrt{9} &= 7 \\ \sqrt{3x+1} + 3 &= 7 \\ \sqrt{3x+1} &= 4 \\ 3x + 1 &= 16 \\ 3x &= 15 \\ x &= \mathbf{5} \end{aligned}$$

17. (5) $5\sqrt{\frac{5}{2}} - 6\sqrt{\frac{2}{5}} + 3\sqrt{40}$

$$= 5\sqrt{\frac{5}{2} \cdot \frac{2}{2}} - 6\sqrt{\frac{2}{5} \cdot \frac{5}{5}} + 3 \cdot 2\sqrt{10}$$

$$= \frac{5}{2}\sqrt{10} - \frac{6}{5}\sqrt{10} + 6\sqrt{10}$$

$$= \frac{25}{10}\sqrt{10} - \frac{12}{10}\sqrt{10} + \frac{60}{10}\sqrt{10} = \mathbf{\frac{73\sqrt{10}}{10}}$$

18. (5) $\left(3x^{\frac{a}{2}} + 2y^{\frac{b}{3}}\right)\left(2x^{\frac{a}{2}} - 2y^{\frac{b}{3}}\right)$

$$= 3x^{\frac{a}{2}}\left(2x^{\frac{a}{2}} - 2y^{\frac{b}{3}}\right) + 2y^{\frac{b}{3}}\left(2x^{\frac{a}{2}} - 2y^{\frac{b}{3}}\right)$$

$$= 6x^{a} - 6x^{\frac{a}{2}}y^{\frac{b}{3}} + 4x^{\frac{a}{2}}y^{\frac{b}{3}} - 4y^{\frac{2b}{3}}$$

$$= \mathbf{6x^{a} - 2x^{\frac{a}{2}}y^{\frac{b}{3}} - 4y^{\frac{2b}{3}}}$$

19. (5) $\dfrac{x^{\left(\frac{a}{4}+1\right)}y^{\left(\frac{c+3}{2}\right)}}{\left(x^2\right)^b y^{-c+1}} = x^{\frac{a}{4}+1}x^{-2b}y^{\frac{c+3}{2}}y^{c-1}$

$$= x^{\frac{a}{4}-2b+1}y^{\left(\frac{c+3}{2}\right)+\left(\frac{2c-2}{2}\right)} = \mathbf{x^{\frac{a-8b+4}{4}}y^{\frac{3c+1}{2}}}$$

20. (5) $\sqrt{3}\sqrt{-3} - \sqrt{-2}\sqrt{-2} + \sqrt{-4} + 5i + \sqrt{-3}$

$$= \sqrt{3}\sqrt{3}i - \sqrt{2}i\sqrt{2}i + \sqrt{4}i + 5i + \sqrt{3}i$$

$$= 3i - 2(ii) + 2i + 5i + \sqrt{3}i$$

$$= 10i + 2 + \sqrt{3}i$$

$$= \mathbf{2 + \left(10 + \sqrt{3}\right)i}$$

21. (5) $7i^6 - 3i^4 + 5i^3 + \sqrt{16}\,i^5$

$$= 7(ii)(ii)(ii) - 3(ii)(ii) + 5(ii)i + 4(ii)(ii)i$$

$$= 7(-1)(-1)(-1) - 3(-1)(-1) + 5(-1)i + 4(-1)(-1)i$$

$$= -7 - 3 - 5i + 4i$$

$$= \mathbf{-10 - i}$$

22. (4)

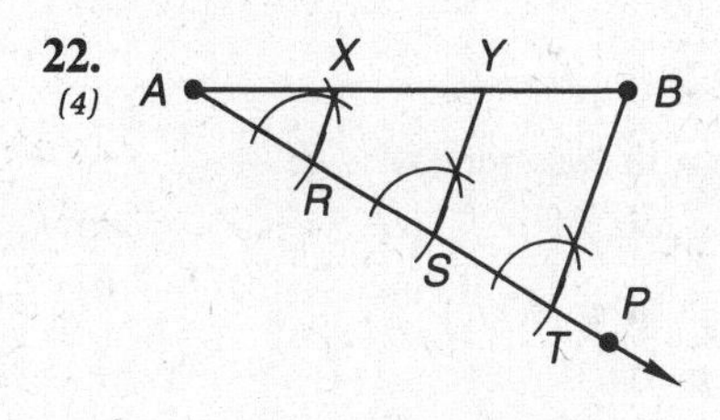

23. (4)

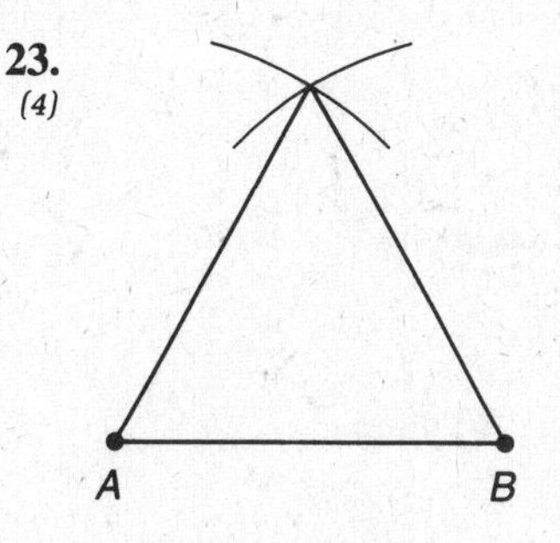

24. (4)

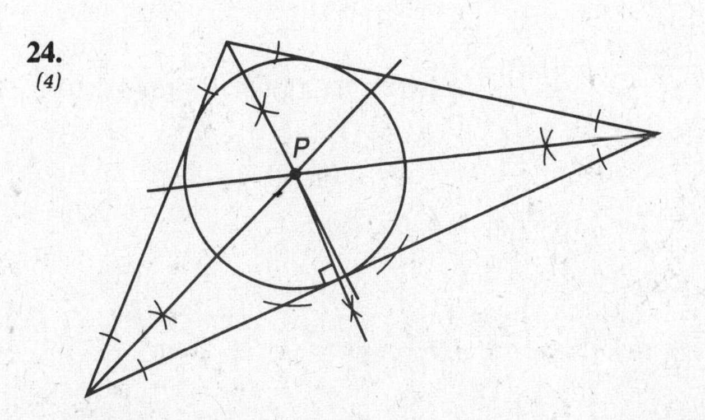

25. (3) First we find the scale factor in both directions.

$$\overrightarrow{SF} \cdot 17 = \frac{85}{7}$$

$$\overrightarrow{SF} = \frac{5}{7} \rightarrow \overleftarrow{SF} = \frac{7}{5}$$

Now we solve for a and b.

$$a = 5 \cdot \overleftarrow{SF} \qquad b = 14 \cdot \overrightarrow{SF}$$

$$a = 5\left(\frac{7}{5}\right) \qquad b = 14\left(\frac{5}{7}\right)$$

$$\mathbf{a = 7} \qquad \mathbf{b = 10}$$

26. (1) First we write two equations.

$$\begin{cases} \text{(a)} & 3x - 2y = 40 \\ \text{(b)} & 40 + 2x + 4y + 60 = 180 \end{cases}$$

Next we simplify (b), then we use elimination to solve.

$$\text{(b)}\quad 40 + 2x + 4y + 60 = 180$$
$$2x + 4y = 80$$
$$x + 2y = 40$$

$$x + 2y = 40$$
$$\text{(a)}\quad 3x - 2y = 40$$
$$4x = 80$$
$$\mathbf{x = 20}$$

Use this value for x in equation (a).

$$\text{(a)}\quad 3x - 2y = 40$$
$$3(20) - 2y = 40$$
$$2y = 20$$
$$\mathbf{y = 10}$$

27. (2) $V = \frac{4}{3}\pi r^3 = \frac{4}{3}\pi(4)^3$

$$\mathbf{V = \frac{256\pi}{3}\ cm^3 = 268.08\ cm^3}$$

$$A_{\text{surface}} = 4\pi r^2 = 4\pi(4^2)$$

$$\mathbf{A_{surface} = 64\pi\ cm^2 = 201.06\ cm^2}$$

28. (2) $V_{\text{cone}} = \frac{1}{3}(A_{\text{base}})(h)$

$$= \left(\frac{1}{3}\right)\left\{\frac{1}{2}[\pi(2)^2] + (4)(4) + \frac{1}{2}(3)(4)\right\}(10)$$

$$= \frac{10}{3}(2\pi + 16 + 6)$$

$$= \frac{10}{3}(2\pi + 22)$$

$$= \mathbf{\left(\frac{220 + 20\pi}{3}\right) cm^3 = 94.28\ cm^3}$$

29. (R) We are given that $a * b = a - 3b$. To evaluate $4 * 3$, we must recognize that a corresponds to 4 and that b corresponds to 3.

If $a * b = a - 3b$, $4 * 3 = 4 - 3(3) = -5$.

Since $4 * 3 = -5$, $(4 * 3) \# 5 = -5 \# 5$.

To evaluate $-5 \# 5$, we see that $a \# b = 2a + 3b$, and recognize that, in this case, a corresponds to -5 and b corresponds to 5.

If $a \# b = 2a + 3b$, $-5 \# 5 = 2(-5) + 3(5) = 5$.

So, $(4 * 3) \# 5 = \mathbf{5}$.

30. (R) We are given that $x * y = 5x + 4y$. To evaluate $4 * 3$, we must recognize that x corresponds to 4 and that y corresponds to 3.

If $x * y = 5x + 4y$, $4 * 3 = 5(4) + 4(3) = 32$.

Since $4 * 3 = 32$, $2 \# (4 * 3) = 2 \# 32$.

To evaluate $2 \# 32$, we see that $x \# y = 2x + 3y$, and recognize that, in this case, x corresponds to 2 and y corresponds to 32.

If $x \# y = 2x + 3y$, $2 \# 32 = 2(2) + 3(32) = 100$.

So, $2 \# (4 * 3) = \mathbf{100}$.

Problem Set 9

1. (R) First we write the two equations.

$$\begin{cases} \text{(a)} & \frac{N_G}{N_B} = \frac{2}{5} \\ \text{(b)} & N_B = 2N_G + 40 \end{cases}$$

Now we can use substitution to solve.

$$\text{(a)}\quad \frac{N_G}{N_B} = \frac{2}{5}$$
$$5N_G = 2N_B$$
$$5N_G = 2(2N_G + 40)$$
$$5N_G = 4N_G + 80$$
$$\mathbf{N_G = 80}$$

$$\text{(b)}\quad N_B = 2N_G + 40$$
$$N_B = 2(80) + 40$$
$$\mathbf{N_B = 200}$$

2. (R) First we write the two equations.

$$\begin{cases} \text{(a)} & 0.14N_B + 0.20N_G = 54 \\ \text{(b)} & 5N_B = 2N_G + 100 \end{cases}$$

Now we can use elimination to solve.

$$\begin{aligned} (100)(a) \quad 14N_B + 20N_G &= 5400 \\ (10)(b) \quad 50N_B &= 20N_G + 1000 \\ \hline 64N_B + 20N_G &= 20N_G + 6400 \\ 64N_B &= 6400 \\ \mathbf{N_B} &= \mathbf{100} \end{aligned}$$

$$\begin{aligned} (b) \quad 5N_B &= 2N_G + 100 \\ 5(100) &= 2N_G + 100 \\ 2N_G &= 400 \\ \mathbf{N_G} &= \mathbf{200} \end{aligned}$$

3. (7) **If two angles of a triangle are congruent, then the sides opposite these angles are congruent.**

4. (7) **Invalid.** The major premise identifies a property about the set of all educated people and the set of people who don't know algebra. The minor premise fails to identify Lao-Tzu as a member of either set.

5. (7) **If the light is on, then the switch is not on. Invalid.** The contrapositive states that the switch is not on if the light is on. The minor premise defines the light as on, so the switch cannot be on, which is contrary to the claim of the conclusion.

6. (7) Consecutive integers: N, $N + 1$, $N + 2$, $N + 3$

$$\begin{aligned} 3[N + (N + 3)] &= -5[N + (N + 1)] - 114 \\ 3(2N + 3) &= -5(2N + 1) - 114 \\ 6N + 9 &= -10N - 5 - 114 \\ 16N &= -128 \\ N &= -8 \end{aligned}$$

−8, −7, −6, −5

7. (5) $\dfrac{\text{Side}_L}{\text{Side}_S} = \dfrac{4}{3}$

$$\frac{\text{Area}_S}{\text{Area}_L} = \left(\frac{\text{Side}_S}{\text{Side}_L}\right)^2 = \left(\frac{3}{4}\right)^2 = \mathbf{\frac{9}{16}}$$

8. (3)

$c^2 \bigcirc a^2 + b^2$

$7^2 \bigcirc 4^2 + 5^2$

$49 \bigcirc 16 + 25$

$49 > 41$

Obtuse triangle

9. (9)
(a) ***AAAS* congruency postulate**
(b) ***SAS* congruency postulate**
(c) ***SSS* congruency postulate**
(d) ***HL* congruency postulate**

10. (9)

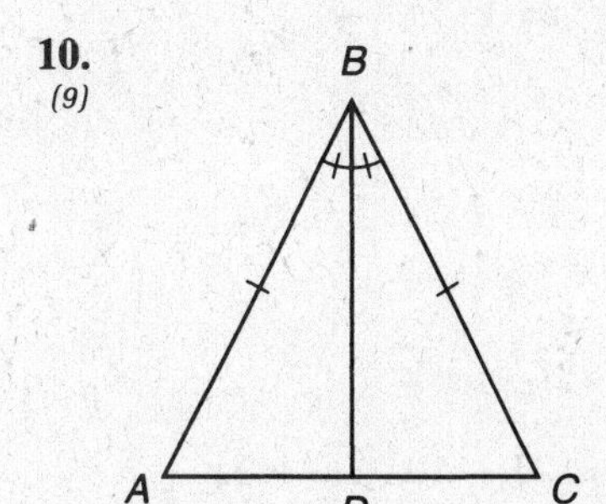

$\mathbf{\Delta ABD \cong \Delta CBD}$ **by SAS congruency postulate.**

11. (9)

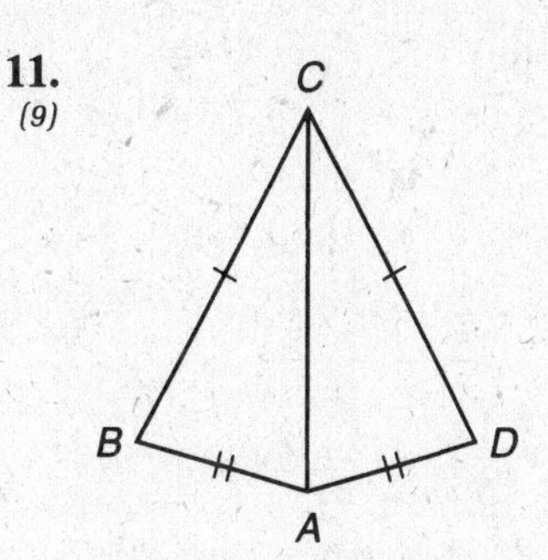

$\mathbf{\Delta ABC \cong \Delta ADC}$ **by *SSS* congruency postulate.**
$\mathbf{\angle ABC \cong \angle ADC}$ **by *CPCTC*.**

12. (5)

$$\begin{aligned} & 7\sqrt{\frac{2}{3}} - 3\sqrt{\frac{3}{2}} + 2\sqrt{24} \\ &= 7\sqrt{\frac{2}{3}\cdot\frac{3}{3}} - 3\sqrt{\frac{3}{2}\cdot\frac{2}{2}} + 2(2)\sqrt{6} \\ &= \frac{7}{3}\sqrt{6} - \frac{3}{2}\sqrt{6} + 4\sqrt{6} \\ &= \frac{14}{6}\sqrt{6} - \frac{9}{6}\sqrt{6} + \frac{24}{6}\sqrt{6} = \mathbf{\frac{29\sqrt{6}}{6}} \end{aligned}$$

13. (5)

$$\begin{aligned} & \sqrt{-2}\sqrt{2} + \sqrt{-4}i - 2i^6 + \sqrt{-2}i^2 \\ &= \sqrt{2}i\sqrt{2} + 2(ii) - 2(ii)(ii)(ii) + \sqrt{2}i(ii) \\ &= 2i + 2(-1) - 2(-1)(-1)(-1) + \sqrt{2}i(-1) \\ &= 2i - 2 + 2 - \sqrt{2}i = \mathbf{(2 - \sqrt{2})i} \end{aligned}$$

14. (5)

$$\begin{aligned} & \left(2x^{\frac{a}{2}} + y^{\frac{b}{2}}\right)\left(2x^{\frac{a}{2}} - y^{\frac{b}{2}}\right) \\ &= \left(2x^{\frac{a}{2}}\right)^2 - \left(y^{\frac{b}{2}}\right)^2 = \mathbf{4x^a - y^b} \end{aligned}$$

15. (5)

$$\frac{x^{\left(\frac{a}{2}-3\right)}y^{\left(\frac{b-3}{2}\right)}}{x^{3a}\left(y^{\frac{1}{3}}\right)^{2b}} = x^{\frac{a}{2}-3}x^{-3a}y^{\frac{b-3}{2}}y^{\frac{-2b}{3}}$$

$$= x^{\frac{a}{2}-3-3a}y^{\frac{b}{2}-\frac{3}{2}-\frac{2b}{3}} = x^{\frac{a-6-6a}{2}}y^{\frac{3b-9-4b}{6}}$$

$$= \mathbf{x^{\frac{-5a-6}{2}}y^{\frac{-b-9}{6}}}$$

16. (6)

$$\sqrt{3x-5} + \sqrt{3x} = 5$$
$$\sqrt{3x-5} = 5 - \sqrt{3x}$$
$$3x - 5 = 25 - 10\sqrt{3x} + 3x$$
$$10\sqrt{3x} = 30$$
$$\sqrt{3x} = 3$$
$$3x = 9$$
$$\mathbf{x = 3}$$

17. (6)

$$\frac{1}{2x-8} + \frac{1}{5} = \frac{1}{7}$$
$$35(2x-8)\left(\frac{1}{2x-8} + \frac{1}{5}\right) = \frac{1}{7}(35)(2x-8)$$
$$35 + 7(2x-8) = 5(2x-8)$$
$$2(2x-8) = -35$$
$$4x - 16 = -35$$
$$4x = -19$$
$$\mathbf{x = \frac{-19}{4}}$$

18. (6) First, label the three equations.

$$\begin{cases} (a)\ 2x + 2y - z = -1 \\ (b)\ x + y - 3z = -8 \\ (c)\ 2x - y + z = 8 \end{cases}$$

Now we can use elimination to solve.

$$\begin{array}{rl} (a) & 2x + 2y - z = -1 \\ -2(b) & -2x - 2y + 6z = 16 \\ \hline & 5z = 15 \\ & \mathbf{z = 3} \end{array}$$

$$\begin{array}{rl} (b) & x + y - 3z = -8 \\ (c) & 2x - y + z = 8 \\ \hline & 3x - 2z = 0 \\ & 3x - 2(3) = 0 \\ & 3x = 6 \\ & \mathbf{x = 2} \end{array}$$

Use the values for x and z in equation (a) to solve for y.

$$\begin{array}{rl} (a) & 2x + 2y - z = -1 \\ & 2(2) + 2y - (3) = -1 \\ & 2y = -2 \\ & \mathbf{y = -1} \end{array}$$

19. (6) First label the three equations.

$$\begin{cases} (a)\ 2x - z = 10 \\ (b)\ y + 2z = -2 \\ (c)\ 3x - 2y = 8 \end{cases}$$

Now we can use elimination to solve.

$$\begin{array}{rl} 2(a) & 4x - 2z = 20 \\ (b) & y + 2z = -2 \\ \hline (d) & 4x + y = 18 \end{array}$$

$$\begin{array}{rl} 2(d) & 8x + 2y = 36 \\ (c) & 3x - 2y = 8 \\ \hline & 11x = 44 \\ & \mathbf{x = 4} \end{array}$$

Use this value for x in equation (c) to solve for y.

$$\begin{array}{rl} (c) & 3x - 2y = 8 \\ & 3(4) - 2y = 8 \\ & 2y = 4 \\ & \mathbf{y = 2} \end{array}$$

Use the value for y in equation (b) to solve for z.

$$\begin{array}{rl} (b) & y + 2z = -2 \\ & (2) + 2z = -2 \\ & 2z = -4 \\ & \mathbf{z = -2} \end{array}$$

20. (3) First separate the triangles.

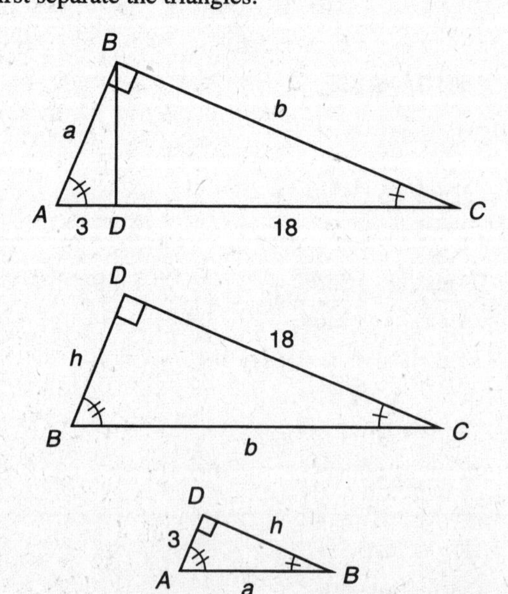

All three triangles are similar. The ratios of sides of similar triangles are equal. We use this fact to solve for a, b, and h.

$$\frac{a}{3} = \frac{21}{a} \qquad \frac{b}{18} = \frac{21}{b} \qquad \frac{3}{h} = \frac{h}{18}$$

$$a^2 = 63 \qquad b^2 = 378 \qquad h^2 = 54$$

$$\boldsymbol{a = 3\sqrt{7}} \qquad \boldsymbol{b = 3\sqrt{42}} \qquad \boldsymbol{h = 3\sqrt{6}}$$

21. (8) First, find similar triangles.

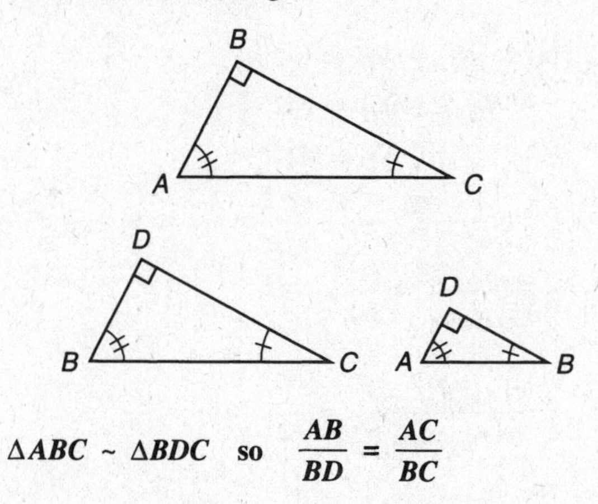

$$\triangle ABC \sim \triangle BDC \quad \text{so} \quad \frac{AB}{BD} = \frac{AC}{BC}$$

22. (8)

$$\frac{10}{8} = \frac{15}{x}$$
$$10x = 120$$
$$x = \mathbf{12}$$

23. (4)

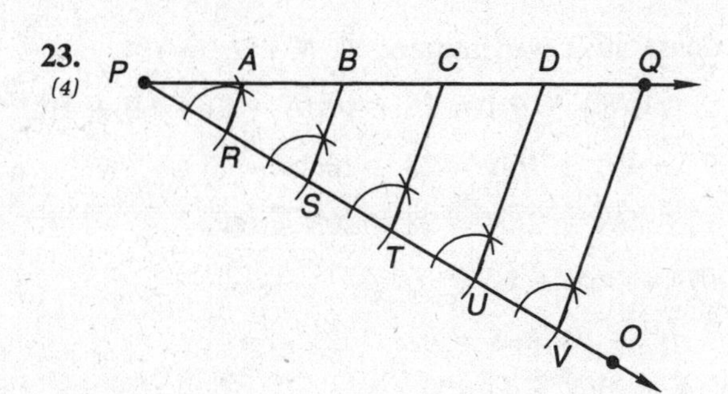

24. (4)

30°

25. (1) First we write two equations.

$$\begin{cases} 2x + 3y = 70 \\ 3x + 5y + 70 = 180 \end{cases}$$

Then we simplify.

$$\begin{matrix} (a) \\ (b) \end{matrix} \begin{cases} 2x + 3y = 70 \\ 3x + 5y = 110 \end{cases}$$

Now we can use elimination to solve.

$$\begin{array}{rrcr} 3(a) & 6x + 9y & = & 210 \\ -2(b) & -6x - 10y & = & -220 \\ \hline & -y & = & -10 \\ & \boldsymbol{y} & = & \mathbf{10} \end{array}$$

$$\begin{aligned} (a) \quad 2x + 3y &= 70 \\ 2x + 3(10) &= 70 \\ 2x &= 40 \\ \boldsymbol{x} &= \mathbf{20} \end{aligned}$$

26. (3) First we find the scale factor in both directions.

$$\overrightarrow{SF} \cdot 5 = 6$$

$$\overrightarrow{SF} = \frac{6}{5} \rightarrow \overleftarrow{SF} = \frac{5}{6}$$

Now we solve for a and b.

$$a = 3 \cdot \overleftarrow{SF} \qquad b = 4 \cdot \overrightarrow{SF}$$

$$a = 3 \cdot \frac{5}{6} \qquad b = 4 \cdot \frac{6}{5}$$

$$\boldsymbol{a = \frac{5}{2}} \qquad \boldsymbol{b = \frac{24}{5}}$$

27. (3) First separate the triangles.

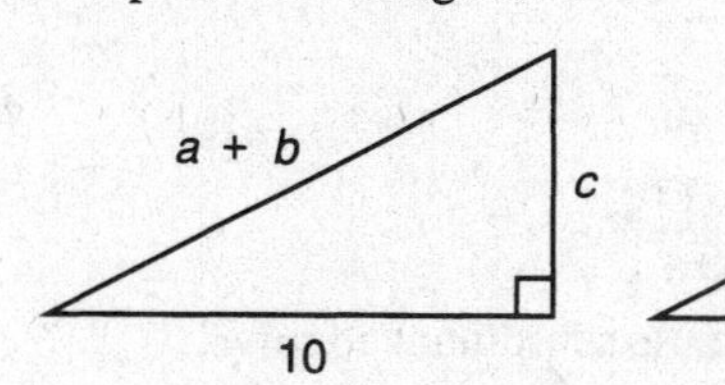

Solve for a using the Pythagorean theorem.

$$a = \sqrt{3^2 + 6^2}$$
$$a = \sqrt{45}$$
$$\boldsymbol{a = 3\sqrt{5}}$$

Now use ratios for similar triangles to solve for b and c.

$$\frac{a}{a + b} = \frac{6}{10}$$

$$\frac{3\sqrt{5}}{3\sqrt{5} + b} = \frac{3}{5}$$

$$15\sqrt{5} = 9\sqrt{5} + 3b$$

$$3b = 6\sqrt{5}$$

$$\boldsymbol{b = 2\sqrt{5}}$$

$$\frac{6}{10} = \frac{3}{c}$$

$$6c = 30$$

$$\boldsymbol{c = 5}$$

28. (2) $V_{\text{pyramid}} = \frac{1}{3}(A_{\text{base}})(h) = \frac{1}{3}\left(\frac{40}{4}\right)^2(6) = \mathbf{200\text{ cm}^3}$

29. (R) $(10 \# 12) * 12 = [2(10) - 3(12)] * 12$
$= (20 - 36) * 12 = (-16) * 12$
$= -16 - 12 = \mathbf{-28}$

30. (1) If $a > 0$ and $b > 0$, then $ab > 0$ and $2ab > 0$.

A^2: $(a + b)^2 = a^2 + 2ab + b^2$

B^2: $a^2 + b^2$

$$a^2 + 2ab + b^2 > a^2 + b^2$$
$$(a + b)^2 > a^2 + b^2$$
$$a + b > \sqrt{a^2 + b^2}$$

Therefore the answer is **A.**

Problem Set 10

1. (R) First we write the two equations.

$$\begin{cases} \text{(a)} & 0.4N_B = 0.6N_S + 128 \\ \text{(b)} & \dfrac{N_B}{N_S} = \dfrac{17}{6} \end{cases}$$

(b) $\dfrac{N_B}{N_S} = \dfrac{17}{6}$

$$N_B = \frac{17}{6}N_S$$

Now we can use substitution to solve.

(a)
$$0.4N_B = 0.6N_S + 128$$
$$0.4\left(\frac{17}{6}N_S\right) = 0.6N_S + 128$$
$$\left(\frac{4}{10}\right)\left(\frac{17}{6}\right)N_S - \frac{6}{10}N_S = 128$$
$$\frac{8}{15}N_S = 128$$
$$\mathbf{N_S = 240}$$

$$N_B = \frac{17}{6}N_S$$
$$N_B = \frac{17}{6}(240)$$
$$\mathbf{N_B = 680}$$

2. (R) First we write the two equations.

$$\begin{cases} \text{(a)} & N_D + N_Q = 142 \\ \text{(b)} & 10N_D + 25N_Q = 2920 \end{cases}$$

(a) $N_D + N_Q = 142$
$$N_D = 142 - N_Q$$

Now we can use substitution to solve.

(b)
$$10N_D + 25N_Q = 2920$$
$$10(142 - N_Q) + 25N_Q = 2920$$
$$1420 - 10N_Q + 25N_Q = 2920$$
$$15N_Q = 1500$$
$$\mathbf{N_Q = 100}$$

(a)
$$N_D + N_Q = 142$$
$$N_D + (100) = 142$$
$$\mathbf{N_D = 42}$$

3. (5) $\dfrac{A_{\text{small circle}}}{A_{\text{big circle}}} = \dfrac{1}{64}$

$$\frac{\pi(r_{\text{small circle}})^2}{\pi(r_{\text{big circle}})^2} = \frac{1}{64}$$
$$64\pi(r_{\text{small circle}})^2 = \pi(r_{\text{big circle}})^2$$
$$8r_{\text{small circle}} = r_{\text{big circle}}$$
$$\frac{r_{\text{small circle}}}{r_{\text{big circle}}} = \mathbf{\frac{1}{8}}$$

4. (7) Consecutive even integers: N, $N + 2$, $N + 4$

$$4[N(N + 4)] = 28 - 10[(N + 2) + (N + 4)]$$
$$4N^2 + 16N = 28 - 10(2N + 6)$$
$$4N^2 + 16N = 28 - 20N - 60$$
$$4N^2 + 36N + 32 = 0$$
$$N^2 + 9N + 8 = 0$$
$$(N + 8)(N + 1) = 0$$

$N + 8 = 0$ $\qquad$ $N + 1 = 0$
$N = -8$ $\qquad$ $N = -1$

We are looking for even integers, so $N = -8$.

−8, −6, −4

5. (10) $m = \dfrac{\text{change in } y}{\text{change in } x} = \dfrac{(1 - 0)}{(0 - 1)} = \dfrac{1}{-1} = -1$

$$y = mx + b$$
$$y = (-1)x + b$$
$$0 = -1(1) + b$$
$$b = 1$$
$$\mathbf{y = -x + 1}$$

6. (10) $y = 2x - 1$

Parallel lines have equal slopes, so $m = 2$.

$y = mx + b$

$y = 2x + b$

$1 = 2(4) + b$

$b = -7$

$\mathbf{y = 2x - 7}$

7. (5) $-2i^5 + 3i^3 - 3i^6$

$= -2(ii)(ii)i + 3(ii)i - 3(ii)(ii)(ii)$

$= -2(-1)(-1)i + 3(-1)i - 3(-1)(-1)(-1)$

$= -2i - 3i + 3 = \mathbf{3 - 5i}$

8. (5) $2i^3 + 4i^5 - 6i^7 + 8i^9$

$= 2(ii)i + 4(ii)(ii)i - 6(ii)(ii)(ii)i + 8(ii)(ii)(ii)(ii)i$

$= 2(-1)i + 4(-1)(-1)i - 6(-1)(-1)(-1)i$
$+ 8(-1)(-1)(-1)(-1)i$

$= -2i + 4i + 6i + 8i = \mathbf{16i}$

9. (10) $\dfrac{2 + i}{3 - 2i} = \dfrac{2 + i}{3 - 2i} \cdot \dfrac{3 + 2i}{3 + 2i}$

$= \dfrac{6 + 4i + 3i + 2i^2}{9 - 4i^2} = \dfrac{4 + 7i}{13} = \mathbf{\dfrac{4}{13} + \dfrac{7}{13}i}$

10. (10) $\dfrac{2 + i - 2i^3}{2i^3 + 4} = \dfrac{2 + i - 2(ii)i}{2(ii)i + 4}$

$= \dfrac{2 + i - 2(-1)i}{2(-1)i + 4} = \dfrac{2 + 3i}{4 - 2i} \cdot \dfrac{4 + 2i}{4 + 2i}$

$= \dfrac{8 + 4i + 12i + 6i^2}{16 - 4i^2} = \dfrac{2 + 16i}{20} = \mathbf{\dfrac{1}{10} + \dfrac{4}{5}i}$

11. (10) $\dfrac{2 + 2\sqrt{3}}{1 - \sqrt{3}} = \dfrac{2 + 2\sqrt{3}}{1 - \sqrt{3}} \cdot \dfrac{1 + \sqrt{3}}{1 + \sqrt{3}}$

$= \dfrac{2 + 2\sqrt{3} + 2\sqrt{3} + 6}{1 - 3} = \dfrac{8 + 4\sqrt{3}}{-2}$

$= \mathbf{-4 - 2\sqrt{3}}$

12. (10) $x^2 - 4x - 12 = 0$

$x^2 - 4x = 12$

$(x^2 - 4x + \quad) = 12$

$(x^2 - 4x + 4) = 12 + 4$

$(x - 2)^2 = 16$

$x - 2 = \pm 4$

$x = \pm 4 + 2$

$x = \mathbf{6, -2}$

13. (10) $x + 3x^2 = -5$

$3x^2 + x = -5$

$x^2 + \dfrac{1}{3}x = -\dfrac{5}{3}$

$\left(x^2 + \dfrac{1}{3}x + \quad\right) = -\dfrac{5}{3}$

$\left(x^2 + \dfrac{1}{3}x + \dfrac{1}{36}\right) = -\dfrac{5}{3} + \dfrac{1}{36}$

$\left(x + \dfrac{1}{6}\right)^2 = -\dfrac{59}{36}$

$x + \dfrac{1}{6} = \pm\sqrt{-\dfrac{59}{36}}$

$x = \mathbf{-\dfrac{1}{6} \pm \dfrac{\sqrt{59}}{6}i}$

14. (7) The contrapositive of the major premise identifies a property about the light when the switch is not on, and the minor premise identifies that the switch is not on. Therefore, the argument is **valid** in concluding that the light has the property that the major premise identifies.

15. (9)
(a) ***SSS* congruency postulate**
(b) ***SAS* congruency postulate**
(c) ***AAAS* congruency postulate**
(d) ***SAS* congruency postulate**

16. (9)

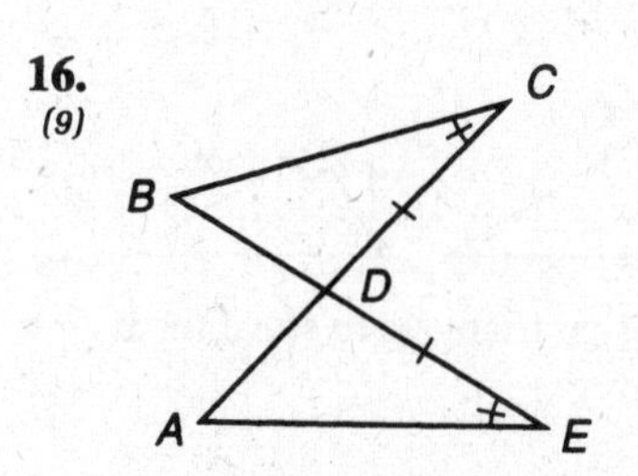

$\triangle BCD \cong \triangle AED$ by *AAAS* congruency postulate.

17. (9)

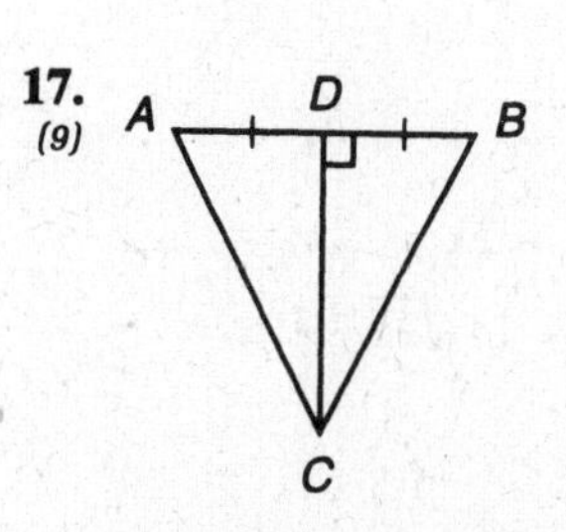

$\triangle ADC \cong \triangle BDC$ by *SAS* congruency postulate.

$\angle A \cong \angle B$ by *CPCTC*.

18. First, find similar triangles.
(8)

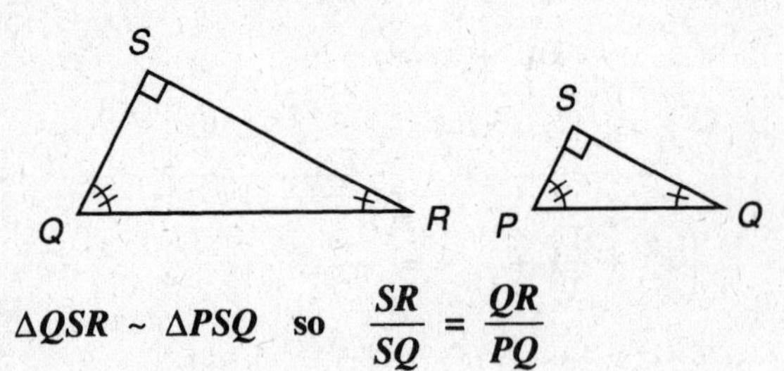

$$\Delta QSR \sim \Delta PSQ \quad \text{so} \quad \frac{SR}{SQ} = \frac{QR}{PQ}$$

19. First, we separate the triangles.
(3)

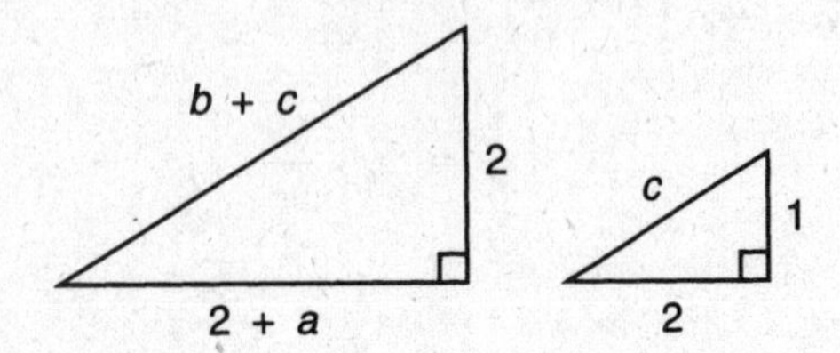

Solve for c using the Pythagorean theorem.

$$c = \sqrt{1^2 + 2^2}$$
$$\boldsymbol{c = \sqrt{5}}$$

Use the ratios for similar triangles to solve for a and b.

$$\frac{2}{2 + a} = \frac{1}{2} \qquad \frac{c}{b + c} = \frac{1}{2}$$
$$2 + a = 4 \qquad \frac{\sqrt{5}}{b + \sqrt{5}} = \frac{1}{2}$$
$$\boldsymbol{a = 2} \qquad 2\sqrt{5} = b + \sqrt{5}$$
$$\boldsymbol{b = \sqrt{5}}$$

20. First we separate the triangles.
(3)

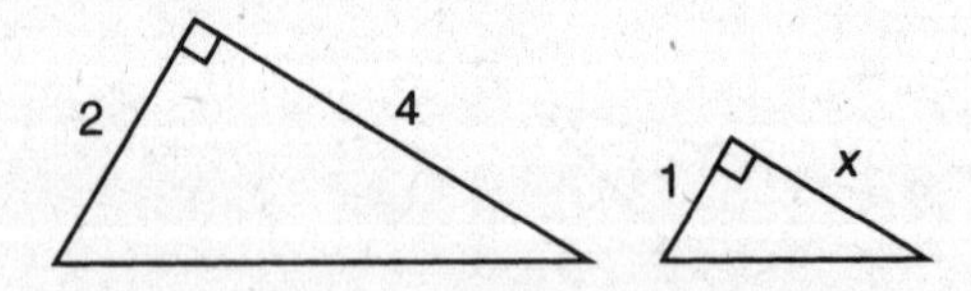

Now use ratios for similar triangles to solve for x.

$$\frac{x}{4} = \frac{1}{2}$$
$$2x = 4$$
$$\boldsymbol{x = 2}$$

21. $\sqrt{2x - 7} + \sqrt{2x} = 7$
(6)

$$\sqrt{2x - 7} = 7 - \sqrt{2x}$$
$$2x - 7 = 49 - 14\sqrt{2x} + 2x$$
$$14\sqrt{2x} = 56$$
$$\sqrt{2x} = 4$$
$$2x = 16$$
$$\boldsymbol{x = 8}$$

22. First, we write the equations.
(6)

$$\begin{cases} \dfrac{3}{4}x + \dfrac{5}{2}y = 8 \\ 0.25y - 0.4z = 0.1 \\ -\dfrac{x}{2} - \dfrac{z}{2} = -\dfrac{5}{2} \end{cases}$$

Then we simplify the equations.

$$\begin{matrix} \text{(a)} \\ \text{(b)} \\ \text{(c)} \end{matrix} \begin{cases} 3x + 10y = 32 \\ 10y - 16z = 4 \\ -x - z = -5 \end{cases}$$

Now we can use elimination to solve.

$$\begin{array}{lrcrcr} \text{(a)} & 3x + 10y & & & = & 32 \\ 3\text{(c)} & -3x & & - 3z & = & -15 \\ \hline & & 10y & - 3z & = & 17 \\ -1\text{(b)} & & -10y & + 16z & = & -4 \\ \hline & & & 13z & = & 13 \\ & & & \boldsymbol{z} & \boldsymbol{=} & \boldsymbol{1} \end{array}$$

Use the value of z in equation (c) to solve for x.

$$\text{(c)} \quad -x - z = -5$$
$$-x - (1) = -5$$
$$\boldsymbol{x = 4}$$

Use the value of x in equation (a) to solve for y.

$$\text{(a)} \quad 3x + 10y = 32$$
$$3(4) + 10y = 32$$
$$10y = 20$$
$$\boldsymbol{y = 2}$$

23.
(5)

$$\frac{x^{-3} + y^{-6}}{x^{-6}y^2} = \frac{\dfrac{1}{x^3} + \dfrac{1}{y^6}}{\dfrac{y^2}{x^6}}$$
$$= \frac{y^6 + x^3}{x^3y^6}\left(\frac{x^6}{y^2}\right) = \boldsymbol{\frac{x^3y^6 + x^6}{y^8}}$$

24.
(5)

$$\frac{x^3\sqrt{x^3}(yx^2)^{\frac{a}{3}}}{x^a y^{2a}} = \frac{x^3x^{\frac{3}{2}}y^{\frac{a}{3}}x^{\frac{2a}{3}}}{x^a y^{2a}}$$
$$= x^3x^{\frac{3}{2}}x^{\frac{2a}{3}}x^{-a}y^{\frac{a}{3}}y^{-2a} = x^{3+\frac{3}{2}+\frac{2a}{3}-a}\,y^{\frac{a}{3}-2a}$$
$$= x^{\frac{18+9+4a-6a}{6}}\,y^{\frac{a-6a}{3}} = \boldsymbol{x^{\frac{27-2a}{6}}\,y^{\frac{-5a}{3}}}$$

25.
(4)

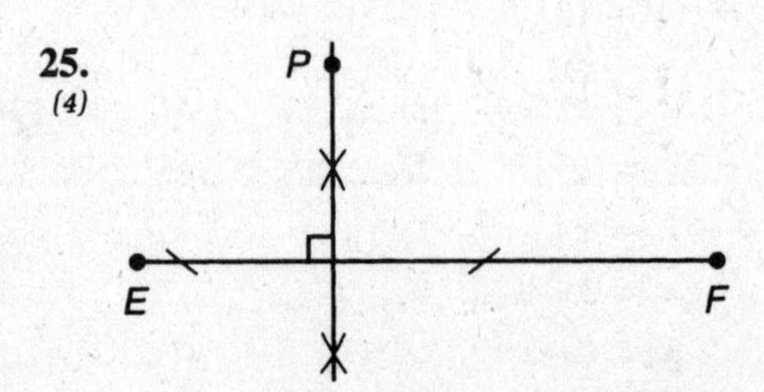

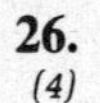

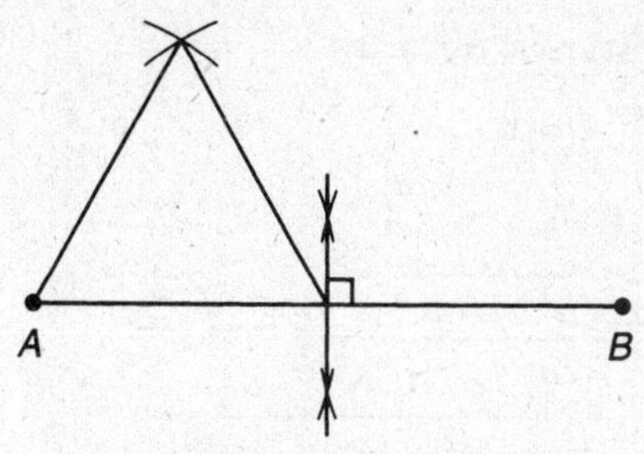

27. (2) $A_{\text{surface}} = 2(A_{\text{base}}) + 12(A_{\text{side}})$

$= 2[5(3)^2] + 12[3(8)]$

$= 2(45) + 12(24) = \mathbf{378\ m^2}$

28. (2) $V = \frac{4}{3}\pi(3)^3$

$\mathbf{V = 36\pi\ in.^3 = 113.10\ in.^3}$

$A_{\text{surface}} = 4\pi(3)^2$

$\mathbf{A_{surface} = 36\pi\ in.^2 = 113.10\ in.^2}$

29. (1) A: $3 * 4 = 5(3) + 4 = 19$

B: $3 \# 4 = 3 + 3(4) = 15$

$19 > 15$

Therefore the answer is **A.**

30. (1) Let v be the unknown angle in the figure.

We know: $x + y + v = 180 = t + z + v$

So: $x + y + v = t + z + v$

$x + y = t + z$

Subtracting y and z from both sides we get:

$x + y - y - z = t + z - y - z$

$x - z = t - y$

Therefore the answer is **C.**

Problem Set 11

1. (R) First we write the two equations.

$$\begin{cases} \text{(a)}\ \dfrac{N_I}{N_E} = \dfrac{4}{7} \\ \text{(b)}\ 5N_E = N_I + 62 \end{cases}$$

(a) $\frac{N_I}{N_E} = \frac{4}{7}$

$N_I = \frac{4}{7}N_E$

Now we can use substitution to solve.

(b) $5N_E = N_I + 62$

$5N_E = \left(\frac{4}{7}N_E\right) + 62$

$5N_E - \frac{4}{7}N_E = 62$

$\frac{31}{7}N_E = 62$

$\mathbf{N_E = 14}$

$N_I = \frac{4}{7}N_E$

$N_I = \frac{4}{7}(14)$

$\mathbf{N_I = 8}$

2. (R) First we write the two equations.

$$\begin{cases} 0.4N_B = N_S + 9 \\ \dfrac{N_B}{N_S} = \dfrac{3}{1} \end{cases}$$

Then we simplify the equations.

$$\begin{cases} \text{(a)}\ 4N_B = 10N_S + 90 \\ \text{(b)}\ N_B = 3N_S \end{cases}$$

Now we can use substitution to solve.

(a) $4N_B = 10N_S + 90$

$4(3N_S) = 10N_S + 90$

$12N_S = 10N_S + 90$

$2N_S = 90$

$\mathbf{N_S = 45}$

(b) $N_B = 3N_S$

$N_B = 3(45)$

$\mathbf{N_B = 135}$

3. (7) Consecutive even integers:

$N, N + 2, N + 4, N + 6$

$6[N + (N + 2)] = 2[N + (N + 6)] + 416$

$6(2N + 2) = 2(2N + 6) + 416$

$12N + 12 = 4N + 12 + 416$

$8N = 416$

$N = 52$

52, 54, 56, 58

4. (7) **Invalid.** The major premise attributes a quality to all members of a certain set. The minor premise does not identify Homer Lee as a member of this set.

5. (11) $x + 16 = 54$

$\mathbf{x = 38}$

$y + 13 = 38$

$\mathbf{y = 25}$

6. (11) $(x + 15)(12) = (24)(x + 5)$

$12x + 180 = 24x + 120$

$-12x = -60$

$\mathbf{x = 5}$

7. (11) $x + 3 = 7$

$\mathbf{x = 4}$

$x + y = 19$

$y = 19 - x$

$y = 19 - (4)$

$\mathbf{y = 15}$

8. (11) $ax^2 + bx + c = 0$

$x^2 + \frac{b}{a}x + \frac{c}{a} = 0$

$\left(x^2 + \frac{b}{a}x \qquad\right) = -\frac{c}{a}$

$\left(x^2 + \frac{b}{a}x + \frac{b^2}{4a^2}\right) = \frac{b^2}{4a^2} - \frac{c}{a}$

$\left(x + \frac{b}{2a}\right)^2 = \frac{b^2}{4a^2} - \frac{4ac}{4a^2}$

$x + \frac{b}{2a} = \pm\sqrt{\frac{b^2 - 4ac}{4a^2}}$

$x = -\frac{b}{2a} \pm \frac{\sqrt{b^2 - 4ac}}{2a}$

$\mathbf{x = \frac{-b \pm \sqrt{b^2 - 4ac}}{2a}}$

9. (11) $3x^2 + 2x + 5 = 0$

$a = 3,\ b = 2,\ c = 5$

$x = \frac{-b \pm \sqrt{b^2 - 4ac}}{2a}$

$x = \frac{-2 \pm \sqrt{2^2 - 4(3)(5)}}{2(3)}$

$x = \frac{-2 \pm \sqrt{-56}}{6} = \frac{-2 \pm 2\sqrt{-14}}{6} = \mathbf{-\frac{1}{3} \pm \frac{\sqrt{14}}{3}i}$

10. (11) $-x = -3x^2 + 4$

$3x^2 - x - 4 = 0$

$a = 3,\ b = -1,\ c = -4$

$x = \frac{-b \pm \sqrt{b^2 - 4ac}}{2a}$

$x = \frac{-(-1) \pm \sqrt{(-1)^2 - 4(3)(-4)}}{2(3)}$

$x = \frac{1 \pm \sqrt{49}}{6} = \frac{1}{6} \pm \frac{7}{6} = \mathbf{\frac{4}{3}, -1}$

11. (10) $2y + 6x = 5$

$y = -3x + \frac{5}{2}$

The slopes of perpendicular lines are negative reciprocals of each other. So, $m = \frac{1}{3}$.

$y = mx + b$

$y = \frac{1}{3}x + b$

$-2 = \left(\frac{1}{3}\right)(3) + b$

$b = -2 - 1 = -3$

$\mathbf{y = \frac{1}{3}x - 3}$

12. (10) $\frac{3i^3 + 2 - 2i^3}{i - 4}$

$= \frac{3(-1)i + 2 - 2(-1)i}{i - 4}$

$= \frac{2 - i}{i - 4} \cdot \frac{i + 4}{i + 4}$

$= \frac{2i + 8 - i^2 - 4i}{i^2 - 16}$

$= \frac{9 - 2i}{-17} = \mathbf{-\frac{9}{17} + \frac{2}{17}i}$

13. (10) $\frac{3 + 2\sqrt{3}}{2 - 3\sqrt{3}}$

$= \frac{3 + 2\sqrt{3}}{2 - 3\sqrt{3}} \cdot \frac{2 + 3\sqrt{3}}{2 + 3\sqrt{3}}$

$= \frac{6 + 4\sqrt{3} + 9\sqrt{3} + 18}{4 - 27}$

$= \frac{24 + 13\sqrt{3}}{-23} = \mathbf{\frac{-24 - 13\sqrt{3}}{23}}$

14. (10)

$$2x^2 = -x - 5$$
$$2x^2 + x = -5$$
$$x^2 + \frac{1}{2}x = -\frac{5}{2}$$
$$\left(x^2 + \frac{1}{2}x + \quad\right) = -\frac{5}{2}$$
$$\left(x^2 + \frac{1}{2}x + \frac{1}{16}\right) = \frac{1}{16} - \frac{5}{2}$$
$$\left(x + \frac{1}{4}\right)^2 = -\frac{39}{16}$$
$$x + \frac{1}{4} = \pm\frac{\sqrt{39}}{4}i$$
$$x = \mathbf{-\frac{1}{4} \pm \frac{\sqrt{39}}{4}i}$$

15. (10)

$$2x + 7 = 3x^2$$
$$3x^2 - 2x = 7$$
$$x^2 - \frac{2}{3}x = \frac{7}{3}$$
$$\left(x^2 - \frac{2}{3}x + \quad\right) = \frac{7}{3}$$
$$\left(x^2 - \frac{2}{3}x + \frac{1}{9}\right) = \frac{7}{3} + \frac{1}{9}$$
$$\left(x - \frac{1}{3}\right)^2 = \frac{22}{9}$$
$$x - \frac{1}{3} = \pm\frac{\sqrt{22}}{3}$$
$$x = \mathbf{\frac{1}{3} \pm \frac{\sqrt{22}}{3}}$$

16. (6) First, label the three equations.

$$\begin{aligned}&(a)\\&(b)\\&(c)\end{aligned}\begin{cases}3x - 2y + z = -1\\ x + 2y - z = 9\\ 2x - y + 2z = 2\end{cases}$$

Now we can use elimination to solve.

$$\begin{array}{lrcl}(a) & 3x - 2y + z &=& -1\\ (b) & x + 2y - z &=& 9\\ \hline & 4x &=& 8\end{array}$$
$$\mathbf{x = 2}$$

$$\begin{array}{lrcl}2(b) & 2x + 4y - 2z &=& 18\\ (c) & 2x - y + 2z &=& 2\\ \hline & 4x + 3y &=& 20\end{array}$$
$$4(2) + 3y = 20$$
$$3y = 12$$
$$\mathbf{y = 4}$$

(b) $x + 2y - z = 9$

$$(2) + 2(4) - z = 9$$
$$\mathbf{z = 1}$$

17. (6)

$$\sqrt{2x + 20} + \sqrt{2x} = 10$$
$$\sqrt{2x + 20} = 10 - \sqrt{2x}$$
$$2x + 20 = 100 - 20\sqrt{2x} + 2x$$
$$20\sqrt{2x} = 80$$
$$\sqrt{2x} = 4$$
$$2x = 16$$
$$x = \mathbf{8}$$

18. (6)

$$\frac{x + 1}{x - 1} - \frac{1}{x} = \frac{2}{x(x - 1)}$$
$$\frac{x(x - 1)(x + 1)}{x - 1} - \frac{x(x - 1)}{x} = \frac{2x(x - 1)}{x(x - 1)}$$
$$x(x + 1) - (x - 1) = 2$$
$$x^2 + x - x + 1 = 2$$
$$x^2 = 1$$

Therefore $x = \pm 1$, but, checking these solutions in the original equation we find $x \neq 1$, so $x = \mathbf{-1}$.

19. (9)
(a) ***HL* congruency postulate**
(b) ***AAAS* congruency postulate**
(c) ***AAAS* congruency postulate**
(d) ***SSS* congruency postulate**

20. (9)

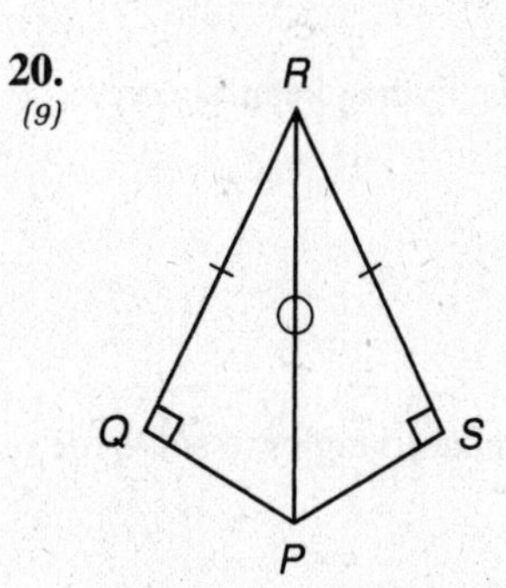

$\Delta PQR \cong \Delta PSR$ by *HL* congruency postulate.

21. (9)

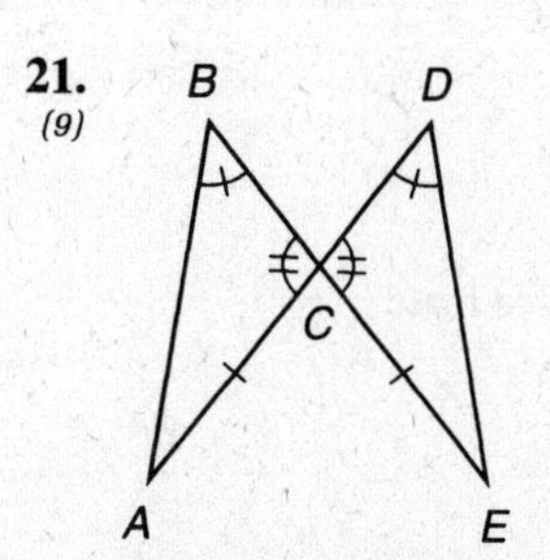

$\Delta ABC \cong \Delta EDC$ by *AAAS* congruency postulate.
$\overline{BC} \cong \overline{DC}$ by *CPCTC*.

22. (8)

$$\frac{10}{15} = \frac{x}{13}$$
$$15x = 13(10)$$
$$x = \frac{13(10)}{15}$$
$$x = \mathbf{\frac{26}{3}}$$

23. (8) First, separate the triangles.

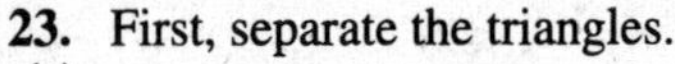

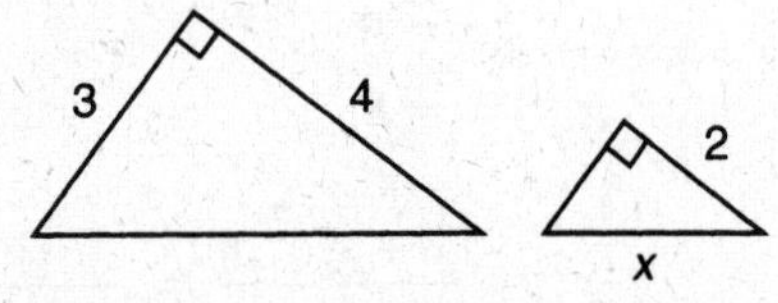

Now use the ratios for similar triangles to solve for x.

$$\frac{x}{2} = \frac{\sqrt{3^2 + 4^2}}{4}$$
$$x = \frac{2(5)}{4}$$
$$x = \mathbf{\frac{5}{2}}$$

24. (3) First, separate the triangles.

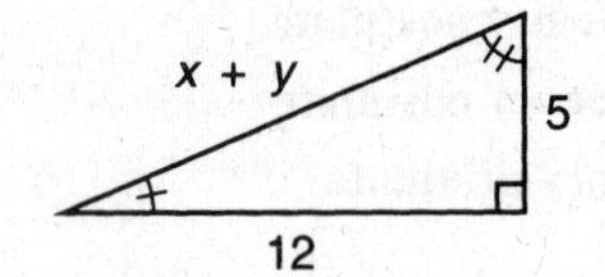

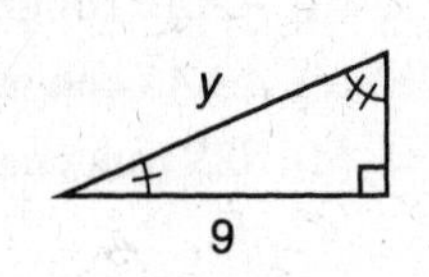

Solve for $(x + y)$ using the Pythagorean Theorem.

$$x + y = \sqrt{12^2 + 5^2}$$
$$x + y = \sqrt{169}$$
$$x + y = 13$$

Now use the ratios for similar triangles to solve for y.

$$\frac{y}{9} = \frac{(x + y)}{12}$$
$$\frac{y}{9} = \frac{(13)}{12}$$
$$12y = 117$$
$$y = \mathbf{\frac{39}{4}}$$

Now use substitution to solve for x.

$$x + y = 13$$
$$x + \left(\frac{39}{4}\right) = 13$$
$$x = \mathbf{\frac{13}{4}}$$

25. (4)

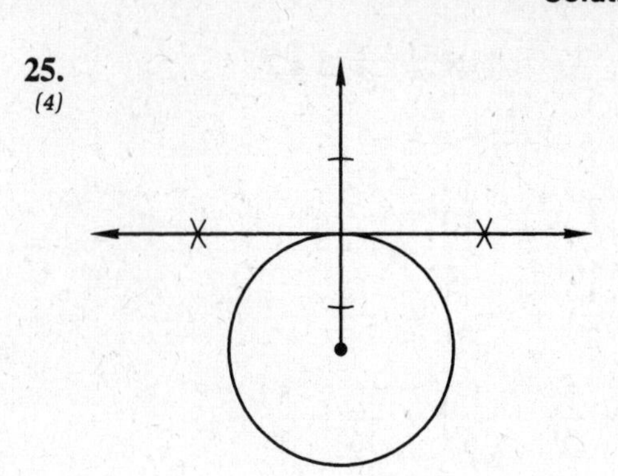

26. (5)

$$\frac{a^{-7} + b^{-3}}{a^4 b^{-5}} = \frac{\frac{1}{a^7} + \frac{1}{b^3}}{\frac{a^4}{b^5}}$$
$$= \left(\frac{b^3 + a^7}{a^7 b^3}\right)\left(\frac{b^5}{a^4}\right) = \mathbf{\frac{b^5 + a^7 b^2}{a^{11}}}$$

27. (5)

$$\frac{x^{3a+2}\left(\sqrt{x^3}\right)^{2a}}{y^{2a-4}\left(\sqrt{x}\right)^{2a+1}} = \frac{x^{3a+2}x^{3a}}{y^{2a-4}x^{a+\frac{1}{2}}}$$
$$= x^{3a+2}x^{3a}x^{-a-\frac{1}{2}}y^{-2a+4}$$
$$= x^{3a+2+3a-a-\frac{1}{2}}y^{-2a+4}$$
$$= x^{5a+\frac{3}{2}}y^{4-2a} = \mathbf{x^{\frac{10a+3}{2}}y^{4-2a}}$$

28. (5)

$$\left(6x^{\frac{3}{2}} - 2y^{\frac{1}{2}}\right)\left(6x^{\frac{3}{2}} + 2y^{\frac{1}{2}}\right)$$
$$= 36x^{\frac{3}{2}+\frac{3}{2}} + 12y^{\frac{1}{2}}x^{\frac{3}{2}} - 12y^{\frac{1}{2}}x^{\frac{3}{2}} - 4y^{\frac{1}{2}+\frac{1}{2}}$$
$$= \mathbf{36x^3 - 4y}$$

29. (1) We know that $x^2 + y^2 = 1$. Since x^2 and y^2 must both be nonnegative, they must be less than or equal to one.

$$x^2 \le 1 \;\rightarrow\; |x| \le 1 \;\rightarrow\; -1 \le x \le 1$$
$$y^2 \le 1 \;\rightarrow\; |y| \le 1 \;\rightarrow\; -1 \le y \le 1$$

Since $|x|$ and $|y|$ are ≤ 1, then $|x||y| \le 1$ which implies that $xy \le 1 < 2$. Therefore the answer is **B.**

30. (1) Triangle ABD is an isosceles triangle, so sides AB and AD are congruent. So we can see that

$$AB + DC = AD + DC = AC$$

Therefore the answer is **C.**

Problem Set 12

1. (R)
N = nonbelievers
T = total

$$N = 85$$
$$N = \frac{5}{7}T$$
$$T = \frac{7}{5}(85)$$
$$T = \mathbf{119}$$

2. (R) First we write the two equations

$$\begin{cases} \text{(a)} & N_G + N_B = 52 \\ \text{(b)} & 3N_B = N_G + 4 \end{cases}$$

$$\text{(a)}\quad N_G + N_B = 52$$
$$N_B = 52 - N_G$$

Now we can use substitution to solve.

$$\text{(b)}\quad 3N_B = N_G + 4$$
$$3(52 - N_G) = N_G + 4$$
$$156 - 3N_G = N_G + 4$$
$$152 = 4N_G$$
$$\mathbf{N_G = 38}$$

$$\text{(a)}\quad N_G + N_B = 52$$
$$(38) + N_B = 52$$
$$\mathbf{N_B = 14}$$

3. (R)
T = tin
A = alloy
Z = zinc

$$0.3A = Z$$
$$0.3A = (183)$$
$$A = 610$$

$T = 0.5A = 0.5(610) =$ **305 grams**

4. (12)
Exterior = 360°
Interior = $(N - 2)180°$
Interior = $(6 - 2)180°$
Interior = 720°

5. (12)
$$\text{Number of diagonals} = \frac{N(N - 3)}{2}$$
$$= \frac{9(9 - 3)}{2} = \mathbf{27}$$

6. (11) First we write two equations.

$$\begin{cases} \text{(a)} & 4x - 3y = 12 \\ \text{(b)} & x + 3y = 8 \end{cases}$$

Now we can use elimination to solve.

$$\begin{aligned} \text{(a)}\quad & 4x - 3y = 12 \\ \text{(b)}\quad & \underline{x + 3y = 8} \\ & 5x = 20 \\ & \mathbf{x = 4} \end{aligned}$$

$$\text{(b)}\quad x + 3y = 8$$
$$(4) + 3y = 8$$
$$3y = 4$$
$$\mathbf{y = \frac{4}{3}}$$

7. (11)
$$x = 2(80)$$
$$\mathbf{x = 160}$$
$$y = 2(30)$$
$$\mathbf{y = 60}$$
$$z = 2(180 - 30 - 80)$$
$$\mathbf{z = 140}$$

8. (11)
$$2(3x + 2x) = 360$$
$$10x = 360$$
$$x = \mathbf{36}$$

9. (11)
$$ax^2 + bx + c = 0$$
$$x^2 + \frac{b}{a}x + \frac{c}{a} = 0$$
$$\left(x^2 + \frac{b}{a}x \qquad\right) = -\frac{c}{a}$$
$$\left(x^2 + \frac{b}{a}x + \frac{b^2}{4a^2}\right) = \frac{b^2}{4a^2} - \frac{c}{a}$$
$$\left(x + \frac{b}{2a}\right)^2 = \frac{b^2}{4a^2} - \frac{4ac}{4a^2}$$
$$x + \frac{b}{2a} = \pm\sqrt{\frac{b^2 - 4ac}{4a^2}}$$
$$x = -\frac{b}{2a} \pm \frac{\sqrt{b^2 - 4ac}}{2a}$$
$$\mathbf{x = \frac{-b \pm \sqrt{b^2 - 4ac}}{2a}}$$

10. (11)

$$2x^2 = x - 5$$
$$2x^2 - x + 5 = 0$$
$$a = 2,\ b = -1,\ c = 5$$
$$x = \frac{-b \pm \sqrt{b^2 - 4ac}}{2a}$$
$$x = \frac{-(-1) \pm \sqrt{(-1)^2 - 4(2)(5)}}{2(2)}$$
$$x = \frac{1 \pm \sqrt{-39}}{4} = \mathbf{\frac{1}{4} \pm \frac{\sqrt{39}}{4}i}$$

11. (11)

$$4 + 3x^2 = -2x$$
$$3x^2 + 2x + 4 = 0$$
$$a = 3,\ b = 2,\ c = 4$$
$$x = \frac{-b \pm \sqrt{b^2 - 4ac}}{2a}$$
$$x = \frac{-(2) \pm \sqrt{(2)^2 - 4(3)(4)}}{2(3)}$$
$$x = \frac{-2 \pm \sqrt{-44}}{6} = -\frac{2}{6} \pm \frac{2\sqrt{11}}{6}i = \mathbf{-\frac{1}{3} \pm \frac{\sqrt{11}}{3}i}$$

12. (10)

$$m = \frac{\text{change in } y}{\text{change in } x} = \frac{(9 - 3)}{(5 - 2)} = \frac{6}{3} = 2$$
$$y = mx + b$$
$$y = 2x + b$$
$$3 = 2(2) + b$$
$$b = -1$$
$$\mathbf{y = 2x - 1}$$

13. (5)

$$3i^3 + 2i^2 - 3i + 4$$
$$= 3(-1)i + 2(-1) - 3i + 4$$
$$= -3i - 2 - 3i + 4 = \mathbf{2 - 6i}$$

14. (10)

$$\frac{3 - \sqrt{2}}{4 + 2\sqrt{2}} = \frac{3 - \sqrt{2}}{4 + 2\sqrt{2}} \cdot \frac{4 - 2\sqrt{2}}{4 - 2\sqrt{2}}$$
$$= \frac{12 - 4\sqrt{2} - 6\sqrt{2} + 4}{4^2 - (2\sqrt{2})^2} = \frac{16 - 10\sqrt{2}}{16 - 8}$$
$$= \mathbf{\frac{8 - 5\sqrt{2}}{4}}$$

15. (10)

$$3x^2 = -6 - x$$
$$3x^2 + x = -6$$
$$x^2 + \frac{x}{3} = -2$$
$$\left(x^2 + \frac{x}{3} \quad \right) = -2$$
$$\left(x^2 + \frac{x}{3} + \frac{1}{36}\right) = -2 + \frac{1}{36}$$
$$\left(x + \frac{1}{6}\right)^2 = -\frac{71}{36}$$
$$x + \frac{1}{6} = \pm\sqrt{-\frac{71}{36}}$$
$$x = \mathbf{-\frac{1}{6} \pm \frac{\sqrt{71}}{6}i}$$

16. (10)

$$4x + 7 = 2x^2$$
$$2x^2 - 4x = 7$$
$$x^2 - 2x = \frac{7}{2}$$
$$(x^2 - 2x \quad) = \frac{7}{2}$$
$$(x^2 - 2x + 1) = \frac{7}{2} + 1$$
$$(x - 1)^2 = \frac{9}{2}$$
$$x - 1 = \pm\sqrt{\frac{9}{2}}$$
$$x = \mathbf{1 \pm \frac{3\sqrt{2}}{2}}$$

17. (6) First, label the three equations.

$$\begin{cases} \text{(a)} & \frac{1}{3}x - \frac{1}{4}y = 2 \\ \text{(b)} & 0.4y + 0.2z = 2 \\ \text{(c)} & \frac{1}{6}x + \frac{1}{2}z = \frac{5}{2} \end{cases}$$

Now we can use elimination to solve for y.

$$\begin{array}{lrcl} (12)(a) & 4x - 3y & = & 24 \\ (-24)(c) & -4x \quad\quad - 12z & = & -60 \\ \hline (d) & -3y - 12z & = & -36 \end{array}$$

$$\begin{array}{lrcl} -\frac{1}{3}(d) & y + 4z & = & 12 \\ (-20)(b) & -8y - 4z & = & -40 \\ \hline & -7y & = & -28 \end{array}$$
$$\mathbf{y = 4}$$

Now use substitution to solve for x and z.

$$12(a) \quad 4x - 3y = 24$$
$$4x - 3(4) = 24$$
$$4x = 36$$
$$\mathbf{x = 9}$$

$$-\frac{1}{3}(d) \quad y + 4z = 12$$
$$(4) + 4z = 12$$
$$4z = 8$$
$$\mathbf{z = 2}$$

18. (6)
$$\sqrt{4x - 12} + \sqrt{4x} = 6$$
$$\sqrt{4x - 12} = 6 - \sqrt{4x}$$
$$4x - 12 = 36 - 12\sqrt{4x} + 4x$$
$$-12 = 36 - 12\sqrt{4x}$$
$$12\sqrt{4x} = 48$$
$$\sqrt{4x} = 4$$
$$4x = 16$$
$$\mathbf{x = 4}$$

19. (5)
$$\sqrt{6}\sqrt{3}\sqrt{-2}\sqrt{-16} + \sqrt{2}\sqrt{-2} - 6\sqrt{-8}\sqrt{2}$$
$$= \sqrt{6}\sqrt{3}\sqrt{2}i\sqrt{16}i + \sqrt{2}\sqrt{2}i - 6\sqrt{8}i\sqrt{2}$$
$$= \sqrt{6}\sqrt{6}(4)(ii) + \sqrt{2}\sqrt{2}i - 12\sqrt{2}\sqrt{2}i$$
$$= 4(6)(-1) + 2i - 24i = \mathbf{-24 - 22i}$$

20. (10)
$$\frac{8i^4 - 4i^3 - 6i}{-3i^3 + 5i - \sqrt{-16}}$$
$$= \frac{8(ii)(ii) - 4(ii)i - 6i}{-3(ii)i + 5i - \sqrt{16}i}$$
$$= \frac{8 + 4i - 6i}{3i + 5i - 4i} = \frac{8 - 2i}{4i}$$
$$= \frac{8}{4i} \cdot \frac{i}{i} - \frac{2i}{4i} = \mathbf{-\frac{1}{2} - 2i}$$

21. (5)
$$\frac{x^{-3}y^{-1} + y^{-5}}{x^{-4}y^5} = \frac{\frac{1}{x^3y} + \frac{1}{y^5}}{\frac{y^5}{x^4}}$$
$$= \left(\frac{y^4 + x^3}{x^3y^5}\right) \cdot \frac{x^4}{y^5} = \mathbf{\frac{xy^4 + x^4}{y^{10}}}$$

22. (9) (a) ***SAS* congruency postulate**

(b) ***HL* congruency postulate**

(c) ***SSS* congruency postulate**

(d) ***AAAS* congruency postulate**

23. (9)

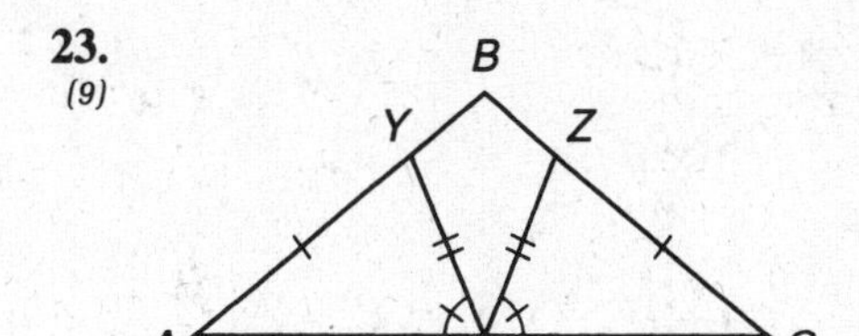

Since $\overline{AB} \cong \overline{CB}$, $\angle A \cong \angle C$;
and by $AA \rightarrow AAA$, $\angle AYX \cong \angle CZX$;
$\triangle AYX \cong \triangle CZX$ by *AAAS* congruency postulate.

24. (9)

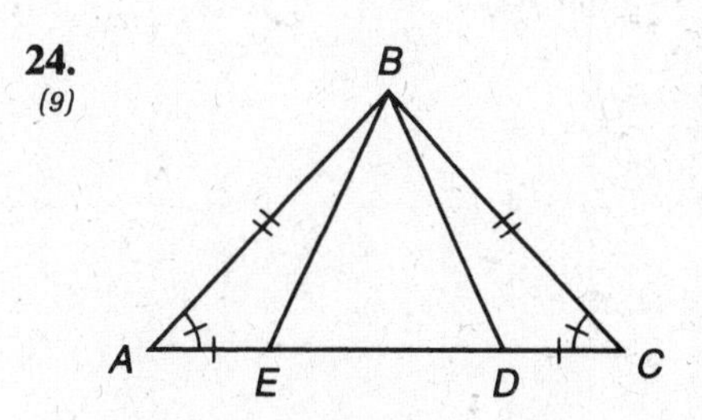

Since $\angle A \cong \angle C$, $\overline{AB} \cong \overline{CB}$.
$\triangle ABE \cong \triangle CBD$ by *SAS* congruency postulate.

25. (3) First, separate the triangles.

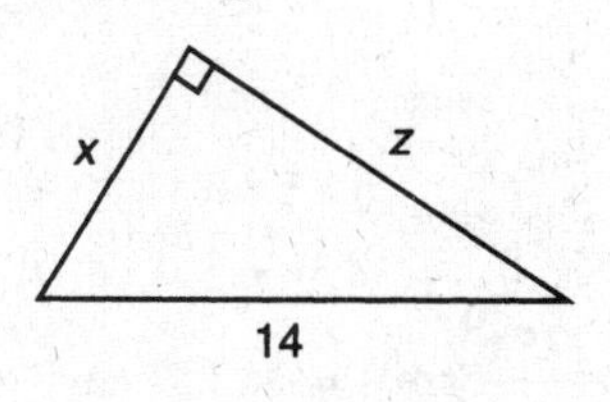

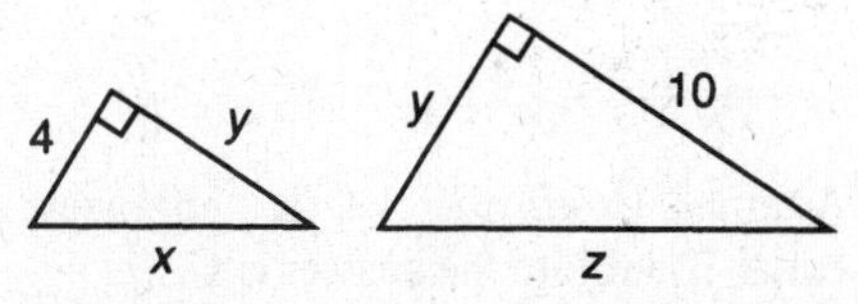

All three triangles are similar. The ratios of the sides of similar triangles are equal. We use this fact to solve for x, y, and z.

$$\frac{x}{14} = \frac{4}{x} \qquad \frac{y}{10} = \frac{4}{y} \qquad \frac{z}{10} = \frac{14}{z}$$
$$x^2 = 56 \qquad y^2 = 40 \qquad z^2 = 140$$
$$\mathbf{x = 2\sqrt{14}} \qquad \mathbf{y = 2\sqrt{10}} \qquad \mathbf{z = 2\sqrt{35}}$$

26. (8)
$$\frac{6}{5} = \frac{8}{x}$$
$$6x = 40$$
$$x = \mathbf{\frac{20}{3}}$$

27. (4)

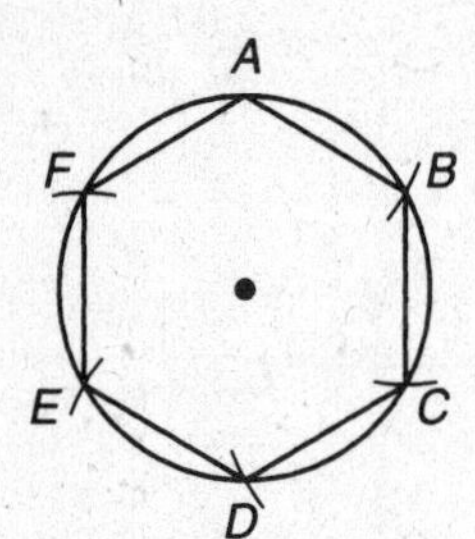

28. (2) The volume of the region in question is the volume of the large sphere minus the volume of the small sphere. To find the volume we first need to find the radii of the two spheres.

$$4\pi(r_{\text{small}})^2 = \pi$$
$$r_{\text{small}}^2 = \frac{1}{4}$$
$$r_{\text{small}} = \frac{1}{2}$$

$$4\pi(r_{\text{large}})^2 = 36\pi$$
$$r_{\text{large}}^2 = 9$$
$$r_{\text{large}} = 3$$

$$V = \frac{4}{3}\pi(r_{\text{large}})^3 - \frac{4}{3}\pi(r_{\text{small}})^3$$
$$= \frac{4}{3}\pi(r_{\text{large}}^3 - r_{\text{small}}^3)$$
$$= \frac{4}{3}\pi\left(27 - \frac{1}{8}\right) = \frac{4}{3}\pi\left(\frac{215}{8}\right)$$
$$= \mathbf{\frac{215\pi}{6}\ cm^3 = 112.57\ cm^3}$$

29. (1) Alternate exterior angles are congruent. A and B are equal. Therefore the answer is **C.**

30. (1) If $x = 0$, $x^4 = 0 = x^2$

If $x \neq 0$, $x^2 > x^4$

Therefore the answer is **D.**

Problem Set 13

1. (R) x = LeAnn's last test score

$$\frac{88 + 100 + 72 + 94 + x}{5} = 90$$
$$88 + 100 + 72 + 94 + x = 90(5)$$
$$x = 450 - 354$$
$$x = \mathbf{96}$$

2. (R) s = successes

f = failures

$$\frac{7}{3} = \frac{s}{f}$$
$$\frac{7}{3} = \frac{s}{1026}$$
$$3s = 7182$$
$$s = \mathbf{2394}$$

3. (7) Consecutive even integers: N, $N + 2$, $N + 4$

$$4[N + (N + 2)] = 2(N + 4)$$
$$4(2N + 2) = 2(N + 4)$$
$$8N + 8 = 2N + 8$$
$$6N = 0$$
$$N = 0$$

0, 2, 4

4. (13) $x = \dfrac{30 + 70}{2} = \mathbf{50}$

5. (13) $x = \dfrac{80 - 50}{2} = \mathbf{15}$

6. (13)
$$5x = 3(9)$$
$$x = \mathbf{\frac{27}{5}}$$

7. (13)
$$5(5 + 7) = 6(6 + x)$$
$$60 = 36 + 6x$$
$$6x = 24$$
$$x = \mathbf{4}$$

8. (12) Number of diagonals $= \dfrac{N(N - 3)}{2}$

$$= \frac{10(10 - 3)}{2}$$
$$= \mathbf{35}$$

9. (12) **Exterior = 360°**

Interior = $(N - 2)180°$

Interior = $(10 - 2)180°$

Interior = 1440°

10. (7) The major premise identified a property of the set of all non-octopi and the minor premise identified that Shannon did not have that propery. That means that Shannon cannot be in the set of all non-octopi, which the conclusion also states. Therefore, the argument is **valid.**

11. (11) $x = \frac{120}{2}$

$\mathbf{x = 60}$

$y = \frac{1}{2}(180 - 120)$

$\mathbf{y = 30}$

12. (11) $85 = \frac{1}{2}(z + 55)$

$170 = z + 55$

$z = 170 - 55$

$\mathbf{z = 115}$

$85 + x = 180$

$\mathbf{x = 95}$

$60 + y = 180$

$\mathbf{y = 120}$

13. (10) $2x - x^2 = 1$

$(x^2 - 2x \quad) = -1$

$(x^2 - 2x + 1) = -1 + 1$

$(x - 1)^2 = 0$

$x - 1 = 0$

$\mathbf{x = 1}$

14. (10) $5x^2 - 2x + 1 = 0$

$\left(x^2 - \frac{2}{5}x \quad\right) = -\frac{1}{5}$

$\left(x^2 - \frac{2}{5}x + \frac{1}{25}\right) = -\frac{1}{5} + \frac{1}{25}$

$\left(x - \frac{1}{5}\right)^2 = -\frac{4}{25}$

$x - \frac{1}{5} = \pm\frac{2}{5}i$

$\mathbf{x = \frac{1}{5} \pm \frac{2}{5}i}$

15. (11) $2x^2 - 3x + 1 = 5$

$2x^2 - 3x - 4 = 0$

$a = 2,\ b = -3,\ c = -4$

$x = \frac{-b \pm \sqrt{b^2 - 4ac}}{2a}$

$x = \frac{-(-3) \pm \sqrt{(-3)^2 - 4(2)(-4)}}{2(2)}$

$x = \frac{3 \pm \sqrt{41}}{4} = \mathbf{\frac{3}{4} \pm \frac{\sqrt{41}}{4}}$

16. (11) $2 + 3x = x^2$

$x^2 - 3x - 2 = 0$

$a = 1,\ b = -3,\ c = -2$

$x = \frac{-b \pm \sqrt{b^2 - 4ac}}{2a}$

$x = \frac{-(-3) \pm \sqrt{(-3)^2 - 4(1)(-2)}}{2(1)}$

$x = \frac{3 \pm \sqrt{17}}{2} = \mathbf{\frac{3}{2} \pm \frac{\sqrt{17}}{2}}$

17. (10) Parallel lines have equal slopes, so $m = 3$.

$y = mx + b$

$y = 3x + b$

$6 = 3(3) + b$

$b = -3$

$\mathbf{y = 3x - 3}$

18. (6) First, label the equations.

$$\begin{cases} \text{(a)} & 2x + 3y - z = 17 \\ \text{(b)} & 3x - y + 2z = 11 \\ \text{(c)} & x - 3y + 3z = -4 \end{cases}$$

Now we can use elimination and substitution to solve.

$$\begin{array}{lr} \text{(a)} & 2x + 3y - z = 17 \\ -2\text{(c)} & -2x + 6y - 6z = 8 \\ \hline \text{(d)} & 9y - 7z = 25 \end{array}$$

$$\begin{array}{lr} \text{(b)} & 3x - y + 2z = 11 \\ (-3)\text{(c)} & -3x + 9y - 9z = 12 \\ \hline \text{(e)} & 8y - 7z = 23 \end{array}$$

$$\begin{array}{lr} \text{(d)} & 9y - 7z = 25 \\ -1\text{(e)} & -8y + 7z = -23 \\ \hline & \mathbf{y = 2} \end{array}$$

(d) $9y - 7z = 25$

$9(2) - 7z = 25$

$-7z = 7$

$\mathbf{z = -1}$

(c) $x - 3y + 3z = -4$

$x - 3(2) + 3(-1) = -4$

$\mathbf{x = 5}$

19. (6)

$$\sqrt{3x+6} + \sqrt{3x-1} = 7$$
$$\sqrt{3x+6} = 7 - \sqrt{3x-1}$$
$$3x + 6 = 49 - 14\sqrt{3x-1} + 3x - 1$$
$$14\sqrt{3x-1} = 49 - 1 - 6$$
$$14\sqrt{3x-1} = 42$$
$$\sqrt{3x-1} = 3$$
$$3x - 1 = 9$$
$$3x = 10$$
$$x = \mathbf{\frac{10}{3}}$$

20. (6)

$$\frac{x-6}{x^2-9x+20} = \frac{2}{x-4} - \frac{x}{x-5}$$
$$\frac{x-6}{(x-4)(x-5)} = \frac{2}{x-4} - \frac{x}{x-5}$$
$$x - 6 = 2(x-5) - x(x-4)$$
$$x - 6 = 2x - 10 - x^2 + 4x$$
$$x^2 - 5x + 4 = 0$$
$$(x-4)(x-1) = 0$$
$$x = \mathbf{1}$$

21. (10)

$$\frac{3-2i+i^3}{1-i} = \frac{3-2i+(-1)i}{1-i}$$
$$= \frac{3-3i}{1-i} = \frac{3(1-i)}{1-i} = \mathbf{3}$$

22. (10)

$$\frac{2-\sqrt{5}}{2+\sqrt{5}} = \frac{2-\sqrt{5}}{2+\sqrt{5}} \cdot \frac{2-\sqrt{5}}{2-\sqrt{5}}$$
$$= \frac{4-4\sqrt{5}+5}{4-5} = \frac{9-4\sqrt{5}}{-1} = \mathbf{-9+4\sqrt{5}}$$

23. (10)

$$\frac{3-2i}{4+6i} = \frac{3-2i}{4+6i} \cdot \frac{4-6i}{4-6i}$$
$$= \frac{12-8i-18i+12i^2}{16-36i^2}$$
$$= \frac{12-26i+12(-1)}{16-36(-1)} = \frac{-26i}{52} = \mathbf{-\frac{1}{2}i}$$

24. (9)

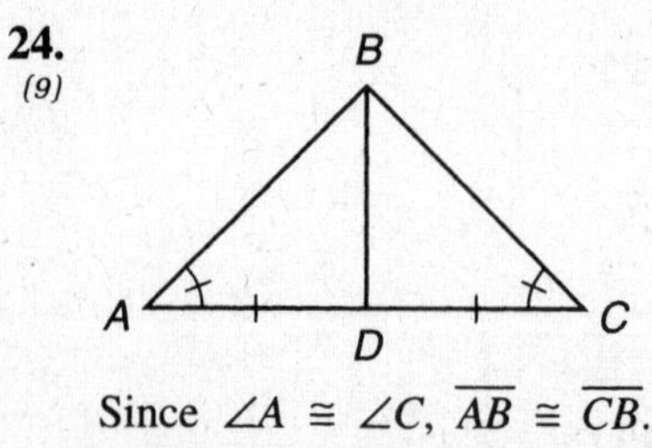

Since $\angle A \cong \angle C$, $\overline{AB} \cong \overline{CB}$.

$\boldsymbol{\triangle ABD \cong \triangle CBD}$ **by *SAS* congruency postulate.**

25. (9)

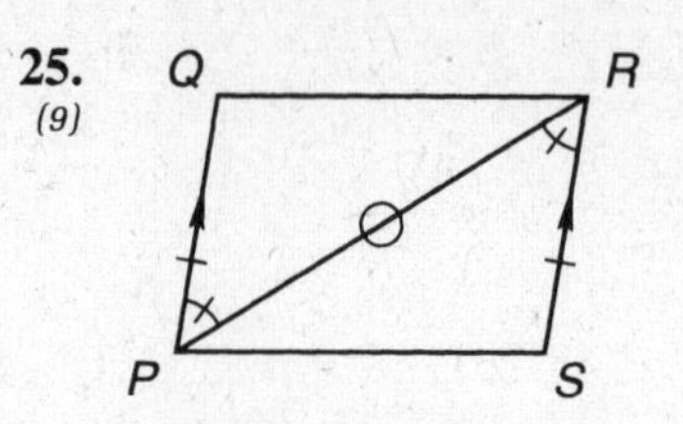

Since $\overline{PQ} \parallel \overline{SR}$, $\angle QPR \cong \angle SRP$

$\boldsymbol{\triangle PQR \cong \triangle RSP}$ **by *SAS* congruency postulate.**

26. (8)

$$\frac{x}{3} = \frac{3}{2}$$
$$2x = 9$$
$$x = \mathbf{\frac{9}{2}}$$

27. (4)

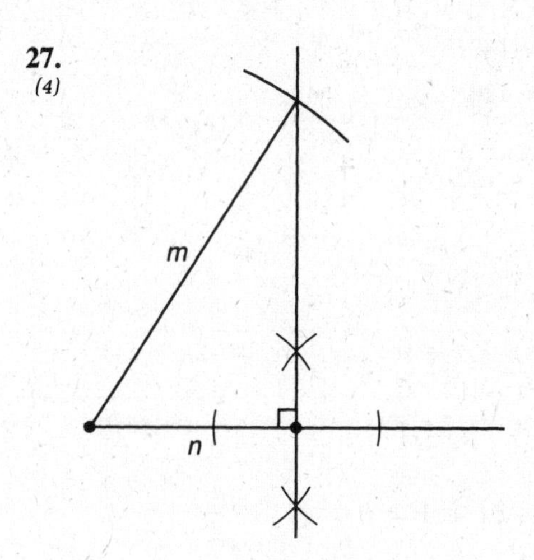

28. (2)

$$V_{\text{cylinder}} = (A_{\text{base}})(h)$$
$$A_{\text{base}} = \frac{1}{2}(\pi r_1^2) + \frac{1}{2}(\pi r_2^2) + \frac{1}{2}(\pi r_3^2) + \left(\frac{1}{2}BH\right)$$

10 10 H 6 6

$$H^2 + 6^2 = 10^2$$
$$H = \sqrt{10^2 - 6^2} = 8$$
$$V_{\text{cylinder}} = \left\{\frac{1}{2}[\pi(5)^2] + \frac{1}{2}[\pi(5)^2] + \frac{1}{2}[\pi(6)^2] + \frac{1}{2}[(12)(8)]\right\}(h)$$
$$= \left[\frac{1}{2}\pi(5^2 + 5^2 + 6^2) + \frac{1}{2}(12)(8)\right](3)$$
$$= (43\pi + 48)(3)$$
$$= \mathbf{(144 + 129\pi)\ m^3 = 549.27\ m^3}$$

29. (1) A: $\dfrac{\frac{1}{2}-\frac{1}{3}}{\frac{1}{2}+\frac{1}{3}} = \dfrac{\frac{3-2}{6}}{\frac{3+2}{6}} = \dfrac{\frac{1}{6}}{\frac{5}{6}} = \dfrac{1}{6}\cdot\dfrac{6}{5} = \dfrac{1}{5}$

B: $\dfrac{1}{5}$

A and B are equal, therefore the answer is **C.**

30. (1) $x > y > 0$

A: $2x - y$
$= x + x - y$
$= (x - y) + x$

B: $x - 2y$
$= x - y - y$
$= (x - y) - y$

If $x > y > 0$, then $x > -y$,

$(x - y) + x > (x - y) - y$

Therefore the answer is **A.**

Problem Set 14

1. (R) First we write the two equations.

$$\begin{cases} \dfrac{N_U}{N_D} = \dfrac{7}{5} \\ 0.1N_U = 0.2N_D - 3 \end{cases}$$

Then we simplify the equations.

$$\begin{cases} \text{(a)}\ 5N_U = 7N_D \\ \text{(b)}\ N_U = 2N_D - 30 \end{cases}$$

Now we can use substitution to solve.

(a)
$$5N_U = 7N_D$$
$$5(2N_D - 30) = 7N_D$$
$$10N_D - 150 = 7N_D$$
$$3N_D = 150$$
$$\mathbf{N_D = 50}$$

(a)
$$5N_U = 7N_D$$
$$5N_U = 7(50)$$
$$\mathbf{N_U = 70}$$

2. (7) Consecutive odd integers: N, $N + 2$, $N + 4$

$$3N = 6(N + 4) + 9$$
$$3N = 6N + 24 + 9$$
$$-3N = 33$$
$$N = -11$$

–11, –9, –7

3. (12) **Exterior = 360°**

Interior $= (N - 2)180°$

Interior $= (17 - 2)180°$

Interior = 2700°

4. (14)

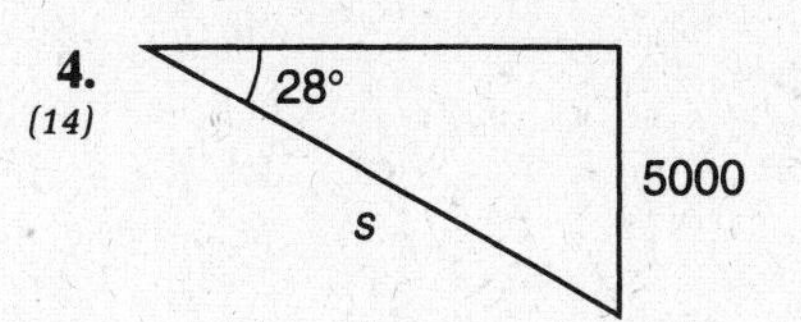

$$\sin 28° = \frac{5000}{s}$$
$$s = \frac{5000}{\sin 28°}$$
$$s = \mathbf{10{,}650.27\ ft}$$

5. (14)

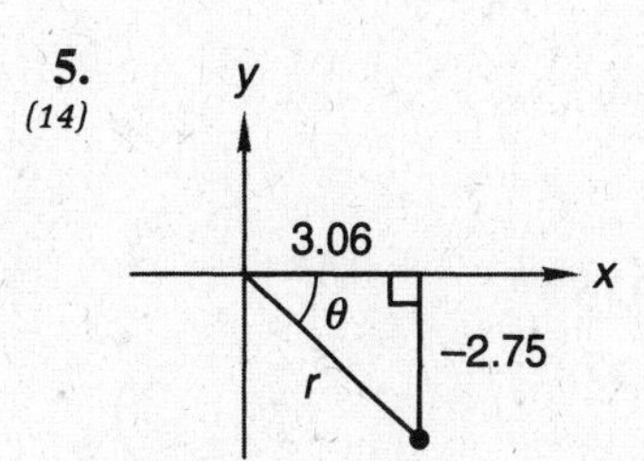

$$r^2 = (3.06)^2 + (2.75)^2$$
$$r^2 = 16.926$$
$$r = 4.11$$

$$\tan\theta = \frac{2.75}{3.06}$$
$$\theta = 41.95°$$

$\mathbf{4.11\ \underline{/318.05°};\ 4.11\ \underline{/-41.95°};}$
$\mathbf{-4.11\ \underline{/138.05°};\ -4.11\ \underline{/-221.95}}$

6. (14)

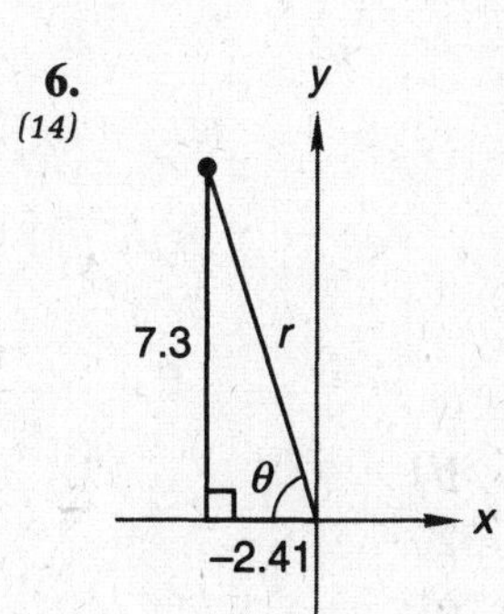

$$r^2 = (2.41)^2 + (7.3)^2$$
$$r^2 = 59.10$$
$$r = 7.69$$

$$\tan\theta = \frac{7.3}{2.41}$$
$$\theta = 71.73°$$

The polar angle is $180° - 71.73° = 108.27°$.

$\mathbf{7.69\ \underline{/108.27°};\ 7.69\ \underline{/-251.73°};}$
$\mathbf{-7.69\ \underline{/288.27°};\ -7.69\ \underline{/-71.73°}}$

7. (14)

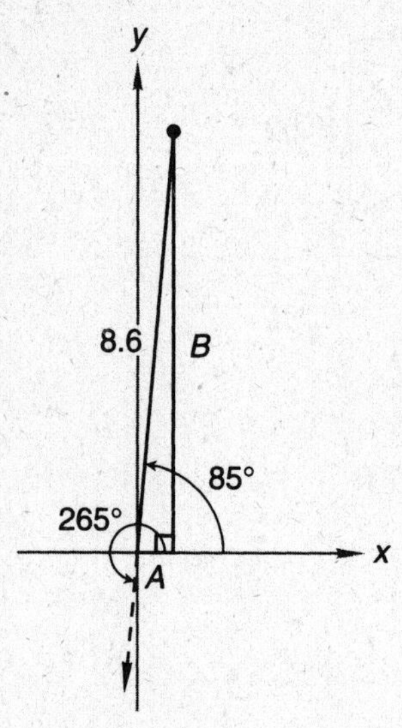

Find A and B.

$A = 8.6 \cos 85° = 0.75$

$B = 8.6 \sin 85° = 8.57$

$\mathbf{0.75\hat{i} + 8.57\hat{j}}$

8. (14)

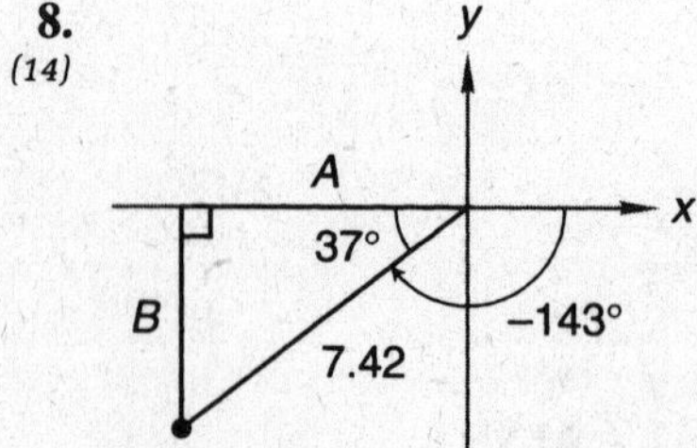

Find A and B.

$A = 7.42 \cos 37° = 5.93$

$B = 7.42 \sin 37° = 4.47$

$\mathbf{-5.93\hat{i} - 4.47\hat{j}}$

9. (11) $y = \frac{1}{2}(40)$

$\mathbf{y = 20}$

$$(4x + 25) + (7x - 13) = 360 - 40$$
$$11x + 12 = 320$$
$$11x = 308$$
$$\mathbf{x = 28}$$

10. (11) $x = 2(70)$

$\mathbf{x = 140}$

$y = 2(40)$

$\mathbf{y = 80}$

$z = 360 - x - y$

$z = 360 - (140) - (80)$

$\mathbf{z = 140}$

11. (13) $x = \frac{1}{2}(60 + 40)$

$\mathbf{x = 50}$

$$y + x = 180$$
$$y + (50) = 180$$
$$\mathbf{y = 130}$$

12. (13) $6(6 + 8) = 5(5 + x)$

$$6(14) = 25 + 5x$$
$$5x = 84 - 25$$
$$5x = 59$$
$$x = \mathbf{\frac{59}{5}}$$

13. (10)

$$4x^2 = -5 - 2x$$
$$4x^2 + 2x = -5$$
$$\left(x^2 + \frac{1}{2}x + \quad\right) = -\frac{5}{4}$$
$$\left(x^2 + \frac{1}{2}x + \frac{1}{16}\right) = -\frac{5}{4} + \frac{1}{16}$$
$$\left(x + \frac{1}{4}\right)^2 = -\frac{19}{16}$$
$$x + \frac{1}{4} = \pm\frac{\sqrt{19}}{4}i$$
$$\mathbf{x = -\frac{1}{4} \pm \frac{\sqrt{19}}{4}i}$$

14. (10)

$$3x + 7 = 4x^2$$
$$4x^2 - 3x = 7$$
$$\left(x^2 - \frac{3}{4}x + \quad\right) = \frac{7}{4}$$
$$\left(x^2 - \frac{3}{4}x + \frac{9}{64}\right) = \frac{7}{4} + \frac{9}{64}$$
$$\left(x - \frac{3}{8}\right)^2 = \frac{121}{64}$$
$$x - \frac{3}{8} = \pm\sqrt{\frac{121}{64}}$$
$$x = \frac{3}{8} \pm \frac{11}{8}$$
$$x = \mathbf{\frac{7}{4}, -1}$$

15. (11)

$$3x^2 = -4 + 2x$$

$$3x^2 - 2x + 4 = 0$$

$$a = 3,\ b = -2,\ c = 4$$

$$x = \frac{-b \pm \sqrt{b^2 - 4ac}}{2a}$$

$$x = \frac{-(-2) \pm \sqrt{(-2)^2 - 4(3)(4)}}{2(3)}$$

$$x = \frac{2 \pm \sqrt{-44}}{6} = \frac{2 \pm 2\sqrt{-11}}{6} = \mathbf{\frac{1}{3} \pm \frac{\sqrt{11}}{3}i}$$

16. (11)

$$5x^2 = -2x + 6$$

$$5x^2 + 2x - 6 = 0$$

$$a = 5,\ b = 2,\ c = -6$$

$$x = \frac{-b \pm \sqrt{b^2 - 4ac}}{2a}$$

$$x = \frac{-(2) \pm \sqrt{(2)^2 - 4(5)(-6)}}{2(5)}$$

$$x = \frac{-2 \pm \sqrt{124}}{10} = \frac{-2 \pm 2\sqrt{31}}{10} = \mathbf{-\frac{1}{5} \pm \frac{\sqrt{31}}{5}}$$

17. (10)

$$2x - 3y - 5 = 0$$

$$3y = 2x - 5$$

$$y = \frac{2}{3}x - \frac{5}{3}$$

The slopes of perpendicular lines are negative reciprocals of each other. So, $m = -\frac{3}{2}$.

$$y = mx + b$$

$$y = -\frac{3}{2}x + b$$

$$1 = -\frac{3}{2}(-3) + b$$

$$b = -\frac{7}{2}$$

$$\mathbf{y = -\frac{3}{2}x - \frac{7}{2}}$$

18. (5)

$$\sqrt{3}\sqrt{3}\sqrt{-3} - \sqrt{-2}\sqrt{-2}\sqrt{-2} - 5\sqrt{-9} + 3i$$

$$= 3\sqrt{3}i - \sqrt{2}i\sqrt{2}i\sqrt{2}i - 5(3i) + 3i$$

$$= 3\sqrt{3}i - 2\sqrt{2}i^3 - 15i + 3i$$

$$= 3\sqrt{3}i - 2\sqrt{2}(-1)i - 12i$$

$$= \mathbf{(3\sqrt{3} + 2\sqrt{2} - 12)i}$$

19. (5)

$$\frac{a^{2x}b^{3x}\left(\sqrt{a^3}\right)^x}{b^{x-y}a} = \frac{a^{2x}b^{3x}a^{\frac{3x}{2}}}{b^{x-y}a}$$

$$= a^{2x}a^{\frac{3x}{2}}a^{-1}b^{3x}b^{-x+y} = a^{2x+\frac{3x}{2}-1}b^{3x-x+y}$$

$$= a^{\frac{7x}{2}-1}b^{2x+y} = \mathbf{a^{\frac{7x-2}{2}}b^{2x+y}}$$

20. (10)

$$\frac{3i^3 - 2i^2 - i^5}{2 - 4i^2 + \sqrt{-4}} = \frac{3i(-1) - 2(-1) - i(-1)(-1)}{2 - 4(-1) + 2i}$$

$$= \frac{-3i - i + 2}{2 + 4 + 2i} = \frac{2 - 4i}{6 + 2i}$$

$$= \frac{2 - 4i}{6 + 2i} \cdot \frac{6 - 2i}{6 - 2i} = \frac{12 - 24i - 4i + 8i^2}{36 - 4i^2}$$

$$= \frac{4 - 28i}{40} = \mathbf{\frac{1}{10} - \frac{7}{10}i}$$

21. (5)

$$\frac{a^{-5} + a^3b^{-2}}{a^4b^{-7}} = \frac{\frac{1}{a^5} + \frac{a^3}{b^2}}{\frac{a^4}{b^7}}$$

$$= \left(\frac{b^2 + a^8}{a^5b^2}\right) \cdot \frac{b^7}{a^4} = \frac{(b^2 + a^8)b^5}{a^9}$$

$$= \mathbf{\frac{b^7 + a^8b^5}{a^9}}$$

22. (9)

Since $BC \parallel DE$, $\angle BCA \cong \angle DEC$ and by $AA \rightarrow AAA$, $\angle B \cong \angle D$.

$\mathbf{\triangle ABC \cong \triangle CDE}$ **by** ***AAAS*** **congruency postulate.**

23. (10)

$$\frac{5 - \sqrt{3}}{4 - 2\sqrt{3}} = \frac{5 - \sqrt{3}}{4 - 2\sqrt{3}} \cdot \frac{4 + 2\sqrt{3}}{4 + 2\sqrt{3}}$$

$$= \frac{20 - 4\sqrt{3} + 10\sqrt{3} - 6}{16 - 12}$$

$$= \frac{14 + 6\sqrt{3}}{4} = \mathbf{\frac{7 + 3\sqrt{3}}{2}}$$

24. (10)

$$\left(2x^{\frac{1}{3}} - 4y^{-\frac{1}{5}}\right)\left(2x^{\frac{1}{3}} + 4y^{-\frac{1}{5}}\right)$$

$$= \left(2x^{\frac{1}{3}}\right)^2 - \left(4y^{-\frac{1}{5}}\right)^2$$

$$= \mathbf{4x^{\frac{2}{3}} - 16y^{-\frac{2}{5}}}$$

25. First, label the three equations.
(6)

$$\begin{cases} \text{(a)} & \dfrac{x}{2} - \dfrac{y}{3} = \dfrac{1}{6} \\ \text{(b)} & \dfrac{x}{3} + \dfrac{2z}{5} = 2\dfrac{3}{5} \\ \text{(c)} & \dfrac{y}{3} + \dfrac{2z}{3} = 4 \end{cases}$$

Now we can use elimination to solve.

$$\text{(a)} \quad \frac{x}{2} - \frac{y}{3} = \frac{1}{6}$$

$$\text{(c)} \quad \frac{y}{3} + \frac{2z}{3} = 4$$

$$\text{(d)} \quad \frac{x}{2} + \frac{2z}{3} = \frac{25}{6}$$

$$6\text{(d)} \quad 3x + 4z = 25$$

$$(-9)\text{(b)} \quad -3x - \frac{18z}{5} = -\frac{117}{5}$$

$$\frac{2}{5}z = \frac{8}{5}$$

$$\boldsymbol{z = 4}$$

$$6\text{(d)} \quad 3x + 4z = 25$$

$$3x + 4(4) = 25$$

$$\boldsymbol{x = 3}$$

$$\text{(a)} \quad \frac{x}{2} - \frac{y}{3} = \frac{1}{6}$$

$$\frac{(3)}{2} - \frac{y}{3} = \frac{1}{6}$$

$$9 - 2y = 1$$

$$\boldsymbol{y = 4}$$

26. $\sqrt{x + 10} - 2 = \sqrt{x}$
(6)

$$\sqrt{x + 10} = \sqrt{x} + 2$$

$$x + 10 = x + 4\sqrt{x} + 4$$

$$4\sqrt{x} = 6$$

$$\sqrt{x} = \frac{3}{2}$$

$$x = \boldsymbol{\frac{9}{4}}$$

27. First, separate the triangles.
(3)

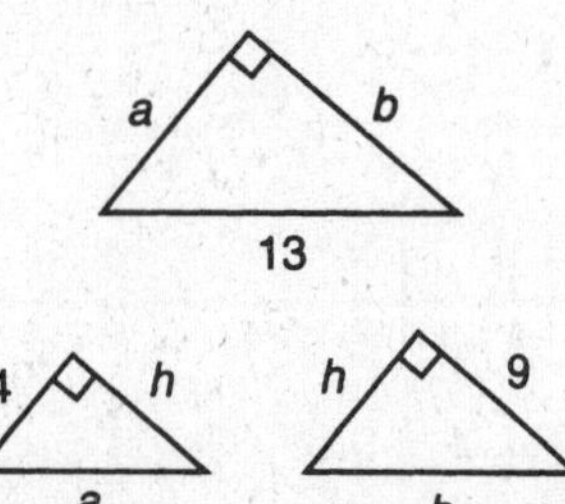

Use the ratios for similar triangles to solve for a and b.

$$\frac{a}{4} = \frac{13}{a} \qquad \frac{b}{9} = \frac{13}{b}$$

$$a^2 = 4(13) \qquad b^2 = 9(13)$$

$$\boldsymbol{a = 2\sqrt{13}} \qquad \boldsymbol{b = 3\sqrt{13}}$$

Now use the Pythagorean Theorem to solve for h.

$$h^2 + 9^2 = b^2$$

$$h = \sqrt{b^2 - 9^2}$$

$$h = \sqrt{(3\sqrt{13})^2 - 81}$$

$$\boldsymbol{h = 6}$$

28.
(2)

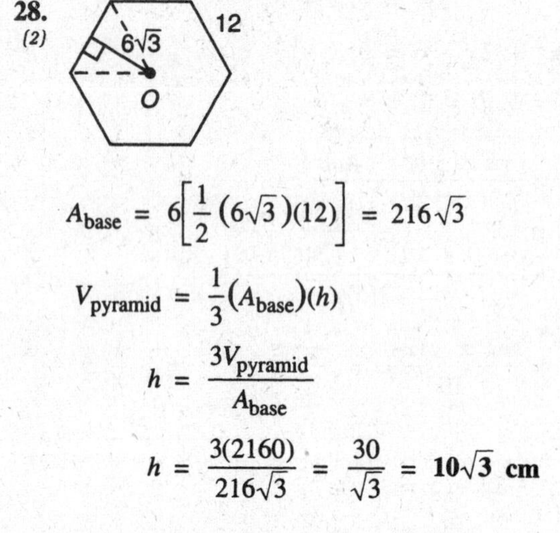

$$A_{\text{base}} = 6\left[\frac{1}{2}(6\sqrt{3})(12)\right] = 216\sqrt{3}$$

$$V_{\text{pyramid}} = \frac{1}{3}(A_{\text{base}})(h)$$

$$h = \frac{3V_{\text{pyramid}}}{A_{\text{base}}}$$

$$h = \frac{3(2160)}{216\sqrt{3}} = \frac{30}{\sqrt{3}} = \boldsymbol{10\sqrt{3}\ \text{cm}}$$

29. We know $\triangle ABC \sim \triangle DEF$ by AAA. Thus,
(1,3)

$$\text{A:} \quad \frac{x}{4} = \frac{3}{2}$$

$$x = 6$$

$$\text{B:} \quad \frac{y}{7} = \frac{2}{3}$$

$$y = \frac{14}{3}$$

$$6 > \frac{14}{3}$$

Therefore the answer is **A**.

30.
(1)

$$PQ > QR$$
$$\theta_1 = 180 - 2x, \ \theta_2 = 180 - 2y$$
$$\theta_1 > \theta_2$$
$$180 - 2x > 180 - 2y$$
$$2y > 2x$$
$$y > x$$

Therefore the answer is **B.**

Problem Set 15

1. (7) Consecutive odd integers:

$N, N + 2, N + 4, N + 6$

$$5[N + (N + 4)] = 22 + 8[(N + 2) + (N + 6)]$$
$$5(2N + 4) = 22 + 8(2N + 8)$$
$$10N + 20 = 22 + 16N + 64$$
$$6N = -66$$
$$N = -11$$

–11, –9, –7, –5

2. (R) c = cobalt, m = mixture

$$\frac{c}{m} = \frac{700 - 430}{700}$$
$$\frac{c}{2800} = \frac{270}{700}$$
$$700c = 270(2800)$$
$$c = \mathbf{1080}$$

3. (10) $m = \dfrac{\text{change in } y}{\text{change in } x} = \dfrac{(5 - 3)}{(-4 + 6)} = \dfrac{2}{2} = 1$

$$y = x + b$$
$$5 = -4 + b$$
$$b = 9$$

$$\mathbf{y = x + 9}$$

4. (14)

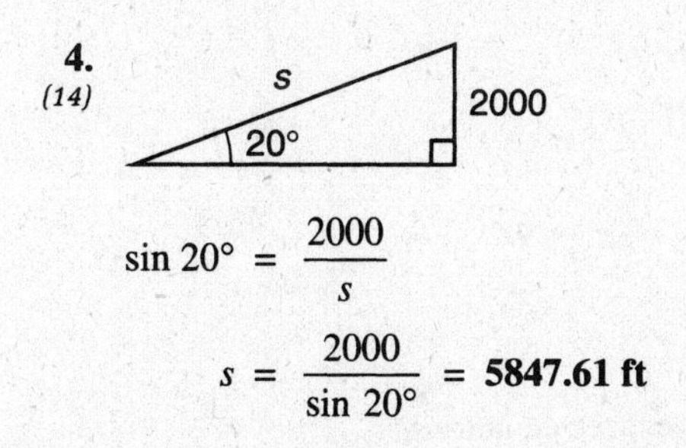

$$\sin 20° = \frac{2000}{s}$$
$$s = \frac{2000}{\sin 20°} = \mathbf{5847.61\ ft}$$

5. (15)

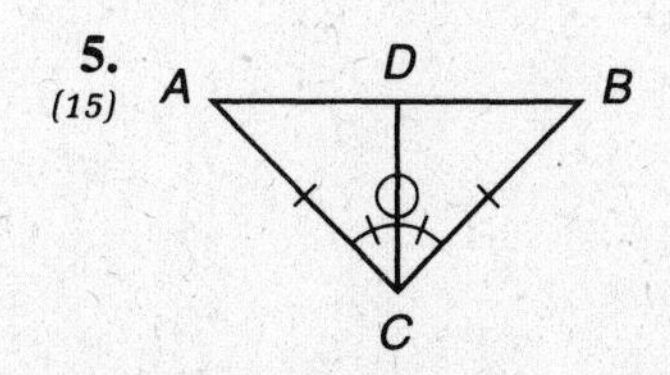

STATEMENTS	REASONS
1. $\overline{AC} \cong \overline{BC}$	1. Given
2. $\angle ACD \cong \angle BCD$	2. Given
3. $\overline{CD} \cong \overline{CD}$	3. Reflexive axiom
4. $\triangle ACD \cong \triangle BCD$	4. *SAS* congruency postulate

6. (15)

STATEMENTS	REASONS
1. $\overline{AB} \cong \overline{AD}$	1. Given
2. $\overline{BC} \cong \overline{DC}$	2. Given
3. $\overline{AC} \cong \overline{AC}$	3. Reflexive axiom
4. $\triangle ABC \cong \triangle ADC$	4. *SSS* congruency postulate

7. (14)

$$r = \sqrt{(7.08)^2 + (4.2)^2} = 8.23$$
$$\tan \theta = \frac{4.2}{7.08}$$
$$\theta = 30.68°$$

The polar angle is $180° - 30.68° = 149.32°$.

8.23 $\underline{/149.32°}$; 8.23 $\underline{/-210.68°}$;
–8.23 $\underline{/329.32°}$; –8.23 $\underline{/-30.68°}$

8. (14)

$$r = \sqrt{(4.2)^2 + 3^2} = 5.16$$
$$\tan \theta = \frac{3}{4.2}$$
$$\theta = 35.54°$$

The polar angle is $360° - 35.54° = 324.46°$.

5.16 $\underline{/324.46°}$; 5.16 $\underline{/-35.54°}$;
–5.16 $\underline{/144.46°}$; –5.16 $\underline{/-215.54°}$

9. (14)

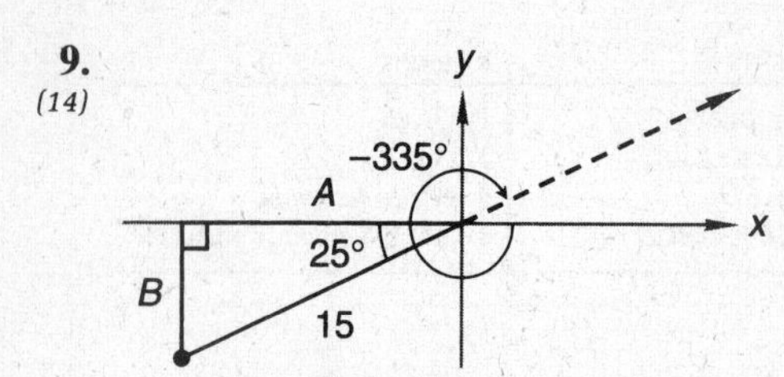

$A = 15 \cos 25° = 13.59$

$B = 15 \sin 25° = 6.34$

$\mathbf{-13.59\hat{i} - 6.34\hat{j}}$

10. (14)

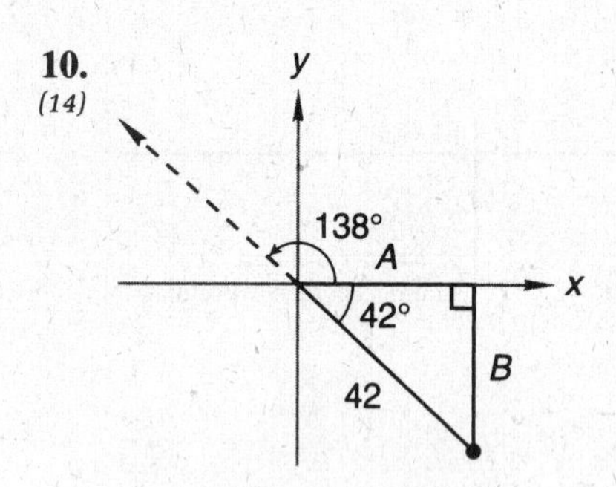

$A = 42 \cos 42° = 31.21$

$B = 42 \sin 42° = 28.10$

$\mathbf{31.21\hat{i} - 28.10\hat{j}}$

11. (13) $7(x) = 5(4)$

$$x = \mathbf{\frac{20}{7}}$$

12. (13) $x = \dfrac{84 - 42}{2} = \mathbf{21}$

13. (10)

$$3x^2 = -4 + 3x$$

$$3x^2 - 3x = -4$$

$$(x^2 - x + \quad) = -\frac{4}{3}$$

$$\left(x^2 - x + \frac{1}{4}\right) = -\frac{4}{3} + \frac{1}{4}$$

$$\left(x - \frac{1}{2}\right)^2 = -\frac{13}{12}$$

$$x - \frac{1}{2} = \pm\frac{\sqrt{13}}{2\sqrt{3}}i \cdot \frac{\sqrt{3}}{\sqrt{3}}$$

$$x - \frac{1}{2} = \pm\frac{\sqrt{39}}{6}i$$

$$x = \mathbf{\frac{1}{2} \pm \frac{\sqrt{39}}{6}i}$$

14. (10)

$$-6x - 9 = 2x^2$$

$$2x^2 + 6x = -9$$

$$(x^2 + 3x + \quad) = -\frac{9}{2}$$

$$\left(x^2 + 3x + \frac{9}{4}\right) = -\frac{9}{2} + \frac{9}{4}$$

$$\left(x + \frac{3}{2}\right)^2 = -\frac{9}{4}$$

$$x + \frac{3}{2} = \pm\frac{3}{2}i$$

$$x = \mathbf{-\frac{3}{2} \pm \frac{3}{2}i}$$

15. (11)

$$-5x - 8 = 3x^2$$

$$3x^2 + 5x + 8 = 0$$

$$a = 3,\ b = 5,\ c = 8$$

$$x = \frac{-b \pm \sqrt{b^2 - 4ac}}{2a}$$

$$x = \frac{-(5) \pm \sqrt{(5)^2 - 4(3)(8)}}{2(3)}$$

$$x = \frac{-5 \pm \sqrt{-71}}{6} = \mathbf{-\frac{5}{6} \pm \frac{\sqrt{71}}{6}i}$$

16. (10) **Line A: $y = 3$**

Use the points $(2, -6)$ and $(6, 0)$ to find the equation of line B.

$$m = \frac{\text{change in } y}{\text{change in } x} = \frac{-6 - 0}{2 - 6} = \frac{3}{2}$$

$$y = \frac{3}{2}x + b$$

$$0 = \frac{3}{2}(6) + b$$

$$b = -9$$

Line B: $y = \frac{3}{2}x - 9$

17. (6) First, label the three equations.

$$\begin{cases} \text{(a)} & \frac{1}{2}x - \frac{1}{4}y = \frac{1}{2} \\ \text{(b)} & 0.2y - 0.2z = 1 \\ \text{(c)} & -\frac{1}{8}x + \frac{1}{4}z = -\frac{1}{4} \end{cases}$$

Now we can use elimination and substitution to solve.

$$\begin{aligned} &\text{(a)} \quad \frac{1}{2}x - \frac{1}{4}y = \frac{1}{2} \\ &\text{4(c)} \quad -\frac{1}{2}x + z = -1 \\ \hline &\quad -\frac{1}{4}y + z = -\frac{1}{2} \\ &\quad z = \frac{1}{4}y - \frac{1}{2} \end{aligned}$$

$$\begin{aligned} \text{(b)} \quad 0.2y - 0.2z &= 1 \\ 0.2y - 0.2\left(\frac{1}{4}y - \frac{1}{2}\right) &= 1 \\ y - \left(\frac{1}{4}y - \frac{1}{2}\right) &= 5 \\ \frac{3}{4}y + \frac{1}{2} &= 5 \\ \frac{3}{4}y &= \frac{9}{2} \\ \mathbf{y} &= \mathbf{6} \end{aligned}$$

$$\begin{aligned} \text{(a)} \quad \frac{1}{2}x - \frac{1}{4}y &= \frac{1}{2} \\ \frac{1}{2}x - \frac{1}{4}(6) &= \frac{1}{2} \\ \frac{1}{2}x &= 2 \\ \mathbf{x} &= \mathbf{4} \end{aligned}$$

$$\begin{aligned} \text{(c)} \quad -\frac{1}{8}x + \frac{1}{4}z &= -\frac{1}{4} \\ -\frac{1}{8}(4) + \frac{1}{4}z &= -\frac{1}{4} \\ \frac{1}{4}z &= \frac{1}{4} \\ \mathbf{z} &= \mathbf{1} \end{aligned}$$

18. (5)
$$\begin{aligned} &4i^3 - 3i^7 + 6i^5 - \sqrt{-4} - 3\sqrt{-9} \\ &= 4(-1)i - 3(-1)(-1)(-1)i + 6(-1)(-1)i \\ &\quad - 2i - 3(3i) \\ &= -4i + 3i + 6i - 2i - 9i \\ &= \mathbf{-6i} \end{aligned}$$

19. (10)
$$\begin{aligned} \frac{2 + 3\sqrt{2}}{2\sqrt{2} - 1} &= \frac{2 + 3\sqrt{2}}{2\sqrt{2} - 1} \cdot \frac{2\sqrt{2} + 1}{2\sqrt{2} + 1} \\ &= \frac{4\sqrt{2} + 12 + 2 + 3\sqrt{2}}{8 - 1} \\ &= \frac{14 + 7\sqrt{2}}{7} = \mathbf{2 + \sqrt{2}} \end{aligned}$$

20. (5)
$$\begin{aligned} \frac{2x^{-1} + 3y^{-1}}{9x^2 - 4y^2} &= \frac{\frac{2}{x} + \frac{3}{y}}{(3x + 2y)(3x - 2y)} \\ &= \frac{\frac{3x + 2y}{xy}}{(3x + 2y)(3x - 2y)} = \mathbf{\frac{1}{xy(3x - 2y)}} \end{aligned}$$

21. (5)
$$\begin{aligned} &\sqrt{2}\sqrt{-2}\sqrt{3}\sqrt{-3} + \sqrt{3}\sqrt{-3} - 7\sqrt{-8}\sqrt{2} \\ &= \sqrt{2}\sqrt{2}i\sqrt{3}\sqrt{3}i + \sqrt{3}\sqrt{3}i - 7(2)\sqrt{2}i\sqrt{2} \\ &= 6(-1) + 3i - 28i \\ &= \mathbf{-6 - 25i} \end{aligned}$$

22. (5)
$$\begin{aligned} &\frac{a^{2z-3}\left(\sqrt[4]{b^4}\right)^{4z+8}}{a^{4z+2}b^{2z-3}} \\ &= a^{2z-3}a^{-4z-2}b^{4z+8}b^{-2z+3} \\ &= \mathbf{a^{-2z-5}b^{2z+11}} \end{aligned}$$

23. (5)
$$\begin{aligned} &\left(3x^{\frac{1}{2}} - 2z^{\frac{1}{4}}\right)\left(3x^{\frac{1}{2}} + 2z^{\frac{1}{4}}\right) \\ &= \left(3x^{\frac{1}{2}}\right)^2 - \left(2z^{\frac{1}{4}}\right)^2 \\ &= \mathbf{9x - 4z^{\frac{1}{2}}} \end{aligned}$$

24. (11) First we write the three equations.

$$\begin{cases} \text{(a)} & x + 2 = 4 \\ \text{(b)} & x + y = z \\ \text{(c)} & 2y - 2 = z \end{cases}$$

We see that we can immediatley solve for x.

$$\begin{aligned} \text{(a)} \quad x + 2 &= 4 \\ \mathbf{x} &= \mathbf{2} \end{aligned}$$

Now we can use substitution to solve.

$$\begin{aligned} \text{(b)} \quad x + y &= z \\ (2) + y &= z \\ 2 + y &= (2y - 2) \\ \mathbf{y} &= \mathbf{4} \end{aligned}$$

$$\begin{aligned} \text{(b)} \quad x + y &= z \\ (2) + (4) &= z \\ \mathbf{z} &= \mathbf{6} \end{aligned}$$

25. (11)
$$8x + 7x = 180$$
$$15x = 180$$
$$x = 12$$
$$m\angle A = 7x$$
$$m\angle A = 7(12)$$
$$\mathbf{m\angle A = 84°}$$
$$m\angle C = 8x$$
$$m\angle C = 8(12)$$
$$\mathbf{m\angle C = 96°}$$

26. (8)
$$\frac{x}{6} = \frac{7}{4}$$
$$4x = 42$$
$$x = \mathbf{\frac{21}{2}}$$

27. (5)
$$BC^2 = 7^2 + x^2$$
$$BC = \sqrt{7^2 + x^2}$$
$$AC^2 = BC^2 + 3^2$$
$$AC^2 = \left(\sqrt{7^2 + x^2}\right)^2 + 3^2$$
$$AC = \sqrt{49 + x^2 + 9}$$
$$AC = \mathbf{\sqrt{58 + x^2}}$$

28. (2)
$$A_{\text{surface}} = 4(A_{\text{side}}) + A_{\text{base}}$$
$$144 = 4\left[\frac{1}{2}\left(\frac{32}{4}\right)l\right] + \left(\frac{32}{4}\right)^2$$
$$144 = 16l + 64$$
$$16l = 80$$
$$\mathbf{l = 5\ m}$$
$$h = \sqrt{5^2 - 4^2}$$
$$\mathbf{h = 3\ m}$$

29. (1)
$a + b > c + b$

By subtracting b from both sides of the inequality we obtain:

$$a > c$$

Therefore the answer is **A.**

30. (1)
$ab > cb$

If $b > 0$, $a > c$.

If $b < 0$, $a < c$.

Therefore the answer is **D.**

Problem Set 16

1. (R) First, write the two equations.

$$\begin{cases} \text{(a)}\ N_C + N_R = 13 \\ \text{(b)}\ 2N_C = \frac{3}{4}N_R + 4 \end{cases}$$

$$\text{(a)}\ N_C + N_R = 13$$
$$N_C = 13 - N_R$$

Now we can use substitution to solve.

$$\text{(b)}\quad 2N_C = \frac{3}{4}N_R + 4$$
$$2(13 - N_R) = \frac{3}{4}N_R + 4$$
$$26 - 2N_R = \frac{3}{4}N_R + 4$$
$$22 = \frac{11}{4}N_R$$
$$\mathbf{N_R = 8}$$

$$\text{(a)}\ N_C + N_R = 13$$
$$N_C + (8) = 13$$
$$\mathbf{N_C = 5}$$

2. (7) Consecutive even integers: N, $N + 2$, $N + 4$
$$7[N + (N + 4)] = 10(N + 2) - 48$$
$$7(2N + 4) = 10N + 20 - 48$$
$$14N + 28 = 10N - 28$$
$$4N = -56$$
$$N = -14$$

–14, –12, –10

3. (10)
$$3y - 2x - 18 = 0$$
$$3y = 2x + 18$$
$$y = \frac{2}{3}x + 6$$

The slopes of perpendicular lines are negative reciprocals of each other, so $m = -\frac{3}{2}$.

$$y = mx + b$$
$$y = -\frac{3}{2}x + b$$
$$2 = -\frac{3}{2}(-4) + b$$
$$b = -4$$
$$\mathbf{y = -\frac{3}{2}x - 4}$$

4. (14)

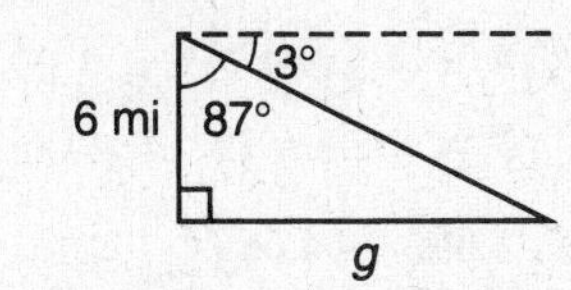

Note: Figure not drawn to scale

$$\tan 87° = \frac{g}{6}$$
$$g = 6\tan 87°$$
$$g = \mathbf{114.49\ mi}$$

5. (16)
$$\frac{\frac{x}{a^2} + \frac{b}{a}}{\frac{1}{a^2} - \frac{k}{ac}} = \frac{\frac{x}{a^2} + \frac{b}{a}}{\frac{1}{a^2} - \frac{k}{ac}} \cdot \frac{\frac{a^2c}{1}}{\frac{a^2c}{1}} = \mathbf{\frac{cx + cab}{c - ka}}$$

6. (16)
$$\frac{p}{m + \frac{m}{1 + \frac{b}{c}}} = \frac{p}{m + \frac{m}{\frac{c+b}{c}}} = \frac{p}{m + \frac{cm}{c+b}}$$
$$= \frac{p}{\frac{m(c+b) + cm}{c+b}} = \frac{p(c+b)}{m(c+b) + cm}$$
$$= \mathbf{\frac{pc + pb}{2mc + mb}}$$

7. (16)
$$\frac{a}{x + \frac{y}{p + \frac{m}{c}}} = \frac{a}{x + \frac{y}{\frac{pc+m}{c}}} = \frac{a}{x + \frac{cy}{pc+m}}$$
$$= \frac{a}{\frac{x(pc+m) + cy}{pc+m}} = \frac{a(pc+m)}{x(pc+m) + cy}$$
$$= \mathbf{\frac{apc + am}{xpc + xm + cy}}$$

8. (16)
$$x = pm\left(\frac{1}{y} + \frac{a}{bd}\right)$$
$$x = \frac{pm}{y} + \frac{pma}{bd}$$
$$bdy \cdot x = \left(\frac{pm}{y} + \frac{pma}{bd}\right)bdy$$
$$bdyx = bdpm + pmay$$
$$bdyx - bdpm = pmay$$
$$d(byx - bpm) = pmay$$
$$\mathbf{d = \frac{pmay}{xyb - pmb}}$$

9. (16)
$$y = m\left(\frac{a}{x} + \frac{b}{mc}\right)$$
$$y = \frac{ma}{x} + \frac{b}{c}$$
$$xc \cdot y = \left(\frac{ma}{x} + \frac{b}{c}\right)xc$$
$$xcy = cma + xb$$
$$xcy - cma = xb$$
$$c(xy - ma) = xb$$
$$\mathbf{c = \frac{bx}{xy - ma}}$$

10. (16)

$$\begin{array}{r} x^2 + 2x + 4 \\ x - 2\overline{)x^3 + 0x^2 + 0x - 1} \\ \underline{x^3 - 2x^2} \qquad\qquad \\ 2x^2 \qquad\qquad \\ \underline{2x^2 - 4x} \qquad \\ 4x - 1 \\ \underline{4x - 8} \\ 7 \end{array}$$

$$\frac{x^3 - 1}{x - 2} = \mathbf{x^2 + 2x + 4 + \frac{7}{x - 2}}$$

Check: $(x^2 + 2x + 4)\left(\frac{x-2}{x-2}\right) + \frac{7}{x-2}$
$$= \frac{x^3 + 2x^2 - 2x^2 + 4x - 4x - 8 + 7}{x - 2}$$
$$= \frac{x^3 - 1}{x - 2}$$

11. (15)

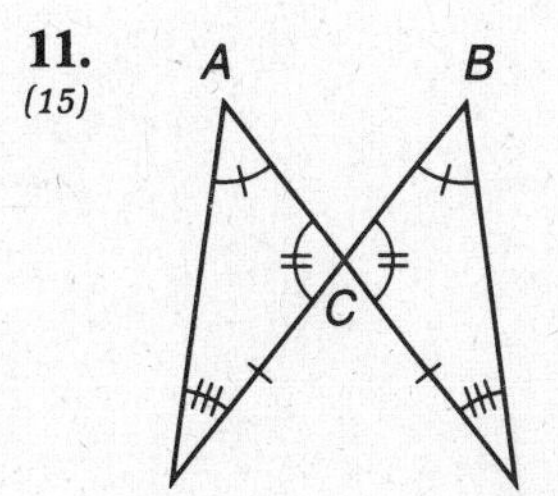

STATEMENTS	REASONS
1. $\angle A \cong \angle B$	1. Given
2. $\angle ACD \cong \angle BCE$	2. Vertical angles are congruent.
3. $\angle D \cong \angle E$	3. If two angles in one triangle are congruent to two angles in a second triangle, then the third angles are congruent.
4. $\overline{CD} \cong \overline{CE}$	4. Given
5. $\Delta ACD \cong \Delta BCE$	5. *AAAS* congruency postulate

12. (15)

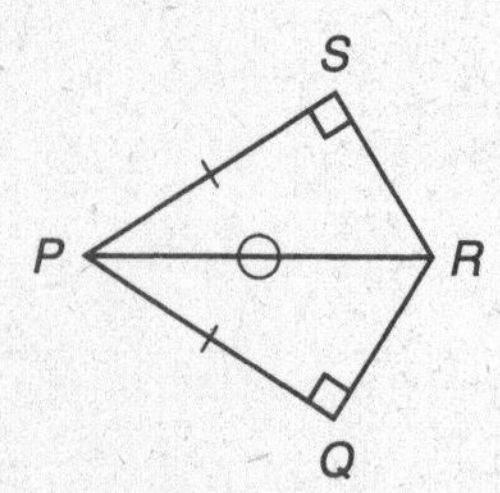

STATEMENTS	REASONS
1. $\angle Q$ and $\angle S$ are right angles.	1. Given
2. ΔPQR and ΔPSR are right triangles.	2. A triangle which contains a right angle is a right triangle.
3. $\overline{PR} \cong \overline{PR}$	3. Reflexive axiom
4. $\overline{PQ} \cong \overline{PS}$	4. Given
5. $\Delta PQR \cong \Delta PSR$	5. *HL* congruency postulate

13. (14)

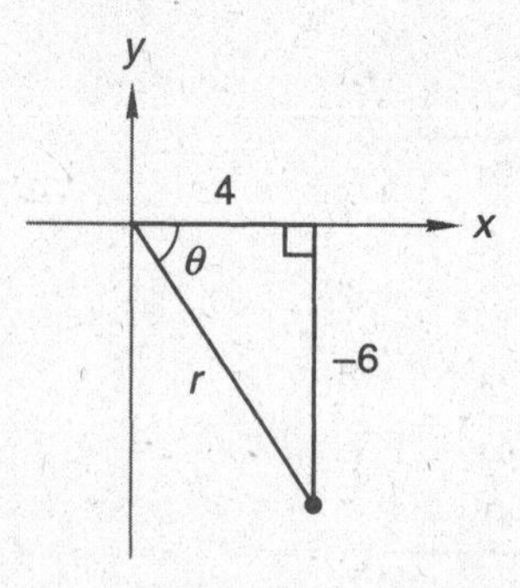

$$r = \sqrt{4^2 + 6^2} = 2\sqrt{13} = 7.21$$

$$\tan\theta = \frac{6}{4}$$

$$\theta = 56.31°$$

The polar angle is $360° - 56.31° = 303.69°$.

$7.21\underline{/303.69°}$; $7.21\underline{/-56.31°}$;
$-7.21\underline{/123.69°}$; $-7.21\underline{/-236.31°}$

14. (14)

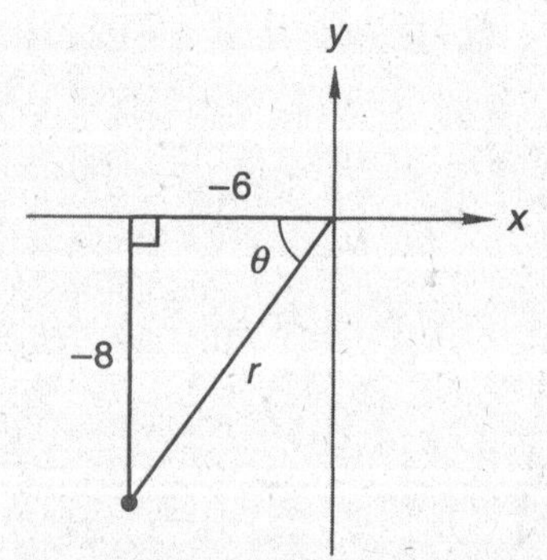

$$r = \sqrt{6^2 + 8^2} = 10$$

$$\tan\theta = \frac{8}{6}$$

$$\theta = 53.13°$$

The polar angle is $180° + 53.13° = 233.13°$.

$10\underline{/233.13°}$; $10\underline{/-126.87°}$;
$-10\underline{/53.13°}$; $-10\underline{/-306.87°}$

15. (14)

$$A = 4.2\cos 75° = 1.09$$

$$B = 4.2\sin 75° = 4.06$$

$$\mathbf{1.09\hat{i} - 4.06\hat{j}}$$

16. (14)

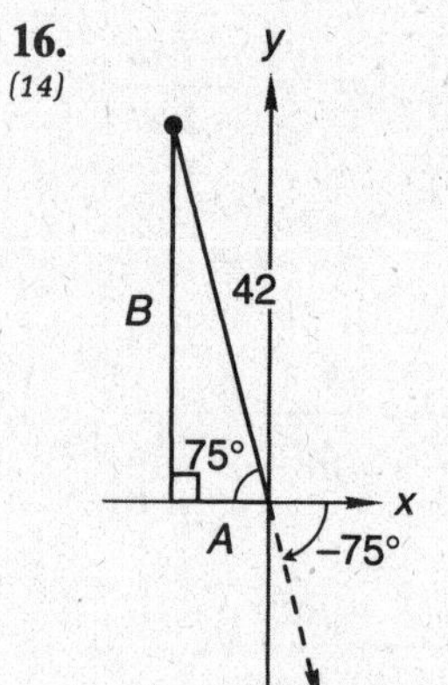

$$A = 42\cos 75° = 10.87$$

$$B = 42\sin 75° = 40.57$$

$$\mathbf{-10.87\hat{i} + 40.57\hat{j}}$$

17. (10)

$$-3x^2 + 2x = 5$$

$$3x^2 - 2x = -5$$

$$x^2 - \frac{2}{3}x = -\frac{5}{3}$$

$$\left(x^2 - \frac{2}{3}x + \quad\right) = -\frac{5}{3}$$

$$\left(x^2 - \frac{2}{3}x + \frac{1}{9}\right) = -\frac{5}{3} + \frac{1}{9}$$

$$\left(x - \frac{1}{3}\right)^2 = -\frac{14}{9}$$

$$x - \frac{1}{3} = \pm\frac{\sqrt{14}}{3}i$$

$$x = \mathbf{\frac{1}{3} \pm \frac{\sqrt{14}}{3}i}$$

18. (11)

$$-5x^2 - x = 7$$

$$5x^2 + x + 7 = 0$$

$$a = 5,\ b = 1,\ c = 7$$

$$x = \frac{-b \pm \sqrt{b^2 - 4ac}}{2a}$$

$$x = \frac{-(1) \pm \sqrt{(1)^2 - 4(5)(7)}}{2(5)}$$

$$x = \frac{-1 \pm \sqrt{-139}}{10} = \mathbf{-\frac{1}{10} \pm \frac{\sqrt{139}}{10}i}$$

19. (10) **Line A: $x = -4$**

Use the points (1, 6) and (4, –6) to find the equation of line B.

$$m = \frac{\text{change in } y}{\text{change in } x} = \frac{6 - (-6)}{1 - 4} = -4$$

$$y = mx + b$$
$$y = -4x + b$$
$$6 = -4(1) + b$$
$$b = 10$$

Line B: $y = -4x + 10$

20. (6) First, label the three equations.

$$\begin{cases} (a) & 2x - z = 5 \\ (b) & 3x + 2y = 13 \\ (c) & y - 2z = 0 \end{cases}$$

(c) $y - 2z = 0$
$$y = 2z$$

Now we can use elimination and substitution to solve.

$$\begin{array}{rrcl} 3(a) & 6x - 3z & = & 15 \\ -2(b) & -6x - 4y & = & -26 \\ \hline & -4y - 3z & = & -11 \\ & -4(2z) - 3z & = & -11 \\ & -8z - 3z & = & -11 \\ & -11z & = & -11 \\ & \mathbf{z} & = & \mathbf{1} \end{array}$$

$$y = 2z$$
$$y = 2(1)$$
$$\mathbf{y = 2}$$

(a) $2x - z = 5$
$$2x - (1) = 5$$
$$2x = 6$$
$$\mathbf{x = 3}$$

21. (6)
$$\frac{x + 2}{x^2 - 3x + 2} - \frac{x - 6}{x^2 - 5x + 6} = \frac{1}{x - 1}$$

$$\frac{x + 2}{(x - 2)(x - 1)} - \frac{x - 6}{(x - 3)(x - 2)} = \frac{1}{(x - 1)}$$

$$(x - 3)(x + 2) - (x - 1)(x - 6) = (x - 2)(x - 3)$$
$$x^2 - x - 6 - x^2 + 7x - 6 = x^2 - 5x + 6$$
$$6x - 12 = x^2 - 5x + 6$$
$$x^2 - 11x + 18 = 0$$
$$(x - 9)(x - 2) = 0$$

Therefore $x = 2$ and 9, but checking these solutions in the original equation we find $x \neq 2$, so $\boldsymbol{x = 9}$.

22. (10)
$$\frac{2i^2 - 3i + 4i^3}{2i^5 - 3i - 2i^2} = \frac{2(-1) - 3i + 4(-1)i}{2(-1)(-1)i - 3i - 2(-1)}$$
$$= \frac{-2 - 3i - 4i}{-i + 2} = \frac{-2 - 7i}{2 - i} \cdot \frac{(2 + i)}{(2 + i)}$$
$$= \frac{-4 - 14i - 2i - 7(-1)}{4 - (-1)} = \frac{3 - 16i}{5}$$
$$= \mathbf{\frac{3}{5} - \frac{16}{5}i}$$

23. (10)
$$\frac{3 + 2\sqrt{3}}{4 - 12\sqrt{3}} = \frac{3 + 2\sqrt{3}}{4 - 12\sqrt{3}} \cdot \frac{4 + 12\sqrt{3}}{4 + 12\sqrt{3}}$$
$$= \frac{12 + 8\sqrt{3} + 36\sqrt{3} + 24(3)}{16 - 144(3)}$$
$$= \frac{84 + 44\sqrt{3}}{-416} = \mathbf{\frac{-21 - 11\sqrt{3}}{104}}$$

24. (5)
$$\sqrt{-2}\sqrt{2} - \sqrt{2}i - \sqrt{-4} + i$$
$$= \sqrt{2}i\sqrt{2} - \sqrt{2}i - 2i + i$$
$$= 2i - \sqrt{2}i - i$$
$$= i - \sqrt{2}i = \mathbf{(1 - \sqrt{2})i}$$

25. (5)
$$\frac{a^3b^{-4} + b^2a^5}{a^{-3}b} = \frac{\frac{a^3}{b^4} + b^2a^5}{\frac{b}{a^3}}$$
$$= \left(\frac{a^3}{b^4} + b^2a^5\right)\frac{a^3}{b}$$
$$= \frac{a^6}{b^5} + a^8b = \mathbf{\frac{a^6 + a^8b^6}{b^5}}$$

26. (1)
$$(6x + 10) + (10x + 10) = 180$$
$$16x = 160$$
$$\mathbf{x = 10}$$

$$z = (10x + 10)$$
$$z = 10(10) + 10$$
$$\mathbf{z = 110}$$

$$\frac{y}{\frac{9}{2}} = \frac{5}{4}$$
$$4y = \frac{45}{2}$$
$$\mathbf{y = \frac{45}{8}}$$

27. First, separate the triangles.
(3)

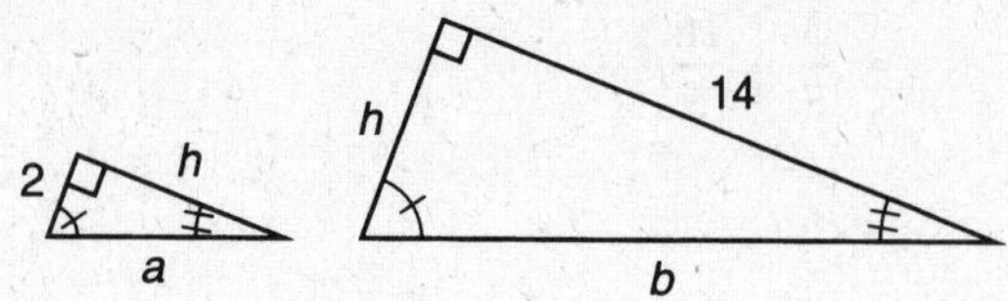

Now use the ratios for similar triangles to solve for a and b.

$$\frac{a}{16} = \frac{2}{a} \qquad \frac{b}{16} = \frac{14}{b}$$

$$a^2 = 32 \qquad b^2 = 224$$

$$\mathbf{a = 4\sqrt{2}} \qquad \mathbf{b = 4\sqrt{14}}$$

Use the Pythagorean Theorem to solve for h.

$$h = \sqrt{a^2 - 2^2}$$

$$h = \sqrt{32 - 4}$$

$$\mathbf{h = 2\sqrt{7}}$$

28. $V = \frac{1}{3}\pi(6)^2(4)$
(2)

$$\mathbf{V = 48\pi \text{ cm}^3 = 150.80 \text{ cm}^3}$$

$$A_{\text{surface}} = \pi(6)^2 + \pi(6)\sqrt{6^2 + 4^2}$$

$$\mathbf{A_{surface} = \left(36\pi + 12\sqrt{13}\pi\right) \text{ cm}^2 = 249.02 \text{ cm}^2}$$

29. We know z is a negative number. Thus,
(1)

$$z^5 < 0 \text{ and } z^4 > 0$$

$$\frac{1}{z^5} < 0 < \frac{1}{z^4}$$

B is greater than A, therefore the answer is **B.**

30. We know that opposite angles of a quadrilateral inscribed in a circle add to make 180°. Thus, $5x + 4y = 180$ and $4x + 5y = 180$, so
(1,11)

$$5x + 4y = 4x + 5y$$

$$x + 4y = 5y$$

$$x = y$$

Therefore the answer is **C.**

Problem Set 17

1. Consecutive even integers: N, $N + 2$, $N + 4$
(7)

$$2(N)(N + 4) = 28 + (N + 2)^2$$

$$2N^2 + 8N = 28 + N^2 + 4N + 4$$

$$N^2 + 4N - 32 = 0$$

$$(N - 4)(N + 8) = 0$$

$$N = 4, -8$$

$$\mathbf{-8, -6, -4 \text{ and } 4, 6, 8}$$

2. First, we write the equations.
(R)

$$\begin{cases} \text{(a)} \ \dfrac{N_S}{N_O} = \dfrac{15}{4} \\ \text{(b)} \ 6N_O = 2N_S - 12 \end{cases}$$

$$\text{(a)} \quad \frac{N_S}{N_O} = \frac{15}{4}$$

$$N_S = \frac{15}{4}N_O$$

Now we can use substitution to solve.

$$\text{(b)} \quad 6N_O = 2N_S - 12$$

$$6N_O = 2\left(\frac{15}{4}N_O\right) - 12$$

$$6N_O = \frac{15}{2}N_O - 12$$

$$-\frac{3}{2}N_O = -12$$

$$\mathbf{N_O = 8}$$

$$N_S = \frac{15}{4}N_O$$

$$N_S = \frac{15}{4}(8)$$

$$\mathbf{N_S = 30}$$

3. $x + 4y = 3$
(10)

$$4y = -x + 3$$

$$y = -\frac{1}{4}x + \frac{3}{4}$$

The slopes of perpendicular lines are negative reciprocals of each other, so $m = 4$.

$$y = mx + b$$

$$y = 4x + b$$

$$-1 = 4(1) + b$$

$$b = -5$$

$$\mathbf{y = 4x - 5}$$

4. (17)
$$\frac{BC}{YZ} = \frac{6}{3} = 2$$
$$\frac{BA}{YX} = \frac{4}{2} = 2$$
$\angle B \cong \angle Y$ given

$\boldsymbol{\Delta ABC \sim \Delta XYZ}$ **by *SAS* similarity postulate.**

5. (17)
$$\frac{HG}{KJ} = \frac{HI}{KL} = \frac{GI}{JL} = \frac{2}{3}$$
$\boldsymbol{\Delta GHI \sim \Delta JKL}$ **by *SSS* similarity postulate.**

6. (17)
$\angle R \cong \angle U$ given

$\angle Q \cong \angle T$ given

$\angle P \cong \angle S$ $AA \rightarrow AAA$

$\boldsymbol{\Delta PQR \sim \Delta STU}$ **by *AAA*.**

7. (17)
$\angle A \cong \angle D$

$\angle B = 180° - (27° + 115°) = 38°$

$\angle B \cong \angle E$

$\angle C \cong \angle F$ $AA \rightarrow AAA$

$\boldsymbol{\Delta ABC \sim \Delta DEF}$ **by *AAA*.**

8. (17)

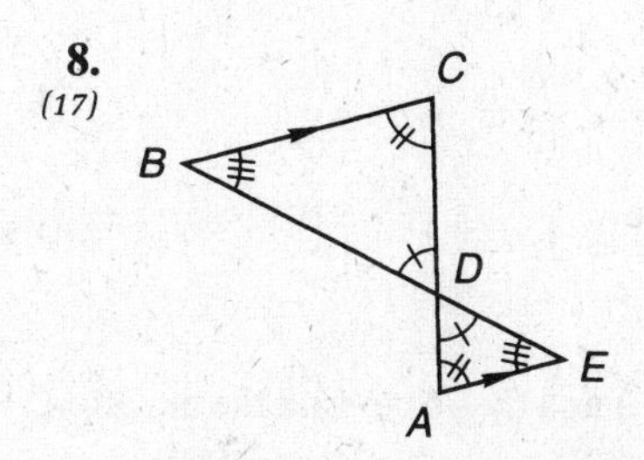

STATEMENTS	REASONS
1. $\overline{BC} \parallel \overline{AE}$	1. Given
2. $\angle ADE \cong \angle CDB$	2. Vertical angles are congruent.
3. $\angle A \cong \angle C$	3. If two parallel lines are cut by a transversal, then each pair of alternate interior angles is congruent.
4. $\angle E \cong \angle B$	4. $AA \rightarrow AAA$
5. $\Delta ADE \sim \Delta CDB$	5. AAA

9. (16)
$$\frac{\frac{a}{bc} - \frac{x}{c^2}}{\frac{c}{a} + \frac{b}{ax}} = \frac{\frac{a}{bc} - \frac{x}{c^2}}{\frac{c}{a} + \frac{b}{ax}} \cdot \frac{\frac{axbc^2}{1}}{\frac{axbc^2}{1}}$$
$$= \boldsymbol{\frac{a^2cx - abx^2}{bxc^3 + b^2c^2}}$$

10. (16)
$$\frac{x}{4 + \frac{x}{1 + \frac{1}{x}}} = \frac{x}{4 + \frac{x}{\frac{x+1}{x}}}$$
$$= \frac{x}{4 + \frac{x^2}{x+1}} = \frac{x}{\frac{4(x+1) + x^2}{x+1}}$$
$$= \frac{x(x+1)}{4(x+1) + x^2} = \boldsymbol{\frac{x^2 + x}{x^2 + 4x + 4}}$$

11. (16)
$$a = b\left(\frac{1+k}{2k}\right)$$
$$a = \frac{b + bk}{2k}$$
$$2ak = b + bk$$
$$2ak - bk = b$$
$$k(2a - b) = b$$
$$\boldsymbol{k = \frac{b}{2a - b}}$$

12. (16)
$$t = ax\left(\frac{b}{x^2} - \frac{2}{br}\right)$$
$$t = \frac{ab}{x} - \frac{2ax}{br}$$
$$brxt = ab^2r - 2ax^2$$
$$brxt - ab^2r = -2ax^2$$
$$r\left(bxt - ab^2\right) = -2ax^2$$
$$\boldsymbol{r = \frac{-2ax^2}{bxt - ab^2}}$$

13. (16)
$$\begin{array}{r} x^2 - 3x + 11 \\ x + 3 \overline{\smash{)}\, x^3 + 0x^2 + 2x - 1} \\ \underline{x^3 + 3x^2} \quad\quad\quad\quad \\ -3x^2 + 2x \quad\;\; \\ \underline{-3x^2 - 9x} \quad\;\; \\ 11x - 1 \\ \underline{11x + 33} \\ -34 \end{array}$$
$$\frac{x^3 + 2x - 1}{x + 3} = \boldsymbol{x^2 - 3x + 11 - \frac{34}{x + 3}}$$
Check:
$$(x^2 - 3x + 11)\left(\frac{x+3}{x+3}\right) - \frac{34}{x+3}$$
$$= \frac{x^3 - 3x^2 + 3x^2 + 11x - 9x + 33 - 34}{x + 3}$$
$$= \frac{x^3 + 2x - 1}{x + 3}$$

14. (15)

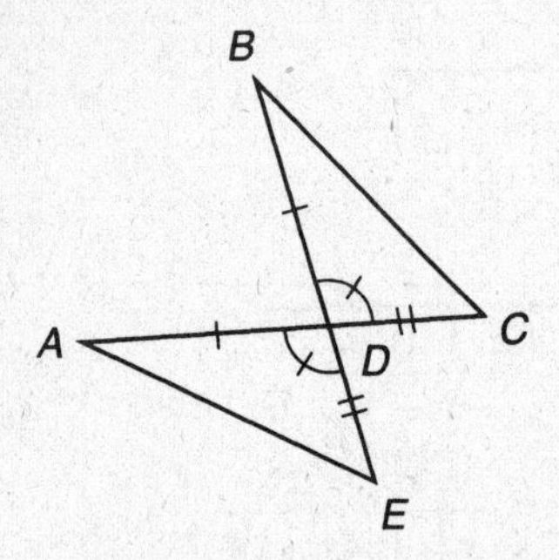

STATEMENTS	REASONS
1. $\overline{AD} \cong \overline{BD}$	1. Given
2. $\angle ADE \cong \angle BDC$	2. Vertical angles are congruent.
3. $\overline{ED} \cong \overline{CD}$	3. Given
4. $\triangle ADE \cong \triangle BDC$	4. *SAS* congruency postulate

15. (14)

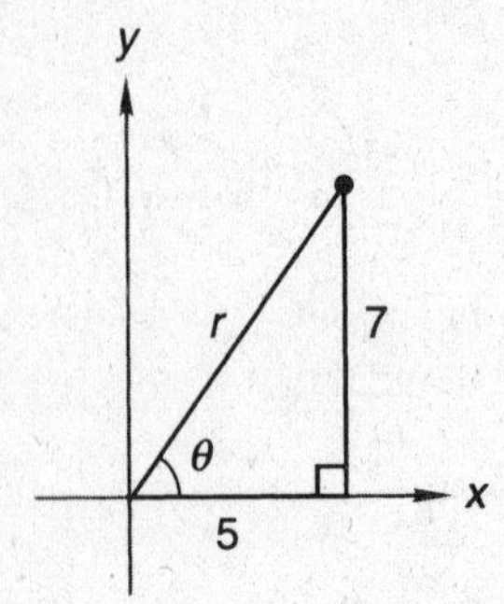

$r = \sqrt{5^2 + 7^2} = 8.60$

$\tan\theta = \frac{7}{5}$

$\theta = 54.46°$

$8.60\underline{/54.46°}$; $8.60\underline{/-305.54°}$;
$-8.60\underline{/234.46°}$; $-8.60\underline{/-125.54°}$

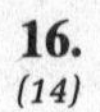

16. (14)

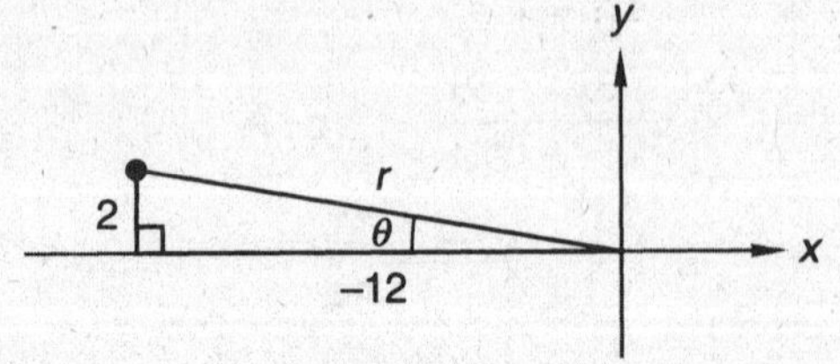

$r = \sqrt{12^2 + 2^2} = 12.17$

$\tan\theta = \frac{2}{12}$

$\theta = 9.46°$

The polar angle is $180° - 9.46° = 170.54°$.

$12.17\underline{/170.54°}$; $12.17\underline{/-189.46°}$;
$-12.17\underline{/350.54°}$; $-12.17\underline{/-9.46°}$

17. (14)

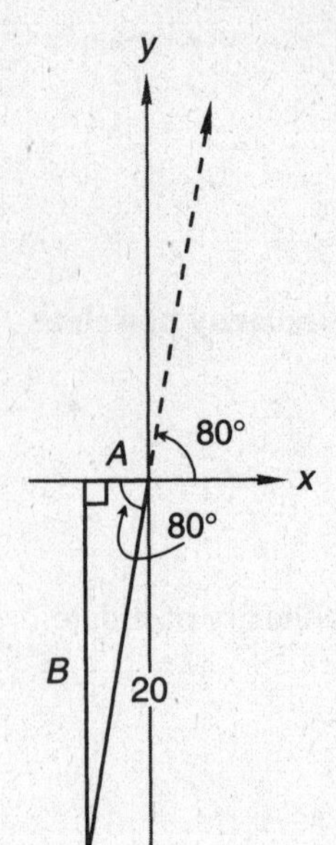

$A = 20\cos 80° = 3.47$

$B = 20\sin 80° = 19.70$

$\mathbf{-3.47\hat{i} - 19.70\hat{j}}$

18. (10)

$$-3 = x - 4x^2$$
$$4x^2 - x = 3$$
$$\left(x^2 - \frac{1}{4}x + \quad\right) = \frac{3}{4}$$
$$\left(x^2 - \frac{1}{4}x + \frac{1}{64}\right) = \frac{3}{4} + \frac{1}{64}$$
$$\left(x - \frac{1}{8}\right)^2 = \frac{49}{64}$$
$$x - \frac{1}{8} = \pm\frac{7}{8}$$
$$x = \frac{1}{8} \pm \frac{7}{8}$$
$$\mathbf{x = 1, -\frac{3}{4}}$$

19. (10) Use the points $(-1, 6)$ and $(3, -6)$ to find the equation of line A.

$$m = \frac{6 + 6}{-1 - 3} = -3$$
$$y = mx + b$$
$$y = -3x + b$$
$$6 = -3(-1) + b$$
$$b = 3$$

Line A: $y = -3x + 3$

Line B: $x = -3$

20. (6) First, label the three equations.

$$\begin{array}{l} \text{(a)} \\ \text{(b)} \\ \text{(c)} \end{array} \begin{cases} x + 2y + z = 0 \\ 3x - y - z = 5 \\ 5x - 2y - 3z = 12 \end{cases}$$

Now we can use elimination and substitution to solve.

$$\begin{aligned} &\text{(a)}\quad x + 2y + z = 0 \\ &\text{(b)}\quad 3x - y - z = 5 \\ \hline &\text{(d)}\quad 4x + y = 5 \\ &\qquad y = 5 - 4x \end{aligned}$$

$$\begin{aligned} -3\text{(b)}\quad -9x + 3y + 3z &= -15 \\ \text{(c)}\quad 5x - 2y - 3z &= 12 \\ \hline -4x + y &= -3 \\ -4x + (5 - 4x) &= -3 \\ -8x &= -8 \\ \mathbf{x} &= \mathbf{1} \end{aligned}$$

$$\begin{aligned} \text{(d)}\quad 4x + y &= 5 \\ 4(1) + y &= 5 \\ \mathbf{y} &= \mathbf{1} \end{aligned}$$

$$\begin{aligned} \text{(a)}\quad x + 2y + z &= 0 \\ (1) + 2(1) + z &= 0 \\ \mathbf{z} &= \mathbf{-3} \end{aligned}$$

21. (6)

$$\begin{aligned} \sqrt{x + 2} &= 2x + 1 \\ x + 2 &= 4x^2 + 4x + 1 \\ 4x^2 + 3x - 1 &= 0 \\ (4x - 1)(x + 1) &= 0 \\ x &= \frac{1}{4}, -1 \\ x &= \mathbf{\frac{1}{4}} \end{aligned}$$

22. (10)

$$\frac{2i^2 + 4i^3 - 1}{i^8 + i^3 - 2} = \frac{2(-1) + 4(-1)i - 1}{(-1)(-1)(-1)(-1) + (-1)i - 2}$$

$$= \frac{-3 - 4i}{-1 - i} = \frac{-3 - 4i}{-1 - i} \cdot \frac{-1 + i}{-1 + i}$$

$$= \frac{3 + 4i - 3i - 4i^2}{1 - i^2} = \frac{3 + i - 4(-1)}{1 - (-1)}$$

$$= \frac{7 + i}{2} = \mathbf{\frac{7}{2} + \frac{1}{2}i}$$

23. (10)

$$\frac{3 - 2\sqrt{5}}{\sqrt{5} - 1} = \frac{3 - 2\sqrt{5}}{\sqrt{5} - 1} \cdot \frac{\sqrt{5} + 1}{\sqrt{5} + 1}$$

$$= \frac{3\sqrt{5} - 2(5) + 3 - 2\sqrt{5}}{5 - 1} = \mathbf{\frac{-7 + \sqrt{5}}{4}}$$

24. (5)

$$\left(3x^{\frac{1}{2}} + y^{\frac{3}{2}}\right)\left(x^{\frac{1}{2}} - 2y^{\frac{3}{2}}\right)$$

$$= 3x^{\frac{1}{2}+\frac{1}{2}} + x^{\frac{1}{2}}y^{\frac{3}{2}} - 6x^{\frac{1}{2}}y^{\frac{3}{2}} - 2y^{\frac{3}{2}+\frac{3}{2}}$$

$$= \mathbf{3x - 5x^{\frac{1}{2}}y^{\frac{3}{2}} - 2y^3}$$

25. (10)

$$\frac{(x^{3a+2})(\sqrt[3]{b})^{2y}}{x^{4a-1}(b^{3a})^{\frac{1}{2}}} = x^{3a+2}x^{-4a+1}b^{\frac{2}{3}y}b^{-\frac{3}{2}a}$$

$$= x^{-a+3}b^{\frac{2}{3}y-\frac{3}{2}a} = \mathbf{x^{-a+3}b^{\frac{4y-9a}{6}}}$$

26. (13)

$$\begin{aligned} x(8 - x) &= 4(3) \\ 8x - x^2 &= 12 \\ x^2 - 8x + 12 &= 0 \\ (x - 6)(x - 2) &= 0 \\ x &= 2, 6 \end{aligned}$$

If $x = 6$, then $8 - x < x$, which is not consistent with the figure. Therefore, $x =$ **2.**

27. (8)

$$\begin{aligned} \frac{x}{10} &= \frac{12 - x}{14} \\ 14x &= 120 - 10x \\ 24x &= 120 \\ x &= 5 \\ \mathbf{AD} &= \mathbf{5} \end{aligned}$$

$$\begin{aligned} DC &= 12 - x \\ DC &= 12 - (5) \\ \mathbf{DC} &= \mathbf{7} \end{aligned}$$

28. (2) Find the radius of the base.

$$\begin{aligned} 2\pi r &= 12\pi \\ r &= 6 \end{aligned}$$

$$A_{\text{base}} + A_{\text{lateral surface}} = A_{\text{cone}}$$

$$\begin{aligned} \pi r^2 + \pi rl &= 96\pi \\ \pi(6)^2 + \pi(6)l &= 96\pi \\ 36 + 6l &= 96 \\ 6l &= 60 \\ \mathbf{l} &= \mathbf{10\ cm} \end{aligned}$$

$$\begin{aligned} h^2 + r^2 &= l^2 \\ h &= \sqrt{l^2 - r^2} \\ h &= \sqrt{10^2 - 6^2} \\ \mathbf{h} &= \mathbf{8\ cm} \end{aligned}$$

29. (1) If $x = -5$ and $y = -3$, $x - 2y > 0$ and $x < y$.

If $x = 5$ and $y = 2$,

then $x - 2y > 0$ but $x > y$.

Therefore the answer is **D.**

30. (1) A: $m = 48$

$$n + 48 = 180$$
$$n = 132$$
$$n - 3m = 132 - 3(48) = -12$$

B: -11

$$-12 < -11$$

Therefore the answer is **B.**

Problem Set 18

1. (18) First, we write the three equations.

$$\begin{cases} N_B = N_W + N_G - 7 \\ N_G = N_B + N_W + 1 \\ 3N_B = N_G \end{cases}$$

Rearrange the equations.

$$\begin{matrix} \text{(a)} \\ \text{(b)} \\ \text{(c)} \end{matrix} \begin{cases} N_B - N_W - N_G = -7 \\ -N_B - N_W + N_G = 1 \\ N_G = 3N_B \end{cases}$$

Now we can use elimination and substitution to solve.

$$\begin{array}{lr} \text{(a)} & N_B - N_W - N_G = -7 \\ \text{(b)} & -N_B - N_W + N_G = 1 \\ \hline & -2N_W = -6 \end{array}$$

$$\mathbf{N_W = 3}$$

$$\text{(a)} \quad N_B - N_W - N_G = -7$$
$$N_B - (3) - (3N_B) = -7$$
$$-2N_B = -4$$
$$\mathbf{N_B = 2}$$

$$\text{(c)} \quad N_G = 3N_B$$
$$N_G = 3(2)$$
$$\mathbf{N_G = 6}$$

2. (18) First, we write the three equations.

$$\begin{matrix} \text{(a)} \\ \text{(b)} \\ \text{(c)} \end{matrix} \begin{cases} N_Q + N_D + N_N = 20 \\ 25N_Q + 10N_D + 5N_N = 205 \\ N_D = 3N_Q \end{cases}$$

Now we can use elimination and substitution to solve.

$$\text{(a)} \quad N_Q + N_D + N_N = 20$$
$$N_Q + (3N_Q) + N_N = 20$$
$$\text{(a')} \quad 4N_Q + N_N = 20$$

$$\text{(b)} \quad 25N_Q + 10N_D + 5N_N = 205$$
$$25N_Q + 10(3N_Q) + 5N_N = 205$$
$$\text{(b')} \quad 55N_Q + 5N_N = 205$$

$$\begin{array}{lr} \text{(b')} & 55N_Q + 5N_N = 205 \\ -5\text{(a')} & -20N_Q - 5N_N = -100 \\ \hline & 35N_Q = 105 \end{array}$$

$$\mathbf{N_Q = 3}$$

$$\text{(c)} \quad N_D = 3N_Q$$
$$N_D = 3(3)$$
$$\mathbf{N_D = 9}$$

$$\text{(a)} \quad N_Q + N_D + N_N = 20$$
$$(3) + (9) + N_N = 20$$
$$\mathbf{N_N = 8}$$

3. (18) $R = \dfrac{kB}{G^2}$

$$5 = \frac{k(2)}{4^2}$$

$$k = \frac{5(4^2)}{2} = 40$$

$$R = \frac{40B}{G^2} = \frac{40(4)}{4^2} = \mathbf{10}$$

4. (10) $5x - 2y + 4 = 0$

$$2y = 5x + 4$$

$$y = \frac{5}{2}x + 2$$

The slopes of perpendicular lines are negative reciprocals of each other, so $m = -\frac{2}{5}$.

$$y = mx + b$$

$$y = -\frac{2}{5}x + b$$

$$3 = \left(-\frac{2}{5}\right)(-2) + b$$

$$b = 3 - \frac{4}{5} = \frac{11}{5}$$

$$\mathbf{y = -\frac{2}{5}x + \frac{11}{5}}$$

5. (18) First, write the two equations.

$$\begin{cases} U + T = 13 & \text{(a)} \\ 10U + T = 45 + 10T + U & \text{(b)} \end{cases}$$

(b) $10U + T = 45 + 10T + U$

$9U - 9T = 45$

Now we can use elimination and substitution to solve.

$$\begin{aligned} 9U - 9T &= 45 \\ 9(a)\quad 9U + 9T &= 117 \\ \hline 18U &= 162 \\ U &= 9 \end{aligned}$$

(a) $U + T = 13$

$(9) + T = 13$

$T = 4$

The number is **49.**

6. (18) P_N = volume of 14% solution

Iodine_1 + iodine added = iodine final

$$0.47(89) + 0.14(P_N) = 0.29(89 + P_N)$$

$$41.83 + 0.14(P_N) = 25.81 + 0.29P_N$$

$$0.15P_N = 16.02$$

$$P_N = \mathbf{106.8\ oz}$$

7. (18) $C = 12 \times 3 = 36$

$H = 1 \times 7 = 7$

$Cl = 35 \times 1 = 35$

$T = 78$

Use ratios to solve.

$$\frac{Cl}{T} = \frac{35}{78}$$

$$\frac{400}{T} = \frac{35}{78}$$

T = **891.43 grams**

8. (17) $\frac{PQ}{ST} = \frac{15}{10} = \frac{3}{2}$

$\frac{PR}{SU} = \frac{12}{8} = \frac{3}{2}$

$\frac{QR}{TU} = \frac{18}{12} = \frac{3}{2}$

$\Delta PQR \sim \Delta STU$ by SSS similarity postulate.

9. (16)

$$\frac{\frac{m^2}{a^2} + \frac{7y}{x}}{\frac{p^2}{ax} - \frac{3}{a^2}} = \frac{\frac{m^2}{a^2} + \frac{7y}{x}}{\frac{p^2}{ax} - \frac{3}{a^2}} \cdot \frac{\frac{a^2x}{1}}{\frac{a^2x}{1}} = \mathbf{\frac{m^2x + 7a^2y}{ap^2 - 3x}}$$

10. (16)

$$\frac{m}{a + \frac{b}{1 + \frac{c}{d}}} = \frac{m}{a + \frac{b}{\frac{d + c}{d}}} = \frac{m}{a + \frac{bd}{d + c}}$$

$$= \frac{m}{\frac{a(d + c) + bd}{d + c}} = \frac{m(d + c)}{ad + ac + bd}$$

$$= \mathbf{\frac{md + mc}{ad + ac + bd}}$$

11. (16)

$$x = kb\left(\frac{1}{c} - \frac{a}{x}\right)$$

$$x = \frac{kb}{c} - \frac{kba}{x}$$

$$x^2c = kbx - kbac$$

$$x^2c + kbac = kbx$$

$$c(x^2 + kba) = kbx$$

$$\mathbf{c = \frac{kbx}{x^2 + kba}}$$

12. (16)

$$mc = p\left(\frac{a}{cm} + \frac{2}{kc}\right)$$

$$mc = \frac{pa}{cm} + \frac{2p}{kc}$$

$$m^2c^2k = pak + 2pm$$

$$m^2c^2k - pak = 2pm$$

$$k(m^2c^2 - pa) = 2pm$$

$$\mathbf{k = \frac{2pm}{m^2c^2 - pa}}$$

13. (16)

$$\begin{array}{r|l} & x^2 + 1 \\ \hline x^2 - 1 & x^4 + 0x^3 + 0x^2 + 0x - 2 \\ & x^4 \qquad\quad - x^2 \\ \hline & x^2 \\ & x^2 \qquad\quad - 1 \\ \hline & -1 \end{array}$$

$$\frac{x^4 - 2}{x^2 - 1} = \mathbf{x^2 + 1 - \frac{1}{x^2 - 1}}$$

Check: $(x^2 + 1)\left(\frac{x^2 - 1}{x^2 - 1}\right) - \frac{1}{x^2 - 1}$

$$= \frac{x^4 - x^2 + x^2 - 1 - 1}{x^2 - 1} = \frac{x^4 - 2}{x^2 - 1}$$

14. (15)

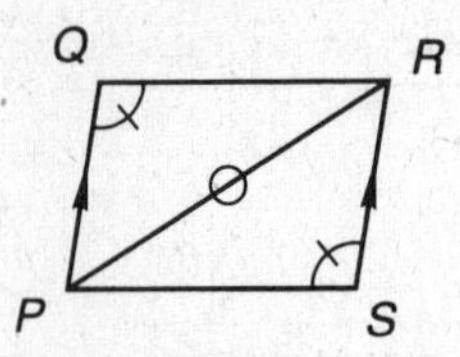

STATEMENTS	REASONS
1. $\angle Q \cong \angle S$	1. Given
2. $\overline{PQ} \parallel \overline{SR}$	2. Given
3. $\angle QPR \cong \angle SRP$	3. If two parallel lines are cut by a transversal, then each pair of alternate interior angles is congruent.
4. $\angle QRP \cong \angle SPR$	4. If two angles in one triangle are congruent to two angles in a second triangle, then the third angles are congruent.
5. $\overline{PR} \cong \overline{PR}$	5. Reflexive axiom
6. $\Delta PQR \cong \Delta RSP$	6. *AAAS* congruency postulate

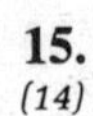

15. (14)

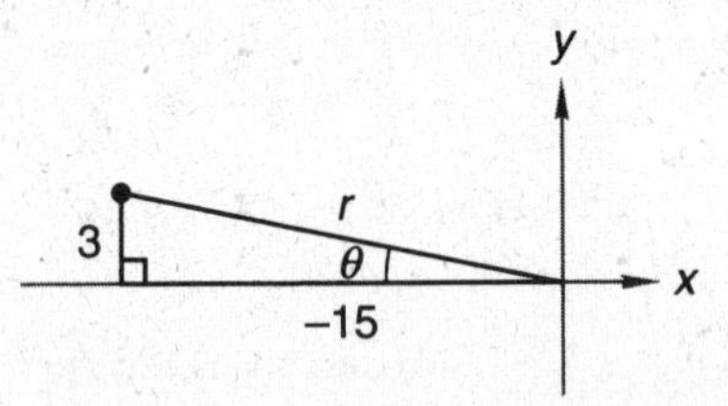

$$r = \sqrt{15^2 + 3^2} = 15.30$$

$$\tan\theta = \frac{3}{15}$$

$$\theta = 11.31°$$

The polar angle is $180° - 11.31° = 168.69°$.

15.30/168.69°; 15.30/–191.31°; –15.30/348.69°; –15.30/–11.31°

16. (14)

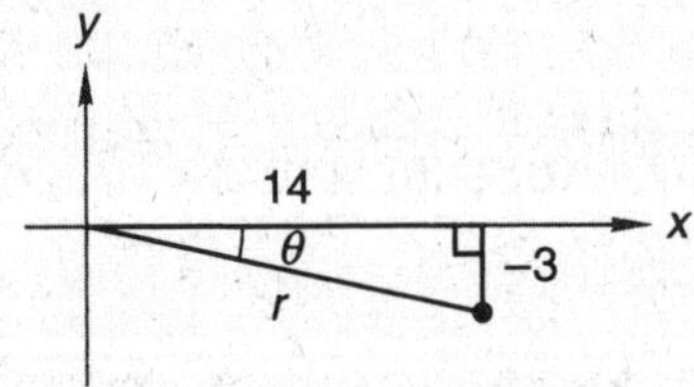

$$r = \sqrt{14^2 + 3^2} = 14.32$$

$$\tan\theta = \frac{3}{14}$$

$$\theta = 12.09°$$

The polar angle is $360° - 12.09° = 347.91°$.

14.32/347.91°; 14.32/–12.09°; –14.32/167.91°; –14.32/–192.09°

17. (14)

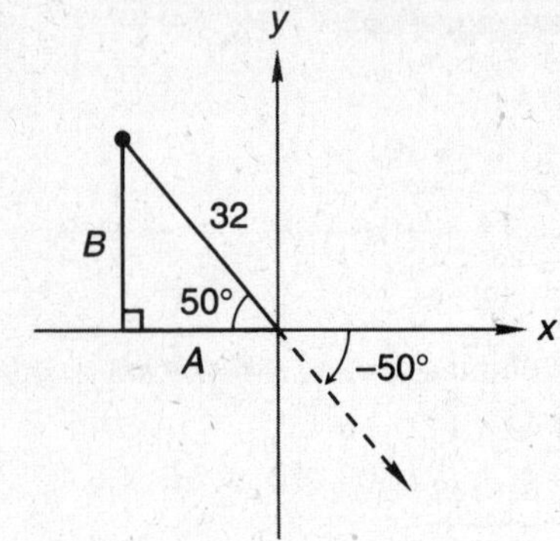

$$A = 32\cos(50°) = 20.57$$

$$B = 32\sin(50°) = 24.51$$

$$\mathbf{-20.57\hat{i} + 24.51\hat{j}}$$

18. (14)

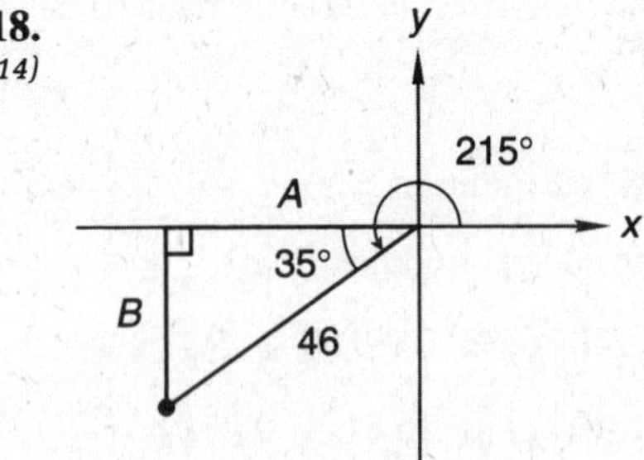

$$A = 46\cos 35° = 37.68$$

$$B = 46\sin 35° = 26.38$$

$$\mathbf{-37.68\hat{i} - 26.38\hat{j}}$$

19. (11)

$$4 = -3x^2 + 5x$$

$$3x^2 - 5x + 4 = 0$$

$$a = 3,\ b = -5,\ c = 4$$

$$x = \frac{-(-5) \pm \sqrt{(-5)^2 - 4(3)(4)}}{2(3)}$$

$$x = \frac{5 \pm \sqrt{-23}}{6} = \mathbf{\frac{5}{6} \pm \frac{\sqrt{23}}{6}i}$$

20. (10)

$$-x + 2x^2 = 3$$

$$2x^2 - x = 3$$

$$\left(x^2 - \frac{1}{2}x + \quad\right) = \frac{3}{2}$$

$$\left(x^2 - \frac{1}{2}x + \frac{1}{16}\right) = \frac{3}{2} + \frac{1}{16}$$

$$\left(x - \frac{1}{4}\right)^2 = \frac{25}{16}$$

$$x - \frac{1}{4} = \pm\sqrt{\frac{25}{16}}$$

$$x = \frac{1}{4} \pm \frac{5}{4} = \mathbf{\frac{3}{2}, -1}$$

21. (6) First, label the three equations.

$$\begin{cases} \text{(a)} & \dfrac{x}{2} - \dfrac{3y}{4} = \dfrac{5}{2} \\ \text{(b)} & 0.4y - 0.05z = -0.95 \\ \text{(c)} & \dfrac{x}{3} - \dfrac{2z}{5} = -\dfrac{8}{15} \end{cases}$$

Now we can use elimination and substitution to solve.

$$\begin{aligned} 20\text{(a)} \quad & 10x - 15y = 50 \\ -30\text{(c)} \quad & -10x + 12z = 16 \\ \hline \text{(d)} \quad & -15y + 12z = 66 \end{aligned}$$

$$\begin{aligned} 8\text{(d)} \quad & -120y + 96z = 528 \\ 300\text{(b)} \quad & 120y - 15z = -285 \\ \hline & 81z = 243 \\ & \mathbf{z = 3} \end{aligned}$$

$$\begin{aligned} 100\text{(b)} \quad 40y - 5z &= -95 \\ 40y - 5(3) &= -95 \\ 40y &= -80 \\ \mathbf{y} &= \mathbf{-2} \end{aligned}$$

$$\begin{aligned} 30\text{(c)} \quad 10x - 12z &= -16 \\ 10x - 12(3) &= -16 \\ 10x &= 20 \\ \mathbf{x} &= \mathbf{2} \end{aligned}$$

22. (5) $\sqrt{5}\sqrt{-5} + \sqrt{-3}\sqrt{-3} + \sqrt{-4} - i$

$= \sqrt{5}\sqrt{5}i + \sqrt{3}i\sqrt{3}i + 2i - i = 5i + 3i^2 + 2i - i$

$= 5i + 3(-1) + 2i - i = \mathbf{-3 + 6i}$

23. (10) $\dfrac{4 + 2\sqrt{12}}{6 - \sqrt{48}} = \dfrac{4 + 4\sqrt{3}}{6 - 4\sqrt{3}} \cdot \dfrac{6 + 4\sqrt{3}}{6 + 4\sqrt{3}}$

$= \dfrac{24 + 24\sqrt{3} + 16\sqrt{3} + 48}{36 - 48}$

$= \dfrac{72 + 40\sqrt{3}}{-12} = \mathbf{\dfrac{-18 - 10\sqrt{3}}{3}}$

24. (5) $b^2\sqrt[3]{ab^4}(ab^3)^{\frac{1}{5}} = b^2a^{\frac{1}{3}}b^{\frac{4}{3}}a^{\frac{1}{5}}b^{\frac{3}{5}}$

$= a^{\frac{1}{3}+\frac{1}{5}}b^{2+\frac{4}{3}+\frac{3}{5}} = a^{\frac{5}{15}+\frac{3}{15}}b^{\frac{30}{15}+\frac{20}{15}+\frac{9}{15}} = \mathbf{a^{\frac{8}{15}}b^{\frac{59}{15}}}$

25. (3) Use the ratios for similar triangles to solve for a.

$$\frac{\frac{2}{4}}{5} = \frac{\frac{a+2}{3}}{2}$$

$$\frac{10}{4} = \frac{2a + 4}{3}$$

$$30 = 8a + 16$$

$$8a = 14$$

$$a = \mathbf{\frac{7}{4}}$$

26. (3) All three triangles are similar. We can use the ratios for similar triangles to solve for x.

$$\frac{6}{x} = \frac{x + 5}{6}$$

$$x^2 + 5x = 36$$

$$x^2 + 5x - 36 = 0$$

$$(x + 9)(x - 4) = 0$$

$$x = 4$$

$$JL = x + (x + 5) = 4 + (4 + 5) = \mathbf{13}$$

27. (3) $A_{\text{shaded region}} = A_{\text{circle}} - A_{\text{square}}$

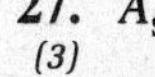

Begin by finding the length of the radius using Pythagorean's Theorem.

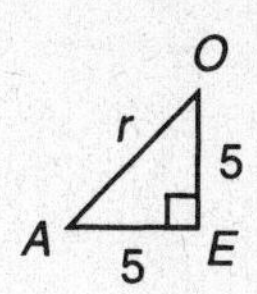

$$r = \sqrt{5^2 + 5^2} = 5\sqrt{2}$$

$$A_{\text{shaded region}} = \pi(5\sqrt{2})^2 - [2(5)]^2$$

$$= \mathbf{(50\pi - 100)\ cm^2 = 57.08\ cm^2}$$

28. (2) $V_{\text{sphere}} = \dfrac{4}{3}\pi r^3 = 2304\pi$

$$r^3 = \left(\frac{3}{4}\right)2304$$

$$r^3 = 1728$$

$$r = 12$$

$$D = 2r = 2(12) = \mathbf{24\ cm}$$

29. (1,3) The sum of the measures of any two sides of a triangle is always greater than the measure of the remaining side. Therefore the answer is **B.**

30. (1) $x^2 - x^4 > 0$

$$x^2 > x^4$$

$$1 > x^2$$

If x^2 must be less than one, then x must be less than one and greater than negative one. Therefore the answer is **B.**

Problem Set 19

1. (5) $\dfrac{\text{Side}_L}{\text{Side}_S} = \dfrac{4}{1}$

$$\frac{\text{Area}_L}{\text{Area}_S} = \left(\frac{\text{Side}_L}{\text{Side}_S}\right)^2 = \left(\frac{4}{1}\right)^2 = \mathbf{\frac{16}{1}}$$

2. First we write the three equations.
(18)

$$\begin{cases} \text{(a)} & 4(N_R + N_B) = 3N_G + 10 \\ \text{(b)} & 4N_B - 3 = N_G + 7 \\ \text{(c)} & N_R + N_B + N_G = 20 \end{cases}$$

(c) $N_R + N_B + N_G = 20$

$$N_R + N_B = 20 - N_G$$

Now we can use substitution to solve.

(a) $4(N_R + N_B) = 3N_G + 10$

$$4(20 - N_G) = 3N_G + 10$$
$$80 - 4N_G = 3N_G + 10$$
$$7N_G = 70$$
$$\mathbf{N_G = 10}$$

(b) $4N_B - 3 = N_G + 7$

$$4N_B - 3 = (10) + 7$$
$$4N_B = 20$$
$$\mathbf{N_B = 5}$$

(c) $N_R + N_B + N_G = 20$

$$N_R + (5) + (10) = 20$$
$$\mathbf{N_R = 5}$$

3. First, write the three equations.
(18)

$$\begin{cases} \text{(a)} & N_N + N_Q + N_D = 50 \\ \text{(b)} & 5N_N + 25N_Q + 10N_D = 640 \\ \text{(c)} & N_Q = N_N + 4 \end{cases}$$

(a) $N_N + N_Q + N_D = 50$

$$N_D = 50 - N_N - N_Q$$

Now we can use substitution to solve.

(b) $5N_N + 25N_Q + 10N_D = 640$

$$5N_N + 25N_Q + 10(50 - N_N - N_Q) = 640$$
$$5N_N + 25N_Q + 500 - 10N_N - 10N_Q = 640$$
$$15N_Q - 5N_N = 140$$
$$3N_Q - N_N = 28$$
$$3(N_N + 4) - N_N = 28$$
$$3N_N + 12 - N_N = 28$$
$$2N_N = 16$$
$$\mathbf{N_N = 8}$$

(c) $N_Q = N_N + 4$

$$N_Q = (8) + 4$$
$$\mathbf{N_Q = 12}$$

(a) $N_N + N_Q + N_D = 50$

$$(8) + (12) + N_D = 50$$
$$\mathbf{N_D = 30}$$

4. $5y - 2x - 30 = 0$
(10)

$$5y = 2x + 30$$
$$y = \frac{2}{5}x + 6$$

Parallel lines have equal slopes, so $m = \frac{2}{5}$.

$$y = mx + b$$
$$y = \frac{2}{5}x + b$$
$$-3 = \frac{2}{5}(5) + b$$
$$b = -5$$
$$\mathbf{y = \frac{2}{5}x - 5}$$

5. $C = \dfrac{kR^2}{S^3}$
(18)

$$6 = \frac{k(4)^2}{(2)^3}$$
$$k = \frac{6 \cdot 8}{16} = 3$$
$$C = \frac{3R^2}{S^3} = \frac{3 \cdot (6)^2}{(1)^3} = \mathbf{108}$$

6. First write the two equations.
(18)

$$\begin{cases} \text{(a)} & T + U = 11 \\ \text{(b)} & 10U + T = 10T + U + 9 \end{cases}$$

(b) $10U + T = 10T + U + 9$

$$9U - 9T = 9$$
$$U - T = 1$$
$$U = T + 1$$

Now we can use substitution to solve.

(a) $T + U = 11$

$$T + (T + 1) = 11$$
$$2T = 10$$
$$T = 5$$

(b) $U = T + 1 = (5) + 1 = 6$

The number is **56.**

7. (18) First, write the two equations.

$$\begin{cases} \text{(a)}\ 0.1(P_N) + 0.4(D_N) = 0.16(50) \\ \text{(b)}\ P_N + D_N = 50 \end{cases}$$

(b) $P_N + D_N = 50$

$$P_N = 50 - D_N$$

Now we can use substitution to solve.

(a) $0.1(P_N) + 0.4(D_N) = 0.16(50)$

$$0.1(50 - D_N) + 0.4D_N = 8$$
$$5 - 0.1D_N + 0.4D_N = 8$$
$$0.3D_N = 3$$
$$\mathbf{D_N = 10 \text{ ml of } 40\%}$$

$P_N = 50 - (10)$

$\mathbf{P_N = 40 \text{ ml of } 10\%}$

8. (19) First, label the two equations.

$$\begin{cases} \text{(a)}\ x^2 + y^2 = 9 \\ \text{(b)}\ y - x = 1 \end{cases}$$

(b) $y - x = 1$

$$y = x + 1$$

Now we can use substitution and the quadratic formula to solve.

(a) $x^2 + y^2 = 9$

$$x^2 + (x + 1)^2 = 9$$
$$2x^2 + 2x + 1 = 9$$
$$2x^2 + 2x - 8 = 0$$
$$x^2 + x - 4 = 0$$

$$x = \frac{-1 \pm \sqrt{(1)^2 - 4(1)(-4)}}{2(1)}$$

$$x = \frac{-1 \pm \sqrt{17}}{2} = -\frac{1}{2} \pm \frac{\sqrt{17}}{2}$$

$y = x + 1$

$$y = \left(-\frac{1}{2} + \frac{\sqrt{17}}{2}\right) + 1 = \frac{1}{2} + \frac{\sqrt{17}}{2}$$

or

$$y = \left(-\frac{1}{2} - \frac{\sqrt{17}}{2}\right) + 1 = \frac{1}{2} - \frac{\sqrt{17}}{2}$$

$$\left(-\frac{1}{2} + \frac{\sqrt{17}}{2}, \frac{1}{2} + \frac{\sqrt{17}}{2}\right), \left(-\frac{1}{2} - \frac{\sqrt{17}}{2}, \frac{1}{2} - \frac{\sqrt{17}}{2}\right)$$

9. (19) First, label the two equations.

$$\begin{cases} \text{(a)}\ x^2 + y^2 = 9 \\ \text{(b)}\ 2x^2 - y^2 = -6 \end{cases}$$

Now we can use elimination and substitution to solve.

(a) $x^2 + y^2 = 9$

(b) $2x^2 - y^2 = -6$

$$3x^2 = 3$$
$$x^2 = 1$$
$$x = \pm 1$$

(a) $x^2 + y^2 = 9$

$$(\pm 1)^2 + y^2 = 9$$
$$1 + y^2 = 9$$
$$y^2 = 8$$
$$y = \pm 2\sqrt{2}$$

$$\mathbf{(1, 2\sqrt{2}), (1, -2\sqrt{2}), (-1, 2\sqrt{2}), (-1, -2\sqrt{2})}$$

10. (19) First, label the two equations.

$$\begin{cases} \text{(a)}\ xy = -4 \\ \text{(b)}\ y = -x - 2 \end{cases}$$

Now we can use substitution and the quadratic formula to solve.

(a) $xy = -4$

$$x(-x - 2) = -4$$
$$-x^2 - 2x = -4$$
$$x^2 + 2x - 4 = 0$$

$$x = \frac{-2 \pm \sqrt{2^2 - 4(1)(-4)}}{2(1)}$$

$$x = \frac{-2 \pm \sqrt{20}}{2}$$

$$x = -1 \pm \sqrt{5}$$

(b) $y = -x - 2$

$$y = -(-1 + \sqrt{5}) - 2 = -1 - \sqrt{5}$$

or

$$y = -(-1 - \sqrt{5}) - 2 = -1 + \sqrt{5}$$

$$\mathbf{(-1 + \sqrt{5}, -1 - \sqrt{5}), (-1 - \sqrt{5}, -1 + \sqrt{5})}$$

11. (19) $4x^{3n+2} - 6x^{4n+1}$
$= 2 \cdot 2 \cdot x^n \cdot x^n \cdot x^n \cdot x^2$
$- 3 \cdot 2 \cdot x^n \cdot x^n \cdot x^n \cdot x^n \cdot x^1$
common factor: $2 \cdot x^n \cdot x^n \cdot x^n \cdot x^1$
$= \mathbf{2x^{3n+1}(2x - 3x^n)}$

12. (19) $\dfrac{x^{4a} - y^{4b}}{x^{2a} + y^{2b}} = \dfrac{(x^{2a})^2 - (y^{2b})^2}{x^{2a} + y^{2b}}$
$= \dfrac{(x^{2a} - y^{2b})(x^{2a} + y^{2b})}{x^{2a} + y^{2b}}$
$= x^{2a} - y^{2b} = (x^a)^2 - (y^b)^2$
$= \mathbf{(x^a + y^b)(x^a - y^b)}$

13. (19) $27x^{12}y^6 - z^9 = (3x^4y^2)^3 - (z^3)^3$
If $F = 3x^4y^2$ and $S = z^3$ and we use the form $F^3 - S^3 = (F - S)(F^2 + FS + S^2)$, then
$27x^{12}y^6 - z^9 = \mathbf{(3x^4y^2 - z^3)(9x^8y^4 + 3x^4y^2z^3 + z^6)}$.

14. (19) $8x^6y^3 + p^3 = (2x^2y)^3 + (p)^3$
If $F = 2x^2y$ and $S = p$ and we use the form $F^3 + S^3 = (F + S)(F^2 - FS + S^2)$, then
$8x^6y^3 + p^3 = \mathbf{(2x^2y + p)(4x^4y^2 - 2x^2yp + p^2)}$.

15. (17) $\dfrac{UV}{XY} = \dfrac{25}{20} = \dfrac{5}{4}$
$\dfrac{VW}{YZ} = \dfrac{15}{12} = \dfrac{5}{4}$
$\angle V \cong \angle Y$
$\mathbf{\Delta UVW \sim \Delta XYZ}$ **by *SAS* similarity postulate.**

16. (16) $\dfrac{\dfrac{3a}{b^2} + \dfrac{2c}{ab}}{\dfrac{a}{c} - \dfrac{b}{d}} = \dfrac{\dfrac{3a}{b^2} + \dfrac{2c}{ab}}{\dfrac{a}{c} - \dfrac{b}{d}} \cdot \dfrac{\dfrac{ab^2cd}{1}}{\dfrac{ab^2cd}{1}}$
$= \mathbf{\dfrac{3a^2cd + 2bc^2d}{a^2b^2d - ab^3c}}$

17. (10) $\dfrac{3i^4 - 2i^2 + i^3}{1 - 3i^2 + \sqrt{-9}} = \dfrac{3(-1)(-1) - 2(-1) + (-1)i}{1 - 3(-1) + 3i}$
$= \dfrac{5 - i}{4 + 3i} \cdot \dfrac{4 - 3i}{4 - 3i} = \dfrac{20 - 4i - 15i + 3i^2}{16 - 9i^2}$
$= \dfrac{20 - 19i + 3(-1)}{16 - 9(-1)} = \dfrac{17 - 19i}{25} = \mathbf{\dfrac{17}{25} - \dfrac{19}{25}i}$

18. (10) $\dfrac{3 - 2\sqrt{12}}{5 + 4\sqrt{27}} = \dfrac{3 - 4\sqrt{3}}{5 + 12\sqrt{3}}$
$= \dfrac{3 - 4\sqrt{3}}{5 + 12\sqrt{3}} \cdot \dfrac{5 - 12\sqrt{3}}{5 - 12\sqrt{3}}$
$= \dfrac{15 - 20\sqrt{3} - 36\sqrt{3} + 144}{25 - 432}$
$= \dfrac{159 - 56\sqrt{3}}{-407} = \mathbf{\dfrac{-159 + 56\sqrt{3}}{407}}$

19. (10) $\dfrac{1 - 3i}{5i + 4} = \dfrac{1 - 3i}{5i + 4} \cdot \dfrac{5i - 4}{5i - 4}$
$= \dfrac{5i - 15i^2 - 4 + 12i}{25i^2 - 16}$
$= \dfrac{11 + 17i}{-41} = \mathbf{-\dfrac{11}{41} - \dfrac{17}{41}i}$

20. (5) $\left(3a^{\frac{x}{3}} - b^{\frac{y}{2}}\right)\left(3a^{\frac{x}{3}} + b^{\frac{y}{2}}\right) = \mathbf{9a^{\frac{2x}{3}} - b^y}$

21. (5) $\dfrac{x^{3c-1}y^{2-4d}}{x^{c+d}y^{c-d}}$
$= x^{3c-1}x^{-c-d}y^{2-4d}y^{-c+d}$
$= \mathbf{x^{2c-1-d}y^{2-3d-c}}$

22. (16)

$$\begin{array}{r} x^3 + 2x^2 + 4x + 6 \\ x - 2 \overline{\smash{)}\, x^4 + 0x^3 + 0x^2 - 2x + 1} \\ \underline{x^4 - 2x^3 } \\ 2x^3 + 0x^2 \\ \underline{2x^3 - 4x^2 } \\ 4x^2 - 2x \\ \underline{4x^2 - 8x } \\ 6x + 1 \\ \underline{6x - 12} \\ 13 \end{array}$$

$\dfrac{x^4 - 2x + 1}{x - 2} = \mathbf{x^3 + 2x^2 + 4x + 6 + \dfrac{13}{x - 2}}$

Check:

$(x^3 + 2x^2 + 4x + 6)\left(\dfrac{x - 2}{x - 2}\right) + \dfrac{13}{x - 2}$
$= \dfrac{x^4 + 2x^3 + 4x^2 + 6x - 2x^3 - 4x^2 - 8x - 12}{x - 2}$
$+ \dfrac{13}{x - 2}$
$= \dfrac{x^4 - 2x + 1}{x - 2}$

23. (16)

$$2t = \frac{1}{3s^2}\left(\frac{5z}{6} - \frac{4m}{n}\right)$$

$$2t = \frac{5z}{18s^2} - \frac{4m}{3s^2n}$$

$$\frac{5z}{18s^2} = 2t + \frac{4m}{3s^2n}$$

$$5z = 36s^2t + \frac{24m}{n}$$

$$z = \frac{36s^2t}{5} + \frac{24m}{5n}$$

$$z = \mathbf{\frac{36s^2nt + 24m}{5n}}$$

24. (15)

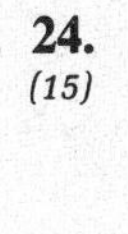

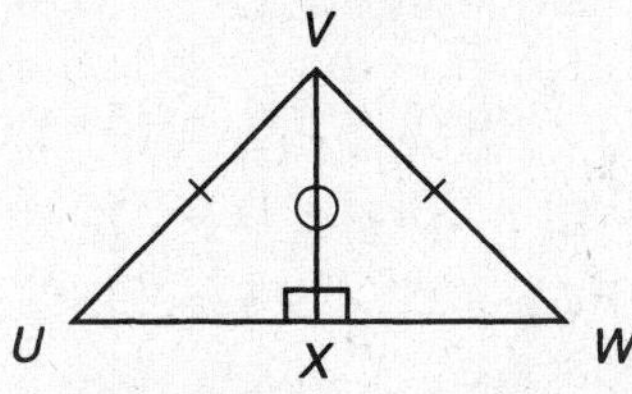

STATEMENTS	REASONS
1. $\overline{VX} \perp \overline{UW}$	1. Given
2. $\angle UXV$ and $\angle WXV$ are right angles.	2. Perpendicular lines intersect to form right angles.
3. ΔUVX and ΔWVX are right triangles.	3. A triangle which contains a right angle is a right triangle.
4. $\overline{UV} \cong \overline{WV}$	4. Given
5. $\overline{VX} \cong \overline{VX}$	5. Reflexive axiom
6. $\Delta UVX \cong \Delta WVX$	6. *HL* congruency postulate

25. (15)

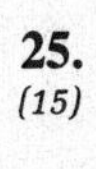

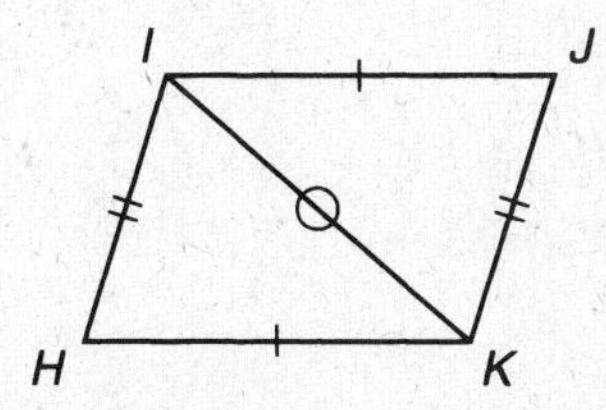

STATEMENTS	REASONS
1. $\overline{HK} \cong \overline{IJ}$	1. Given
2. $\overline{HI} \cong \overline{KJ}$	2. Given
3. $\overline{IK} \cong \overline{IK}$	3. Reflexive axiom
4. $\Delta HIK \cong \Delta JKI$	4. *SSS* congruency postulate

26. (13) $m\angle AOB = 50°$

$$m\angle ABO = 90°$$

$$x = 180 - m\angle ABO - m\angle AOB$$

$$x = 180 - (90) - (50)$$

$$x = \mathbf{40}$$

27. (13)

$$30 = \frac{1}{2}(120 - x)$$

$$30 = 60 - \frac{1}{2}x$$

$$\frac{1}{2}x = 30$$

$$x = \mathbf{60}$$

28. (2) (a) First, find the radius of the sphere.

$$4\pi r^2 = 2916\pi$$

$$r^2 = 729$$

$$r = 27$$

Then find the volume.

$$V_{\text{sphere}} = \frac{4}{3}\pi r^3$$

$$= \frac{4}{3}\pi(27)^3$$

$$= \mathbf{26{,}244\pi\ m^3 = 82{,}447.96\ m^3}$$

(b) $4\pi r^2 = \frac{4}{3}\pi r^3$

$$r = \frac{4\pi}{\frac{4}{3}\pi}$$

$$r = \mathbf{3\ cm}$$

29. (1,11) A: $m\angle A = 30°$

Since $\overline{AB}$ is a diameter, $m\angle C = 90°$

$m\angle B = 180° - 90° - 30° = 60°$

B: $m\angle C - m\angle A = 90° - 30° = 60°$

A and B are equal. Therefore the answer is **C.**

30. (1,3) Since $\angle DBC \cong \angle BDC$, $\overline{DC} \cong \overline{BC}$

$AC = AD + DC$

Since $\overline{DC} \cong \overline{BC}$, $AC = AD + BC$

Therefore the answer is **C.**

Problem Set 20

1. (14)

$$\sin 65° = \frac{h}{250}$$

$$h = 250 \sin 65°$$

$$h = \mathbf{226.58\ m}$$

2. (10) $5x + 3y + 2 = 0$

$$3y = -5x - 2$$

$$y = -\frac{5}{3}x - \frac{2}{3}$$

The slopes of perpendicular lines are negative reciprocals of each other, so $m = \frac{3}{5}$.

$$y = mx + b$$

$$y = \frac{3}{5}x + b$$

$$0 = \frac{3}{5}(-2) + b$$

$$b = \frac{6}{5}$$

$$\mathbf{y = \frac{3}{5}x + \frac{6}{5}}$$

3. (18) First we write the two equations.

$$\begin{cases} \text{(a)} & 0.4N_R + 0.6N_G = 56 \\ \text{(b)} & \dfrac{N_R}{N_G} = \dfrac{1}{4} \end{cases}$$

Rearrange the equations and use substitution to solve.

$$\text{(b)} \quad \frac{N_R}{N_G} = \frac{1}{4}$$

$$N_G = 4N_R$$

$$\text{(a)} \quad 0.4N_R + 0.6N_G = 56$$

$$0.4N_R + 0.6(4N_R) = 56$$

$$2.8N_R = 56$$

$$\mathbf{N_R = 20}$$

$$N_G = 4N_R$$

$$N_G = 4(20)$$

$$\mathbf{N_G = 80}$$

4. (18) First we write the two equations.

$$\begin{cases} 2N_R = 3N_B + 8 \\ \dfrac{N_R}{N_R + N_B} = \dfrac{5}{7} \end{cases}$$

Rearrange the equations.

$$\begin{cases} \text{(a)} & 2N_R - 3N_B = 8 \\ \text{(b)} & 2N_R - 5N_B = 0 \end{cases}$$

Now we can use elimination and substitution to solve.

$$\text{(b)} \quad 2N_R - 5N_B = 0$$

$$-1\text{(a)} \quad -2N_R + 3N_B = -8$$

$$-2N_B = -8$$

$$\mathbf{N_B = 4}$$

$$\text{(a)} \quad 2N_R - 3N_B = 8$$

$$2N_R - 3(4) = 8$$

$$2N_R = 20$$

$$\mathbf{N_R = 10}$$

5. (18) First, write the three equations.

$$\begin{cases} \text{(a)} & N_Q + N_D + N_N = 12 \\ \text{(b)} & 25N_Q + 10N_D + 5N_N = 155 \\ \text{(c)} & N_N = N_Q + 1 \end{cases}$$

Now we can use elimination and substitution to solve.

$$\text{(a)} \quad N_Q + N_D + N_N = 12$$

$$N_Q + N_D + (N_Q + 1) = 12$$

$$\text{(a}'\text{)} \quad 2N_Q + N_D = 11$$

$$\text{(b)} \quad 25N_Q + 10N_D + 5N_N = 155$$

$$25N_Q + 10N_D + 5(N_Q + 1) = 155$$

$$25N_Q + 10N_D + 5N_Q + 5 = 155$$

$$\text{(b}'\text{)} \quad 30N_Q + 10N_D = 150$$

$$-10\text{(a}'\text{)} \quad -20N_Q - 10N_D = -110$$

$$\text{(b}'\text{)} \quad 30N_Q + 10N_D = 150$$

$$10N_Q = 40$$

$$\mathbf{N_Q = 4}$$

$$\text{(c)} \quad N_N = N_Q + 1$$

$$N_N = (4) + 1$$

$$\mathbf{N_N = 5}$$

$$\text{(a)} \quad N_Q + N_D + N_N = 12$$

$$(4) + N_D + (5) = 12$$

$$\mathbf{N_D = 3}$$

6. (18) First, write the two equations.

$$\begin{cases} \text{(a)} & T + U = 12 \\ \text{(b)} & \dfrac{U}{T} = \dfrac{1}{2} \end{cases}$$

$$\text{(b)} \quad \frac{U}{T} = \frac{1}{2}$$

$$T = 2U$$

Now we can use substitution to solve.

$$\text{(a)} \quad T + U = 12$$

$$(2U) + U = 12$$

$$3U = 12$$

$$U = 4$$

$T = 2U$

$T = 2(4)$

$T = 8$

The number is **84.**

7. (18) P_N = copper

$\text{Copper}_1 + \text{copper added} = \text{copper final}$

$$\begin{aligned} 0.2(20) + P_N &= 0.5(20 + P_N) \\ 4 + P_N &= 10 + 0.5P_N \\ 0.5P_N &= 6 \\ P_N &= \mathbf{12\ lb} \end{aligned}$$

8. (18) P_N = alcohol

Iodine initial = Iodine final

$$\begin{aligned} 0.035(100) &= 0.015(100 + P_N) \\ 3.5 &= 1.5 + 0.015P_N \\ 0.015P_N &= 2 \\ P_N &= \mathbf{133.33\ oz} \end{aligned}$$

9. (20)

2, 60°, 1, 30°, $\sqrt{3}$

$$\cos\theta = \frac{\text{adjacent}}{\text{hypotenuse}}$$

$$\cos 30° = \frac{\sqrt{3}}{2}$$

$$6\sqrt{2}\cos 30° = 6\sqrt{2}\left(\frac{\sqrt{3}}{2}\right) = \frac{6}{2}\sqrt{6} = \mathbf{3\sqrt{6}}$$

10. (20)

$\sqrt{2}$, 45°, 1, 45°, 1

$$\sin\theta = \frac{\text{opposite}}{\text{hypotenuse}}$$

$$\sin 45° = \frac{1}{\sqrt{2}} = \frac{\sqrt{2}}{2}$$

$$6\sqrt{3}\sin 45° = 6\sqrt{3}\left(\frac{\sqrt{2}}{2}\right) = \frac{6}{2}\sqrt{6} = \mathbf{3\sqrt{6}}$$

11. (20)

2, 60°, 1, 30°, $\sqrt{3}$

$$\tan\theta = \frac{\text{opposite}}{\text{adjacent}}$$

$$\tan 60° = \frac{\sqrt{3}}{1} = \sqrt{3}$$

$$4\tan 60° = 4 \cdot \sqrt{3} = \mathbf{4\sqrt{3}}$$

12. (17)

$$\frac{VW}{YX} = \frac{35}{20} = \frac{7}{4}$$

$$\frac{WZ}{XZ} = \frac{28}{28 - 12} = \frac{28}{16} = \frac{7}{4}$$

$$\frac{VZ}{YZ} = \frac{21}{12} = \frac{7}{4}$$

$\Delta VWZ \sim \Delta YXZ$ by SSS similarity postulate

13. (17)

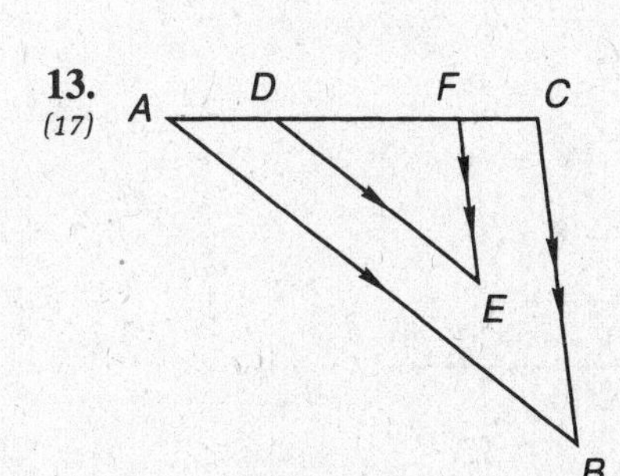

STATEMENTS	REASONS
1. $\overline{AB} \parallel \overline{DE}$	1. Given
2. $\angle BAC \cong \angle EDF$	2. If two parallel lines are cut by a transversal, then each pair of corresponding angles is congruent.
3. $\overline{BC} \parallel \overline{EF}$	3. Given
4. $\angle BCA \cong \angle EFD$	4. If two parallel lines are cut by a transversal, then each pair of corresponding angles is congruent.
5. $\angle B \cong \angle E$	5. $AA \rightarrow AAA$
6. $\Delta ABC \sim \Delta DEF$	6. AAA

14. (6) First, label the three equations.

$$\begin{cases} \text{(a)} & \frac{3}{4}x - \frac{5}{2}z = -2 \\ \text{(b)} & 0.4x + 0.3z = 2.2 \\ \text{(c)} & -\frac{1}{2}x + \frac{3}{2}y = \frac{5}{2} \end{cases}$$

Now we can use elimination and substitution to solve.

$$\begin{aligned} 12(a) \quad & 9x - 30z = -24 \\ 100(b) \quad & 40x + 30z = 220 \\ \hline & 49x = 196 \\ & \mathbf{x = 4} \end{aligned}$$

$$\begin{aligned} 12(a) \quad 9x - 30z &= -24 \\ 9(4) - 30z &= -24 \\ -30z &= -60 \\ \mathbf{z} &= \mathbf{2} \end{aligned}$$

$$\begin{aligned} 2(c) \quad -x + 3y &= 5 \\ -(4) + 3y &= 5 \\ 3y &= 9 \\ \mathbf{y} &= \mathbf{3} \end{aligned}$$

15. (19) First, label the two equations.

$$\begin{array}{l}(a)\\(b)\end{array}\begin{cases}x^2 + y^2 = 9\\ y - 2x = 1\end{cases}$$

(b) $y - 2x = 1$

$$y = 2x + 1$$

Now we can use substitution and the quadratic formula to solve.

(a) $x^2 + y^2 = 9$

$$x^2 + (2x + 1)^2 = 9$$

$$5x^2 + 4x - 8 = 0$$

$$x = \frac{-4 \pm \sqrt{(4)^2 - 4(5)(-8)}}{2(5)}$$

$$x = \frac{-4 \pm \sqrt{176}}{10} = -\frac{2}{5} \pm \frac{2\sqrt{11}}{5}$$

$$y = 2x + 1$$

$$y = 2\left(-\frac{2}{5} + \frac{2\sqrt{11}}{5}\right) + 1 = \frac{1}{5} + \frac{4\sqrt{11}}{5}$$

or

$$y = 2\left(-\frac{2}{5} - \frac{2\sqrt{11}}{5}\right) + 1 = \frac{1}{5} - \frac{4\sqrt{11}}{5}$$

$$\left(-\frac{2}{5} + \frac{2\sqrt{11}}{5}, \frac{1}{5} + \frac{4\sqrt{11}}{5}\right),$$

$$\left(-\frac{2}{5} - \frac{2\sqrt{11}}{5}, \frac{1}{5} - \frac{4\sqrt{11}}{5}\right)$$

16. (6)

$$\frac{x}{x - 3} - \frac{2}{x + 1} = \frac{x + 9}{x^2 - 2x - 3}$$

$$x(x + 1) - 2(x - 3) = x + 9$$

$$x^2 + x - 2x + 6 - x - 9 = 0$$

$$x^2 - 2x - 3 = 0$$

$$(x - 3)(x + 1) = 0$$

$$x = 3, -1$$

If we check these values for x in the original equation, we find that they would give a zero in a denominator. Therefore, there is **no solution;** $\boldsymbol{x \neq 3, -1}$.

17. (19) $27x^3y^6 + 8p^3 = \left(3xy^2\right)^3 + (2p)^3$

$$= \boldsymbol{\left(3xy^2 + 2p\right)\left(9x^2y^4 - 6xy^2p + 4p^2\right)}$$

18. (19) $8x^3y^{12} - 27z^9 = \left(2xy^4\right)^3 - \left(3z^3\right)^3$

$$= \boldsymbol{\left(2xy^4 - 3z^3\right)\left(4x^2y^8 + 6xy^4z^3 + 9z^6\right)}$$

19. (16)

$$\frac{x}{2y - \dfrac{6z}{3 + \dfrac{s}{t}}} = \frac{x}{2y - \dfrac{6z}{\dfrac{3t + s}{t}}}$$

$$= \frac{x}{2y - \dfrac{6zt}{3t + s}} = \frac{x}{\dfrac{2y(3t + s) - 6zt}{3t + s}}$$

$$= \frac{x(3t + s)}{6yt + 2ys - 6zt} = \boldsymbol{\frac{3tx + xs}{6yt + 2ys - 6zt}}$$

20. (5) $\left(2x^5y\right)^{-2}\left(\dfrac{2x^3y^2}{xy^{-1}}\right)^{-1}$

$$= (2)^{-2}x^{-10}y^{-2}(2)^{-1}x^{-3}y^{-2}x^1y^{-1}$$

$$= (2)^{-3}x^{-10-3+1}y^{-2-2-1}$$

$$= 2^{-3}x^{-12}y^{-5} = \boldsymbol{\frac{1}{8x^{12}y^5}}$$

21. (5) $\sqrt{7}\sqrt{7}\sqrt{-7} + \sqrt{-5}\sqrt{-5}\sqrt{-5} - 3\sqrt{-4} + 2i$

$$= 7\sqrt{7}i + \sqrt{5}i\sqrt{5}i\sqrt{5}i - 3(2i) + 2i$$

$$= 7\sqrt{7}i + 5\sqrt{5}i^3 - 6i + 2i$$

$$= 7\sqrt{7}i + 5\sqrt{5}(-1)i - 6i + 2i$$

$$= \boldsymbol{\left(7\sqrt{7} - 5\sqrt{5} - 4\right)i}$$

22. (16)

$$s^2 = \frac{1}{df}\left(\frac{4y}{3} - \frac{7x^3}{g}\right)$$

$$s^2 = \frac{4y}{3df} - \frac{7x^3}{gdf}$$

$$\frac{4y}{3df} = s^2 + \frac{7x^3}{gdf}$$

$$4y = 3dfs^2 + \frac{21x^3}{g}$$

$$y = \frac{3dfs^2}{4} + \frac{21x^3}{4g} = \boldsymbol{\frac{3dfgs^2 + 21x^3}{4g}}$$

23. (16)

$$\begin{array}{r} x^2 + 1 \\ x^2 - 1\overline{)\,x^4 + 0x^3 + 0x^2 + 0x - 6} \\ \underline{x^4 - x^2 } \\ x^2 - 6 \\ \underline{x^2 - 1} \\ -5 \end{array}$$

$$\frac{x^4 - 6}{x^2 - 1} = \boldsymbol{x^2 + 1 - \frac{5}{x^2 - 1}}$$

Check: $(x^2 + 1)\left(\dfrac{x^2 - 1}{x^2 - 1}\right) - \dfrac{5}{x^2 - 1}$

$$= \frac{x^4 - 1 - 5}{x^2 - 1} = \frac{x^4 - 6}{x^2 - 1}$$

24. (14)

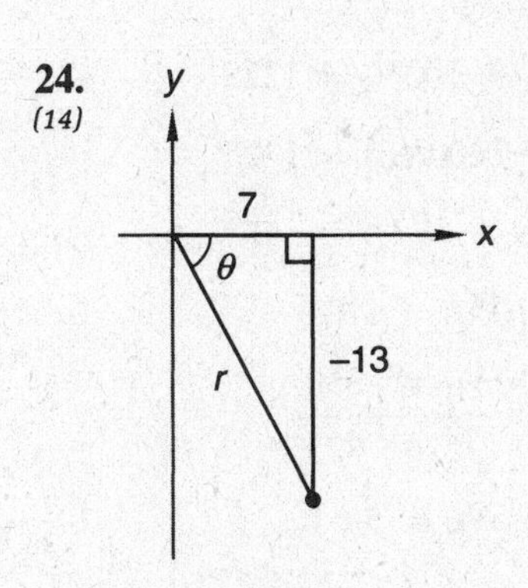

$$r = \sqrt{7^2 + 13^2} = 14.76$$

$$\tan\theta = \frac{13}{7}$$

$$\theta = 61.70°$$

The polar angle is $360° - 61.70° = 298.30°$.

$\mathbf{14.76\underline{/298.30°}};\ \mathbf{14.76\underline{/-61.70°}};$
$\mathbf{-14.76\underline{/118.30°}};\ \mathbf{-14.76\underline{/-241.70°}}$

25. (10)

$$-5x - 8 = 3x^2$$

$$3x^2 + 5x = -8$$

$$\left(x^2 + \frac{5}{3}x + \quad\right) = -\frac{8}{3}$$

$$\left(x^2 + \frac{5}{3}x + \frac{25}{36}\right) = -\frac{8}{3} + \frac{25}{36}$$

$$\left(x + \frac{5}{6}\right)^2 = -\frac{71}{36}$$

$$x + \frac{5}{6} = \pm\sqrt{-\frac{71}{36}}$$

$$\mathbf{x = -\frac{5}{6} \pm \frac{\sqrt{71}}{6}i}$$

26. (15)

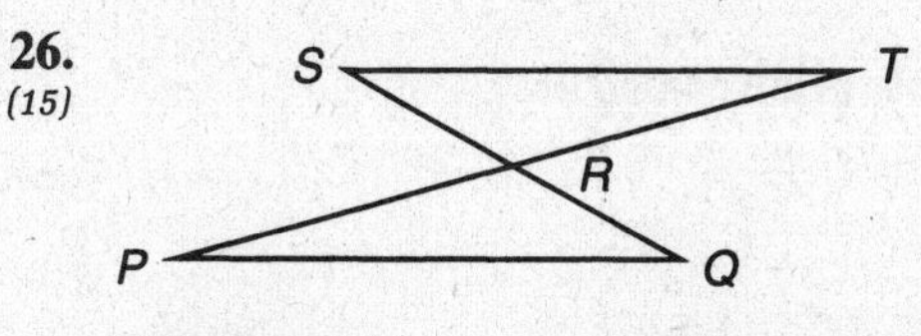

STATEMENTS	REASONS
1. $\overline{PQ} \parallel \overline{ST}$	1. Given
2. $\angle P \cong \angle T$ and $\angle Q \cong \angle S$	2. If two parallel lines are cut by a transversal, then each pair of alternate interior angles is congruent.
3. $\angle PRQ \cong \angle TRS$	3. Vertical angles are congruent.
4. $\overline{PQ} \cong \overline{ST}$	4. Given
5. $\triangle PRQ \cong \triangle TRS$	5. *AAAS* congruency postulate

27. (13) $3x = 7 \cdot 9$

$$x = 21$$

$$r = \frac{21 + 3}{2} = \mathbf{12\ m}$$

28. (2)

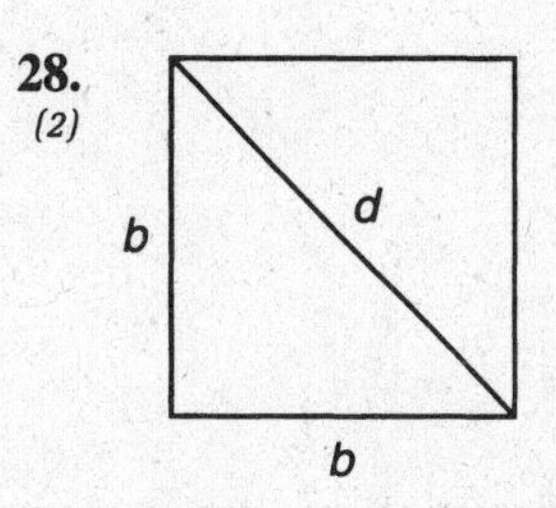

$$b^2 + b^2 = d^2$$

$$2b^2 = (12\sqrt{2})^2$$

$$b^2 = 144$$

$$b = 12$$

$$A_{\text{surface}} = A_{\text{base}} + A_{\text{sides}}$$

$$A_{\text{surface}} = b^2 + 4\left(\frac{1}{2}bl\right)$$

$$384 = (12)^2 + 4\left[\frac{1}{2}(12)l\right]$$

$$384 = 144 + 24l$$

$$24l = 240$$

$$\mathbf{l = 10\ cm}$$

$$h^2 + \left(\frac{1}{2}b\right)^2 = l^2$$

$$h^2 + (6)^2 = 10^2$$

$$h^2 = 64$$

$$\mathbf{h = 8\ cm}$$

29. (1) $x > 0,\ y < 0$

A: $x - y > 0$

B: $y - x < 0$

$x - y > y - x$

We know that x is positive and y is negative. Thus $x - y$ is positive and $y - x$ is negative. Therefore the answer is **A.**

30. (1)

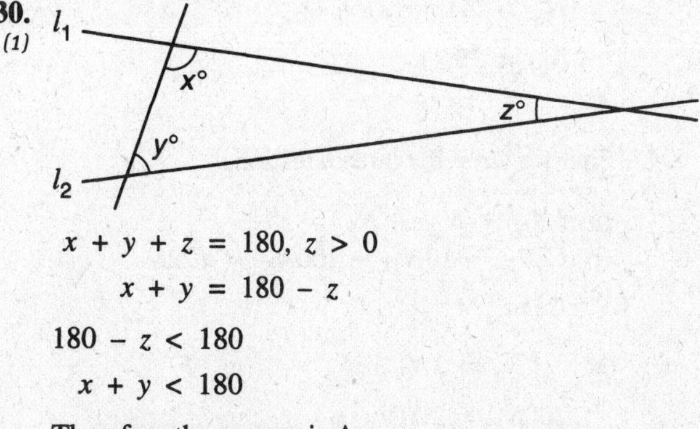

$$x + y + z = 180,\ z > 0$$

$$x + y = 180 - z$$

$$180 - z < 180$$

$$x + y < 180$$

Therefore the answer is **A.**

Problem Set 21

1. w = wanted
(R)

$$7\frac{1}{8} = \frac{2}{7}w$$

$$\frac{57}{8} = \frac{2}{7}w$$

$$w = \frac{57}{8} \cdot \frac{7}{2} = \mathbf{\frac{399}{16}}$$

2. $m_{\text{original}} = \frac{[7 - (-2)]}{[5 - (-4)]} = \frac{9}{9} = 1$
(10)

The slopes of perpendicular lines are negative reciprocals of each other, so $m = -1$.

$y = -x + b$

$5 = -(-2) + b$

$b = 5 - 2 = 3$

$\mathbf{y = -x + 3}$

3. First we write the three equations.
(18)

$$\begin{cases} \text{(a)} & N_G = N_R + N_B + 9 \\ \text{(b)} & N_B + N_G = 4N_R + 1 \\ \text{(c)} & N_B = N_R - 2 \end{cases}$$

Use substitution to solve.

(b) $N_B + N_G = 4N_R + 1$

$(N_R - 2) + (N_R + N_B + 9) = 4N_R + 1$

$(N_R - 2) + [N_R + (N_R - 2) + 9] = 4N_R + 1$

$N_R - 2 + N_R + N_R - 2 + 9 = 4N_R + 1$

$3N_R + 5 = 4N_R + 1$

$\mathbf{N_R = 4}$

(c) $N_B = N_R - 2$

$N_B = (4) - 2$

$\mathbf{N_B = 2}$

(a) $N_G = N_R + N_B + 9$

$N_G = (4) + (2) + 9$

$\mathbf{N_G = 15}$

4. First we write the three equations.
(18)

$$\begin{cases} \text{(a)} & N_N + N_D + N_S = 35 \\ \text{(b)} & 5N_N + 10N_D + 100N_S = 1225 \\ \text{(c)} & N_S = 2N_N \end{cases}$$

(c) $N_S = 2N_N$

$100N_S = 200N_N$

Now we can use substitution and elimination to solve.

(a) $N_N + N_D + N_S = 35$

$N_N + N_D + (2N_N) = 35$

(a′) $3N_N + N_D = 35$

(b) $5N_N + 10N_D + 100N_S = 1225$

$5N_N + 10N_D + (200N_N) = 1225$

(b′) $205N_N + 10N_D = 1225$

$$\begin{array}{rrcrcl} \text{(b')} & 205N_N & + & 10N_D & = & 1225 \\ -10\text{(a')} & -30N_N & - & 10N_D & = & -350 \\ \hline & 175N_N & & & = & 875 \end{array}$$

$\mathbf{N_N = 5}$

(c) $N_S = 2N_N$

$N_S = 2(5)$

$\mathbf{N_S = 10}$

(a) $N_N + N_D + N_S = 35$

$(5) + N_D + (10) = 35$

$\mathbf{N_D = 20}$

5. First, write the two equations.
(18)

$$\begin{cases} \text{(a)} & T + U = 9 \\ \text{(b)} & 10U + T = 10T + U - 9 \end{cases}$$

(a) $T + U = 9$

$U = 9 - T$

Now we can use substitution to solve.

(b) $10U + T = 10T + U - 9$

$9U - 9T = -9$

$9(9 - T) - 9T = -9$

$-18T = -90$

$T = 5$

$U = 9 - T = 9 - 5 = 4$

The numbers are **54** and **45.**

6. P_N = key ingredient
(18)

Ingredient$_1$ + ingredient added = ingredient final

$0.73(143) + 0.37(P_N) = 0.51(143 + P_N)$

$104.39 + 0.37P_N = 72.93 + 0.51P_N$

$0.14P_N = 31.46$

$P_N = \mathbf{224.71\ liters}$

7. (21) **(a), (d)**

Sets (a) and (d) are functions because for each x value (the domain) there is only one corresponding y value. Set (b) is not a function because it contains the points (2, 3) and (2, –3). The x value of 2 in the domain corresponds to both 3 and –3. Set (c) is not a function because the x value of 1 corresponds to both –1 and 2.

8. (21) (a) This is **not a function** because a in the domain corresponds to both 2 and 4 in the range.

(b) This is a **function** because although two members of the domain correspond to 1, each member only corresponds to a single value.

9. (21) (a) $\{x \in \mathbb{R}\}$

(b) $\{x \in \mathbb{R} \mid x \geq -5\}$

If $x < -5$, then $g(x)$ will be the square root of a negative number, which is not allowed.

(c) $\{x \in \mathbb{R} \mid x \leq 6\}$

If $x > 6$, then $h(x)$ will be the square root of a negative number, which is not allowed.

10. (21) $f(x) = x^2 - x + 3$

$f(4) = (4)^2 - (4) + 3 = \mathbf{15}$

11. (20,21) $g(\theta) = 4\sqrt{3}\cos\theta$

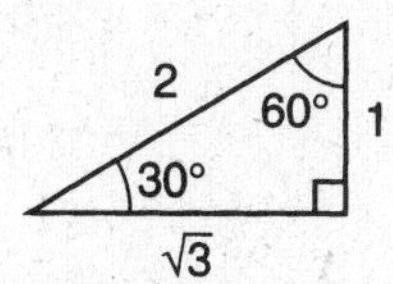

$g(60°) = 4\sqrt{3}\cos 60° = 4\sqrt{3}\left(\frac{1}{2}\right) = \mathbf{2\sqrt{3}}$

12. (20,21) $h(\theta) = 8\tan\theta$

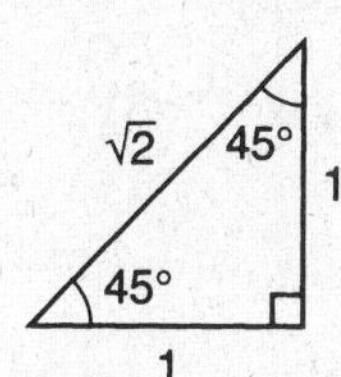

$h(45°) = 8\tan 45° = 8(1) = \mathbf{8}$

13. (17) $\frac{RT}{QT} = \frac{RT}{RT - RQ} = \frac{49}{49 - 21} = \frac{7}{4}$

$\frac{ST}{PT} = \frac{42}{24} = \frac{7}{4}$

$\angle PTQ \cong \angle STR$

$\mathbf{\triangle PQT \sim \triangle SRT}$ **by** ***SAS*** **similarity postulate.**

14. (17)

STATEMENTS	REASONS
1. $\overline{ED} \parallel \overline{BC}$	1. Given
2. $\angle ADE \cong \angle ACB$ $\angle AED \cong \angle ABC$	2. If two parallel lines are cut by a transversal, then each pair of corresponding angles is congruent.
3. $\angle A \cong \angle A$	3. Reflexive axiom
4. $\triangle AED \sim \triangle ABC$	4. *AAA*

15. (19) First, label the equations.

$$\begin{matrix}\text{(a)}\\ \text{(b)}\end{matrix}\begin{cases} xy = 6 \\ x - y = 3 \end{cases}$$

(b) $x - y = 3$

$y = x - 3$

Now we can use substitution and the quadratic formula to solve.

(a) $xy = 6$

$x(x - 3) = 6$

$x^2 - 3x - 6 = 0$

$$x = \frac{-(-3) \pm \sqrt{(-3)^2 - 4(1)(-6)}}{2(1)}$$

$$x = \frac{3 \pm \sqrt{33}}{2}$$

$y = x - 3$

$$y = \left(\frac{3 + \sqrt{33}}{2}\right) - 3 \quad \text{or} \quad \left(\frac{3 - \sqrt{33}}{2}\right) - 3$$

$$y = \frac{-3 + \sqrt{33}}{2} \quad \text{or} \quad \frac{-3 - \sqrt{33}}{2}$$

$$\mathbf{\left(\frac{3 + \sqrt{33}}{2}, \frac{-3 + \sqrt{33}}{2}\right), \left(\frac{3 - \sqrt{33}}{2}, \frac{-3 - \sqrt{33}}{2}\right)}$$

16. (19) $64a^3b^9 - 8p^3 = 8\left[\left(2ab^3\right)^3 - p^3\right]$

$= \mathbf{8(2ab^3 - p)(4a^2b^6 + 2ab^3p + p^2)}$

17. (19) $27b^9a^6 - 64c^3 = \left(3b^3a^2\right)^3 - (4c)^3$

$= \mathbf{(3b^3a^2 - 4c)(9b^6a^4 + 12b^3a^2c + 16c^2)}$

18. (16)
$$\frac{\frac{a^3}{x^2y} - \frac{6m^2}{y^2x}}{\frac{l^2}{y^2} - \frac{6t}{x^2}} = \frac{\frac{a^3}{x^2y} - \frac{6m^2}{y^2x}}{\frac{l^2}{y^2} - \frac{6t}{x^2}} \cdot \frac{\frac{y^2x^2}{1}}{\frac{y^2x^2}{1}}$$
$$= \mathbf{\frac{a^3y - 6xm^2}{x^2l^2 - 6ty^2}}$$

19. (5) $(5x^{a-b} + y^{b-a})(x^{b-a} - y^{a-b})$
$$= 5x^{a-b+b-a} + x^{b-a}y^{b-a} - 5x^{a-b}y^{a-b} - y^{b-a+a-b}$$
$$= 5x^0 + x^{b-a}y^{b-a} - 5x^{a-b}y^{a-b} - y^0$$
$$= \mathbf{4 - 5x^{a-b}y^{a-b} + x^{b-a}y^{b-a}}$$

20. (5) $\sqrt{3}\sqrt{2}\sqrt{-3}\sqrt{-2} + \sqrt{3}\sqrt{-9} + 5\sqrt{-16} - 3i$
$$= \sqrt{3}\sqrt{2}\sqrt{3}i\sqrt{2}i + \sqrt{3}(3i) + 5(4i) - 3i$$
$$= 6i^2 + 3\sqrt{3}i + 20i - 3i = \mathbf{-6 + (3\sqrt{3} + 17)i}$$

21. (10)
$$\frac{3 - 2\sqrt{12}}{1 - 3\sqrt{3}} = \frac{3 - 4\sqrt{3}}{1 - 3\sqrt{3}} \cdot \frac{1 + 3\sqrt{3}}{1 + 3\sqrt{3}}$$
$$= \frac{3 + 5\sqrt{3} - 36}{1 - 27} = \frac{-33 + 5\sqrt{3}}{-26} = \mathbf{\frac{33 - 5\sqrt{3}}{26}}$$

22. (10)
$$\frac{5i^2 - 2i^3 + i^4}{-2i^3 + 4i - \sqrt{-16}} = \frac{-5 + 2i + 1}{2i + 4i - 4i} = \frac{-4 + 2i}{2i}$$
$$= \frac{-2 + i}{i} \cdot \frac{i}{i} = \frac{-2i + i^2}{i^2} = \frac{-2i - 1}{-1} = \mathbf{1 + 2i}$$

23. (16)
$$r = \frac{d + 1}{2d - w}$$
$$r(2d - w) = d + 1$$
$$2dr - wr = d + 1$$
$$2dr - d = wr + 1$$
$$d(2r - 1) = wr + 1$$
$$\mathbf{d = \frac{wr + 1}{2r - 1}}$$

24. (16)
$$\begin{array}{r} x^2 - 1 \\ x^2 + 2\overline{)x^4 + 0x^3 + x^2 - x + 1} \\ \underline{x^4 \quad\quad + 2x^2} \\ -x^2 - x + 1 \\ \underline{-x^2 \quad\quad - 2} \\ -x + 3 \end{array}$$

$$\frac{x^4 + x^2 - x + 1}{x^2 + 2} = \mathbf{x^2 - 1 - \frac{x - 3}{x^2 + 2}}$$

Check: $(x^2 - 1)\left(\frac{x^2 + 2}{x^2 + 2}\right) - \frac{x - 3}{x^2 + 2}$
$$= \frac{x^4 + 2x^2 - x^2 - 2 - x + 3}{x^2 + 2}$$
$$= \frac{x^4 + x^2 - x + 1}{x^2 + 2}$$

25. (14)

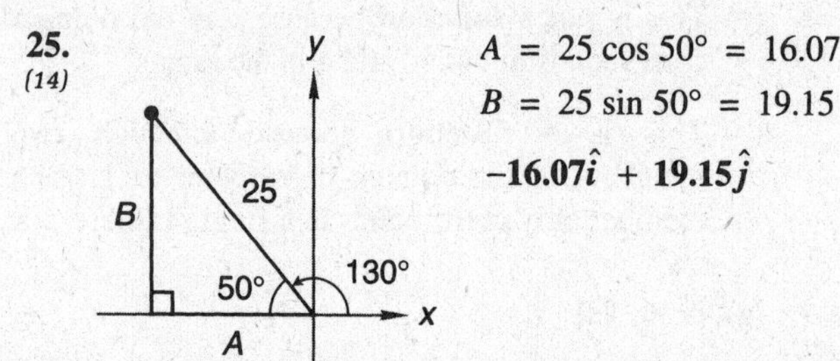

$A = 25 \cos 50° = 16.07$
$B = 25 \sin 50° = 19.15$
$\mathbf{-16.07\hat{i} + 19.15\hat{j}}$

26. (15)

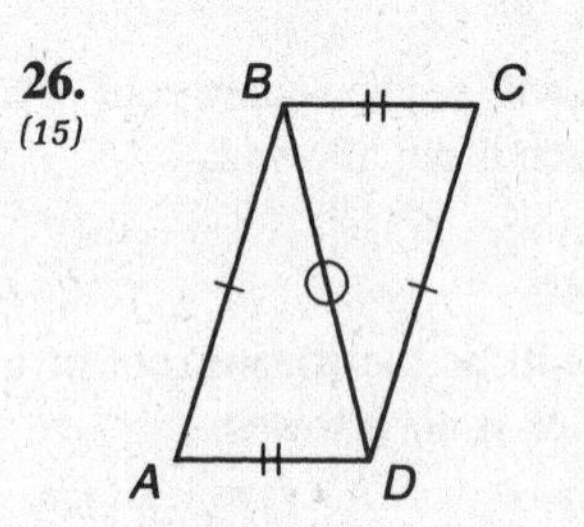

STATEMENTS	REASONS
1. $\overline{AB} \cong \overline{DC}$	1. Given
2. $\overline{AD} \cong \overline{BC}$	2. Given
3. $\overline{BD} \cong \overline{BD}$	3. Reflexive axiom
4. $\triangle ABD \cong \triangle CDB$	4. *SSS* congruency postulate

27. (13)
$$31 = \frac{1}{2}[(180 - x) - x]$$
$$31 = \frac{1}{2}(180 - 2x)$$
$$31 = 90 - x$$
$$\mathbf{x = 59}$$

28. (2) $V_{\text{solid}} = V_{\text{cone}} + V_{\text{cylinder}} + V_{\text{hemisphere}}$
$$V_{\text{solid}} = \frac{1}{3}\pi(3)^2(9) + \pi(3)^2(18) + \frac{2}{3}\pi(3)^3$$
$$= 27\pi + 162\pi + 18\pi$$
$$= \mathbf{207\pi\ m^3 = 650.31\ m^3}$$

29. (1) $(n + 1) - (n - 1) = 2$

$n + 1 > n - 1$

Therefore the answer is **A.**

30. (1) $AC + BC > AB$

$7 + 4 > AB$

$11 > AB$

Therefore the answer is **B.**

Problem Set 22

1. $m_{\text{original}} = \dfrac{(-2 - 7)}{[-8 - (-5)]} = \dfrac{-9}{-3} = 3$
(10)

The slopes of perpendicular lines are negative reciprocals of each other, so $m = -\frac{1}{3}$.

$$y = -\frac{1}{3}x + b$$

$$-2 = \left(-\frac{1}{3}\right)(5) + b$$

$$b = -2 + \frac{5}{3} = -\frac{1}{3}$$

$$\mathbf{y = -\frac{1}{3}x - \frac{1}{3}}$$

2. Consecutive even integers: N, $N + 2$, $N + 4$
(7)

$$(N + 2)(N + 4) = 4 + 10N$$

$$N^2 + 2N + 4N + 8 - 10N - 4 = 0$$

$$N^2 - 4N + 4 = 0$$

$$(N - 2)(N - 2) = 0$$

$$N = 2$$

2, 4, 6

3.
(14)

43°
85 m
43°
l

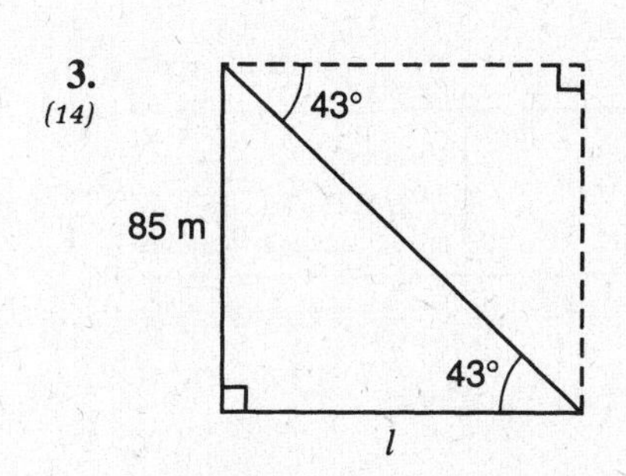

$$\tan 43° = \frac{85}{l}$$

$$l = \frac{85}{\tan 43°}$$

$$l = \mathbf{91.15\ m}$$

4. First, write the three equations.
(18)

$$\begin{cases} \text{(a)} & N_P + N_N + N_Q = 24 \\ \text{(b)} & N_P + 5N_N + 25N_Q = 160 \\ \text{(c)} & N_N = N_P \end{cases}$$

Now we can use substitution and elimination to solve.

(a) $N_P + N_N + N_Q = 24$

$(N_N) + N_N + N_Q = 24$

(a′) $2N_N + N_Q = 24$

(b) $N_P + 5N_N + 25N_Q = 160$

$(N_N) + 5N_N + 25N_Q = 160$

(b′) $6N_N + 25N_Q = 160$

(b′) $6N_N + 25N_Q = 160$

-3(a′) $-6N_N - 3N_Q = -72$

$22N_Q = 88$

$\mathbf{N_Q = 4}$

(a) $N_P + N_N + N_Q = 24$

$(N_N) + N_N + (4) = 24$

$2N_N = 20$

$\mathbf{N_N = 10}$

(c) $N_N = N_P$

$\mathbf{N_P = 10}$

5. First, write the two equations.
(18)

$$\begin{cases} \text{(a)} & T + U = 8 \\ \text{(b)} & 10T + U = 10U + T + 36 \end{cases}$$

(a) $T + U = 8$

$U = 8 - T$

Now we can use substitution to solve.

(b) $10T + U = 10U + T + 36$

$9T - 9U = 36$

$9T - 9(8 - T) = 36$

$18T = 108$

$T = 6$

$U = 8 - T = 8 - (6) = 2$

The numbers are **62, 26.**

6. (a) $\{x \in \mathbb{R} \mid |x - 3| < 4\}$
(22)

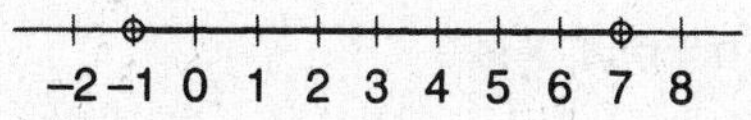

−2 −1 0 1 2 3 4 5 6 7 8

(b) $\{x \in \mathbb{R} \mid |x - 3| \geq 4\}$

−2 −1 0 1 2 3 4 5 6 7 8

7. $f(x) = |x|$
(22)

y
3
2
x
−3 −2 −1
1 2 3
−2
−3

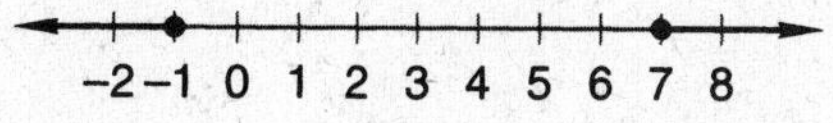

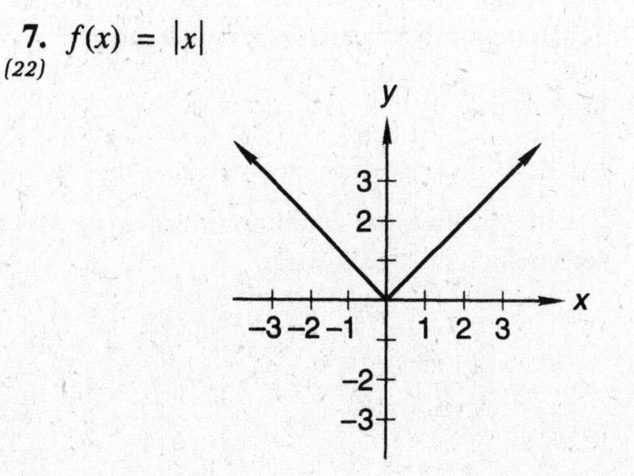

8. (22) $g(x) = \dfrac{1}{x + 3}$

The original function crosses the x-axis at -3. Therefore the reciprocal function will have an asymptote at -3.

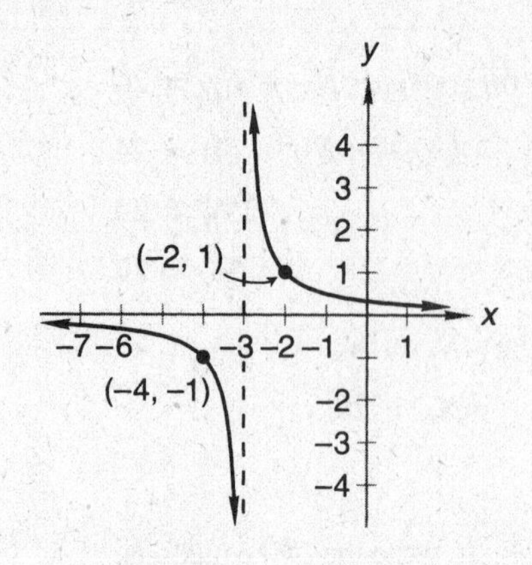

9. (21) (a) This is a **function** because all members of the domain correspond to only one member of the range.

(b) This is **not a function** because y in the domain corresponds to both 1 and 2 in the range.

10. (21) (a) **Function, 1 to 1**

This is a function because it passes the vertical line test. It is 1 to 1 because it passes the horizontal line test.

(b) **Function, not 1 to 1**

This is a function because it passes the vertical line test. It is not 1 to 1 because it fails the horizontal line test.

(c) **Not a function**

This is a not a function because it fails the vertical line test.

(d) **Function, 1 to 1**

This is a function because it passes the vertical line test. It is 1 to 1 because it passes the horizontal line test.

11. (21,22) (a) $\{x \in \mathbb{R} \mid x \geq 8\}$

If $x < 8$, then $f(x)$ will be the square root of a negative number, which is not allowed.

(b) $\{x \in \mathbb{R}\}$

(c) $\{x \in \mathbb{R} \mid x \neq -3\}$

If $x = -3$, then the denominator of $h(x)$ will be zero, which is not allowed.

12. (21) $f(x) = x^2 + 2x - 3$

$f(5) = 5^2 + 2(5) - 3 = \mathbf{32}$

13. (20,21) $g(\theta) = 2\sqrt{2}\sin\theta$

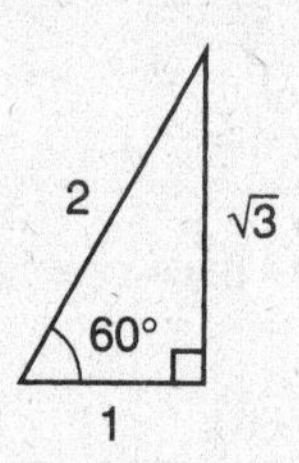

$$g(60°) = 2\sqrt{2}\left(\frac{\sqrt{3}}{2}\right) = \boldsymbol{\sqrt{6}}$$

14. (20,21) $h(\theta) = 3\sqrt{3}\cos\theta$

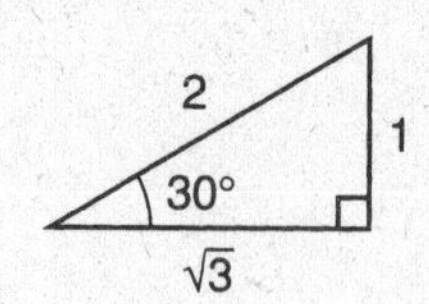

$$h(30°) = 3\sqrt{3}\left(\frac{\sqrt{3}}{2}\right) = \boldsymbol{\frac{9}{2}}$$

15. (17)

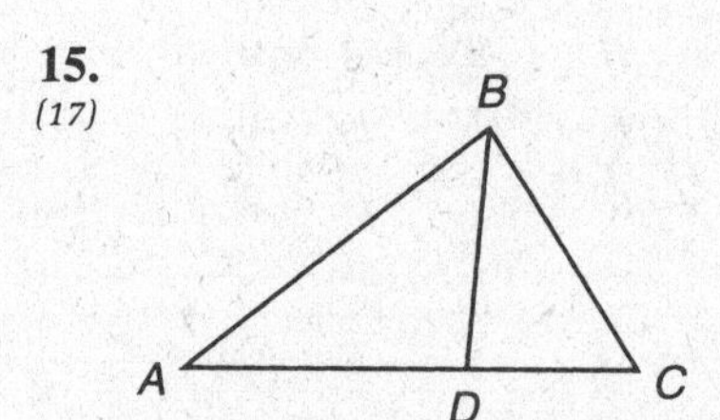

STATEMENTS	REASONS
1. $\frac{AC}{BC} = \frac{BC}{DC}$	1. Given
2. $\angle C \cong \angle C$	2. Reflexive axiom
3. $\triangle ABC \sim \triangle BDC$	3. *SAS* similarity postulate

16. (19) First, label the two equations.

$$\begin{aligned} &(a) \quad \begin{cases} x^2 - 2y^2 = -9 \\ x^2 + y^2 = 18 \end{cases} \\ &(b) \end{aligned}$$

Now we can use elimination and substitution to solve.

$$\begin{array}{rl} (a) & x^2 - 2y^2 = -9 \\ 2(b) & \underline{2x^2 + 2y^2 = 36} \\ & 3x^2 \qquad = 27 \\ & x^2 = 9 \\ & x = \pm 3 \end{array}$$

$$\begin{aligned} (b)\ x^2 + y^2 &= 18 \\ y &= \pm\sqrt{18 - x^2} \\ y &= \pm\sqrt{18 - (\pm 3)^2} \\ y &= \pm 3 \end{aligned}$$

(3, 3), (3, −3), (−3, 3), (−3, −3)

17. (19)
$$\frac{x^4 - y^4}{x - y} = \frac{(x^2 - y^2)(x^2 + y^2)}{x - y}$$
$$= \frac{(x + y)(x - y)(x^2 + y^2)}{x - y} = \mathbf{(x^2 + y^2)(x + y)}$$

18. (19)
$$8x^3b^6 - 27p^3 = (2xb^2)^3 - (3p)^3$$
$$= \mathbf{(2xb^2 - 3p)(4x^2b^4 + 6xb^2p + 9p^2)}$$

19. (16)
$$\frac{3s}{2m - \dfrac{z}{1 + \dfrac{k}{l}}} = \frac{3s}{2m - \dfrac{z}{\dfrac{l + k}{l}}}$$
$$= \frac{3s}{2m - \dfrac{zl}{l + k}} = \frac{3s}{\dfrac{2m(l + k) - zl}{l + k}}$$
$$= \frac{3s(l + k)}{2m(l + k) - zl} = \mathbf{\frac{3sl + 3sk}{2ml + 2mk - zl}}$$

20. (10)
$$\frac{6i^3 - 4i^4 - 6i}{-3i^4 - 5i + \sqrt{-25}} = \frac{-6i - 4 - 6i}{-3 - 5i + 5i}$$
$$= \frac{-4 - 12i}{-3} = \mathbf{\frac{4}{3} + 4i}$$

21. (5)
$$\frac{x^{3a+2}\left(\sqrt{y^3}\right)^{2a}}{y^{2a-4}\left(\sqrt{x}\right)^{2a+1}} = x^{3a+2}x^{-a-\frac{1}{2}}y^{3a}y^{-2a+4}$$
$$= x^{3a+2-a-\frac{1}{2}}y^{3a-2a+4} = \mathbf{x^{2a+\frac{3}{2}}y^{a+4}}$$

22. (5)
$$\frac{a^{\frac{5}{2}}b^{-\frac{2}{3}} + a^{-\frac{1}{2}}b^{\frac{1}{3}}}{a^{-\frac{1}{2}}b^{\frac{1}{3}}} = \frac{a^{\frac{5}{2}}b^{-\frac{2}{3}}}{a^{-\frac{1}{2}}b^{\frac{1}{3}}} + \frac{a^{-\frac{1}{2}}b^{\frac{1}{3}}}{a^{-\frac{1}{2}}b^{\frac{1}{3}}}$$
$$= \frac{a^3}{b} + 1 = \frac{a^3}{b} + \frac{b}{b} = \mathbf{\frac{a^3 + b}{b}}$$

23. (14)

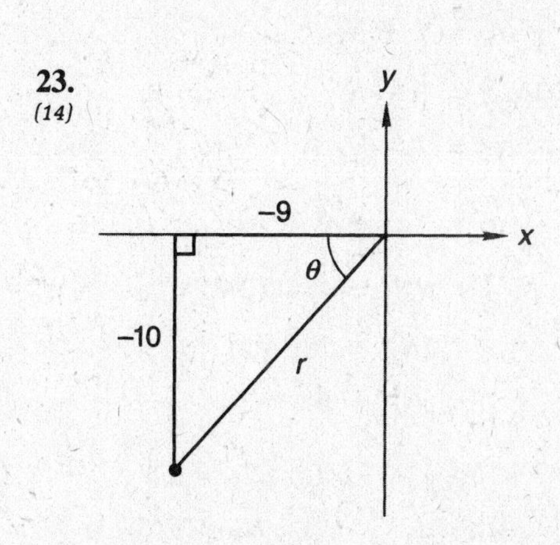

$$r = \sqrt{9^2 + 10^2} = \sqrt{81 + 100} = 13.45$$

$$\tan\theta = \frac{10}{9}$$
$$\theta = 48.01°$$

The polar angle is $180° + 48.01° = 228.01°$.

$\mathbf{13.45\underline{/228.01°}}$; $\mathbf{13.45\underline{/-131.99°}}$;
$\mathbf{-13.45\underline{/48.01°}}$; $\mathbf{-13.45\underline{/-311.99°}}$

24. (10)
$$4x^2 = -5 + 6x$$
$$4x^2 - 6x = -5$$
$$\left(x^2 - \frac{3}{2}x \qquad\right) = -\frac{5}{4}$$
$$\left(x^2 - \frac{3}{2}x + \frac{9}{16}\right) = -\frac{5}{4} + \frac{9}{16}$$
$$\left(x - \frac{3}{4}\right)^2 = -\frac{11}{16}$$
$$x - \frac{3}{4} = \pm\sqrt{-\frac{11}{16}}$$
$$x = \mathbf{\frac{3}{4} \pm \frac{\sqrt{11}\,i}{4}}$$

25. (15)

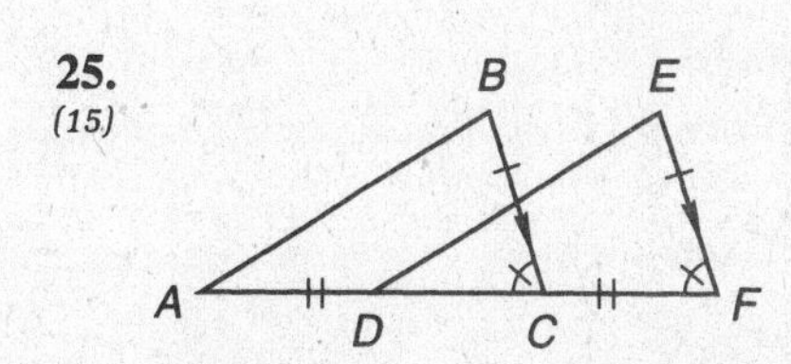

STATEMENTS	REASONS
1. $\overline{BC} \cong \overline{EF}$	1. Given
2. $\overline{BC} \parallel \overline{EF}$	2. Given
3. $\angle ACB \cong \angle DFE$	3. If two parallel lines are cut by a transversal, then each pair of corresponding angles is congruent.
4. $\overline{AC} \cong \overline{DF}$	4. Given
5. $\triangle ABC \cong \triangle DEF$	5. *SAS* congruency postulate

26. (15)

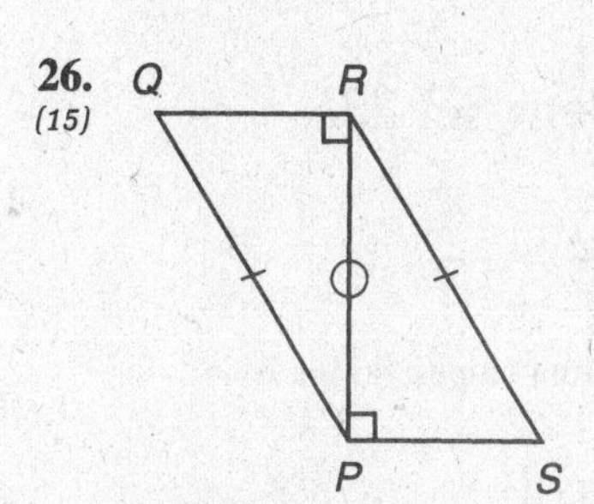

STATEMENTS	REASONS
1. $\overline{RP} \perp \overline{PS}, \overline{RP} \perp \overline{QR}$	1. Given
2. $\angle PRQ$ and $\angle RPS$ are right angles.	2. Perpendicular lines intersect to form right angles.
3. $\triangle PRQ$ and $\triangle RPS$ are right triangles.	3. A triangle which contains a right angle is a right triangle.
4. $\overline{PQ} \cong \overline{SR}$	4. Given
5. $\overline{RP} \cong \overline{RP}$	5. Reflexive axiom
6. $\triangle PRQ \cong \triangle RPS$	6. *HL* congruency postulate

27. (13) $m\widehat{AB} = m\widehat{BC}$

$m\widehat{AB} = 2(26°) = 52°$

$m\widehat{BC} = 52°$

$x = \frac{1}{2}[(180 - 52) - 52] = \frac{1}{2}(76) = \mathbf{38}$

28. (2) $V = V_{\text{cone}} + V_{\text{cylinder}} + V_{\text{hemisphere}}$

$= \frac{1}{3}\pi(5)^2\left[\sqrt{(5\sqrt{2})^2 - 5^2}\right] + \pi(5)^2(7)$

$+ \frac{1}{2}\left(\frac{4}{3}\right)\pi(5)^3$

$= \frac{125}{3}\pi + 175\pi + \frac{250}{3}\pi$

$= \mathbf{300\pi \text{ cm}^3 = 942.48 \text{ cm}^3}$

29. (1) If a is less than c, and c is less than 5, then a is less than 5.

$0 < a < c < 5$

$a < 5$

Therefore the answer is **A.**

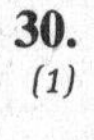

30. (1)

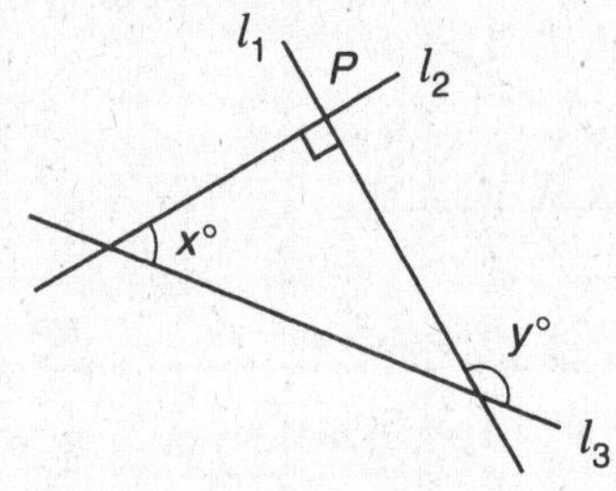

$x + 90 + (180 - y) = 180$

$x + 90 - y = 0$

$x = y - 90$

Therefore the answer is **C.**

Problem Set 23

1. (10) $m_{\text{original}} = \frac{(-1 - 2)}{(7 - 5)} = -\frac{3}{2}$

Parallel lines have equal slopes, so $m = -\frac{3}{2}$.

$y = -\frac{3}{2}x + b$

$1 = -\frac{3}{2}(1) + b$

$b = 1 + \frac{3}{2} = \frac{5}{2}$

$\mathbf{y = -\frac{3}{2}x + \frac{5}{2}}$

2. (18) w = reduction wanted

$\frac{2}{9}w = 3\frac{1}{3}$

$\frac{2}{9}w = \frac{10}{3}$

$w = \frac{10}{3}\left(\frac{9}{2}\right) = \mathbf{15}$

3. (18) First, write the three equations.

$$\begin{cases} \text{(a)} & N_N + N_D + N_Q = 21 \\ \text{(b)} & 5N_N + 10N_D + 25N_Q = 285 \\ \text{(c)} & N_Q = 2N_D \end{cases}$$

Now we can use substitution to solve.

(a) $N_N + N_D + N_Q = 21$

$N_N + N_D + (2N_D) = 21$

(a′) $N_N = 21 - 3N_D$

(b) $5N_N + 10N_D + 25N_Q = 285$

$5(21 - 3N_D) + 10N_D + 25(2N_D) = 285$

$105 - 15N_D + 10N_D + 50N_D = 285$

$45N_D = 180$

$\mathbf{N_D = 4}$

(a′) $N_N = 21 - 3N_D$

$N_N = 21 - 3(4)$

$\mathbf{N_N = 9}$

(c) $N_Q = 2N_D$

$N_Q = 2(4)$

$\mathbf{N_Q = 8}$

4. (18) First, write the three equations.

$$\begin{cases} \text{(a)} & 2(N_B + N_G) = 4 + N_R \\ \text{(b)} & N_B + N_R = 7 + N_G \\ \text{(c)} & N_R = 2N_G \end{cases}$$

Now we can use substitution to solve.

(b) $N_B + N_R = 7 + N_G$

$N_B + (2N_G) = 7 + N_G$

(b′) $N_B = 7 - N_G$

(a) $2(N_B + N_G) = 4 + N_R$

$2[(7 - N_G) + N_G] = 4 + (2N_G)$

$2N_G = 10$

$\mathbf{N_G = 5}$

(b′) $N_B = 7 - N_G$

$N_B = 7 - (5)$

$\mathbf{N_B = 2}$

(c) $N_R = 2N_G$

$N_R = 2(5)$

$\mathbf{N_R = 10}$

5. First, write the two equations.
(18)

$$\begin{cases} \text{(a)} & T + U = 6 \\ \text{(b)} & 10T + U = 10U + T + 18 \end{cases}$$

(a) $T + U = 6$

$U = 6 - T$

Now we can use substitution to solve.

(b)
$$10T + U = 10U + T + 18$$
$$9T - 9U = 18$$
$$9T - 9(6 - T) = 18$$
$$18T = 72$$
$$T = 4$$

$U = 6 - T = 6 - (4) = 2$

The numbers are **42** and **24.**

6. Acid initial = acid final
(18)

$$9 = 0.25(27 + W)$$
$$9 = 6.75 + 0.25W$$
$$0.25W = 2.25$$
$$W = \mathbf{9\ gallons}$$

7. $f(x) = 2^x$
(23)

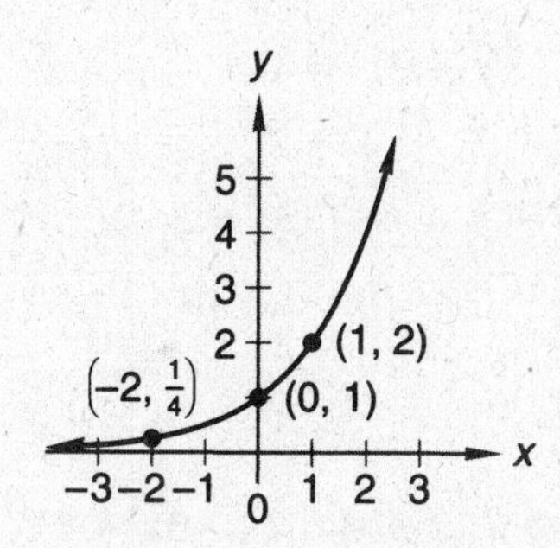

8. $g(x) = \left(\frac{1}{2}\right)^x$
(23)

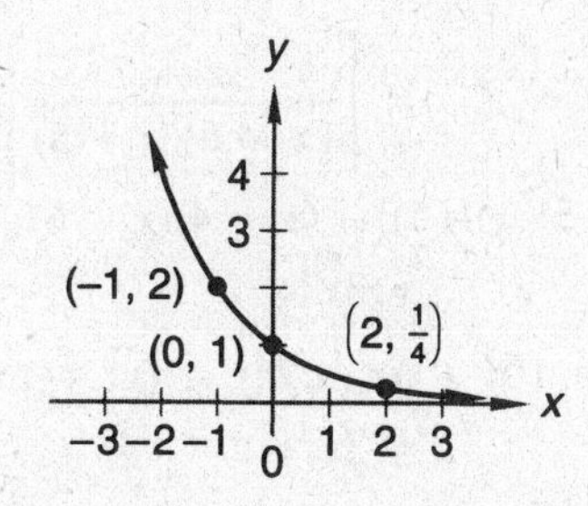

9. (a) $|x + 2| > 5$
(22)

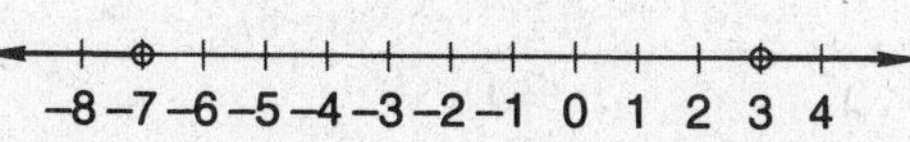

(b) $|x + 2| \leq 5$

−8 −7 −6 −5 −4 −3 −2 −1 0 1 2 3 4

10. $g(x) = \frac{1}{x - 2}$
(22)

When $x < 2, y < 0$. When $x > 2, y > 0$.

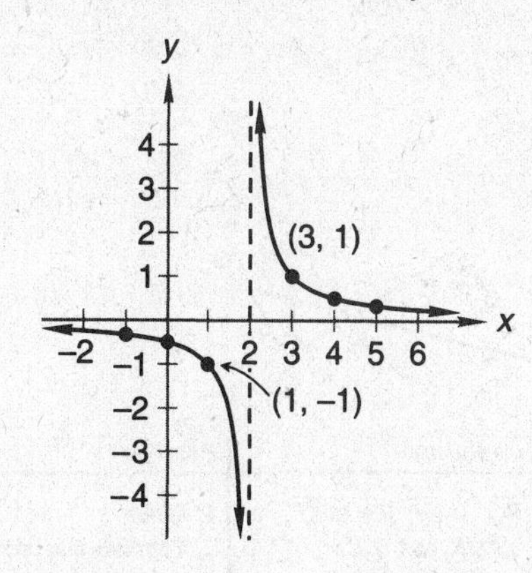

11. **(b), (d)**
(21)

Sets (b) and (d) are functions because for each x value (the domain) there is only one corresponding y value. Set (a) is not a function because it contains points (2, −1) and (2, 1). The x value of 2 in the domain corresponds to both −1 and 1. Set (c) is not a function because the x value of 1 corresponds to both 6 and −3.

12. (a) $\{x \in \mathbb{R}\}$
(21,22)

(b) $\{x \in \mathbb{R}\}$

(c) $\{x \in \mathbb{R} \mid x \neq 2\}$

If x is 2 then we have division by zero, which is not allowed.

13. $f(\theta) = 2\sqrt{2} \sin \theta$
(20,21)

$\sqrt{2}$, 45°, 1, 45°, 1

$$f(45°) = 2\sqrt{2} \sin (45°) = 2\sqrt{2}\left(\frac{1}{\sqrt{2}}\right) = \mathbf{2}$$

14. $g(\theta) = \tan \theta$
(20,21)

2, 60°, 1, 30°, $\sqrt{3}$

$$8g(60°) = 8 \tan (60°) = 8\left(\frac{\sqrt{3}}{1}\right) = \mathbf{8\sqrt{3}}$$

15. (8)
$$\frac{x}{x + 22} = \frac{18}{30} = \frac{3}{5}$$
$$5x = 3(x + 22)$$
$$2x = 66$$
$$x = \mathbf{33}$$

16. (17)

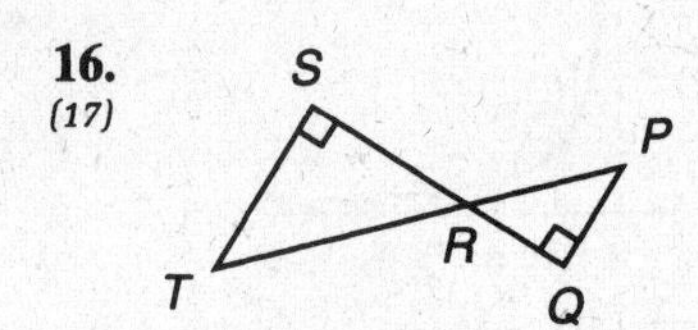

STATEMENTS	REASONS
1. $\overline{RQ} \perp \overline{QP}, \overline{RS} \perp \overline{ST}$	1. Given
2. $\angle PQR$ and $\angle TSR$ are right angles.	2. Perpendicular lines intersect to form right angles.
3. $\angle PQR \cong \angle TSR$	3. All right angles are congruent.
4. $\angle PRQ \cong \angle TRS$	4. Vertical angles are congruent.
5. $\angle P \cong \angle T$	5. $AA \rightarrow AAA$
6. $\Delta PQR \sim \Delta TSR$	6. AAA

17. (6) First, label the three equations.

$$\begin{cases} \text{(a)} & \frac{1}{2}x - \frac{1}{3}y + z = 1 \\ \text{(b)} & x - y + z = -2 \\ \text{(c)} & x + \frac{2}{3}z = 0 \end{cases}$$

(c) $x + \frac{2}{3}z = 0$
$$x = -\frac{2}{3}z$$

Now we can use substitution and elimination to solve.

(a) $\frac{1}{2}x - \frac{1}{3}y + z = 1$
$$\frac{1}{2}\left(-\frac{2}{3}z\right) - \frac{1}{3}y + z = 1$$
(a′) $-\frac{1}{3}y + \frac{2}{3}z = 1$

(b) $x - y + z = -2$
$$\left(-\frac{2}{3}z\right) - y + z = -2$$
(b′) $-y + \frac{1}{3}z = -2$

(b′) $-y + \frac{1}{3}z = -2$
−3(a′) $y - 2z = -3$
$$-\frac{5}{3}z = -5$$
$$\mathbf{z = 3}$$

$$x = -\frac{2}{3}z$$
$$x = -\frac{2}{3}(3)$$
$$\mathbf{x = -2}$$

(b) $x - y + z = -2$
$$y = x + z + 2$$
$$y = (-2) + (3) + 2$$
$$\mathbf{y = 3}$$

18. (19) First we write the two equations.

$$\begin{cases} \text{(a)} & xy = 9 \\ \text{(b)} & x + y = 6 \end{cases}$$

(b) $x + y = 6$
$$x = 6 - y$$

Now we can use substitution to solve.

(a) $xy = 9$
$$(6 - y)y = 9$$
$$y^2 - 6y + 9 = 0$$
$$(y - 3)^2 = 0$$
$$y = 3$$
$$x = 6 - y = 6 - (3) = 3$$
(3, 3)

19. (6)
$$\frac{2x - 5}{x - 6} - \frac{x + 4}{x + 5} = \frac{3x + 7}{x^2 - x - 30}$$
$$(x - 6)(x + 5)\left(\frac{2x - 5}{x - 6} - \frac{x + 4}{x + 5}\right) = \left[\frac{3x + 7}{(x - 6)(x + 5)}\right](x - 6)(x + 5)$$
$$(2x - 5)(x + 5) - (x + 4)(x - 6) = 3x + 7$$
$$(2x^2 + 10x - 5x - 25) - (x^2 - 6x + 4x - 24) = 3x + 7$$
$$x^2 + 4x - 8 = 0$$
$$x = \frac{-(4) \pm \sqrt{(4)^2 - 4(1)(-8)}}{2(1)}$$
$$x = \frac{-4 \pm \sqrt{48}}{2}$$
$$\mathbf{x = -2 \pm 2\sqrt{3}}$$

20. (19) $8a^3b^6 + 27a^9b^3$

$= a^3b^3(8b^3 + 27a^6)$

$= a^3b^3\left[(2b)^3 + (3a^2)^3\right]$

$= \mathbf{a^3b^3(2b + 3a^2)(4b^2 - 6ba^2 + 9a^4)}$

21. (16) $\dfrac{2}{1 + \dfrac{3x}{y - \dfrac{z}{k}}} = \dfrac{2}{1 + \dfrac{3x}{\left(\dfrac{yk - z}{k}\right)}}$

$= \dfrac{2}{1 + \dfrac{3xk}{(yk - z)}} = \dfrac{2}{\dfrac{1(yk - z) + 3xk}{(yk - z)}}$

$= \dfrac{\mathbf{2(yk - z)}}{\mathbf{yk - z + 3xk}}$

22. (5) $\sqrt{5}\sqrt{2}\sqrt{-5}\sqrt{-2} - 3\sqrt{2}i + \sqrt{5}\sqrt{2} - \sqrt{-5}i$

$= \sqrt{5}\sqrt{2}\sqrt{5}i\sqrt{2}i - 3\sqrt{2}i + \sqrt{5}\sqrt{2} - \sqrt{5}i^2$

$= (5)(2)(-1) - 3\sqrt{2}i + \sqrt{10} + \sqrt{5}$

$= \mathbf{\sqrt{10} + \sqrt{5} - 10 - 3\sqrt{2}i}$

23. (10) $\dfrac{4 + \sqrt{27}}{5 - 2\sqrt{3}} = \dfrac{(4 + 3\sqrt{3})}{(5 - 2\sqrt{3})} \cdot \dfrac{(5 + 2\sqrt{3})}{(5 + 2\sqrt{3})}$

$= \dfrac{20 + 8\sqrt{3} + 15\sqrt{3} + 6(3)}{25 - 4(3)} = \dfrac{\mathbf{38 + 23\sqrt{3}}}{\mathbf{13}}$

24. (5) $(m + n)(m^{-1} + n^{-1})^{-1} = \dfrac{m + n}{m^{-1} + n^{-1}}$

$= \dfrac{m + n}{\dfrac{1}{m} + \dfrac{1}{n}} = \dfrac{m + n}{\dfrac{n + m}{mn}} = \mathbf{mn}$

25. (16)

$$
\begin{array}{r}
x^2 - x + 4 \\
x + 3\overline{\smash{)}x^3 + 2x^2 + x} \\
\underline{x^3 + 3x^2} \\
-x^2 + x \\
\underline{-x^2 - 3x} \\
4x \\
\underline{4x + 12} \\
-12
\end{array}
$$

$\dfrac{x^3 + 2x^2 + x}{x + 3} = \mathbf{x^2 - x + 4 - \dfrac{12}{x + 3}}$

Check: $(x^2 - x + 4)\left(\dfrac{x + 3}{x + 3}\right) - \dfrac{12}{x + 3}$

$= \dfrac{x^3 - x^2 + 3x^2 + 4x - 3x + 12 - 12}{x + 3}$

$= \dfrac{x^3 + 2x^2 + x}{x + 3}$

26. (15)

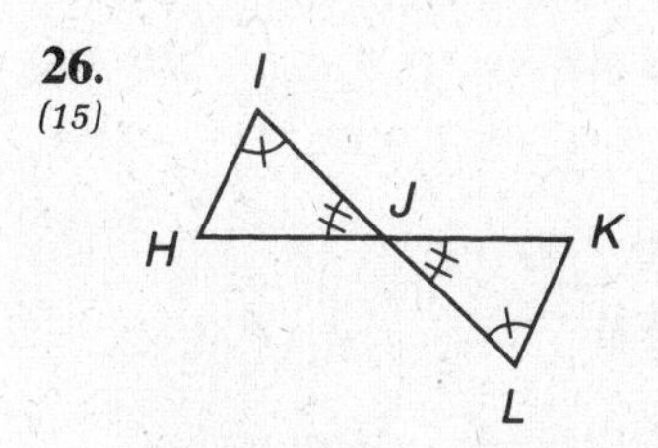

STATEMENTS	REASONS
1. $\angle I \cong \angle L$	1. Given
2. J is the midpoint of $\overline{IL}$	2. Given
3. $\overline{IJ} \cong \overline{LJ}$	3. A midpoint divides a segment into two congruent segments.
4. $\angle IJH \cong \angle LJK$	4. Vertical angles are congruent.
5. $\angle H \cong \angle K$	5. If two angles in one triangle are congruent to two angles in a second triangle, then the third angles are congruent.
6. $\Delta HIJ \cong \Delta KLJ$	6. *AAAS* congruency postulate

27. (11,13)

$2x(2x + 3) = (4x + 3)(2x - 1)$

$4x^2 + 6x = 8x^2 - 4x + 6x - 3$

$4x^2 - 4x - 3 = 0$

$(2x - 3)(2x + 1) = 0$

$x = -\dfrac{1}{2}, \dfrac{3}{2}$

$x = \dfrac{3}{2}$

$AB = (2x + 3) + 2x$

$AB = 2\left(\dfrac{3}{2}\right) + 3 + 2\left(\dfrac{3}{2}\right)$

$\mathbf{AB = 9}$

$CD = (4x + 3) + (2x - 1)$

$CD = 4\left(\dfrac{3}{2}\right) + 3 + 2\left(\dfrac{3}{2}\right) - 1$

$\mathbf{CD = 11}$

28. (2) $A_{\text{surface}} = \dfrac{1}{2}4\pi(12)^2 + 18\left[2\pi(12)\right] + 15\pi(12)$

$= 288\pi + 432\pi + 180\pi$

$= \mathbf{900\pi\ m^2 = 2827.43\ m^2}$

29. (1) A: If $a * b = a + 3b$, $2 * 0 = 2 + 3(0) = 2$.

Since $2 * 0 = 2$, $(2 * 0) * 5 = 2 * 5$, and $2 * 5 = 2 + 3(5) = 17$.

So, $(2 * 0) * 5 = 17$.

B: If $a * b = a + 3b$, $0 * 5 = 0 + 3(5) = 15$.

Since $0 * 5 = 15$, $2 * (0 * 5) = 2 * 15$, and $2 * 15 = 2 + 3(15) = 47$.

So, $2 * (0 * 5) = 47$.

$17 < 47$

$(2 * 0) * 5 < 2 * (0 * 5)$

Therefore the answer is **B.**

30. (1)

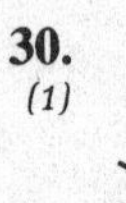

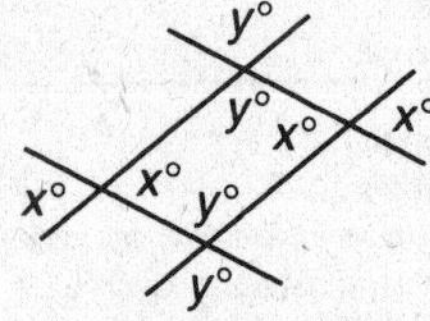

The angles vertical to the $x°$ and $y°$ angles form the interior angles of a quadrilateral.

$2x + 2y = 360$ and $x + y = 180$.

$160 < 180$

Therefore the answer is **B.**

Problem Set 24

1. (10) $m_{\text{original}} = \dfrac{(-3 - 2)}{(8 - 5)} = \dfrac{-5}{3}$

The slopes of perpendicular lines are negative reciprocals of each other, so $m = \frac{3}{5}$.

$$y = \frac{3}{5}x + b$$

$$-3 = \frac{3}{5}(-2) + b$$

$$b = -\frac{9}{5}$$

$$\mathbf{y = \frac{3}{5}x - \frac{9}{5}}$$

2. (7) Consecutive even integers:

$N,\ N + 2,\ N + 4,\ N + 6$

$$N + (N + 2) = (N + 4)(N + 6) - 70$$

$$2N + 2 = N^2 + 6N + 4N + 24 - 70$$

$$N^2 + 8N - 48 = 0$$

$$(N - 4)(N + 12) = 0$$

$$N = 4, -12$$

4, 6, 8, 10 and −12, −10, −8, −6

3. (18) First we write the three equations.

$$\begin{cases} \text{(a)}\ N_D + N_Q + N_H = 14 \\ \text{(b)}\ 10N_D + 25N_Q + 50N_H = 250 \\ \text{(c)}\ N_Q = N_H \end{cases}$$

Now we can use substitution to solve.

(a) $N_D + N_Q + N_H = 14$

$N_D + (N_H) + N_H = 14$

(a′) $N_D = 14 - 2N_H$

(b) $10N_D + 25N_Q + 50N_H = 250$

$10(14 - 2N_H) + 25(N_H) + 50N_H = 250$

$140 + 55N_H = 250$

$55N_H = 110$

$\mathbf{N_H = 2}$

(a′) $N_D = 14 - 2N_H$

$N_D = 14 - 2(2)$

$\mathbf{N_D = 10}$

(c) $N_Q = N_H$

$\mathbf{N_Q = 2}$

4. (18) First, write the three equations.

$$\begin{cases} \text{(a)}\ N_R + N_B = N_G + 1 \\ \text{(b)}\ 3N_R + 2N_B = N_G + 9 \\ \text{(c)}\ 3N_B = N_G + 2 \end{cases}$$

(c) $3N_B = N_G + 2$

$N_G = 3N_B - 2$

Now use elimination and substitution to solve.

$$\begin{array}{lrcl} \text{(b)} & 3N_R + 2N_B & = & N_G + 9 \\ -3\text{(a)} & -3N_R - 3N_B & = & -3N_G - 3 \\ \hline & -N_B & = & -2N_G + 6 \end{array}$$

$N_B = 2N_G - 6$

$N_B = 2(3N_B - 2) - 6$

$N_B = 6N_B - 4 - 6$

$-5N_B = -10$

$\mathbf{N_B = 2}$

$N_G = 3N_B - 2$

$N_G = 3(2) - 2$

$\mathbf{N_G = 4}$

(a) $N_R + N_B = N_G + 1$

$N_R + (2) = (4) + 1$

$\mathbf{N_R = 3}$

5. (18) First, write the two equations.

$$\begin{cases} \text{(a)}\ T + U = 10 \\ \text{(b)}\ 10T + U = 10U + T + 36 \end{cases}$$

(a) $T + U = 10$

$U = 10 - T$

Now we can use substitution to solve.

(b)
$$10T + U = 10U + T + 36$$
$$9T - 9U = 36$$
$$9T - 9(10 - T) = 36$$
$$18T = 126$$
$$T = 7$$
$$U = 10 - T = 10 - (7) = 3$$

The numbers are **73, 37.**

6. (18) P_N = salt

$$\text{Salt}_1 + \text{salt added} = \text{salt final}$$
$$0.0475(176) + 0.035P_N = 0.04(176 + P_N)$$
$$8.36 + 0.035P_N = 7.04 + 0.04P_N$$
$$0.005P_N = 1.32$$
$$P_N = \mathbf{264\ ml}$$

7. (23) $f(x) = 3^x$

8. (23) $g(x) = \left(\frac{1}{3}\right)^x$

9. (22) $g(x) = \dfrac{1}{x + 2}$

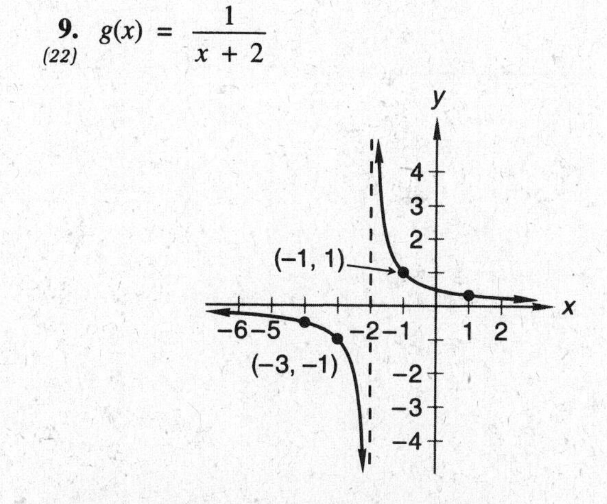

10. (21) (a) **Not a function**

(b) **Not a function**

11. (21) (a) **Not a function**

(b) **Function, not 1 to 1**

(c) **Not a function**

(d) **Function, not 1 to 1**

12. (21,22) (a) $\{x \in \mathbb{R} \mid x \geq -6\}$

(b) $\{x \in \mathbb{R}\}$

(c) $\{x \in \mathbb{R} \mid x \neq 0\}$

13. (24)

$$\tan 30° = \frac{1}{\sqrt{3}}$$
$$\cos 60° = \frac{1}{2}$$

$$\frac{\sqrt{3}}{2}\tan 30° + \sqrt{2}\cos 60°$$
$$= \frac{\sqrt{3}}{2}\left(\frac{1}{\sqrt{3}}\right) + \sqrt{2}\left(\frac{1}{2}\right) = \mathbf{\frac{1 + \sqrt{2}}{2}}$$

14. (24) (a) $(f + g)(x) = x + 3 + x^2 + 1 = x^2 + x + 4$

$(f + g)(3) = (3)^2 + (3) + 4 = \mathbf{16}$

(b) $(f - g)(x) = (x + 3) - (x^2 + 1)$

$= -x^2 + x + 2$

$(f - g)(3) = -(3)^2 + (3) + 2 = \mathbf{-4}$

(c) $(f \circ g)(x) = (x^2 + 1) + 3 = x^2 + 4$

$(f \circ g)(3) = (3)^2 + 4 = \mathbf{13}$

15. (20,21)

$$\tan 45° = \frac{1}{1} = 1$$

$$\tan 30° = \frac{1}{\sqrt{3}}$$

$$f(45°) + 2f(30°) = \tan 45° + 2\tan 30°$$
$$= 1 + 2\left(\frac{1}{\sqrt{3}}\right) = \mathbf{1 + \frac{2\sqrt{3}}{3}}$$

16. (3) First we solve for the scale factor.

$$4 \cdot \overrightarrow{SF} = 6$$

$$\overrightarrow{SF} = \frac{3}{2}$$

Now use the scale factor to solve for the missing sides.

$$DE = 8 \cdot \overrightarrow{SF} = 8 \cdot \frac{3}{2} = 12$$

$$DF = 10 \cdot \overrightarrow{SF} = 10 \cdot \frac{3}{2} = 15$$

$$\text{Perimeter} = 6 + 12 + 15 = \mathbf{33}$$

17. (19) First, label the two equations.

$$\begin{matrix}\text{(a)} \\ \text{(b)}\end{matrix} \begin{cases} x^2 + y^2 = 4 \\ 2x + y = 1 \end{cases}$$

$$\text{(b)}\quad 2x + y = 1$$

$$y = 1 - 2x$$

Now we can use substitution and the quadratic formula to solve.

$$\text{(a)}\quad x^2 + y^2 = 4$$

$$x^2 + (1 - 2x)^2 = 4$$

$$x^2 + 1 - 4x + 4x^2 = 4$$

$$5x^2 - 4x - 3 = 0$$

$$x = \frac{-(-4) \pm \sqrt{(-4)^2 - 4(5)(-3)}}{2(5)}$$

$$x = \frac{4 \pm \sqrt{76}}{10} = \frac{2 \pm \sqrt{19}}{5}$$

$$y = 1 - 2x$$

$$y = 1 - 2\left(\frac{2 + \sqrt{19}}{5}\right) \quad \text{or} \quad 1 - 2\left(\frac{2 - \sqrt{19}}{5}\right)$$

$$y = \frac{1 - 2\sqrt{19}}{5} \quad \text{or} \quad \frac{1 + 2\sqrt{19}}{5}$$

$$\mathbf{\left(\frac{2 + \sqrt{19}}{5}, \frac{1 - 2\sqrt{19}}{5}\right), \left(\frac{2 - \sqrt{19}}{5}, \frac{1 + 2\sqrt{19}}{5}\right)}$$

18. (19) $4x^{5N+2} + 3x^{8N+3}$

$$= 4\left(x^{5N}\right)\left(x^2\right) + 3\left(x^{8N}\right)\left(x^3\right)$$

$$= x^{5N}x^2\left(4 + 3x^{3N}x\right)$$

$$= \mathbf{x^{5N+2}\left(4 + 3x^{3N+1}\right)}$$

19. (16)

$$\frac{\dfrac{ac^3}{x^3y^2} - \dfrac{6m^3}{x^2y}}{\dfrac{d^2f}{x^2f} - \dfrac{g}{y}} = \frac{\dfrac{ac^3 - 6m^3xy}{x^3y^2}}{\dfrac{d^2y - gx^2}{x^2y}}$$

$$= \frac{x^2y\left(ac^3 - 6m^3xy\right)}{x^3y^2\left(d^2y - gx^2\right)} = \frac{ac^3 - 6m^3xy}{xy\left(d^2y - gx^2\right)}$$

$$= \mathbf{\frac{ac^3 - 6m^3xy}{xy^2d^2 - gx^3y}}$$

20. (10)

$$\frac{12i^6 + 3i^5 - 2i^4}{1 - 5i^2 + 3i^3} = \frac{-12 + 3i - 2}{1 + 5 - 3i}$$

$$= \left(\frac{-14 + 3i}{6 - 3i}\right)\left(\frac{6 + 3i}{6 + 3i}\right)$$

$$= \frac{-84 - 42i + 18i + 9i^2}{36 - 9i^2}$$

$$= \frac{-93 - 24i}{45} = \mathbf{-\frac{31}{15} - \frac{8}{15}i}$$

21. (10)

$$\frac{\sqrt[3]{a}\sqrt{b}\left(\sqrt[4]{ab}\right)^3}{a^{-2}b^3\left(\sqrt{a}\right)^3b^{-2}} = \frac{a^{\frac{1}{3}}b^{\frac{1}{2}}a^{\frac{3}{4}}b^{\frac{3}{4}}a^2b^2}{a^{\frac{3}{2}}b^3}$$

$$= a^{\frac{4}{12}+\frac{9}{12}+\frac{24}{12}-\frac{18}{12}}b^{\frac{2}{4}+\frac{3}{4}+\frac{8}{4}-\frac{12}{4}} = \mathbf{a^{\frac{19}{12}}b^{\frac{1}{4}}}$$

22. (5)

$$\left(\frac{1}{x + y}\right)^{-1}\left(x^{-1} + y^{-1}\right)^{-1} = \left(\frac{1}{\dfrac{1}{x + y}}\right)\left(\frac{1}{\dfrac{1}{x} + \dfrac{1}{y}}\right)$$

$$= (x + y)\left(\frac{1}{\dfrac{x + y}{xy}}\right) = \frac{xy(x + y)}{(x + y)} = \mathbf{xy}$$

23. (14)

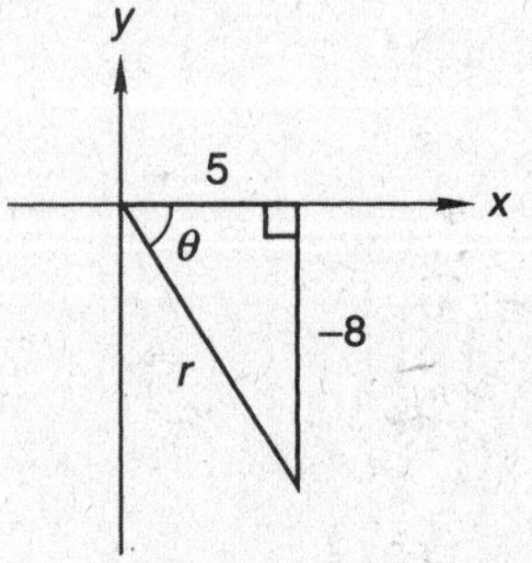

y, x, 5, θ, −8, r

$$r = \sqrt{5^2 + 8^2} = 9.43$$

$$\tan\theta = \frac{8}{5}$$

$$\theta = 57.99°$$

The polar angle is $360° - 57.99° = 302.01°$.

9.43/302.01°; 9.43/–57.99°;
–9.43/122.01°; –9.43/–237.99°

24. (10)

$$9x^2 = -3x + 8$$

$$x^2 + \frac{1}{3}x = \frac{8}{9}$$

$$\left(x^2 + \frac{1}{3}x + \frac{1}{36}\right) = \frac{8}{9} + \frac{1}{36}$$

$$\left(x + \frac{1}{6}\right)^2 = \frac{33}{36}$$

$$x + \frac{1}{6} = \pm\sqrt{\frac{33}{36}}$$

$$x = -\frac{\mathbf{1}}{\mathbf{6}} \pm \frac{\sqrt{\mathbf{33}}}{\mathbf{6}}$$

25. (21) $f(x + 1) = 2(x + 1) = \mathbf{2x + 2}$

26. (15)

B C E A D

STATEMENTS	REASONS
1. $\overline{BE} \cong \overline{CE}$	1. Given
2. $\overline{AE} \cong \overline{DE}$	2. Given
3. $\angle BEA \cong \angle CED$	3. Vertical angles are congruent.
4. $\triangle ABE \cong \triangle DCE$	4. *SAS* congruency postulate

27. (13)

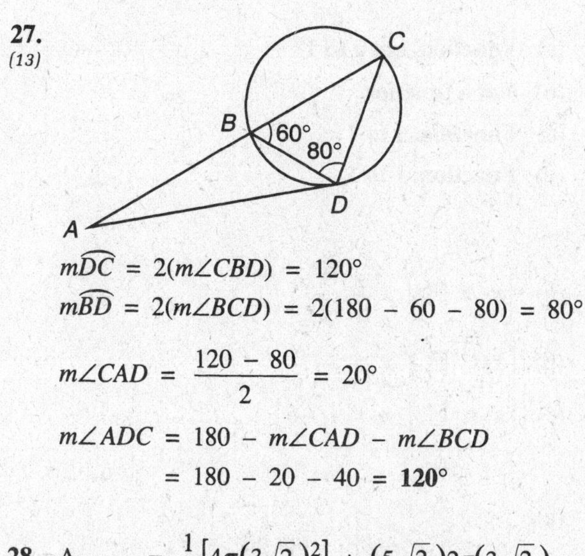

$$m\widehat{DC} = 2(m\angle CBD) = 120°$$

$$m\widehat{BD} = 2(m\angle BCD) = 2(180 - 60 - 80) = 80°$$

$$m\angle CAD = \frac{120 - 80}{2} = 20°$$

$$m\angle ADC = 180 - m\angle CAD - m\angle BCD$$
$$= 180 - 20 - 40 = \mathbf{120°}$$

28. (2)

$$A_{\text{surface}} = \frac{1}{2}\left[4\pi(3\sqrt{2})^2\right] + (5\sqrt{2})2\pi(3\sqrt{2}) + \sqrt{12^2 + (3\sqrt{2})^2}\,\pi(3\sqrt{2})$$
$$= 36\pi + 60\pi + 54\pi$$
$$= \mathbf{150\pi\ cm^2} = \mathbf{471.24\ cm^2}$$

29. (1) $a > b$

If we divide both sides of the inequality by –1, we get

$$-a < -b$$

Therefore the answer is **B.**

30. (1,3) The length of a side of a triangle is less than the sum of the lengths of the two remaining sides.

$$AC < 5 + 12$$
$$17 > AC$$

Therefore the answer is **A.**

Problem Set 25

1. (25)

$O_N - 5$ = Orville's age 5 years ago

$O_N + 5$ = Orville's age 5 years from now

$W_N - 5$ = Wilbur's age 5 years ago

$W_N + 5$ = Wilbur's age 5 years from now

First, write the two equations.

$$\begin{cases}(O_N - 5) = 2(W_N - 5)\\ (O_N + 5) = \frac{3}{2}(W_N + 5)\end{cases}$$

Simplify the equations.

$$\begin{cases}\text{(a)}\ O_N - 2W_N = -5\\ \text{(b)}\ 2O_N - 3W_N = 5\end{cases}$$

Now use substitution and elimination to solve.

$$\begin{array}{rr}\text{(b)} & 2O_N - 3W_N = 5\\ -2\text{(a)} & -2O_N + 4W_N = 10\\ \hline & W_N = 15\end{array}$$

$$\text{(a)}\ O_N - 2W_N = -5$$
$$O_N - 2(15) = -5$$
$$O_N = 25$$

Orville = 25 yr; Wilbur = 15 yr

2. (25)

$$R_ET_E + R_JT_J = \text{holes}$$

$$\left(\frac{2}{3}\right)(1 + T) + \left(\frac{3}{2}\right)(T) = 6$$

$$\frac{2}{3} + \frac{2}{3}T + \frac{3}{2}T = 6$$

$$\frac{13}{6}T = \frac{16}{3}$$

$$T = \frac{\mathbf{32}}{\mathbf{13}}\ \mathbf{hr}$$

3. (25)
$$R_ST_S + R_FT_F = \text{sandcastles}$$
$$\left(\frac{3}{5}\right)(3) + R_F(3) = 3$$
$$\frac{3}{5} + R_F = 1$$
$$R_F = \mathbf{\frac{2}{5}\ \frac{sandcastles}{hr}}$$

4. (25)
$$RMT = J$$
$$R(6)(9) = 1$$
$$R = \frac{1}{54}\ \frac{\text{job}}{\text{men-days}}$$
$$RMT = J$$
$$\frac{1}{54}(M)(5) = 15$$
$$M = \mathbf{162\ men}$$

5. (18) First, write the two equations.

(a) $T + U = 14$
(b) $\dfrac{T}{U} = 0.75$

(b) $\dfrac{T}{U} = 0.75$
$$T = 0.75U$$

Now we can use substitution to solve.

(a)
$$T + U = 14$$
$$(0.75U) + U = 14$$
$$1.75U = 14$$
$$U = 8$$

(a)
$$T + U = 14$$
$$T + (8) = 14$$
$$T = 6$$

The number is **68.**

6. (10) $m_{\text{original}} = \dfrac{(4 - 2)}{[3 - (-1)]} = \dfrac{1}{2}$

The slopes of parallel lines are equal, so $m = \frac{1}{2}$.

$$y = \frac{1}{2}x + b$$
$$1 = \frac{1}{2}(3) + b$$
$$b = -\frac{1}{2}$$
$$\mathbf{y = \frac{1}{2}x - \frac{1}{2}}$$

7. (23) $f(x) = 4^x$

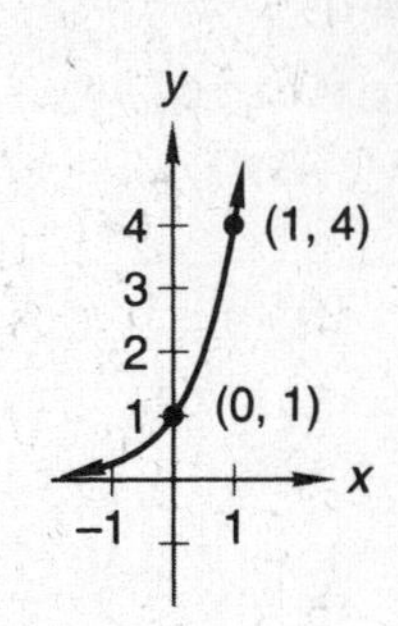

8. (23) $f(x) = \left(\dfrac{1}{4}\right)^x = (4)^{-x}$

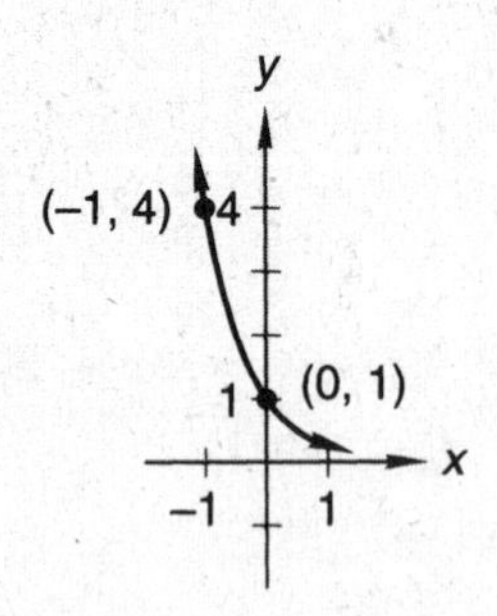

9. (22) (a) $|x + 3| > 6$

–9 0 3

(b) $|x - 2| \le 5$

–4 –3 0 7 8

10. (21)
(a) **Function, not 1 to 1**
(b) **Not a function**
(c) **Function, 1 to 1**
(d) **Function, 1 to 1**

11. (21)
(a) $\mathbf{\left\{x \in \mathbb{R} \mid x \ge \frac{1}{2}\right\}}$
(b) $\mathbf{\{x \in \mathbb{R} \mid x > 3\}}$
(c) $\mathbf{\{x \in \mathbb{R} \mid x \ne 1, -6\}}$

12. (24) (a)

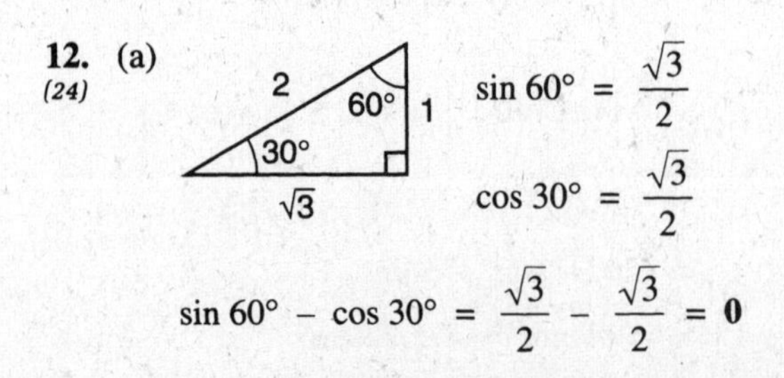

$$\sin 60° = \frac{\sqrt{3}}{2}$$
$$\cos 30° = \frac{\sqrt{3}}{2}$$
$$\sin 60° - \cos 30° = \frac{\sqrt{3}}{2} - \frac{\sqrt{3}}{2} = \mathbf{0}$$

(b) $\cos 45° = \frac{1}{\sqrt{2}} = \frac{\sqrt{2}}{2}$

$\sin 60° = \frac{\sqrt{3}}{2}$

$\frac{\sqrt{2}}{2}\cos 45° - 2\sin 60°$

$= \frac{\sqrt{2}}{2}\left(\frac{\sqrt{2}}{2}\right) - 2\left(\frac{\sqrt{3}}{2}\right) = \mathbf{\frac{1}{2} - \sqrt{3}}$

13. (24) (a) $(fg)(x) = (2x)(1 - x^2) = 2x - 2x^3$

$(fg)(2) = 2(2) - 2(2)^3 = \mathbf{-12}$

(b) $\left(\frac{f}{g}\right)(x) = \frac{2x}{(1 - x^2)}$

$\left(\frac{f}{g}\right)(2) = \frac{2(2)}{1 - 2^2} = \mathbf{-\frac{4}{3}}$

(c) $(g \circ f)(x) = 1 - (2x)^2 = 1 - 4x^2$

$(g \circ f)(2) = 1 - 4(2)^2 = \mathbf{-15}$

14. (21,24)

$\cos 30° = \frac{\sqrt{3}}{2}$

$\sin 30° = \frac{1}{2}$

$f(30°) = \frac{\sqrt{3}}{2}\cos 30° - \frac{1}{2}\sin 30°$

$= \frac{\sqrt{3}}{2}\left(\frac{\sqrt{3}}{2}\right) - \frac{1}{2}\left(\frac{1}{2}\right) = \frac{3}{4} - \frac{1}{4} = \mathbf{\frac{1}{2}}$

15. (24)

$\sin 30° = \frac{1}{2}$

$\tan 30° = \frac{1}{\sqrt{3}}$

$\sqrt{2}m(30°) - 2n(30°) = \sqrt{2}\sin 30° - 2\tan 30°$

$= \sqrt{2}\left(\frac{1}{2}\right) - 2\left(\frac{1}{\sqrt{3}}\right) = \mathbf{\frac{\sqrt{2}}{2} - \frac{2\sqrt{3}}{3}}$

16. (3) First we solve for the scale factor.

$\frac{22}{5} \cdot \overrightarrow{SF} = \frac{176}{25}$

$\overrightarrow{SF} = \frac{8}{5}$

Then we use the scale factor to find the unknown perimeter.

$P = \left(\frac{22}{5} + \frac{48}{5} + 6\right)\overrightarrow{SF} = 20\left(\frac{8}{5}\right) = \mathbf{32}$

17. (19) First we label the equations.

$$\begin{cases} \text{(a)}\ x^2 + y^2 = 9 \\ \text{(b)}\ 2x^2 - y^2 = 6 \end{cases}$$

(b) $2x^2 - y^2 = 6$

$y^2 = 2x^2 - 6$

Now we can use substitution to solve.

(a) $x^2 + y^2 = 9$

$x^2 + (2x^2 - 6) = 9$

$3x^2 = 15$

$x^2 = 5$

$x = \pm\sqrt{5}$

$y^2 = 2x^2 - 6$

$y^2 = 2(\pm\sqrt{5})^2 - 6$

$y^2 = 4$

$y = \pm 2$

$\mathbf{(\sqrt{5}, 2), (\sqrt{5}, -2), (-\sqrt{5}, 2), (-\sqrt{5}, -2)}$

18. (19) $\frac{x^{4a} - y^{4a}}{x^a + y^a} = \frac{(x^{2a} + y^{2a})(x^{2a} - y^{2a})}{x^a + y^a}$

$= \frac{(x^{2a} + y^{2a})(x^a + y^a)(x^a - y^a)}{x^a + y^a}$

$= \mathbf{(x^{2a} + y^{2a})(x^a - y^a)}$

19. (19) $8x^3y^6 - 27a^6b^9 = (2xy^2)^3 - (3a^2b^3)^3$

$= \mathbf{(2xy^2 - 3a^2b^3)(4x^2y^4 + 6xy^2a^2b^3 + 9a^4b^6)}$

20. (16) $\frac{3s}{2 - \frac{6l}{m - \frac{q}{r}}} = \frac{3s}{2 - \frac{6l}{\frac{mr - q}{r}}}$

$= \frac{3s}{\frac{2(mr - q) - 6rl}{mr - q}} = \mathbf{\frac{3smr - 3sq}{2mr - 2q - 6rl}}$

21. (10)

$$\frac{3 - 2\sqrt{12}}{4 + 3\sqrt{3}} = \left(\frac{3 - 4\sqrt{3}}{4 + 3\sqrt{3}}\right)\left(\frac{4 - 3\sqrt{3}}{4 - 3\sqrt{3}}\right)$$

$$= \frac{12 - 9\sqrt{3} - 16\sqrt{3} + 12(3)}{16 - 9(3)}$$

$$= \frac{48 - 25\sqrt{3}}{-11} = \mathbf{\frac{-48 + 25\sqrt{3}}{11}}$$

22. (5)

$$\left(3x^{\frac{1}{2}} - 2z^{\frac{1}{4}}\right)\left(3x^{\frac{1}{2}} + 2z^{\frac{1}{4}}\right)$$

$$= \left(3x^{\frac{1}{2}}\right)^2 - \left(2z^{\frac{1}{4}}\right)^2 = \mathbf{9x - 4z^{\frac{1}{2}}}$$

23. (16)

$$\begin{array}{r} x^3 - 4x^2 - x - 2 \\ x - 2\overline{)x^4 - 6x^3 + 7x^2 + 0x - 5} \\ \underline{x^4 - 2x^3} \qquad\qquad\qquad \\ -4x^3 + 7x^2 \qquad\quad \\ \underline{-4x^3 + 8x^2} \qquad\quad \\ -x^2 + 0x \quad \\ \underline{-x^2 + 2x} \quad \\ -2x - 5 \\ \underline{-2x + 4} \\ -9 \end{array}$$

$$\frac{x^4 - 6x^3 + 7x^2 - 5}{x - 2}$$

$$= \mathbf{x^3 - 4x^2 - x - 2 - \frac{9}{x - 2}}$$

Check: $\left(x^3 - 4x^2 - x - 2\right)\left(\frac{x - 2}{x - 2}\right) - \frac{9}{x - 2}$

$$= \frac{(x^3 - 4x^2 - x - 2)(x - 2) - 9}{x - 2}$$

$$= \frac{x^4 - 6x^3 + 7x^2 - 5}{x - 2}$$

24. (16)

$$5s = \frac{3k}{m}\left(\frac{6t}{z} - \frac{4k}{m}\right)$$

$$5s = \frac{18kt}{mz} - \frac{12k^2}{m^2}$$

$$5s + \frac{12k^2}{m^2} = \frac{18kt}{mz}$$

$$z\left(\frac{5sm^2 + 12k^2}{m^2}\right) = \frac{18kt}{m}$$

$$z = \mathbf{\frac{18ktm}{5sm^2 + 12k^2}}$$

25. (21)

$$f(x + 2) = (x + 2)^2 + 2(x + 2)$$

$$= x^2 + 4x + 4 + 2x + 4$$

$$= \mathbf{x^2 + 6x + 8}$$

26. (15)

STATEMENTS	REASONS
1. $\overline{AB} \cong \overline{CB}$	1. Given
2. $\angle ABD \cong \angle CBD$	2. Given
3. $\overline{BD} \cong \overline{BD}$	3. Reflexive axiom
4. $\triangle ABD \cong \triangle CBD$	4. *SAS* congruency postulate
5. $\overline{AD} \cong \overline{CD}$	5. *CPCTC*

27. (13)

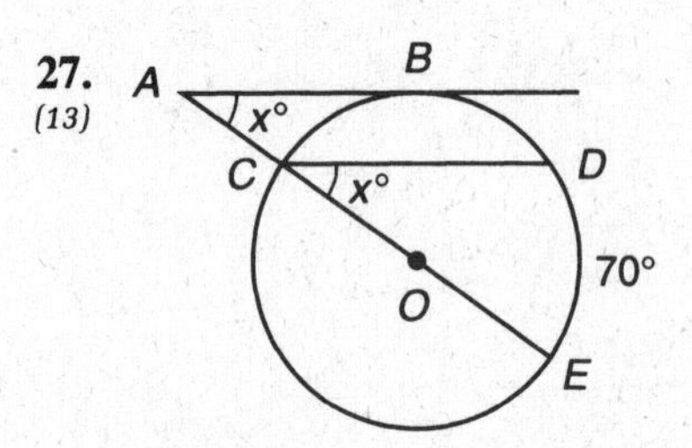

$m\angle DCE = x°$

$m\widehat{DE} = 360° - m\widehat{DCE} = 70°$

$x = \frac{1}{2}(70) = \mathbf{35}$

28. (2)

$$V_{\text{pyramid}} = \frac{1}{3}(A_{\text{base}})(h) = 13{,}824$$

$$\frac{1}{3}\left\{6\left[\frac{1}{2}(16\sqrt{3})(24)\right]\right\}h = 13{,}824$$

$$384\sqrt{3}\,h = 13{,}824$$

$$h = \mathbf{12\sqrt{3}\text{ m}}$$

29. (1) $ac^2 > bc^2$

We divide both sides of this inequality by c^2 which is a positive number, and we get

$$a > b$$

Therefore the answer is **A.**

30. (1) Since $BC = AC$, $m\angle A = m\angle B$.

So $m\angle A = 60°$.

$m\angle C = 180° - 60° - 60° = 60°$

ΔABC is equilateral, so $x = 3\frac{1}{2}$, and $\frac{x}{2} = 1\frac{3}{4}$.

Therefore the answer is **C.**

Problem Set 26

1. (25) First, write the two equations.

$$\begin{cases} C_N + 10 = 2(E_N + 10) \\ C_N - 5 = 5(E_N - 5) \end{cases}$$

Now we can rearrange the equations to solve.

$$\begin{aligned} &\text{(a)} \\ &\text{(b)} \end{aligned} \begin{cases} C_N = 2E_N + 10 \\ C_N = 5E_N - 20 \end{cases}$$

$$5E_N - 20 = 2E_N + 10$$
$$3E_N = 30$$
$$E_N = 10$$

(a) $C_N = 2E_N + 10 = 2(10) + 10 = 30$

Charlotte = 30 yr; Emily = 10 yr

2. (25)

$$R_DT_D + R_FT_F = 1 \text{ lawn}$$
$$\frac{1}{30}(10 + T) + \frac{1}{40}(T) = 1$$
$$\frac{1}{3} + \frac{1}{30}T + \frac{1}{40}T = 1$$
$$40 + 4T + 3T = 120$$
$$7T = 80$$
$$T = \mathbf{\frac{80}{7}\ min}$$

3. (25)

$$RMT = F$$
$$R(20)(10) = 360$$
$$R = 1.8\ \frac{\text{lb}}{\text{man-days}}$$
$$RMT = F$$
$$(1.8)(20 + 5)T = 360$$
$$T = \frac{360}{(1.8)(25)} = \mathbf{8\ days}$$

4. (25) First, write the two equations.

$$\begin{aligned} &\text{(a)} \\ &\text{(b)} \end{aligned} \begin{cases} R_DT_D = D_D \\ R_ST_S = D_S \end{cases}$$

$$D_D = 325 \text{ mi}, \ D_S = 450 \text{ mi}$$

$$R_D = R_S + 20, \ T_D = \frac{1}{2}T_S$$

Now we can use substitution to solve.

(a)
$$R_DT_D = D_D$$
$$(R_S + 20)\left(\frac{1}{2}T_S\right) = 325$$
$$\frac{1}{2}R_ST_S + 10T_S = 325$$
$$\frac{1}{2}(450) + 10T_S = 325$$
$$10T_S = 100$$
$$T_S = 10 \text{ hr}$$

$$T_D = \frac{1}{2}T_S = \frac{1}{2}(10) = 5 \text{ hr}$$

(a)
$$R_DT_D = D_D$$
$$R_D(5) = 325$$
$$R_D = 65 \text{ mph}$$

$$R_D = R_S + 20$$
$$(65) = R_S + 20$$
$$R_S = 45 \text{ mph}$$

Donnie = 65 mph; time = 5 hr
Sarah = 45 mph; time = 10 hr

5. (18)

$$\frac{P_1V_1}{T_1} = \frac{P_2V_2}{T_2}$$
$$\frac{(450)(400)}{1000} = \frac{(450)V_2}{2000}$$
$$V_2 = \mathbf{800\ liters}$$

6. (26) (a) $\mathbf{\log_k 7 = p}$

(b) $\mathbf{k^p = 7}$

7. (26) (a) $\mathbf{b^a = 12}$

(b) $\mathbf{\log_b 12 = a}$

8. (26)

$$\log_b 27 = 3$$
$$b^3 = 27$$
$$(b^3)^{\frac{1}{3}} = (27)^{\frac{1}{3}}$$
$$b = 3$$

9. (26)

$$\log_2 \frac{1}{8} = m$$

$$2^m = \frac{1}{8}$$

$$2^m = \frac{1}{2^3}$$

$$2^m = 2^{-3}$$

$$m = \mathbf{-3}$$

10. (26)

$$\log_{\frac{1}{2}} c = -4$$

$$c = \left(\frac{1}{2}\right)^{-4} = 2^4 = \mathbf{16}$$

11. (23) (a) $f(x) = 2.5^x$

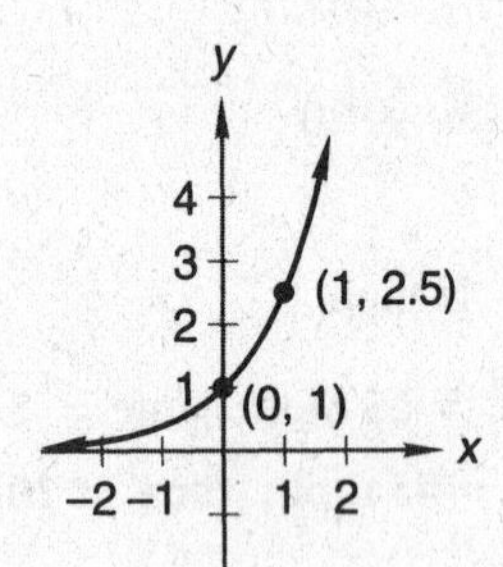

(b) $g(x) = \left(\frac{2}{5}\right)^x$

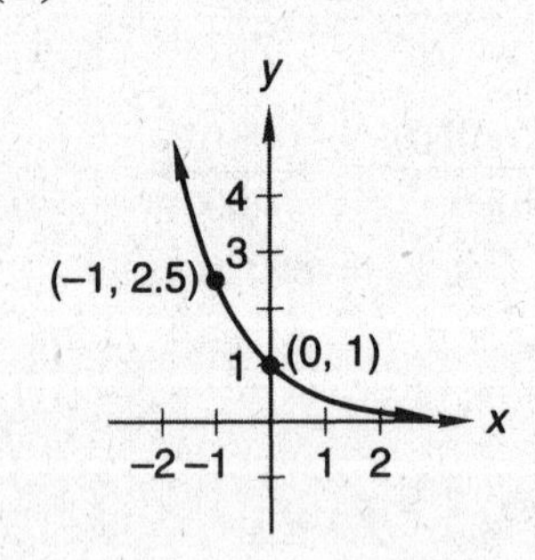

12. (22) $g(x) = \dfrac{1}{x - 3}$

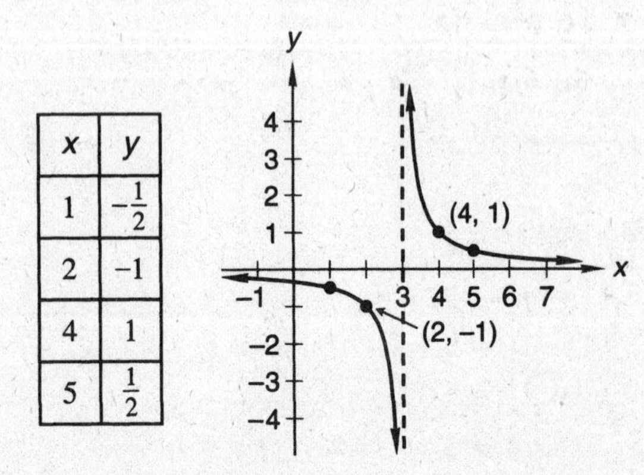

x	y
1	$-\frac{1}{2}$
2	-1
4	1
5	$\frac{1}{2}$

13. (21)

(a) **Not a function**

(b) **Function, 1 to 1**

(c) **Function, not 1 to 1**

(d) **Function, not 1 to 1**

14. (21)

(a) Domain of f: $\left\{\boldsymbol{x \in \mathbb{R} \mid x \geq -\frac{1}{2}}\right\}$

(b) Domain of g: $\{\boldsymbol{x \in \mathbb{R} \mid x \geq 0}\}$

(c) Domain of h: $\left\{\boldsymbol{x \in \mathbb{R} \mid x \neq -\frac{1}{2}, 3}\right\}$

15. (24)

2, 60°, 1, 30°, $\sqrt{3}$

$$\sin 30° = \frac{1}{2}$$

$$\cos 30° = \frac{\sqrt{3}}{2}$$

$$\sin 60° = \frac{\sqrt{3}}{2}$$

$$2 \sin 30° \cos 30° - \sin 60°$$

$$= 2\left(\frac{1}{2}\right)\left(\frac{\sqrt{3}}{2}\right) - \frac{\sqrt{3}}{2} = \mathbf{0}$$

16. (24)

2, 30°, $\sqrt{3}$, 60°, 1

$$\sin 60° = \frac{\sqrt{3}}{2}$$

$$\cos 60° = \frac{1}{2}$$

$$\tan 60° = \sqrt{3}$$

$$\frac{\sin 60°}{\cos 60°} - \tan 60° = \frac{\frac{\sqrt{3}}{2}}{\frac{1}{2}} - \sqrt{3} = \mathbf{0}$$

17. (24)

$\sqrt{2}$, 45°, 1, 45°, 1

$$\cos 45° = \frac{1}{\sqrt{2}} = \frac{\sqrt{2}}{2}$$

2, 30°, $\sqrt{3}$, 60°, 1

$$\sin 60° = \frac{\sqrt{3}}{2}$$

$$\frac{\sqrt{3}}{2} \cos 45° - \frac{\sqrt{2}}{2} \sin 60°$$

$$= \frac{\sqrt{3}}{2}\left(\frac{\sqrt{2}}{2}\right) - \frac{\sqrt{2}}{2}\left(\frac{\sqrt{3}}{2}\right) = \mathbf{0}$$

18. (24)

$\sqrt{2}$, 45°, 45°, 1, 1 $\tan 45° = 1$

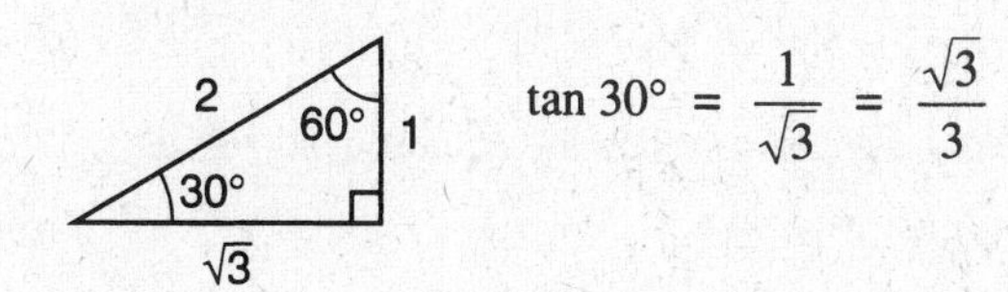

$$\tan 30° = \frac{1}{\sqrt{3}} = \frac{\sqrt{3}}{3}$$

$$\frac{\sqrt{2}}{2}r(45)° - \sqrt{3}r(30)°$$

$$= \frac{\sqrt{2}}{2}\tan 45° - \sqrt{3}\tan 30°$$

$$= \left(\frac{\sqrt{2}}{2}\right)(1) - \left(\sqrt{3}\right)\left(\frac{\sqrt{3}}{3}\right) = \mathbf{\frac{\sqrt{2}}{2} - 1}$$

19. (24) (a) $(f + g)(x) = (1 + x) + (x - x^2)$

$$= -x^2 + 2x + 1$$

$$(f + g)(-1) = -(-1)^2 + 2(-1) + 1 = \mathbf{-2}$$

(b) $(fg)(x) = (1 + x)(x - x^2) = -x^3 + x$

$$(fg)(-1) = -(-1)^3 + (-1) = 1 - 1 = \mathbf{0}$$

(c) $(g \circ f)(x) = 1 + x - (1 + x)^2 = -x^2 - x$

$$(g \circ f)(-1) = -(-1)^2 - (-1) = \mathbf{0}$$

20. (17)

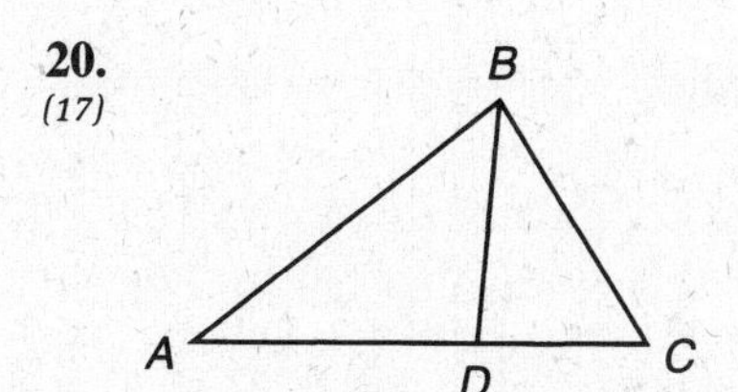

STATEMENTS	REASONS
1. $AC \cdot DC = BC \cdot BC$	1. Given
2. $\frac{AC}{BC} = \frac{BC}{DC}$	2. Division
3. $\angle C \cong \angle C$	3. Reflexive axiom
4. $\Delta ABC \sim \Delta BDC$	4. *SAS* similarity postulate

21. (19) First, we label the two equations.

$$\begin{cases} \text{(a)} & 2x^2 - y^2 = 1 \\ \text{(b)} & y + 2x = 1 \end{cases}$$

(b) $y + 2x = 1$

$$y = 1 - 2x$$

Now we can use substitution to solve.

(a)
$$2x^2 - y^2 = 1$$
$$2x^2 - (1 - 2x)^2 - 1 = 0$$
$$2x^2 - 1 + 4x - 4x^2 - 1 = 0$$
$$-2x^2 + 4x - 2 = 0$$
$$-2(x^2 - 2x + 1) = 0$$
$$(x - 1)^2 = 0$$
$$x - 1 = 0$$
$$\mathbf{x = 1}$$

$$y = 1 - 2x$$
$$y = 1 - 2(1)$$
$$\mathbf{y = -1}$$

22. (6)
$$\sqrt{x + 6} - x = 4$$
$$\sqrt{x + 6} = 4 + x$$
$$x + 6 = 16 + 8x + x^2$$
$$x^2 + 7x + 10 = 0$$
$$(x + 2)(x + 5) = 0$$

Therefore $x = -2, -5$ but, checking these solutions in the original equation we find $x \neq -5$, so $\mathbf{x = -2}$.

23. (19)
$$\frac{x^3 - y^3}{x - y} = \frac{(x - y)(x^2 + xy + y^2)}{x - y}$$
$$= \mathbf{x^2 + xy + y^2}$$

24. (19)
$$64x^{12}y^6 - 27a^6b^9 = (4x^4y^2)^3 - (3a^2b^3)^3$$
$$= \mathbf{(4x^4y^2 - 3a^2b^3)(16x^8y^4 + 12x^4y^2a^2b^3 + 9a^4b^6)}$$

25. (14)

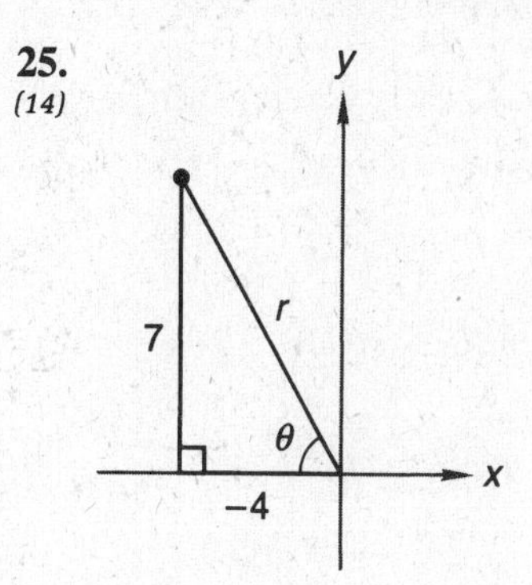

$$r = \sqrt{4^2 + 7^2} = 8.06$$

$$\tan\theta = \frac{7}{4}$$

$$\theta = 60.26°$$

The polar angle is $180° - 60.26° = 119.74°$.

8.06/119.74°; 8.06/−240.26°; −8.06/299.74°; −8.06/−60.26°

26. (21) $f(x - 5) = \dfrac{(x - 5) + 1}{(x - 5) - 2} = \mathbf{\dfrac{x - 4}{x - 7}}$

27. (13) $x = \dfrac{(360 - 115) - 115}{2} = \mathbf{65}$

28. (2)

$$4\pi(r_1)^2 = 144\pi \text{ m}^2$$
$$(r_1)^2 = 36$$
$$r_1 = 6 \text{ m}$$
$$4\pi(r_2)^2 = 144\pi + 880\pi$$
$$4\pi(r_2)^2 = 1024\pi$$
$$(r_2)^2 = 256$$
$$r_2 = 16 \text{ m}$$
$$r_2 - r_1 = \mathbf{10 \text{ m}}$$

29. (1)

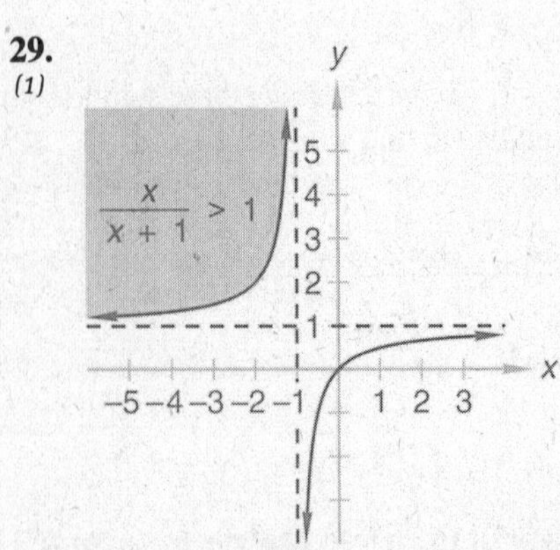

This is the graph of $y = \frac{x}{x+1}$. The shaded portion is those values which satisfy the inequality $\frac{x}{x+1} > 1$. All such values occur only when $x < 0$. Therefore the answer is **B.**

30. (1,12)

$$m\angle E = \frac{1}{2}(m\widehat{AB} + m\widehat{BC} + m\widehat{CD})$$
$$m\angle B = \frac{1}{2}(m\widehat{CD} + m\widehat{DE} + m\widehat{AE})$$
$$m\angle E + m\angle B$$
$$= \frac{1}{2}(m\widehat{AB} + m\widehat{BC} + 80 + m\widehat{CD} + m\widehat{DE} + m\widehat{EA})$$
$$= 40 + \frac{1}{2}(m\widehat{AB} + m\widehat{BC} + m\widehat{CD} + m\widehat{DE} + m\widehat{EA})$$
$$= 40 + \frac{1}{2}(360) = 220$$

$230 > 220$

Therefore the answer is **A.**

Problem Set 27

1. (25) First, write the two equations.

$$\begin{cases} G_N + 1 = 2(M_N + 1) \\ G_N - 5 = 8(M_N - 5) \end{cases}$$

Now we can simplify the equations to solve.

$$\begin{matrix} \text{(a)} \\ \text{(b)} \end{matrix} \begin{cases} G_N = 2M_N + 1 \\ G_N = 8M_N - 35 \end{cases}$$

$$2M_N + 1 = 8M_N - 35$$
$$6M_N = 36$$
$$M_N = 6$$

(a) $G_N = 2M_N + 1 = 2(6) + 1 = 13$

Marshall = 6 yr; George = 13 yr

2. (25)

$$R_M T_M + R_A T_A = \text{cake}$$
$$\frac{1}{10}(3 + 4) + R_A(4) = 1$$
$$\frac{7}{10} + 4R_A = 1$$
$$4R_A = \frac{3}{10}$$
$$R_A = \frac{3}{40}\,\frac{\text{cake}}{\text{min}}$$
$$R_A T_A = \text{cake}$$
$$\left(\frac{3}{40}\right)T_A = 1$$
$$T_A = \mathbf{\frac{40}{3} \textbf{ min}}$$

3. (25)

$$RWT = J$$
$$R(2)(3) = 6$$
$$R = 1\,\frac{\text{job}}{\text{worker-day}}$$
$$RWT = J$$
$$(1)(2 + 4)(T) = 6$$
$$6(T) = 6$$
$$T = \mathbf{1 \textbf{ day}}$$

4. (18) First, write the three equations.

$$\begin{matrix} \text{(a)} \\ \text{(b)} \\ \text{(c)} \end{matrix} \begin{cases} N_R = N_B + N_W - 11 \\ N_W = N_B + N_R - 3 \\ N_W = N_B + 1 \end{cases}$$

Now we can use substitution to solve.

(b) $N_W = N_B + N_R - 3$

$(N_B + 1) = N_B + N_R - 3$

$\mathbf{N_R = 4}$

(a) $N_R = N_B + N_W - 11$

$(4) = N_B + (N_B + 1) - 11$

$14 = 2N_B$

$\mathbf{N_B = 7}$

(c) $N_W = N_B + 1$

$N_W = (7) + 1$

$\mathbf{N_W = 8}$

5. (18) $a = \dfrac{kw}{s^2}$

$28 = \dfrac{k(7)}{(3)^2}$

$k = 36$

$a = \dfrac{36w}{s^2} = \dfrac{36(4)}{(2)^2} = \mathbf{36\ acorns}$

6. (10) $3y - 2x + 1 = 0$

$y = \dfrac{2}{3}x - \dfrac{1}{3}$

Parallel lines have equal slopes, so $m = \frac{2}{3}$.

$y = \dfrac{2}{3}x + b$

$-1 = \dfrac{2}{3}(2) + b$

$b = -\dfrac{7}{3}$

$\mathbf{y = \dfrac{2}{3}x - \dfrac{7}{3}}$

7. (26) $\mathbf{\log_3 7 = k}$

8. (26) $\mathbf{m^n = 8}$

9. (26) $\log_b 64 = 3$

$b^3 = 64$

$b = (64)^{\frac{1}{3}} = \mathbf{4}$

10. (26) $\log_3 \dfrac{1}{27} = n$

$3^n = \dfrac{1}{27}$

$3^n = \dfrac{1}{3^3}$

$3^n = 3^{-3}$

$\mathbf{n = -3}$

11. (26) $\log_{\frac{1}{2}} a = -2$

$a = \left(\dfrac{1}{2}\right)^{-2} = 2^2 = \mathbf{4}$

12. (23) (a) $f(x) = 5^x$

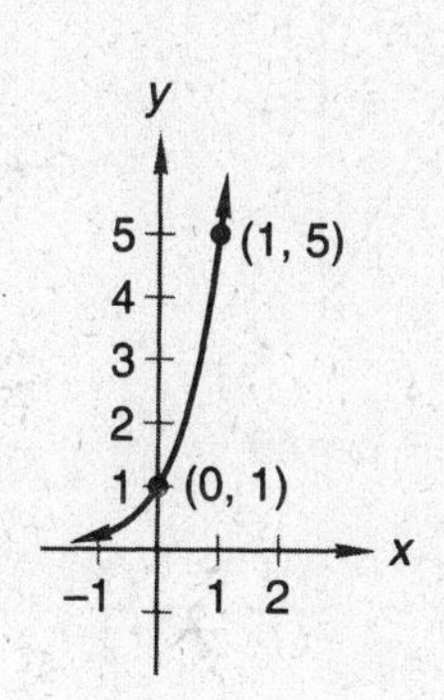

(b) $g(x) = \left(\dfrac{1}{5}\right)^x$

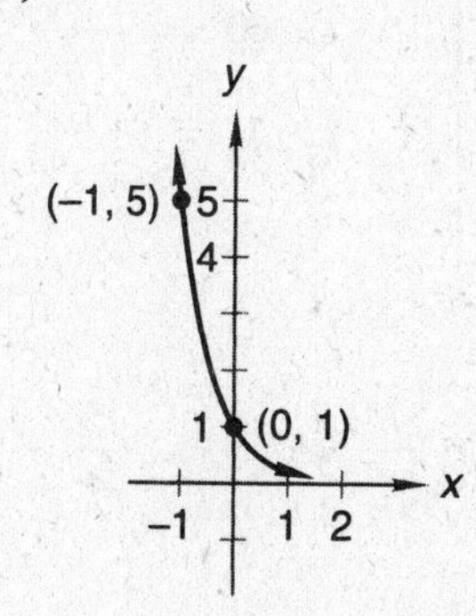

13. (22) $g(x) = \dfrac{1}{\frac{1}{2}x + 1} = \dfrac{2}{x + 2}$

x	y
−4	−1
−3	−2
−1	2
0	1

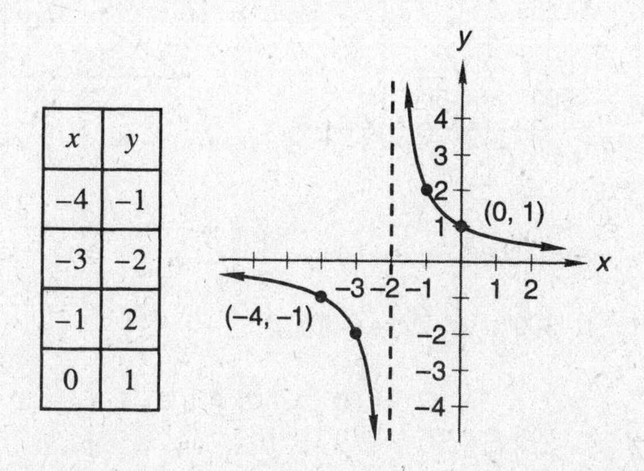

14. (21)
(a) **Not a function**
(b) **Not a function**
(c) **Function, 1 to 1**
(d) **Function, not 1 to 1**

15. (21)
(a) Domain of f: $\left\{\boldsymbol{x \in \mathbb{R} \mid x \leq \frac{1}{3}}\right\}$

(b) Domain of g: $\left\{\boldsymbol{x \in \mathbb{R} \mid x \geq -10, x \neq 2}\right\}$

(c) Domain of h: $\left\{\boldsymbol{x \in \mathbb{R} \mid x \neq -3, 1}\right\}$

16. (27)

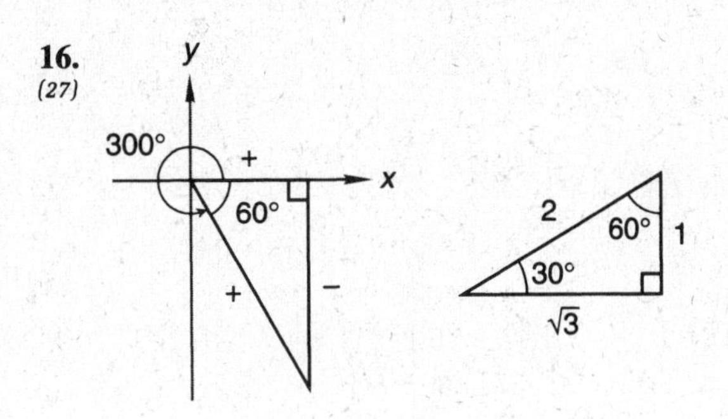

$$\cos 300° = \cos 60° = \frac{1}{2}$$

$$3 \cos 300° = 3\left(\frac{1}{2}\right) = \mathbf{\frac{3}{2}}$$

17. (24)

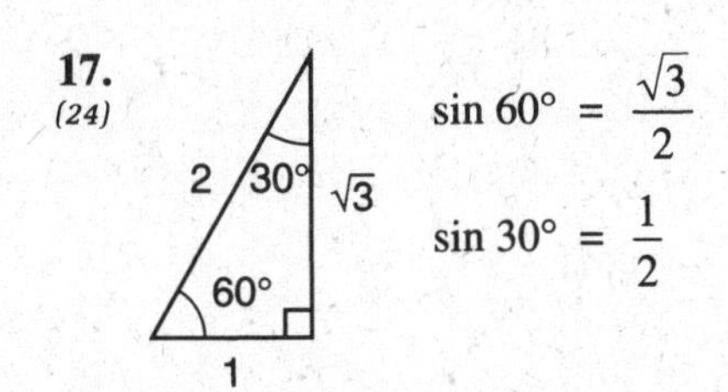

$$\sin 60° = \frac{\sqrt{3}}{2}$$

$$\sin 30° = \frac{1}{2}$$

$$\sin 60° + \frac{\sqrt{3}}{2} \sin 30° = \frac{\sqrt{3}}{2} + \frac{\sqrt{3}}{2}\left(\frac{1}{2}\right) = \mathbf{\frac{3\sqrt{3}}{4}}$$

18. (27)

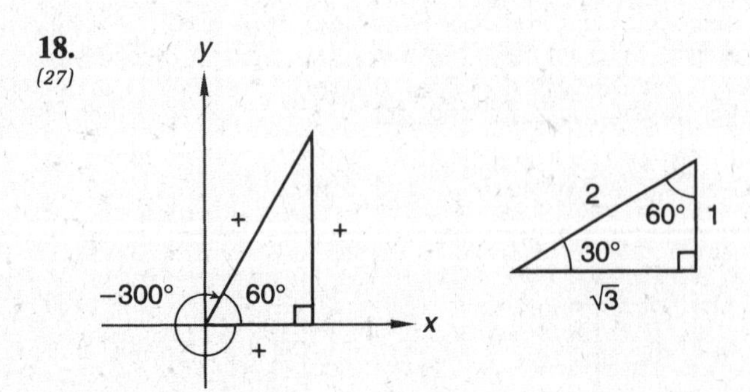

$$\cos(-300°) = \cos 60° = \frac{1}{2}$$

$$f(-300°) = -2\cos(-300°)$$

$$= -2\cos 60° = -2\left(\frac{1}{2}\right) = \mathbf{-1}$$

19. (24,26)
(a)
$$(f - g)(x) = 3 - \log_{10} x - x^2 - 2$$
$$= 1 - \log_{10} x - x^2$$
$$(f - g)(10) = 1 - \log_{10} 10 - (10)^2 = \mathbf{-100}$$

(b)
$$\left(\frac{f}{g}\right)(x) = \frac{3 - \log_{10} x}{x^2 + 2}$$
$$\left(\frac{f}{g}\right)(10) = \frac{3 - \log_{10} 10}{(10)^2 + 2} = \mathbf{\frac{1}{51}}$$

(c)
$$(g \circ f)(x) = (3 - \log_{10} x)^2 + 2$$
$$(g \circ f)(10) = (3 - \log_{10} 10)^2 + 2 = \mathbf{6}$$

20. (19) First, label the two equations.

$$\begin{cases} x^2 + y^2 = 16 & \text{(a)} \\ y - 3x = 4 & \text{(b)} \end{cases}$$

(b)
$$y - 3x = 4$$
$$y = 4 + 3x$$

Now we can use substitution to solve.

(a)
$$x^2 + y^2 = 16$$
$$x^2 + (4 + 3x)^2 = 16$$
$$10x^2 + 24x + 16 = 16$$
$$10x^2 + 24x = 0$$
$$2x(5x + 12) = 0$$
$$x = 0, -\frac{12}{5}$$

(b) $y = 4 + 3x = 4 + 3(0) = 4$

or

$$y = 4 + 3\left(-\frac{12}{5}\right) = -\frac{16}{5}$$

$$\mathbf{(0, 4), \left(-\frac{12}{5}, -\frac{16}{5}\right)}$$

21. (19)
$$3y^{2n+1} + 12y^{3n+2} = 3y^{2n+1} + 3y^{2n+1}(4y^{n+1})$$
$$= \mathbf{3y^{2n+1}(1 + 4y^{n+1})}$$

22. (10)
$$\frac{\sqrt{-16}\,i^3 - 4i^2}{1 - \sqrt{-4}} = \frac{4i^4 - 4i^2}{1 - 2i} = \frac{4 + 4}{1 - 2i}$$
$$= \frac{8}{1 - 2i} \cdot \frac{1 + 2i}{1 + 2i} = \frac{8 + 16i}{1 - 4i^2} = \frac{8 + 16i}{5}$$
$$= \mathbf{\frac{8}{5} + \frac{16}{5}i}$$

23. (16)

$$2r = \frac{1}{25}\left(\frac{3z}{a} - \frac{k}{p}\right)$$

$$50r = \frac{3z}{a} - \frac{k}{p}$$

$$50r + \frac{k}{p} = \frac{3z}{a}$$

$$\frac{50pr + k}{p} = \frac{3z}{a}$$

$$\mathbf{a = \frac{3zp}{50pr + k}}$$

24. (10)

$$4x^2 - 2x = -2$$

$$\left(x^2 - \frac{1}{2}x + \quad\right) = -\frac{1}{2}$$

$$\left(x^2 - \frac{1}{2}x + \frac{1}{16}\right) = -\frac{1}{2} + \frac{1}{16}$$

$$\left(x - \frac{1}{4}\right)^2 = -\frac{7}{16}$$

$$x - \frac{1}{4} = \pm\sqrt{-\frac{7}{16}}$$

$$x = \mathbf{\frac{1}{4} \pm \frac{\sqrt{7}}{4}i}$$

25. (21) $f(x + h) = \mathbf{\frac{2}{x + h}}$

26. (15)

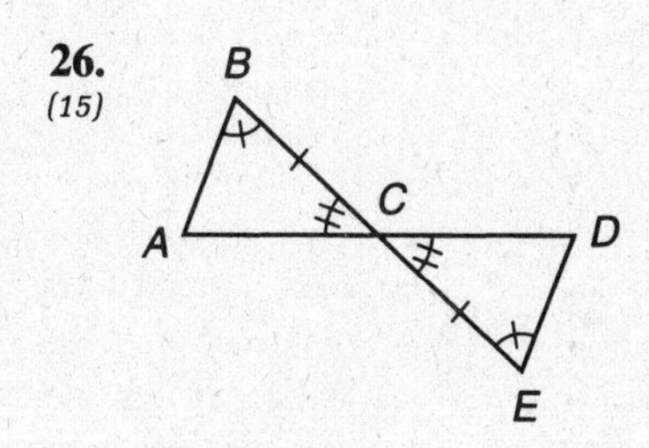

STATEMENTS	REASONS
1. C is the midpoint of $\overline{BE}$	1. Given
2. $\overline{BC} \cong \overline{EC}$	2. A midpoint divides a segment into two congruent segments.
3. $\angle B \cong \angle E$	3. Given
4. $\angle BCA \cong \angle ECD$	4. Vertical angles are congruent.
5. $\angle A \cong \angle D$	5. If two angles in one triangle are congruent to two angles in a second triangle, then the third angles are congruent.
6. $\triangle ABC \cong \triangle DEC$	6. *AAAS* congruency postulate
7. $\overline{AC} \cong \overline{DC}$	7. *CPCTC*

27. (13)

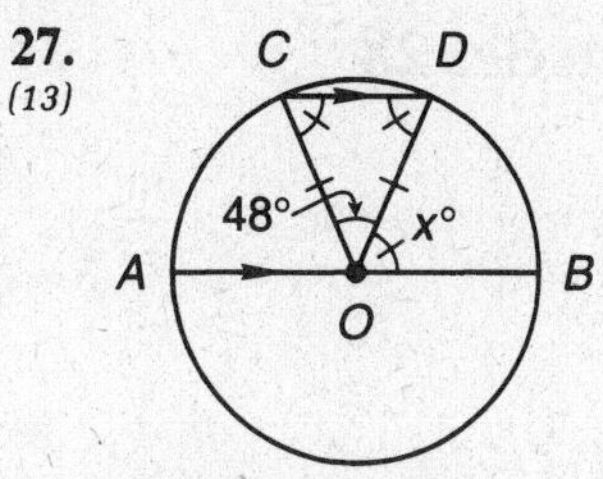

ΔOCD is isosceles

$\angle D \cong \angle C \cong \angle DOB$

$m\angle COD = m\overset{\frown}{CD} = 48°$

$$m\angle C + m\angle D + m\angle COD = 180$$

$$2m\angle C + m\angle COD = 180$$

$$2x + 48 = 180$$

$$x = \frac{180 - 48}{2} = \mathbf{66}$$

28. (2) $V_{\text{sphere}} = \frac{4}{3}\pi r^3$

$$\frac{4}{3}\pi(r_1)^3 = 4500\pi \text{ cm}^3$$

$$(r_1)^3 = 3375$$

$$r_1 = 15 \text{ cm}$$

$$\frac{4}{3}\pi(r_2)^3 = 4500\pi - 3528\pi$$

$$\frac{4}{3}\pi(r_2)^3 = 972\pi$$

$$(r_2)^3 = 729$$

$$r_2 = 9 \text{ cm}$$

$$r_1 - r_2 = 15 - 9 = \mathbf{6 \text{ cm}}$$

29. (1) $x = (-8)^{\frac{1}{3}} = -2$

$y = (-16)^{\frac{1}{3}} = -2\sqrt[3]{2}$

$x > y$

Therefore the answer is **A.**

30. (1)

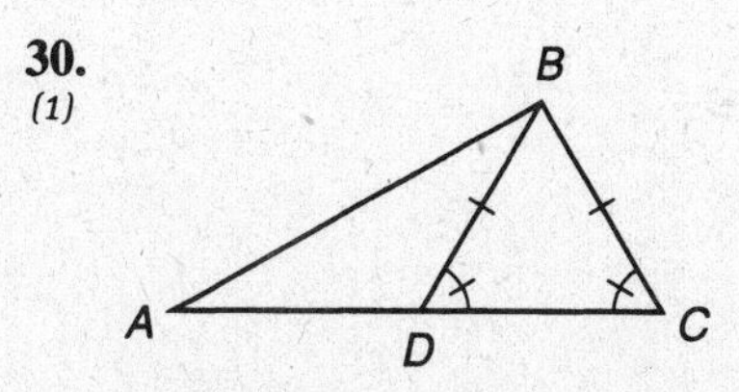

Since $\angle BDC \cong \angle BCD$, $BC \cong BD$.

If $BC = BD$ and $BC < AB$, then $BD < AB$.

Therefore the answer is **A.**

Problem Set 28

1. (28) Distance = x mi, time = $\frac{x}{p}$ hr, rate = $p\,\frac{\text{mi}}{\text{hr}}$

New distance = x mi

New time = $\left(\frac{x}{p} - 2\right)$ hr

New rate = $\dfrac{x}{\frac{x}{p} - 2} = \mathbf{\dfrac{xp}{x - 2p}\,\dfrac{mi}{hr}}$

2. (28) Distance = RT mi, rate = $R\,\frac{\text{mi}}{\text{hr}}$, time = T hr

New distance = $(RT + 20)$ mi

New rate = $(R + 5)\,\frac{\text{mi}}{\text{hr}}$

New time = $\mathbf{\dfrac{RT + 20}{R + 5}}$ **hr**

3. (25) First we write the two equations.

$$\begin{aligned}&\text{(a)} \; \begin{cases} D_N + 7 = 3T_N \\ D_N = 2T_N \end{cases} \\ &\text{(b)}\end{aligned}$$

Now we can use substitution to solve.

(a) $D_N + 7 = 3T_N$

$(2T_N) + 7 = 3T_N$

$T_N = 7$

(b) $D_N = 2T_N = 2(7) = 14$

Thomas = 7 yr; Dylan = 14 yr

4. (25) $R_J T_J + R_I T_I = \text{mission}$

$\frac{1}{20}(3 + 4) + R_I(4) = 1$

$\frac{7}{20} + 4R_I = 1$

$4R_I = \frac{13}{20}$

$R_I = \frac{13}{80}\,\frac{\text{mission}}{\text{hr}}$

$R_I T_I = \text{mission}$

$\frac{13}{80}T_I = 1$

$T_I = \mathbf{\frac{80}{13}}$ **hr**

5. (25) $RWT = J$

$R(7)(5) = 3$

$R = \frac{3}{35}\,\frac{\text{jobs}}{\text{worker-day}}$

$RWT = J$

$\left(\frac{3}{35}\right)(7 + 3)(T) = 4$

$\frac{30}{35}(T) = 4$

$T = \mathbf{\frac{14}{3}}$ **days**

6. (28) $\frac{8!}{2!\,4!} = \frac{8 \cdot 7 \cdot 6 \cdot 5 \cdot 4!}{2 \cdot 1 \cdot 4!}$
$= 4 \cdot 7 \cdot 6 \cdot 5 = \mathbf{840}$

7. (28) $\frac{9!}{3!\,3!} = \frac{9 \cdot 8 \cdot 7 \cdot 6 \cdot 5 \cdot 4 \cdot 3!}{3 \cdot 2 \cdot 1 \cdot 3!}$
$= 3 \cdot 4 \cdot 7 \cdot 6 \cdot 5 \cdot 4 = \mathbf{10{,}080}$

8. (26) $\mathbf{\log_2 9 = k}$

9. (26) $\mathbf{m^n = 6}$

10. (26) $\log_c 81 = 4$

$c^4 = 81$

$c = 81^{\frac{1}{4}} = \mathbf{3}$

11. (26) $\log_2 \frac{1}{16} = m$

$2^m = \frac{1}{16}$

$2^m = \frac{1}{2^4}$

$2^m = 2^{-4}$

$m = \mathbf{-4}$

12. (26) $\log_{\frac{1}{4}} b = -2$

$b = \left(\frac{1}{4}\right)^{-2} = 4^2 = \mathbf{16}$

13. (23) (a) $f(x) = \left(\frac{5}{4}\right)^x$

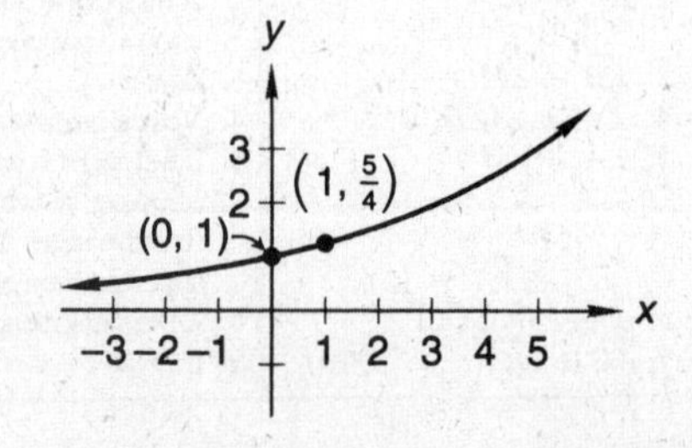

(b) $g(x) = \left(\dfrac{4}{5}\right)^x$

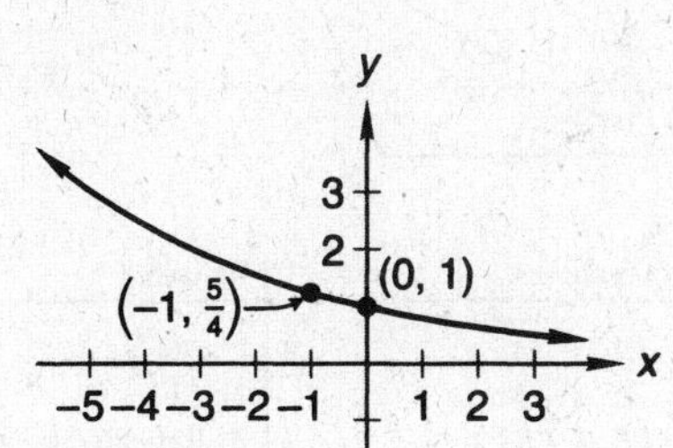

14. $g(x) = \dfrac{1}{\frac{1}{2}x - 1} = \dfrac{2}{x - 2}$
(22)

x	y
0	−1
1	−2
3	2
4	1

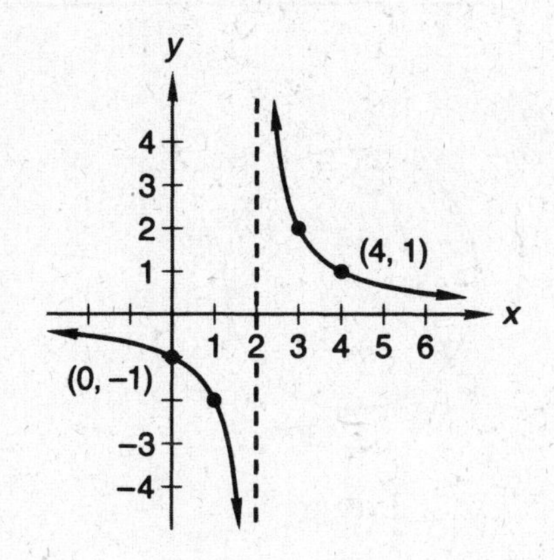

15. (a) Domain of f: $\{x \in \mathbb{R} \mid x \geq -4\}$
(21)

(b) $g(x) = \dfrac{1}{x(x - 1)}$

Domain of g: $\{x \in \mathbb{R} \mid x \neq 0, 1\}$

(c) Domain of h: $\{x \in \mathbb{R} \mid x \geq 4\}$

16.
(27)

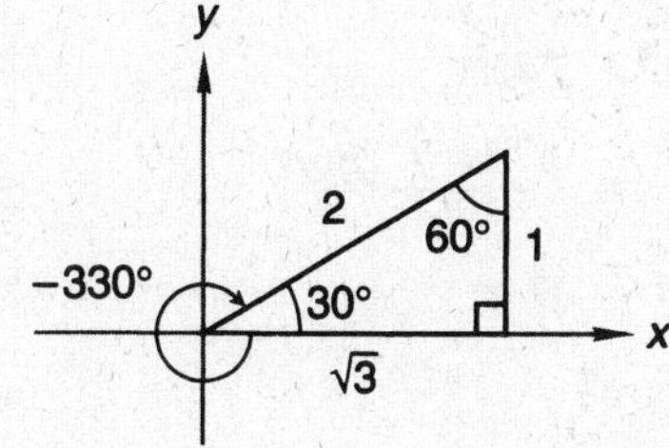

$\cos(-330°) = \cos 30° = \dfrac{\sqrt{3}}{2}$

$2\cos(-330°) = 2\left(\dfrac{\sqrt{3}}{2}\right) = \sqrt{3}$

17.
(24)

$\cos 60° = \dfrac{1}{2}$

$\sin 45° = \dfrac{1}{\sqrt{2}} = \dfrac{\sqrt{2}}{2}$

$s(60°) + \sqrt{2}\,t(45°) = \cos 60° + \sqrt{2} \sin 45°$

$= \dfrac{1}{2} + \sqrt{2}\left(\dfrac{\sqrt{2}}{2}\right) = \mathbf{\dfrac{3}{2}}$

18. (a) $(f + g)(x) = 2x + 3 + \dfrac{1}{2}x - \dfrac{3}{2}$
(24)

$= \dfrac{5}{2}x + \dfrac{3}{2}$

$(f + g)(-3) = \dfrac{5}{2}(-3) + \dfrac{3}{2} = \mathbf{-6}$

(b) $(f \circ g)(x) = 2\left(\dfrac{1}{2}x - \dfrac{3}{2}\right) + 3 = x$

$(f \circ g)(-3) = \mathbf{-3}$

(c) $(g \circ f)(x) = \dfrac{1}{2}(2x + 3) - \dfrac{3}{2} = x$

$(g \circ f)(-3) = \mathbf{-3}$

19. First, label the two equations.
(19)

$$\begin{cases} \text{(a)}\ x^2 + y^2 = 4 \\ \text{(b)}\ x^2 - y^2 = 4 \end{cases}$$

(a) $x^2 + y^2 = 4$

$x^2 = 4 - y^2$

Now we can use substitution to solve.

(b) $x^2 - y^2 = 4$

$(4 - y^2) - y^2 = 4$

$-2y^2 = 0$

$y = 0$

(b) $x^2 - y^2 = 4$

$x^2 - (0)^2 = 4$

$x = \pm 2$

(2, 0), (−2, 0)

20. $\dfrac{\dfrac{x^2y}{ca^3} - \dfrac{y^3z}{a^2}}{\dfrac{s^2t}{a^2} - \dfrac{r^2z}{a^3c}} \cdot \dfrac{a^3c}{a^3c} = \mathbf{\dfrac{x^2y - y^3zca}{s^2tca - r^2z}}$
(16)

21. (10)
$$\frac{\sqrt{2}\sqrt{-3}\sqrt{6} + \sqrt{-16} + \sqrt{5}\sqrt{-5}}{1 + \sqrt{-4}i^4}$$
$$= \frac{\sqrt{2}\sqrt{3}i\sqrt{6} + 4i + 5i}{1 + 2i^5} = \frac{6i + 4i + 5i}{1 + 2i}$$
$$= \frac{15i}{1 + 2i} \cdot \frac{1 - 2i}{1 - 2i} = \frac{15i - 30i^2}{1 - 4i^2}$$
$$= \frac{15i + 30}{5} = \mathbf{6 + 3i}$$

22. (10)
$$\frac{4 - 3\sqrt{2}}{2 - 2\sqrt{2}} = \frac{4 - 3\sqrt{2}}{2 - 2\sqrt{2}} \cdot \frac{2 + 2\sqrt{2}}{2 + 2\sqrt{2}}$$
$$= \frac{8 + 2\sqrt{2} - 12}{4 - 8} = \frac{-4 + 2\sqrt{2}}{-4} = \mathbf{\frac{2 - \sqrt{2}}{2}}$$

23. (5)
$$\frac{x^{a+2}\left(\sqrt{y^4}\right)^{2a-1}}{y^{3a+2}} = x^{a+2}\left(y^2\right)^{2a-1}y^{-3a-2}$$
$$= x^{a+2}y^{4a-2}y^{-3a-2} = \mathbf{x^{a+2}y^{a-4}}$$

24. (16)
$$\begin{array}{r} x^3 - 2x^2 - 2x \\ x - 1 \overline{)\, x^4 - 3x^3 + 0x^2 + 2x - 6} \\ \underline{x^4 - x^3} \\ -2x^3 + 0x^2 \\ \underline{-2x^3 + 2x^2} \\ -2x^2 + 2x \\ \underline{-2x^2 + 2x} \\ 0 - 6 \end{array}$$

$$\frac{x^4 - 3x^3 + 2x - 6}{x - 1} = \mathbf{x^3 - 2x^2 - 2x - \frac{6}{x - 1}}$$

Check:
$$\left(x^3 - 2x^2 - 2x\right)\left(\frac{x - 1}{x - 1}\right) - \frac{6}{x - 1}$$
$$= \frac{x^4 - 2x^3 - x^3 - 2x^2 + 2x^2 + 2x - 6}{x - 1}$$
$$= \frac{x^4 - 3x^3 + 2x - 6}{x - 1}$$

25. (21)
$$f(x - 3) = -(x - 3)^2 - 6(x - 3) - 8$$
$$= -x^2 + 6x - 9 - 6x + 18 - 8$$
$$= \mathbf{1 - x^2}$$

26. (15)

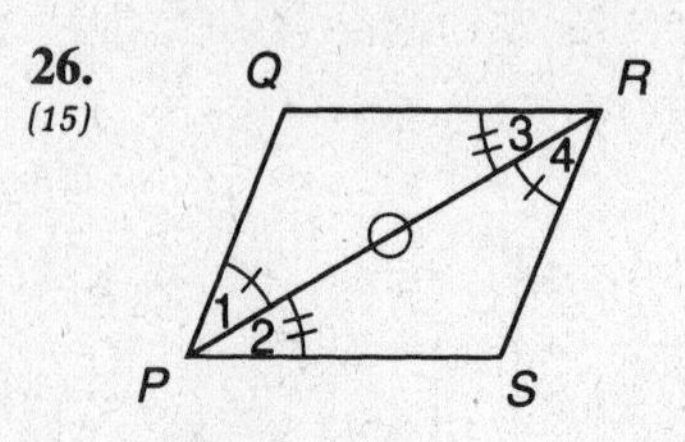

STATEMENTS	REASONS
1. $\angle 1 \cong \angle 4$	1. Given
2. $\angle 2 \cong \angle 3$	2. Given
3. $\overline{PR} \cong \overline{PR}$	3. Reflexive axiom
4. $\angle Q \cong \angle S$	4. If two angles in one triangle are congruent to two angles in a second triangle, then the third angles are congruent.
5. $\Delta PQR \cong \Delta RSP$	5. AAAS Congruency postulate
6. $\overline{QR} \cong \overline{SP}$	6. CPCTC

27. (11,13)

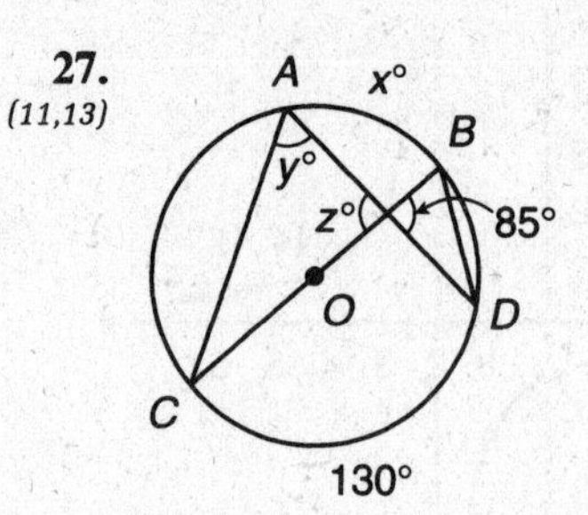

$$y = \frac{1}{2}(130)$$
$$\mathbf{y = 65}$$
$$\mathbf{z = 85}$$
$$m\angle C = 180° - 85° - 65° = 30°$$
$$x = 2(30)$$
$$\mathbf{x = 60}$$

28. (2)
$$32\pi = 2\pi r$$
$$r = 16$$
$$\pi(16)^2 + \pi(16)l = 576\pi$$
$$l = \frac{576\pi - 256\pi}{16\pi}$$
$$\mathbf{l = 20\ m}$$
$$h^2 + (16)^2 = (20)^2$$
$$h^2 = 144$$
$$\mathbf{h = 12\ m}$$

29. (1)
$$x^2 + y^2 = 25$$
$$3^2 + y^2 = 25$$
$$y^2 = 16$$
$$y = \pm 4$$

We have been given insufficient information. Therefore the answer is **D.**

30. (1)

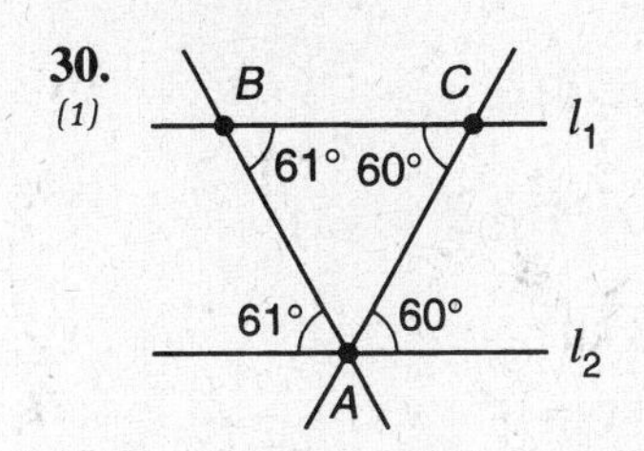

$m\angle ABC = 61° > 60° = m\angle ACB$

The side of the triangle opposite the larger angle is longer, therefore the answer is **B.**

Problem Set 29

1. (28) Distance $= g$ mi, rate $= \frac{g}{x}\frac{\text{mi}}{\text{hr}}$, time $= x$ hr

New rate $= \left(\frac{g}{x} + p\right)\frac{\text{mi}}{\text{hr}}$

New time $= ax$ hr

New distance $= \left(\frac{g}{x} + p\right)ax = \mathbf{(ag + pax)\ mi}$

2. (25)

$$R_DT_D + R_LT_L = \text{henways}$$

$$\frac{1}{4}(1 + 13) + R_L(13) = 10$$

$$\frac{14}{4} + 13R_L = 10$$

$$13R_L = \frac{26}{4}$$

$$R_L = \mathbf{\frac{1}{2}\ \frac{henway}{day}}$$

3. (25)

$$RWT = J$$

$$R(20)(6) = 10$$

$$R = \frac{1}{12}\ \frac{\text{jobs}}{\text{woman-day}}$$

$$RWT = J$$

$$\frac{1}{12}(30)(T) = 20$$

$$\frac{30}{12}(T) = 20$$

$$T = \mathbf{8\ days}$$

4. (25) First, write the two equations.

$$\begin{cases} \text{(a)}\ O_N - 10 = 2(N_N - 10) \\ \text{(b)}\ 2(O_N + 10) = 3(N_N + 10) - 10 \end{cases}$$

(a)
$$O_N - 10 = 2(N_N - 10)$$
$$O_N - 10 = 2N_N - 20$$
$$O_N = 2N_N - 10$$

Now we can use substitution to solve.

(b)
$$2(O_N + 10) = 3(N_N + 10) - 10$$
$$2O_N + 20 = 3N_N + 30 - 10$$
$$2O_N = 3N_N$$
$$2(2N_N - 10) = 3N_N$$
$$4N_N - 20 = 3N_N$$
$$N_N = 20$$

$$O_N = 2N_N - 10 = 2(20) - 10 = 30$$

$$O_N + 45 = \textbf{Odessa} = \mathbf{75\ yr}$$

$$N_N + 45 = \textbf{Nat} = \mathbf{65\ yr}$$

5. (18) First, we write the three equations.

$$\begin{cases} \text{(a)}\ N_N + N_D + N_P = 30 \\ \text{(b)}\ 5N_N + 10N_D + N_P = 135 \\ \text{(c)}\ 2N_D = N_P \end{cases}$$

Now we can use substitution to solve.

(a)
$$N_N + N_D + N_P = 30$$
$$N_N + N_D + (2N_D) = 30$$
$$N_N = 30 - 3N_D$$

(b)
$$5N_N + 10N_D + N_P = 135$$
$$5(30 - 3N_D) + 10N_D + (2N_D) = 135$$
$$150 - 15N_D + 10N_D + 2N_D = 135$$
$$3N_D = 15$$
$$\mathbf{N_D = 5}$$

(c)
$$2N_D = N_P$$
$$2(5) = N_P$$
$$\mathbf{N_P = 10}$$

(a)
$$N_N + N_D + N_P = 30$$
$$N_N + (5) + (10) = 30$$
$$\mathbf{N_N = 15}$$

6. (28)
$$\frac{8!}{2!2!} = \frac{8 \cdot 7 \cdot 6 \cdot 5 \cdot 4 \cdot 3 \cdot 2!}{2 \cdot 1 \cdot 2!}$$
$$= 4 \cdot 7 \cdot 6 \cdot 5 \cdot 4 \cdot 3 = \mathbf{10{,}080}$$

7. (28)
$$\frac{15!}{8!5!} = \frac{15 \cdot 14 \cdot 13 \cdot 12 \cdot 11 \cdot 10 \cdot 9 \cdot 8!}{8! \cdot 5 \cdot 4 \cdot 3 \cdot 2 \cdot 1}$$
$$= 3 \cdot 14 \cdot 13 \cdot 3 \cdot 11 \cdot 5 \cdot 3 = \mathbf{270{,}270}$$

8. $\log_3 12 = k$
(26)

9. $b^3 = 5$
(26)

10. $\log_b 49 = 2$
(26)

$b^2 = 49$

$b = 49^{\frac{1}{2}} = \mathbf{7}$

11. $\log_5 \frac{1}{125} = c$
(26)

$5^c = \frac{1}{125}$

$5^c = \frac{1}{5^3}$

$5^c = 5^{-3}$

$c = \mathbf{-3}$

12. $\log_{\frac{1}{4}} k = -3$
(26)

$k = \left(\frac{1}{4}\right)^{-3} = 4^3 = \mathbf{64}$

13. $f(x) = 3.2^x$
(23)

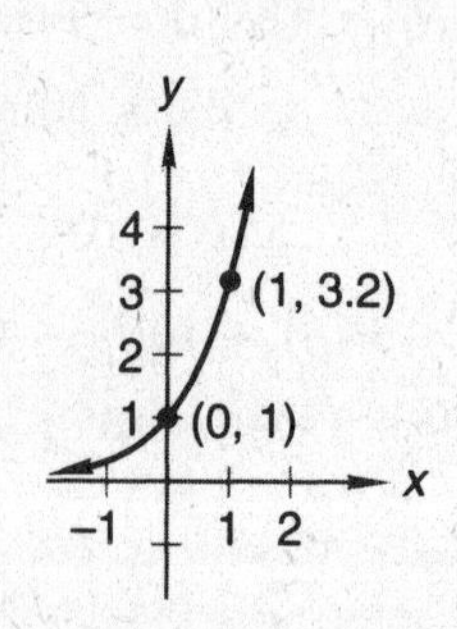

14. $g(x) = \frac{1}{x^2}$
(22)

x	y
−2	$\frac{1}{4}$
−1	1
1	1
2	$\frac{1}{4}$

15. (a) **Function, not 1 to 1**
(21)
(b) **Not a function**
(c) **Function, 1 to 1**
(d) **Not a function**

16. (a) Domain of f: $\left\{x \in \mathbb{R} \mid x \geq \frac{5}{4}\right\}$
(21)

(b) Domain of g: $\{x \in \mathbb{R} \mid x > -3\}$

(c) $h(x) = \frac{\sqrt{5x + 10}}{(2x + 1)(x - 1)}$

Domain of h:

$\left\{x \in \mathbb{R} \mid x \geq -2, x \neq -\frac{1}{2}, 1\right\}$

17.
(24,29)

$\sin 30° = \frac{1}{2}$

$\cos 30° = \frac{\sqrt{3}}{2}$

$\sin 60° = \frac{\sqrt{3}}{2}$

$2 \sin 30° \cos 30° - \sin 60° + \cos 90°$

$= 2\left(\frac{1}{2}\right)\left(\frac{\sqrt{3}}{2}\right) - \frac{\sqrt{3}}{2} + 0 = \mathbf{0}$

18. $\sin 180° = 0$
(29)

$\cos 90° = 0$

$\cos 270° = 0$

$2 \sin 180° - 3 \cos 90° + \cos 270°$

$= 0 - 0 + 0 = \mathbf{0}$

19.
(24,29)

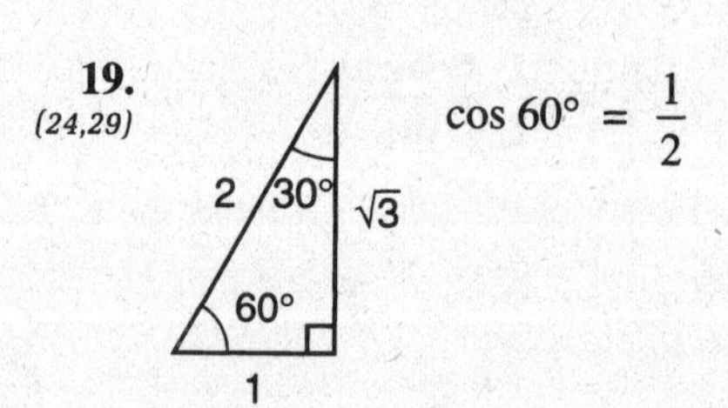

$\cos 60° = \frac{1}{2}$

$\sin 180° = 0$

$3f(60°) + g(180°) = 3 \cos 60° + \sin 180°$

$= 3\left(\frac{1}{2}\right) + 0 = \mathbf{\frac{3}{2}}$

20.
(24,29)

$\cos 30° = \frac{\sqrt{3}}{2}$

$\sin 180° = 0$

$\cos 180° = -1$

$\sin (30° - 300°) = \sin (-270°) = \sin 90° = 1$

$\frac{\sqrt{3}}{2}r(180°) + s(30°)$

$= \frac{\sqrt{3}}{2}\left(\frac{\sin 180°}{\cos 180°}\right) + \cos 30° + \sin(30° - 300°)$

$= \frac{\sqrt{3}}{2}\left(\frac{0}{-1}\right) + \frac{\sqrt{3}}{2} + 1 = \mathbf{\frac{\sqrt{3}}{2} + 1}$

21. (24) (a) $(f - g)(x) = \log_4 x - 4^x$

$(f - g)(2) = \log_4 2 - 4^2 = \frac{1}{2} - 16 = \mathbf{-15\frac{1}{2}}$

(b) $(fg)(x) = (\log_4 x)4^x$

$(fg)(2) = (\log_4 2)4^2 = \left(\frac{1}{2}\right)16 = \mathbf{8}$

(c) $(f \circ g)(x) = \log_4(4^x)$

$(f \circ g)(2) = \log_4(4^2) = \mathbf{2}$

22. (19) (a) $4a^{3m+2} - 16a^{3m} = 4a^{3m}(a^2 - 4)$

$= \mathbf{4a^{3m}(a + 2)(a - 2)}$

(b) $8a^3b^3 - 27c^6d^6 = (2ab)^3 - (3c^2d^2)^3$

$= \mathbf{(2ab - 3c^2d^2)(4a^2b^2 + 6abc^2d^2 + 9c^4d^4)}$

23. (16) $\frac{3}{2c + \frac{t}{1 + \frac{4}{z}}} = \frac{3}{2c + \frac{t}{\frac{z+4}{z}}} = \frac{3}{2c + \frac{tz}{z+4}}$

$= \frac{3}{\frac{2c(z+4) + tz}{z+4}} = \mathbf{\frac{3z + 12}{2cz + 8c + tz}}$

24. (10) $\frac{\sqrt{5}\sqrt{-5}\sqrt{6}\sqrt{-6} + \sqrt{-25} - \sqrt{-16}}{-3i^2 - \sqrt{-9}i}$

$= \frac{5i \cdot 6i + 5i - 4i}{-3i^2 - 3i^2} = \frac{30i^2 + i}{-6i^2}$

$= \frac{-30 + i}{6} = \mathbf{-5 + \frac{1}{6}i}$

25. (14)

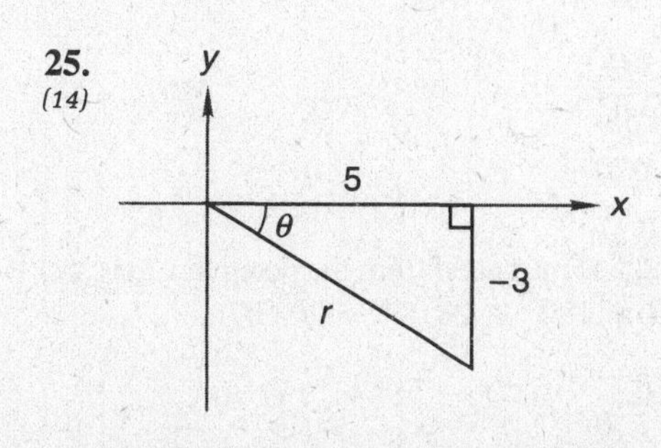

$r = \sqrt{5^2 + 3^2} = 5.83$

$\tan\theta = \frac{3}{5}$

$\theta = 30.96$

The polar angle is $360° - 30.96° = 329.04°$.

$\mathbf{5.83\underline{/329.04°}; 5.83\underline{/-30.96};}$

$\mathbf{-5.83\underline{/149.04°}; -5.83\underline{/-210.96°}}$

26. (21) $f(x + h) = \sqrt{x + h} + (x + h)^2$

$= \mathbf{\sqrt{x + h} + x^2 + 2xh + h^2}$

27. (15)

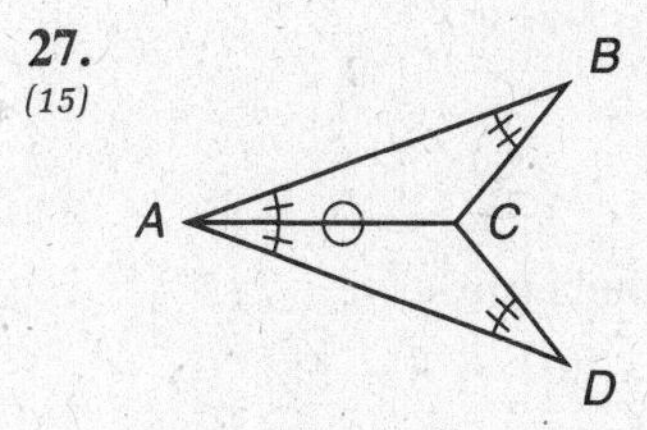

STATEMENTS	REASONS
1. $\overline{AC}$ bisects $\angle BAD$	1. Given
2. $\angle BAC \cong \angle DAC$	2. A bisector divides an angle into two congruent angles.
3. $\angle ABC \cong \angle ADC$	3. Given
4. $\angle ACB \cong \angle ACD$	4. If two angles in one triangle are congruent to two angles in a second triangle, then the third angles are congruent.
5. $\overline{AC} \cong \overline{AC}$	5. Reflexive axiom
6. $\Delta ABC \cong \Delta ADC$	6. *AAAS* congruency postulate
7. $\overline{BC} \cong \overline{DC}$	7. *CPCTC*

28. (13) $y + 2y + y + \frac{1}{2}y = 360$

$\frac{9}{2}y = 360$

$\mathbf{y = 80}$

$x = \frac{1}{2}(160 - 40)$

$\mathbf{x = 60}$

29. (1)

$x + y = 7$

$x - y = 5$

$2x = 12$

$x = 6$

$y = 1$

$x > y$

Therefore the answer is **A.**

30. (1) In any triangle, the side opposite the larger angle is longer. Therefore the answer is **B.**

Problem Set 30

1. (18)

$$b = \frac{kg^2}{t}$$

$$200 = \frac{k(10)^2}{(20)}$$

$$k = 40$$

$$b = \frac{40g^2}{t} = \frac{40(8)^2}{2} = \mathbf{1280\ boys}$$

2. (18) First write the two equations.

$$\begin{cases} \text{(a)}\ T + U = 5 \\ \text{(b)}\ 10T + U = 10U + T - 27 \end{cases}$$

(a) $T + U = 5$

$$U = 5 - T$$

Now we can use substitution to solve.

(b) $10T + U = 10U + T - 27$

$$9T - 9U = -27$$

$$9T - 9(5 - T) = -27$$

$$18T = 18$$

$$T = 1$$

$$U = 5 - T = 5 - (1) = 4$$

The numbers are **14** and **41.**

3. (18) First write the two equations.

$$\begin{cases} \text{(a)}\ P_N + D_N = 20 \\ \text{(b)}\ 0.9P_N + 0.58D_N = 0.78(20) \end{cases}$$

(a) $P_N + D_N = 20$

$$P_N = 20 - D_N$$

Now we can use substitution to solve.

(b) $0.9P_N + 0.58D_N = 0.78(20)$

$$0.9(20 - D_N) + 0.58D_N = 15.6$$

$$0.32D_N = 2.4$$

$$D_N = \mathbf{7.5\ liters\ of\ 58\%\ sol.}$$

(a) $P_N + D_N = 20$

$$P_N + (7.5) = 20$$

$$P_N = \mathbf{12.5\ liters\ of\ 90\%\ sol.}$$

4. (18) $\text{Iodine}_1 + \text{iodine added} = \text{iodine final}$

$$0.06(370) + 0.135I = 0.1(370 + I)$$

$$22.2 + 0.135I = 37 + 0.1I$$

$$0.035I = 14.8$$

$$I = \mathbf{422.86\ ml}$$

5. (10) $3x + 2y - 4 = 0$

$$2y = -3x + 4$$

$$y = -\frac{3}{2}x + 2$$

Parallel lines have equal slopes, so $m = -\frac{3}{2}$.

$$y = -\frac{3}{2}x + b$$

$$4 = -\frac{3}{2}(-2) + b$$

$$b = 1$$

$$\mathbf{y = -\frac{3}{2}x + 1}$$

6. (30)

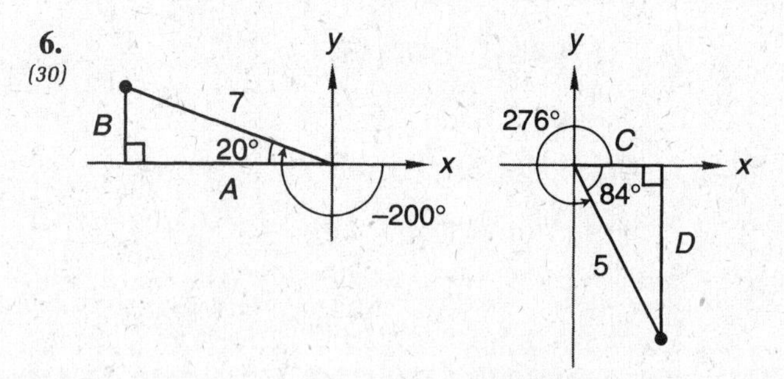

Note:
Figure not drawn to scale

$$A = 7\cos 20° = 6.578$$

$$B = 7\sin 20° = 2.394$$

$$C = 5\cos 84° = 0.523$$

$$D = 5\sin 84° = 4.973$$

$$\text{Resultant} = (0.523 - 6.578)\hat{i} + (2.394 - 4.973)\hat{j}$$

$$= -6.055\hat{i} - 2.579\hat{j}$$

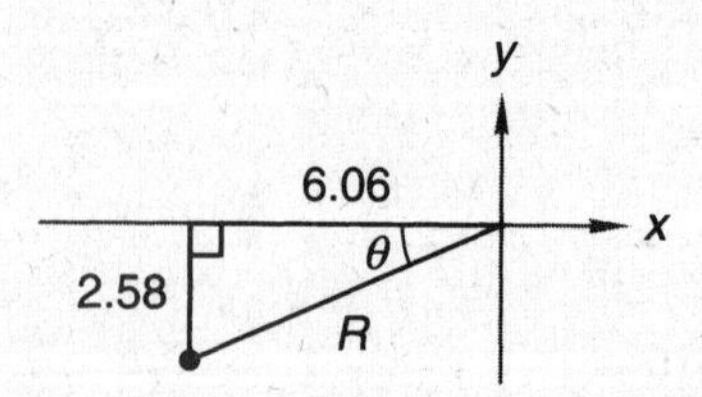

$$R = \sqrt{(6.055)^2 + (2.579)^2} = 6.58$$

$$\tan\theta = \frac{2.579}{6.055}$$

$$\theta = 23.07°$$

The polar angle is measured from the positive x axis and is $180° + \theta = 180° + 23.07° = 203.07°$.

6.58/203.07°

7. (30)

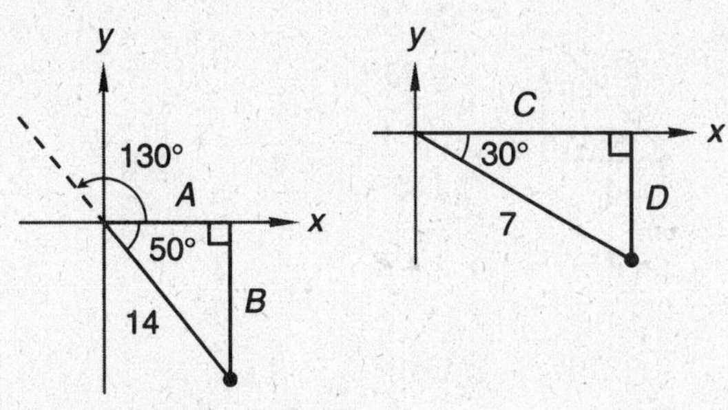

$A = 14 \cos 50° = 8.999$

$B = 14 \sin 50° = 10.725$

$C = 7 \cos 30° = 6.062$

$D = 7 \sin 30° = 3.50$

$$\text{Resultant} = (8.999 + 6.062)\hat{i} + (-10.725 - 3.50)\hat{j}$$
$$= 15.061\hat{i} - 14.225\hat{j}$$

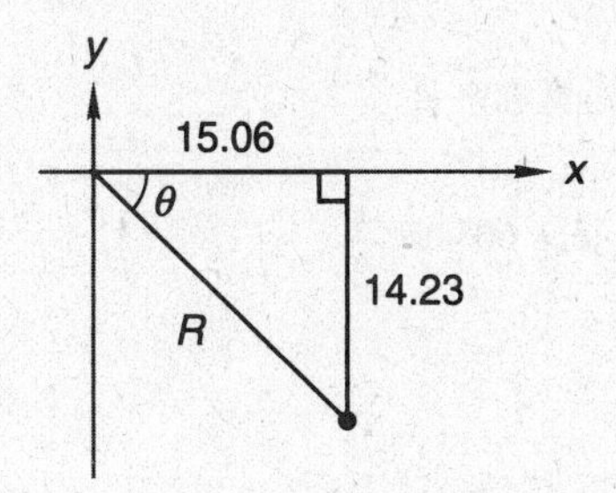

$R = \sqrt{(15.061)^2 + (14.225)^2} = 20.72$

$$\tan\theta = \frac{14.225}{15.061}$$
$$\theta = 43.36°$$

The polar angle is measured from the positive x axis and is $-43.36°$.

Resultant = $20.72\underline{/-43.36°}$, so

Equilibrant = $\mathbf{20.72\underline{/136.64°}}$

8. (30)

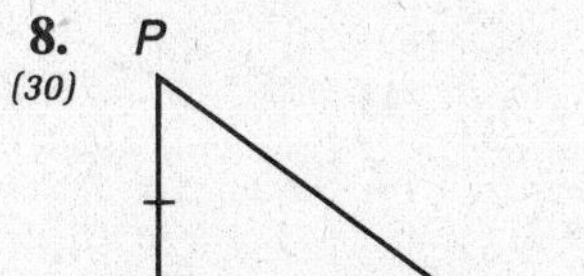

Statements	Reasons
1. $\overline{PQ} \cong \overline{SR}$	1. Given
2. $\angle PQR$ and $\angle SRQ$ are right angles.	2. Given
3. $\angle PQR \cong \angle SRQ$	3. All right angles are congruent.
4. $\overline{QR} \cong \overline{QR}$	4. Reflexive axiom
5. $\Delta PQR \cong \Delta SRQ$	5. *SAS* congruency postulate

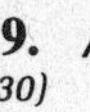

9. (30)

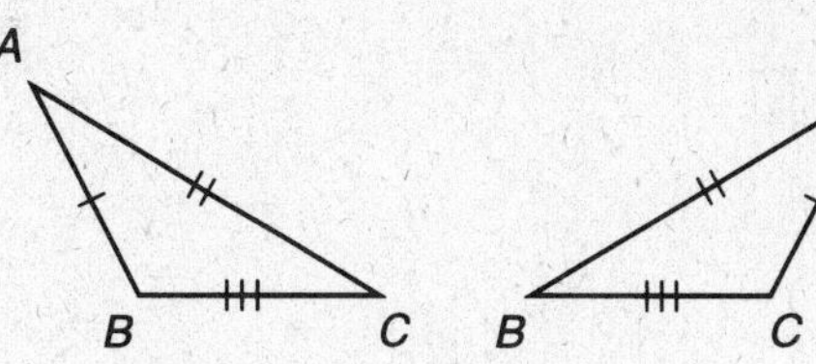

Statements	Reasons
1. $\overline{AB} \cong \overline{DC}$	1. Given
2. $\overline{AC} \cong \overline{DB}$	2. Given
3. $\overline{BC} \cong \overline{BC}$	3. Reflexive axiom
4. $\Delta ABC \cong \Delta DCB$	4. *SSS* congruency postulate

10. (28) $\frac{7!}{5!2!} = \frac{7 \cdot 6 \cdot 5!}{5! \cdot 2 \cdot 1} = 7 \cdot 3 = \mathbf{21}$

11. (26) $\log_b \frac{1}{9} = 2$

$$b^2 = \frac{1}{9}$$
$$b = \left(\frac{1}{9}\right)^{\frac{1}{2}} = \mathbf{\frac{1}{3}}$$

12. (26) $\log_9 729 = c$

$$9^c = 729$$
$$9^c = 9^3$$
$$c = \mathbf{3}$$

13. (26) $\log_{\frac{1}{5}} k = -2$

$$k = \left(\frac{1}{5}\right)^{-2} = 5^2 = \mathbf{25}$$

14. (23) $f(x) = 0.7^x$

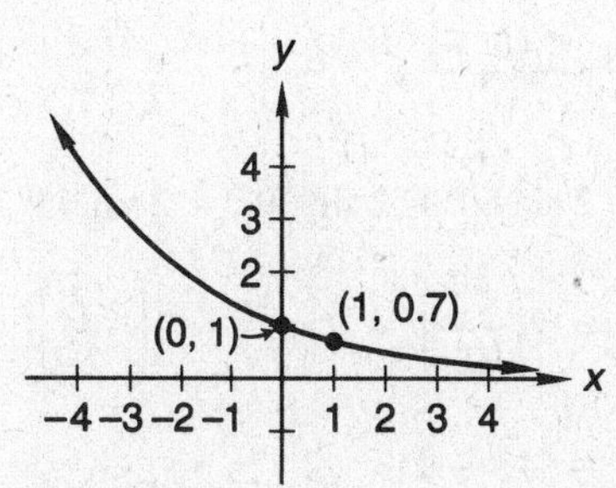

15. (22) $y = \dfrac{1}{|x|}$

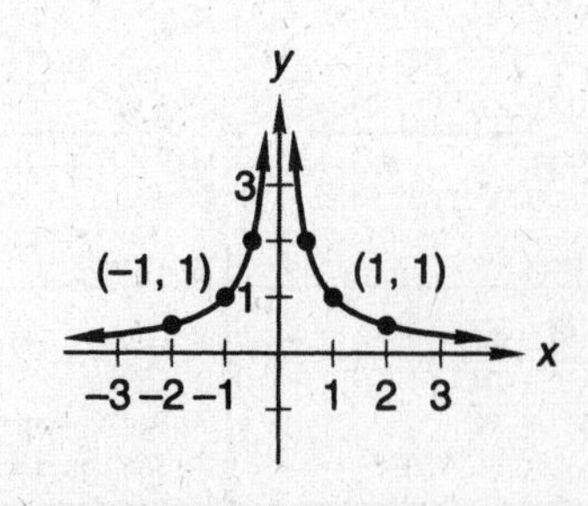

x	-2	-1	$-\frac{1}{2}$	$\frac{1}{2}$	1	2
y	$\frac{1}{2}$	1	2	2	1	$\frac{1}{2}$

16. (21) (a) Domain of f: $\{x \in \mathbb{R} \mid x \neq 0\}$

(b) Domain of g: $\left\{x \in \mathbb{R} \mid x > -\dfrac{7}{2}\right\}$

(c) Domain of h: $\{x \in \mathbb{R} \mid x \geq 3\}$

17. (24,29) (a)

$$\sin 30° = \frac{1}{2}$$

$$\cos 60° = \frac{1}{2}$$

$$\cos 180° = -1$$

$$\sin 180° = 0$$

$$3 \sin 30° \cos 180° - 2 \sin 180° \cos 60°$$

$$= 3\left(\frac{1}{2}\right)(-1) - 2(0)\left(\frac{1}{2}\right) = -\frac{3}{2} - 0 = \mathbf{-\frac{3}{2}}$$

(b)

$$\sin 60° = \frac{\sqrt{3}}{2}$$

$$\sin 30° = \frac{1}{2}$$

$$\sin 45° = \frac{1}{\sqrt{2}} \text{ or } \frac{\sqrt{2}}{2}$$

$$2 \sin 60 + 2 \sin 30° + 2 \sin 45°$$

$$= 2\left(\frac{\sqrt{3}}{2}\right) + 2\left(\frac{1}{2}\right) + 2\left(\frac{\sqrt{2}}{2}\right)$$

$$= \mathbf{\sqrt{3} + 1 + \sqrt{2}}$$

18. (24,29)

$$\cos 45° = \frac{\sqrt{2}}{2}$$

$$\sin 45° = \frac{\sqrt{2}}{2}$$

$$\sin 270° = -1$$

$$\frac{1}{2}r(270°) + s(45°)$$

$$= \frac{1}{2}(1 + 5 \sin 270°) + \cos 45° \sin 45°$$

$$= \frac{1}{2}[1 + 5(-1)] + \left(\frac{\sqrt{2}}{2}\right)\left(\frac{\sqrt{2}}{2}\right)$$

$$= \frac{1}{2}(-4) + \frac{2}{4} = -2 + \frac{1}{2} = \mathbf{-\frac{3}{2}}$$

19. (24,29)

$$\sin 60° = \frac{\sqrt{3}}{2}$$

$$\cos 60° = \frac{1}{2}$$

$$\sin 45° = \frac{\sqrt{2}}{2}$$

$$\cos 45° = \frac{\sqrt{2}}{2}$$

$$5f(60°) + 2g(45°)$$

$$= 5(2 \sin 60° - 3 \cos 60°) + 2\left(\frac{\sin 45°}{\cos 45°}\right)$$

$$= 5\left[2\left(\frac{\sqrt{3}}{2}\right) - 3\left(\frac{1}{2}\right)\right] + 2\left(\frac{\frac{\sqrt{2}}{2}}{\frac{\sqrt{2}}{2}}\right)$$

$$= 5\left(\sqrt{3} - \frac{3}{2}\right) + 2(1)$$

$$= 5\sqrt{3} - \frac{15}{2} + 2 = \mathbf{5\sqrt{3} - \frac{11}{2}}$$

20. (24) (a) $(f \circ g)(x) = 2\left[\dfrac{1}{2}(x + 1)\right] - 1$

$$= x + 1 - 1 = \mathbf{x}$$

(b) $(g \circ f)(x) = \dfrac{1}{2}(2x - 1 + 1) = \mathbf{x}$

21. (19) First write the two equations.

$$\begin{cases} \text{(a)}\ x^2 - y^2 = 8 \\ \text{(b)}\ 2x^2 + y^2 = 19 \end{cases}$$

Now we can use elimination and substitution to solve.

$$\begin{aligned} \text{(a)} \quad x^2 - y^2 &= 8 \\ \text{(b)} \quad 2x^2 + y^2 &= 19 \\ \hline 3x^2 \quad\quad &= 27 \\ x^2 &= 9 \\ x &= \pm 3 \end{aligned}$$

$$\begin{aligned} \text{(a)} \quad x^2 - y^2 &= 8 \\ (\pm 3)^2 - y^2 &= 8 \\ y^2 &= 1 \\ y &= \pm 1 \end{aligned}$$

(3, 1), (3, –1), (–3, 1), (–3, –1)

22. (19) $\dfrac{x^{2a} - y^{2b}}{x^a - y^b} = \dfrac{(x^a + y^b)(x^a - y^b)}{(x^a - y^b)} = \mathbf{x^a + y^b}$

23. (19) $2x^{3N+1} + 6x^{5N+2}$

$= 2 \cdot x^N \cdot x^N \cdot x^N \cdot x + 6 \cdot x^N \cdot x^N \cdot x^N \cdot x^N \cdot x^N \cdot x^2$

$= \mathbf{2x^{3N+1}(1 + 3x^{2N+1})}$

24. (16)
$$\frac{k}{a + \dfrac{b}{x + \dfrac{c}{d}}} = \frac{k}{a + \dfrac{b}{\dfrac{xd + c}{d}}}$$

$$= \frac{k}{a + \dfrac{bd}{xd + c}} = \frac{k}{\dfrac{a(xd + c) + bd}{xd + c}}$$

$$= \frac{k(xd + c)}{axd + ac + bd} = \mathbf{\frac{kxd + kc}{axd + ac + bd}}$$

25. (10)
$$\frac{2i - 4}{3 + 6i} \cdot \frac{3 - 6i}{3 - 6i}$$

$$= \frac{6i - 12 - 12i^2 + 24i}{9 - 36i^2}$$

$$= \frac{30i - 12 + 12}{9 + 36} = \frac{30i}{45} = \mathbf{\frac{2}{3}i}$$

26. (21) $f(x + h) = (x + h)^2 - (x + h)$

$= \mathbf{x^2 + 2hx + h^2 - x - h}$

27. (11)
$$\begin{aligned} 2 \cdot m\angle DAC &= 70° \\ m\angle DAC &= 35° \end{aligned}$$

$$\begin{aligned} y + 30 + 35 + 40 &= 180 \\ y &= \mathbf{75} \end{aligned}$$

$$\begin{aligned} x + y + m\angle DAC &= 180 \\ x + (75) + (35) &= 180 \\ x &= \mathbf{70} \end{aligned}$$

28. (2) $V_{\text{pyramid}} = \dfrac{1}{3}(A_{\text{base}})(h)$

$$4096 = \frac{1}{3}\left(\frac{128}{4}\right)^2 (h)$$

$$h = 12$$

$$A_{\text{surface}} = A_{\text{base}} + A_{\text{lateral}}$$

$$= \left(\frac{128}{4}\right)^2 + 4\left(\frac{1}{2}\right)\left(\frac{128}{4}\right)\left\{\sqrt{12^2 + \left[\frac{1}{2}\left(\frac{128}{4}\right)\right]^2}\right\}$$

$$= 1024 + 1280 = \mathbf{2304\ cm^2}$$

29. (1,5) $x - x^{\frac{1}{2}} = x^{\frac{1}{2}}\left(x^{\frac{1}{2}} - 1\right)$

Since $x > 0$, we can take the square root of the entire inequality $0 < x < 1$ without changing the inequality.

$$0 < x < 1$$

$$0^{\frac{1}{2}} < x^{\frac{1}{2}} < 1^{\frac{1}{2}}$$

$$x^{\frac{1}{2}} < 1$$

$$x^{\frac{1}{2}} - 1 < 0$$

Since $x^{\frac{1}{2}} > 0$ and $x^{\frac{1}{2}} - 1 < 0$,

$$x^{\frac{1}{2}}\left(x^{\frac{1}{2}} - 1\right) < 0$$

$$x - x^{\frac{1}{2}} < 0$$

$$x < x^{\frac{1}{2}}$$

Therefore the answer is **A.**

30. (1,20)

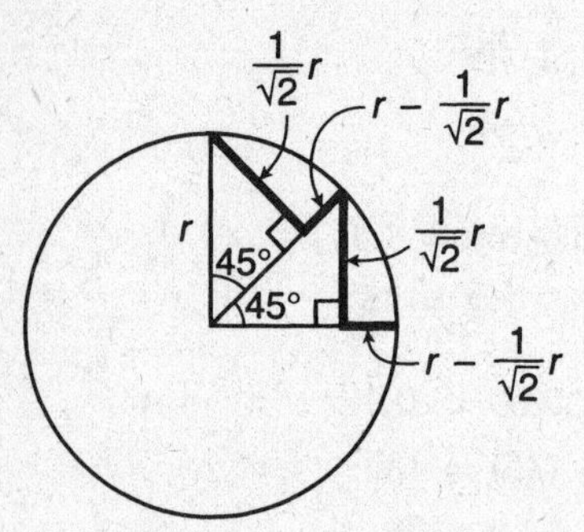

A: sum of segments

$$2\left(\frac{1}{\sqrt{2}}r\right) + 2\left(r - \frac{1}{\sqrt{2}}r\right)$$

$$= \frac{2}{\sqrt{2}}r + 2r - \frac{2}{\sqrt{2}}r = 2r$$

B: $2r$

A and B are equal. Therefore the answer is **C.**

Problem Set 31

1. (28) Distance $= m$

Rate $= p$

Time $= \frac{m}{p}$

New distance $= m + 5$

New time $= \frac{m}{p} + 1$

$$\text{New rate} = \frac{\text{new distance}}{\text{new time}}$$

$$= \frac{m + 5}{\frac{m}{p} + 1}$$

$$= \frac{m + 5}{\frac{m + p}{p}} = \mathbf{\frac{pm + 5p}{m + p}\ \frac{mi}{hr}}$$

2. (25) First we write the two equations.

$$\begin{aligned}&\text{(a)} \\ &\text{(b)}\end{aligned}\begin{cases} D_N - 10 = 3(L_N - 10) \\ D_N = 2L_N \end{cases}$$

Now we can use substitution to solve.

(a) $D_N - 10 = 3(L_N - 10)$

$(2L_N) - 10 = 3L_N - 30$

$L_N = \textbf{Lannes} = \mathbf{20\ yr}$

(b) $D_N = 2L_N = 2(20) = \textbf{Davout} = \mathbf{40\ yr}$

3. (7) $N, N + 1, N + 2, N + 3$

$$N(N + 2) = -13(N + 3) + 25$$

$$N^2 + 2N = -13N - 39 + 25$$

$$N^2 + 15N + 14 = 0$$

$$(N + 14)(N + 1) = 0$$

$$N = -14, -1$$

If $N = -1$, we do not have four consecutive negative integers, so $N = -14$.

The integers are: **−14, −13, −12, −11.**

4. (18) First we write the three equations.

$$\begin{aligned}&\text{(a)} \\ &\text{(b)} \\ &\text{(c)}\end{aligned}\begin{cases} N_B = N_R + N_G + 2 \\ 2N_B = 3N_R + 2N_G \\ N_B + N_R + N_G = 22 \end{cases}$$

(a) $N_B = N_R + N_G + 2$

$N_B - N_R - N_G = 2$

Now we can use elimination to solve.

$$\begin{aligned} N_B - N_R - N_G &= 2 \\ \text{(c)}\quad N_B + N_R + N_G &= 22 \\ \hline 2N_B \qquad\qquad &= 24 \\ \mathbf{N_B} &= \mathbf{12} \end{aligned}$$

(a) $N_B = N_R + N_G + 2$

$(12) = N_R + N_G + 2$

(a′) $N_R + N_G = 10$

(b) $2N_B = 3N_R + 2N_G$

$2(12) = 3N_R + 2N_G$

(b′) $24 = 3N_R + 2N_G$

$$\begin{aligned} \text{(b′)}\quad 3N_R + 2N_G &= 24 \\ -2\text{(a′)}\quad -2N_R - 2N_G &= -20 \\ \hline \mathbf{N_R} &= \mathbf{4} \end{aligned}$$

(a′) $N_R + N_G = 10$

$(4) + N_G = 10$

$\mathbf{N_G = 6}$

5. (10) $m = \frac{y_2 - y_1}{x_2 - x_1} = \frac{(-1) - (1)}{(3) - (1)} = -1$

Parallel lines have equal slopes, so $m = -1$.

$y = -x + b$

$7 = -(2) + b$

$b = 9$

$\mathbf{y = -x + 9}$

6. (31) (a) ***x* axis, no.**

$(-y) = x^2$ is not equivalent to $y = x^2$.

***y* axis, yes.**

$y = (-x)^2$ is equivalent to $y = x^2$.

Origin, no.

$(-y) = (-x)^2$ is not equivalent to $y = x^2$.

(b) ***x* axis, yes.**

$x^2 + (-y)^2 = 9$ is equivalent to $x^2 + y^2 = 9$.

***y* axis, yes.**

$(-x)^2 + y^2 = 9$ is equivalent to $x^2 + y^2 = 9$.

Origin, yes.

$(-x)^2 + (-y)^2 = 9$ is equivalent to $x^2 + y^2 = 9$.

7. (31) Replace x with $x - 2$.

$\mathbf{y = (x - 2)^2}$

8. (31) $g(x) = |x - 1|$

This is $f(x)$ translated 1 unit to the right.

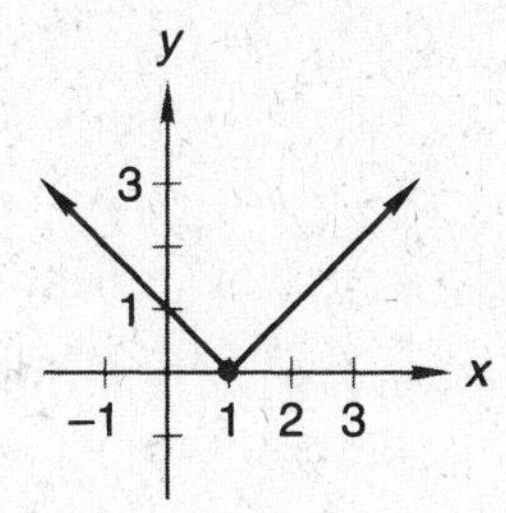

9. (30)

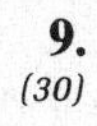

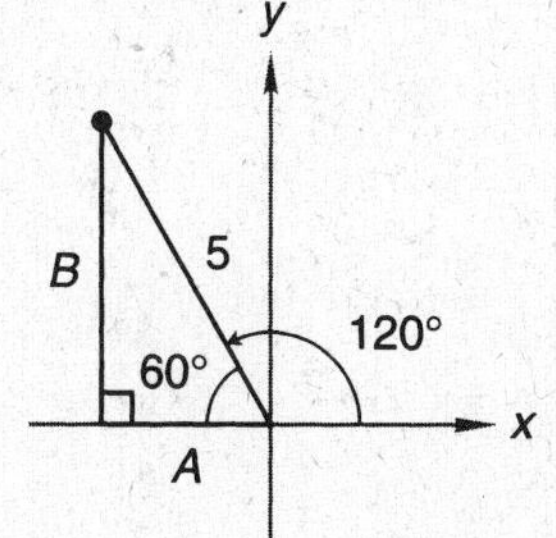

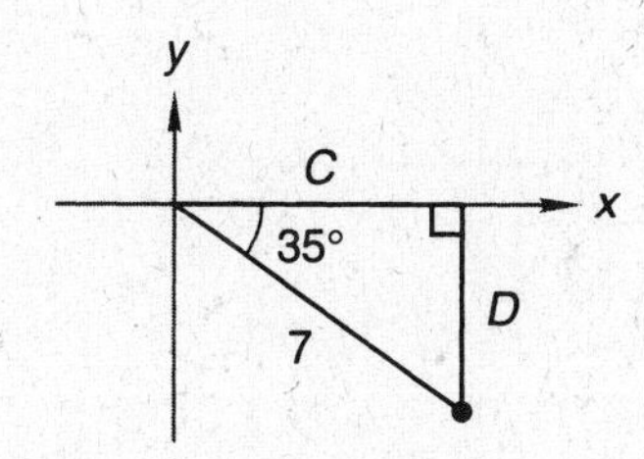

$A = 5 \cos 60° = 2.5$

$B = 5 \sin 60° = 4.3301$

$C = 7 \cos 35° = 5.7341$

$D = 7 \sin 35° = 4.0150$

$$\text{Resultant} = (5.7341 - 2.5000)\,\hat{i} + (4.3301 - 4.0150)\,\hat{j} = 3.2341\hat{i} + 0.3151\hat{j}$$

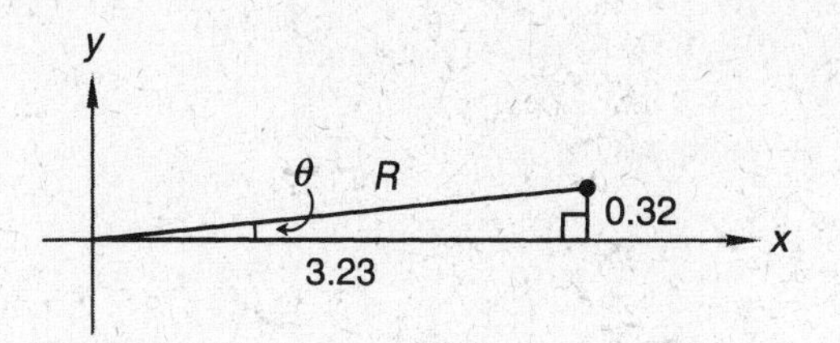

$R = \sqrt{(3.2341)^2 + (0.3151)^2} = 3.2494$

$\tan \theta = \dfrac{0.3151}{3.2341}$

$\theta = 5.56°$

3.25$\underline{/5.56°}$

10. (30)

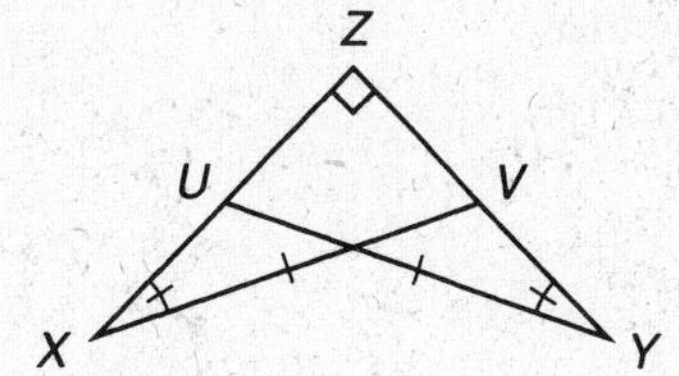

STATEMENTS	REASONS
1. $\angle ZXV \cong \angle ZYU$	1. Given
2. $\overline{XV} \cong \overline{YU}$	2. Given
3. $\angle Z \cong \angle Z$	3. Reflexive axiom
4. $\angle ZVX \cong \angle ZUY$	4. If two angles in one triangle are congruent to two angles in a second triangle, then the third angles are congruent.
5. $\Delta XZV \cong \Delta YZU$	5. *AAAS* congruency postulate

11. (28) $\dfrac{7!}{4!3!} = \dfrac{7 \cdot 6 \cdot 5 \cdot 4!}{3 \cdot 2 \cdot 1 \cdot 4!} = 7 \cdot 5 = \mathbf{35}$

12. (26) $\log_a 8 = 3$

$a^3 = 8$

$a = 8^{\frac{1}{3}} = \mathbf{2}$

13. (26) $\log_3 \dfrac{1}{9} = b$

$3^b = \dfrac{1}{9}$

$3^b = 3^{-2}$

$b = \mathbf{-2}$

14. (26) $\log_3 c = -3$

$c = 3^{-3} = \dfrac{1}{3^3} = \mathbf{\dfrac{1}{27}}$

15. (22) Graph $y = \dfrac{1}{\sqrt{x}}$. Plot point (1, 1).

Asymptote at $x = 0$.

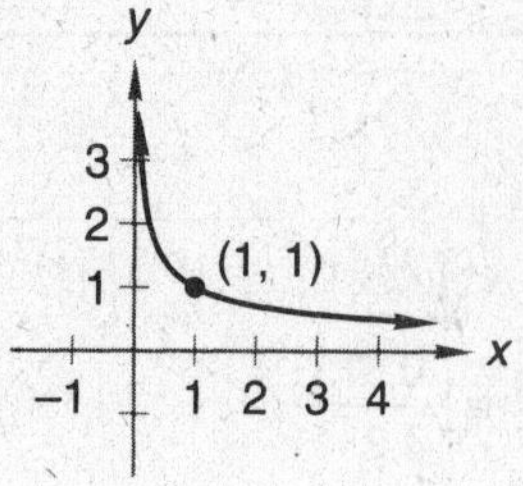

16. (21) (a) **Function, not 1 to 1**

(b) **Function, not 1 to 1**

(c) **Not a function**

(d) **Function, 1 to 1**

17. (21) $2x + 6 \geq 0$

$x \geq -3$

$(2x - 3)(2x + 1) \neq 0$

$x \neq -\dfrac{1}{2}, \dfrac{3}{2}$

$\left\{\mathbf{x \in \mathbb{R} \mid x \geq -3, x \neq -\dfrac{1}{2}, \dfrac{3}{2}}\right\}$

18. (24)

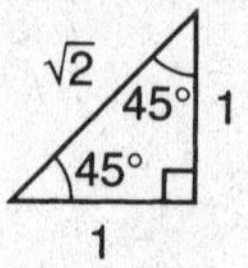

$\sin 45° = \dfrac{1}{\sqrt{2}} = \dfrac{\sqrt{2}}{2}$

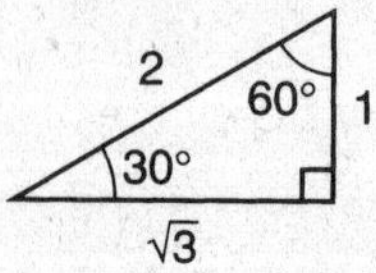

$\tan 30° = \dfrac{1}{\sqrt{3}} = \dfrac{\sqrt{3}}{3}$

$\sqrt{2}\, f(45°) - \sqrt{3}\, g(30°) = \sqrt{2} \sin 45° - \sqrt{3} \tan 30°$

$= \sqrt{2}\left(\dfrac{\sqrt{2}}{2}\right) - \sqrt{3}\left(\dfrac{\sqrt{3}}{3}\right) = 1 - 1 = \mathbf{0}$

19. (24,29)

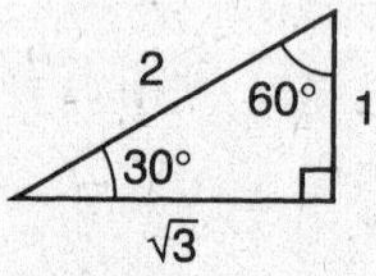

$\cos 30° = \dfrac{\sqrt{3}}{2}$

$\tan 180° = 0$

$3r(180°) + 2s(30°) = 3 \tan 180° + 2 \cos 30°$

$= 3(0) + 2\left(\dfrac{\sqrt{3}}{2}\right) = \mathbf{\sqrt{3}}$

20. (24,29) $\cos 180° = -1, \ \sin 180° = 0$

(a) $(f + g)(180°) = \cos 180° + \sin 180°$
$= -1 + 0 = \mathbf{-1}$

(b) $(f - g)(180°) = \cos 180° - \sin 180°$
$= -1 - 0 = \mathbf{-1}$

(c) $fg(180°) = \cos 180° \sin 180° = (-1)(0) = \mathbf{0}$

21. (19) First we label the equations.

$$\begin{aligned} &(a) \\ &(b) \end{aligned} \begin{cases} x^2 - y^2 = 9 \\ x + y = 9 \end{cases}$$

(b) $x + y = 9$

$x = 9 - y$

Now we can use substitution to solve.

(a)
$$\begin{aligned} x^2 - y^2 &= 9 \\ (9 - y)^2 - y^2 &= 9 \\ 81 - 18y + y^2 - y^2 &= 9 \\ 18y &= 72 \\ \mathbf{y} &\mathbf{= 4} \end{aligned}$$

(b)
$$\begin{aligned} x + y &= 9 \\ x + (4) &= 9 \\ \mathbf{x} &\mathbf{= 5} \end{aligned}$$

22. (19) $49a^{5n+2} - 7a^{6n+3} = 7^2a^{5n}a^2 - 7a^{6n}a^3$

$= 7a^{5n}a^2(7 - a^na) = \mathbf{7a^{5n+2}(7 - a^{n+1})}$

23. (16) $\dfrac{\dfrac{3a}{b} - \dfrac{2b}{a}}{\dfrac{ab}{c} + \dfrac{c}{a}} = \dfrac{\dfrac{3a^2 - 2b^2}{ab}}{\dfrac{a^2b + c^2}{ac}} \cdot \dfrac{\dfrac{abc}{1}}{\dfrac{abc}{1}} = \mathbf{\dfrac{3a^2c - 2b^2c}{a^2b^2 + bc^2}}$

24. (5) $\dfrac{x^{-3}y + y^{-2}}{x^{-2}y} = \dfrac{\dfrac{y}{x^3} + \dfrac{1}{y^2}}{\dfrac{y}{x^2}} \cdot \dfrac{\dfrac{x^2}{y}}{\dfrac{x^2}{y}}$

$= \dfrac{x^2}{y}\left(\dfrac{y^3 + x^3}{x^3y^2}\right) = \mathbf{\dfrac{y^3 + x^3}{xy^3}}$

25. (16)

$$\begin{array}{r} 2x^2 - 2x + 3 \\ x + 1 \overline{\smash{)}\, 2x^3 + 0x^2 + x - 1} \\ \underline{2x^3 + 2x^2} \qquad\quad \\ -2x^2 + x \quad\; \\ \underline{-2x^2 - 2x} \quad\; \\ 3x - 1 \\ \underline{3x + 3} \\ -4 \end{array}$$

$$\frac{2x^3 + x - 1}{x + 1} = \mathbf{2x^2 - 2x + 3 - \frac{4}{x + 1}}$$

Check:

$$(2x^2 - 2x + 3)\left(\frac{x+1}{x+1}\right) - \frac{4}{(x+1)}$$

$$= \frac{2x^3 + 2x^2 - 2x^2 - 2x + 3x + 3 - 4}{x + 1}$$

$$= \frac{2x^3 + x - 1}{x + 1}$$

26. (16)

$$y = 3\left(\frac{px + 2}{r + 5x}\right)$$

$$y = \frac{3px + 6}{r + 5x}$$

$$ry + 5xy = 3px + 6$$

$$5xy - 3px = 6 - ry$$

$$x(5y - 3p) = 6 - ry$$

$$x = \mathbf{\frac{6 - yr}{5y - 3p}}$$

27. (10)

$$4x^2 + 3x - 2 = 0$$

$$x^2 + \frac{3}{4}x = \frac{1}{2}$$

$$\left(x^2 + \frac{3}{4}x + \frac{9}{64}\right) = \frac{1}{2} + \frac{9}{64}$$

$$\left(x + \frac{3}{8}\right)^2 = \frac{41}{64}$$

$$x + \frac{3}{8} = \pm\sqrt{\frac{41}{64}}$$

$$x = \mathbf{-\frac{3}{8} \pm \frac{\sqrt{41}}{8}}$$

28. (21) $f(x) = 1 + x^2$

$$f(x + h) = 1 + (x + h)^2 = 1 + x^2 + 2xh + h^2$$

$$f(x + h) - f(x)$$

$$= (1 + x^2 + 2xh + h^2) - (1 + x^2)$$

$$= \mathbf{2xh + h^2}$$

29. (1)

$x < -1$ and $x < -1$

$x^2 > (-1)^2$ $\quad$ $x^3 < (-1)^3$

$x^2 > 1$ $\quad$ $x^3 < -1$

$x^2 > x^3$

Therefore the answer is **B.**

30. (1,5) A: $\sqrt[6]{9(\sqrt[3]{3})} = (3^2)^{\frac{1}{6}}\left(3^{\frac{1}{3}}\right)^{\frac{1}{6}} = 3^{\frac{7}{18}}$

B: $\sqrt[3]{3(\sqrt[6]{9})} = \left(3^{\frac{1}{3}}\right)\left[(3^2)^{\frac{1}{6}}\right]^{\frac{1}{3}} = 3^{\frac{4}{9}} = 3^{\frac{8}{18}}$

$$3^{\frac{7}{18}} < 3^{\frac{8}{18}}$$

Therefore the answer is **B.**

Problem Set 32

1. (28) Distance $= x$, rate $= p$, time $= \frac{x}{p}$

New distance $= x + 10$

New time $= \frac{x}{p} - 2$

New rate $= \frac{\text{new distance}}{\text{new time}}$

$$= \frac{x + 10}{\frac{x}{p} - 2} = \frac{x + 10}{\frac{x - 2p}{p}} = \mathbf{\frac{xp + 10p}{x - 2p}\ \frac{mi}{hr}}$$

2. (25)

$$R_T T_T + R_A T_A = \text{tasks}$$

$$\frac{1}{40}(T + 60) + \frac{1}{60}T = 5$$

$$\frac{T}{40} + \frac{T}{60} = \frac{200}{40} - \frac{60}{40}$$

$$\frac{5T}{120} = \frac{140}{40}$$

$$T = \mathbf{84\ min}$$

3. (25)

$$RWT = \text{jobs}$$

$$R(30)(4) = 6$$

$$R = \frac{1}{20}\ \frac{\text{jobs}}{\text{worker-day}}$$

$$RWT = \text{jobs}$$

$$\left(\frac{1}{20}\right)(20)T = 4$$

$$T = \mathbf{4\ days}$$

4. (R) The only even number between 6 and 24 that is evenly divisible by $5\frac{1}{2}$ is 22. Therefore, the builder should buy boards **22 ft** long to avoid throwing away any wood.

5. (25)

O_N = Ophelia now

$O_N - 10$ = Ophelia 10 years ago

L_N = Laertes now

$L_N - 10$ = Laertes 10 years ago

$$\begin{cases} \text{(a)} & O_N = \frac{4}{5}L_N \\ \text{(b)} & O_N - 10 = \frac{3}{5}(L_N - 10) \end{cases}$$

(b) $$O_N - 10 = \frac{3}{5}(L_N - 10)$$

$$\left(\frac{4}{5}L_N\right) - 10 = \frac{3}{5}L_N - 6$$

$$\frac{1}{5}L_N = 4$$

$$L_N = 20$$

(a) $O_N = \frac{4}{5}L_N = \frac{4}{5}(20) = 16$

$O_N + 20 =$ **Ophelia = 36 yr**

$L_N + 20 =$ **Laertes = 40 yr**

6. (32)

$$y = 3x + 4$$

$$-3x = -y + 4$$

$$x = \frac{1}{3}y - \frac{4}{3}$$

Interchange x and y.

$$\mathbf{y = \frac{1}{3}x - \frac{4}{3}}$$

7. (32) (a)

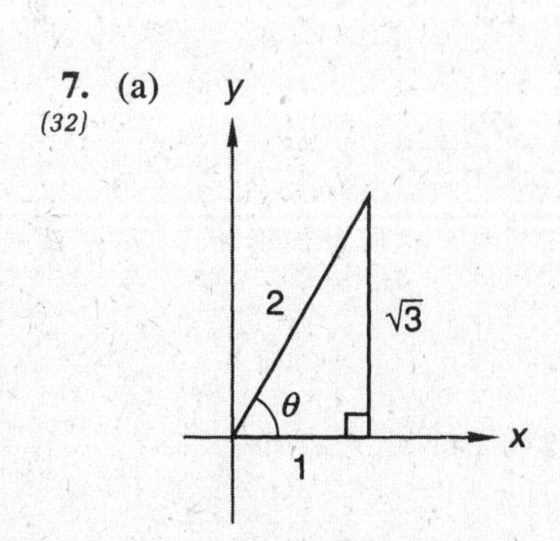

$-90° \le \theta \le 90°$

$\text{Arcsin } \frac{\sqrt{3}}{2} = \mathbf{60°}$

(b)

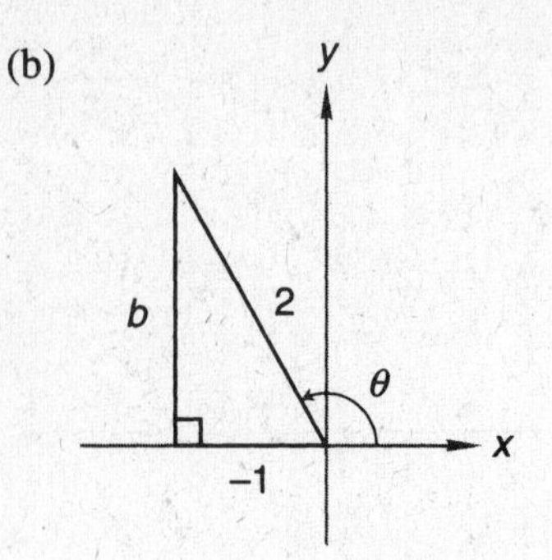

$0° \le \theta \le 180°$

$b = \sqrt{2^2 - (-1)^2} = \sqrt{3}$

$\sin\left[\text{Arccos}\left(-\frac{1}{2}\right)\right] = \sin\theta = \mathbf{\frac{\sqrt{3}}{2}}$

(c)

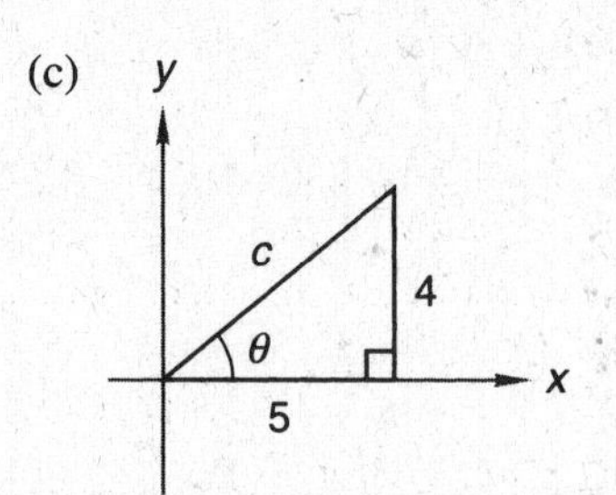

$-90° < \theta < 90°$

$c = \sqrt{4^2 + 5^2} = \sqrt{41}$

$\cos\left(\text{Arctan } \frac{4}{5}\right) = \cos\theta = \frac{5}{\sqrt{41}} = \mathbf{\frac{5\sqrt{41}}{41}}$

8. (31) (a) ***x* axis, yes.**
$(-y)^2 = x$ is equivalent to $y^2 = x$.

***y* axis, no.**
$y^2 = (-x)$ is not equivalent to $y^2 = x$.

Origin, no.
$(-y)^2 = (-x)$ is not equivalent to $y^2 = x$.

(b) ***x* axis, no.**
$(-y) = |x|$ is not equivalent to $y = |x|$.

***y* axis, yes.**
$y = |-x|$ is equivalent to $y = |x|$.

Origin, no.
$(-y) = |-x|$ is not equivalent to $y = |x|$.

9. (31) Replace x with $x + 2$.

$$\mathbf{y = (x + 2)^2}$$

10. (31) $g(x) = |x + 1|$

This is $f(x)$ translated 1 unit to the left.

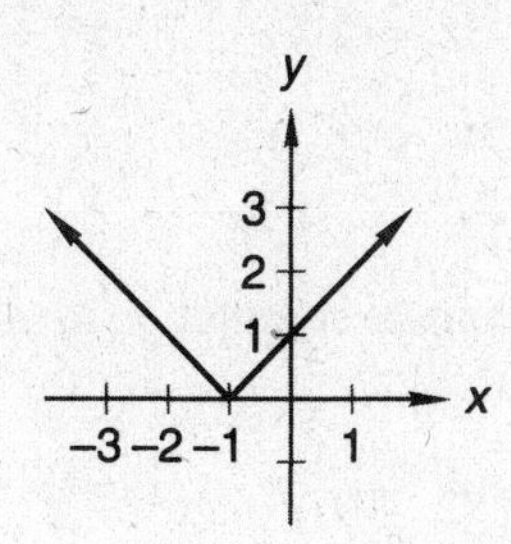

11. (30) $-2\underline{/120°} + 6\underline{/-130°}$

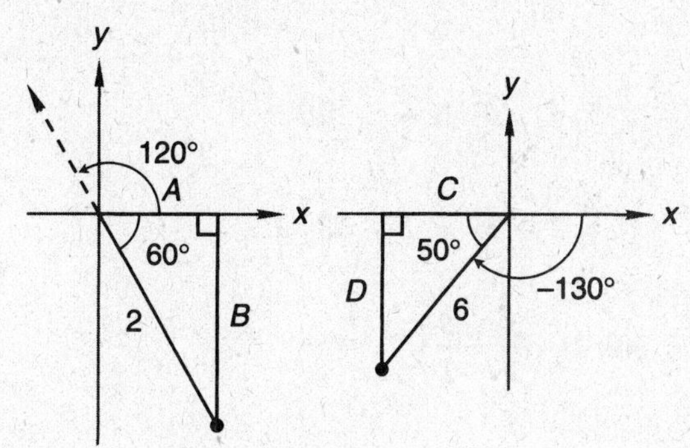

$A = 2 \cos 60° = 1$

$B = 2 \sin 60° = 1.7321$

$C = 6 \cos 50° = 3.8567$

$D = 6 \sin 50° = 4.5963$

$$\text{Resultant} = (1 - 3.8567)\hat{i} + (-1.7321 - 4.5963)\hat{j}$$

$$= -2.8567\hat{i} - 6.3284\hat{j}$$

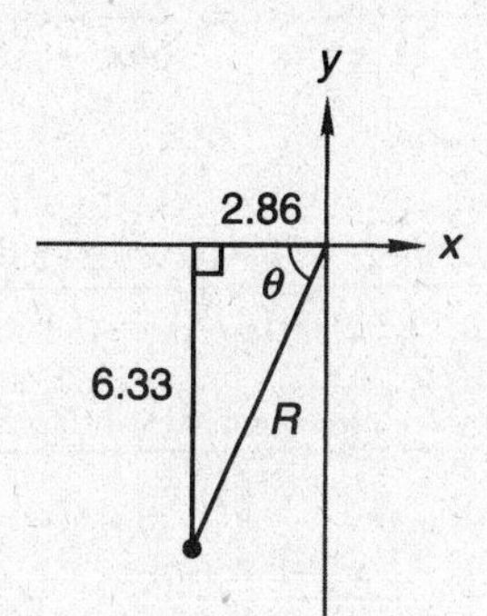

$$R = \sqrt{(2.8567)^2 + (6.3284)^2} = 6.94$$

$$\tan\theta = \frac{6.328}{2.857}$$

$$\theta = 65.70°$$

The polar angle is measured from the positive x axis and is $180° + \theta = 180° + 65.70° = 245.70°$.

Resultant = $6.94\underline{/245.70°}$, so

Equilibrant = $\mathbf{6.94\underline{/65.70°}}$

12. (30)

STATEMENTS	REASONS
1. $\angle BCD \cong \angle EDC$	1. Given
2. $\angle CDB \cong \angle DCE$	2. Given
3. $\overline{CD} \cong \overline{CD}$	3. Reflexive axiom
4. $\angle CBD \cong \angle DEC$	4. If two angles in one triangle are congruent to two angles in a second triangle, then the third angles are congruent.
5. $\triangle BCD \cong \triangle EDC$	5. *AAAS* congruency postulate
6. $\overline{BD} \cong \overline{EC}$	6. *CPCTC*

13. (26) $\log_a 36 = 2$

$$a^2 = 36$$

$$a = 36^{\frac{1}{2}}$$

$$a = \mathbf{6}$$

14. (26) $\log_3 \frac{1}{81} = n$

$$3^n = \frac{1}{81}$$

$$3^n = \frac{1}{3^4}$$

$$3^n = 3^{-4}$$

$$n = \mathbf{-4}$$

15. (26) $\log_{\frac{1}{3}} p = -4$

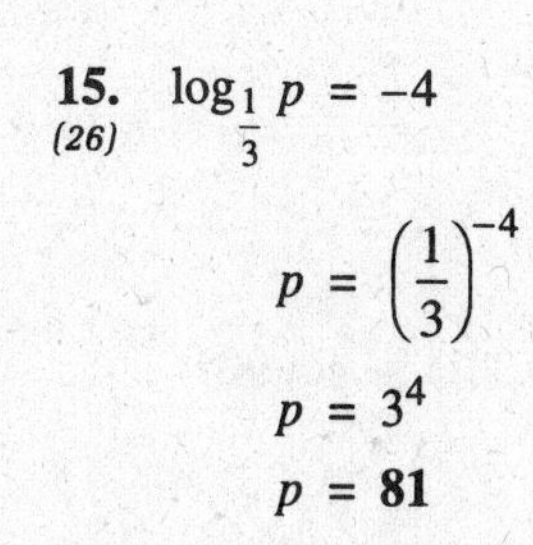

$$p = \left(\frac{1}{3}\right)^{-4}$$

$$p = 3^4$$

$$p = \mathbf{81}$$

16. (23,31) $f(x) = 2 + \left(\frac{1}{4}\right)^x$

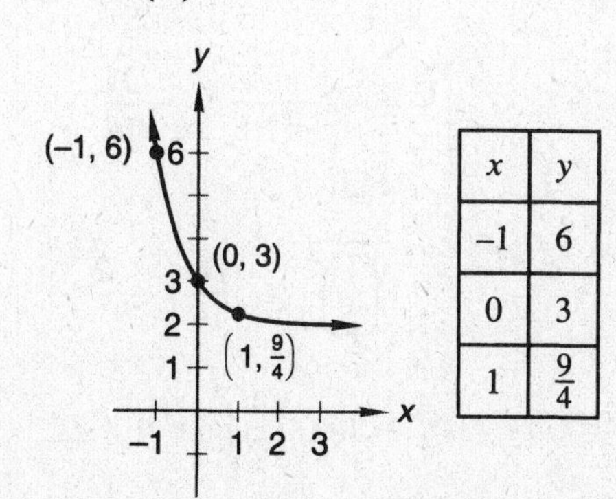

x	y
-1	6
0	3
1	$\frac{9}{4}$

17. (22) $g(x) = \dfrac{1}{|x - 2|}$

Plot points (3, 1) and (1, 1).

Asymptote: $x = 2$

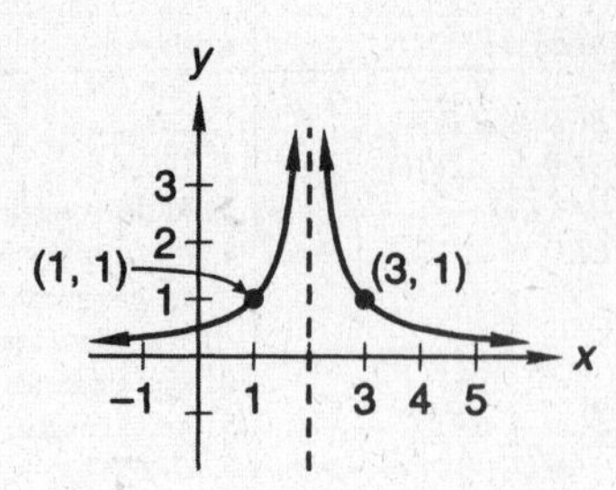

18. (21,22) Domain of $f = \{x \in \mathbb{R} \mid x \geq 0, x \neq 1\}$

19. (24)

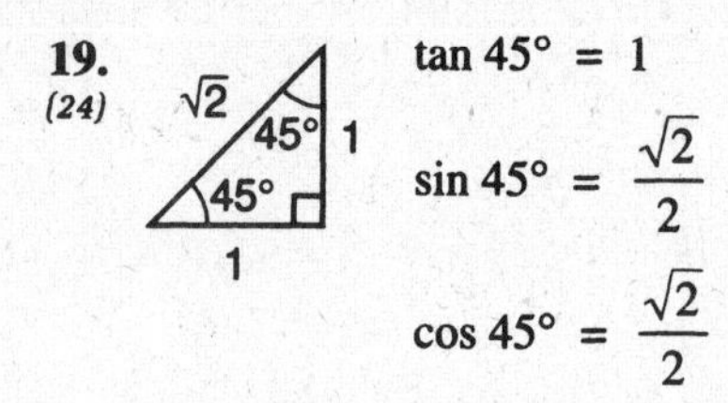

$\tan 45° = 1$

$\sin 45° = \dfrac{\sqrt{2}}{2}$

$\cos 45° = \dfrac{\sqrt{2}}{2}$

$-2\sqrt{2}\tan 45° - \sin 45° \cos 45°$

$= -2\sqrt{2}(1) - \left(\dfrac{\sqrt{2}}{2}\right)\left(\dfrac{\sqrt{2}}{2}\right) = \mathbf{-2\sqrt{2} - \dfrac{1}{2}}$

20. (24,29)

y, √3, –30°, x, –1, 2

$\sin(-30°) = -\dfrac{1}{2}$

$\cos 90° = 0$

$2f(-30°) + g(90°) = 2\sin(-30°) + \cos 90°$

$= 2\left(-\dfrac{1}{2}\right) + (0) = \mathbf{-1}$

21. (24,26) (a) $(f + g)(4) = \log_{\frac{1}{2}} 4 + \left(\dfrac{1}{2}\right)^4$

$= -2 + \dfrac{1}{16} = \mathbf{-\dfrac{31}{16}}$

(b) $(f/g)(4) = \dfrac{\log_{\frac{1}{2}} 4}{\left(\frac{1}{2}\right)^4} = \dfrac{-2}{\frac{1}{16}} = \mathbf{-32}$

(c) $(f \circ g)(4) = \log_{\frac{1}{2}}\left(\dfrac{1}{2}\right)^4 = \mathbf{4}$

22. (19) First we label the equations.

$$\begin{cases} \text{(a)} & y^2 - x^2 = 4 \\ \text{(b)} & y + 3x = 2 \end{cases}$$

(b) $y + 3x = 2$

$y = 2 - 3x$

Now we can use substitution to solve.

(a)
$$\begin{aligned} y^2 - x^2 &= 4 \\ (2 - 3x)^2 - x^2 &= 4 \\ 4 - 12x + 9x^2 - x^2 &= 4 \\ 8x^2 - 12x &= 0 \\ x(2x - 3) &= 0 \\ x &= 0, \frac{3}{2} \end{aligned}$$

$y = 2 - 3x = 2 - 3(0) = 2$

or

$y = 2 - 3x = 2 - 3\left(\dfrac{3}{2}\right) = -\dfrac{5}{2}$

$\mathbf{(0, 2), \left(\dfrac{3}{2}, -\dfrac{5}{2}\right)}$

23. (16) $\dfrac{2}{6b + \dfrac{3t}{1 + \frac{3}{x}}} = \dfrac{2}{6b + \dfrac{3t}{\frac{x + 3}{x}}}$

$= \dfrac{2}{6b + \dfrac{3tx}{x + 3}} = \dfrac{2}{\dfrac{6b(x + 3) + 3tx}{x + 3}}$

$= \dfrac{2(x + 3)}{6b(x + 3) + 3tx} = \mathbf{\dfrac{2x + 6}{6bx + 18b + 3tx}}$

24. (10) $\dfrac{\sqrt{3}\sqrt{-3}\sqrt{2}\sqrt{-2} - \sqrt{-16} + \sqrt{-5}\sqrt{5}}{1 - \sqrt{-16}i^2}$

$= \dfrac{\sqrt{3}\sqrt{3}\sqrt{2}\sqrt{2}ii - \sqrt{16}i + \sqrt{5}\sqrt{5}i}{1 - \sqrt{16}i^3}$

$= \dfrac{-6 + i}{1 + 4i} \cdot \dfrac{1 - 4i}{1 - 4i}$

$= \dfrac{-6 + i + 24i - 4i^2}{1 - 16i^2} = \mathbf{\dfrac{-2}{17} + \dfrac{25}{17}i}$

25. (5) $\dfrac{z^{3+b}\left(\sqrt{z^3}\right)^{b+1}}{z^b} = z^{3+b} \cdot z^{\frac{3}{2}(b+1)} \cdot z^{-b}$

$= z^{3+b-b} \cdot z^{\frac{3}{2}b + \frac{3}{2}} = \mathbf{z^{\frac{3}{2}b + \frac{9}{2}}}$

26. (10)

$$6x^2 - 2x + 1 = 0$$

$$x^2 - \frac{1}{3}x = -\frac{1}{6}$$

$$\left(x^2 - \frac{1}{3}x + \frac{1}{36}\right) = -\frac{1}{6} + \frac{1}{36}$$

$$\left(x - \frac{1}{6}\right)^2 = -\frac{5}{36}$$

$$x - \frac{1}{6} = \pm\frac{\sqrt{5}i}{6}$$

$$\mathbf{x = \frac{1}{6} \pm \frac{\sqrt{5}i}{6}}$$

27. (21) $f(x + h) - f(x)$

$$= \left[(x + h)^2 + 3(x + h) + 1\right] - \left(x^2 + 3x + 1\right)$$

$$= x^2 + 2xh + h^2 + 3x + 3h + 1 - x^2 - 3x - 1$$

$$= \mathbf{h^2 + 2xh + 3h}$$

28. (2)

$$C = 2\pi r$$

$$30\pi = 2\pi r$$

$$r = 15$$

$$V_{\text{cone}} = \frac{1}{3}(A_{\text{base}})(h)$$

$$1500\pi = \frac{1}{3}\left[\pi(15)^2\right](h)$$

$$h = 20$$

$$l = \sqrt{15^2 + 20^2} = 25$$

$$A_{\text{surface}} = A_{\text{base}} + \pi rl$$

$$= \pi(15)^2 + \pi(15)(25)$$

$$= \mathbf{600\pi\ m^2 = 1884.96\ m^2}$$

29. (1) Since $ab = 10$, a and b can be any number except 0. If $a = 1$, then $b = 10$, and if $a = 10$, then $b = 1$. We do not have enough information to compare a to b. Therefore the answer is **D.**

30. (1) If $x > 1$, then $\frac{1}{x} < 1$.

So $x > 1 > \frac{1}{x}$.

Therefore the answer is **A.**

Problem Set 33

1. (18) First, we write the three equations.

$$\begin{cases} \text{(a)}\ N_N + N_D + N_Q = 11 \\ \text{(b)}\ 5N_N + 10N_D + 25N_Q = 155 \\ \text{(c)}\ 2N_Q = N_D + N_N + 1 \end{cases}$$

(c) $2N_Q = N_D + N_N + 1$

$$-N_N - N_D + 2N_Q = 1$$

Now we can use elimination to solve.

$$-N_N - N_D + 2N_Q = 1$$

(a) $N_N + N_D + N_Q = 11$

$$3N_Q = 12$$

$$\mathbf{N_Q = 4}$$

−5(a) $-5N_N - 5N_D - 5N_Q = -55$

(b) $5N_N + 10N_D + 25N_Q = 155$

$$5N_D + 20N_Q = 100$$

$$5N_D + 20(4) = 100$$

$$5N_D = 20$$

$$\mathbf{N_D = 4}$$

(a) $N_N + N_D + N_Q = 11$

$$N_N + (4) + (4) = 11$$

$$\mathbf{N_N = 3}$$

2. (25)

$$RGT = \text{pounds}$$

$$R(15)(2) = 45$$

$$R = \frac{3}{2}\,\frac{\text{lb}}{\text{gourmand-hr}}$$

$$RGT = \text{pounds}$$

$$\frac{3}{2}(25)(10) = \mathbf{375\ lb}$$

3. (10) $4x - 2y + 3 = 0$

$$-2y = -4x - 3$$

$$y = 2x + \frac{3}{2}$$

Slopes of perpendicular lines are negative reciprocals of each other, so $m = -\frac{1}{2}$.

$$y = -\frac{1}{2}x + b$$

$$2 = -\frac{1}{2}(2) + b$$

$$b = 3$$

$$\mathbf{y = -\frac{1}{2}x + 3}$$

4. (33) We are not told anything about the angles, so we do not know if the figure is a square. Therefore, the answer is **C.**

5. (33) Area $= \frac{1}{2}H(B_1 + B_2) = \frac{1}{2}(8)(5 + 7) = \mathbf{48\ m^2}$

6. (32)
$$y = -4x + 7$$
$$4x = -y + 7$$
$$x = -\frac{1}{4}y + \frac{7}{4}$$

Interchange x and y.

$$\mathbf{y = -\frac{1}{4}x + \frac{7}{4}}$$

7. (32) (a)

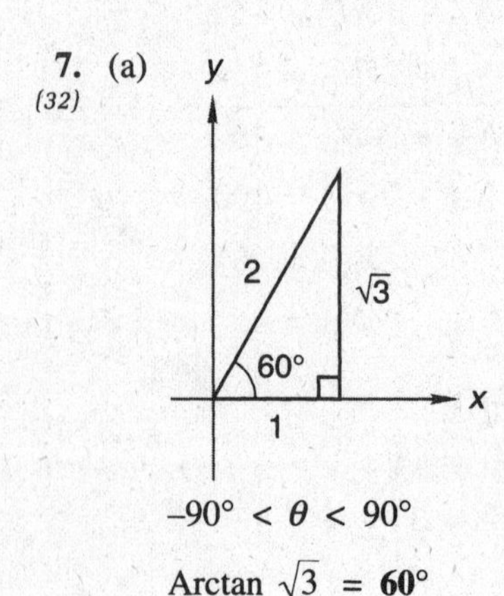

$-90° < \theta < 90°$

Arctan $\sqrt{3} = \mathbf{60°}$

(b)

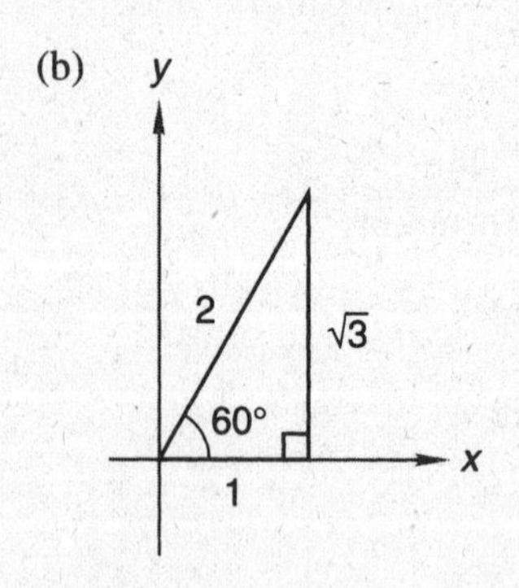

$0° \leq \theta \leq 180°$

$$\sin\left(\text{Arccos}\ \frac{1}{2}\right) = \sin 60° = \mathbf{\frac{\sqrt{3}}{2}}$$

(c)

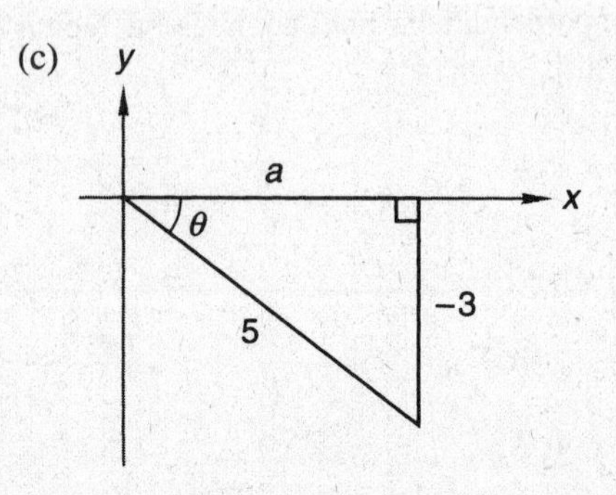

$-90° \leq \theta \leq 90°$

$$a = \sqrt{5^2 - (-3)^2} = 4$$

$$\tan\left[\text{Arcsin}\left(-\frac{3}{5}\right)\right] = \tan\theta = \mathbf{-\frac{3}{4}}$$

8. (31) (a) ***x* axis, no.**

$(-y)x = 1$ is not equivalent to $yx = 1$.

***y* axis, no.**

$y(-x) = 1$ is not equivalent to $yx = 1$.

Origin, yes.

$(-y)(-x) = 1$ is equivalent to $yx = 1$.

(b) ***x* axis, yes.**

$4x^2 + 9(-y)^2 = 36$ is equivalent to $4x^2 + 9y^2 = 36$.

***y* axis, yes.**

$4(-x)^2 + 9y^2 = 36$ is equivalent to $4x^2 + 9y^2 = 36$.

Origin, yes.

$4x^2 + 9y^2 = 36$ is symmetric about the x and y axes.

9. (31) Replace y with $y - 3$.

$$y - 3 = |x|$$
$$\mathbf{y = |x| + 3}$$

10. (31) $g(x) = x^2 + 2$

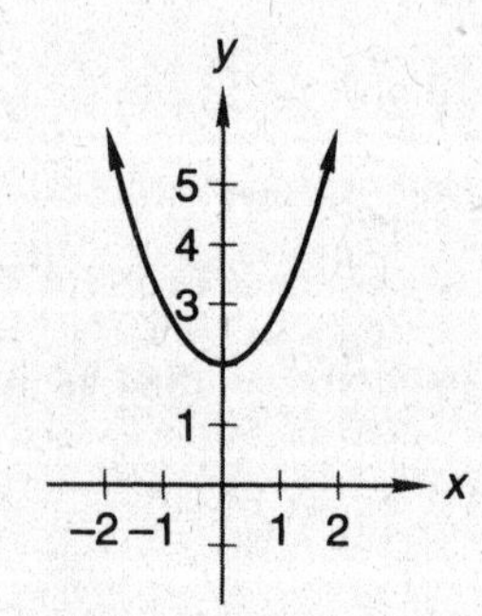

11. (30) $-2\underline{/140°} + 3\underline{/-120°}$

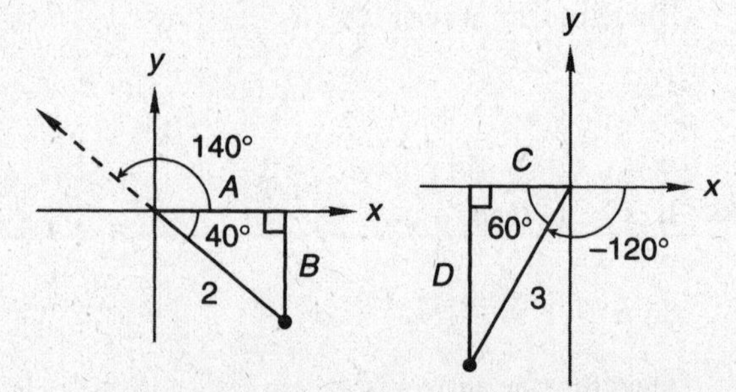

$A = 2\cos 40° = 1.5321$

$B = 2\sin 40° = 1.2856$

$C = 3\cos 60° = 1.5$

$D = 3\sin 60° = 2.5981$

$\text{Resultant} = (1.5321 - 1.5)\hat{i}$

$+ (-1.2856 - 2.5981)\hat{j}$

$= 0.0321\hat{i} - 3.8837\hat{j}$

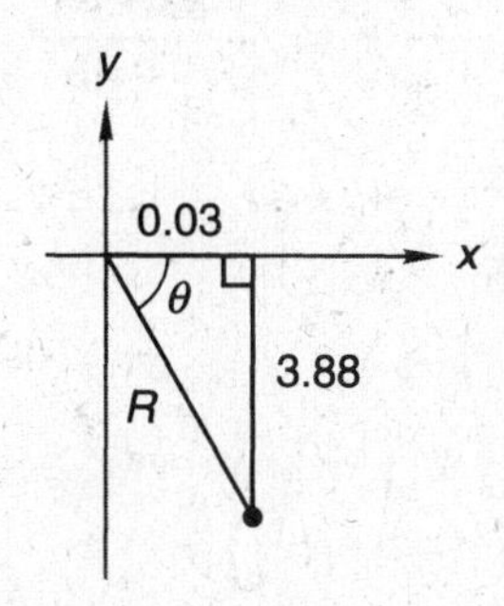

Note: Figure not drawn to scale

$R = \sqrt{(0.0321)^2 + (3.8837)^2} = 3.88$

$\tan\theta = \dfrac{3.8837}{0.0321}$

$\theta = 89.53°$

The polar angle is measured from the positive x axis and is $360° - \theta = 360° - 89.53° = 270.47°$.

3.88 $\underline{/270.47°}$

12. (28) $\dfrac{12!}{4!8!} = \dfrac{12 \cdot 11 \cdot 10 \cdot 9 \cdot 8!}{4 \cdot 3 \cdot 2 \cdot 1 \cdot 8!}$

$= 11 \cdot 5 \cdot 9 = \mathbf{495}$

13. (26) $\log_d 216 = 3$

$d^3 = 216$

$d = 216^{\frac{1}{3}}$

$d = \mathbf{6}$

14. (26) $\log_2 \dfrac{1}{64} = m$

$2^m = \dfrac{1}{64}$

$2^m = \dfrac{1}{2^6}$

$m = \mathbf{-6}$

15. (26) $\log_{\frac{1}{2}} p = -3$

$p = \left(\dfrac{1}{2}\right)^{-3}$

$p = \dfrac{1}{2^{-3}}$

$p = \mathbf{8}$

16. (23) $y = \left(\dfrac{1}{3}\right)^x - 2$

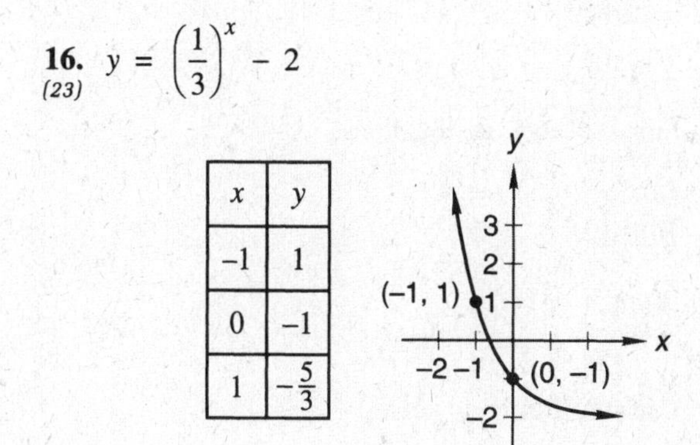

x	y
-1	1
0	-1
1	$-\frac{5}{3}$

17. (22) $y = \dfrac{1}{(x-2)^2}$

Asymptote: $x = 2$

Plot points (1, 1), (3, 1).

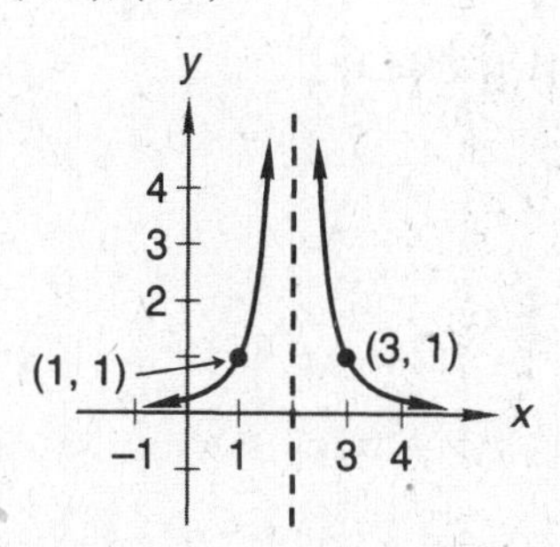

18. (21,22) $\{x \in \mathbb{R} \mid x \geq 2, x \leq -2, x \neq 3\}$

19. (24,29)

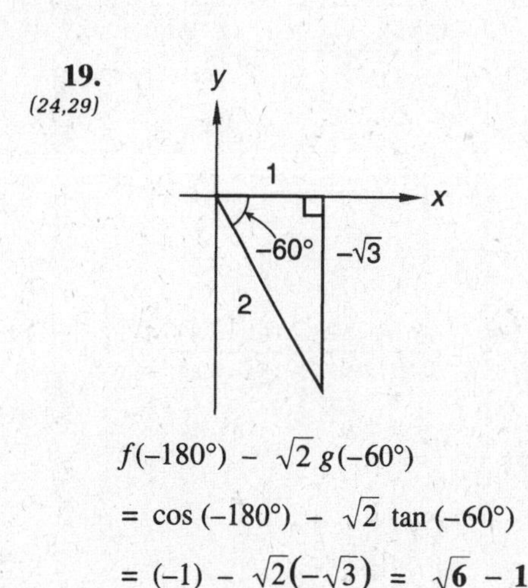

$f(-180°) - \sqrt{2}\, g(-60°)$

$= \cos(-180°) - \sqrt{2}\tan(-60°)$

$= (-1) - \sqrt{2}\left(-\sqrt{3}\right) = \mathbf{\sqrt{6} - 1}$

20. (24,26)

(a) $(f - g)(x) = 3^{x+1} - \log_3 x$

$(f - g)(3) = 3^{(3+1)} - \log_3 3 = 81 - 1 = \mathbf{80}$

(b) $(f \circ g)(x) = 3^{\log_3 x + 1}$

$(f \circ g)(3) = 3^{\log_3 3 + 1} = 3^{1+1} = \mathbf{9}$

(c) $(g \circ f)(x) = \log_3 3^{x+1} = x + 1$

$(g \circ f)(3) = 3 + 1 = \mathbf{4}$

21. (19) First we label the equations.

$$\begin{cases} x^2 - y^2 = 4 & \text{(a)} \\ x + y = 2 & \text{(b)} \end{cases}$$

(b) $x + y = 2$

$x = 2 - y$

Use substitution to solve.

(a)
$$x^2 - y^2 = 4$$
$$(2 - y)^2 - y^2 = 4$$
$$4 - 4y + y^2 - y^2 = 4$$
$$4y = 0$$
$$\mathbf{y = 0}$$

$x = 2 - y$

$x = 2 - (0)$

$\mathbf{x = 2}$

22. (19) $4a^{x+2} - 12a^{x+3} = 4a^x a^2 - 12a^x a^3$

$= \mathbf{4a^{x+2}(1 - 3a)}$

23. (16)
$$\frac{\dfrac{a^2b}{c^2d^3} - \dfrac{fg^2}{cd^3}}{\dfrac{x^2y}{c^2d} + \dfrac{z^3}{cd^3}} = \frac{\dfrac{a^2b - fg^2c}{c^2d^3}}{\dfrac{x^2yd^2 + z^3c}{c^2d^3}} = \mathbf{\frac{a^2b - fg^2c}{x^2yd^2 + cz^3}}$$

24. (10)
$$\frac{\sqrt{2}\sqrt{-3}\sqrt{-2}\sqrt{3} - \sqrt{-16}}{-4i^3 - \sqrt{-16i}}$$

$$= \frac{(2)(3)i^2 - 4i}{-4i^3 - 4i^2} = \frac{-6 - 4i}{4i + 4} \cdot \frac{(4i - 4)}{(4i - 4)}$$

$$= \frac{-24i - 16i^2 + 24 + 16i}{16i^2 - 16}$$

$$= \frac{-8i + 40}{-32} = \mathbf{-\frac{5}{4} + \frac{1}{4}i}$$

25. (16)
$$h = \frac{2d}{f}\left(\frac{4s}{g} + \frac{2s}{r}\right)$$
$$h = \frac{8sd}{fg} + \frac{4sd}{fr}$$
$$hfgr = 8sdr + 4sdg$$
$$hfgr - 4sdg = 8sdr$$
$$g(hfr - 4sd) = 8sdr$$
$$\mathbf{g = \frac{8dsr}{hfr - 4ds}}$$

26. (21) $f(x + h) - f(x)$

$= (x + h + 1)^2 - (x + 1)^2$

$= x^2 + xh + x + xh + h^2 + h + x + h + 1 - (x^2 + 2x + 1)$

$= \mathbf{h^2 + 2h + 2xh}$

27. (33)

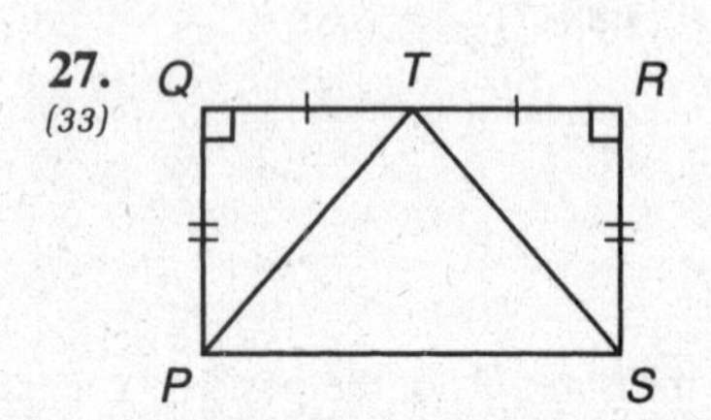

Statements	Reasons
1. $PQRS$ is a rectangle.	1. Given
2. $\angle PQT$ and $\angle SRT$ are right angles.	2. A rectangle contains four right angles.
3. $\angle PQT \cong \angle SRT$	3. All right angles are congruent.
4. T is the midpoint of $\overline{QR}$.	4. Given
5. $\overline{QT} \cong \overline{RT}$	5. A midpoint divides a segment into two congruent segments.
6. $\overline{PQ} \cong \overline{SR}$	6. Opposite sides of a rectangle are congruent.
7. $\Delta PQT \cong \Delta SRT$	7. *SAS* congruency postulate

28. (13)

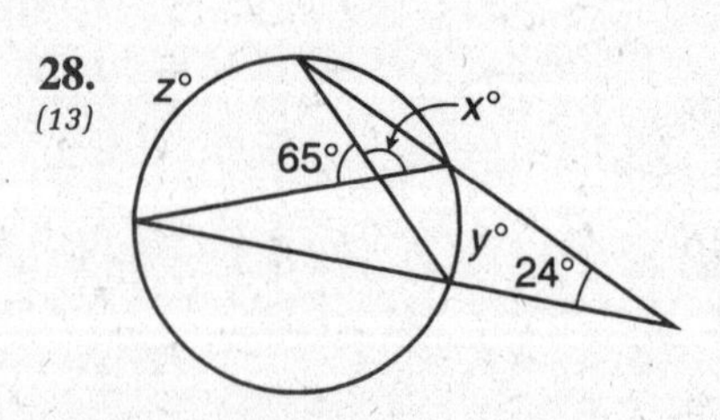

$x = 180 - 65$

$\mathbf{x = 115}$

We write two equations.

$$\begin{cases} \dfrac{y + z}{2} = 65 & \text{(a)} \\ \dfrac{z - y}{2} = 24 & \text{(b)} \end{cases}$$

(a) $\dfrac{y + z}{2} = 65$

$y + z = 130$

$y = 130 - z$

Use substitution to solve.

(b) $\dfrac{z - y}{2} = 24$

$z - y = 48$

$z - (130 - z) = 48$

$2z = 178$

$\mathbf{z = 89}$

$y = 130 - z$

$y = 130 - (89)$

$\mathbf{y = 41}$

29. (1) $\dfrac{a}{b} = 4$

$a = 4b$

A and B are equal. Therefore the answer is **C.**

30. (1)

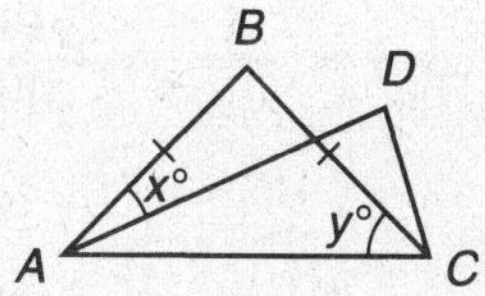

Since $AB = BC$, $m\angle BAC = m\angle BCA = y°$.

$x° = m\angle BAC - m\angle DAC$

$= y° - m\angle DAC$

$x° < y°$

Therefore the answer is **B.**

Problem Set 34

1. (28) Distance $= m$

Rate $= z$

Time $= \dfrac{m}{z}$

New distance $= m$

New time $= \dfrac{m}{z} - 3 = \dfrac{m - 3z}{z}$

New rate $= \dfrac{m}{\frac{m - 3z}{z}} = \mathbf{\dfrac{mz}{m - 3z}\ \dfrac{mi}{hr}}$

2. (25) $RET = \text{concoctions}$

$R = \dfrac{\text{concoctions}}{ET}$

$R = \dfrac{20}{(3)(5)} = \dfrac{4}{3}\ \dfrac{\text{concoctions}}{\text{eater-hr}}$

$RET = \text{concoctions}$

$T = \dfrac{\text{concoctions}}{RE}$

$T = \dfrac{14}{\left(\frac{4}{3}\right)(5)} = \mathbf{\dfrac{21}{10}\ hr}$

3. (25) $R_ST + R_BT = \text{jobs}$

$\left(\dfrac{2}{5}\right)T + \left(\dfrac{1}{6}\right)T = 10$

$\dfrac{17}{30}T = 10$

$T = \mathbf{\dfrac{300}{17}\ hr}$

4. (18) $B = \dfrac{\$120}{16} = \$7.50 \text{ per call}$

$A = \dfrac{3}{2}B = \dfrac{3}{2}(\$7.50) = \$11.25 \text{ per call}$

$12A = 12(\$11.25) = \mathbf{\$135}$

5. (34) $\displaystyle\sum_{i=1}^{7} 3 = 3 + 3 + 3 + 3 + 3 + 3 + 3 = \mathbf{21}$

6. (34) $\displaystyle\sum_{j=0}^{2} \frac{3^j}{1 - 2j} = \frac{3^{(0)}}{1 - 2(0)} + \frac{3^{(1)}}{1 - 2(1)} + \frac{3^{(2)}}{1 - 2(2)}$

$= \dfrac{1}{1} + \dfrac{3}{-1} + \dfrac{9}{-3}$

$= 1 - 3 - 3 = \mathbf{-5}$

7. (34) From the graph of the line we pick two ordered pairs: (70, 850) and (90, 700).

$m = \dfrac{850 - 700}{70 - 90} = -7.5$

$\text{Mo} = -7.5\text{Zr} + b$

$700 = -7.5(90) + b$

$b = 1375$

$\mathbf{Mo = -7.5Zr + 1375}$

8. (34) From the graph of the line we pick two ordered pairs: (102, 100) and (105, 120).

$$m = \frac{120 - 100}{105 - 102} = 6.67$$

$$H = 6.67C + b$$

$$100 = 6.67(102) + b$$

$$b = -580$$

$$\mathbf{H = 6.67C - 580}$$

9. (34) $\mathbf{f(x) = \sqrt{x}}$

$\mathbf{g(x) = x + 1}$

10. (33) A property of a rectangle is that it has 4 right angles. The square also has 4 right angles. However, we are not told anything about the sides or diagonals. Therefore the answer is **A.**

11. (33) Median $= \frac{1}{2}(B_1 + B_2)$

$$3x + 5 = \frac{1}{2}[(2x - 1) + (6x - 9)]$$

$$3x + 5 = \frac{1}{2}(8x - 10)$$

$$3x + 5 = 4x - 5$$

$$x = \mathbf{10}$$

12. (32) $y = \frac{1}{3}x - \frac{5}{6}$

$$-\frac{1}{3}x = -y - \frac{5}{6}$$

$$x = 3y + \frac{5}{2}$$

Interchange x and y.

$$\mathbf{y = 3x + \frac{5}{2}}$$

13. (32) (a)

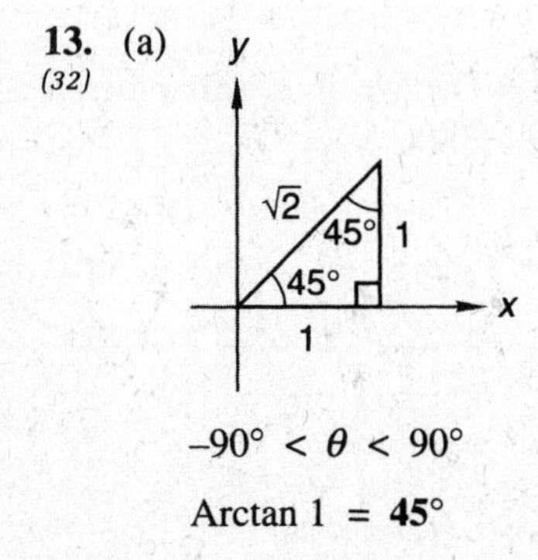

$-90° < \theta < 90°$

Arctan 1 = **45°**

(b)

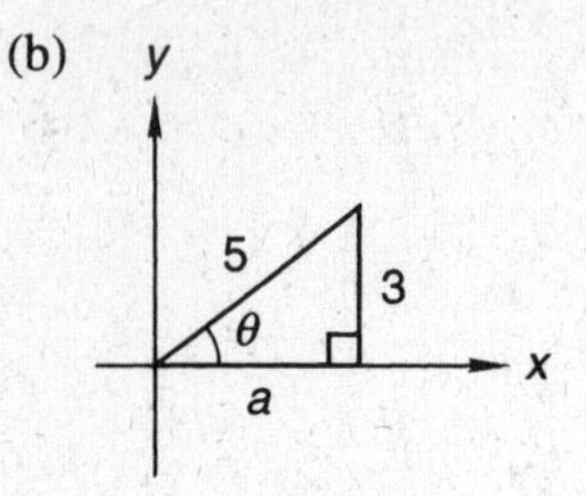

$-90° \le \theta \le 90°$

$$a = \sqrt{5^2 - 3^2} = 4$$

$$\tan\left(\text{Arcsin}\ \frac{3}{5}\right) = \tan\theta = \mathbf{\frac{3}{4}}$$

(c)

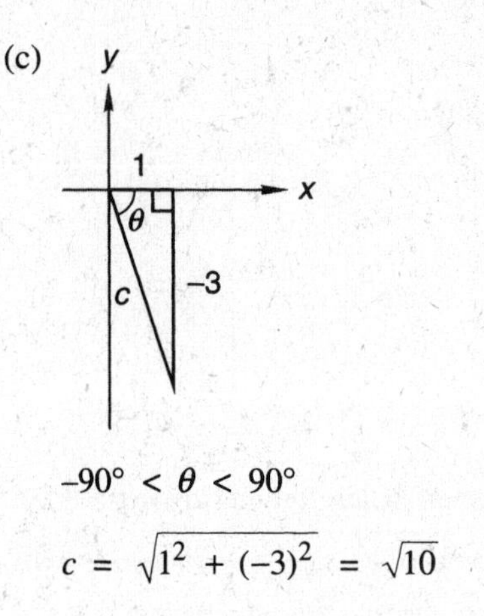

$-90° < \theta < 90°$

$$c = \sqrt{1^2 + (-3)^2} = \sqrt{10}$$

$$\cos[\text{Arctan}(-3)] = \cos\theta = \frac{1}{\sqrt{10}} = \mathbf{\frac{\sqrt{10}}{10}}$$

14. (31) (a) **x axis, yes.**

$9x^2 + (-y)^2 = 9$ is equivalent to $9x^2 + y^2 = 9$.

y axis, yes.

$9(-x)^2 + y^2 = 9$ is equivalent to $9x^2 + y^2 = 9$.

Origin, yes.

$9(-x)^2 + (-y)^2 = 9$ is equivalent to $9x^2 + y^2 = 9$.

(b) **x axis, no.**

$-y = x^3$ is not equivalent to $y = x^3$.

y axis, no.

$y = (-x)^3$ is not equivalent to $y = x^3$.

Origin, yes.

$-y = (-x)^3$ is equivalent to $y = x^3$.

15. (31) Replace y with $y + 3$.

$$y + 3 = |x|$$

$$\mathbf{y = |x| - 3}$$

16. (31) $g(x) = x^2 - 2$

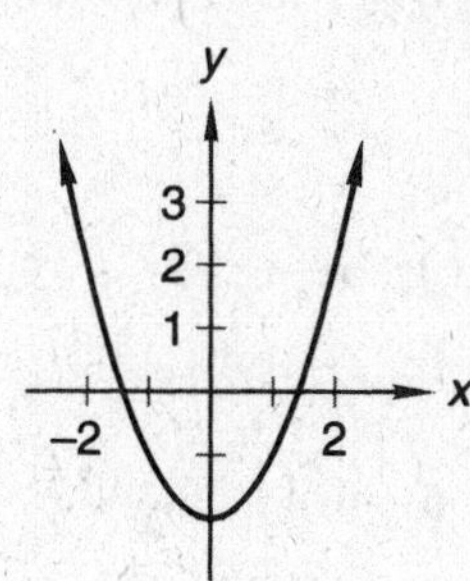

17. (30) $-3\underline{/-135°} - 2\underline{/-140°}$

$= -3\underline{/-135°} + (-2\underline{/-140°})$

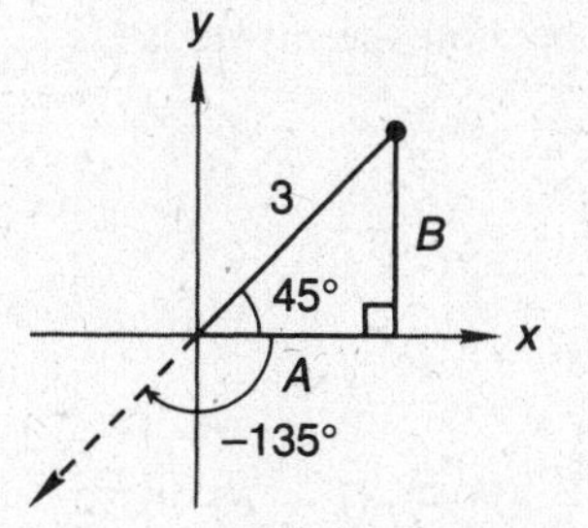

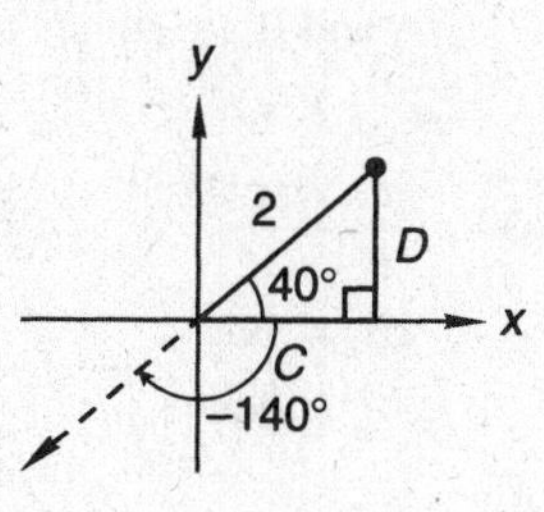

$A = 3 \cos 45° = 2.1213$

$B = 3 \sin 45° = 2.1213$

$C = 2 \cos 40° = 1.5321$

$D = 2 \sin 40° = 1.2856$

Resultant $= (2.1213 + 1.5321)\hat{i}$

$+ (2.1213 + 1.2856)\hat{j}$

$= 3.6534\hat{i} + 3.4069\hat{j}$

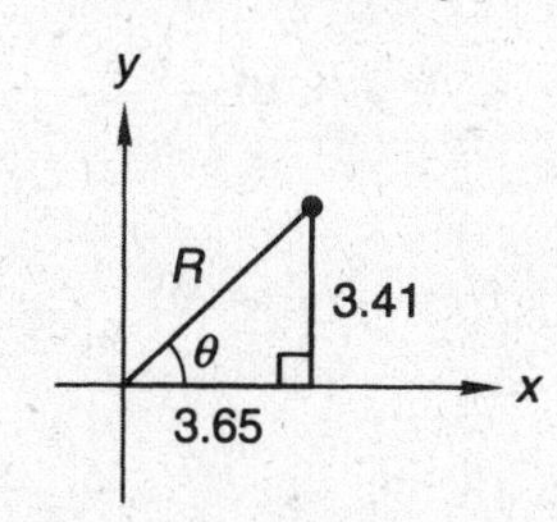

$R = \sqrt{(3.6534)^2 + (3.4069)^2} = 5$

$\tan\theta = \dfrac{3.4069}{3.6534}$

$\theta = 43°$

Resultant $= 5\underline{/43°}$, so

Equilibrant $=$ $\mathbf{5\underline{/223°}}$

18. (26) $\log_b 125 = 3$

$b^3 = 125$

$b = 125^{\frac{1}{3}} = \mathbf{5}$

19. (26) $\log_2 \dfrac{1}{32} = a$

$2^a = \dfrac{1}{32}$

$2^a = \dfrac{1}{2^5}$

$a = \mathbf{-5}$

20. (26) $\log_{\frac{1}{5}} p = -2$

$p = \left(\dfrac{1}{5}\right)^{-2} = \dfrac{1}{5^{-2}} = \mathbf{25}$

21. (21) (a) **Not a function**

(b) **Not a function**

(c) **Function, 1 to 1**

(d) **Function, not 1 to 1**

22. (21) $x + 11 \geq 0$

$x \geq -11$

$x^2 - x - 12 \neq 0$

$(x - 4)(x + 3) \neq 0$

$x \neq 4, -3$

$\{\mathbf{x \in \mathbb{R} \mid x \geq -11, x \neq 4, -3}\}$

23. (24,29)

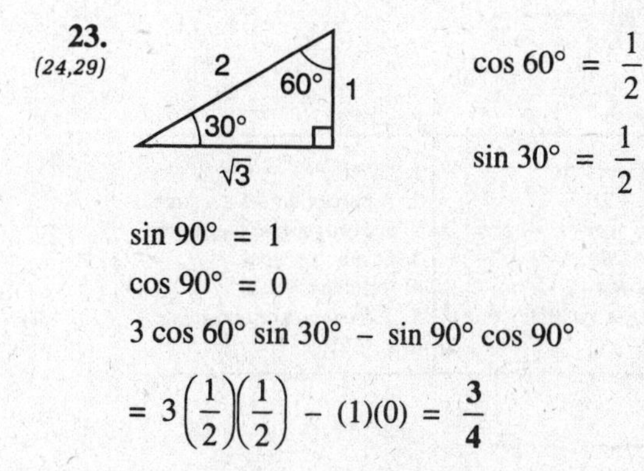

$\cos 60° = \dfrac{1}{2}$

$\sin 30° = \dfrac{1}{2}$

$\sin 90° = 1$

$\cos 90° = 0$

$3 \cos 60° \sin 30° - \sin 90° \cos 90°$

$= 3\left(\dfrac{1}{2}\right)\left(\dfrac{1}{2}\right) - (1)(0) = \mathbf{\dfrac{3}{4}}$

24. (24) (a) $(fg)(x) = (3x - 5)(2x^2) = 6x^3 - 10x^2$

$(fg)(2) = 6(2)^3 - 10(2)^2 = 48 - 40 = \mathbf{8}$

(b) $(gf)(x) = (2x^2)(3x - 5) = 6x^3 - 10x^2$

$(gf)(2) = 6(2)^3 - 10(2)^2 = 48 - 40 = \mathbf{8}$

(c) $(f \circ g)(x) = 3(2x^2) - 5 = 6x^2 - 5$

$(f \circ g)(2) = 6(2)^2 - 5 = 24 - 5 = \mathbf{19}$

25. First we label the equations.
(19)

$$\begin{cases} \text{(a)} & x^2 + y^2 = 10 \\ \text{(b)} & 2x^2 - y^2 = 17 \end{cases}$$

Next we use elimination to solve.

$$\begin{aligned} \text{(a)} \quad x^2 + y^2 &= 10 \\ \text{(b)} \quad 2x^2 - y^2 &= 17 \\ \hline 3x^2 &= 27 \\ x^2 &= 9 \\ x &= \pm 3 \end{aligned}$$

$$\begin{aligned} \text{(a)} \quad x^2 + y^2 &= 10 \\ y^2 &= 10 - x^2 \\ y^2 &= 10 - (\pm 3)^2 \\ y^2 &= 1 \\ y &= \pm 1 \end{aligned}$$

(3, 1), (3, –1), (–3, 1), (–3, –1)

26. $f(x + h) - f(x) = \dfrac{1}{x + h} - \dfrac{1}{x}$
(21)

$$= \frac{x - (x + h)}{x(x + h)} = \boldsymbol{\frac{-h}{x^2 + hx}}$$

27.
(15)

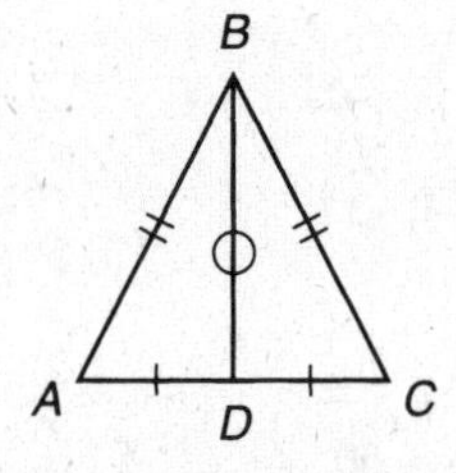

STATEMENTS	REASONS
1. $\overline{BD}$ bisects $\overline{AC}$	1. Given
2. $\overline{AD} \cong \overline{CD}$	2. A bisector divides a segment into two congruent segments.
3. $\overline{AB} \cong \overline{CB}$	3. Given
4. $\overline{BD} \cong \overline{BD}$	4. Reflexive axiom
5. $\triangle ABD \cong \triangle CBD$	5. *SSS* congruency postulate
6. $\angle A \cong \angle C$	6. *CPCTC*

28.
(2)

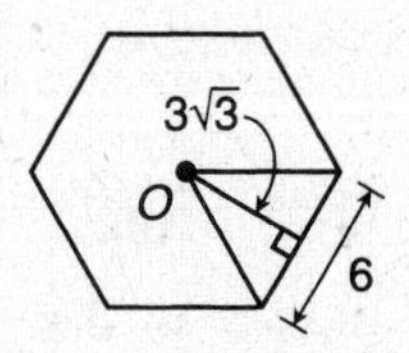

$$V = \frac{1}{3}(A_{\text{base}})(h)$$

$$216 = \frac{1}{3}\left[6\left(\frac{1}{2}\right)(6)\left(3\sqrt{3}\right)\right](h)$$

$$h = 4\sqrt{3}$$

l
h = 4√3
3√3

$$l = \sqrt{\left(3\sqrt{3}\right)^2 + \left(4\sqrt{3}\right)^2} = 5\sqrt{3}$$

$$A_{\text{surface}} = A_{\text{base}} + 6A_{\text{side}}$$

$$= 6\left[\frac{1}{2}(6)\left(3\sqrt{3}\right)\right] + 6\left[\frac{1}{2}(6)\left(5\sqrt{3}\right)\right]$$

$$= \mathbf{144\sqrt{3}\ cm^2}$$

29. $\dfrac{x + 2y}{x} = \dfrac{x}{x} + \dfrac{2y}{x} = 1 + \dfrac{2y}{x}$
(1)

A and B are equal. Therefore the answer is **C.**

30. $\dfrac{a}{b} = \dfrac{c}{d}$
(1)

$$\begin{aligned} ad &= bc \\ 0 &= bc - ad \end{aligned}$$

$$\frac{a + c}{b + d} - \frac{a}{b} = \frac{(a + c)b - a(b + d)}{b(b + d)}$$

$$= \frac{ab + bc - ab - ad}{b(b + d)} = \frac{bc - ad}{b(b + d)} = 0$$

Therefore the answer is **C.**

Problem Set 35

1. First we write three equations.
(18)

$$\begin{cases} \text{(a)} & 4N_W = 9N_R + 10 \\ \text{(b)} & \dfrac{N_B}{N_R} = \dfrac{3}{1} \\ \text{(c)} & N_B + N_R + N_W = 65 \end{cases}$$

$$\text{(b)} \quad \frac{N_B}{N_R} = \frac{3}{1}$$

$$N_B = 3N_R$$

$$\text{(a)} \quad 4N_W = 9N_R + 10$$

$$N_W = \frac{9}{4}N_R + \frac{5}{2}$$

Now we can use substitution to solve.

$$\text{(c)} \quad N_B + N_R + N_W = 65$$

$$(3N_R) + N_R + \left(\frac{9}{4}N_R + \frac{5}{2}\right) = 65$$

$$\frac{25}{4}N_R = \frac{125}{2}$$

$$\mathbf{N_R = 10}$$

$$N_B = 3N_R$$
$$N_B = 3(10)$$
$$\mathbf{N_B = 30}$$

(c) $N_B + N_R + N_W = 65$

$$(30) + (10) + N_W = 65$$
$$\mathbf{N_W = 25}$$

2. (18)
$$B = \frac{kWD^2}{L}$$
$$2000 = \frac{k(2)(8)^2}{16}$$
$$k = 250$$
$$B = \frac{250WD^2}{L} = \frac{250(8)(2)^2}{16} = \mathbf{500\ lb}$$

3. (25)
$$R_BT_B + R_ZT_Z = \text{jobs}$$
$$\frac{3}{8}(4) + R_Z(4) = 6$$
$$4R_Z = \frac{9}{2}$$
$$R_Z = \mathbf{\frac{9}{8}\frac{jobs}{hr}}$$

4. (6)
$$T = \sqrt{\frac{L}{4}}$$
$$3T = 3\sqrt{\frac{L}{4}} = \sqrt{\frac{3^2L}{4}} = \sqrt{\frac{9L}{4}}$$

She must **multiply it by 9.**

5. (35)
$$-3 + \frac{3}{5}(\Delta C) = -3 + \frac{3}{5}(C_F - C_I)$$
$$= -3 + \frac{3}{5}[9 - (-3)] = -3 + \frac{36}{5} = \mathbf{\frac{21}{5}}$$

6. (35)
$$-3\frac{5}{6} + \frac{7}{8}\left[2\frac{1}{6} - \left(-3\frac{5}{6}\right)\right] = -\frac{23}{6} + \frac{7}{8}(6)$$
$$= -\frac{23}{6} + \frac{21}{4} = \mathbf{\frac{17}{12}}$$

7. (35)
$$D = \sqrt{(x_2 - x_1)^2 + (y_2 - y_1)^2}$$
$$= \mathbf{\sqrt{(x - 3)^2 + (y + 2)^2}}$$

8. (34)
$$\sum_{j=5}^{6} \frac{2}{j} = \frac{2}{5} + \frac{2}{6} = \frac{12}{30} + \frac{10}{30} = \mathbf{\frac{11}{15}}$$

9. (34)
$$\sum_{n=0}^{3} \frac{3^n}{n+1} = \frac{3^0}{0+1} + \frac{3^1}{1+1} + \frac{3^2}{2+1} + \frac{3^3}{3+1}$$
$$= 1 + \frac{3}{2} + 3 + \frac{27}{4} = \mathbf{\frac{49}{4}}$$

10. (34) From the graph of the line we pick two ordered pairs: (90, 500) and (100, 50).

$$m = \frac{500 - 50}{90 - 100} = -45$$
$$Y = -45B + b$$
$$500 = -45(90) + b$$
$$b = 4550$$
$$\mathbf{Y = -45B + 4550}$$

11. (34) $\mathbf{f(x) = x^2}$

12. (33) A property of a rhombus is that it has congruent consecutive sides. We are not told anything about the angles or diagonals, so we do not know if the parallelogram is a square. Therefore the answer is **C.**

13. (33)
$$A_{\text{trapezoid}} = \frac{1}{2}(H)(B_1 + B_2)$$
$$100 = \frac{1}{2}(10)(2x + 2 + x)$$
$$100 = 10x + 10 + 5x$$
$$90 = 15x$$
$$x = \mathbf{6}$$

14. (32)
$$y = \frac{5}{6}x + \frac{4}{3}$$
$$-\frac{5}{6}x = -y + \frac{4}{3}$$
$$x = \frac{6}{5}y - \frac{8}{5}$$

Interchange x and y.

$$\mathbf{y = \frac{6}{5}x - \frac{8}{5}}$$

15. (32)

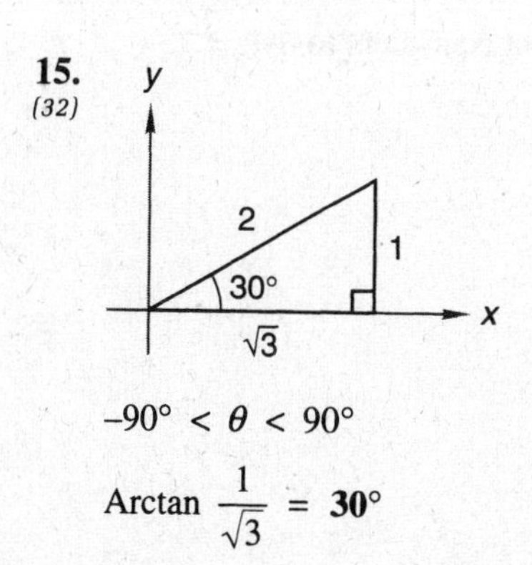

$$-90° < \theta < 90°$$
$$\text{Arctan}\ \frac{1}{\sqrt{3}} = \mathbf{30°}$$

16. (32)

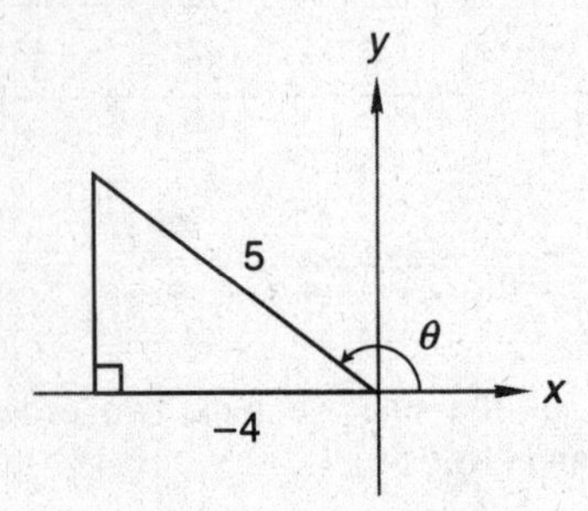

$0° \le \theta \le 180°$

$$\cos\left[\text{Arccos}\left(-\frac{4}{5}\right)\right] = \cos\theta = \mathbf{-\frac{4}{5}}$$

17. (32)

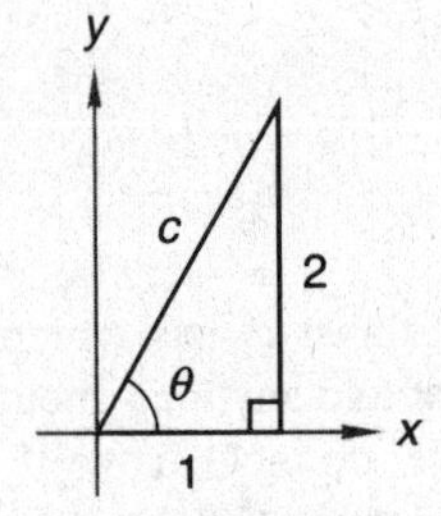

$-90° < \theta < 90°$

$$c = \sqrt{1^2 + 2^2} = \sqrt{5}$$

$$\sin(\text{Arctan } 2) = \sin\theta = \frac{2}{\sqrt{5}} = \mathbf{\frac{2\sqrt{5}}{5}}$$

18. (31) (a) **x axis, no.**

$-y = -x^2$ is not equivalent to $y = -x^2$.

y axis, yes.

$y = -(-x)^2$ is equivalent to $y = -x^2$.

Origin, no.

$-y = -(-x)^2$ is not equivalent to $y = -x^2$.

(b) **x axis, yes.**

$|-y| = x$ is equivalent to $|y| = x$.

y axis, no.

$|y| = -x$ is not equivalent to $|y| = x$.

Origin, no.

$|-y| = -x$ is not equivalent to $|y| = x$.

19. (31) Replace y with $-y$.

$$-y = \sqrt{x}$$

$$\mathbf{y = -\sqrt{x}}$$

20. (31) Replace x with $x - 3$.

$$\mathbf{y = (x - 3)^3}$$

21. (30) $-2\underline{/120°} - 3\underline{/135°} = -2\underline{/120°} + (-3\underline{/135°})$

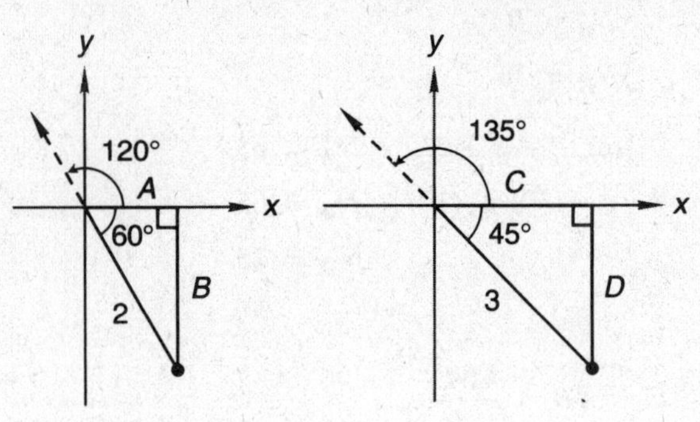

$A = 2\cos 60° = 1$

$B = 2\sin 60° = 1.7321$

$C = 3\cos 45° = 2.1213$

$D = 3\sin 45° = 2.1213$

$$\text{Resultant} = (1 + 2.1213)\hat{i} + (-1.7321 - 2.1213)\hat{j}$$

$$= 3.1213\hat{i} - 3.8534\hat{j}$$

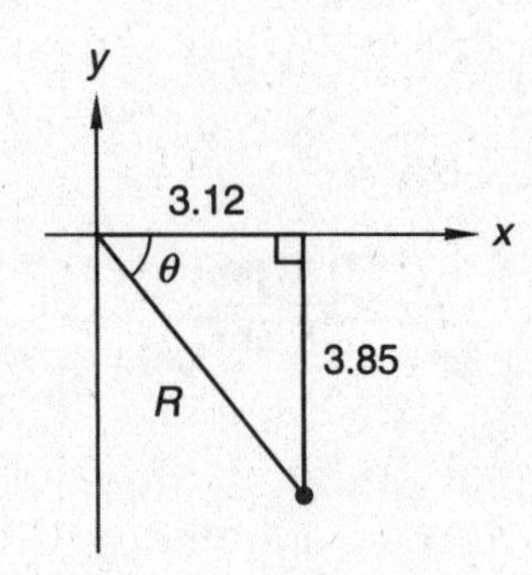

$$R = \sqrt{(3.1213)^2 + (3.8534)^2} = 4.96$$

$$\tan\theta = \frac{3.8534}{3.1213}$$

$$\theta = 50.99°$$

The polar angle is $360° - 50.99° = 309.01°$.

$\mathbf{4.96\underline{/309.01°}}$

22. (26) (a) $\log_a \frac{64}{27} = 3$

$$a^3 = \frac{64}{27}$$

$$a^3 = \frac{4^3}{3^3}$$

$$a^3 = \left(\frac{4}{3}\right)^3$$

$$a = \mathbf{\frac{4}{3}}$$

(b) $\log_9 \dfrac{1}{27} = n$

$$9^n = \frac{1}{27}$$

$$9^n = \frac{1}{3^3}$$

$$9^n = \frac{1}{9^{\frac{3}{2}}}$$

$$n = -\frac{3}{2}$$

(c) $\log_{\frac{1}{3}} m = -3$

$$\left(\frac{1}{3}\right)^{-3} = m$$

$$\frac{1}{3^{-3}} = m$$

$$m = 27$$

23. (21) (a) **Domain** $= \{x \in \mathbb{R} \mid -4 \leq x \leq 6\}$;

Range $= \{y \in \mathbb{R} \mid -2 \leq y \leq 3\}$

(b) **Domain** $= \{x \in \mathbb{R} \mid -4 \leq x \leq 4\}$;

Range $= \{y \in \mathbb{R} \mid -2 \leq y \leq 2\}$

24. (24,29)

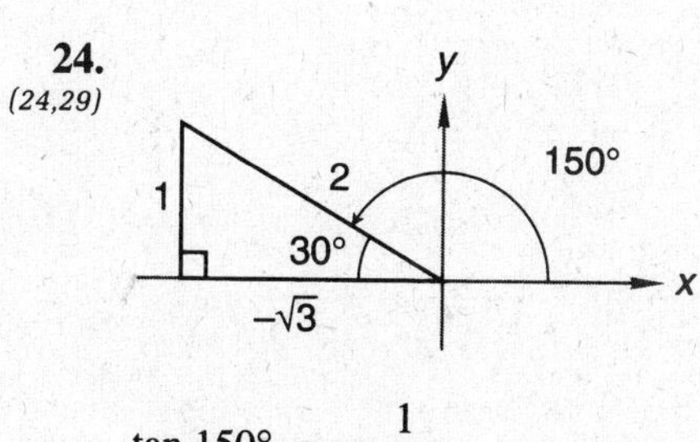

$$\tan 150° = -\frac{1}{\sqrt{3}}$$

$$\cos(-90°) = 0$$

$$f(150°) - g(-90°) = \tan 150° - \cos(-90°)$$

$$= -\frac{1}{\sqrt{3}} - 0 = -\frac{\sqrt{3}}{3}$$

25. (24) (a) $(f \circ g)(x) = 4\left[\frac{1}{4}(x - 5)\right] + 5$

$$= x - 5 + 5 = x$$

(b) $(g \circ f)(x) = \frac{1}{4}(4x + 5 - 5) = \frac{1}{4}(4x) = x$

26. (21) $f(x + h) - f(x)$

$$= \frac{x + h + 1}{x + h} - \frac{x + 1}{x}$$

$$= \frac{x(x + h + 1) - (x + 1)(x + h)}{x(x + h)}$$

$$= \frac{x^2 + xh + x - x^2 - x - xh - h}{x(x - h)}$$

$$= \frac{-h}{x(x + h)}$$

27. (11)

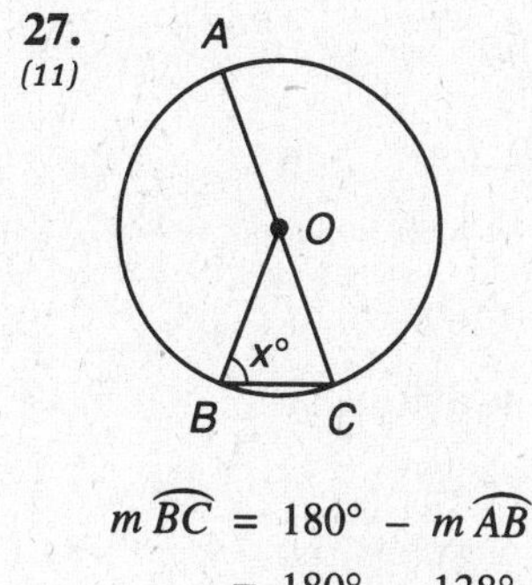

$$m\widehat{BC} = 180° - m\widehat{AB}$$

$$= 180° - 138° = 42°$$

$$m\angle BOC = m\widehat{BC} = 42°$$

$$42 + x + x = 180$$

$$2x = 138$$

$$x = 69$$

28. (15)

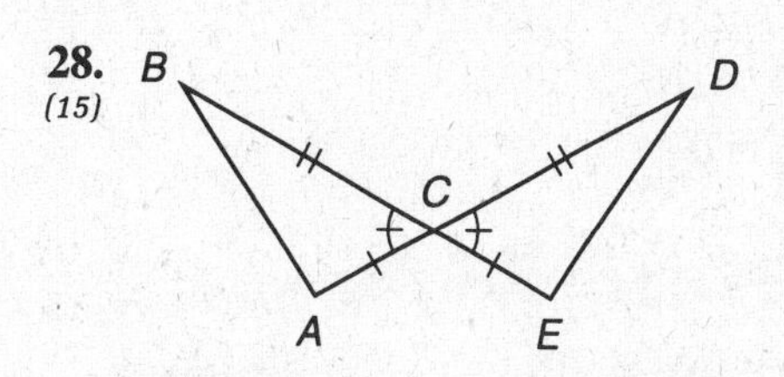

STATEMENTS	REASONS
1. $\overline{AC} \cong \overline{EC}$	1. Given
2. $\overline{BC} \cong \overline{DC}$	2. Given
3. $\angle BCA \cong \angle DCE$	3. Vertical angles are congruent.
4. $\triangle ABC \cong \triangle EDC$	4. *SAS* congruency postulate
5. $\overline{AB} \cong \overline{ED}$	5. *CPCTC*

29. (1) $(a - b) - (a + b) = -2b$

Since $b < 0$, $-2b > 0$.

Therefore the answer is **A**.

30. (1)

$$x > 1$$

$$x^4 > x^3$$

$$\frac{1}{x^4} < \frac{1}{x^3}$$

$$B < A$$

Therefore the answer is **A**.

Problem Set 36

1. (36) First we write the equations.

$$\begin{cases} \text{(a)}\ (P + W)T_D = D_D \\ \text{(b)}\ (P - W)T_U = D_U \end{cases}$$

$P = 4W,\ T_D = T_U + 1$

Next we use substitution to solve.

Upwind:

(b) $$\begin{aligned} (P - W)T_U &= D_U \\ (4W - W)T_U &= 300 \\ 3WT_U &= 300 \\ WT_U &= 100 \end{aligned}$$

Downwind:

(a) $$\begin{aligned} (P + W)T_D &= D_D \\ (4W + W)(T_U + 1) &= 800 \\ 5WT_U + 5W &= 800 \\ 5(100) + 5W &= 800 \\ 5W &= 300 \\ W &= 60 \end{aligned}$$

$P = 4W = 4(60) = 240$

Plane = 240 mph; Wind = 60 mph

2. (18) First write the three equations.

$$\begin{cases} \text{(a)}\ 0.20N_W = N_R + 10 \\ \text{(b)}\ \frac{3}{5}N_B = 3N_R \\ \text{(c)}\ 0.10(N_B + N_R) + 94 = N_W \end{cases}$$

(a) $$\begin{aligned} 0.20N_W &= N_R + 10 \\ N_W &= 5N_R + 50 \end{aligned}$$

(b) $$\begin{aligned} \frac{3}{5}N_B &= 3N_R \\ N_B &= 5N_R \end{aligned}$$

Now we can use substitution to solve.

(c) $$\begin{aligned} 0.10(N_B + N_R) + 94 &= N_W \\ N_B + N_R + 940 &= 10N_W \\ N_B + N_R - 10N_W &= -940 \\ (5N_R) + N_R - 10(5N_R + 50) &= -940 \\ -44N_R &= -440 \\ \mathbf{N_R} &= \mathbf{10} \end{aligned}$$

$$\begin{aligned} N_W &= 5N_R + 50 \\ N_W &= 5(10) + 50 \\ \mathbf{N_W} &= \mathbf{100} \end{aligned} \qquad \begin{aligned} N_B &= 5N_R \\ N_B &= 5(10) \\ \mathbf{N_B} &= \mathbf{50} \end{aligned}$$

3. (5,10) $$\frac{2}{\sqrt{6}}x = 3\sqrt{2}$$

$$x = \frac{3\sqrt{12}}{2} = \frac{3 \cdot 2\sqrt{3}}{2} = \mathbf{3\sqrt{3}}$$

4. (28) Distance $= m$, rate $= p$, time $= \dfrac{m}{p}$

$$\text{New rate} = \frac{m}{\frac{m}{p} + 2} = \mathbf{\frac{mp}{m + 2p}\ \frac{mi}{hr}}$$

5. (5) $A_1 = l_1w_1$

$A_2 = l_2w_2 = (1.2l_1)(0.8w_1) = .96l_1w_1 = 0.96A_1$

$A_2 - A_1 = 0.96A_1 - 1.00A_1 = -0.04A_1$

Decreased by 4%.

6. (35) $$-3\frac{3}{7} + \frac{2}{9}(C_F - C_I) = -3\frac{3}{7} + \frac{2}{9}\left[4\frac{2}{7} - \left(-3\frac{3}{7}\right)\right]$$

$$= -\frac{24}{7} + \frac{2}{9}\left(\frac{30}{7} + \frac{24}{7}\right) = -\frac{24}{7} + \frac{2}{9}\left(\frac{54}{7}\right)$$

$$= -\frac{24}{7} + \frac{12}{7} = \mathbf{-\frac{12}{7}}$$

7. (35) $\mathbf{D = \sqrt{(x + 2)^2 + (y - 5)^2}}$

8. (34) $$\sum_{x=3}^{5} \frac{3}{x + 1} = \frac{3}{3 + 1} + \frac{3}{4 + 1} + \frac{3}{5 + 1}$$

$$= \frac{3}{4} + \frac{3}{5} + \frac{1}{2} = \frac{15}{20} + \frac{12}{20} + \frac{10}{20} = \mathbf{\frac{37}{20}}$$

9. (34) From the graph of the line we pick two ordered pairs: (52, 100) and (54, 80).

$$m = \frac{80 - 100}{54 - 52} = \frac{-20}{2} = -10$$

R = −10W + b

100 = −10(52) + b

b = 620

R = −10W + 620

10. (34) $\mathbf{f(x) = \sqrt{x}}$

$\mathbf{g(x) = x - 1}$

11. (33)

$$\text{Median} = \frac{1}{2}(B_1 + B_2)$$

$$5x - 5 = \frac{1}{2}[(2x + 6) + (6x - 2)]$$

$$5x - 5 = \frac{1}{2}(8x + 4)$$

$$5x - 5 = 4x + 2$$

$$x = \mathbf{7}$$

12. (32)

$$y = \frac{1}{3}x - \frac{7}{3}$$

$$-\frac{1}{3}x = -y - \frac{7}{3}$$

$$x = 3y + 7$$

Interchange x and y.

$$\mathbf{y = 3x + 7}$$

13. (32)

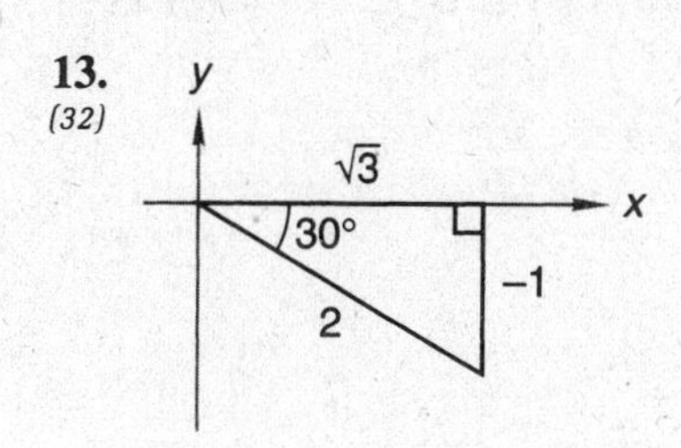

$$-90° < \theta < 90°$$

$$\text{Arctan}\left(-\frac{1}{3}\sqrt{3}\right) = \text{Arctan}\left(-\frac{1}{\sqrt{3}}\right) = \mathbf{-30°}$$

14. (32)

$$0° \le \theta \le 180°$$

$$b = \sqrt{7^2 - 3^2} = \sqrt{40} = 2\sqrt{10}$$

$$\sin\left(\text{Arccos}\,\frac{3}{7}\right) = \sin\theta = \frac{\mathbf{2\sqrt{10}}}{\mathbf{7}}$$

15. (31) (a) ***x* axis, no.**

$(-y)x = -1$ is not equivalent to $yx = -1$.

***y* axis, no.**

$y(-x) = -1$ is not equivalent to $yx = -1$.

Origin, yes.

$(-y)(-x) = -1$ is equivalent to $yx = -1$.

(b) ***x* axis, no.**

$(-y) = -|x|$ is not equivalent to $y = -|x|$.

***y* axis, yes.**

$y = -|-x|$ is equivalent to $y = -|x|$.

Origin, no.

$-y = -|-x|$ is not equivalent to $y = -|x|$.

16. (31) Replace x with $-x$.

$$\mathbf{y = \sqrt{-x}}$$

17. (31) Replace x with $x + 3$.

$$\mathbf{y = (x + 3)^3}$$

18. (30) $-3\underline{/-135°} + 4\underline{/140°}$

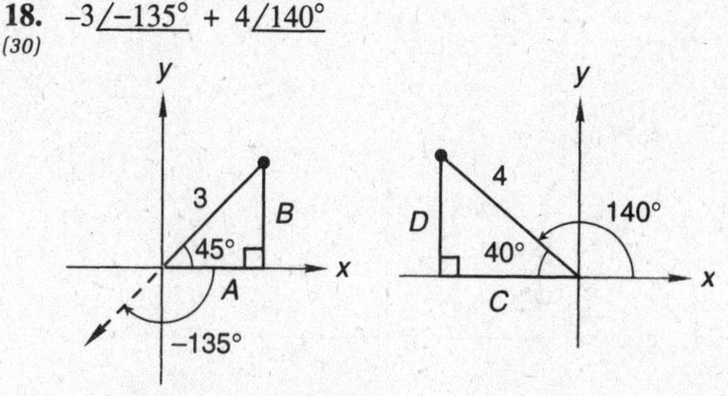

$$A = 3\cos 45° = 2.1213$$
$$B = 3\sin 45° = 2.1213$$
$$C = 4\cos 40° = 3.0642$$
$$D = 4\sin 40° = 2.5712$$

$$\text{Resultant} = (2.1213 - 3.0642)\hat{i} + (2.1213 + 2.5712)\hat{j} = -0.9429\hat{i} + 4.6925\hat{j}$$

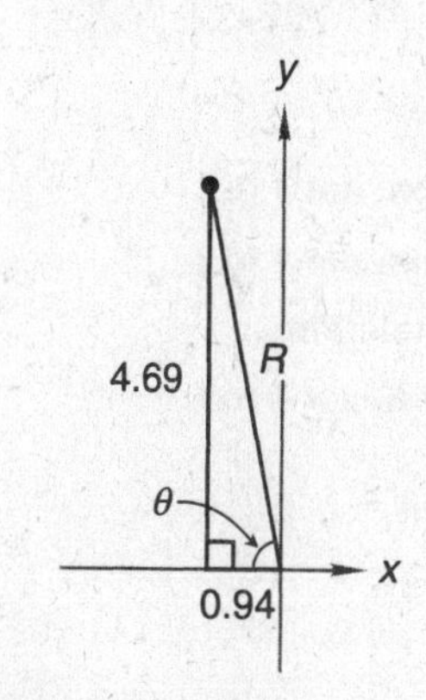

$$R = \sqrt{(0.9429)^2 + (4.6925)^2} = 4.79$$

$$\tan\theta = \frac{4.6925}{0.9429}$$

$$\theta = 78.64°$$

Since θ is in the second quadrant, the polar angle is $180° - 78.64° = 101.36°$.

Resultant $= 4.79\underline{/101.36°}$, so
Equilibrant $=$ $\mathbf{4.79\underline{/281.36°}}$

19. (28) $\dfrac{10!}{2!7!} = \dfrac{10 \cdot 9 \cdot 8 \cdot 7!}{2 \cdot 1 \cdot 7!} = 5 \cdot 9 \cdot 8 = \mathbf{360}$

20. (26) (a) $\log_n 27 = 2$

$n^2 = 27$

$n = \sqrt{27} = \mathbf{3\sqrt{3}}$

(b) $\log_2 \dfrac{\sqrt{2}}{64} = t$

$2^t = \dfrac{2^{\frac{1}{2}}}{2^6}$

$2^t = 2^{-\frac{11}{2}}$

$t = \mathbf{-\dfrac{11}{2}}$

(c) $\log_3 k = -2$

$k = 3^{-2} = \mathbf{\dfrac{1}{9}}$

21. (22) $g(x) = \dfrac{1}{\frac{1}{2}x^2 - 2} = \dfrac{2}{x^2 - 4}$

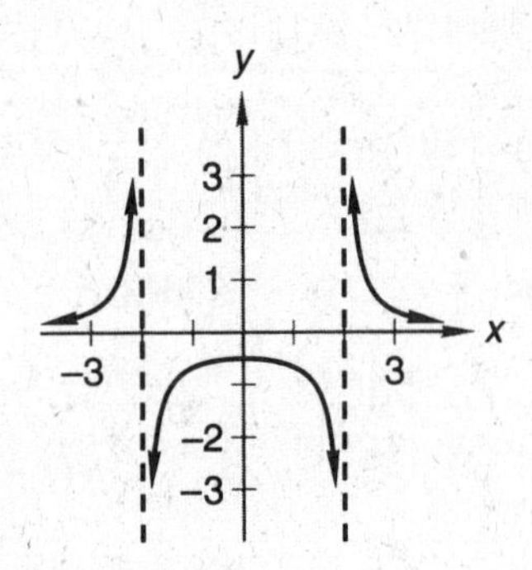

22. (21) (a) **Function, not 1 to 1**

(b) **Function, not 1 to 1**

(c) **Not a function**

(d) **Function, not 1 to 1**

23. (21) $9 - 3x \geq 0$

$9 \geq 3x$

$x \leq 3$

$2x^2 + 9x + 10 \neq 0$

$(2x + 5)(x + 2) \neq 0$

$2x + 5 \neq 0 \qquad x + 2 \neq 0$

$x \neq -\dfrac{5}{2} \qquad x \neq -2$

$\left\{\mathbf{x \in \mathbb{R} \mid x \leq 3, x \neq -\dfrac{5}{2}, -2}\right\}$

24. (36) (a)

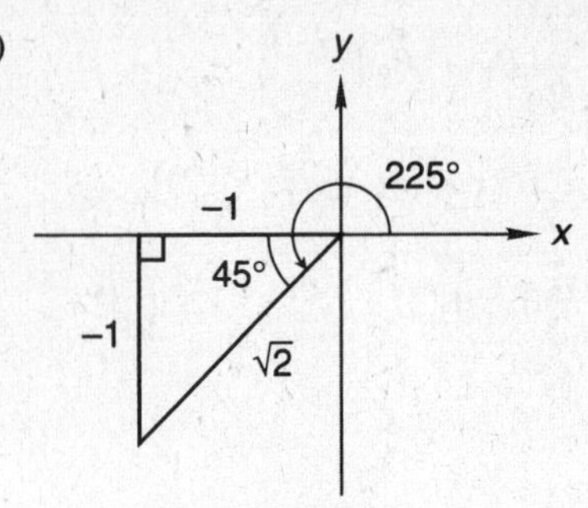

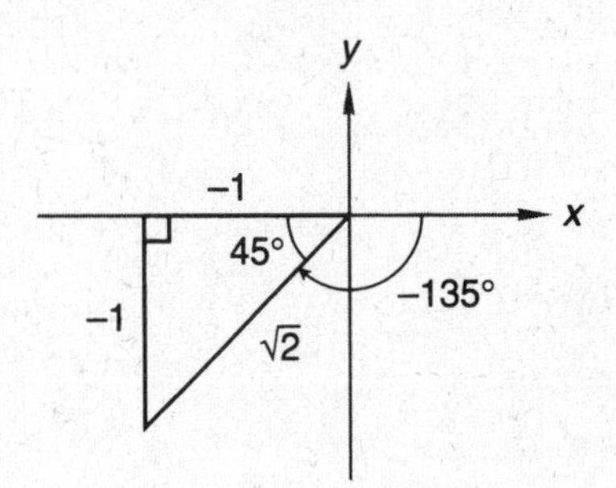

$\cos 225° + \tan(-135°) = -\cos 45° + \tan 45°$

$= -\dfrac{\sqrt{2}}{2} + 1 = \mathbf{1 - \dfrac{\sqrt{2}}{2}}$

(b)

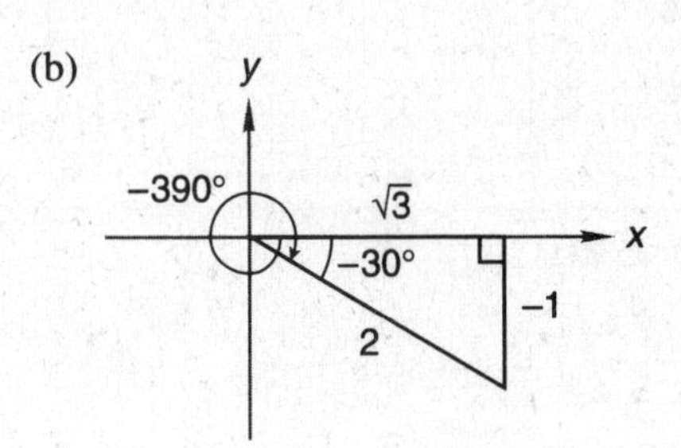

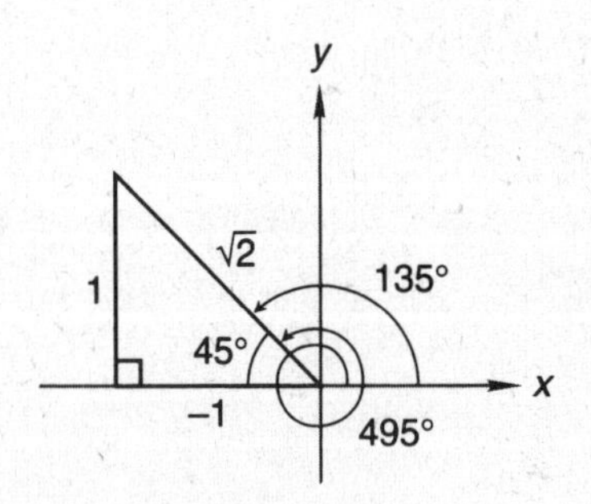

$\sin(-390°) - \cos 495°$

$= \sin(-30°) - \cos 135°$

$= -\sin 30° - (-\cos 45°)$

$= -\dfrac{1}{2} - \left(-\dfrac{\sqrt{2}}{2}\right) = \mathbf{\dfrac{-1 + \sqrt{2}}{2}}$

25. (29,36) $f(90°) - g(420°) = \cos 90° - \tan 420°$

$= \cos 90° - \tan 60° = 0 - \sqrt{3} = \mathbf{-\sqrt{3}}$

26. (24) (a) $\left(\frac{f}{g}\right)(x) = \frac{\frac{1}{x}}{x^3 + 1}$

$$\left(\frac{f}{g}\right)(2) = \frac{\frac{1}{2}}{2^3 + 1} = \frac{1}{2(9)} = \mathbf{\frac{1}{18}}$$

(b) $(f \circ g)(x) = \frac{1}{x^3 + 1}$

$$(f \circ g)(2) = \frac{1}{2^3 + 1} = \mathbf{\frac{1}{9}}$$

(c) $(g \circ f)(x) = \left(\frac{1}{x}\right)^3 + 1$

$$(g \circ f)(2) = \left(\frac{1}{2}\right)^3 + 1 = \frac{1}{8} + 1 = \mathbf{\frac{9}{8}}$$

27. (21) $f(x + h) - f(x)$

$$= [(x + h) - 1]^2 - (x - 1)^2$$
$$= (x^2 + 2xh + h^2) - 2(x + h) + 1 - (x^2 - 2x + 1)$$
$$= x^2 + 2xh + h^2 - 2x - 2h + 1 - x^2 + 2x - 1$$
$$= \mathbf{h^2 + 2xh - 2h}$$

28. (30)

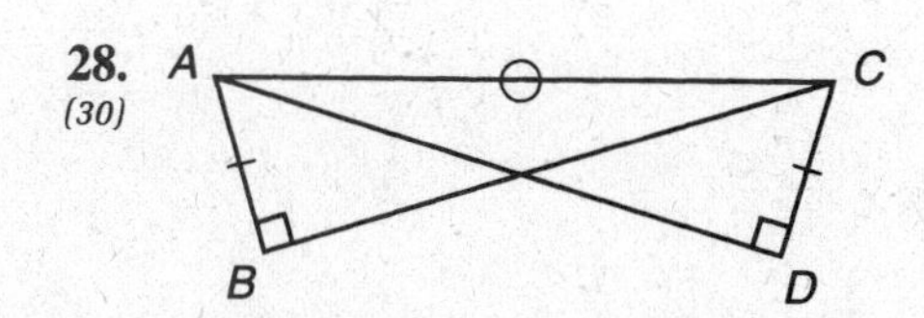

STATEMENTS	REASONS
1. $\overline{AB} \perp \overline{BC}$, $\overline{CD} \perp \overline{DA}$	1. Given
2. $\angle ABC$ and $\angle CDA$ are right angles.	2. Perpendicular lines intersect to form right angles.
3. $\triangle ABC$ and $\triangle CDA$ are right triangles.	3. A triangle which contains a right angle is a right triangle.
4. $\overline{AC} \cong \overline{AC}$	4. Reflexive axiom
5. $\overline{AB} \cong \overline{CD}$	5. Given
6. $\triangle ABC \cong \triangle CDA$	6. *HL* congruency postulate
7. $\overline{BC} \cong \overline{DA}$	7. *CPCTC*

29. (1) $(x + y)^2 = x^2 + 2xy + y^2$

$$= (x^2 + y^2) + 2(xy)$$
$$= 8 + 2(4) = 16$$

Therefore the answer is **C.**

30. (1,33)

$m\angle AEB = a°$

$AB = AE$

$BC = ED$

$AD = AE + ED = AB + BC$

Therefore the answer is **C.**

Problem Set 37

1. (36) First we write the equations.

$$\begin{cases} \text{(a)}\ (P - W)T_U = D_U \\ \text{(b)}\ (P + W)T_D = D_D \end{cases}$$

$P = 6W$, $T_U = T_D + 3$

Next we use substitution to solve.

Downwind:

(b) $(P + W)T_D = D_D$
$$(6W + W)T_D = 560$$
$$7WT_D = 560$$
$$WT_D = 80$$

Upwind:

(a) $(P - W)T_U = D_U$
$$(6W - W)(T_D + 3) = 700$$
$$5WT_D + 15W = 700$$
$$5(80) + 15W = 700$$
$$15W = 300$$
$$W = 20$$

$P = 6W = 6(20) = 120$

Plane = 120 mph; Wind = 20 mph

2. (25) $R = \frac{2000}{2000 + 3000 + 5000} = \frac{2000}{10{,}000} = 0.2$

$x = R(1920) = (0.2)(1920) = \mathbf{\$384}$

3. (18) 50% of $\frac{3}{4} = \frac{1}{2}\left(\frac{3}{4}\right) = \mathbf{\frac{3}{8}}$

4. (R) $RWT = \text{dollars}$

$$R(10)(15) = 6000$$
$$R = 40\ \frac{\text{dollars}}{\text{people-days}}$$
$$RWT = \text{dollars}$$
$$(40)(20)(5) = \mathbf{\$4000}$$

5. (28) Distance $= k$, rate $= m$, time $= \frac{k}{m}$

$$\text{New rate} = \frac{\text{new distance}}{\text{new time}}$$

$$= \frac{k + 10}{\frac{k}{m} - 2} = \mathbf{\frac{mk + 10m}{k - 2m}\ \frac{mi}{hr}}$$

6. (37) $\sqrt{(x - 3)^2 + (y - 2)^2} = \sqrt{(x + 4)^2 + (y + 3)^2}$

$$(x - 3)^2 + (y - 2)^2 = (x + 4)^2 + (y + 3)^2$$

$$(x^2 - 6x + 9) + (y^2 - 4y + 4)$$
$$= (x^2 + 8x + 16) + (y^2 + 6y + 9)$$
$$-6x - 4y + 13 = 8x + 6y + 25$$
$$10y = -14x - 12$$
$$\mathbf{y = -\frac{7}{5}x - \frac{6}{5}}$$

7. (37) $x = \frac{x_1 + x_2}{2} = \frac{-4 + 3}{2} = -\frac{1}{2}$

$$y = \frac{y_1 + y_2}{2} = \frac{7 + 2}{2} = \frac{9}{2}$$

$$\mathbf{\left(-\frac{1}{2}, \frac{9}{2}\right)}$$

8. (35) $-2\frac{2}{5} + \frac{3}{7}\left[-3\frac{1}{5} - \left(-2\frac{2}{5}\right)\right]$

$$= -\frac{12}{5} + \frac{3}{7}\left(-\frac{16}{5} + \frac{12}{5}\right)$$

$$= -\frac{12}{5} - \frac{12}{35} = \mathbf{-\frac{96}{35}}$$

9. (34) $\sum_{k=2}^{5}(k^2 - 2)$

$$= (2^2 - 2) + (3^2 - 2) + (4^2 - 2) + (5^2 - 2)$$
$$= 2 + 7 + 14 + 23 = \mathbf{46}$$

10. (34) From the graph of the line we pick two ordered pairs: (28, 80) and (34, 120).

$$m = \frac{120 - 80}{34 - 28} = \frac{40}{6} = 6.67$$

$$O = 6.67I + b$$
$$120 = 6.67(34) + b$$
$$b = -106.7$$
$$\mathbf{O = 6.67I - 106.7}$$

11. (34) $f(x) = \sqrt{x}$

12. (33) A property of rectangles and squares is that all angles are right angles. We are not told anything about the sides or diagonals so we do not know if the figure is a square. Therefore the answer is **A.**

13. (33) $A_{\text{trapezoid}} = \frac{1}{2}(H)(B_1 + B_2)$

$$180 = \frac{1}{2}(15)[(x + 3) + (2x + 3)]$$
$$180 = \frac{15}{2}(3x + 6)$$
$$24 = 3x + 6$$
$$3x = 18$$
$$\mathbf{x = 6}$$

14. (32) $y = 4x + \frac{2}{3}$

$$-4x = -y + \frac{2}{3}$$
$$x = \frac{1}{4}y - \frac{1}{6}$$

Interchange x and y.

$$\mathbf{y = \frac{1}{4}x - \frac{1}{6}}$$

15. (32) (a) $-90° < \theta < 90°$

$$\tan 0° = 0$$
$$\text{Arctan } 0 = \mathbf{0°}$$

(b)

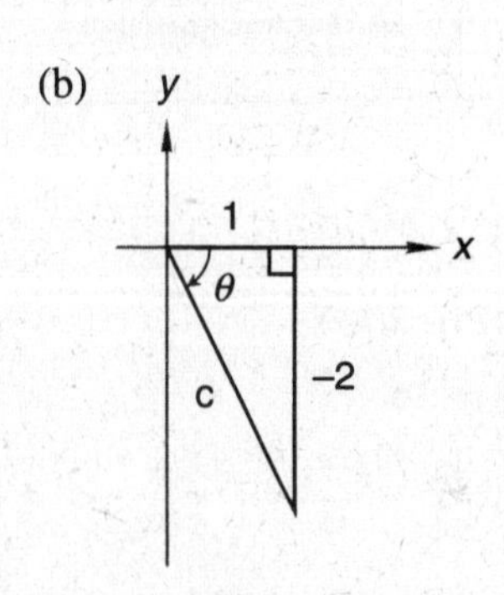

$$-90° < \theta < 90°$$
$$c = \sqrt{1^2 + (-2)^2} = \sqrt{5}$$
$$\cos[\text{Arctan}(-2)] = \cos\theta = \frac{1}{\sqrt{5}} = \mathbf{\frac{\sqrt{5}}{5}}$$

16. (31) (a) **x axis, no.**

$(-y) = x^3 - x$ is not equivalent to $y = x^3 - x$.

y axis, no.

$y = (-x)^3 - (-x)$ is not equivalent to $y = x^3 - x$.

Origin, yes.

$(-y) = (-x)^3 - (-x)$ is equivalent to $y = x^3 - x$.

(b) **x axis, yes.**

$(-y)^2 = -x$ is equivalent to $y^2 = -x$.

y axis, no.

$y^2 = -(-x)$ is not equivalent to $y^2 = -x$.

Origin, no.

$(-y)^2 = -(-x)$ is not equivalent to $y^2 = -x$.

17. (31) Replace y with $-y$.

$$-y = 3x$$
$$\mathbf{y = -(3^x)}$$

18. (31) Replace y with $y - 3$.

$$y - 3 = 4^x$$
$$\mathbf{y = 4^x + 3}$$

19. (26) (a) $\log_m 32 = 5$

$$m^5 = 32$$
$$m = \sqrt[5]{32} = \mathbf{2}$$

(b) $\log_3 \frac{\sqrt{3}}{81} = n$

$$3^n = \frac{3^{\frac{1}{2}}}{3^4}$$
$$3^n = 3^{-\frac{7}{2}}$$
$$n = \mathbf{-\frac{7}{2}}$$

(c) $\log_{\frac{1}{2}} t = -\frac{3}{2}$

$$t = \left(\frac{1}{2}\right)^{-\frac{3}{2}}$$
$$t = \frac{1}{2^{-\frac{3}{2}}} = 2^{\frac{3}{2}} = 8^{\frac{1}{2}} = \mathbf{2\sqrt{2}}$$

20. (21) (a) **Domain** $= \{x \in \mathbb{R} \mid -3 \le x \le 5\}$

Range $= \{y \in \mathbb{R} \mid 1 \le y \le 7\}$

(b) **Domain** $= \{x \in \mathbb{R} \mid -2 \le x \le 2\}$

Range $= \{y \in \mathbb{R} \mid -8 \le y \le 8\}$

21. (36) (a) $\sin(-390°) + \cos 495°$

$$= \sin(-30°) + \cos 135°$$
$$= (-\sin 30°) + (-\cos 45°)$$
$$= -\frac{1}{2} - \frac{\sqrt{2}}{2} = \mathbf{\frac{-1 - \sqrt{2}}{2}}$$

(b) $\tan(-495°) - \sin 225°$

$$= \tan(-135°) - \sin 225°$$
$$= \tan 45° - (-\sin 45°)$$
$$= 1 + \frac{\sqrt{2}}{2} = \mathbf{\frac{2 + \sqrt{2}}{2}}$$

22. (24,26) $(g \circ f)(x) = \log_{10} 10^x$

(a) $(g \circ f)(2) = \log_{10} 10^2 = \mathbf{2}$

(b) $(g \circ f)(3) = \log_{10} 10^3 = \mathbf{3}$

(c) $(g \circ f)(4) = \log_{10} 10^4 = \mathbf{4}$

23. (19) First we label the equations.

$$\begin{cases} x^2 + y^2 = 4 & \text{(a)} \\ 2x + y = 2 & \text{(b)} \end{cases}$$

(b) $2x + y = 2$

$$y = 2 - 2x$$

Next we use substitution to solve.

(a)
$$x^2 + y^2 = 4$$
$$x^2 + (2 - 2x)^2 = 4$$
$$x^2 + (4 - 8x + 4x^2) = 4$$
$$5x^2 - 8x = 0$$
$$x(5x - 8) = 0$$
$$x = 0, \frac{8}{5}$$

If $x = 0$, then $y = 2 - 2x = 2 - 2(0) = 2$.

If $x = \frac{8}{5}$, then $y = 2 - 2\left(\frac{8}{5}\right) = -\frac{6}{5}$.

$\mathbf{(0, 2), \left(\frac{8}{5}, -\frac{6}{5}\right)}$

24. (6)

$$\frac{x}{x^2 - 1} - \frac{1}{x^2 - x} = \frac{1}{x^3 - x}$$

$$\frac{x}{(x + 1)(x - 1)} - \frac{1}{x(x - 1)} = \frac{1}{x(x - 1)(x + 1)}$$

$$x(x + 1)(x - 1)\left[\frac{x}{(x + 1)(x - 1)} - \frac{1}{x(x - 1)}\right]$$

$$= x(x + 1)(x - 1)\left[\frac{1}{x(x - 1)(x + 1)}\right]$$

$$x^2 - x - 1 = 1$$

$$x^2 - x - 2 = 0$$

$$(x - 2)(x + 1) = 0$$

Therefore $x = 2, -1$ but, checking these solutions in the original equation we find $x \neq -1$, so $x =$ **2.**

25. (16)

$$\begin{array}{r} x^3 + x + 1 \\ x - 1 \overline{)\, x^4 - x^3 + x^2 + 0x - 1} \\ \underline{x^4 - x^3} \\ 0x^3 + x^2 + 0x \\ \underline{x^2 - x} \\ x - 1 \\ \underline{x - 1} \\ 0 \end{array}$$

$$\frac{x^4 - x^3 + x^2 - 1}{x - 1} = \mathbf{x^3 + x + 1}$$

Check: $(x - 1)(x^3 + x + 1)$

$$= x^4 + x^2 + x - x^3 - x - 1$$

$$= x^4 - x^3 + x^2 - 1$$

26. (21)

$$\frac{f(x + h) - f(x)}{h} = \frac{(x + h)^2 + 1 - (x^2 + 1)}{h}$$

$$= \frac{x^2 + 2xh + h^2 + 1 - x^2 - 1}{h} = \frac{2xh + h^2}{h}$$

$$= \mathbf{2x + h}$$

27. (30)

Q R P S

STATEMENTS	REASONS
1. $\overline{PQ} \cong \overline{SR}$	1. Given
2. $\overline{QS} \cong \overline{RP}$	2. Given
3. $\overline{PS} \cong \overline{PS}$	3. Reflexive axiom
4. $\Delta PQS \cong \Delta SRP$	4. *SSS* congruency postulate
5. $\angle PQS \cong \angle SRP$	5. *CPCTC*

28. (2) $A_{\text{surface}} = 4\pi r^2$

$$4\pi(r_1)^2 = 1764\pi$$

$$(r_1)^2 = 441$$

$$r_1 = 21$$

$$4\pi(r_2)^2 = 1764\pi - 468\pi$$

$$4\pi(r_2)^2 = 1296\pi$$

$$(r_2)^2 = 324$$

$$r_2 = 18$$

$$V_{\text{sphere}} = \frac{4}{3}\pi r^3$$

$$V_1 - V_2 = \frac{4}{3}\pi(r_1)^3 - \frac{4}{3}\pi(r_2)^3$$

$$= \frac{4}{3}\pi(21)^3 - \frac{4}{3}\pi(18)^3$$

$$= 12{,}348\pi - 7776\pi$$

$$= \mathbf{4572\pi\ cm^3 = 14{,}363.36\ cm^3}$$

29. (1) Since $0 < x < 1$,

$$x^2 > x^3$$

$$\frac{1}{x^2} < \frac{1}{x^3}$$

Therefore the answer is **B.**

30. (1,12) Interior angles $= (N - 2)180° = (5 - 2)180° = 540°$

$$m\angle 1 + m\angle 2 + m\angle 3 + m\angle 4 + m\angle 5 = 540°$$

$$m\angle 1 + m\angle 2 + m\angle 3 + m\angle 4 + 90° = 540°$$

$$m\angle 1 + m\angle 2 + m\angle 3 + m\angle 4 = 450°$$

Therefore the answer is **A.**

Problem Set 38

1. (36) First we write the equations.

$$\begin{cases} \text{(a)} & (B + C)T_D = D_D \\ \text{(b)} & (B - C)T_U = D_U \end{cases}$$

$$B = 4C,\ T_U = T_D - 1$$

Next we use substitution to solve.

Downstream:

(a) $(B + C)T_D = D_D$

$$(4C + C)T_D = 60$$

$$5CT_D = 60$$

$$CT_D = 12$$

Upstream:

(b) $(B - C)T_U = D_U$

$$(4C - C)(T_D - 1) = 27$$
$$3CT_D - 3C = 27$$
$$3(12) - 3C = 27$$
$$3C = 9$$
$$C = 3$$

$B = 4C = 4(3) = 12$

Boat = 12 mph; Current = 3 mph

2. (28) $\dfrac{\text{Total distance}}{\text{Total time}} = \mathbf{\dfrac{RT + 100}{T + P}\ \dfrac{mi}{hr}}$

3. (38) $\text{Overall overage rate} = \dfrac{\text{overall distance}}{\text{overall time}}$

$$40 \text{ mph} = \frac{300 \text{ mi} + 200 \text{ mi} + 500 \text{ mi}}{\text{overall time}}$$
$$40 \text{ mph} = \frac{1000 \text{ mi}}{\text{overall time}}$$
$$\text{Overall time} = 25 \text{ hr}$$

$$T_1 = \frac{D_1}{R_1} = \frac{300 \text{ mi}}{60 \text{ mph}} = 5 \text{ hr}$$

$$T_2 = \frac{D_2}{R_2} = \frac{200 \text{ mi}}{20 \text{ mph}} = 10 \text{ hr}$$

$$T_3 = \text{overall time} - T_2 - T_1$$
$$= 25 \text{ hr} - 10 \text{ hr} - 5 \text{ hr} = 10 \text{ hr}$$

$$\text{Rate of third leg} = R_3 = \frac{D_3}{T_3} = \frac{500 \text{ mi}}{10 \text{ hr}} = \mathbf{50 \text{ mph}}$$

4. (18) First we write the equations.

(a) $\begin{cases} P_N + D_N = 20 \\ 0.9P_N + 0.75D_N = 0.78(20) \end{cases}$ (b)

(a) $P_N + D_N = 20$

$$P_N = 20 - D_N$$

Next we use substitution to solve.

(b) $0.9P_N + 0.75D_N = 0.78(20)$

$$0.9(20 - D_N) + 0.75D_N = 15.6$$
$$0.15D_N = 2.4$$
$$D_N = \mathbf{16 \text{ liters of } 75\%}$$

(a) $P_N + D_N = 20$

$$P_N + (16) = 20$$
$$P_N = \mathbf{4 \text{ liters of } 90\%}$$

5. (25) $R_W = \dfrac{1}{4 + 6} = \dfrac{1}{10}$

$$R_W T_W + R_C T_C = \text{jobs}$$
$$\frac{1}{10}(4 + 2) + R_C(2) = 1$$
$$R_C = \frac{1}{5}\ \frac{\text{job}}{\text{hr}}$$

It would take Sally **5 hr** to do one job.

6. (38) $6 \cdot 5 \cdot 4 = \mathbf{120}$

7. (38) $5 \cdot 5 \cdot 5 \cdot 5 \cdot 5 = \mathbf{3125}$

8. (38) $3 \cdot 3 \cdot 3 \cdot 3 \cdot 3 = \mathbf{243}$

9. (38) $\left(x + \dfrac{2}{5}\right)\left(x - \dfrac{4}{3}\right) = 0$

$$x^2 + \frac{2}{5}x - \frac{4}{3}x - \frac{8}{15} = 0$$
$$x^2 + \frac{6}{15}x - \frac{20}{15}x - \frac{8}{15} = 0$$
$$\mathbf{x^2 - \frac{14}{15}x - \frac{8}{15} = 0}$$

10. (37) $\sqrt{(x - 4)^2 + (y - 2)^2} = \sqrt{(x + 4)^2 + (y - 3)^2}$

$$(x - 4)^2 + (y - 2)^2 = (x + 4)^2 + (y - 3)^2$$
$$(x^2 - 8x + 16) + (y^2 - 4y + 4)$$
$$= (x^2 + 8x + 16) + (y^2 - 6y + 9)$$
$$-8x - 4y + 20 = 8x - 6y + 25$$
$$2y = 16x + 5$$
$$\mathbf{y = 8x + \frac{5}{2}}$$

11. (37) $x = \dfrac{x_1 + x_2}{2} = \dfrac{1 + 6}{2} = \dfrac{7}{2}$

$$y = \frac{y_1 + y_2}{2} = \frac{3 + 5}{2} = 4$$

$$\mathbf{\left(\frac{7}{2}, 4\right)}$$

12. (34) $\displaystyle\sum_{k=-1}^{3}\left(\frac{k^2}{4} - k\right)$

$$= \left(\frac{(-1)^2}{4} - (-1)\right) + \left(\frac{0^2}{4} - 0\right) + \left(\frac{1^2}{4} - 1\right)$$

$$+ \left(\frac{2^2}{4} - 2\right) + \left(\frac{3^2}{4} - 3\right)$$

$$= \frac{5}{4} + 0 - \frac{3}{4} - 1 - \frac{3}{4} = \mathbf{-\frac{5}{4}}$$

13. (34) From the graph of the line we choose two ordered pairs: (100, 19.7) and (94.6, 28).

$$m = \frac{28 - 19.7}{94.6 - 100} = \frac{8.3}{-5.4} = -1.54$$

$$\text{V} = -1.54\text{K} + b$$
$$19.7 = (-1.54)(100) + b$$
$$b = 173.7$$

$$\mathbf{V = -1.54K + 173.7}$$

14. (34) $\mathbf{f(x) = x^3}$

$\mathbf{g(x) = 2x + 3}$

15. (33) Both the square and the rhombus have diagonals that are perpendicular bisectors of each other. Since we know nothing about sides or angles, we must choose the rhombus. Therefore the answer is **C.**

16. (33) $A_{\text{trapezoid}} = \frac{1}{2}(H)(B_1 + B_2)$

$$= \frac{1}{2}(4)(6 + 10) = \frac{(4)(16)}{2} = \mathbf{32\ cm^2}$$

17. (32) (a)

$-90° < \theta < 90°$

$\text{Arctan}(-1) = \mathbf{-45°}$

(b)

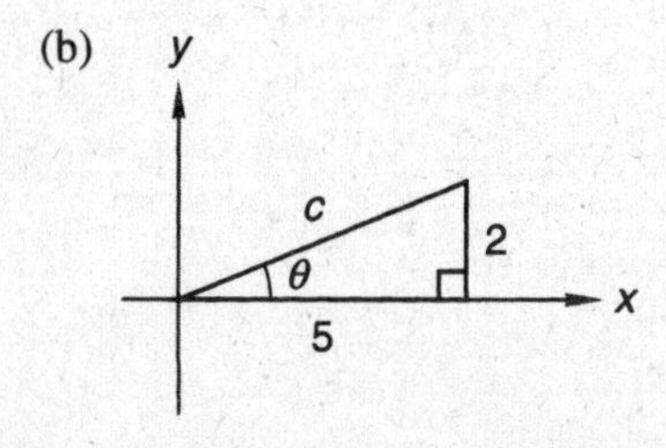

$-90° < \theta < 90°$

$c = \sqrt{5^2 + 2^2} = \sqrt{29}$

$$\sin\left(\text{Arctan}\,\frac{2}{5}\right) = \sin\theta = \frac{2}{\sqrt{29}} = \mathbf{\frac{2\sqrt{29}}{29}}$$

18. (31) Replace x with $-x$.

$\mathbf{y = 3^{-x}}$

19. (31) Replace y with $y + 3$.

$y + 3 = 4^x$

$\mathbf{y = 4^x - 3}$

20. (30) $4\underline{/-45°} + 16\underline{/330°}$

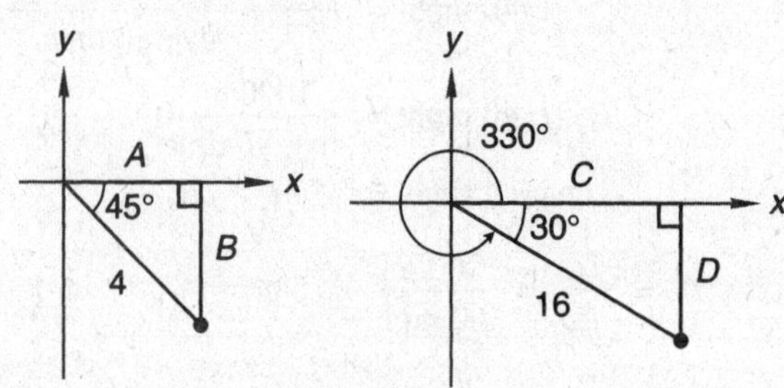

$A = 4\cos 45° = 2.8284$
$B = 4\sin 45° = 2.8284$
$C = 16\cos 30° = 13.8564$
$D = 16\sin 30° = 8$

$$\text{Resultant} = (2.8284 + 13.8564)\hat{i} + (-2.8284 - 8)\hat{j}$$
$$= 16.6848\hat{i} - 10.8284\hat{j}$$

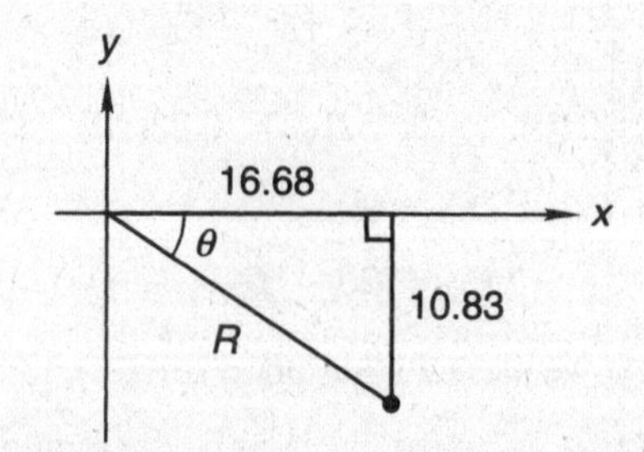

$$R = \sqrt{(16.6848)^2 + (10.8284)^2} = 19.89$$

$$\tan\theta = \frac{10.8284}{16.6848}$$
$$\theta = 32.98°$$

Since θ is in the fourth quadrant, the polar angle is $360° - 32.98° = 327.02°$.

19.89$\underline{/327.02°}$

21. (26) (a) $\log_k \frac{1}{8} = -3$

$$\frac{1}{8} = k^{-3}$$

$$\frac{1}{8} = \frac{1}{k^3}$$

$$k^3 = 8$$

$$k = \mathbf{2}$$

(b) $\log_2 \frac{\sqrt[3]{2}}{16} = s$

$$2^s = \frac{2^{\frac{1}{3}}}{2^4}$$

$$2^s = 2^{-\frac{11}{3}}$$

$$s = -\frac{\mathbf{11}}{\mathbf{3}}$$

(c) $\log_{\frac{1}{3}} r = -2$

$$r = \left(\frac{1}{3}\right)^{-2} = \frac{1}{3^{-2}} = 3^2 = \mathbf{9}$$

22. (21) $100 + x \geq 0$

$$x \geq -100$$

$$x^2 + 7x + 10 \neq 0$$

$$(x + 5)(x + 2) \neq 0$$

$$x \neq -5, -2$$

$$\{x \in \mathbb{R} \mid x \geq -100, x \neq -5, -2\}$$

23. (29,36) (a) $\sin(-390°) - \tan 495°$

$$= \sin(-30°) - \tan 135°$$

$$= -\sin 30° - (-\tan 45°)$$

$$= -\frac{1}{2} - (-1) = -\frac{1}{2} + 1 = \frac{\mathbf{1}}{\mathbf{2}}$$

(b) $\sin 90° \cos 135° - \tan 30°$

$$= \sin 90°(-\cos 45°) - \tan 30°$$

$$= (1)\left(-\frac{\sqrt{2}}{2}\right) - \frac{\sqrt{3}}{3} = -\frac{\sqrt{2}}{2} - \frac{\sqrt{3}}{3}$$

24. (24) (a) $(f \circ g)(x) = \frac{1}{7}[(7x + 2) - 2] = \frac{1}{7}(7x) = \mathbf{x}$

(b) $(g \circ f)(x) = 7\left[\frac{1}{7}(x - 2)\right] + 2$

$$= 7\left(\frac{1}{7}x\right) = \mathbf{x}$$

25. (19) First we label the equations.

$$\begin{cases} \text{(a)}\ 2y^2 - x^2 = 5 \\ \text{(b)}\ 2y^2 + x^2 = 11 \end{cases}$$

Then use elimination to solve.

$$\begin{array}{ll} \text{(a)} & 2y^2 - x^2 = 5 \\ \text{(b)} & \underline{2y^2 + x^2 = 11} \\ & 4y^2 \quad\quad = 16 \end{array}$$

$$y^2 = 4$$

$$y = \pm 2$$

(a) $2y^2 - x^2 = 5$

$$2(\pm 2)^2 - x^2 = 5$$

$$8 - x^2 = 5$$

$$x^2 = 3$$

$$x = \pm\sqrt{3}$$

$$(\sqrt{3}, 2), (\sqrt{3}, -2), (-\sqrt{3}, 2), (-\sqrt{3}, -2)$$

26. (16) $x = x_0 + v_0 t + \frac{1}{2}gt^2$

$$\frac{1}{2}gt^2 = x - x_0 - v_0 t$$

$$g = \frac{2(x - x_0 - v_0 t)}{t^2}$$

27. (21) $\frac{f(x + h) - f(x)}{h}$

$$= \frac{[(x + h)^2 - 2(x + h)] - (x^2 - 2x)}{h}$$

$$= \frac{x^2 + 2xh + h^2 - 2x - 2h - x^2 + 2x}{h}$$

$$= \frac{2xh + h^2 - 2h}{h} = \mathbf{h + 2x - 2}$$

28. (30,33)

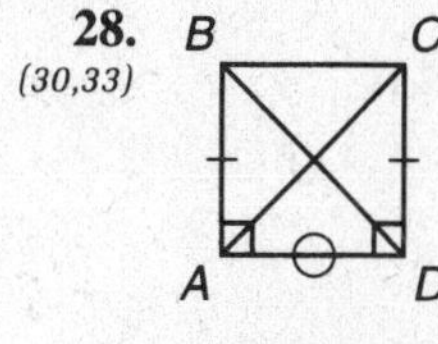

STATEMENTS	REASONS
1. *ABCD* is a square.	1. Given
2. $\angle BAD$ and $\angle CDA$ are right angles.	2. A square contains four right angles.
3. $\angle BAD \cong \angle CDA$	3. All right angles are congruent.
4. $\overline{AB} \cong \overline{DC}$	4. All sides of a square are congruent.
5. $\overline{AD} \cong \overline{AD}$	5. Reflexive axiom
6. $\triangle BAD \cong \triangle CDA$	6. *SAS* congruency postulate
7. $\overline{BD} \cong \overline{CA}$	7. *CPCTC*

29. If $x < 0$, then
(1)

$$\frac{1}{x} < 0 \text{ and } \frac{1}{x^2} > 0, \text{ so } \frac{1}{x} < \frac{1}{x^2}.$$

Therefore the answer is **B.**

30. First we label the equations.
(1)

$$\begin{aligned}&(a)\ \begin{cases}6x - y - 8 = 0\\ 4x - 3y - 7 = 0\end{cases}\end{aligned}$$
(b)

Then we use elimination to solve.

$$\begin{array}{lrcl}(b) & 4x - 3y - 7 &=& 0\\ -3(a) & -18x + 3y + 24 &=& 0\\ \hline & -14x \quad + 17 &=& 0\end{array}$$

$$-14x = -17$$

$$x = \frac{17}{14}$$

(a) $6x - y - 8 = 0$

$$6\left(\frac{17}{14}\right) - y - 8 = 0$$

$$y = 6\left(\frac{17}{14}\right) - 8$$

$$y = \frac{51}{7} - \frac{56}{7} = -\frac{5}{7}$$

$$x + y = \frac{17}{14} - \frac{5}{7} = \frac{17}{14} - \frac{10}{14} = \frac{7}{14} = \frac{1}{2}$$

$$\frac{1}{2} > \frac{1}{4}$$

A > B

Therefore the answer is **A.**

Problem Set 39

1. $35.5°\left(\frac{\pi}{180°}\right)\left(\frac{7920}{2}\right) = 781\pi = \mathbf{2453.58\ mi}$
(39)

2. $1.2°\left(\frac{\pi}{180°}\right)(1500) = 10\pi = \mathbf{31.42\ ft}$
(39)

3. $4 \cdot 3 \cdot 2 \cdot 1 = \mathbf{24}$
(38)

4. $7 \cdot 6 \cdot 5 = \mathbf{210}$
(38)

5. Overall average rate $= \dfrac{\text{overall distance}}{\text{overall time}}$
(38)

$$20 = \frac{60 + 20 + 20}{\text{overall time}}$$

Overall time = 5 hr

$$T_1 = \frac{D_1}{R_1} = \frac{60 \text{ mi}}{30 \text{ mph}} = 2 \text{ hr}$$

$$T_2 = \frac{D_2}{R_2} = \frac{20 \text{ mi}}{10 \text{ mph}} = 2 \text{ hr}$$

$$T_3 = \text{overall time} - T_2 - T_1$$
$$= 5 \text{ hr} - 2 \text{ hr} - 2 \text{ hr} = 1 \text{ hr}$$

$$R_3 = \frac{D_3}{T_3} = \frac{20}{1} = \mathbf{20\ mph}$$

6. $\dfrac{\text{Total distance}}{\text{Total time}} = \mathbf{\dfrac{2y}{s + 10}\ \dfrac{yd}{s}}$
(28)

7. $RWT = \text{jobs}$
(25)

$$R(5)(10) = 1$$

$$R = \frac{1}{50} \frac{\text{jobs}}{\text{men-days}}$$

$$RWT = \text{jobs}$$

$$\left(\frac{1}{50}\right)(2)T = 1$$

$$T = \mathbf{25\ days}$$

8. $y - y_1 = m(x - x_1)$
(39)

$$y - 3 = \frac{1}{2}(x + 2)$$

$$y - 3 = \frac{1}{2}x + 1$$

$$\frac{1}{2}x - y + 4 = 0$$

$$\mathbf{x - 2y + 8 = 0}$$

9. $3x - 2y + 6 = 0$
(39)

$$3x - 2y = -6$$

$$\frac{3x}{-6} - \frac{2y}{-6} = \frac{-6}{-6}$$

$$\mathbf{\frac{x}{-2} + \frac{y}{3} = 1}$$

10. $\cos\left(-\frac{13\pi}{6}\right) + \sin\frac{13\pi}{6}$
(39)

$$= \cos\left(-\frac{13(180°)}{6}\right) + \sin\frac{13(180°)}{6}$$

$$= \cos(-390°) + \sin(390°)$$

$$= \cos 30° + \sin 30° = \frac{\sqrt{3}}{2} + \frac{1}{2} = \mathbf{\frac{\sqrt{3} + 1}{2}}$$

11. (39) $\sin\frac{\pi}{3}\cos\frac{\pi}{3} - \sin\frac{\pi}{3}$

$= \sin\frac{180°}{3}\cos\frac{180°}{3} - \sin\frac{180°}{3}$

$= \sin 60°\cos 60° - \sin 60°$

$= \left(\frac{\sqrt{3}}{2}\right)\left(\frac{1}{2}\right) - \frac{\sqrt{3}}{2} = \frac{\sqrt{3}}{4} - \frac{2\sqrt{3}}{4} = \mathbf{-\frac{\sqrt{3}}{4}}$

12. (39) $\tan\left(-\frac{\pi}{4}\right)\cos\frac{\pi}{2} + \sin\frac{\pi}{6}$

$= \tan\left(-\frac{180°}{4}\right)\cos\frac{180°}{2} + \sin\frac{180°}{6}$

$= \tan(-45°)\cos 90° + \sin 30°$

$= -\tan 45°\cos 90° + \sin 30° = (-1)(0) + \frac{1}{2} = \mathbf{\frac{1}{2}}$

13. (38) $(x + \sqrt{3})(x - \sqrt{3}) = 0$

$\mathbf{x^2 - 3 = 0}$

14. (37) $\sqrt{(x - 5)^2 + (y - 1)^2} = \sqrt{(x + 4)^2 + (y + 3)^2}$

$(x - 5)^2 + (y - 1)^2 = (x + 4)^2 + (y + 3)^2$

$x^2 - 10x + 25 + y^2 - 2y + 1$

$= x^2 + 8x + 16 + y^2 + 6y + 9$

$-10x + 25 - 2y + 1 = 8x + 16 + 6y + 9$

$-10x - 2y + 26 = 8x + 6y + 25$

$8y = -18x + 1$

$\mathbf{y = -\frac{9}{4}x + \frac{1}{8}}$

15. (37) $x = \frac{x_1 + x_2}{2} = \frac{-2 + 6}{2} = 2$

$y = \frac{y_1 + y_2}{2} = \frac{6 + 3}{2} = \frac{9}{2}$

$\mathbf{\left(2, \frac{9}{2}\right)}$

16. (35) $-3\frac{2}{3} + \frac{5}{6}\left[6\frac{1}{3} - \left(-3\frac{2}{3}\right)\right] = -\frac{11}{3} + \frac{5}{6}\left(\frac{30}{3}\right)$

$= -\frac{11}{3} + \frac{25}{3} = \mathbf{\frac{14}{3}}$

17. (34) $\sum_{p=-1}^{3} \frac{p^2 - 2}{4}$

$= \frac{(-1)^2 - 2}{4} + \frac{(0)^2 - 2}{4} + \frac{(1)^2 - 2}{4}$

$+ \frac{(2)^2 - 2}{4} + \frac{(3)^2 - 2}{4}$

$= -\frac{1}{4} + \left(-\frac{1}{2}\right) + \left(-\frac{1}{4}\right) + \frac{1}{2} + \frac{7}{4} = \mathbf{\frac{5}{4}}$

18. (34) From the graph of the line we choose two ordered pairs: (13, 260) and (20, 220).

$m = \frac{220 - 260}{20 - 13} = \frac{-40}{7} = -5.71$

$S = -5.71P + b$

$220 = -5.71(20) + b$

$b = 334.2$

$\mathbf{S = -5.71P + 334.2}$

19. (34) $f(x) = \sqrt{x}$

20. (33) **No. Isosceles trapezoid.**

21. (33) Median $= \frac{1}{2}(B_1 + B_2)$

$20x - 3 = \frac{1}{2}[(8x + 1) + (20x - 1)]$

$20x - 3 = \frac{1}{2}(28x)$

$20x - 3 = 14x$

$6x = 3$

$\mathbf{x = \frac{1}{2}}$

22. (32) (a)

y
1
x
−45°
−1
√2

$-90° \le \theta \le 90°$

$\text{Arcsin}\left(-\frac{\sqrt{2}}{2}\right) = \text{Arcsin}\left(-\frac{1}{\sqrt{2}}\right) = \mathbf{-45°}$

(b)

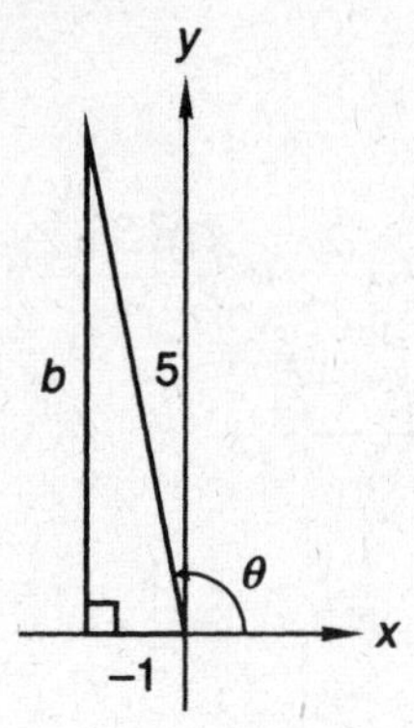

$0° \le \theta \le 180°$

$b = \sqrt{5^2 - (-1)^2} = \sqrt{24} = 2\sqrt{6}$

$\sin\left[\text{Arccos}\left(-\frac{1}{5}\right)\right] = \sin\theta = \mathbf{\frac{2\sqrt{6}}{5}}$

23. (31) (a) ***x* axis, yes.**

$|x| + |-y| = 2$ is equivalent to $|x| + |y| = 2$.

***y* axis, yes.**

$|-x| + |y| = 2$ is equivalent to $|x| + |y| = 2$.

Origin, yes.

$|-x| + |-y| = 2$ is equivalent to $|x| + |y| = 2$.

(b) ***x* axis, no.**

$(-y) = x^4$ is not equivalent to $y = x^4$.

***y* axis, yes.**

$y = (-x)^4$ is equivalent to $y = x^4$.

Origin, no.

$(-y) = (-x)^4$ is not equivalent to $y = x^4$.

24. (26) (a)
$$\log_b \frac{1}{27} = -3$$
$$\frac{1}{27} = b^{-3}$$
$$\frac{1}{27} = \frac{1}{b^3}$$
$$b = \sqrt[3]{27}$$
$$b = \mathbf{3}$$

(b)
$$\log_4 \frac{1}{64} = k$$
$$\frac{1}{64} = 4^k$$
$$\frac{1}{4^3} = 4^k$$
$$4^{-3} = 4^k$$
$$k = \mathbf{-3}$$

(c)
$$\log_{\frac{1}{3}} m = -\frac{2}{3}$$
$$m = \left(\frac{1}{3}\right)^{-\frac{2}{3}}$$
$$m = \frac{1}{3^{-\frac{2}{3}}}$$
$$m = 3^{\frac{2}{3}}$$
$$m = \mathbf{\sqrt[3]{9}}$$

25. (21) (a) **Domain** = $\{x \in \mathbb{R} \mid -4 \le x \le 4\}$

Range = $\{y \in \mathbb{R} \mid 0 \le y \le 3\}$

(b) **Domain** = $\{x \in \mathbb{R} \mid -5 \le x \le 2\}$

Range = $\{y \in \mathbb{R} \mid -4 \le y \le 5\}$

26. (24) (a)
$$(f \circ g)(x) = \frac{2}{3}\left[\frac{3}{2}\left(x + \frac{4}{3}\right)\right] - \frac{4}{3}$$
$$= x + \frac{4}{3} - \frac{4}{3}$$
$$= \mathbf{x}$$

(b)
$$(g \circ f)(x) = \frac{3}{2}\left[\left(\frac{2}{3}x - \frac{4}{3}\right) + \frac{4}{3}\right]$$
$$= \frac{3}{2}\left(\frac{2}{3}x\right)$$
$$= \mathbf{x}$$

27. (33)

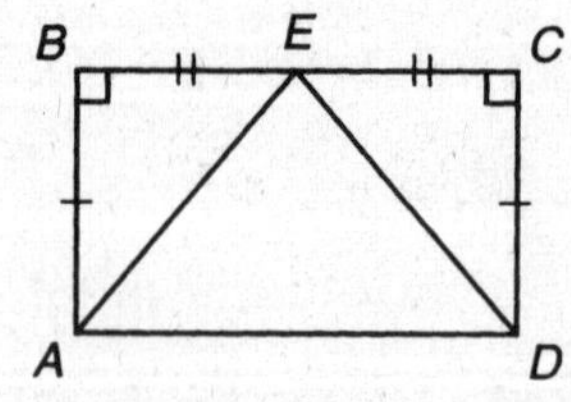

STATEMENTS	REASONS
1. *ABCD* is a rectangle.	1. Given
2. ∠*B* and ∠*C* are right angles.	2. A rectangle contains four right angles.
3. ∠*B* ≅ ∠*C*	3. All right angles are congruent.
4. $\overline{AB} \cong \overline{CD}$	4. Opposite sides of a rectangle are congruent.
5. *E* is the midpoint of $\overline{BC}$.	5. Given
6. $\overline{BE} \cong \overline{CE}$	6. A midpoint divides a segment into two congruent segments.
7. △*ABE* ≅ △*DCE*	7. *SAS* congruency postulate
8. $\overline{AE} \cong \overline{DE}$	8. *CPCTC*

28. (2)
$$V_{\text{sphere}} = \frac{4}{3}\pi r^3$$

$$\frac{4}{3}\pi(r_1)^3 = 18{,}432\pi$$
$$(r_1)^3 = 13{,}824$$
$$r_1 = 24$$

$$\frac{4}{3}\pi(r_2)^3 = 29{,}484\pi + 18{,}432\pi$$
$$(r_2)^3 = 35{,}937$$
$$r_2 = 33$$
$$A_{\text{surface}} = 4\pi r^2$$
$$A_2 - A_1 = 4\pi(r_2)^2 - 4\pi(r_1)^2$$
$$= 4\pi(33)^2 - 4\pi(24)^2$$
$$= 4356\pi - 2304\pi$$
$$= \mathbf{2052\pi\ m^2} = \mathbf{6446.55\ m^2}$$

29. (1)
$$(a + b)^2 - (a^2 + b^2) = 2ab > 0$$
$$(a + b)^2 > a^2 + b^2$$

Therefore the answer is **B.**

30. (1)
$$xy - \frac{x}{y} = x\left(y - \frac{1}{y}\right)$$

Since $y > 1$, $\left(y - \frac{1}{y}\right) > 0$, and

since $x < 0$, $x\left(y - \frac{1}{y}\right) < 0$, so $xy < \frac{x}{y}$.

Therefore the answer is **B.**

Problem Set 40

1. (39)
$$35°\left(\frac{\pi}{180°}\right)30 = \frac{35\pi}{6} = \mathbf{18.33\ ft}$$

2. (38)
$$\text{Overall average rate} = \frac{\text{overall distance}}{\text{overall time}}$$
$$18 = \frac{45 + 50 + 85}{\text{overall time}}$$
$$18 = \frac{180\text{ mi}}{\text{overall time}}$$
$$\text{Overall time} = 10\text{ hr}$$

$$T_1 = \frac{D_1}{R_1} = \frac{45\text{ mi}}{15\text{ mph}} = 3\text{ hr}$$
$$T_2 = \frac{D_2}{R_2} = \frac{50\text{ mi}}{25\text{ mph}} = 2\text{ hr}$$
$$T_3 = \text{overall time} - T_1 - T_2$$
$$= 10\text{ hr} - 3\text{ hr} - 2\text{ hr} = 5\text{ hr}$$

$$R_3 = \frac{D_3}{T_3} = \frac{85}{5} = \mathbf{17\ mph}$$

3. (36) First we write the equations.

$$\begin{cases} \text{(a)}\ (P + W)T_D = D_D \\ \text{(b)}\ (P - W)T_U = D_U \end{cases}$$

$$P = 6W,\ T_U = T_D + 1$$

Next we use substitution to solve.

Downwind:

$$\text{(a)}\quad (P + W)T_D = D_D$$
$$(6W + W)T_D = 700$$
$$(7W)T_D = 700$$
$$WT_D = 100$$

Upwind:

$$\text{(b)}\quad (P - W)T_U = D_U$$
$$(6W - W)(T_D + 1) = 600$$
$$5WT_D + 5W = 600$$
$$5(100) + 5W = 600$$
$$5W = 100$$
$$W = 20$$
$$P = 6W = 6(20) = 120$$

Plane = 120 mph; Wind = 20 mph

4. (18)
$$\text{Tin}_1 + \text{tin added} = \text{tin final}$$
$$0.16(410) + P_N = 0.18(410 + P_N)$$
$$65.6 + P_N = 73.8 + 0.18P_N$$
$$0.82P_N = 8.2$$
$$P_N = \mathbf{10\ lb}$$

5. (38) $5 \cdot 5 \cdot 5 \cdot 5 = \mathbf{625}$

6. (38) $20 \cdot 19 \cdot 18 \cdot 17 \cdot 16 = \mathbf{1{,}860{,}480}$

7. (40)
$$\log_a 3 + \log_a 4 = \log_a (x - 2)$$
$$\log_a (3 \cdot 4) = \log_a (x - 2)$$
$$12 = x - 2$$
$$x = \mathbf{14}$$

8. *(40)*
$$\log_3 (x+2) - \log_3 (x-1) = \log_3 4$$
$$\log_3 \left(\frac{x+2}{x-1}\right) = \log_3 4$$
$$\frac{x+2}{x-1} = 4$$
$$x + 2 = 4x - 4$$
$$3x = 6$$
$$x = \mathbf{2}$$

9. *(40)*
$$3 \log_c x = \log_c 27$$
$$\log_c x^3 = \log_c 27$$
$$x^3 = 27$$
$$x = \mathbf{3}$$

10. *(40)*
$$\log_{12} x = \frac{2}{3} \log_{12} 27$$
$$\log_{12} x = \log_{12} 27^{\frac{2}{3}}$$
$$x = 27^{\frac{2}{3}}$$
$$x = \mathbf{9}$$

11. *(40)*
$$(f \circ g)(x) = 2\left(\frac{1}{2}x - \frac{3}{2}\right) + 3$$
$$= x - 3 + 3 = x$$
$$(g \circ f)(x) = \frac{1}{2}(2x + 3) - \frac{3}{2}$$
$$= x + \frac{3}{2} - \frac{3}{2} = x$$
$$(f \circ g)(x) = (g \circ f)(x) = x$$

Yes, f and g are inverse functions.

12. *(39)*
$$y - y_1 = m(x - x_1)$$
$$y - 4 = -\frac{1}{3}(x + 3)$$
$$\frac{1}{3}x + 1 + y - 4 = 0$$
$$\mathbf{x + 3y - 9 = 0}$$

13. *(39)*
$$2x - 4y = -5$$
$$\frac{2x}{-5} - \frac{4y}{-5} = 1$$
$$\mathbf{\frac{x}{-\frac{5}{2}} + \frac{y}{\frac{5}{4}} = 1}$$

14. *(39)*
$$\sin \frac{2\pi}{3} \cos \frac{2\pi}{3} - \sin \frac{4\pi}{3}$$
$$= \sin \frac{2(180°)}{3} \cos \frac{2(180°)}{3} - \sin \frac{4(180°)}{3}$$
$$= \sin 120° \cos 120° - \sin 240°$$
$$= \sin 60°(-\cos 60°) - (-\sin 60°)$$
$$= \left(\frac{\sqrt{3}}{2}\right)\left(-\frac{1}{2}\right) - \left(-\frac{\sqrt{3}}{2}\right)$$
$$= -\frac{\sqrt{3}}{4} + \frac{\sqrt{3}}{2} = \mathbf{\frac{\sqrt{3}}{4}}$$

15. *(39)*
$$\tan\left(-\frac{3\pi}{4}\right) \cos \frac{\pi}{2} - \sin \frac{\pi}{2}$$
$$= \tan\left[-\frac{3(180°)}{4}\right] \cos \frac{180°}{2} - \sin \frac{180°}{2}$$
$$= \tan(-135°) \cos 90° - \sin 90°$$
$$= \tan 45° \cos 90° - \sin 90°$$
$$= 1(0) - 1 = \mathbf{-1}$$

16. *(39)*
$$\sin\left(-\frac{19\pi}{6}\right) + \sin \frac{13\pi}{6}$$
$$= \sin\left[-\frac{19(180°)}{6}\right] + \sin\left[\frac{13(180°)}{6}\right]$$
$$= \sin(-570°) + \sin 390°$$
$$= \sin 30° + \sin 30°$$
$$= \frac{1}{2} + \frac{1}{2} = \mathbf{1}$$

17. *(38)*
$$\left(x + \frac{5}{3}\right)\left(x - \frac{2}{5}\right) = 0$$
$$x^2 - \frac{2}{5}x + \frac{5}{3}x - \frac{10}{15} = 0$$
$$\mathbf{x^2 + \frac{19}{15}x - \frac{2}{3} = 0}$$

18. *(37)*
$$\sqrt{(x+4)^2 + (y-2)^2} = \sqrt{(x-5)^2 + (y-3)^2}$$
$$(x+4)^2 + (y-2)^2 = (x-5)^2 + (y-3)^2$$
$$x^2 + 8x + 16 + y^2 - 4y + 4$$
$$= x^2 - 10x + 25 + y^2 - 6y + 9$$
$$18x + 2y - 14 = 0$$
$$\mathbf{9x + y - 7 = 0}$$

19. (34) From the graph of the line we choose two ordered pairs: (180, 0) and (100, 90).

$$m = \frac{90 - 0}{100 - 180} = -\frac{9}{8}$$

$$N = -\frac{9}{8}R + b$$

$$0 = -\frac{9}{8}(180) + b$$

$$b = 202.5$$

$$\mathbf{N = -\frac{9}{8}R + 202.5}$$

20. (34) $\mathbf{f(x) = x + 1;\ g(x) = x^2}$

21. (a)

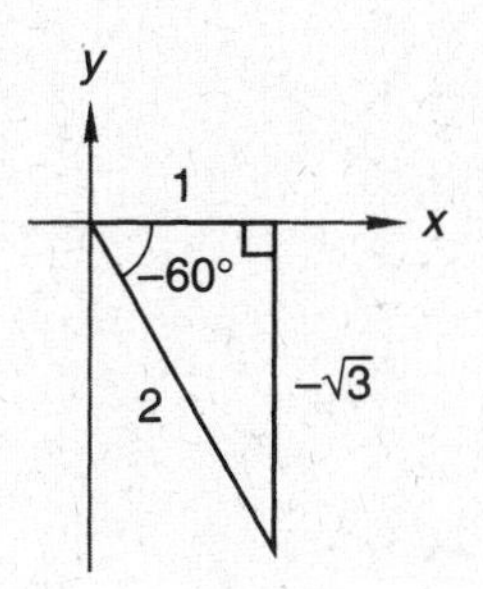

$-90° < \theta < 90°$

$\text{Arctan}\left(-\sqrt{3}\right) = \mathbf{-60°}$

(b)

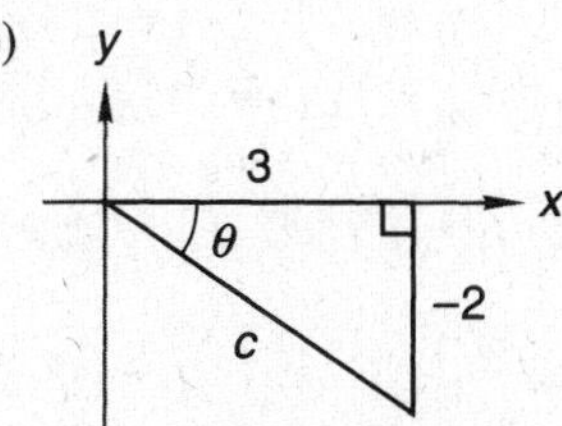

$-90° < \theta < 90°$

$$c = \sqrt{3^2 + (-2)^2} = \sqrt{13}$$

$$\cos\left[\text{Arctan}\left(-\frac{2}{3}\right)\right] = \cos\theta = \frac{3}{\sqrt{13}} = \mathbf{\frac{3\sqrt{13}}{13}}$$

22. (31) Replace y with $-y$.

$$-y = x^2$$

$$\mathbf{y = -x^2}$$

23. (31) Replace x with $x - 3$.

$$\mathbf{y = \frac{1}{x - 3}}$$

24. (21) $$\frac{\sqrt{x}}{x(x^2 + x - 2)} = \frac{\sqrt{x}}{x(x + 2)(x - 1)}$$

$x \geq 0,\ x \neq 0, 1, -2$

$\mathbf{\{x \in \mathbb{R} \mid x > 0, x \neq 1\}}$

25. (19) $125x^3y^6 - 216a^3y^9$

$$= y^6(125x^3 - 216a^3y^3) = y^6[(5x)^3 - (6ay)^3]$$

$$= \mathbf{y^6(5x - 6ay)(25x^2 + 30xay + 36a^2y^2)}$$

26. (21) $$\frac{f(x + h) - f(x)}{h}$$

$$= \frac{[3(x + h)^2 - 4(x + h)] - (3x^2 - 4x)}{h}$$

$$= \frac{3(x^2 + 2xh + h^2) - 4x - 4h - 3x^2 + 4x}{h}$$

$$= \frac{6xh + 3h^2 - 4h}{h} = \mathbf{6x + 3h - 4}$$

27. (15)

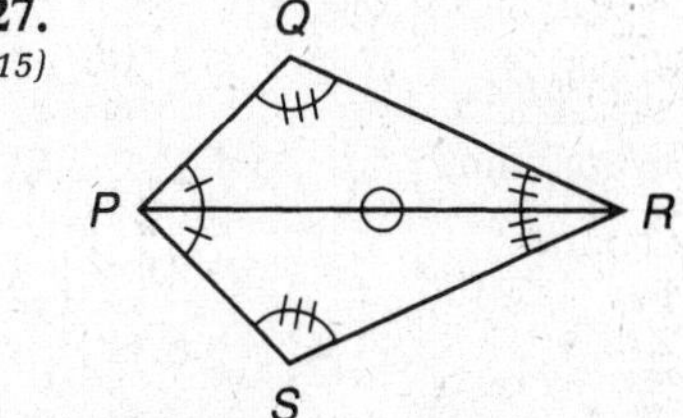

STATEMENTS	REASONS
1. $\overline{PR}$ bisects $\angle QPS$	1. Given
2. $\overline{PR}$ bisects $\angle QRS$	2. Given
3. $\angle QPR \cong \angle SPR$ $\angle QRP \cong \angle SRP$	3. A bisector divides an angle into two congruent angles.
4. $\overline{PR} \cong \overline{PR}$	4. Reflexive axiom
5. $\angle Q \cong \angle S$	5. If two angles in one triangle are congruent to two angles in a second triangle, then the third angles are congruent.
6. $\triangle PQR \cong \triangle PSR$	6. *AAAS* congruency postulate

28. (2) $$V_{\text{cone}} = \frac{1}{3}V_{\text{cylinder}}$$

$$V_{\text{discarded}} = V_{\text{cylinder}} - V_{\text{cone}}$$

$$= V_{\text{cylinder}} - \frac{1}{3}V_{\text{cylinder}}$$

$$= \mathbf{\frac{2}{3}}V_{\text{cylinder}}$$

29. (1,40) A: $2\log_3 4 = \log_3 4^2 = \log_3 16$

B: $4\log_3 2 = \log_3 2^4 = \log_3 16$

A and B are equal. Therefore the answer is **C.**

30. (1) Since $AB < BC < AC$,

$$m\angle C < m\angle A < m\angle B.$$
$$m\angle C + m\angle A + m\angle B = 180°$$
$$m\angle C + m\angle C + m\angle C < 180°$$
$$3m\angle C < 180°$$
$$m\angle C < 60°$$

Therefore the answer is **B.**

Problem Set 41

1. (39) $50°\left(\frac{\pi}{180°}\right)35 = \frac{175\pi}{18} = \mathbf{30.54\ ft}$

2. (38) Overall average rate $= \frac{\text{overall distance}}{\text{overall time}}$

$$15 \text{ mph} = \frac{36 \text{ miles} + 45 \text{ miles} + 54 \text{ miles}}{\text{overall time}}$$
$$15 \text{ mph} = \frac{135 \text{ miles}}{\text{overall time}}$$

Overall time = 9 hours

$$T_1 = \frac{D_1}{R_1} = \frac{36 \text{ mi}}{12 \text{ mph}} = 3 \text{ hr}$$
$$T_2 = \frac{D_2}{R_2} = \frac{45 \text{ mi}}{15 \text{ mph}} = 3 \text{ hr}$$
$$T_3 = \text{overall time} - T_1 - T_2$$
$$= 9 \text{ hr} - 3 \text{ hr} - 3 \text{ hr} = 3 \text{ hr}$$
$$R_3 = \frac{D_3}{T_3} = \frac{54 \text{ mi}}{3 \text{ hr}} = \mathbf{18\ mph}$$

3. (25)

$$RWT = \text{jobs}$$
$$R(400)(10) = 1$$
$$R = \frac{1}{4000} \frac{\text{job}}{\text{men-days}}$$
$$RWT = \text{jobs}$$
$$\frac{1}{4000}[(600)(5) + (x)5] = 1 \text{ job}$$
$$3000 + 5x = 4000$$
$$x = \frac{1000}{5} = \mathbf{200\ men}$$

4. (25)

$$R_BT_B + R_DT_D = \text{jobs}$$
$$\frac{1}{3}(3 + T) + \frac{1}{6}T = 4$$
$$1 + \frac{1}{3}T + \frac{1}{6}T = 4$$
$$\frac{1}{2}T = 3$$
$$T = 6 \text{ hr}$$

Time = noon + 3 hours + 6 hours

Time = **9:00 P.M.**

5. (28) New rate $= \frac{\text{distance}}{\text{new time}} = \mathbf{\frac{m}{h - 3} \frac{mi}{hr}}$

6. (41) $\csc(-330°) - \sec 390°$

$$= \csc 30° - \sec 30°$$
$$= \frac{1}{\sin 30°} - \frac{1}{\cos 30°}$$
$$= \frac{1}{\frac{1}{2}} - \frac{1}{\frac{\sqrt{3}}{2}}$$
$$= 2 - \frac{2}{\sqrt{3}} = \mathbf{\frac{6 - 2\sqrt{3}}{3}}$$

7. (41) $\sec\left(-\frac{\pi}{3}\right) + \cot\left(-\frac{13\pi}{6}\right)$

$$= \sec\left(\frac{\pi}{3}\right) + \cot\left(-\frac{\pi}{6}\right)$$
$$= \frac{1}{\cos\frac{\pi}{3}} + \frac{1}{\left(-\tan\frac{\pi}{6}\right)}$$
$$= \frac{1}{\cos 60°} - \frac{1}{\tan 30°}$$
$$= \frac{1}{\frac{1}{2}} - \frac{1}{\frac{1}{\sqrt{3}}} = \mathbf{2 - \sqrt{3}}$$

8. (41) (a) ${}_nP_r = \frac{n!}{(n - r)!}$

$${}_{10}P_4 = \frac{10!}{(10 - 4)!} = \frac{10!}{6!} = \mathbf{5040}$$

(b) ${}_nP_r = \frac{n!}{(n - r)!}$

$${}_{13}P_5 = \frac{13!}{(13 - 5)!} = \frac{13!}{8!} = \mathbf{154{,}440}$$

9. (40)
$$\log_4 2 + \log_4 8 = \log_4 (x + 3)$$
$$\log_4 (2 \cdot 8) = \log_4 (x + 3)$$
$$16 = x + 3$$
$$x = \mathbf{13}$$

10. (40)
$$\log_3 (x - 2) - \log_3 (x + 1) = \log_3 5$$
$$\log_3 \left(\frac{x - 2}{x + 1}\right) = \log_3 5$$
$$\frac{(x - 2)}{(x + 1)} = 5$$
$$x - 2 = 5x + 5$$
$$x = -\frac{7}{4}$$

No solution. You cannot take the log of a negative number.

11. (40)
$$4 \log_c x = \log_c 81$$
$$\log_c x^4 = \log_c 81$$
$$x^4 = 81$$
$$x = \mathbf{3}$$

12. (40)
$$\log_{13} x = \frac{3}{4} \log_{13} 16$$
$$\log_{13} x = \log_{13} 16^{\frac{3}{4}}$$
$$x = 16^{\frac{3}{4}} = \mathbf{8}$$

13. (40)
$$(f \circ g)(x) = 4\left(\frac{1}{4}x + 3\right) - 3$$
$$= x + 12 - 3 = x + 9$$
$$(g \circ f)(x) = \frac{1}{4}(4x - 3) + 3$$
$$= x - \frac{3}{4} + 3 = x + 2\frac{1}{4}$$

$(f \circ g)(x) \neq x$

$(g \circ f)(x) \neq x$

No, f and g are not inverse functions.

14. (39)
$$y - y_1 = m(x - x_1)$$
$$y - 3 = -\frac{1}{3}(x + 2)$$
$$-3y + 9 = x + 2$$
$$\mathbf{x + 3y - 7 = 0}$$

15. (39)
$$5x - 3y = -6$$
$$\frac{5x}{-6} - \frac{3y}{-6} = \frac{-6}{-6}$$
$$\frac{5x}{-6} + \frac{y}{2} = 1$$
$$\mathbf{\frac{x}{-\frac{6}{5}} + \frac{y}{2} = 1}$$

16. (38)
$$\left(x + \frac{1}{2}\right)\left(x - \frac{2}{3}\right) = 0$$
$$x^2 - \frac{2}{3}x + \frac{1}{2}x - \frac{2}{6} = 0$$
$$\mathbf{x^2 - \frac{1}{6}x - \frac{1}{3} = 0}$$

17. (37,39)
$$\sqrt{(x - 2)^2 + (y - 4)^2} = \sqrt{(x + 4)^2 + (y - 6)^2}$$
$$x^2 - 4x + 4 + y^2 - 8y + 16$$
$$= x^2 + 8x + 16 + y^2 - 12y + 36$$
$$-12x + 4y = 32$$
$$-\frac{12x}{32} + \frac{4y}{32} = \frac{32}{32}$$
$$-\frac{3x}{8} + \frac{y}{8} = 1$$
$$\mathbf{\frac{x}{-\frac{8}{3}} + \frac{y}{8} = 1}$$

18. (35)
$$-\frac{15}{4} + \frac{4}{13}\left(6 + \frac{15}{4}\right) = -\frac{15}{4} + \frac{4}{13}\left(\frac{39}{4}\right)$$
$$= -\frac{15}{4} + 3 = \mathbf{-\frac{3}{4}}$$

19. (34)
$$\sum_{z=-2}^{1}\left(\frac{z^3 - 1}{3}\right)$$
$$= \left(\frac{(-2)^3 - 1}{3}\right) + \left(\frac{(-1)^3 - 1}{3}\right) + \left(\frac{(0)^3 - 1}{3}\right) + \left(\frac{(1)^3 - 1}{3}\right)$$
$$= \left(\frac{-8 - 1}{3}\right) + \left(\frac{-1 - 1}{3}\right) + \left(\frac{-1}{3}\right) + \left(\frac{0}{3}\right)$$
$$= -3 - \frac{2}{3} - \frac{1}{3} = \mathbf{-4}$$

20. (34) From the graph of the line we choose two ordered pairs: (0, 0) and (10, 10).

$$m = \frac{10 - 0}{10 - 0} = 1$$
$$H = 1S + b$$
$$0 = 1(0) + b$$
$$b = 0$$
$$\mathbf{H = S}$$

21. (33) Since the diagonal bisects the angles of two opposite vertices, all sides must be congruent. Therefore, the answer is **C.**

22. (33) $A_{\text{trapezoid}} = \frac{1}{2}(H)(B_1 + B_2)$

$$60 = \frac{1}{2}(H)(8 + 12)$$
$$H = \mathbf{6\text{ cm}}$$

23. (32) (a)

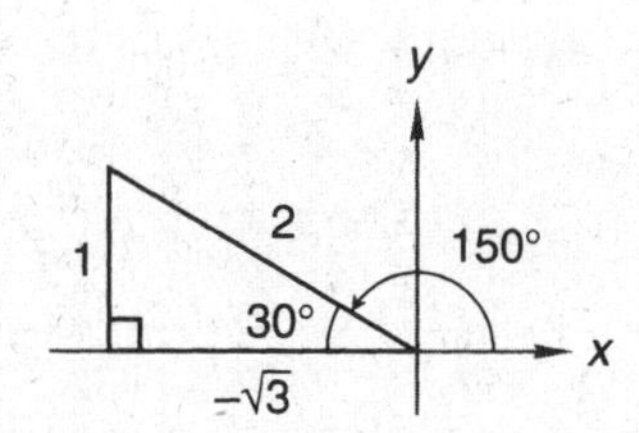

$$0° \le \theta \le 180°$$
$$\text{Arccos}\left(-\frac{\sqrt{3}}{2}\right) = \mathbf{150°}$$

(b)

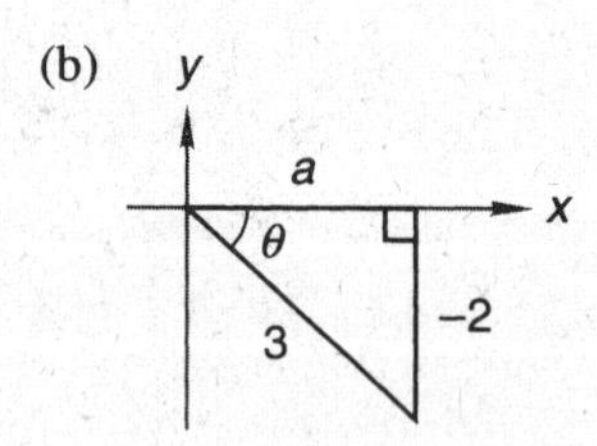

$$-90° \le \theta \le 90°$$
$$a = \sqrt{3^2 - (-2)^2} = \sqrt{5}$$
$$\cos\left[\text{Arcsin}\left(-\frac{2}{3}\right)\right] = \cos\theta = \mathbf{\frac{\sqrt{5}}{3}}$$

24. (31) Replace x with $x + 3$.

$$\mathbf{y = \frac{1}{x + 3}}$$

25. (31) Replace y with $y - 2$ and x with $x - 3$.

$$y - 2 = |x - 3|$$
$$\mathbf{y = |x - 3| + 2}$$

26. (30) $-2\underline{/140°} + 8\underline{/150°}$

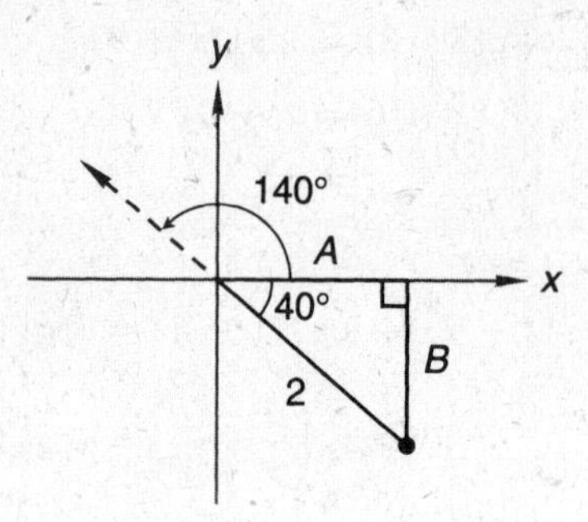

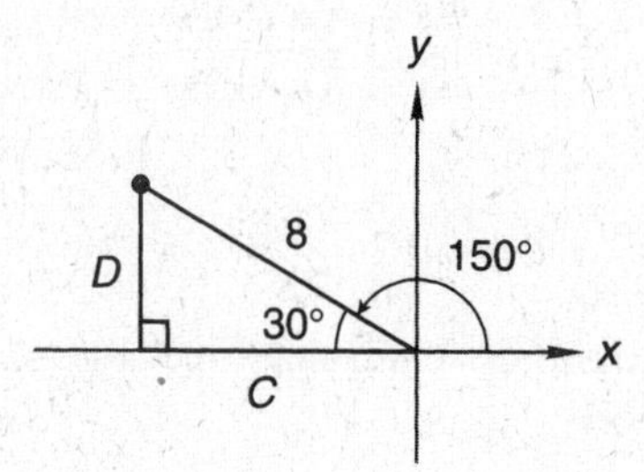

$$A = 2\cos 40° = 1.5321$$
$$B = 2\sin 40° = 1.2856$$
$$C = 8\cos 30° = 6.9282$$
$$D = 8\sin 30° = 4$$

$$\text{Resultant} = (1.5321 - 6.9282)\hat{i} + (-1.2856 + 4)\hat{j}$$
$$= -5.3961\hat{i} + 2.7144\hat{j}$$

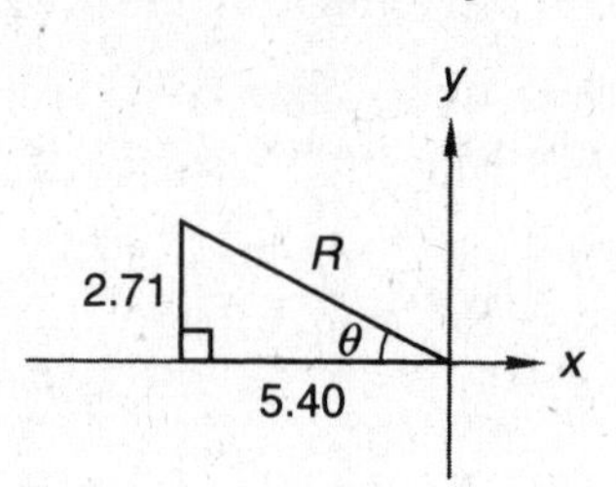

$$R = \sqrt{(5.3961)^2 + (2.7144)^2} = 6.04$$

$$\tan\theta = \frac{2.7144}{5.3961}$$
$$\theta = 26.70°$$

Since θ is in the second quadrant, the polar angle is $180° - 26.70° = 153.30°$.

$\mathbf{6.04\underline{/153.30°}}$

27. (21) (a) **Domain** $= \{x \in \mathbb{R} \mid \mathbf{-6 \le x \le 5}\}$

Range $= \{y \in \mathbb{R} \mid \mathbf{-3 \le y \le 4}\}$

(b) **Domain** $= \{x \in \mathbb{R} \mid \mathbf{-4 \le x \le 4}\}$

Range $= \{y \in \mathbb{R} \mid \mathbf{-2 \le y \le 6}\}$

28. (21) $$\frac{f(x + h) - f(x)}{h} = \frac{\frac{1}{x + h} - \frac{1}{x}}{h}$$

$$= \frac{\frac{x - (x + h)}{x(x + h)}}{h} = \frac{-h}{xh(x + h)} = \mathbf{\frac{-1}{x(x + h)}}$$

29. (1,40) A: $\log_2 6 + \log_2 5 = \log_2 30$

B: $\log_2 3 + \log_2 10 = \log_2 30$

A and B are equal. Therefore the answer is **C.**

30. (1)
$$AC = BC$$
$$3x + 16 = 2x + 22$$
$$x = 6$$
$$6 > 5$$
$$x > 5$$

Therefore the answer is **A.**

Problem Set 42

1. (39) $$40°\left(\frac{\pi}{180°}\right)1500 = \frac{3000\pi}{9} = \mathbf{1047.20\ ft}$$

2. (36) First, we write the equations.

$$\begin{aligned} &(a) \\ &(b) \end{aligned} \begin{cases} (B + C)T_D = D_D \\ (B - C)T_U = D_U \end{cases}$$

$$D_D = D_U = 75$$
$$T_D = 3,\ T_U = 5$$

Next, we use substitution to solve.

$$D_D = D_U$$
$$(B + C)T_D = (B - C)T_U$$
$$(B + C)3 = (B - C)5$$
$$3B + 3C = 5B - 5C$$
$$2B = 8C$$
$$B = 4C$$

(a) $(B + C)T_D = D_D$
$$(B + C)3 = 75$$
$$(4C + C)3 = 75$$
$$15C = 75$$
$$\mathbf{C = 5\ mph}$$

$$B = 4C$$
$$B = 4(5)$$
$$\mathbf{B = 20\ mph}$$

3. (25) $$\frac{\left(\frac{440}{49}\right)}{\left(\frac{100}{9.8}\right)} = \frac{9.8(440)}{49(100)} = \mathbf{0.88}$$

4. (25)
$$R_eT_e + R_aT_a = \text{jobs}$$
$$\left(\frac{1}{12}\right)(5) + R_a(5) = 1\frac{1}{24}$$
$$5R_a = \frac{15}{24}$$
$$R_a = \frac{1}{8}$$
$$R_aT_a = \text{jobs}$$
$$\left(\frac{1}{8}\right)T = 2$$
$$\mathbf{T = 16\ hr}$$

5. (25)
$$RWT = \text{jobs}$$
$$R(5)(3) = 7$$
$$R = \frac{7}{15}\ \frac{\text{jobs}}{\text{worker-day}}$$
$$RWT = \text{jobs}$$
$$\left(\frac{7}{15}\right)(10)T = 14$$
$$\mathbf{T = 3\ days}$$

6. (25) $$\text{Rate} = \frac{\$160}{40\text{ hr}} = 4\ \frac{\text{dollars}}{\text{hr}}$$

$$\text{New rate} = 1.25 \times \text{rate} = 1.25(4) = 5\ \frac{\text{dollars}}{\text{hr}}$$

$$\text{New rate} \times T = \text{dollars}$$
$$5(60) = \mathbf{\$300}$$

7. (42)
$$\sqrt{(x - h)^2 + (y - k)^2} = r$$
$$\sqrt{(x - 3)^2 + (y - 4)^2} = 3$$
$$\mathbf{(x - 3)^2 + (y - 4)^2 = 3^2}$$

8. (42) $$y = \left(\frac{1}{2}\right)^{2x} = \left(\frac{1}{4}\right)^x$$

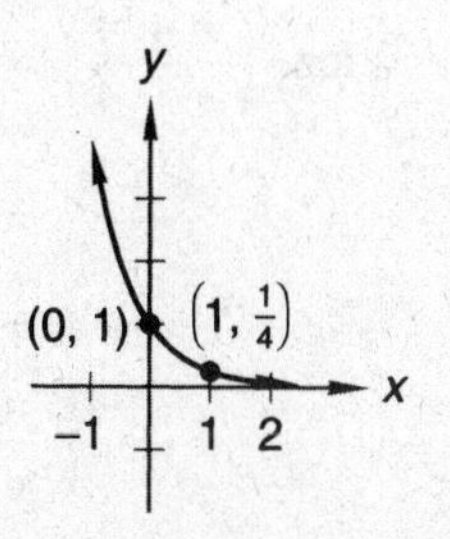

9. (42) $y = \left(\frac{1}{2}\right)^{x-2} = \left(\frac{1}{2}\right)^{x}\left(\frac{1}{2}\right)^{-2} = 4\left(\frac{1}{2}\right)^{x}$

(0, 4)
(1, 2)

10. (41) $\sec\frac{\pi}{4} - \sin\frac{13\pi}{6} - \cos\left(-\frac{8\pi}{3}\right)$

$= \frac{1}{\cos\frac{\pi}{4}} - \sin\frac{\pi}{6} - \left(-\cos\frac{\pi}{3}\right)$

$= \frac{1}{\cos 45°} - \sin 30° + \cos 60°$

$= \frac{1}{\frac{\sqrt{2}}{2}} - \frac{1}{2} + \frac{1}{2} = \frac{2}{\sqrt{2}} = \mathbf{\sqrt{2}}$

11. (41) $\csc\left(\frac{-\pi}{6}\right) - \cot\left(\frac{-13\pi}{6}\right)$

$= \frac{1}{-\sin\frac{\pi}{6}} - \frac{1}{-\tan\frac{\pi}{6}}$

$= \frac{1}{-\sin 30°} + \frac{1}{\tan 30°}$

$= \frac{1}{-\frac{1}{2}} + \frac{1}{\frac{1}{\sqrt{3}}} = \mathbf{-2 + \sqrt{3}}$

12. (41) ${}_{10}P_6 = \frac{10!}{(10-6)!} = \frac{10!}{4!} = \mathbf{151{,}200}$

13. (40) $\log_2(x+4) - \log_2(x-4) = \log_2 5$

$\log_2\left(\frac{x+4}{x-4}\right) = \log_2 5$

$\frac{x+4}{x-4} = 5$

$x + 4 = 5(x-4)$

$4x = 24$

$x = \mathbf{6}$

14. (40) $2\log_7 x = 4$

$\log_7 x^2 = 4$

$x^2 = 7^4$

$x = 7^2 = \mathbf{49}$

15. (40) $4\log_6 x = \log_6 256$

$\log_6 x^4 = \log_6 256$

$x^4 = 256$

$x = \mathbf{4}$

16. (40) $(f \circ g)(x) = \frac{\left(\frac{2}{x}\right)}{2} = \frac{1}{x}$

$(g \circ f)(x) = \frac{2}{\left(\frac{x}{2}\right)} = \frac{4}{x}$

$(f \circ g)(x) \neq x$

$(g \circ f)(x) \neq x$

No, ***f*** **and** ***g*** **are not inverse functions.**

17. (39) $y - y_1 = m(x - x_1)$

$y - 3 = -\frac{2}{5}(x+1)$

$\frac{2}{5}x + y - \frac{13}{5} = 0$

$\mathbf{2x + 5y - 13 = 0}$

18. (39) $6x - 5y = 3$

$\frac{6x}{3} - \frac{5y}{3} = \frac{3}{3}$

$2x - \frac{5y}{3} = 1$

$\mathbf{\frac{x}{\frac{1}{2}} + \frac{y}{-\frac{3}{5}} = 1}$

19. (38) $\left(x + \frac{1}{2}\right)\left(x - \frac{4}{5}\right) = 0$

$x^2 - \frac{4}{5}x + \frac{1}{2}x - \frac{4}{10} = 0$

$\mathbf{x^2 - \frac{3}{10}x - \frac{2}{5} = 0}$

20. (37,39) $\sqrt{(x-1)^2 + (y-4)^2} = \sqrt{(x+3)^2 + (y-2)^2}$

$x^2 - 2x + 1 + y^2 - 8y + 16$

$= x^2 + 6x + 9 + y^2 - 4y + 4$

$-8x - 4y + 4 = 0$

$\mathbf{2x + y - 1 = 0}$

21. *(34)* From the graph of the line we choose two ordered pairs: (14, 10) and (55, 60).

$$m = \frac{60 - 10}{55 - 14} = \frac{50}{41} = 1.22$$

$$\text{I} = 1.22\text{A} + b$$
$$60 = 1.22(55) + b$$
$$b = -7.1$$
$$\mathbf{I = 1.22A - 7.1}$$

22. *(33)* Since the diagonal does not bisect angles of vertices, all sides of the quadrilateral are not congruent. Therefore the answer is **A.**

23. *(33)* $A_{\text{trapezoid}} = \frac{1}{2}(H)(B_1 + B_2)$

$$50 = \frac{1}{2}(5)[(x - 3) + (2x - 4)]$$
$$20 = 3x - 7$$
$$27 = 3x$$
$$x = \mathbf{9\text{ cm}}$$

24. *(32)* (a)

$-90° \le \theta \le 90°$

$$\text{Arcsin}\left(-\frac{\sqrt{3}}{2}\right) = \mathbf{-60°}$$

(b)

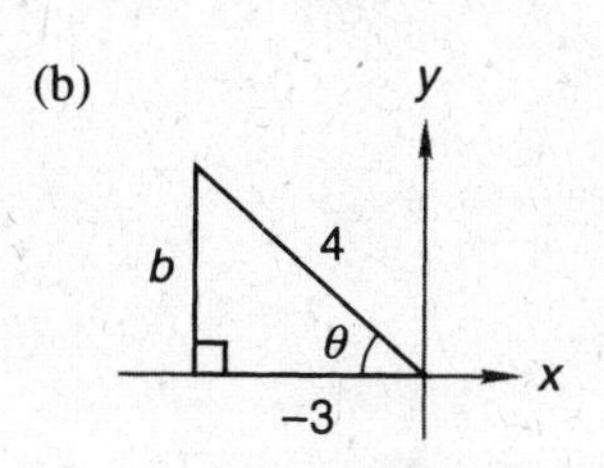

$0° \le \theta \le 180°$

$$b = \sqrt{4^2 - (-3)^2} = \sqrt{7}$$

$$\sin\left[\text{Arccos}\left(-\frac{3}{4}\right)\right] = \sin\theta = \mathbf{\frac{\sqrt{7}}{4}}$$

25. *(31)* Replace y with $y - 2$ and x with $x + 3$.

$$y - 2 = |x + 3|$$
$$\mathbf{y = |x + 3| + 2}$$

26. *(31)* Replace $g(x)$ with $g(x) - 3$ and x with $x - 2$.

$$g(x) - 3 = |x - 2|$$
$$\mathbf{g(x) = |x - 2| + 3}$$

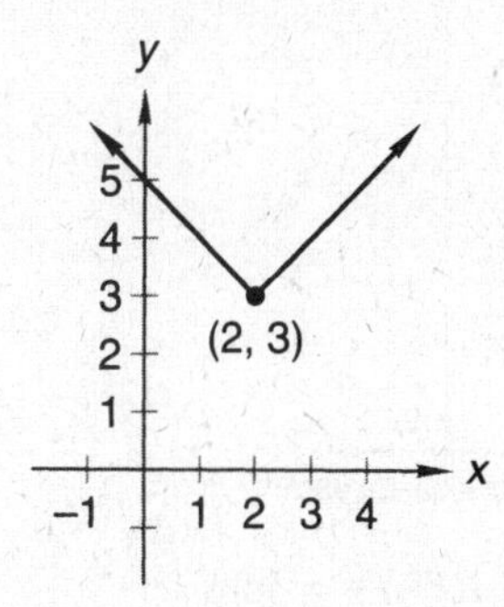

27. *(21,22)* $\left\{\mathbf{x \in \mathbb{R} \mid x \ge 0, x \ne \frac{1}{2}}\right\}$

28. *(15)*

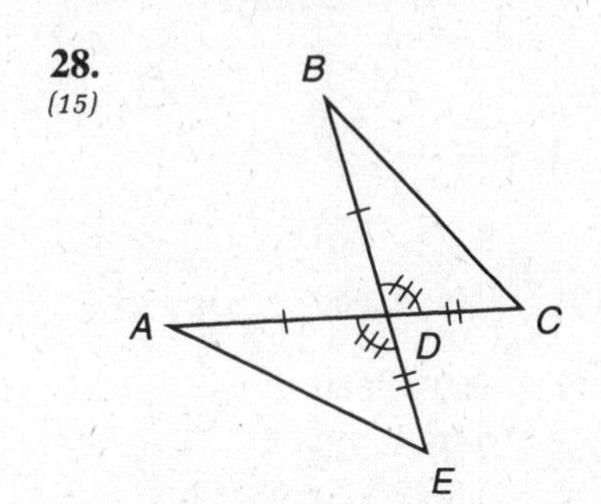

STATEMENTS	REASONS
1. $\overline{AD} \cong \overline{BD}$	1. Given
2. $\overline{ED} \cong \overline{CD}$	2. Given
3. $\angle BDC \cong \angle ADE$	3. Vertical angles are congruent
4. $\triangle BDC \cong \triangle ADE$	4. SAS congruency postulate
5. $\angle A \cong \angle B$	5. CPCTC

29. *(1,5)*

$$\sqrt{x} + \sqrt{x - 1} = 1$$
$$\sqrt{x - 1} = 1 - \sqrt{x}$$
$$\left(\sqrt{x - 1}\right)^2 = \left(1 - \sqrt{x}\right)^2$$
$$x - 1 = 1 - 2\sqrt{x} + x$$
$$\sqrt{x} = 1$$
$$x = 1$$

Therefore the answer is **C.**

30. *(1)* $x^2 = 4$: $x = 2$ or $x = -2$

$y^2 = 9$: $y = 3$ or $y = -3$

Therefore the answer is **D.**

Problem Set 43

1. (39) $H = (\tan 0.5°)5000 = \mathbf{43.63\ ft}$

2. (39) $1000\left(\frac{\pi}{6}\right) = \mathbf{523.60\ ft}$

3. (25)
$$RWT = \text{jobs}$$
$$R(10)(12) = 1$$
$$R = \frac{1}{120}\ \frac{\text{job}}{\text{worker-days}}$$
$$RW_1T_1 + RW_2T_2 = \text{jobs}$$
$$\left(\frac{1}{120}\right)(8)(5) + \left(\frac{1}{120}\right)(4)(T) = 1$$
$$\frac{1}{30}T = \frac{80}{120}$$
$$T = \mathbf{20\ days}$$

4. (18) $\frac{\$560}{35\text{ hr}} = \frac{\$16}{\text{hr}}$
$$40(16) + (T - 40)\left(\frac{3}{2}\right)(16) = 880$$
$$640 + 24T - 960 = 880$$
$$24T = 1200$$
$$T = \mathbf{50\ hr}$$

5. (28) $G(130 + f + s) = 2000$
$$G = \mathbf{\frac{2000}{130 + f + s}\ gal}$$

6. (25) First, we write two equations.

$$\begin{cases} \text{(a)}\ 5S_N = 3J_N + 30 \\ \text{(b)}\ 2(S_N + 10) = (J_N + 10) + 26 \end{cases}$$

(b) $2(S_N + 10) = (J_N + 10) + 26$
$$J_N = 2S_N - 16$$

Next, we use substitution to solve.

(a) $5S_N = 3J_N + 30$
$$5S_N = 3(2S_N - 16) + 30$$
$$5S_N = 6S_N - 18$$
$$S_N = 18$$

$$J_N = 2S_N - 16 = 2(18) - 16 = 20$$
$$S_N + 13 = \textbf{Sally} = \mathbf{31\ yr}$$
$$J_N + 13 = \textbf{John} = \mathbf{33\ yr}$$

7. (43) Function $= -\cos\theta$

Amplitude $= 4$

$\mathbf{y = -4\cos\theta}$

8. (43) Function $= \sin x$

Amplitude $= 5$

$\mathbf{y = 5\sin x}$

9. (42)
$$\sqrt{(x - h)^2 + (y - k)^2} = r$$
$$\sqrt{[x - (-2)]^2 + (y - 3)^2} = 4$$
$$\mathbf{(x + 2)^2 + (y - 3)^2 = 4^2}$$

10. (42) $y = \left(\frac{1}{3}\right)^{x-1} = \left(\frac{1}{3}\right)^x\left(\frac{1}{3}\right)^{-1} = 3\left(\frac{1}{3}\right)^x$

y
3 (0, 3)
2
1 (1, 1)
1 2
x

11. (41) $\sec\left(-\frac{7\pi}{6}\right) + \csc\frac{\pi}{3} + \tan\frac{8\pi}{3}$
$$= \frac{1}{\cos\left(-\frac{7\pi}{6}\right)} + \frac{1}{\sin\left(\frac{\pi}{3}\right)} + \tan\frac{2\pi}{3}$$
$$= \frac{1}{-\cos\frac{\pi}{6}} + \frac{1}{\sin\frac{\pi}{3}} - \tan\frac{\pi}{3}$$
$$= \frac{1}{-\cos 30°} + \frac{1}{\sin 60°} - \tan 60°$$
$$= \frac{1}{-\frac{\sqrt{3}}{2}} + \frac{1}{\frac{\sqrt{3}}{2}} - \frac{\sqrt{3}}{1} = \mathbf{-\sqrt{3}}$$

12. (41) $\sec\left(-\frac{3\pi}{4}\right) - \cot\frac{\pi}{2} + \csc\frac{3\pi}{2}$
$$= \frac{1}{\cos\left(-\frac{3\pi}{4}\right)} - 0 + \frac{1}{\sin\left(\frac{3\pi}{2}\right)}$$
$$= \frac{1}{-\cos 45°} + \frac{1}{\sin 270°}$$
$$= \frac{1}{-\frac{1}{\sqrt{2}}} + \frac{1}{-1} = \mathbf{-\sqrt{2} - 1}$$

13. (41)
$$_8P_6 - {_8P_5} = \frac{8!}{(8-6)!} - \frac{8!}{(8-5)!}$$
$$= \frac{8!}{2!} - \frac{8!}{3!} = \mathbf{13{,}440}$$

14. (40)
$$\log_3 3^2 + \log_5 5^3 - \log_2 2^4$$
$$= 2 + 3 - 4 = \mathbf{1}$$

15. (40)
$$\log_3 6 + \log_3 3 = \log_3 (4x + 2)$$
$$\log_3 (6 \cdot 3) = \log_3 (4x + 2)$$
$$18 = 4x + 2$$
$$x = \mathbf{4}$$

16. (40)
$$\log_7 (x + 2) - \log_7 (x - 4) = \log_7 2$$
$$\log_7 \left(\frac{x+2}{x-4}\right) = \log_7 2$$
$$\frac{x+2}{x-4} = 2$$
$$x + 2 = 2x - 8$$
$$x = \mathbf{10}$$

17. (40)
$$2 \log_x 4 = 2$$
$$\log_x 4^2 = 2$$
$$x^2 = 16$$
$$x = \mathbf{4}$$

18. (40)
$$(f \circ g)(x) = \left(\sqrt[3]{x}\right)^3 = x$$
$$(g \circ f)(x) = \sqrt[3]{x^3} = x$$
$$(g \circ f)(x) = (f \circ g)(x) = x$$
Yes, f and g are inverse functions.

19. (38)
$$\left[x - \left(1 + \sqrt{3}\right)\right]\left[x - \left(1 - \sqrt{3}\right)\right] = 0$$
$$x^2 - x\left(1 - \sqrt{3}\right) - x\left(1 + \sqrt{3}\right) + (1 - 3) = 0$$
$$x^2 - x + \sqrt{3}x - x - \sqrt{3}x - 2 = 0$$
$$\mathbf{x^2 - 2x - 2 = 0}$$

20. (37)
$$\sqrt{(x+6)^2 + (y-4)^2} = \sqrt{x^2 + (y-8)^2}$$
$$x^2 + 12x + 36 + y^2 - 8y + 16$$
$$= x^2 + y^2 - 16y + 64$$
$$12x - 12 = -8y$$
$$y = \mathbf{-\frac{3}{2}x + \frac{3}{2}}$$

21. (37)
$$\frac{x_1 + x_2}{2} = \frac{7 - 6}{2} = \frac{1}{2}$$
$$\frac{y_1 + y_2}{2} = \frac{4 + 2}{2} = 3$$
$$\mathbf{\left(\frac{1}{2}, 3\right)}$$

22. (35)
$$\frac{8}{3} + \frac{1}{3}\left(\frac{19}{3} - \frac{8}{3}\right) = \frac{8}{3} + \frac{1}{3}\left(\frac{11}{3}\right)$$
$$= \frac{24}{9} + \frac{11}{9} = \mathbf{\frac{35}{9}}$$

23. (34)
$$\sum_{i=-1}^{3} i^2 - 3$$
$$= \left[(-1)^2 - 3\right] + \left[(0)^2 - 3\right] + \left[(1)^2 - 3\right]$$
$$+ \left[(2)^2 - 3\right] + \left[(3)^2 - 3\right]$$
$$= -2 - 3 - 2 + 1 + 6 = \mathbf{0}$$

24. (34) From the graph of the line we choose two ordered pairs: (101.5, 110) and (99, 130).
$$m = \frac{130 - 110}{99 - 101.5} = -8$$
$$\text{H} = -8\text{C} + b$$
$$110 = -8(101.5) + b$$
$$b = 922$$
$$\mathbf{H = -8C + 922}$$

25. (32) (a)

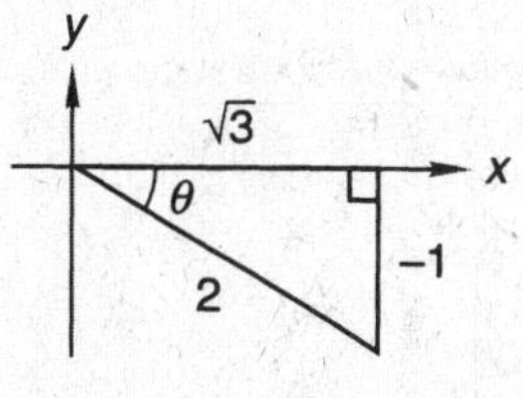

$$-90° \le \theta \le 90°$$
$$\text{Arcsin}\left(-\frac{1}{2}\right) = \mathbf{-30°}$$

(b) $\sin\left[\text{Arcsin}\left(-\frac{3}{5}\right)\right] = \mathbf{-\frac{3}{5}}$

26. (21) (a) **Domain** $= \{x \in \mathbb{R} \mid -6 \le x \le 2\}$

Range $= \{y \in \mathbb{R} \mid -2 \le y \le 6\}$

(b) **Domain** $= \{x \in \mathbb{R} \mid -5 \le x \le 5\}$

Range $= \{y \in \mathbb{R} \mid -4 \le y \le 4\}$

27. (31) Replace y with $y + 1$ and x with $x - 1$.

$$y + 1 = (x - 1)^2$$

$$\mathbf{y = (x - 1)^2 - 1}$$

28. (31)

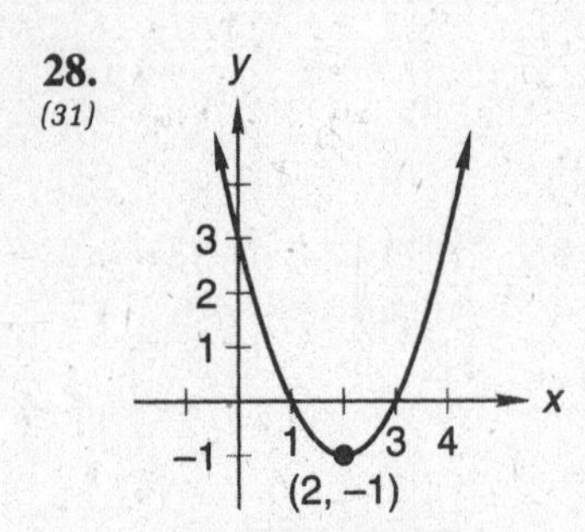

Replace y with $y + 1$ and x with $x - 2$.

$$y + 1 = (x - 2)^2$$

$$y = (x - 2)^2 - 1$$

$$\mathbf{g(x) = (x - 2)^2 - 1}$$

29. (1,5) $\sqrt{x^2} = |x| \geq x$

Therefore the answer is **D.**

30. (1,16) A: $\dfrac{1}{1 + \dfrac{1}{x}} = \dfrac{1}{\dfrac{x + 1}{x}} = \dfrac{x}{x + 1}$

Since $x > 1$, $\dfrac{x}{x + 1} > \dfrac{1}{1 + x}$.

Therefore the answer is **A.**

Problem Set 44

1. (44) $\text{Rate} = \dfrac{x}{d} - 5 = \dfrac{x - 5d}{d}\ \dfrac{\text{dollars}}{\text{drum}}$

$$\text{Rate} \times N = \text{price}$$

$$\frac{x - 5d}{d} \times N = 100$$

$$N = \frac{100}{\dfrac{x - 5d}{d}} = \mathbf{\frac{100d}{x - 5d}\ drums}$$

2. (44) $\text{Rate} = \dfrac{c}{hw}\ \dfrac{\text{cookies}}{\text{worker-hr}}$

$$RWT = \text{cookies}$$

$$\frac{c}{hw}(w + m)T = c$$

$$T = \frac{hwc}{c(w + m)} = \mathbf{\frac{hw}{w + m}\ hr}$$

3. (44) $\text{Rate} = \dfrac{p}{mc}\ \dfrac{\text{pounds}}{\text{child-min}}$

$$RWT = \text{pounds}$$

$$\left(\frac{p}{mc}\right)(c + n)T = p$$

$$T = \frac{pmc}{p(c + n)} = \mathbf{\frac{mc}{c + n}\ min}$$

4. (25) $R_T = 3R_A$

$$R_AT_A + R_TT_T = \text{miles}$$

$$R_A(1) + 3R_A(10) = 3100$$

$$R_A = 100\ \frac{\text{mi}}{\text{yr}}$$

$$R_AT_A = \text{miles}$$

$$(100)1 = \mathbf{100\ mi}$$

5. (38) $5 \cdot 4 \cdot 3 \cdot 2 = \mathbf{120}$

6. (39) $60°\left(\dfrac{\pi}{180°}\right)300 = 100\pi = \mathbf{314.16\ yd}$

7. (43) $\text{Function} = -\cos x$

$$\text{Amplitude} = 10$$

$$\mathbf{y = -10 \cos x}$$

8. (43) $\text{Function} = \sin x$

$$\text{Amplitude} = 11$$

$$\mathbf{y = 11 \sin x}$$

9. (42) $\sqrt{(x - h)^2 + (y - k)^2} = 5$

$$\mathbf{(x - h)^2 + (y - k)^2 = 5^2}$$

10. (42) $f(x) = 2^{-x-2} = 2^{-x}2^{-2} = \dfrac{1}{4}\left(\dfrac{1}{2}\right)^x$

11. (41)

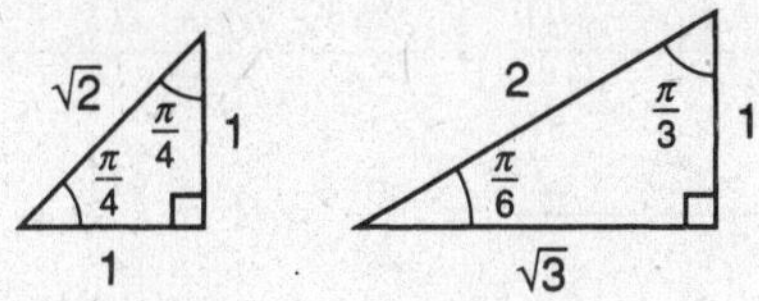

$$\csc\left(-\frac{3\pi}{4}\right) + \cos\left(-\frac{13\pi}{3}\right)$$

$$= -\csc\frac{\pi}{4} + \cos\frac{\pi}{3}$$

$$= \frac{1}{-\sin\frac{\pi}{4}} + \cos\frac{\pi}{3} = \mathbf{-\sqrt{2} + \frac{1}{2}}$$

12. (41)

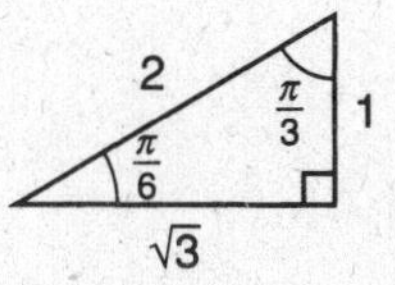

$$\sec\left(-\frac{19\pi}{6}\right) + \csc\left(-\frac{\pi}{3}\right) + \cos\frac{7\pi}{2}$$

$$= \frac{1}{\left(-\cos\frac{\pi}{6}\right)} + \frac{1}{\left(-\sin\frac{\pi}{3}\right)} + \cos\frac{3\pi}{2}$$

$$= -\frac{2}{\sqrt{3}} + \left(-\frac{2}{\sqrt{3}}\right) + 0 = -\frac{4}{\sqrt{3}} = \mathbf{-\frac{4\sqrt{3}}{3}}$$

13. (40) $\log_7 7^3 + \log_{10} 10^4 + \log_5 5 - \log_2 2$

$$= 3 + 4 + 1 - 1 = \mathbf{7}$$

14. (40) $\log_4 8 + \log_4 6 = \log_4 (3x + 2)$

$$\log_4 48 = \log_4 (3x + 2)$$

$$48 = 3x + 2$$

$$3x = 46$$

$$x = \mathbf{\frac{46}{3}}$$

15. (40) $\log_3 (x + 3) - \log_3 (x - 2) = \log_3 12$

$$\log_3 \frac{x+3}{x-2} = \log_3 12$$

$$\frac{x+3}{x-2} = 12$$

$$12x - 24 = x + 3$$

$$11x = 27$$

$$x = \mathbf{\frac{27}{11}}$$

16. (26) $\log_x 16 = 4$

$$x^4 = 16$$

$$x = 16^{\frac{1}{4}}$$

$$x = \mathbf{2}$$

17. (40) $\log_{15} x = \frac{2}{3}\log_{15} 8$

$$x = 8^{\frac{2}{3}} = \mathbf{4}$$

18. (40) $(f \circ g)(x) = \frac{1}{3}(3x + 9) - 3 = x + 3 - 3 = x$

$$(g \circ f)(x) = 3\left(\frac{1}{3}x - 3\right) + 9 = x - 9 + 9 = x$$

$$(g \circ f)(x) = (f \circ g)(x) = x$$

Yes, *f* and *g* are inverse functions.

19. (38)

$$(x - 3)\left(x + \frac{1}{4}\right) = 0$$

$$x^2 - 3x + \frac{1}{4}x - \frac{3}{4} = 0$$

$$x^2 - \frac{12}{4}x + \frac{1}{4}x - \frac{3}{4} = 0$$

$$\mathbf{x^2 - \frac{11}{4}x - \frac{3}{4} = 0}$$

20. (39)

$$y - y_1 = m(x - x_1)$$

$$y - 3 = -\frac{1}{4}(x + 2)$$

$$4y - 12 = -x - 2$$

$$\mathbf{x + 4y - 10 = 0}$$

21. (37,39) $\sqrt{(x + 3)^2 + (y + 2)^2} = \sqrt{(x - 2)^2 + (y - 3)^2}$

$$x^2 + 6x + 9 + y^2 + 4y + 4$$

$$= x^2 - 4x + 4 + y^2 - 6y + 9$$

$$10x + 10y = 0$$

$$\mathbf{x + y = 0}$$

22. (37) $x = \frac{x_1 + x_2}{2} = \frac{-3 + 8}{2} = \frac{5}{2}$

$$y = \frac{y_1 + y_2}{2} = \frac{4 - 3}{2} = \frac{1}{2}$$

$$\mathbf{\left(\frac{5}{2}, \frac{1}{2}\right)}$$

23. (35) $\frac{5}{4} + \frac{4}{5}\left(-\frac{20}{3} - \frac{5}{4}\right) = \frac{5}{4} + \frac{4}{5}\left(-\frac{95}{12}\right)$

$= \frac{5}{4} - \frac{19}{3} = \mathbf{\frac{-61}{12}}$

24. (34) From the graph of the line we choose two ordered pairs: (28, 0) and (37, 500).

$m = \frac{500 - 0}{37 - 28} = \frac{500}{9} = 55.56$

$F = 55.56D + b$

$0 = 55.56(28) + b$

$b = -1555.68$

$\mathbf{F = 55.56D - 1555.68}$

25. (34) $\mathbf{f(x) = \log_3 x}$

$\mathbf{g(x) = x + 2}$

26. (32) (a)

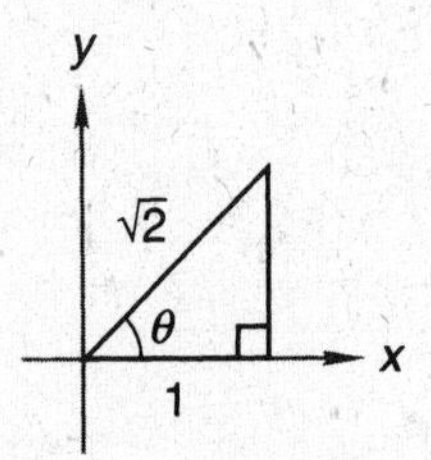

$0° \le \theta \le 180°$

$\text{Arccos}\left(\frac{\sqrt{2}}{2}\right) = \mathbf{45°}$

(b)

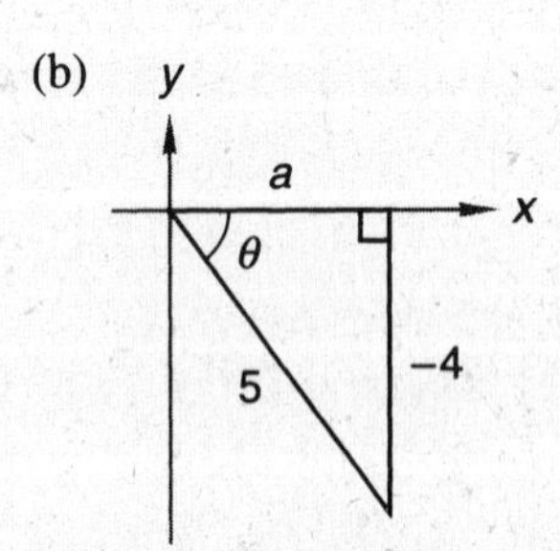

$-90° \le \theta \le 90°$

$a = \sqrt{5^2 - (-4)^2} = 3$

$\cos\left[\text{Arcsin}\left(-\frac{4}{5}\right)\right] = \cos\theta = \mathbf{\frac{3}{5}}$

27. (31) Replace y with $y + 1$ and x with $x + 3$.

$y + 1 = (x + 3)^2$

$\mathbf{y = (x + 3)^2 - 1}$

28. (21) $\frac{f(x + h) - f(x - h)}{2h} = \frac{2(x + h) - 2(x - h)}{2h}$

$= \frac{2x + 2h - 2x + 2h}{2h} = \frac{4h}{2h} = \mathbf{2}$

29. (1) $(x - y)^2 = x^2 - 2xy + y^2$

$= (x^2 + y^2) - 2xy = 40 - 2(16) = 8$

$12 > 8$

Therefore the answer is **B.**

30. (1,12) $m\angle 1 + m\angle 2 + m\angle 3 = 360°$

$360° > 180°$

Therefore the answer is **B.**

Problem Set 45

1. (45) $1 \cdot 5 \cdot 5 = \mathbf{25}$

2. (45)

MATH						ENGLISH		
6	5	4	3	2	1	3	2	1

→ $6! \times 3! = 4320$

ENGLISH			MATH					
3	2	1	6	5	4	3	2	1

→ $3! \times 6! = 4320$

$4320 + 4320 = \mathbf{8640}$

3. (44) $\text{Rate} = \frac{c}{fk}\ \frac{\text{articles}}{\text{worker-hr}}$

$RWT = \text{articles}$

$\frac{c}{fk}(k - x)T = c + 10$

$T = \mathbf{\frac{fk(c + 10)}{c(k - x)}\ hr}$

4. (39) $40.5°\left(\frac{\pi}{180°}\right)3960 = \mathbf{2799.16\ mi}$

5. (25,38) First, we write the equations.

(a) $R_O T_O = D_O$
(b) $R_B T_B = D_B$

$R_B = 2R_O,\ T_O + T_B = 9$

$D_O = D_B = 24\ \text{mi}$

$$T_O + T_B = 9$$
$$T_B = 9 - T_O$$

Next, we use substitution to solve.

(b) $R_B T_B = D_B$

$$2R_O(9 - T_O) = 24$$
$$18R_O - 2R_O T_O = 24$$
$$18R_O - 2(24) = 24$$
$$18R_O = 72$$
$$\mathbf{R_O = 4\ mph}$$

$$R_B = 2R_O$$
$$R_B = 2(4)$$
$$\mathbf{R_B = 8\ mph}$$

(a) $R_O T_O = D_O$

$$R_O T_O = 24$$
$$(4)T_O = 24$$
$$\mathbf{T_O = 6\ hr}$$

(b) $R_B T_B = D_B$

$$R_B T_B = 24$$
$$(8)T_B = 24$$
$$\mathbf{T_B = 3\ hr}$$

6. *(18)* $\text{Alcohol}_1 - \text{alcohol extracted} = \text{alcohol final}$

$$0.92(4000) - A = 0.8(4000 - A)$$
$$3680 - A = 3200 - 0.8A$$
$$0.2A = 480$$
$$A = \mathbf{2400\ liters}$$

7. *(18)* First, we write three equations.

$$\begin{cases} \text{(a)}\ \dfrac{N_G}{N_B} = \dfrac{2}{1} \\ \text{(b)}\ 2(N_B + N_W) = N_G + 10 \\ \text{(c)}\ N_G + N_B + N_W = 35 \end{cases}$$

Next, we can rearrange and use substitution to solve.

(a) $\dfrac{N_G}{N_B} = \dfrac{2}{1}$

$$N_G = 2N_B$$

(b) $2(N_B + N_W) = N_G + 10$

$$2N_B + 2N_W = (2N_B) + 10$$
$$2N_W = 10$$
$$\mathbf{N_W = 5}$$

(c) $N_G + N_B + N_W = 35$

$$(2N_B) + N_B + (5) = 35$$
$$3N_B = 30$$
$$\mathbf{N_B = 10}$$

$$N_G = 2N_B$$
$$N_G = 2(10)$$
$$\mathbf{N_G = 20}$$

8. *(45)* Use a calculator to get:

$\text{Cu} = a\text{Pb} + b$

$a = -0.2317$

$b = 55.2056$

$\mathbf{r = -0.8109}$

$\mathbf{Cu = -0.2317Pb + 55.2056}$

This is **not a good correlation** since r is not between 0.9 and 1.0 or between –0.9 and –1.0.

9. *(45)* Use a calculator to obtain:

$\text{Rh} = a\text{Dy} + b$

$a = 0.7686$

$b = -2.2571$

$\mathbf{r = 0.9944}$

$\mathbf{Rh = 0.7686Dy - 2.2571}$

This is a **good correlation** since $0.9 < |r| < 1.0$.

10. *(43)* Function = $-\sin\theta$

Amplitude = 10

$\mathbf{y = -10\sin\theta}$

11. *(43)* Function = $\cos x$

Amplitude = 8

$\mathbf{y = 8\cos x}$

12. *(42)* $\sqrt{[x - (-2)]^2 + (y - 5)^2} = 6$

$$\mathbf{(x + 2)^2 + (y - 5)^2 = 6^2}$$

13. (42) $f(x) = \left(\frac{1}{2}\right)^{-x+1} = \left[\left(\frac{1}{2}\right)^{-1}\right]^{x}\left(\frac{1}{2}\right)^{1} = \frac{1}{2} \cdot 2^{x}$

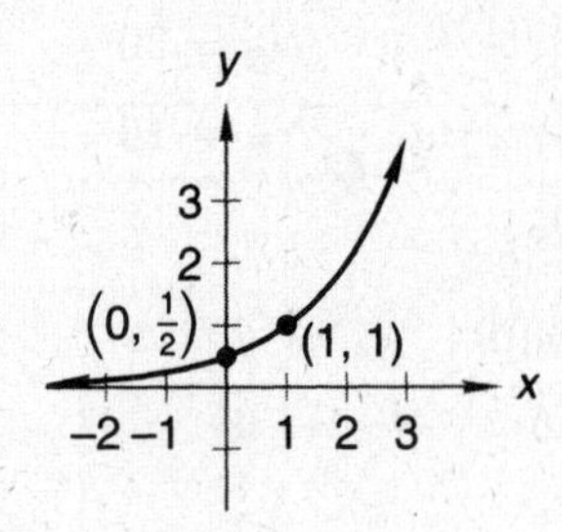

14. (41) $\csc \frac{3\pi}{4} - \sec\left(-\frac{5\pi}{6}\right) + \cos \frac{9\pi}{4}$

$= \frac{1}{\sin \frac{\pi}{4}} - \frac{1}{-\cos \frac{\pi}{6}} + \cos \frac{\pi}{4}$

$= \sqrt{2} - \frac{2}{\sqrt{3}} + \frac{1}{\sqrt{2}}$

$= \sqrt{2} - \frac{2\sqrt{3}}{3} + \frac{\sqrt{2}}{2} = \mathbf{\frac{9\sqrt{2} + 4\sqrt{3}}{6}}$

15. (41) $\sec\left(-\frac{19\pi}{6}\right) + \cos \frac{7\pi}{2} - \sin \frac{10\pi}{3}$

$= \frac{1}{-\cos \frac{\pi}{6}} + \cos \frac{3\pi}{2} - \left(-\sin \frac{\pi}{3}\right)$

$= -\frac{2}{\sqrt{3}} + 0 + \frac{\sqrt{3}}{2} = \mathbf{-\frac{\sqrt{3}}{6}}$

16. (41) ${}_6P_3 - {}_6P_2 = \frac{6!}{(6-3)!} - \frac{6!}{(6-2)!}$

$= \frac{6!}{3!} - \frac{6!}{4!} = \mathbf{90}$

17. (40) $\log_3 9 - \log_5 5^3 + \log_7 7^2 - \log_{11} 1$

$= \log_3 3^2 - \log_5 5^3 + \log_7 7^2 - \log_{11} 1$

$= 2 - 3 + 2 - 0 = \mathbf{1}$

18. (40) $\log_5 7 + \log_5 8 = \log_5 (2x - 4)$

$2x - 4 = 7 \cdot 8$

$2x = 60$

$x = \mathbf{30}$

19. (40) $\log_3 (x + 1) - \log_3 x = \log_3 15$

$\frac{x+1}{x} = 15$

$x + 1 = 15x$

$x = \mathbf{\frac{1}{14}}$

20. (40) $\frac{3}{4} \log_{10} 10{,}000 = x$

$\log_{10} 10^{4\left(\frac{3}{4}\right)} = x$

$\log_{10} 10^3 = x$

$x = \mathbf{3}$

21. (40) $(f \circ g)(x) = \frac{1}{\left(\frac{1}{x}\right)} = x$

$(g \circ f)(x) = \frac{1}{\left(\frac{1}{x}\right)} = x$

$(g \circ f)(x) = (f \circ g)(x) = x$

Yes, *f* and *g* are inverse functions.

22. (38) $\left[x - (2 + \sqrt{5})\right]\left[x - (2 - \sqrt{5})\right] = 0$

$x^2 - (2 + \sqrt{5})x - (2 - \sqrt{5})x$

$+ (2 + \sqrt{5})(2 - \sqrt{5}) = 0$

$x^2 - 2x - \sqrt{5}x - 2x + \sqrt{5}x + 4 - 5 = 0$

$\mathbf{x^2 - 4x - 1 = 0}$

23. (37) $\sqrt{(x+4)^2 + (y+3)^2} = \sqrt{(x-4)^2 + (y-6)^2}$

$x^2 + 8x + 16 + y^2 + 6y + 9$

$= x^2 - 8x + 16 + y^2 - 12y + 36$

$16x + 18y = 27$

$18y = -16x + 27$

$\mathbf{y = -\frac{8}{9}x + \frac{3}{2}}$

24. (34) $\mathbf{f(x) = 2x + 3}$

25. (21,22) Domain of $f(x) = \mathbf{\left\{x \in \mathbb{R} \mid x \geq 0, x \neq \frac{1}{4}\right\}}$

26. (31) Replace *y* with $y - 3$ and *x* with $x - 2$.

$\mathbf{y = \sqrt{x - 2} + 3}$

27. *(31)*

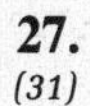

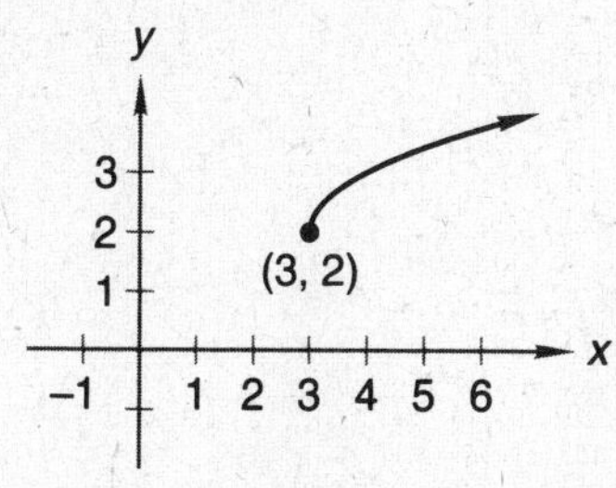

Replace $f(x)$ with $g(x) - 2$ and x with $x - 3$.

$\mathbf{g(x) = \sqrt{x - 3} + 2}$

28. *(15)*

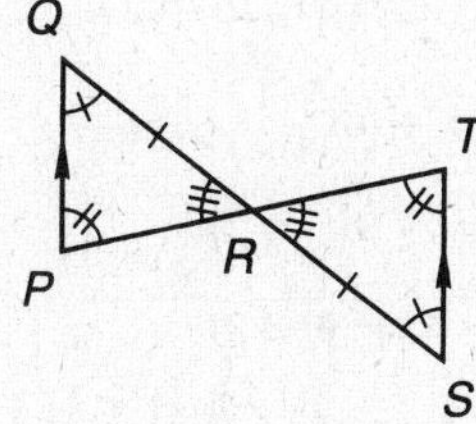

STATEMENTS	REASONS
1. $\overline{PQ} \parallel \overline{ST}$	1. Given
2. $\angle Q \cong \angle S$ $\angle P \cong \angle T$	2. If parallel lines are intersected by a transversal, then each pair of alternate interior angles is congruent.
3. R is the midpoint of $\overline{QS}$.	3. Given
4. $\overline{QR} \cong \overline{RS}$	4. A midpoint divides a segment into two congruent segments.
5. $\angle QRP \cong \angle SRT$	5. Vertical angles are congruent.
6. $\triangle QRP \cong \triangle SRT$	6. *AAAS* congruency postulate
7. $\overline{PR} \cong \overline{TR}$	7. *CPCTC*

29. *(1)*

$$x < 1 \qquad\qquad y > 3$$
$$x + 1 < 2 \qquad\qquad y - 1 > 2$$

$x + 1 < y - 1$

Therefore the answer is **B.**

30. *(1,2)*

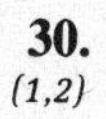

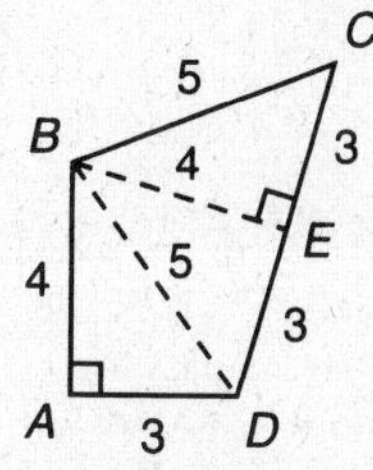

Draw $\overline{BD}$.

$BD = \sqrt{4^2 + 3^2} = 5$

Since $BD = BC$, if we draw $BE \perp CD$, BE will bisect CD, so $DE = CE = 3$ and $BE = \sqrt{5^2 - 3^2} = 4$.

$$\text{Area of } ABCD = \text{Area } \Delta BAD + \text{Area } \Delta BDE + \text{Area } \Delta BEC$$
$$= \frac{1}{2}(4)(3) + \frac{1}{2}(4)(3) + \frac{1}{2}(4)(3)$$
$$= 6 + 6 + 6 = 18$$

$18 > 16$

Therefore the answer is **A.**

Problem Set 46

1. *(45)* $1 \cdot 5 \cdot 5 = \mathbf{25}$

2. *(45)*

LITERATURE						MATH			
6	5	4	3	2	1	4	3	2	1

→ $6! \times 4! = 17{,}280$

MATH				LITERATURE					
4	3	2	1	6	5	4	3	2	1

→ $4! \times 6! = 17{,}280$

$17{,}280 + 17{,}280 = \mathbf{34{,}560}$

3. *(44)* $\text{Rate} = \dfrac{x}{dy}\,\dfrac{\text{food}}{\text{worker-day}}$

$RWT = \text{food}$

$\dfrac{x}{dy}(y + 50)T = x$

$\mathbf{T = \dfrac{dy}{y + 50}}$ **days**

4. *(44)* $\text{Rate} = \dfrac{d}{b}\,\dfrac{\text{dollars}}{\text{ball}}$

$\text{New rate} = \dfrac{d}{b} + 2 = \dfrac{d + 2b}{b}\,\dfrac{\text{dollars}}{\text{ball}}$

$\text{New rate} \times N = \text{price}$

$\left(\dfrac{d + 2b}{b}\right)N = 200$

$\mathbf{N = \dfrac{200b}{d + 2b}}$ **balls**

5. *(39)* $60°\left(\dfrac{\pi}{180°}\right)600 = 200\pi = \mathbf{628.32\ m}$

6. (38)
$$\begin{cases} \text{(a)} & R_O T_O = D_O \\ \text{(b)} & R_B T_B = D_B \end{cases}$$

$R_B = 2R_O$

$T_O + T_B = 6$

$D_B = D_O = 36 \text{ mi}$

$$T_O + T_B = 6$$
$$T_B = 6 - T_O$$

(b)
$$R_B T_B = D_B$$
$$(2R_O)(6 - T_O) = 36$$
$$12R_O - 2R_O T_O = 36$$
$$12R_O - 2(36) = 36$$
$$12R_O = 108$$
$$\mathbf{R_O = 9 \text{ mph}}$$

$R_B = 2R_O$

$R_B = 2(9)$

$\mathbf{R_B = 18 \text{ mph}}$

7. (18) Alcohol$_1$ − alcohol extracted = alcohol final
$$0.9(1000) - A = 0.8(1000 - A)$$
$$900 - A = 800 - 0.8A$$
$$0.2A = 100$$
$$\mathbf{A = 500 \text{ liters}}$$

8. (18)
$$\begin{cases} \text{(a)} & \dfrac{N_R}{N_E} = \dfrac{3}{1} \\ \text{(b)} & \dfrac{N_E}{N_D} = \dfrac{2}{1} \\ \text{(c)} & N_R + N_E + N_D = 18 \end{cases}$$

(b) $\dfrac{N_E}{N_D} = \dfrac{2}{1}$

$$N_D = \frac{1}{2}N_E$$

(a) $\dfrac{N_R}{N_E} = \dfrac{3}{1}$

$$N_R = 3N_E$$

(c)
$$N_R + N_E + N_D = 18$$
$$(3N_E) + N_E + \left(\frac{1}{2}N_E\right) = 18$$
$$\frac{9}{2}N_E = 18$$
$$\mathbf{N_E = 4}$$

$N_R = 3N_E$

$N_R = 3(4)$

$\mathbf{N_R = 12}$

$N_D = \dfrac{1}{2}N_E$

$N_D = \dfrac{1}{2}(4)$

$\mathbf{N_D = 2}$

9. (46)
$$[x - (2 - 3i)][x - (2 + 3i)] = 0$$
$$(x - 2 + 3i)(x - 2 - 3i) = 0$$
$$x^2 - 2x - 3ix - 2x + 4 + 6i$$
$$+ 3ix - 6i - 9i^2 = 0$$
$$x^2 - 4x + 4 + 9 = 0$$
$$\mathbf{x^2 - 4x + 13 = 0}$$

10. (46) $x^2 + 3x + 6 = 0$

$$x = \frac{-3 \pm \sqrt{9 - 4(1)(6)}}{2(1)}$$

$$x = \frac{-3 \pm \sqrt{-15}}{2} = -\frac{3}{2} \pm \frac{\sqrt{15}}{2}i$$

$$\left(\boldsymbol{x + \frac{3}{2} - \frac{\sqrt{15}}{2}i}\right)\left(\boldsymbol{x + \frac{3}{2} + \frac{\sqrt{15}}{2}i}\right)$$

11. (45) Use a calculator to get:

C = aH + b

$a = -0.1138$

$b = 50.1680$

$\mathbf{r = -0.9439}$

C = −0.1138H + 50.1680

This is a **good correlation** since $0.9 < |r| < 1.0$.

12. (43) Function = $-\cos x$

Amplitude = 12

$\mathbf{y = -12 \cos x}$

13. (43) Function = $-\sin \theta$

Amplitude = 9

$\mathbf{y = -9 \sin \theta}$

14. (42) $\sqrt{(x-h)^2 + (y-k)^2} = 6$

$$\mathbf{(x - h)^2 + (y - k)^2 = 6^2}$$

15. (42) $f(x) = \left(\frac{1}{3}\right)^{-x+2} = \left[\left(\frac{1}{3}\right)^{-1}\right]^x \left(\frac{1}{3}\right)^2 = \frac{1}{9} \cdot 3^x$

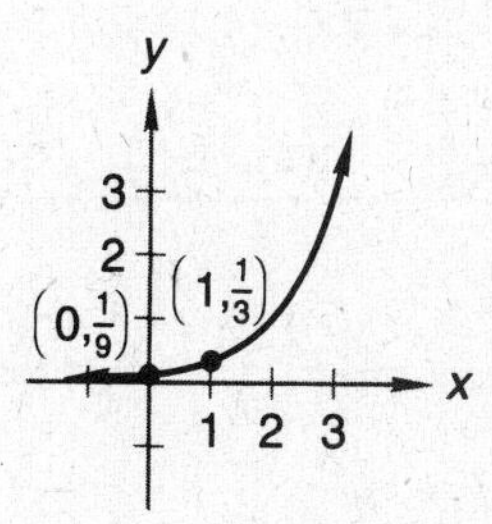

16. (41) (a) $\cos\left(-\frac{13\pi}{6}\right) + \sin\frac{13\pi}{4} - \cot\frac{9\pi}{2}$

$$= \cos\frac{\pi}{6} + \left(-\sin\frac{\pi}{4}\right) - \cot\frac{\pi}{2}$$

$$= \frac{\sqrt{3}}{2} - \frac{\sqrt{2}}{2} - 0$$

$$= \mathbf{\frac{\sqrt{3} - \sqrt{2}}{2}}$$

(b) $\csc\frac{7\pi}{4} - \sec\left(-\frac{\pi}{4}\right) + \tan 3\pi$

$$= \frac{1}{\left(-\sin\frac{\pi}{4}\right)} - \frac{1}{\cos\frac{\pi}{4}} + \tan\pi$$

$$= -\frac{2}{\sqrt{2}} - \frac{2}{\sqrt{2}} + 0 = -\frac{4}{\sqrt{2}} = \mathbf{-2\sqrt{2}}$$

17. (41) ${}_8P_4 - {}_8P_2 = \frac{8!}{(8-4)!} - \frac{8!}{(8-2)!}$

$$= \frac{8!}{4!} - \frac{8!}{6!} = \mathbf{1624}$$

18. (40) $3\log_7 7 - \log_5 25 + \log_3\frac{1}{3} + \log_{10} 1$

$$= \log_7 7^3 - \log_5 5^2 + \log_3 3^{-1} + \log_{10} 1$$

$$= 3 - 2 - 1 + 0 = \mathbf{0}$$

19. (40) $\frac{5}{2}\log_4 4 = \log_4(2x - 3)$

$$\log_4 4^{\frac{5}{2}} = \log_4(2x - 3)$$

$$4^{\frac{5}{2}} = 2x - 3$$

$$32 = 2x - 3$$

$$35 = 2x$$

$$x = \mathbf{\frac{35}{2}}$$

20. (40) $\log_3(x + 2) - \log_3 x = \log_3 10$

$$\log_3\frac{x+2}{x} = \log_3 10$$

$$\frac{x+2}{x} = 10$$

$$10x = x + 2$$

$$9x = 2$$

$$x = \mathbf{\frac{2}{9}}$$

21. (40) $\log_8 x = \frac{3}{4}\log_8 81$

$$\log_8 x = \log_8 81^{\frac{3}{4}}$$

$$x = 81^{\frac{3}{4}}$$

$$x = \mathbf{27}$$

22. (40) $(f \circ g)(x) = 3\left(\frac{1}{3}x - \frac{1}{3}\right) + 1 = x - 1 + 1 = x$

$$(g \circ f)(x) = \frac{1}{3}(3x + 1) - \frac{1}{3} = x + \frac{1}{3} - \frac{1}{3} = x$$

$$(f \circ g)(x) = (g \circ f)(x) = x$$

Yes, f **and** g **are inverse functions.**

23. (33) $\text{Median} = \frac{1}{2}(B_1 + B_2)$

$$12x + 4 = \frac{1}{2}[(6x + 2) + (45x - 3)]$$

$$24x + 8 = 51x - 1$$

$$-27x = -9$$

$$x = \mathbf{\frac{1}{3}}$$

24. (32) (a)

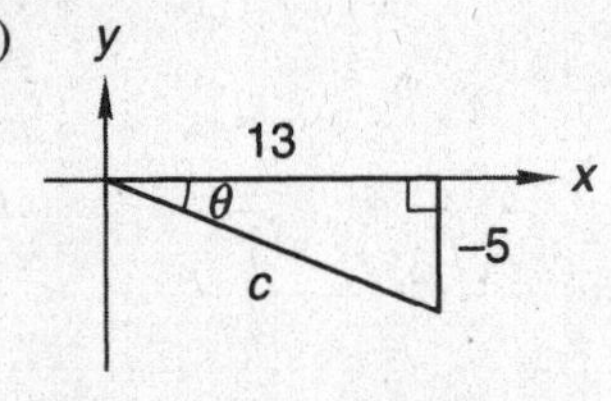

$-90° < \theta < 90°$

$c = \sqrt{13^2 + (-5)^2} = \sqrt{194}$

$\sin\left[\text{Arctan}\left(-\frac{5}{13}\right)\right]$

$= \sin\theta = \frac{-5}{\sqrt{194}} = \mathbf{\frac{-5\sqrt{194}}{194}}$

(b)

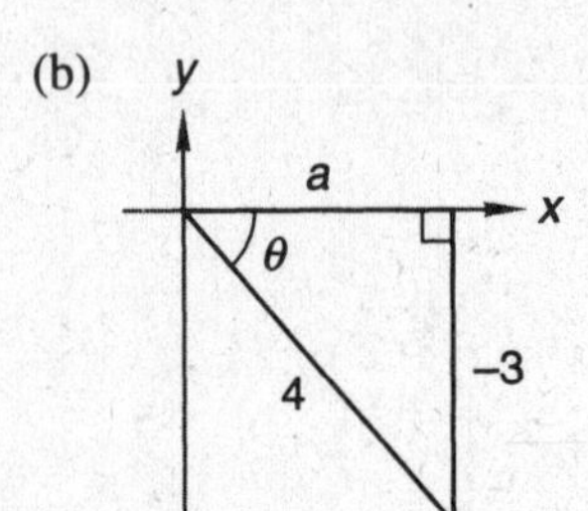

$-90° \le \theta \le 90°$

$a = \sqrt{4^2 - (-3)^2} = \sqrt{7}$

$\cos\left[\text{Arcsin}\left(-\frac{3}{4}\right)\right] = \cos\theta = \mathbf{\frac{\sqrt{7}}{4}}$

25. (31) Replace y with $y + 4$ and x with $x - 1$.

$y + 4 = \sqrt{x - 1}$

$\mathbf{y = \sqrt{x - 1} - 4}$

26. (31) Replace $f(x)$ with $g(x) + 5$ and x with $x - 4$.

$g(x) + 5 = \sqrt{x - 4}$

$\mathbf{g(x) = \sqrt{x - 4} - 5}$

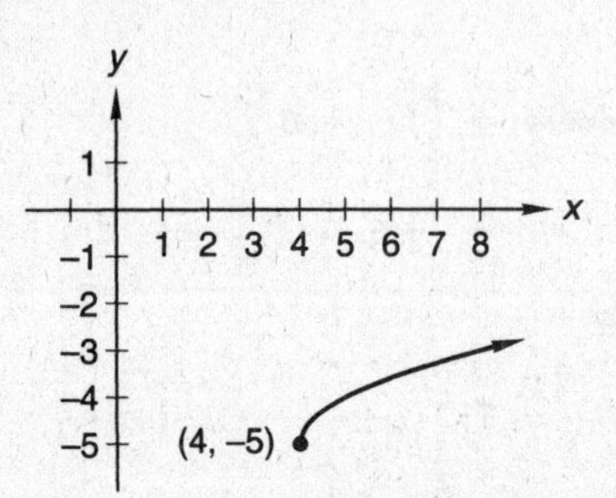

27. (21) (a) **Domain** $= \{x \in \mathbb{R} \mid -4 \le x \le 6\}$

Range $= \{y \in \mathbb{R} \mid -2 \le y \le 3\}$

(b) **Domain** $= \{x \in \mathbb{R} \mid -3 \le x \le 2\}$

Range $= \{y \in \mathbb{R} \mid -4 \le y \le 4\}$

28. (21) $\frac{f(x + h) - f(x)}{h}$

$= \frac{(x + h)[1 - (x + h)] - x(1 - x)}{h}$

$= \frac{x + h - x^2 - 2xh - h^2 - x + x^2}{h}$

$= \frac{h - 2xh - h^2}{h} = \mathbf{1 - 2x - h}$

29. (1) $\frac{x^2 + 2xy + y^2}{x + y} = \frac{(x + y)(x + y)}{x + y} = x + y$

$x + y < x + y + 2$

Therefore the answer is **B.**

30. (1,20) $3x = 180$

$x = 60$

Since $x = 60$, $m\angle A = 30°$.

So, $\sin m\angle A = \frac{BC}{AB}$

$BC = AB \sin 30° = 8\left(\frac{1}{2}\right) = 4$

Therefore the answer is **C.**

Problem Set 47

1. (45)

2	**3**	2	1	→	$2 \cdot 1 = 2$
3	**2**	2	1	→	$2 \cdot 1 = 2$
2	**2**	**3**	1	→	$2 \cdot 1 = 2$
2	**3**	**2**	1	→	$2 \cdot 1 = 2$
2	1	**2**	**3**	→	$2 \cdot 1 = 2$
2	1	**3**	**2**	→	$2 \cdot 1 = 2$

12

2. (44) $\text{Rate} = \dfrac{1}{dm}\dfrac{\text{job}}{\text{worker-day}}$

$RWT = \text{jobs}$

$\dfrac{1}{dm}(40)T = 1$

$T = \dfrac{dm}{40}$ **days**

3. (44) $\text{Rate} = \dfrac{x}{dk}\dfrac{\text{items}}{\text{worker-hr}}$

$RWT = \text{items}$

$\dfrac{x}{dk}(k - y)(T) = x$

$T = \dfrac{dk}{k - y}$ **hr**

4. (44) $\text{Rate} = \dfrac{d}{12}\dfrac{\text{dollars}}{\text{pencil}}$

$\text{Rate} \times N = \text{price}$

$\dfrac{d}{12} \times m = \dfrac{dm}{12}$ **dollars**

5. (39) $40°\left(\dfrac{\pi}{180°}\right)20 = \dfrac{40\pi}{9} = \mathbf{13.96\ ft}$

6. (36) (a) $\begin{cases}(B + W)T_D = D_D \\ (B - W)T_U = D_U\end{cases}$ (b)

$B = 4W,\ T_D = T_U + 1$

(b) $(B - W)T_U = D_U$

$(4W - W)T_U = 30$

$3WT_U = 30$

$WT_U = 10$

(a) $(B + W)T_D = D_D$

$5W(T_U + 1) = 100$

$5WT_U + 5W = 100$

$5(10) + 5W = 100$

$5W = 50$

$\mathbf{W = 10\ mph}$

$B = 4W$

$B = 4(10)$

$\mathbf{B = 40\ mph}$

7. (47) Function $= -\sin\theta$

Centerline $= -5$

Amplitude $= 25$

$\mathbf{y = -5 - 25\sin\theta}$

8. (47) Function $= \cos x$

Centerline $= -3$

Amplitude $= 7$

$\mathbf{y = -3 + 7\cos x}$

9. (47) (a)

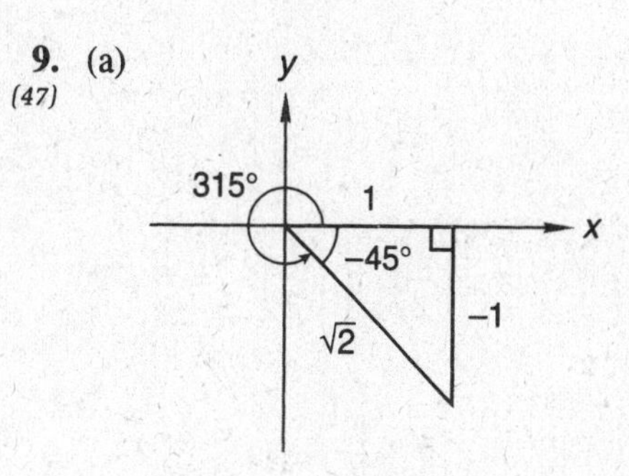

$-90° < \theta < 90°$

$\text{Arctan}(\tan 315°) = \text{Arctan}(-1) = \mathbf{-45°}$

(b)

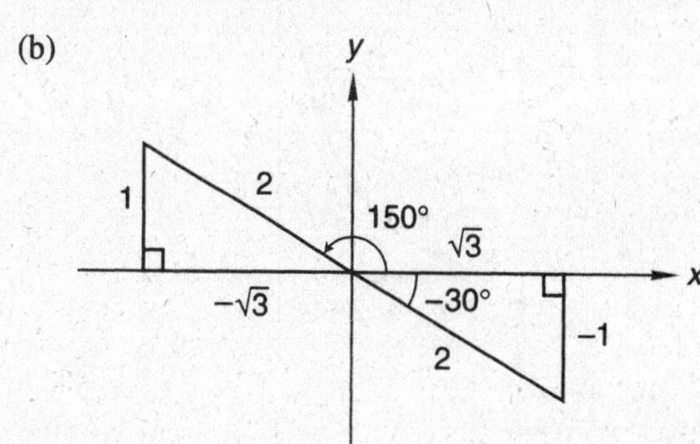

$-90° < \theta < 90°$

$\text{Arctan}(\tan 150°) = \text{Arctan}\left(-\dfrac{1}{\sqrt{3}}\right) = \mathbf{-30°}$

10. (46)

$[x - (3 + 2i)][x - (3 - 2i)] = 0$

$(x - 3 - 2i)(x - 3 + 2i) = 0$

$x^2 - 3x + 2ix - 3x + 9 - 6i - 2ix + 6i - 4i^2 = 0$

$x^2 - 6x + 9 + 4 = 0$

$\mathbf{x^2 - 6x + 13 = 0}$

11. (46) $2(x^2 + 2x + 2) = 0$

$x = \dfrac{-2 \pm \sqrt{4 - 4 \cdot 2}}{2} = \dfrac{-2 \pm \sqrt{-4}}{2} = -1 \pm i$

$\mathbf{2(x + 1 - i)(x + 1 + i)}$

12. Use a calculator to get:
(45)

$O = aN + b$

$a = 0.2434$

$b = -1.1762$

$\mathbf{r = 0.9596}$

$\mathbf{O = 0.2434N - 1.1762}$

This is a **good correlation** since $0.9 < |r| < 1$.

13. $\sqrt{(x-2)^2 + (y-1)^2} = 3$
(42)

$$\mathbf{(x-2)^2 + (y-1)^2 = 3^2}$$

14. $f(x) = \left(\frac{1}{2}\right)^{-x-3} = \left[\left(\frac{1}{2}\right)^{-1}\right]^x \left(\frac{1}{2}\right)^{-3} = 8 \cdot 2^x$
(42)

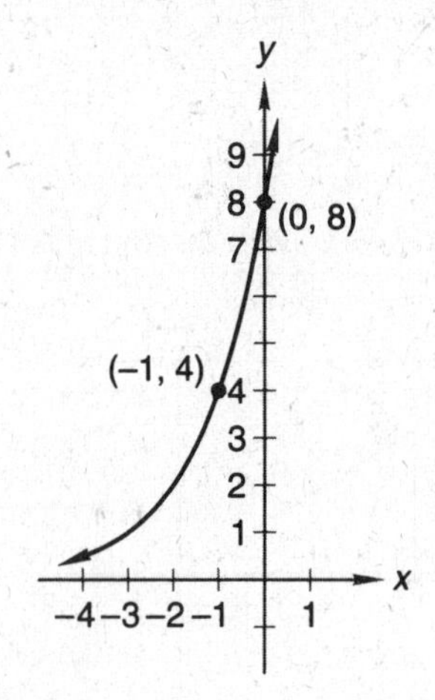

15. (a) $\cot\left(-\frac{3\pi}{4}\right) - \csc\frac{5\pi}{6} + \sin\frac{7\pi}{6} - \cos\left(-\frac{\pi}{4}\right)$
(41)

$$= \frac{1}{\tan\frac{\pi}{4}} - \frac{1}{\sin\frac{\pi}{6}} + \left(-\sin\frac{\pi}{6}\right) - \cos\frac{\pi}{4}$$

$$= 1 - 2 - \frac{1}{2} - \frac{\sqrt{2}}{2} = \mathbf{\frac{-3-\sqrt{2}}{2}}$$

(b) $\sec\frac{13\pi}{6}\cos\frac{13\pi}{6} = \frac{1}{\cos\frac{\pi}{6}}\cos\frac{\pi}{6} = \mathbf{1}$

16. $\log_2 8 + 3\log_3 3 - \log_2 1 + \log_4 \frac{1}{4}$
(40)

$= \log_2 2^3 + \log_3 3^3 - \log_2 1 + \log_4 4^{-1}$

$= 3 + 3 - 0 - 1 = \mathbf{5}$

17. $\log_3 4 + \log_3 5 = \log_3(4x + 5)$
(40)

$$\log_3(4 \cdot 5) = \log_3(4x+5)$$

$$20 = 4x + 5$$

$$4x = 15$$

$$x = \mathbf{\frac{15}{4}}$$

18. $\log_7 6 - \log_7(x - 1) = \log_7 3$
(40)

$$\log_7 \frac{6}{x-1} = \log_7 3$$

$$\frac{6}{x-1} = 3$$

$$3x - 3 = 6$$

$$3x = 9$$

$$x = \mathbf{3}$$

19. $3\log_x 4 = 2$
(40)

$$\log_x 4^3 = 2$$

$$x^2 = 4^3$$

$$x^2 = 64$$

$$x = \mathbf{8}$$

20. $2x - 6y = 3$
(39)

$$\frac{2}{3}x - 2y = 1$$

$$\mathbf{\frac{x}{\frac{3}{2}} + \frac{y}{-\frac{1}{2}} = 1}$$

21. $(x, y) = \left(\frac{-4+6}{2}, \frac{-2+4}{2}\right) = \mathbf{(1, 1)}$
(37)

22. $-3\frac{2}{3} + \frac{4}{5}\left(4\frac{1}{4} + 3\frac{2}{3}\right) = -\frac{11}{3} + \frac{4}{5}\left(\frac{17}{4} + \frac{11}{3}\right)$
(35)

$$= -\frac{11}{3} + \frac{4}{5}\left(\frac{51+44}{12}\right) = -\frac{11}{3} + \frac{19}{3} = \mathbf{\frac{8}{3}}$$

23. $\sum_{k=-1}^{2} \frac{k^2+1}{2}$
(34)

$$= \frac{(-1)^2+1}{2} + \frac{0+1}{2} + \frac{1^2+1}{2} + \frac{2^2+1}{2}$$

$$= 1 + \frac{1}{2} + 1 + \frac{5}{2} = \mathbf{5}$$

24. (a)
(32)

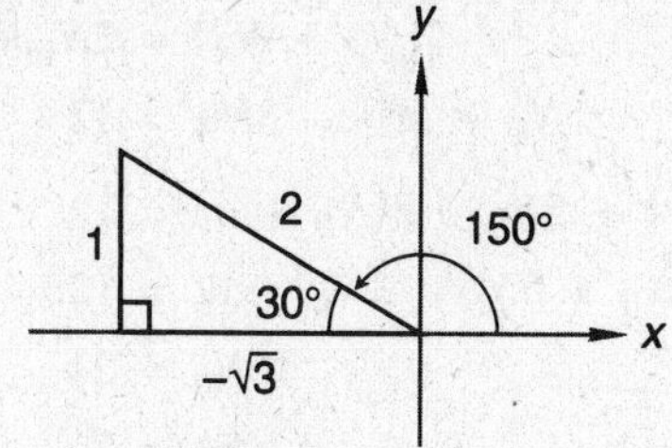

$0° \le \theta \le 180°$

$\text{Arccos}\left(-\frac{\sqrt{3}}{2}\right) = \mathbf{150°}$

(b)

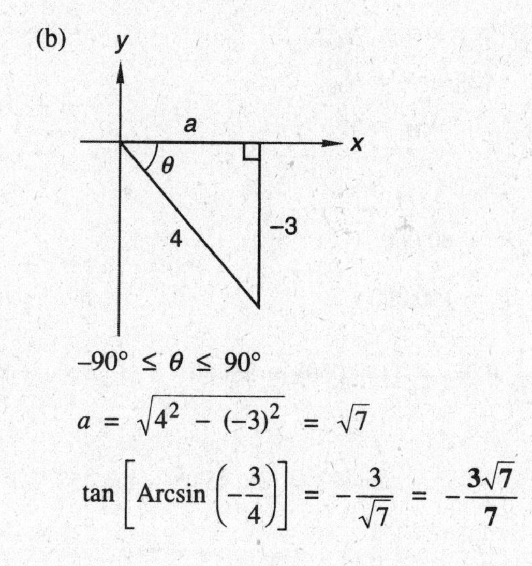

$-90° \le \theta \le 90°$

$a = \sqrt{4^2 - (-3)^2} = \sqrt{7}$

$\tan\left[\text{Arcsin}\left(-\frac{3}{4}\right)\right] = -\frac{3}{\sqrt{7}} = -\frac{\mathbf{3\sqrt{7}}}{\mathbf{7}}$

25. (a) ***x* axis, yes; *y* axis, no; origin, no**
(31)
(b) ***x* axis, no; *y* axis, yes; origin, no**

26. Replace y with $y - 2$ and x with $x + 4$.
(31)

$y - 2 = \sqrt{x + 4}$

$\mathbf{y = \sqrt{x + 4} + 2}$

27. Replace $f(x)$ with $g(x) - 1$ and x with $x + 4$.
(31)

$g(x) - k = \sqrt{x - h}$

$g(x) - 1 = \sqrt{x + 4}$

$\mathbf{g(x) = \sqrt{x + 4} + 1}$

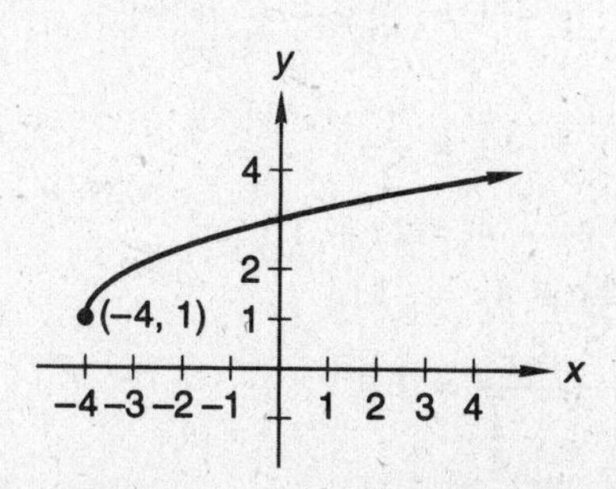

28. (a) $(f - g)(x) = 3x^2 - \left(x - \frac{2}{x}\right)$
(24)

$= 3x^2 - x + \frac{2}{x}$

$(f - g)(3) = 3(3)^2 - 3 + \frac{2}{3}$

$= 27 - 2\frac{1}{3} = \mathbf{\frac{74}{3}}$

(b) $\left(\frac{f}{g}\right)(x) = \frac{3x^2}{x - \frac{2}{x}}$

$\left(\frac{f}{g}\right)(3) = \frac{3(3)^2}{3 - \frac{2}{3}} = \frac{27}{\frac{7}{3}} = \mathbf{\frac{81}{7}}$

(c) $(g \circ f)(x) = (3x^2) - \frac{2}{(3x^2)}$

$(g \circ f)(3) = 3(3)^2 - \frac{2}{3(3)^2} = 27 - \frac{2}{27}$

$= \frac{729}{27} - \frac{2}{27} = \mathbf{\frac{727}{27}}$

29. A: $(x - y)^2 = x^2 - 2xy + y^2$
(1)

B: $(y - x)^2 = y^2 - 2yx + x^2$

$x^2 - 2xy + y^2 = y^2 - 2yx + x^2$

A and B are equal. Therefore the answer is **C.**

30. (a) $\begin{cases} x - y = 60 \\ x + y = 180 \end{cases}$ (b)
(1)

(a) $x - y = 60$

$x = 60 + y$

(b) $x + y = 180$

$(60 + y) + y = 180$

$2y = 120$

$y = 60$

$60 < 65$

A < B

Therefore the answer is **B.**

Problem Set 48

1. $2 \cdot 8 \cdot 5 = \mathbf{80}$
(45)

2. $9 \cdot 9 \cdot 4 = \mathbf{324}$
(45)

3. (39) $42.4°\left(\frac{\pi}{180°}\right)3960 = 932.8\pi = \mathbf{2930.48\text{ mi}}$

4. (25) $\text{Cottage} = 12\text{ oz}\cdot\frac{216\text{ cal}}{8\text{ oz}} = 324\text{ cal}$

$\text{Cream} = 6\text{ oz}\cdot\frac{106\text{ cal}}{\text{oz}} = 636\text{ cal}$

$36\text{ oz}\cdot\frac{324\text{ cal} + 636\text{ cal}}{12\text{ oz} + 6\text{ oz}} = \mathbf{1920\text{ cal}}$

5. (44) $R_m = \frac{1}{m},\ R_1 = \frac{1}{W_1},\ R_2 = \frac{1}{W_2}$

$$R_m = R_1 + R_2$$

$$\frac{1}{m} = \frac{1}{W_1} + \frac{1}{W_2}$$

$$\frac{1}{m} = \frac{W_2 + W_1}{W_1W_2}$$

$$\mathbf{m = \frac{W_1W_2}{W_1 + W_2}}$$

6. (18) $\begin{cases}\text{(a) } 0.4P_N + 0.8D_N = 0.5(600)\\ \text{(b) } P_N + D_N = 600\end{cases}$

(b) $P_N + D_N = 600$

$$P_N = 600 - D_N$$

(a) $0.4P_N + 0.8D_N = 0.5(600)$

$$0.4(600 - D_N) + 0.8D_N = 300$$

$$0.40D_N = 60$$

$$D_N = \mathbf{150\text{ ml of }80\%}$$

$$P_N = 600 - D_N$$
$$P_N = 600 - (150)$$
$$P_N = \mathbf{450\text{ ml of }40\%}$$

7. (18) $\begin{cases}\text{(a) } \frac{N_R}{N_B} = \frac{2}{1}\\ \text{(b) } 5(N_R + N_B) = 3N_W + 12\\ \text{(c) } N_B + 4 = N_W\end{cases}$

(a) $\frac{N_R}{N_B} = \frac{2}{1}$

$$N_R = 2N_B$$

(b) $5(N_R + N_B) = 3N_W + 12$

$$5N_R + 5N_B - 3N_W = 12$$
$$5(2N_B) + 5N_B - 3(N_B + 4) = 12$$
$$10N_B + 5N_B - 3N_B - 12 = 12$$
$$12N_B = 24$$
$$\mathbf{N_B = 2}$$

$$N_R = 2N_B$$
$$N_R = 2(2)$$
$$\mathbf{N_R = 4}$$

(c) $N_B + 4 = N_W$

$$(2) + 4 = N_W$$
$$\mathbf{N_W = 6}$$

8. (25) $\frac{3}{7}P = 60{,}000$

$$P = 140{,}000$$

$$\frac{1}{28}P = \frac{1}{28}(140{,}000) = \mathbf{\$5000}$$

9. (48) $\csc^2 405° - \tan 45° = (\csc 45°)^2 - \tan 45°$

$$= \left(\frac{\sqrt{2}}{1}\right)^2 - 1 = 2 - 1 = \mathbf{1}$$

10. (48) $\sec^2(-30°) + \sin^3 90° = (\sec 30°)^2 + (1)^3$

$$= \left(\frac{2}{\sqrt{3}}\right)^2 + 1 = \frac{4}{3} + 1 = \mathbf{\frac{7}{3}}$$

11. (39,48) $m_0 = -\frac{10}{5} = -2$

$$(x_m, y_m) = \left(\frac{3+8}{2}, \frac{6-4}{2}\right) = \left(\frac{11}{2}, 1\right)$$

The slope of the perpendicular bisector is the negative reciprocal so, $m = \frac{1}{2}$.

$$y - y_m = m(x - x_m)$$
$$y - 1 = \frac{1}{2}\left(x - \frac{11}{2}\right)$$
$$y - 1 = \frac{1}{2}x - \frac{11}{4}$$
$$4y - 4 = 2x - 11$$
$$-2x + 4y = -7$$
$$\mathbf{\frac{x}{\frac{7}{2}} + \frac{y}{-\frac{7}{4}} = 1}$$

12. (39,48) $m_0 = \frac{-8}{6} = -\frac{4}{3}$

$$(x_m, y_m) = \left(\frac{-2+4}{2}, \frac{6-2}{2}\right) = (1, 2)$$

The slope of the perpendicular bisector is the negative reciprocal so, $m = \frac{3}{4}$.

$$y - y_m = m(x - x_m)$$
$$y - 2 = \frac{3}{4}(x - 1)$$
$$y - 2 = \frac{3}{4}x - \frac{3}{4}$$
$$\mathbf{y = \frac{3}{4}x + \frac{5}{4}}$$

13. (47) Function = $-\cos x$

Centerline = -5

Amplitude = 20

$\mathbf{y = -5 - 20\cos x}$

14. (47) Function = $-\sin\theta$

Centerline = 6

Amplitude = 14

$\mathbf{y = 6 - 14\sin\theta}$

15. (47) (a) $-90° < \theta < 90°$

Arctan (tan 240°) = Arctan (tan 60°) = **60°**

(b) $-90° < \theta < 90°$

Arctan (tan 330°) = Arctan [tan (−30°)] = **−30°**

16. (46)
$$[x - (1 + 3i)][x - (1 - 3i)] = 0$$
$$(x - 1 - 3i)(x - 1 + 3i) = 0$$
$$x^2 - x + 3ix - x + 1 - 3i - 3ix + 3i - 9i^2 = 0$$
$$\mathbf{x^2 - 2x + 10 = 0}$$

17. (46) $x^2 - 2x + 3 = 0$

$$x = \frac{-(-2) \pm \sqrt{(-2)^2 - 4(1)(3)}}{2(1)}$$
$$x = \frac{2 \pm \sqrt{-8}}{2}$$
$$x = \frac{2 \pm 2\sqrt{2}i}{2} = 1 \pm \sqrt{2}i$$
$$\mathbf{(x - 1 - \sqrt{2}i)(x - 1 + \sqrt{2}i)}$$

18. (45) Use a calculator to get:

I = aE + b

a = 4.2727

b = −19.4935

$\mathbf{r = 0.8253}$

I = 4.2727E − 19.4935

Since $0.7 < |r| < 1.0$ this is a **good correlation** for an experiment in the social sciences.

19. (42) $(x - h)^2 + (y - k)^2 = r$

$\mathbf{(x - 3)^2 + (y - 2)^2 = 4^2}$

20. (42) $f(x) = \left(\frac{1}{2}\right)^{x-3} = \left(\frac{1}{2}\right)^x\left(\frac{1}{2}\right)^{-3} = 8\left(\frac{1}{2}\right)^x$

y
9
8 (0, 8)
7
6
5
4 (1, 4)
3
2
1
−1 1 2 3 4 x

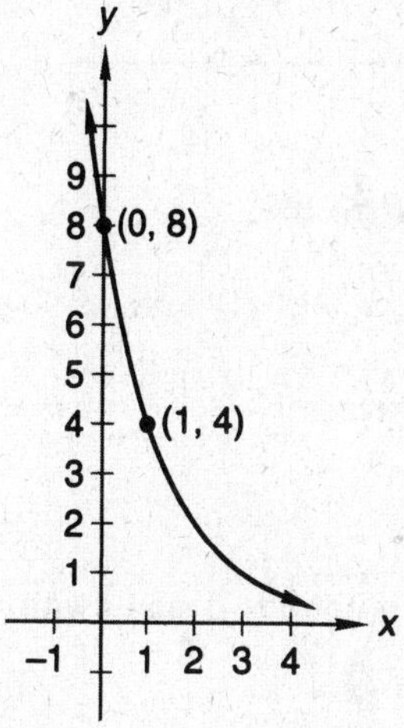

21. (41) ${}_9P_3 - {}_9P_2 = \frac{9!}{(9-3)!} - \frac{9!}{(9-2)!}$

$$= \frac{9!}{6!} - \frac{9!}{7!} = \mathbf{432}$$

22. (40) $2\log_4 4 + 3\log_3 3^2 - \log_4 \frac{1}{16} + \log_{10} 10$

$= \log_4 4^2 + \log_3 3^6 - \log_4 4^{-2} + 1$

$= 2 + 6 - (-2) + 1 = \mathbf{11}$

23. (40) $\log_4 5 + \log_4 \frac{1}{5} = \log_4 (2x + 1)$

$$\log_4\left(5 \cdot \frac{1}{5}\right) = \log_4 (2x + 1)$$
$$1 = 2x + 1$$
$$2x = 0$$
$$\mathbf{x = 0}$$

24. (40) $\log_8 (x - 2) - \log_8 3 = \log_8 16$

$$\log_8 \frac{x-2}{3} = \log_8 16$$

$$\frac{x-2}{3} = 16$$

$$x = \mathbf{50}$$

25. (32) (a) $-90° \le \theta \le 90°$

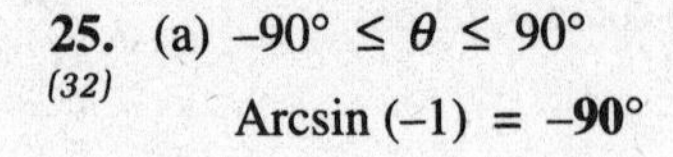

$$\text{Arcsin}(-1) = \mathbf{-90°}$$

(b)

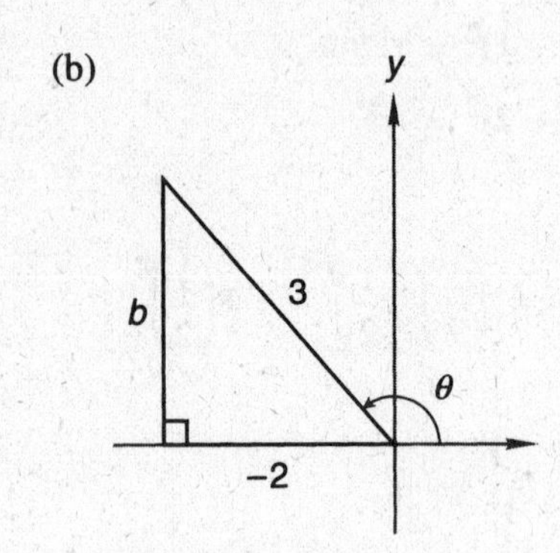

$$0° \le \theta \le 180°$$

$$b = \sqrt{3^2 - (-2)^2} = \sqrt{5}$$

$$\sin\left[\text{Arccos}\left(-\frac{2}{3}\right)\right] = \sin\theta = \mathbf{\frac{\sqrt{5}}{3}}$$

26. (31) Replace y with $y + 3$ and x with $x + 1$.

$$y + 3 = \sqrt{x+1}$$

$$\mathbf{y = \sqrt{x+1} - 3}$$

27. (31) Replace $f(x)$ with $g(x) + 2$ and x with $x + 9$.

$$g(x) + 2 = \sqrt{x - h}$$

$$\mathbf{g(x) = \sqrt{x+9} - 2}$$

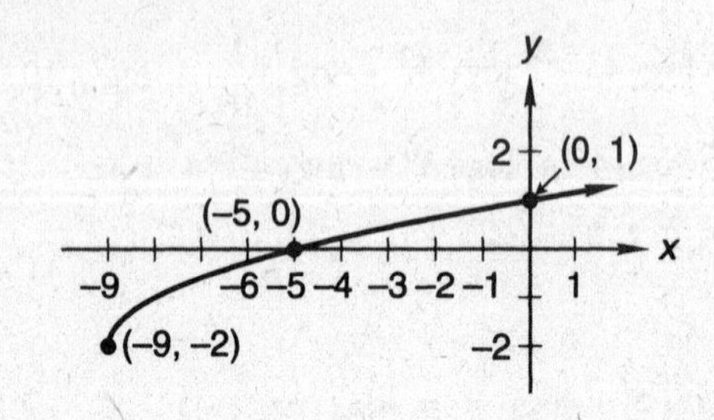

28. (15)

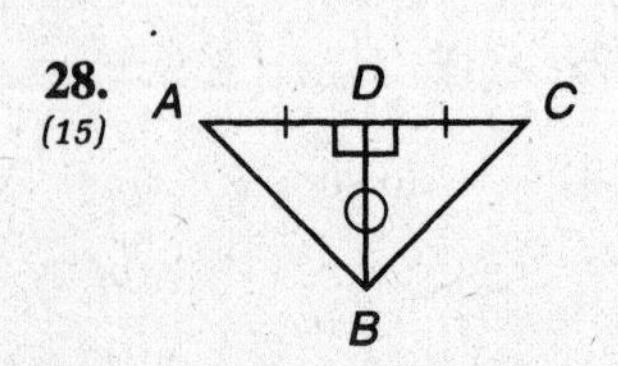

STATEMENTS	REASONS
1. $\overline{BD}$ is perpendicular bisector of $\overline{AC}$	1. Given
2. $\overline{AD} \cong \overline{DC}$ $\angle ADB \cong \angle CDB$	2. Perpendicular bisector bisects segment into two congruent segments and forms congruent right angles.
3. $\overline{DB} \cong \overline{DB}$	3. Reflexive axiom
4. $\Delta ADB \cong \Delta CDB$	4. *SAS* congruency postulate
5. $\overline{AB} \cong \overline{CB}$	5. *CPCTC*

29. (1,40) A: $2\log_2 3 - \log_2 5 = \log_2 3^2 - \log_2 5 = \log_2 \frac{9}{5}$

B: $\log_2 9 - \frac{1}{2}\log_2 25 = \log_2 9 - \log_2 5 = \log_2 \frac{9}{5}$

A and B are equal. Therefore the answer is **C.**

30. (1)

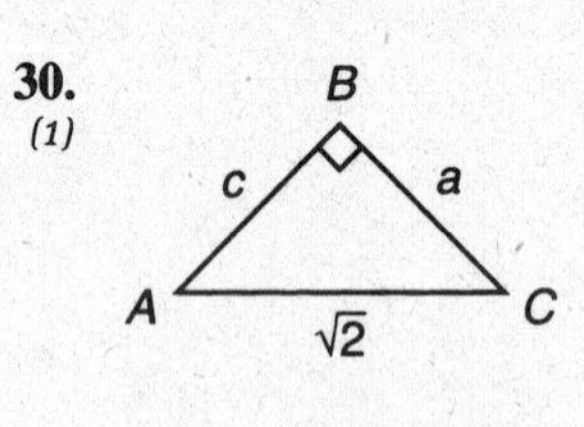

$$\text{Area} = \frac{1}{2}(AB)(BC)$$

Depending on the length of AB and BC, the area of ΔABC can be anywhere from almost zero to $\frac{1}{2}$. Therefore the answer is **D.**

Problem Set 49

1. (45) $2 \cdot 4! \cdot 3! = \mathbf{288}$

2. (45) $4 \cdot 6 \cdot 4 \cdot 5 = \mathbf{480}$

3. (39) $43.6°\left(\frac{\pi}{180°}\right)3960 = 959.20\pi = \mathbf{3013.42\ mi}$

4. (25)

$$R_R T_R + R_B T_B = \text{jobs}$$

$$\frac{2}{3}T + \frac{5}{6}(T - 3) = 74$$

$$\frac{2}{3}T + \frac{5}{6}T - \frac{5}{2} = 74$$

$$\frac{9}{6}T = \frac{153}{2}$$

$$T = \mathbf{51\ days}$$

5. (25) Rate $= \frac{240}{40} = 6\ \frac{\text{dollars}}{\text{hr}}$

New rate $= 1.2 \times \text{rate} = 1.2(6) = 7.2\ \frac{\text{dollars}}{\text{hr}}$

New rate $\times\ T$ = dollars

$7.2(60) = \mathbf{\$432}$

6. (44) $R_c = \frac{d}{c}\ \frac{\text{dollars}}{\text{cat}}$

$R_p = \frac{7}{4}R_c = \frac{7}{4}\left(\frac{d}{c}\right) = \frac{7d}{4c}\ \frac{\text{dollars}}{\text{parrot}}$

Rate $\times\ N$ = price

$\left(\frac{7d}{4c}\right)p = \mathbf{\frac{7dp}{4c}}$ **dollars**

7. (44) RIT = gallons

$R(S)11 = G$

$R = \frac{G}{11S}\ \frac{\text{gallons}}{\text{infant-day}}$

RIT = gallons

$\frac{G}{11S}(S + 14)T = K$

$T = \mathbf{\frac{11KS}{G(S + 14)}}$ **days**

8. (49) $M = b^a$ so $a = \log_b M$

$N = b^c$ so $c = \log_b N$

$MN = b^a b^c$

$MN = b^{a+c}$

$\log_b MN = \log_b b^{a+c}$

$\log_b MN = (a + c)$

$\log_b MN = \log_b M + \log_b N$

9. (48) $\sec^2\left(-\frac{13\pi}{6}\right) - \tan^3\frac{\pi}{4} = \frac{1}{\left(\cos\frac{\pi}{6}\right)^2} - 1^3$

$= \left(\frac{2}{\sqrt{3}}\right)^2 - 1 = \frac{4}{3} - 1 = \mathbf{\frac{1}{3}}$

10. (48) $\csc^2\left(-\frac{\pi}{3}\right) + \sec^2\left(\frac{2\pi}{3}\right) - \tan^2\frac{5\pi}{6}$

$= \frac{1}{\left(-\sin\frac{\pi}{3}\right)^2} + \frac{1}{\left(-\cos\frac{\pi}{3}\right)^2} - \left(-\tan\frac{\pi}{6}\right)^2$

$= \left(-\frac{2}{\sqrt{3}}\right)^2 + (-2)^2 - \left(-\frac{1}{\sqrt{3}}\right)^2$

$= \frac{4}{3} + 4 - \frac{1}{3} = \mathbf{5}$

11. (39,48) $m_0 = \frac{4 + 4}{6 + 2} = \frac{8}{8} = 1$

$(x_m, y_m) = \left(\frac{-2 + 6}{2}, \frac{-4 + 4}{2}\right) = (2, 0)$

The slope of the perpendicular bisector is the negative reciprocal so, $m = -1$.

$y - y_m = m(x - x_m)$

$y - 0 = -(x - 2)$

$y = -x + 2$

$\mathbf{x + y - 2 = 0}$

12. (39,48) $m_0 = \frac{4 + 5}{3 + 2} = \frac{9}{5}$

$(x_m, y_m) = \left(\frac{-2 + 3}{2}, \frac{-5 + 4}{2}\right) = \left(\frac{1}{2}, -\frac{1}{2}\right)$

The slope of the perpendicular bisector is the negative reciprocal so, $m = -\frac{5}{9}$.

$y - y_m = m(x - x_m)$

$y + \frac{1}{2} = -\frac{5}{9}\left(x - \frac{1}{2}\right)$

$y + \frac{1}{2} = -\frac{5}{9}x + \frac{5}{18}$

$18y + 9 = -10x + 5$

$10x + 18y = -4$

$\mathbf{\frac{x}{-\frac{2}{5}} + \frac{y}{-\frac{2}{9}} = 1}$

13. (37,39) $\sqrt{(x - 1)^2 + (y + 1)^2} = \sqrt{(x - 2)^2 + (y + 6)^2}$

$x^2 - 2x + 1 + y^2 + 2y + 1$

$= x^2 - 4x + 4 + y^2 + 12y + 36$

$2x - 10y - 38 = 0$

$\mathbf{x - 5y - 19 = 0}$

14. (47) Function = $\sin x$

Centerline = 1

Amplitude = 5

$\mathbf{y = 1 + 5 \sin x}$

15. (47) Function = $-\cos\theta$

Centerline = 5

Amplitude = 15

$\mathbf{y = 5 - 15\cos\theta}$

16. (47) (a) $-90° < \theta < 90°$

Arctan (tan 135°) = Arctan [tan (−45°)] = **−45°**

(b) $-90° < \theta < 90°$

Arctan (tan 120°) = Arctan [tan (−60°)] = **−60°**

17. (46)

$$[x - (2 + i)][x - (2 - i)] = 0$$
$$(x - 2 - i)(x - 2 + i) = 0$$
$$x^2 - 2x + xi - 2x + 4 - 2i - xi + 2i - i^2 = 0$$
$$\mathbf{x^2 - 4x + 5 = 0}$$

18. (46) $3x^2 - 12x + 24 = 0$

$3(x^2 - 4x + 8) = 0$

$$x = \frac{-(-4) \pm \sqrt{(-4)^2 - 4(1)(8)}}{2(1)}$$

$$x = \frac{4 \pm \sqrt{-16}}{2} = 2 \pm 2i$$

$\mathbf{3(x - 2 - 2i)(x - 2 + 2i)}$

19. (45) Use a calculator to get:

Ca = aK + b

$a = 0.5176$

$b = 6.6444$

$\mathbf{r = 0.9352}$

Ca = 0.5176K + 6.6444

This is a **good correlation.**

20. (42) $\sqrt{[x - (-2)]^2 + [y - (-3)]^2} = 3$

$$\mathbf{(x + 2)^2 + (y + 3)^2 = 3^2}$$

21. (42) $f(x) = \left(\frac{1}{2}\right)^{-x+2} = \left(\frac{1}{2}\right)^{-x}\left(\frac{1}{2}\right)^{2} = \frac{1}{4} \cdot 2^x$

y
2
(0, 1/4)
(1, 1/2)
x
−1
1 2

22. (40) $4\log_2 2 + \log_{10} 100 - 5\log_5 5 + \log_3 \frac{1}{3}$

$= \log_2 2^4 + \log_{10} 10^2 - \log_5 5^5 + \log_3 3^{-1}$

$= 4 + 2 - 5 + (-1) = \mathbf{0}$

23. (40) (a) $\log_{10}(x - 2) - 2\log_{10} 3 = \log_{10} 4$

$$\log_{10}\frac{(x - 2)}{3^2} = \log_{10} 4$$

$$\frac{x - 2}{9} = 4$$

$$x - 2 = 36$$

$$\mathbf{x = 38}$$

(b) $2\log_x 9 = 4$

$\log_x 9^2 = 4$

$9^2 = x^4$

$3^4 = x^4$

$\mathbf{x = 3}$

24. (33) $A = \frac{1}{2}(h)(B_1 + B_2)$

$150 = \frac{1}{2}(x)(x + 2x)$

$150 = \frac{3}{2}x^2$

$100 = x^2$

$x =$ **10 ft**

25. (31) Replace y with $y + 4$ and x with $x - 1$.

$y + 4 = |x - 1|$

$\mathbf{y = |x - 1| - 4}$

26. (31) Replace $f(x)$ with $g(x) + 3$ and x with $x - 4$.

$g(x) + 3 = |x - 4|$

$\mathbf{g(x) = |x - 4| - 3}$

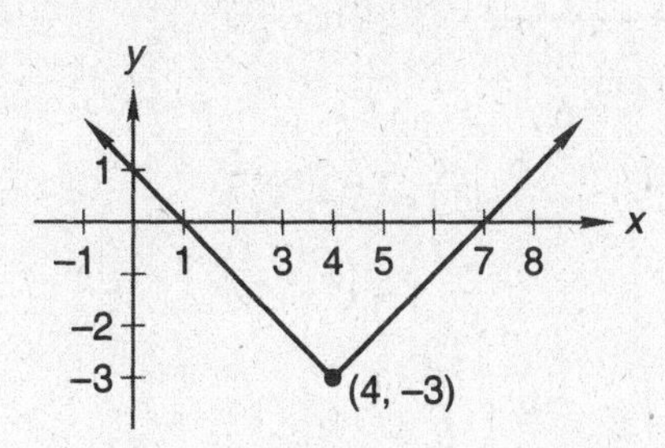

27. (21) (a) **Domain** $= \{x \in \mathbb{R} \mid -2 < x \leq 4\}$

Range $= \{y \in \mathbb{R} \mid -3 < y \leq 3\}$

(b) **Domain** $= \{x \in \mathbb{R} \mid -3 \leq x < 6\}$

Range $= \{y \in \mathbb{R} \mid -4 \leq y \leq 5\}$

28. (21) $\dfrac{f(x+h) - f(x-h)}{2h} = \dfrac{(x+h)^2 - (x-h)^2}{2h}$

$= \dfrac{x^2 + 2xh + h^2 - x^2 + 2xh - h^2}{2h}$

$= \dfrac{4xh}{2h} = \mathbf{2x}$

29. (1) $A = \dfrac{x^2 - 1}{x - 1} = \dfrac{(x+1)(x-1)}{x-1} = x + 1$

$x + 1 > x$

Therefore the answer is **A.**

30. (1,5)

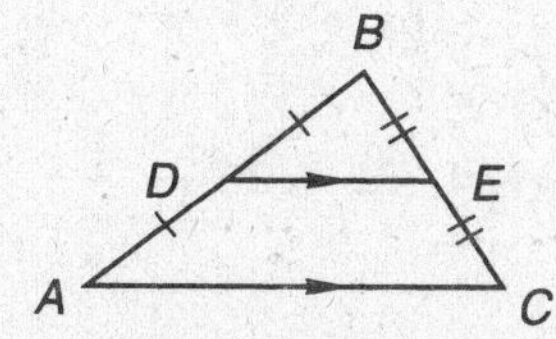

$\Delta DBE \sim \Delta ABC$ by AAA

Since $AD = DB$ and $BE = EC$ then, $AB = AD + DB = 2DB$ and $BE + EC = 2BE$.

Therefore the ratio of the sides is 2:1, so

$\dfrac{\text{Area of } \Delta ABC}{\text{Area of } \Delta DBE} = \left(\dfrac{2}{1}\right)^2$ and

$\dfrac{1}{4}$Area of ΔABC = area of ΔDBE.

Therefore the answer is **C.**

Problem Set 50

1. (45) $5 \cdot 4 \cdot 2 = \mathbf{40}$

2. (45) $5 + (5 \cdot 5) + (1 \cdot 2 \cdot 5)$

$= 5 + 25 + 10 = \mathbf{40}$

3. (39) $39^\circ\left(\dfrac{\pi}{180^\circ}\right)3960 = 858\pi = \mathbf{2695.49\ mi}$

4. (25) (a) $3S_N = 5J_N + 15$
(b) $2(S_N + 10) = 4(J_N + 10) - 20$

(b) $2(S_N + 10) = 4(J_N + 10) - 20$

$2SN = 4J_N$

$S_N = 2J_N$

(a) $3S_N = 5J_N + 15$

$3(2J_N) = 5J_N + 15$

$J_N = 15$

(a) $3S_N = 5J_N + 15$

$3S_N = 5(15) + 15$

$S_N = 30$

$J_N + 15 =$ **John = 30 yr**

$S_N + 15 =$ **Sally = 45 yr**

5. (25) $R_BT_B + R_LT_L = \text{jobs}$

$\dfrac{\left(\frac{1}{3}\right)}{4}(T) + \dfrac{\left(\frac{1}{2}\right)}{2}(T + 2) = 30$

$\dfrac{1}{12}T + \dfrac{1}{4}T + \dfrac{1}{2} = 30$

$\dfrac{1}{3}T = \dfrac{59}{2}$

$T = \mathbf{\dfrac{177}{2}\ hr}$

6. (25) Children's milk + adults' milk + leftover milk = 16 qt

$40\left(\dfrac{1}{4}\right) + 12\left(\dfrac{1}{3}\right) + M = 16$

$10 + 4 + M = 16$

$\mathbf{M = 2\ qt}$

7. (28) New rate $= \dfrac{\text{distance}}{\text{new time}} = \mathbf{\dfrac{d}{h - 4}\ \dfrac{mi}{hr}}$

8. (50)

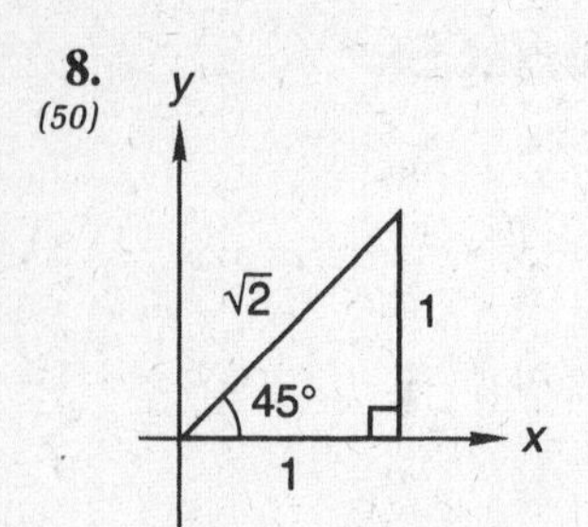

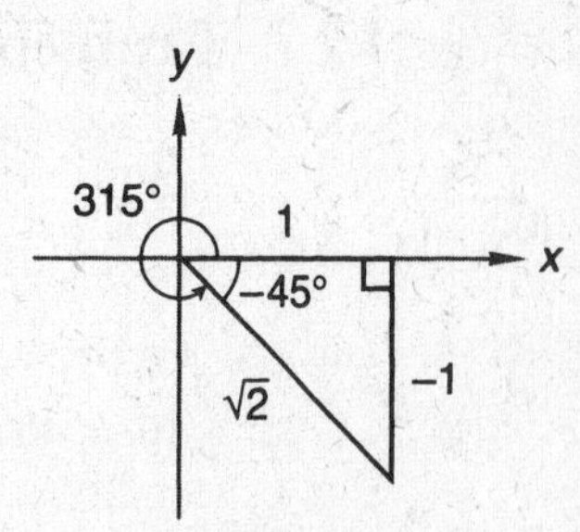

$$\cos\theta = \frac{1}{\sqrt{2}}$$

$$\theta = \mathbf{45°, 315°}$$

9. (50)

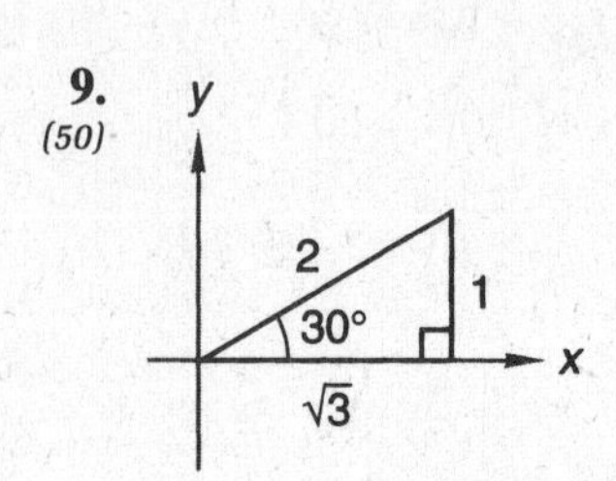

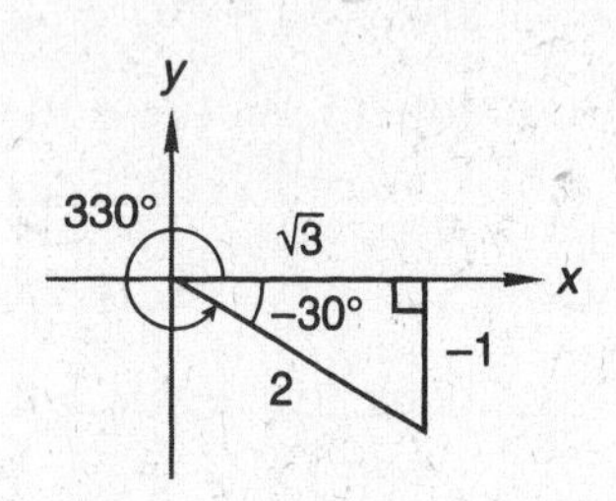

$$-2 + \sqrt{3}\sec\theta = 0$$

$$\sec\theta = \frac{2}{\sqrt{3}}$$

$$\frac{1}{\cos\theta} = \frac{2}{\sqrt{3}}$$

$$\cos\theta = \frac{\sqrt{3}}{2}$$

$$\theta = \mathbf{30°, 330°}$$

10. (49) $M = b^a$ so $\log_b M = a$

$N = b^c$ so $\log_b N = c$

$$\frac{M}{N} = \frac{b^a}{b^c}$$

$$\log_b \frac{M}{N} = \log_b b^{a-c}$$

$$\log_b \frac{M}{N} = a - c$$

$$\log_b \frac{M}{N} = \log_b M - \log_b N$$

11. (48) (a) $\sec^2 \frac{3\pi}{4} - \csc^2 \frac{\pi}{4} - \cos^2 4\pi$

$$= \frac{1}{\left(-\cos\frac{\pi}{4}\right)^2} - \frac{1}{\left(\sin\frac{\pi}{4}\right)^2} - \cos^2(0)$$

$$= (-\sqrt{2})^2 - (\sqrt{2})^2 - 1^2 = \mathbf{-1}$$

(b) $\cot^2\left(-\frac{13\pi}{6}\right) - 3\tan\left(-\frac{7\pi}{3}\right)$

$$= \frac{1}{\left(-\tan\frac{\pi}{6}\right)^2} - 3\left(-\tan\frac{\pi}{3}\right)$$

$$= (\sqrt{3})^2 - 3(\sqrt{3}) = \mathbf{3 + 3\sqrt{3}}$$

12. (48) $\sec^2\left(\frac{3\pi}{4}\right) - \csc^2\left(\frac{7\pi}{6}\right)$

$$= \frac{1}{\left(-\cos\frac{\pi}{4}\right)^2} - \frac{1}{\left(-\sin\frac{\pi}{6}\right)^2}$$

$$= (-\sqrt{2})^2 - (-2)^2 = \mathbf{-2}$$

13. (39,48) $m_0 = \frac{4 - 5}{-2 - 3} = \frac{1}{5}$

$$(x_m, y_m) = \left(\frac{-2 + 3}{2}, \frac{4 + 5}{2}\right) = \left(\frac{1}{2}, \frac{9}{2}\right)$$

The slope of the perpendicular bisector is the negative reciprocal, so $m = -5$.

$$y - y_m = m(x - x_m)$$

$$y - \frac{9}{2} = -5\left(x - \frac{1}{2}\right)$$

$$y - \frac{9}{2} = -5x + \frac{5}{2}$$

$$\mathbf{5x + y - 7 = 0}$$

14. (37) $\sqrt{(x - 6)^2 + (y - 3)^2} = \sqrt{(x - 3)^2 + (y - 2)^2}$

$x^2 - 12x + 36 + y^2 - 6y + 9$

$$= x^2 - 6x + 9 + y^2 - 4y + 4$$

$$-2y = 6x - 32$$

$$\mathbf{y = -3x + 16}$$

15. (47) Function $= \cos x$

Centerline $= 5$

Amplitude $= 25$

$\mathbf{y = 5 + 25\cos x}$

16. (47) Function $= -\cos\theta$

Centerline $= 15$

Amplitude $= 25$

$\mathbf{y = 15 - 25\cos\theta}$

17. (46)
$$\left[x - (1 + \sqrt{3}i)\right]\left[x - (1 - \sqrt{3}i)\right] = 0$$
$$(x - 1 - \sqrt{3}i)(x - 1 + \sqrt{3}i) = 0$$
$$x^2 - x + \sqrt{3}ix - x + 1 - \sqrt{3}i - \sqrt{3}ix + \sqrt{3}i - (\sqrt{3}i)^2 = 0$$
$$\mathbf{x^2 - 2x + 4 = 0}$$

18. (46) $x^2 - 2x + 4 = 0$
$$x = \frac{-(-2) \pm \sqrt{(-2)^2 - 4(1)(4)}}{2(1)}$$
$$x = \frac{2 \pm \sqrt{-12}}{2} = 1 \pm \sqrt{3}i$$
$$\mathbf{(x - 1 - \sqrt{3}i)(x - 1 + \sqrt{3}i)}$$

19. (45) Use a calculator to get:

$C = a\text{H} + b$

$a = 0.1178$

$b = -4.1948$

$\mathbf{r = 0.9112}$

$\mathbf{C = 0.1178H - 4.1948}$

This is a **good correlation.**

20. (42)
$$\sqrt{[x - (-2)]^2 + (y - 4)^2} = 4$$
$$\mathbf{(x + 2)^2 + (y - 4)^2 = 4^2}$$

21. (42)
$$f(x) = \left(\frac{1}{2}\right)^{-x-4} = \left[\left(\frac{1}{2}\right)^{-1}\right]^{x+4} = 2^{x+4}$$
$$= 2^4 \cdot 2^x = 16 \cdot 2^x$$

22. (40) (a)
$$\log_3 (x - 3) + \log_3 10 = \log_3 22$$
$$\log_3 (x - 3)(10) = \log_3 22$$
$$(x - 3)(10) = 22$$
$$10x - 30 = 22$$
$$10x = 52$$
$$x = \mathbf{\frac{26}{5}}$$

(b)
$$4 \log_5 3 - \log_5 x = 0$$
$$\log_5 3^4 = \log_5 x$$
$$3^4 = x$$
$$x = \mathbf{81}$$

23. (34)
$$\sum_{p=1}^{3} \frac{p^3 - 2}{p + 1}$$
$$= \frac{(1)^3 - 2}{1 + 1} + \frac{(2)^3 - 2}{2 + 1} + \frac{(3)^3 - 2}{3 + 1}$$
$$= \frac{-1}{2} + \frac{6}{3} + \frac{25}{4}$$
$$= \mathbf{\frac{31}{4}}$$

24. (32) (a) $0° \le \theta \le 180°$

Arccos (cos 150°) = **150°**

(b)

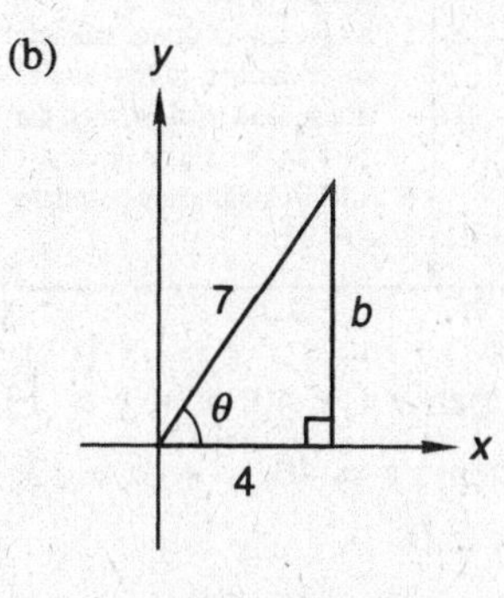

$0° \le \theta \le 180°$
$$b = \sqrt{7^2 - 4^2} = \sqrt{33}$$
$$\sin\left(\text{Arccos } \frac{4}{7}\right) = \sin\theta = \mathbf{\frac{\sqrt{33}}{7}}$$

25. (31) Replace y with $y + 3$ and x with $x + 3$.
$$y + 3 = |x + 3|$$
$$\mathbf{y = |x + 3| - 3}$$

26. (31) Replace $f(x)$ with $g(x) + 5$ and x with $x + 4$.

$g(x) + 5 = |x + 4|$

$\mathbf{g(x) = |x + 4| - 5}$

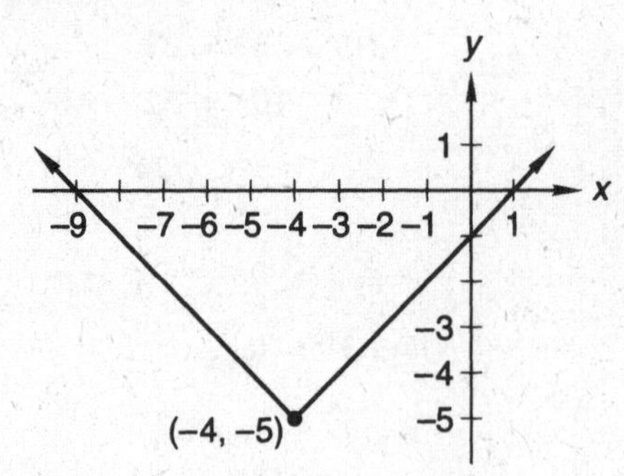

27. (21) (a) **Domain** $= \{x \in \mathbb{R} \mid -5 \leq x < 4\}$

Range $= \{y \in \mathbb{R} \mid -2 \leq y \leq 7\}$

(b) **Domain** $= \{x \in \mathbb{R} \mid -4 < x < 3\}$

Range $= \{y \in \mathbb{R} \mid -4 < y \leq 4\}$

28. (15)

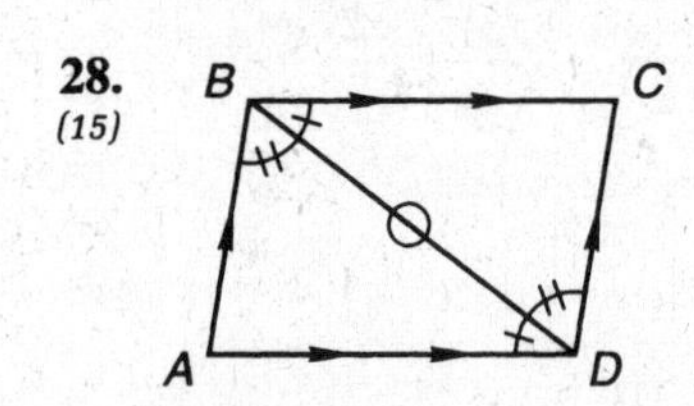

STATEMENTS	REASONS
1. $\overline{AB} \parallel \overline{DC}$	1. Given
2. $\overline{AD} \parallel \overline{BC}$	2. Given
3. $\angle DBC \cong \angle BDA$ $\angle ABD \cong \angle CDB$	3. If two parallel lines are intersected by a transversal, then alternate interior angles are congruent.
4. $\overline{BD} \cong \overline{BD}$	4. Reflexive axiom
5. $\angle BAD \cong \angle BCD$	5. If two angles in one triangle are congruent to two angles in a second triangle, then the third angles are congruent.
6. $\triangle ABD \cong \triangle CDB$	6. *AAAS* Congruency postulate
7. $\overline{AB} \cong \overline{DC}$	7. *CPCTC*

29. (1) If $x = 4$, $y = 10$, then $xy = 40$, $x + y > 13$.

If $x = 8$, $y = 5$, then $xy = 40$, $x + y = 13$.

Therefore the answer is ***D***.

30. (1,3) If side length $= 2$, $r_S = 1$. So, $A_S = \pi(r_S)^2 = \pi$.

Diagonal of square $= \sqrt{2^2 + 2^2} = 2\sqrt{2}$

Diagonal of square = diameter of large circle

If $D_L = 2\sqrt{2}$, $r_L = \sqrt{2}$. So, $A_L = \pi r_L^2 = 2\pi$.

$$\frac{1}{\sqrt{2}}A_L = \frac{1}{\sqrt{2}}(2\pi) = \sqrt{2}\pi$$

$$\sqrt{2}\pi > \pi$$

$$B > A$$

Therefore, the answer is ***B***.

Problem Set 51

1. (45)

B	3	2	1

$\rightarrow 3 \cdot 2 \cdot 1 = 6$

3	2	*B*	1

$\rightarrow 3 \cdot 2 \cdot 1 = 6$

3	2	1	*B*

$\rightarrow 3 \cdot 2 \cdot 1 = 6$

Total: **18**

2. (45)

$$4 \cdot 5 \cdot 5 = 100$$
$$5 \cdot 5 = 25$$
$$5 = 5$$

Total: **130**

3. (39) $\frac{\pi}{3}(300) = 100\pi = \mathbf{314.16\ m}$

4. (25) $R_F T_F + R_S T_S = \text{jobs}$

$$\frac{7}{3}T + \frac{8}{5}T = 59$$

$$\left(\frac{59}{15}\right)T = 59$$

$$T = \mathbf{15\ hr}$$

5. (44) $d \text{ dollars} \times \frac{100 \text{ cents}}{1 \text{ dollar}} = 100d \text{ cents}$

$$\text{Rate} \times N = \text{price}$$

$$3\frac{\text{cents}}{\text{stamp}} \times N = 100d \text{ cents}$$

$$N = \frac{100d \text{ cents}}{3\frac{\text{cents}}{\text{stamp}}} = \mathbf{\frac{100d}{3}\ stamps}$$

6. (25) $\left(\frac{16 \text{ oz}}{3 \text{ cups}}\right)15 \text{ cups} = 80 \text{ oz of raisins}$

$$\frac{80 \text{ oz}}{\frac{10 \text{ oz}}{\text{package}}} = \mathbf{8\ packages}$$

7. (28) $\text{Rate} = \frac{d}{m}\left(\frac{60 \text{ min}}{1 \text{ hr}}\right) = \frac{60d}{m} \text{ mph}$

$$\text{Time} = \frac{d}{\frac{60d}{m}} = \frac{m}{60} \text{ hr}$$

$$\text{New time} = \frac{m}{60} - 1 = \frac{m - 60}{60} \text{ hr}$$

$$\text{New rate} = \frac{d}{\frac{m - 60}{60}} = \mathbf{\frac{60d}{m - 60}\ mph}$$

8. (51) Use a calculator to get:

(a) $\log 3.5 = \mathbf{0.5441}$

(b) $\ln 3.5 = \mathbf{1.2528}$

9. (51) Use a calculator to get:

$\ln 3600 = 8.188689$

$3600 = \mathbf{e^{8.1887}}$

10. (50) $\sin\theta = -\frac{\sqrt{3}}{2}$

$\theta = \mathbf{240°, 300°}$

11. (50) $2 + \sqrt{3}\csc\theta = 0$

$\csc\theta = -\frac{2}{\sqrt{3}}$

$\frac{1}{\sin\theta} = -\frac{2}{\sqrt{3}}$

$\sin\theta = -\frac{\sqrt{3}}{2}$

$\theta = \mathbf{240°, 300°}$

12. (49) $N = b^c, \log_b N = c.$

$\log_b N^x = \log_b (b^c)^x = \log_b b^{cx} = cx.$

Therefore, $\log_b N^x = cx = x\log_b N.$

13. (48) $\sec^3 0 - \csc^2\frac{5\pi}{6} + \cos^4 9\pi$

$= \frac{1}{(\cos 0)^3} - \frac{1}{\left(\sin\frac{\pi}{6}\right)^2} + (\cos\pi)^4$

$= 1^3 - (2)^2 + (-1)^4 = \mathbf{-2}$

14. (37) $(x, y) = \left(\frac{-3+6}{2}, \frac{4+2}{2}\right) = \mathbf{\left(\frac{3}{2}, 3\right)}$

15. (37,39) $\sqrt{(x+3)^2 + (y-4)^2} = \sqrt{(x-6)^2 + (y+2)^2}$

$x^2 + 6x + 9 + y^2 - 8y + 16$

$= x^2 - 12x + 36 + y^2 + 4y + 4$

$18x - 12y - 15 = 0$

$\mathbf{6x - 4y - 5 = 0}$

16. (37,39) $\sqrt{(x+2)^2 + (y-1)^2} = \sqrt{(x-3)^2 + (y-2)^2}$

$x^2 + 4x + 4 + y^2 - 2y + 1$

$= x^2 - 6x + 9 + y^2 - 4y + 4$

$10x + 2y - 8 = 0$

$\mathbf{5x + y - 4 = 0}$

17. (47) Function = $\sin x$

Centerline = 5

Amplitude = 25

$\mathbf{y = 5 + 25\sin x}$

18. (47) Function = $-\cos\theta$

Centerline = 10

Amplitude = 20

$\mathbf{y = 10 - 20\cos\theta}$

19. (46) $[x - (3 - i)][x - (3 + i)] = 0$

$(x - 3 + i)(x - 3 - i) = 0$

$x^2 - 3x + ix - 3x + 9 - 3i - ix + 3i + 1 = 0$

$\mathbf{x^2 - 6x + 10 = 0}$

20. (46) $4x^2 + 8x + 28 = 0$

$4(x^2 + 2x + 7) = 0$

$x = \frac{-2 \pm \sqrt{2^2 - 4(1)(7)}}{2(1)}$

$x = \frac{-2 \pm \sqrt{-24}}{2}$

$x = \frac{-2 \pm 2\sqrt{6}i}{2} = -1 \pm \sqrt{6}i$

$\mathbf{4(x + 1 - \sqrt{6}i)(x + 1 + \sqrt{6}i)}$

21. (45) Use a calculator to get:

$H = aP + b$

$a = 0.8072$

$b = 55.9193$

$\mathbf{r = 0.9111}$

$\mathbf{H = 0.8072P + 55.9193}$

This is a **good correlation.**

22. (42) $\sqrt{(x+3)^2 + (y-2)^2} = 5$

$\mathbf{(x + 3)^2 + (y - 2)^2 = 5^2}$

23. (42) $f(x) = \left(\frac{1}{2}\right)^{x+2} = \left(\frac{1}{2}\right)^x\left(\frac{1}{2}\right)^2 = \frac{1}{4}\left(\frac{1}{2}\right)^x$

24. (41) ${}_8P_5 - {}_8P_3 = \frac{8!}{(8-5)!} - \frac{8!}{(8-3)!}$

$= \frac{8!}{3!} - \frac{8!}{5!} = \mathbf{6384}$

25. (40) (a) $\log_5(x + 2) - 2\log_5 6 = \log_5 1$

$\log_5(x + 2) - \log_5 6^2 = \log_5 1$

$\log_5 \frac{x + 2}{36} = \log_5 1$

$\frac{x + 2}{36} = 1$

$x + 2 = 36$

$x = \mathbf{34}$

(b) $3\log_7 2 = 2\log_7 3 + \log_7 x$

$\log_7 2^3 = \log_7 3^2 + \log_7 x$

$\log_7 8 = \log_7 9x$

$8 = 9x$

$x = \mathbf{\frac{8}{9}}$

26. (40) $(f \circ g)(x) = \dfrac{1}{\frac{3x+1}{x} - 3} = \dfrac{1}{\frac{3x + 1 - 3x}{x}} = x$

$(g \circ f)(x) = \dfrac{3\left(\frac{1}{x-3}\right) + 1}{\frac{1}{x-3}} = \dfrac{\frac{3 + x - 3}{x - 3}}{\frac{1}{x-3}} = x$

$(f \circ g)(x) = (g \circ f)(x) = x$

Yes, f and g are inverse functions.

27. (32,47) (a)

$-90° \leq \theta \leq 90°$

$a = \sqrt{4^2 - 3^2} = \sqrt{7}$

$\cos\left(\text{Arcsin } \frac{3}{4}\right) = \cos\theta = \mathbf{\frac{\sqrt{7}}{4}}$

(b)

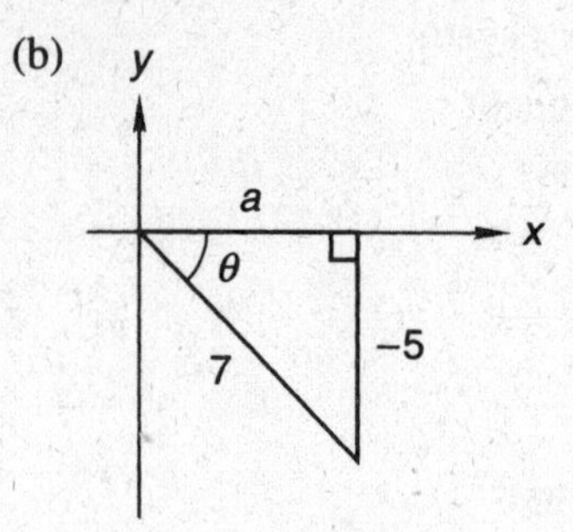

$-90° \leq \theta \leq 90°$

$a = \sqrt{7^2 - (-5)^2} = 2\sqrt{6}$

$\tan\left[\text{Arcsin } -\frac{5}{7}\right] = \tan\theta = -\frac{5}{2\sqrt{6}} = \mathbf{-\frac{5\sqrt{6}}{12}}$

(c) $-90° < \theta < 90°$

Arctan (tan 210°) = Arctan (tan 30°) = **30°**

28. (31) $-y = \sqrt{x - 2}$

$\mathbf{y = -\sqrt{x - 2}}$

29. (31) The graph of g is the graph of f translated six units to the left. Therefore the answer is **B.**

30. (21) $\dfrac{f(x + h) - f(x - h)}{2h}$

$= \dfrac{[(x + h)^2 + (x + h)] - [(x - h)^2 + (x - h)]}{2h}$

$= \dfrac{(x^2 + 2xh + h^2 + x + h)}{2h} - \dfrac{(x^2 - 2xh + h^2 + x - h)}{2h}$

$= \dfrac{4xh + 2h}{2h} = \mathbf{2x + 1}$

Problem Set 52

1. (45)

6	6	5	6	1

→ $6^3 \cdot 5 = \mathbf{1080}$

2. (45) $2 \cdot 5 \cdot 5 = 50$

$5 \cdot 5 = 25$

$5 = 5$

80

3. (39) $38.8°\left(\frac{\pi}{180°}\right)3960 = 853.6\pi = \mathbf{2681.66\ mi}$

4. (25)
$$R_RT_R + R_BT_B = \text{jobs}$$
$$\frac{5}{4}(7) + R_B(7) = 12$$
$$7R_B = \frac{13}{4}$$
$$R_B = \frac{13}{28}$$
$$R_BT = 2$$
$$\left(\frac{13}{28}\right)T = 2$$
$$T = \mathbf{\frac{56}{13}\ hr}$$

5. (25)
$$\text{Rate} = \frac{200}{40} = 5\ \frac{\text{dollars}}{\text{hr}}$$
$$\text{New rate} = 1.4 \times \text{rate} = 1.4(5) = 7\ \frac{\text{dollars}}{\text{hr}}$$
$$\text{New rate} \times T = \text{dollars}$$
$$7(30) = \mathbf{\$210}$$

6. (38)
$$\text{Average Price} = \frac{\text{Total Price}}{\text{Total number of shares}}$$
$$58 = \frac{60(50) + 56(25) + x(25)}{50 + 25 + 25}$$
$$100(58) = 3000 + 1400 + 25x$$
$$25x = 1400$$
$$x = \mathbf{\$56}$$

7. (28)
$$\text{Rate} = \frac{D}{S}$$
$$\text{Time} = S$$
$$\text{New time} = S - 120$$
$$\text{New rate} = \mathbf{\frac{D}{S - 120}\ \frac{ft}{s}}$$

8. (36)
$$\begin{cases}(a)\ (B + W)T_D = D_D \\ (b)\ (B - W)T_U = D_U\end{cases}$$
$$B = 2W,\ T_D = T_U + 1$$

(b)
$$(B - W)T_U = D_U$$
$$(2W - W)T_U = 10$$
$$WT_U = 10$$

(a)
$$(B + W)T_D = D_D$$
$$(2W + W)(T_U + 1) = 45$$
$$(3W)(T_U + 1) = 45$$
$$3WT_U + 3W = 45$$
$$3(10) + 3W = 45$$
$$3W = 15$$
$$W = \mathbf{5\ \frac{mi}{hr}}$$

$$B = 2W$$
$$B = 2(5)$$
$$B = \mathbf{10\ \frac{mi}{hr}}$$

9. (52)
$$\cos 2\theta = -\frac{1}{2}$$
$$2\theta = 120°, 240°, 480°, 600°$$
$$\theta = \mathbf{60°, 120°, 240°, 300°}$$

10. (52)
$$\tan 3\theta = 1$$
$$3\theta = 45°, 225°, 405°, 585°, 765°, 945°$$
$$\theta = \mathbf{15°, 75°, 135°, 195°, 255°, 315°}$$

11. (52)
$$\sin\frac{\theta}{2} = \frac{\sqrt{3}}{2}$$
$$\frac{\theta}{2} = 60°, 120°$$
$$\theta = \mathbf{120°, 240°}$$

12. (51)
(a) $\log 16.3 = \mathbf{1.2122}$
(b) $\ln 16.3 = \mathbf{2.7912}$

13. (51)
$$\ln 3800 = 8.2428$$
$$3800 = \mathbf{e^{8.2428}}$$

14. (48)
$$\sin^2\left(-\frac{\pi}{3}\right) - \cos^2\left(\frac{2\pi}{3}\right) + \csc^2\left(\frac{4\pi}{3}\right)$$
$$= \left(-\sin\frac{\pi}{3}\right)^2 - \left(-\cos\frac{\pi}{3}\right)^2 + \frac{1}{\left(-\sin\frac{\pi}{3}\right)^2}$$
$$= \left(-\frac{\sqrt{3}}{2}\right)^2 - \left(-\frac{1}{2}\right)^2 + \left(-\frac{2}{\sqrt{3}}\right)^2$$
$$= \frac{3}{4} - \frac{1}{4} + \frac{4}{3} = \mathbf{\frac{11}{6}}$$

15. (48) $\csc^2\left(-\frac{13\pi}{6}\right) - \sec^2 \frac{19\pi}{6} - \cot^2 \frac{5\pi}{4}$

$= \frac{1}{\left(-\sin \frac{\pi}{6}\right)^2} - \frac{1}{\left(-\cos \frac{\pi}{6}\right)^2} - \frac{1}{\left(\tan \frac{\pi}{4}\right)^2}$

$= (-2)^2 - \left(-\frac{2}{\sqrt{3}}\right)^2 - \frac{1}{(1)^2}$

$= 4 - \frac{4}{3} - 1 = \mathbf{\frac{5}{3}}$

16. (48) $m_0 = \frac{8 - (-2)}{-4 - (-6)} = \frac{10}{2} = 5$

$(x_m, y_m) = \left(\frac{-6 - 4}{2}, \frac{8 - 2}{2}\right) = (-5, 3)$

The slope of the perpendicular line is the negative reciprocal, so $m = -\frac{1}{5}$.

$y = -\frac{1}{5}x + b$

$3 = -\frac{1}{5}(-5) + b$

$b = 2$

$y = -\frac{1}{5}x + 2$

$5y = -x + 10$

$x + 5y = 10$

$\mathbf{\frac{x}{10} + \frac{y}{2} = 1}$

17. (47) Function $= -\sin x$

Centerline $= 4$

Amplitude $= 7$

$\mathbf{y = 4 - 7 \sin x}$

18. (47) Function $= -\cos \theta$

Centerline $= 2$

Amplitude $= 8$

$\mathbf{y = 2 - 8 \cos \theta}$

19. (46)

$[x - (-2 + i)][x - (-2 - i)] = 0$

$(x + 2 - i)(x + 2 + i) = 0$

$x^2 + 2x - ix + 2x + 4 - 2i + ix + 2i + 1 = 0$

$\mathbf{x^2 + 4x + 5 = 0}$

20. (46) $x^2 + 4x + 5 = 0$

$x = \frac{-4 \pm \sqrt{4^2 - 4(1)(5)}}{2(1)} = \frac{-4 \pm \sqrt{-4}}{2} = -2 \pm i$

$\mathbf{(x + 2 - i)(x + 2 + i)}$

21. (45) Use a calculator to get:

$W = aH + b$

$a = 13.4787$

$b = -761.8723$

$\mathbf{r = 0.7815}$

$\mathbf{W = 13.4787H - 761.8723}$

This is **not a good correlation.**

22. (42) $\sqrt{(x - h)^2 + (y - k)^2} = r$

$\mathbf{(x - h)^2 + (y - k)^2 = r^2}$

23. (42) $f(x) = \left(\frac{1}{4}\right)^{-2x+1} = \left(\frac{1}{4}\right)^1\left(\frac{1}{4}\right)^{-2x} = \frac{1}{4} \cdot 16^x$

y
4
3
2
1
x
(1, 4)
(0, 1/4)

24. (40) $\log_6 (x - 2) + 2 \log_6 3 = \log_6 2$

$\log_6 (x - 2) + \log_6 3^2 = \log_6 2$

$\log_6 9(x - 2) = \log_6 2$

$9(x - 2) = 2$

$x - 2 = \frac{2}{9}$

$x = \mathbf{\frac{20}{9}}$

25. (40) $2 \log_{\frac{1}{2}} 3 - \log_{\frac{1}{2}} (2x - 3) = 1$

$\log_{\frac{1}{2}} 3^2 - \log_{\frac{1}{2}} (2x - 3) = 1$

$\log_{\frac{1}{2}} \frac{9}{2x - 3} = 1$

$\frac{9}{2x - 3} = \left(\frac{1}{2}\right)^1$

$2x - 3 = 18$

$2x = 21$

$x = \mathbf{\frac{21}{2}}$

26. (34) $\sum_{k=-1}^{3}(k^3 - 2)$

$= [(-1)^3 - 2] + [(0)^3 - 2] + [(1)^3 - 2] + [(2)^3 - 2] + [(3)^3 - 2]$

$= -3 + (-2) + (-1) + 6 + 25 = \mathbf{25}$

27. (40) $(f \circ g)(x) = \dfrac{\dfrac{x}{x+1}}{\dfrac{x}{x+1} - 1} = \dfrac{\dfrac{x}{x+1}}{\dfrac{x - x - 1}{x+1}} = -x$

$(g \circ f)(x) = \dfrac{\dfrac{x}{x-1}}{\dfrac{x}{x-1} + 1} = \dfrac{\dfrac{x}{x-1}}{\dfrac{x + x - 1}{x-1}} = \dfrac{x}{2x - 1}$

$(f \circ g)(x) \neq x$

$(g \circ f)(x) \neq x$

No, f and g are not inverse functions.

28. (33) $A_{\text{trapezoid}} = \frac{1}{2}(H)(B_1 + B_2)$

$20 = \frac{1}{2}(x)[(x - 3) + (2x - 4)]$

$40 = x(3x - 7)$

$40 = 3x^2 - 7x$

$3x^2 - 7x - 40 = 0$

$(3x + 8)(x - 5) = 0$

$x = -\frac{8}{3}, 5$

$x = \mathbf{5}$

29. (31) Replace y with $-y$ and x with $x + 4$.

$-y = \sqrt{x + 4}$

$\mathbf{y = -\sqrt{x + 4}}$

30. (31) The graph of g is the graph of f translated five units to the right. Therefore the answer is **A.**

Problem Set 53

1. (45) $5 \cdot 4 \cdot 3 \cdot 3 = \mathbf{180}$

2. (39) $270°\left(\dfrac{\pi}{180°}\right)2000 = 3000\pi = \mathbf{9424.78\ m}$

3. (25) $R_M T_M + R_J T_J = \text{jobs}$

$\frac{8}{3}(T + 3) + \frac{5}{2}T = 39$

$\frac{8}{3}T + 8 + \frac{5}{2}T = 39$

$\frac{31}{6}T = 31$

$T = \mathbf{6\ days}$

4. (44) $RWT = \text{jobs}$

$Rmh = j$

$R = \dfrac{j}{mh}\ \dfrac{\text{jobs}}{\text{worker-hr}}$

$RWT = \text{jobs}$

$\left(\dfrac{j}{mh}\right)(m + p)T = j$

$T = \mathbf{\dfrac{hm}{m + p}\ hr}$

5. (28,44) Rate $= \dfrac{Y}{S}\ \dfrac{\text{yd}}{\text{s}}$

New distance $= 3Y$ yd

New time $= (S + 20)$ s

New rate $= \dfrac{\text{New distance}}{\text{New time}} = \mathbf{\dfrac{3Y}{S + 20}\ \dfrac{yd}{s}}$

6. (53) $2800\text{ francs}\left(\dfrac{1\text{ dollar}}{350\text{ francs}}\right) = 8\text{ dollars}$

$8\text{ dollars}\left(\dfrac{400\text{ francs}}{1\text{ dollar}}\right) = 3200\text{ francs}$

Profit $= 3200 - 2800 =$ **400 francs**

7. (53) $\left(50\ \dfrac{\text{mi}}{\text{hr}}\right)\left(\dfrac{5280\text{ ft}}{1\text{ mi}}\right)\left(\dfrac{12\text{ in.}}{1\text{ ft}}\right)\left(\dfrac{2.54\text{ cm}}{1\text{ in.}}\right)\left(\dfrac{1\text{ hr}}{60\text{ min}}\right)$

$= \mathbf{\dfrac{(50)(5280)(12)(2.54)}{60}\ \dfrac{cm}{min} = 134{,}112\ \dfrac{cm}{min}}$

8. (53) $(30\text{ liters})\left(\dfrac{1000\text{ cm}^3}{1\text{ liter}}\right)\left(\dfrac{1\text{ in}^3}{(2.54)^3\text{ cm}^3}\right)$

$= \dfrac{(30)(1000)}{(2.54)^3}\text{ in.}^3 = \mathbf{\dfrac{(30)(10)^3}{(2.54)^3}\ in.^3 = 1830.71\ in.^3}$

9. (53) $\omega = \frac{v}{r}$

$$\omega = \frac{50\ \frac{\text{mi}}{\text{hr}}}{15\text{ in.}} = \frac{50\text{ mi}}{15\text{ hr-in.}}$$

$$\omega = \left(\frac{50\text{ mi}}{15\text{ hr-in.}}\right)\left(\frac{5280\text{ ft}}{1\text{ mi}}\right)\left(\frac{12\text{ in.}}{1\text{ ft}}\right) \times \left(\frac{1\text{ hr}}{60\text{ min}}\right)\left(\frac{1\text{ min}}{60\text{ s}}\right)$$

$$\omega = \frac{50(5280)(12)\text{ rad}}{15(60)(60)\text{ s}} \times \frac{1\text{ rev}}{2\pi\text{ rad}}$$

$$\omega = \mathbf{\frac{(50)(5280)(12)}{(15)(60)(60)(2\pi)}\ \frac{rev}{s} = 9.34\ \frac{rev}{s}}$$

10. (50) $\sin\theta = \frac{\sqrt{3}}{2}$

$\theta = \mathbf{60°, 120°}$

11. (52) $\cos 3\theta = 1$

$3\theta = 0°, 360°, 720°$

$\theta = \mathbf{0°, 120°, 240°}$

12. (52) $\sin\frac{\theta}{2} = 1$

$\frac{\theta}{2} = 90°$

$\theta = \mathbf{180°}$

13. (51) $\log 6200 = 3.7924$

$6200 = \mathbf{10^{3.7924}}$

14. (51) $\ln 5400 = 8.5942$

$5400 = \mathbf{e^{8.5942}}$

15. (48) $\cot^2\left(-\frac{\pi}{3}\right) - \sec^2 0$

$$= \frac{1}{\left(-\tan\frac{\pi}{3}\right)^2} - \frac{1}{(\cos 0)^2}$$

$$= \frac{1}{3} - 1 = \mathbf{-\frac{2}{3}}$$

16. (48) $\sec^2\left(-\frac{3\pi}{4}\right) + \cos^2\left(\frac{13\pi}{6}\right)$

$$= \frac{1}{\left(-\cos\frac{\pi}{4}\right)^2} + \left(\cos\frac{\pi}{6}\right)^2$$

$$= 2 + \frac{3}{4} = \mathbf{\frac{11}{4}}$$

17. (37,39) $\sqrt{(x-8)^2 + (y-4)^2} = \sqrt{(x+3)^2 + (y-4)^2}$

$x^2 - 16x + 64 + y^2 - 8y + 16$

$= x^2 + 6x + 9 + y^2 - 8y + 16$

$22x - 55 = 0$

$\mathbf{2x - 5 = 0}$

18. (47) Function = $\cos x$

Centerline = 1

Amplitude = 5

$\mathbf{y = 1 + 5\cos x}$

19. (47) Function = $-\cos\theta$

Centerline = -1

Amplitude = 7

$\mathbf{y = -1 - 7\cos\theta}$

20. (46) $\left[x - \left(-1 + 2\sqrt{2}i\right)\right]\left[x - \left(-1 - 2\sqrt{2}i\right)\right] = 0$

$\left(x + 1 - 2\sqrt{2}i\right)\left(x + 1 + 2\sqrt{2}i\right) = 0$

$x^2 + x - 2\sqrt{2}ix + x + 1 - 2\sqrt{2}i$

$+ 2\sqrt{2}ix + 2\sqrt{2}i + 8 = 0$

$\mathbf{x^2 + 2x + 9 = 0}$

21. (45) Use a calculator to get:

$V = aW + b$

$a = 59.8788$

$b = 323.7576$

$\mathbf{r = 0.9815}$

$\mathbf{V = 59.88W + 323.76}$

This is a **good correlation.**

22. (42) $\sqrt{\left(x - \sqrt{3}\right)^2 + [y - (-2)]^2} = 5^2$

$\mathbf{\left(x - \sqrt{3}\right)^2 + (y + 2)^2 = 5^2}$

23. (42)
$$f(x) = \left(\frac{2}{3}\right)^{3x-2} = \left(\frac{2}{3}\right)^{3x}\left(\frac{2}{3}\right)^{-2}$$
$$= \left[\left(\frac{2}{3}\right)^3\right]^x\left(\frac{3}{2}\right)^2 = \frac{9}{4}\left(\frac{8}{27}\right)^x$$

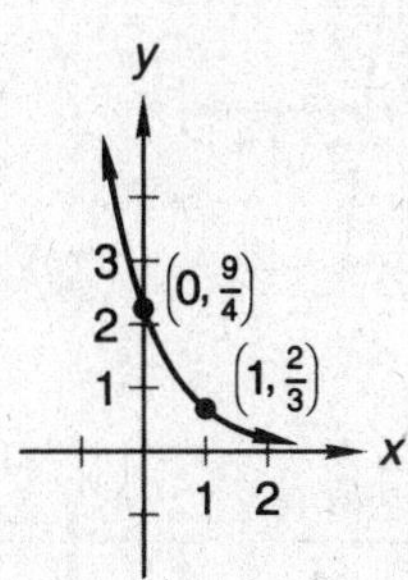

24. (40)
$$\log_{15} 4 - \log_{15} \frac{2}{x} = 2$$
$$\log_{15} \frac{4x}{2} = 2$$
$$\frac{4x}{2} = 15^2$$
$$2x = 225$$
$$x = \mathbf{\frac{225}{2}}$$

25. (40)
$$2 \log_7 4 - \log_7 x = \frac{2}{3} \log_7 8$$
$$\log_7 4^2 - \log_7 x = \log_7 8^{\frac{2}{3}}$$
$$\log_7 \frac{4^2}{x} = \log_7 4$$
$$\frac{4^2}{x} = 4$$
$$16 = 4x$$
$$x = \mathbf{4}$$

26. (40)
$$(f \circ g)(x) = \frac{1}{1 - \frac{x-1}{x}} = \frac{1}{\frac{x - x + 1}{x}} = \frac{1}{\frac{1}{x}} = x$$
$$(g \circ f)(x) = \frac{\frac{1}{1-x} - 1}{\frac{1}{1-x}}$$
$$= \frac{\frac{1 - 1 + x}{1-x}}{\frac{1}{1-x}} = \frac{\frac{x}{1-x}}{\frac{1}{1-x}} = x$$
$$(f \circ g)(x) = (g \circ f)(x) = x$$

Yes, f and g are inverse functions.

27. (33)
$$A_{\text{trapezoid}} = \frac{1}{2}(H)(B_1 + B_2)$$
$$6 = \frac{1}{2}(x)[(2x - 3) + (3x - 1)]$$
$$12 = x(5x - 4)$$
$$12 = 5x^2 - 4x$$
$$5x^2 - 4x - 12 = 0$$
$$(5x + 6)(x - 2) = 0$$
$$x = 2, -\frac{6}{5}$$
$$x = \mathbf{2}$$

28. (31) Replace y with $-y$, then replace y with $y - 4$.
$$-(y - 4) = x^2$$
$$-y = x^2 - 4$$
$$\mathbf{y = -x^2 + 4}$$

29. (31) The graph of g is the graph of f translated three units down. Therefore the answer is **D.**

30. (21)
$$\frac{(x + h)^2 - (x + h) - [(x - h)^2 - (x - h)]}{2h}$$
$$= \frac{x^2 + 2xh + h^2 - x - h}{2h}$$
$$+ \frac{-x^2 + 2xh - h^2 + x - h}{2h}$$
$$= \frac{4xh - 2h}{2h} = \mathbf{2x - 1}$$

Problem Set 54

1. (53) $v = r\omega$
$$v = (1.5 \text{ ft})\left(30 \frac{\text{rev}}{\text{min}}\right)\left(\frac{1 \text{ mi}}{5280 \text{ ft}}\right)\left(\frac{60 \text{ min}}{1 \text{ hr}}\right)\left(\frac{2\pi \text{ rad}}{1 \text{ rev}}\right)$$
$$v = \frac{(1.5)(30)(60)(2\pi)}{5280} \frac{\text{mi}}{\text{hr}} = \mathbf{3.21 \frac{mi}{hr}}$$

2. (53) $\omega = \frac{v}{r}$
$$\omega = \frac{30 \frac{\text{km}}{\text{hr}}}{70 \text{ cm}} = \frac{30 \text{ km}}{70 \text{ cm-hr}}$$
$$\omega = \left(\frac{30 \text{ km}}{70 \text{ cm-hr}}\right)\left(\frac{1 \text{ hr}}{60 \text{ min}}\right)\left(\frac{1000 \text{ m}}{1 \text{ km}}\right)\left(\frac{100 \text{ cm}}{1 \text{ m}}\right)$$
$$\omega = \frac{(30)(1000)(100)}{(70)(60)} \frac{\text{rad}}{\text{min}} = \mathbf{714.29 \frac{rad}{min}}$$

3. *(45)* $3 \cdot 3 \cdot 4 = \mathbf{36}$

4. *(39)* $80°\left(\dfrac{\pi}{180°}\right)300 = \dfrac{400\pi}{3} = \mathbf{418.88\ m}$

5. *(28)* $\text{Rate} = \dfrac{R}{m}\dfrac{\text{mi}}{\text{hr}}$

$\text{New rate} = \mathbf{\dfrac{R}{m - 2}\dfrac{mi}{hr}}$

6. *(25)* $R_H T_H + R_J T_J = \text{jobs}$

$\dfrac{1}{4}T + \dfrac{1}{8}(T + 2) = \dfrac{25}{4}$

$\dfrac{3}{8}T = 6$

$T = \mathbf{16\ hr}$

7. *(18)* $\text{Salt}_1 + \text{salt added} = \text{salt final}$

$(0.05)40 + 0.20P_N = 0.10(40 + P_N)$

$2 + 0.20P_N = 4 + 0.1P_N$

$0.1P_N = 2$

$P_N = \mathbf{20\ liters}$

8. *(54)* $y = (x^2 - 6x \quad) + 4$

$y = (x^2 - 6x + 9) + 4 - 9$

$\mathbf{y = (x - 3)^2 - 5}$

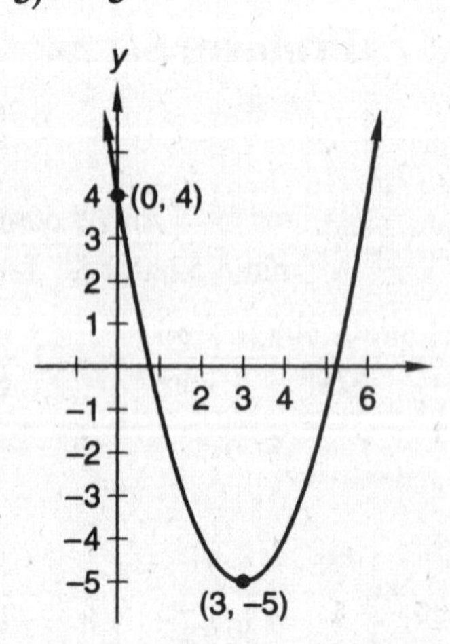

9. *(54)* $y = -(x^2 + 4x \quad) + 6$

$y = -(x^2 + 4x + 4) + 6 + 4$

$\mathbf{y = -(x + 2)^2 + 10}$

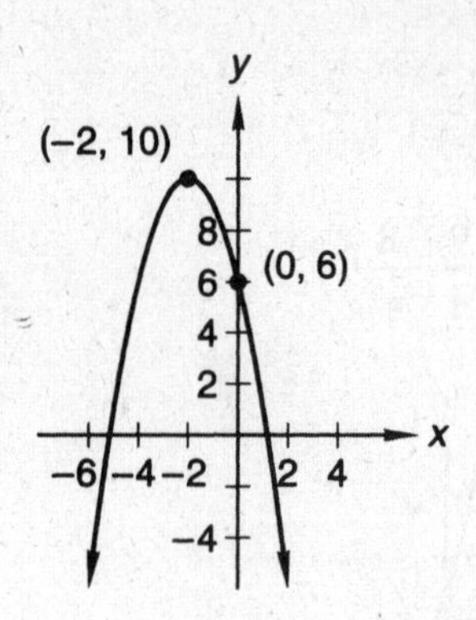

10. *(53)* $\left(40\ \dfrac{\text{mi}}{\text{hr}}\right)\left(\dfrac{5280\text{ ft}}{1\text{ mi}}\right)\left(\dfrac{12\text{ in.}}{1\text{ ft}}\right)\left(\dfrac{2.54\text{ cm}}{1\text{ in.}}\right)$

$\times \left(\dfrac{1\text{ hr}}{60\text{ min}}\right)\left(\dfrac{1\text{ min}}{60\text{ s}}\right)$

$= \mathbf{\dfrac{(40)(5280)(12)(2.54)}{(60)(60)}\ \dfrac{cm}{s} = 1788.16\ \dfrac{cm}{s}}$

11. *(53)* $\left(12\ \dfrac{\text{L}}{\text{s}}\right)\left(\dfrac{1000\text{ cm}^3}{1\text{ L}}\right)\left(\dfrac{60\text{ s}}{1\text{ min}}\right)\left(\dfrac{60\text{ min}}{1\text{ hr}}\right)$

$= \mathbf{(12)(10)^3(60)(60)\ \dfrac{cm^3}{hr} = 4.32 \times 10^7\ \dfrac{cm^3}{hr}}$

12. *(50)* $\cos\theta = \dfrac{-\sqrt{2}}{2}$

$= \dfrac{-1}{\sqrt{2}}$

$\theta = \mathbf{135°, 225°}$

13. *(52)* $\sin 3\theta = \dfrac{1}{2}$

$3\theta = 30°, 150°, 390°, 510°, 750°, 870°$

$\theta = \mathbf{10°, 50°, 130°, 170°, 250°, 290°}$

14. *(52)* $\sec 2\theta = 1$

$2\theta = 0°, 360°$

$\theta = \mathbf{0°, 180°}$

15. *(51)* $\log 3100 = 3.4914$

$3100 = \mathbf{10^{3.4914}}$

16. *(51)* $\ln 3100 = 8.0392$

$3100 = \mathbf{e^{8.0392}}$

17. (48) $\tan^2 \frac{19\pi}{6} - \csc^2 \frac{\pi}{2} + \sec^2 3\pi$

$= \left(\tan \frac{\pi}{6}\right)^2 - \frac{1}{\left(\sin \frac{\pi}{2}\right)^2} + \frac{1}{(\cos \pi)^2}$

$= \left(\frac{1}{\sqrt{3}}\right)^2 - 1 + (-1)^2 = \mathbf{\frac{1}{3}}$

18. (48) $\sec^2\left(-\frac{3\pi}{4}\right) - \cot^2\left(\frac{7\pi}{6}\right)$

$= \frac{1}{\left(-\cos \frac{\pi}{4}\right)^2} - \frac{1}{\left(\tan \frac{\pi}{6}\right)^2}$

$= 2 - 3 = \mathbf{-1}$

19. (37) $(x, y) = \left(\frac{6 + (-2)}{2}, \frac{7 + 5}{2}\right) = \mathbf{(2, 6)}$

20. (37,39) $\sqrt{(x + 4)^2 + (y - 8)^2} = \sqrt{(x - 4)^2 + (y - 6)^2}$

$x^2 + 8x + 16 + y^2 - 16y + 64$

$= x^2 - 8x + 16 + y^2 - 12y + 36$

$16x - 4y = -28$

$\frac{16x}{-28} - \left(\frac{4y}{-28}\right) = \frac{-28}{-28}$

$\mathbf{\frac{x}{-\frac{7}{4}} + \frac{y}{7} = 1}$

21. (47) Function $= -\sin x$

Centerline $= 9$

Amplitude $= 1$

$\mathbf{y = 9 - \sin x}$

22. (47) Function $= \cos \theta$

Centerline $= 6$

Amplitude $= 6$

$\mathbf{y = 6 + 6\cos \theta}$

23. (46) $5(x^2 + 3x + 9) = 0$

$x = \frac{-3 \pm \sqrt{9 - 36}}{2}$

$x = -\frac{3}{2} \pm \frac{\sqrt{-27}}{2} = -\frac{3}{2} \pm \frac{3\sqrt{3}i}{2}$

$\mathbf{5\left(x + \frac{3}{2} - \frac{3\sqrt{3}}{2}i\right)\left(x + \frac{3}{2} + \frac{3\sqrt{3}}{2}i\right)}$

24. (45) Use a calculator to get:

$S = aT + b$

$a = -0.089$

$b = 14.47$

$\mathbf{r = -0.9047}$

$\mathbf{S = -0.089T + 14.47}$

This is a **good correlation.**

25. (42) $\mathbf{(x - 3)^2 + (y - 2)^2 = 6^2}$

26. (42) $f(x) = 2^{-x-3} = 2^{-x}2^{-3} = \left(\frac{1}{2}\right)^x\left(\frac{1}{2}\right)^3 = \frac{1}{8}\left(\frac{1}{2}\right)^x$

27. (40) (a) $\log_5 6 - \log_5 \frac{1}{x + 1} = \log_5 7$

$\log_5\left(\frac{6}{\frac{1}{x + 1}}\right) = \log_5 7$

$\log_5 6(x + 1) = \log_5 7$

$6x + 6 = 7$

$x = \mathbf{\frac{1}{6}}$

(b) $2 \ln 5 - \ln x = \frac{3}{4} \ln 16$

$\ln 5^2 - \ln x = \ln 16^{\frac{3}{4}}$

$\ln \frac{5^2}{x} = \ln 16^{\frac{3}{4}}$

$\frac{25}{x} = 8$

$x = \mathbf{\frac{25}{8}}$

28. (34) $\sum_{k=-3}^{-1} \frac{(k - 1)(k + 1)}{k} = \sum_{k=-3}^{-1} \frac{k^2 - 1}{k}$

$= \frac{(-3)^2 - 1}{-3} + \frac{(-2)^2 - 1}{-2} + \frac{(-1)^2 - 1}{-1}$

$= -\frac{8}{3} + \left(-\frac{3}{2}\right) + 0 = \mathbf{-\frac{25}{6}}$

29. (31) Replace y with $-y$, then replace y with $y + 2$.

$$-(y + 2) = |x|$$
$$-y - 2 = |x|$$
$$-y = |x| + 2$$
$$\mathbf{y = -|x| - 2}$$

30. (31) The graph of g is the graph of f translated six units up. Therefore the answer is **C.**

Problem Set 55

1. (55)

R	W	R
W	R	W

2 possiblities

2. (55) $(N - 1)! = (4 - 1)! = 3! = 3 \cdot 2 \cdot 1 = \mathbf{6}$

3. (55) $\dfrac{N!}{a!b!} = \dfrac{8!}{2!2!} = 8 \cdot 7 \cdot 6 \cdot 5 \cdot 3 \cdot 2 \cdot 1$
$= \mathbf{10{,}080}$

4. (53) $\omega = \dfrac{v}{r}$

$$\omega = \frac{10\ \frac{\text{km}}{\text{hr}}}{3\text{ m}} = \frac{10\text{ km}}{3\text{ m-hr}}$$

$$\omega = \left(\frac{10\text{ km}}{3\text{ m-hr}}\right)\left(\frac{1000\text{ m}}{1\text{ km}}\right)\left(\frac{1\text{ hr}}{60\text{ min}}\right)\left(\frac{1\text{ min}}{60\text{ s}}\right)$$

$$\omega = \mathbf{\frac{(10)(1000)}{(60)(60)(3)}\ \frac{rad}{s} = 0.93\ \frac{rad}{s}}$$

5. (53) $v = r\omega$

$$v = (0.5\text{ in.})\left(40\ \frac{\text{rad}}{\text{s}}\right)\left(\frac{1\text{ yd}}{36\text{ in.}}\right)\left(\frac{60\text{ s}}{1\text{ min}}\right)$$

$$v = \mathbf{\frac{(0.5)(40)(60)}{36}\ \frac{yd}{min} = 33.33\ \frac{yd}{min}}$$

6. (38) (a) $\begin{cases} R_O T_O = 400 \\ R_B T_B = 400 \end{cases}$ (b)

$R_B = 2R_O,\ T_B = 6 - T_O$

(b)
$$R_B T_B = 400$$
$$(2R_O)(6 - T_O) = 400$$
$$12R_O - 2R_O T_O = 400$$
$$12R_O - 2(400) = 400$$
$$12R_O = 1200$$
$$R_O = \mathbf{100\ mph}$$

7. (18) (a) $N_R = N_W + 2$; (b) $N_R = N_G + 1$; (c) $N_R + N_W + N_G = 15$

(a) $N_R = N_W + 2$
$N_W = N_R - 2$

(b) $N_R = N_G + 1$
$N_G = N_R - 1$

(c)
$$N_R + N_W + N_G = 15$$
$$N_R + (N_R - 2) + (N_R - 1) = 15$$
$$3N_R = 18$$
$$\mathbf{N_R = 6}$$

$N_W = N_R - 2$
$N_W = (6) - 2$
$\mathbf{N_W = 4}$

$N_G = N_R - 1$
$N_G = (6) - 1$
$\mathbf{N_G = 5}$

8. (54) $y = (x^2 - 8x\quad) + 12$
$y = (x^2 - 8x + 16) + 12 - 16$
$\mathbf{y = (x - 4)^2 - 4}$

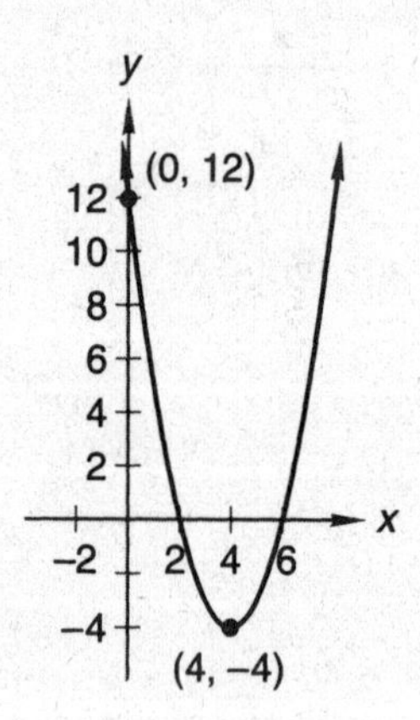

9. (54) $y = -(x^2 + 2x\quad) - 2$
$y = -(x^2 + 2x + 1) - 2 + 1$
$\mathbf{y = -(x + 1)^2 - 1}$

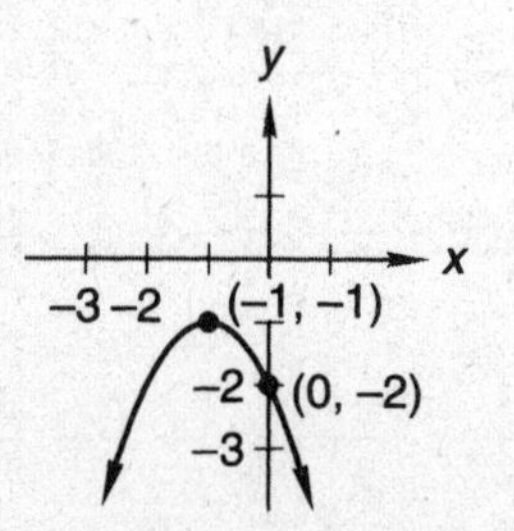

10. (53)
$$\left(20\ \frac{\text{m}}{\text{s}}\right)\left(\frac{60\text{ s}}{1\text{ min}}\right)\left(\frac{60\text{ min}}{1\text{ hr}}\right)\left(\frac{100\text{ cm}}{1\text{ m}}\right)\left(\frac{1\text{ in.}}{2.54\text{ cm}}\right)$$
$$\times\left(\frac{1\text{ ft}}{12\text{ in.}}\right)\left(\frac{1\text{ mi}}{5280\text{ ft}}\right)$$
$$= \frac{\mathbf{(20)(100)(60)(60)}}{\mathbf{(2.54)(12)(5280)}}\ \frac{\mathbf{mi}}{\mathbf{hr}} = \mathbf{44.74\ \frac{mi}{hr}}$$

11. (53)
$$\left(12\ \frac{\text{liter}}{\text{hr}}\right)\left(\frac{1\text{ hr}}{60\text{ min}}\right)\left(\frac{10^3\text{ cm}^3}{1\text{ liter}}\right)\left(\frac{1\text{ in.}}{2.54\text{ cm}}\right)^3$$
$$= \frac{\mathbf{(12)(10)^3}}{\mathbf{(2.54)^3(60)}}\ \frac{\mathbf{in.^3}}{\mathbf{min}} = \mathbf{12.20\ \frac{in.^3}{min}}$$

12. (50)
$$3\tan\theta = \sqrt{3}$$
$$\tan\theta = \frac{\sqrt{3}}{3}$$
$$\tan\theta = \frac{1}{\sqrt{3}}$$
$$\theta = \mathbf{30°, 210°}$$

13. (52)
$$\tan 3\theta = \frac{\sqrt{3}}{3}$$
$$\tan 3\theta = \frac{1}{\sqrt{3}}$$
$$3\theta = 30°, 210°, 390°, 570°, 750°, 930°$$
$$\theta = \mathbf{10°, 70°, 130°, 190°, 250°, 310°}$$

14. (52)
$$\csc 2\theta = 1$$
$$\sin 2\theta = 1$$
$$2\theta = 90°, 450°$$
$$\theta = \mathbf{45°, 225°}$$

15. (51)
$$\log_{10} 5800 = 3.7634$$
$$5800 = \mathbf{10^{3.7634}}$$

16. (51)
$$\ln 5800 = 8.6656$$
$$5800 = \mathbf{e^{8.6656}}$$

17. (48)
$$\sec^2\left(-\frac{3\pi}{4}\right) - \cot^2\frac{7\pi}{6} + 2\tan^2\left(-\frac{7\pi}{6}\right)$$
$$= \frac{1}{\left(-\cos\frac{\pi}{4}\right)^2} - \frac{1}{\left(\tan\frac{\pi}{6}\right)^2} + 2\left(-\tan\frac{\pi}{6}\right)^2$$
$$= 2 - 3 + \frac{2}{3} = \mathbf{-\frac{1}{3}}$$

18. (48)
$$\sec^2\left(\frac{3\pi}{4}\right) - \csc^2\left(-\frac{4\pi}{3}\right)$$
$$= \frac{1}{\left(-\cos\frac{\pi}{4}\right)^2} - \frac{1}{\left(\sin\frac{\pi}{3}\right)^2}$$
$$= 2 - \frac{4}{3} = \mathbf{\frac{2}{3}}$$

19. (48)
$$m_0 = \frac{-3-3}{-2-4} = \frac{-6}{-6} = 1$$
$$(x_m, y_m) = \left(\frac{-2+4}{2}, \frac{-3+3}{2}\right) = (1, 0)$$

The slope of the perpendicular line is the negative reciprocal, so $m = -1$.

$$y = -x + b$$
$$0 = -1(1) + b$$
$$b = 1$$
$$y = -x + 1$$
$$\mathbf{x + y - 1 = 0}$$

20. (47)
Function = $\sin\theta$
Centerline = 7
Amplitude = 3
$$\mathbf{y = 7 + 3\sin\theta}$$

21. (47)
Function = $-\cos\theta$
Centerline = 8
Amplitude = 9
$$\mathbf{y = 8 - 9\cos\theta}$$

22. (46)
$$x^2 - 5x + 8 = 0$$
$$x = \frac{5 \pm \sqrt{25-32}}{2} = \frac{5}{2} \pm \frac{\sqrt{7}}{2}i$$
$$\mathbf{\left(x - \frac{5}{2} - \frac{\sqrt{7}}{2}i\right)\left(x - \frac{5}{2} + \frac{\sqrt{7}}{2}i\right)}$$

23. (45)
Use a calculator to get:
$P = aH + b$
$a = 1.057$
$b = 60.43$
$\mathbf{r = 0.1018}$
$\mathbf{P = 1.057H + 60.43}$
This is **not a good correlation.**

24. (42)
$$\mathbf{(x-2)^2 + (y-5)^2 = 7^2}$$

25. (41) $_7P_4 - {}_7P_3 = \frac{7!}{(7-4)!} - \frac{7!}{(7-3)!}$

$= \frac{7!}{3!} - \frac{7!}{4!} = \mathbf{630}$

26. (40,51) (a) $\log_{18} 12x - 2\log_{18} 2 = 1$

$\log_{18} 12x - \log_{18} 2^2 = 1$

$\log_{18} \frac{12x}{4} = 1$

$\log_{18} 3x = 1$

$3x = 18^1$

$x = \mathbf{6}$

(b) $\frac{2}{3}\ln 8 + 2\ln 3 = \ln(x+5)$

$\ln 8^{\frac{2}{3}} + \ln 3^2 = \ln(x+5)$

$\ln(4)(9) = \ln(x+5)$

$36 = x + 5$

$x = \mathbf{31}$

27. (40) $(f \circ g)(x) = \frac{2\left(\frac{2x+1}{3x-2}\right)+1}{3\left(\frac{2x+1}{3x-2}\right)-2}$

$= \frac{2(2x+1)+(3x-2)}{3(2x+1)-2(3x-2)} = \frac{7x}{7} = x$

$(g \circ f)(x) = \frac{2\left(\frac{2x+1}{3x-2}\right)+1}{3\left(\frac{2x+1}{3x-2}\right)-2}$

$= \frac{2(2x+1)+(3x-2)}{3(2x+1)-2(3x-2)} = \frac{7x}{7} = x$

$(g \circ f)(x) = (f \circ g)(x) = x$

Yes, f and g are inverse functions.

28. (32) (a)

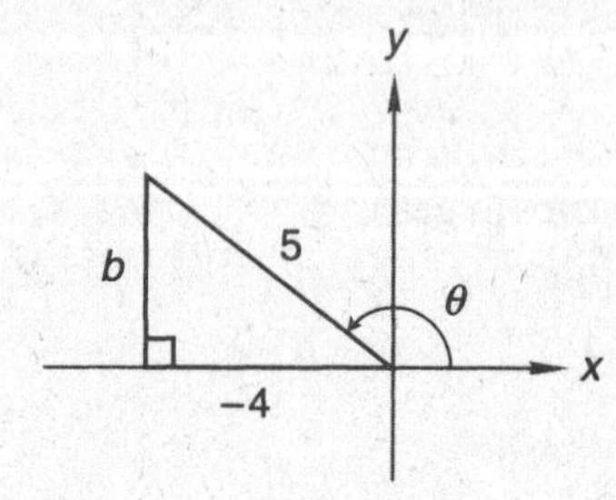

$0° \le \theta \le 180°$

$b = \sqrt{5^2 - (-4)^2} = 3$

$\tan\left[\text{Arccos}\left(-\frac{4}{5}\right)\right] = \tan\theta = \mathbf{-\frac{3}{4}}$

(b)

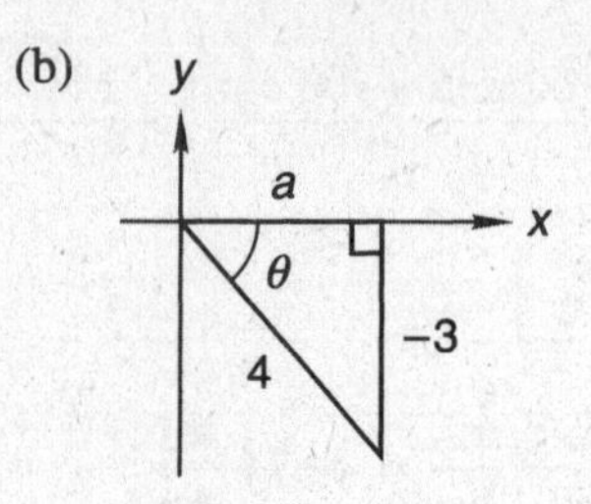

$-90° \le \theta \le 90°$

$a = \sqrt{4^2 - (-3)^2} = \sqrt{7}$

$\cos\left[\text{Arcsin}\left(-\frac{3}{4}\right)\right] = \cos\theta = \mathbf{\frac{\sqrt{7}}{4}}$

29. (31) Replace x with $-x$ and y with $y - 3$.

$y - 3 = \sqrt{-x}$

$\mathbf{y = \sqrt{-x} + 3}$

30. (31) The graph of g is the graph of f translated seven units to the left. Therefore the answer is **B.**

Problem Set 56

1. (45) $5 \cdot 5 \cdot 5 = \mathbf{125}$

2. (55) $(12-1)! = 11! = \mathbf{39{,}916{,}800}$

3. (55) $\frac{9!}{2!3!2!} = \mathbf{15{,}120}$

4. (53) $v = r\omega$

$v = (10 \text{ in.})\left(\frac{40 \text{ rev}}{1 \text{ min}}\right)\left(\frac{2\pi \text{ rad}}{1 \text{ rev}}\right)\left(\frac{2.54 \text{ cm}}{1 \text{ in.}}\right)\left(\frac{1 \text{ min}}{60 \text{ s}}\right)$

$v = \mathbf{\frac{(40)(2\pi)(10)(2.54)}{60} \frac{cm}{s} = 106.40 \frac{cm}{s}}$

5. (53) $\omega = \frac{v}{r}$

$\omega = \frac{260 \frac{\text{km}}{\text{hr}}}{1 \text{ ft}} = \frac{260 \text{ km}}{\text{hr-ft}}$

$\omega = \left(\frac{260 \text{ km}}{\text{hr-ft}}\right)\left(\frac{1 \text{ hr}}{60 \text{ min}}\right)\left(\frac{1000 \text{ m}}{1 \text{ km}}\right)\left(\frac{100 \text{ cm}}{1 \text{ m}}\right)$

$\times \left(\frac{1 \text{ in.}}{2.54 \text{ cm}}\right)\left(\frac{1 \text{ ft}}{12 \text{ in.}}\right)$

$\omega = \frac{(260)(1000)(100)}{(60)(2.54)(12)} \frac{\text{rad}}{\text{min}} \cdot \frac{1 \text{ rev}}{2\pi \text{ rad}}$

$\omega = \mathbf{\frac{(260)(1000)(100)}{(60)(2.54)(12)(2\pi)} \frac{rev}{min} = 2262.70 \frac{rev}{min}}$

6. (25) $\text{Rate} = \frac{450}{30} = 15\ \frac{\text{dollars}}{\text{hr}}$

$$\text{New Rate} = 1.1(15) = 16.50\ \frac{\text{dollars}}{\text{hr}}$$

$$\text{New Rate} \times T = \text{dollars}$$

$$16.5(36) = \mathbf{\$594}$$

7. (18) $\begin{cases} \text{(a)}\ T + U = 11 \\ \text{(b)}\ 10U + T = 10T + U - 27 \end{cases}$

(a) $T + U = 11$

$$T = 11 - U$$

(b)

$$10U + T = 10T + U - 27$$
$$9U - 9T = -27$$
$$9U - 9(11 - U) = -27$$
$$18U = 72$$
$$U = 4$$

$$T = 11 - U = 11 - (4) = 7$$

The original number was **74.**

8. (25) $\begin{cases} \text{(a)}\ A_N = Y_N + 40 \\ \text{(b)}\ A_N - 10 = 7(Y_N - 10) + 22 \end{cases}$

(b)

$$A_N - 10 = 7Y_N - 70 + 22$$
$$A_N - 7Y_N = -38$$
$$(Y_N + 40) - 7Y_N = -38$$
$$-6Y_N = -78$$
$$Y_N = 13$$

(a) $A_N = Y_N + 40 = (13) + 40 = 53$

Youngster = 13 yr; Ancient one = 53 yr

9. (56)

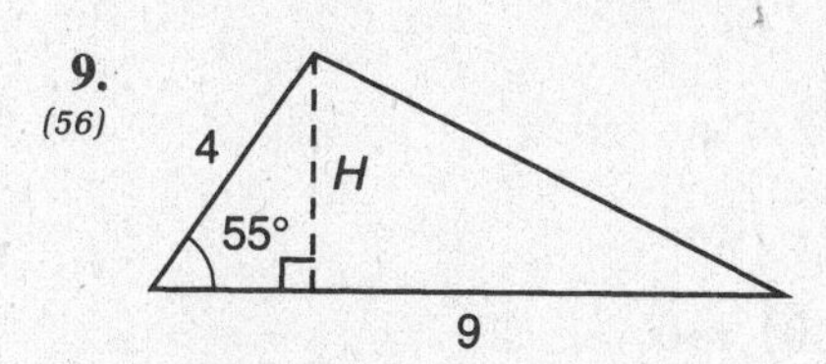

$$\sin 55° = \frac{H}{4}$$

$$H = 4 \sin 55° = 3.276 \text{ m}$$

$$\text{Area} = \frac{1}{2}BH = \frac{1}{2}(9)(3.276) = \mathbf{14.74\ m^2}$$

10. (56)

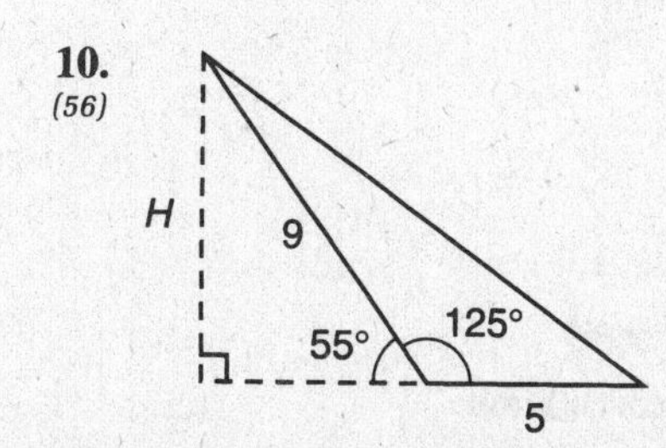

$$\sin 55° = \frac{H}{9}$$

$$H = 9 \sin 55° = 7.37 \text{ m}$$

$$\text{Area} = \frac{1}{2}BH = \frac{1}{2}(5)(7.37) = \mathbf{18.43\ cm^2}$$

11. (56)

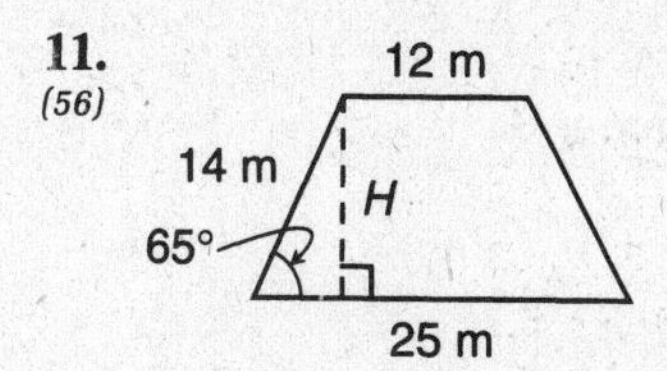

$$\sin 65° = \frac{H}{14}$$

$$H = 14 \sin 65° = 12.688 \text{ m}$$

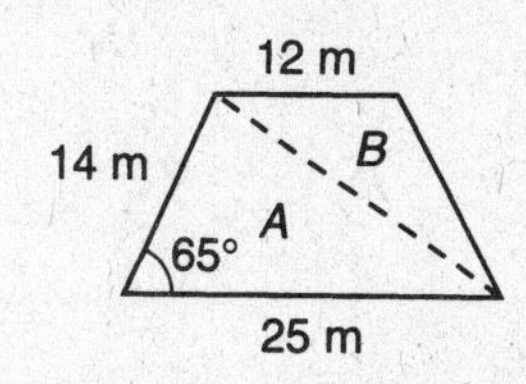

$$\text{Total Area} = \text{Area } A + \text{Area } B$$

$$= \frac{1}{2}(25)(12.688) + \frac{1}{2}(12)(12.688)$$

$$= 158.60 + 76.13 = \mathbf{234.73\ m^2}$$

12. (56)

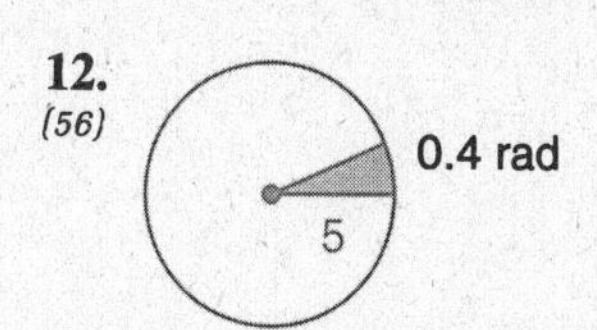

$$A_{\text{sector}} = \pi r^2\left(\frac{0.4}{2\pi}\right) = \pi(5)^2\left(\frac{0.4}{2\pi}\right) = \mathbf{5\ m^2}$$

13. (56)

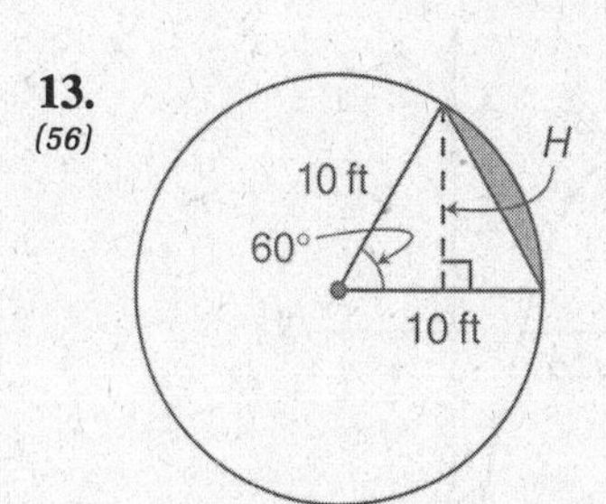

$$H = 10 \sin 60° = 8.66 \text{ ft}$$

$$A_{\text{segment}} = A_{\text{sector}} - A_{\text{triangle}}$$

$$= \pi r^2\left(\frac{60°}{360°}\right) - \frac{1}{2}rH$$

$$= \pi(10)^2\left(\frac{60°}{360°}\right) - \frac{1}{2}(10)(8.66)$$

$$= 52.36 - 43.3 = \mathbf{9.06\ ft^2}$$

14. (56) $\begin{cases} y < x^2 - 3 \\ y \geq x + 1 \end{cases}$

This system of inequalities designates the region below the parabola and above or on the line. Therefore the answer is **B.**

Test point: $(-4, 0)$

Parabola: $0 < (-4)^2 - 3$

$0 < 13$ True

Line: $0 \geq -4 + 1$

$0 \geq -3$ True

15. (54) $y = (x^2 - 4x \quad) + 6$

$y = (x^2 - 4x + 4) + 6 - 4$

$\mathbf{y = (x - 2)^2 + 2}$

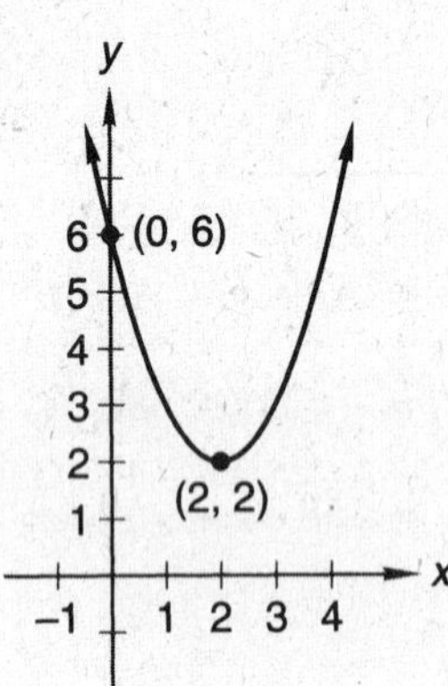

16. (54) $y = -(x^2 + 6x \quad) - 6$

$y = -(x^2 + 6x + 9) - 6 + 9$

$\mathbf{y = -(x + 3)^2 + 3}$

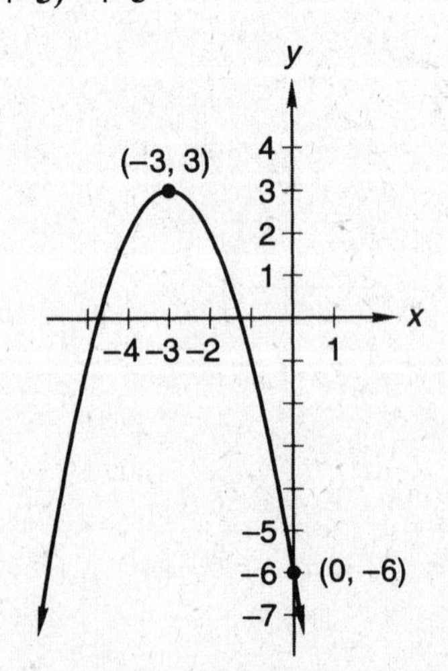

17. (50) $\tan \theta = -1$

$\theta = \mathbf{135°, 315°}$

18. (52) $\sin 3\theta = -\dfrac{\sqrt{3}}{2}$

$3\theta = 240°, 300°, 600°, 660°, 960°, 1020°$

$\theta = \mathbf{80°, 100°, 200°, 220°, 320°, 340°}$

19. (52) $\csc 2\theta = -1$

$2\theta = 270°, 630°$

$\theta = \mathbf{135°, 315°}$

20. (51) $\log_{10} 65{,}000 = 4.8129$

$65{,}000 = \mathbf{10^{4.8129}}$

21. (51) $\ln 65{,}000 = 11.0821$

$65{,}000 = \mathbf{e^{11.0821}}$

22. (48) $\sec^2\left(-\dfrac{2\pi}{3}\right) - \tan^2\left(\dfrac{9\pi}{4}\right)$

$= \left(-\sec\dfrac{\pi}{3}\right)^2 - \left(\tan\dfrac{\pi}{4}\right)^2 = (-2)^2 - 1^2 = \mathbf{3}$

23. (37,39) $\sqrt{(x + 2)^2 + (y - 4)^2} = \sqrt{(x + 4)^2 + (y + 4)^2}$

$x^2 + 4x + 4 + y^2 - 8y + 16$

$= x^2 + 8x + 16 + y^2 + 8y + 16$

$4x + 16y + 12 = 0$

$\mathbf{x + 4y + 3 = 0}$

24. (37) $(x, y) = \left(\dfrac{4 + 12}{2}, \dfrac{7 + (-3)}{2}\right) = \mathbf{(8, 2)}$

25. (47) Function $= -\cos x$

Centerline $= 3$

Amplitude $= 9$

$\mathbf{y = 3 - 9 \cos x}$

26. (46) $3(x^2 + 6x + 10) = 0$

$x = \dfrac{-6 \pm \sqrt{36 - 40}}{2} = -\dfrac{6}{2} \pm \dfrac{\sqrt{4}}{2}i = -3 \pm i$

$\mathbf{3(x + 3 - i)(x + 3 + i)}$

27. (45) Use a calculator to get:

$P = aI + b$

$a = 0.0996$

$b = 47.9949$

$\mathbf{r = 0.1503}$

$\mathbf{P = 0.0996I + 47.9949}$

This is **not a good correlation.**

28. (42) $(x - 4)^2 + (y - 5)^2 = 5^2$

29. (40,51) (a) $\log_2 128 = x$

$$2^x = 128$$
$$2^x = 2^7$$
$$\mathbf{x = 7}$$

(b) $\frac{2}{3}\log 27 + \log x = \log 12$

$$\log 27^{\frac{2}{3}} + \log x = \log 12$$
$$\log 27^{\frac{2}{3}} x = \log 12$$
$$27^{\frac{2}{3}} x = 12$$
$$9x = 12$$
$$\mathbf{x = \frac{4}{3}}$$

30. (21) $\frac{f(x + h) - f(x - h)}{2h}$

$$= \frac{(x + h)(x + h - 2) - (x - h)(x - h - 2)}{2h}$$
$$= \frac{x^2 + 2hx - 2x + h^2 - 2h}{2h}$$
$$- \frac{x^2 + 2hx + 2x - h^2 - 2h}{2h}$$
$$= \frac{4hx - 4h}{2h} = \mathbf{2x - 2}$$

Problem Set 57

1. (55) $\frac{12!}{6!4!2!} = \mathbf{13{,}860}$

2. (55) $\frac{11!}{5!2!2!} = \mathbf{83{,}160}$

3. (53) $v = r\omega$

$$v = (4\text{ cm})\left(\frac{400\text{ rad}}{1\text{ min}}\right)\left(\frac{60\text{ min}}{1\text{ hr}}\right)\left(\frac{1\text{ in.}}{2.54\text{ cm}}\right)$$
$$\times\left(\frac{1\text{ ft}}{12\text{ in}}\right)\left(\frac{1\text{ mile}}{5280\text{ ft}}\right)$$
$$v = \frac{\mathbf{(4)(400)(60)}}{\mathbf{(2.54)(12)(5280)}}\frac{\textbf{mi}}{\textbf{hr}} = \mathbf{0.60}\ \frac{\textbf{mi}}{\textbf{hr}}$$

4. (53) $\omega = \frac{v}{r} = \frac{40\ \frac{\text{km}}{\text{hr}}}{5\text{ cm}} = \frac{40\text{ km}}{5\text{ cm-hr}}$

$$\omega = \left(\frac{40\text{ km}}{5\text{cm-hr}}\right)\left(\frac{1\text{ hr}}{60\text{ min}}\right)\left(\frac{1\text{ min}}{60\text{ s}}\right)$$
$$\times\left(\frac{1000\text{ m}}{1\text{ km}}\right)\left(\frac{100\text{ cm}}{1\text{ m}}\right)$$
$$\omega = \frac{\mathbf{(40)(1000)(100)}}{\mathbf{(5)(60)(60)}}\frac{\textbf{rad}}{\textbf{s}} = \mathbf{222.22}\ \frac{\textbf{rad}}{\textbf{s}}$$

5. (36) (a) $\begin{cases}(B + W)T_D = D_D\\(B - W)T_U = D_U\end{cases}$ (b)

$B = 3W,\ T_D = T_U + 1$

(a)
$$(B + W)T_D = D_D$$
$$(3W + W)T_D = 24$$
$$(4W)T_D = 24$$
$$WT_D = 6$$

$$T_D = T_U + 1$$
$$T_U = T_D - 1$$

(b)
$$(B - W)T_U = D_U$$
$$(3W - W)(T_D - 1) = 8$$
$$(2W)(T_D - 1) = 8$$
$$2WT_D - 2W = 8$$
$$2(6) - 2W = 8$$
$$2W = 4$$
$$\mathbf{W = 2\ mph}$$

$$B = 3W$$
$$B = 3(2)$$
$$\mathbf{B = 6\ mph}$$

6. (39) $38.3°\left(\frac{\pi}{180°}\right)6000 = \mathbf{4010.77\ km}$

7. (18) $B = kf$

$$k = \frac{B}{f} = \frac{100}{2} = 50$$
$$B = 50f = 50(40) = \mathbf{2000\ boys}$$

8. (57) Function = $\sin\theta$

Amplitude = 6

Period = 360°

Phase angle = 90°

$\mathbf{y = 6\sin(\theta - 90°)}$

9. *(57)* Function = $\sin\theta$

Amplitude = 10

Period = 180°

Phase angle = 0

$$\text{Coefficient} = \frac{360°}{180°} = 2$$

$$\mathbf{y = 10 \sin 2\theta}$$

10. *(56)*

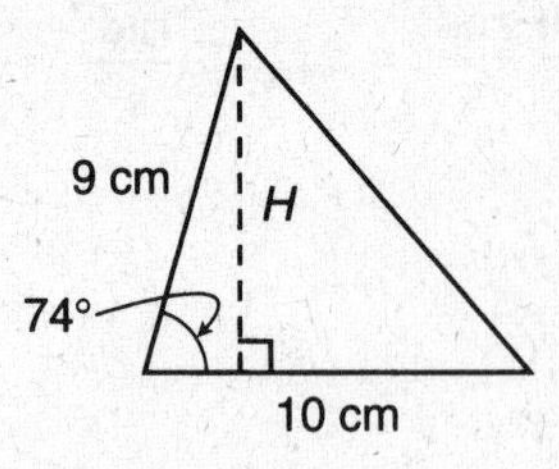

$$\sin 74° = \frac{H}{9}$$

$$H = 9 \sin 74° = 8.651 \text{ cm}$$

$$\text{Area} = \frac{1}{2}BH = \frac{1}{2}(10)(8.651) = \mathbf{43.26\ cm^2}$$

11. *(56)*

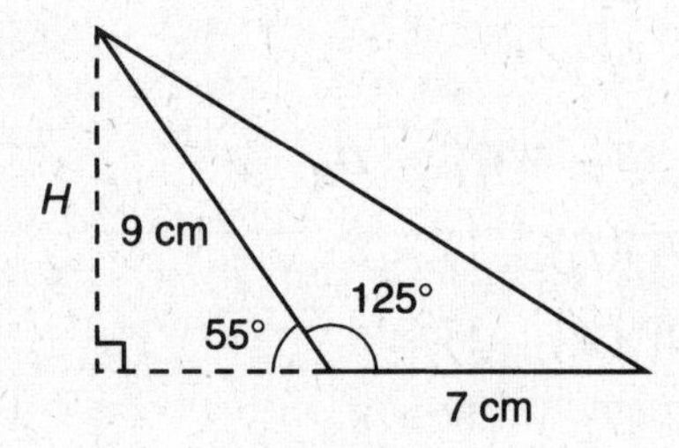

$$\sin 55° = \frac{H}{9}$$

$$H = 9 \sin 55° = 7.372 \text{ cm}$$

$$\text{Area} = \frac{1}{2}BH = \frac{1}{2}(7)(7.372) = \mathbf{25.80\ cm^2}$$

12. *(56)*

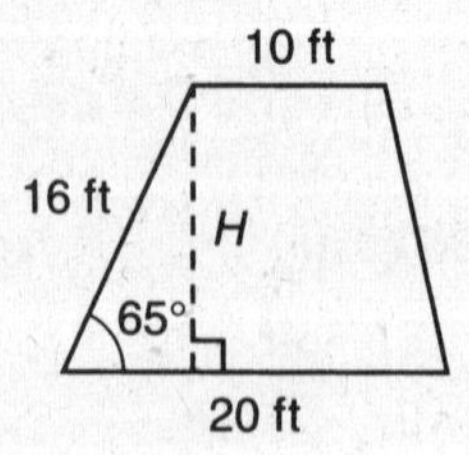

$$\sin 65° = \frac{H}{16}$$

$$H = 16 \sin 65° = 14.5009 \text{ ft}$$

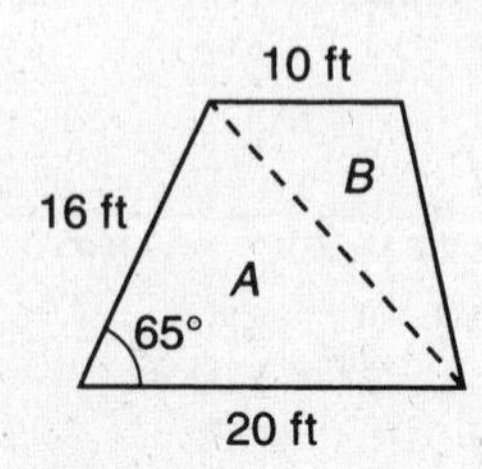

$$\text{Total Area} = \text{Area } A + \text{Area } B$$

$$= \frac{1}{2}(20)(14.5009) + \frac{1}{2}(10)(14.5009)$$

$$= 145.009 + 72.50 = \mathbf{217.51\ ft^2}$$

13. *(56)*

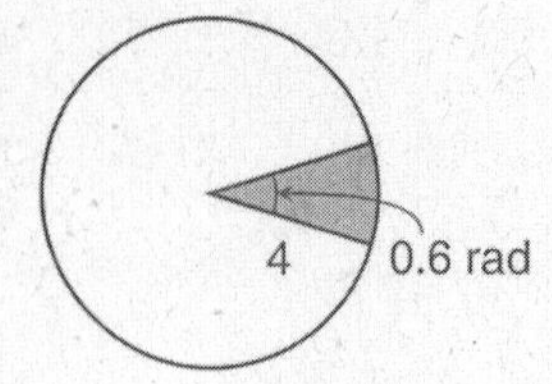

$$\text{Area} = \pi r^2\left(\frac{0.6}{2\pi}\right) = \pi(4)^2\left(\frac{0.6}{2\pi}\right) = \mathbf{4.80\ m^2}$$

14. *(56)*

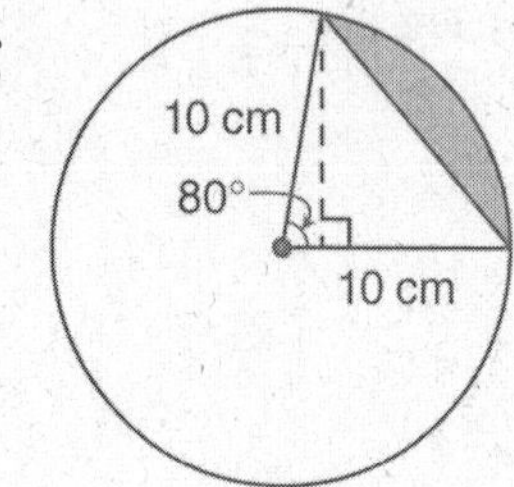

$$H = 10 \sin 80° = 9.848 \text{ cm}$$

$$A_{\text{segment}} = A_{\text{sector}} - A_{\text{triangle}}$$

$$= \pi r^2\left(\frac{80°}{360°}\right) - \frac{1}{2}rH$$

$$= \pi(10)^2\left(\frac{80°}{360°}\right) - \frac{1}{2}(10)(9.848)$$

$$= 69.81 - 49.24 = \mathbf{20.57\ cm^2}$$

15. *(56)*
$$\begin{cases} y \le -x + 1 & \text{(line)} \\ y \ge (x-2)^2 - 3 & \text{(parabola)} \end{cases}$$

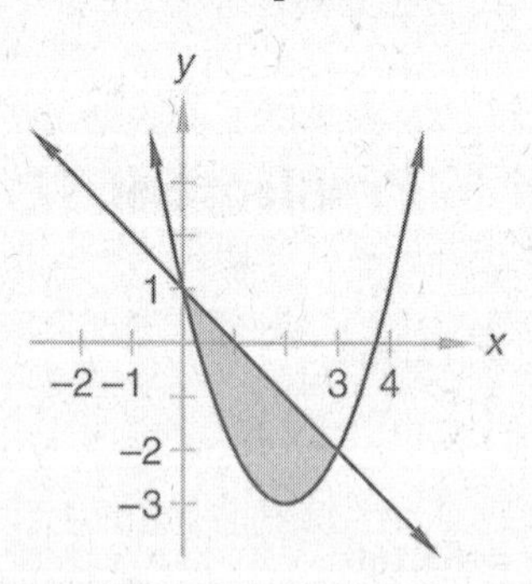

The region must be on or above the parabola and on or below the line.

Test point: (1, –1)

Line: $-1 \le -(1) + 1$

$-1 \le 0$ True

Parabola: $-1 \ge (1 - 2)^2 - 3$

$-1 \ge (-1)^2 - 3$

$-1 \ge 1 - 3$

$-1 \ge -2$ True

16. (54) $y = -(x^2 + 4x \quad\quad) + 4$

$y = -(x^2 + 4x + 4) + 4 + 4$

$\mathbf{y = -(x + 2)^2 + 8}$

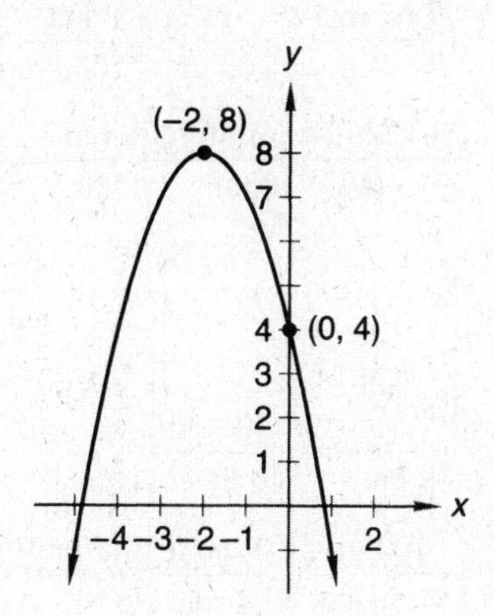

17. (50) $2\cos\theta = 1$

$\cos\theta = \dfrac{1}{2}$

$\theta = \mathbf{60°, 300°}$

18. (52) $2\cos 3\theta = 1$

$\cos 3\theta = \dfrac{1}{2}$

$3\theta = 60°, 300°, 420°, 660°, 780°, 1020°$

$\theta = \mathbf{20°, 100°, 140°, 220°, 260°, 340°}$

19. (52) $\cot 3\theta = 0$

$3\theta = 90°, 270°, 450°, 630°, 810°, 990°$

$\theta = \mathbf{30°, 90°, 150°, 210°, 270°, 330°}$

20. (51) $\log_{10} 10{,}000 = 4$

$10{,}000 = \mathbf{10^4}$

21. (51) $\ln 10{,}000 = 9.2103$

$10{,}000 = \mathbf{e^{9.2103}}$

22. (39,48) $m_0 = \dfrac{8 - (-6)}{4 - (-4)} = \dfrac{14}{8} = \dfrac{7}{4}$

$(x_m, y_m) = \left(\dfrac{-4 + 4}{2}, \dfrac{-6 + 8}{2}\right) = (0, 1)$

The slope of a perpendicular line is the negative reciprocal, so $m = -\frac{4}{7}$.

$y = -\dfrac{4}{7}x + b$

$1 = -\dfrac{4}{7}(0) + b$

$b = 1$

$y = -\dfrac{4}{7}x + 1$

$\mathbf{\dfrac{x}{\frac{7}{4}} + \dfrac{y}{1} = 1}$

23. (47) $\text{Arctan}\left(\tan\dfrac{7\pi}{6}\right) = \text{Arctan}\left(\tan\dfrac{\pi}{6}\right) = \mathbf{\dfrac{\pi}{6}}$

24. (48) $\cot^3\dfrac{\pi}{4} + \tan^2\dfrac{8\pi}{3} = \left(\cot\dfrac{\pi}{4}\right)^3 + \left(-\tan\dfrac{\pi}{3}\right)^2$

$= 1^3 + (-\sqrt{3})^2 = \mathbf{4}$

25. (46) $x^2 - 3x + 4 = 0$

$x = \dfrac{3 \pm \sqrt{3^2 - 16}}{2} = \dfrac{3}{2} \pm \dfrac{\sqrt{7}}{2}i$

$\mathbf{\left(x - \dfrac{3}{2} - \dfrac{\sqrt{7}}{2}i\right)\left(x - \dfrac{3}{2} + \dfrac{\sqrt{7}}{2}i\right)}$

26. (40,51) (a) $\dfrac{3}{4}\log_{\frac{1}{2}} 16 - \log_{\frac{1}{2}}(x - 1) = 2$

$\log_{\frac{1}{2}}\dfrac{16^{\frac{3}{4}}}{(x - 1)} = 2$

$\dfrac{8}{x - 1} = \left(\dfrac{1}{2}\right)^2$

$\dfrac{8}{x - 1} = \dfrac{1}{4}$

$32 = x - 1$

$x = \mathbf{33}$

(b) $\dfrac{1}{2}\ln 25 + \ln(x - 2) = \ln(2x + 3)$

$\ln 25^{\frac{1}{2}}(x - 2) = \ln(2x + 3)$

$5x - 10 = 2x + 3$

$3x = 13$

$x = \mathbf{\dfrac{13}{3}}$

27. (40)

$$(f \circ g)(x) = \frac{-8}{\left(\frac{-4x - 8}{x}\right) + 4}$$

$$= \frac{-8}{\frac{-4x - 8 + 4x}{x}} = \frac{-8x}{-8} = x$$

$$(g \circ f)(x) = \frac{-4\left(\frac{-8}{x + 4}\right) - 8}{\frac{-8}{x + 4}}$$

$$= \frac{\frac{32 - 8x - 32}{x + 4}}{\frac{-8}{x + 4}} = \frac{-8x}{-8} = x$$

$(f \circ g)(x) = (g \circ f)(x) = x$

Yes, f **and** g **are inverse functions.**

28. (33)

$$75 = \frac{1}{2}(2x)(x + 2 + 3x - 7)$$

$$75 = 4x^2 - 5x$$

$$4x^2 - 5x - 75 = 0$$

$$x = \frac{5 \pm \sqrt{25 + 1200}}{8}$$

$$x = \frac{5}{8} \pm \frac{35}{8}$$

$$x = 5, -\frac{15}{4}$$

$$x = \mathbf{5}$$

29. (31) Replace x with $-x$ then replace x with $x - 4$.

$$y = \sqrt{-(x - 4)}$$

30. (31) The graph of g is the graph of f translated eight units down. Therefore the answer is **D.**

Problem Set 58

1. (55) $\frac{9!}{5!4!} = \mathbf{126}$

2. (55) $\frac{5!}{2!2!} = \mathbf{30}$

3. (55) $(N - 1)! = (5 - 1)! = 4! = \mathbf{24}$

4. (53) $v = r\omega$

$$v = (10 \text{ in.})\left(\frac{525 \text{ rad}}{1 \text{ s}}\right)\left(\frac{2.54 \text{ cm}}{1 \text{ in.}}\right)\left(\frac{60 \text{ s}}{1 \text{ min}}\right) \times \left(\frac{60 \text{ min}}{1 \text{ hr}}\right)\left(\frac{\text{m}}{100 \text{ cm}}\right)\left(\frac{\text{km}}{1000 \text{ m}}\right)$$

$$v = \frac{(10)(525)(2.54)(60)(60)}{(100)(1000)} \frac{\text{km}}{\text{hr}} = \mathbf{480.06} \, \frac{\textbf{km}}{\textbf{hr}}$$

5. (53) $\omega = \frac{v}{r}$

$$\omega = \frac{100 \frac{\text{mi}}{\text{hr}}}{40 \text{ cm}} = \frac{100 \text{ mi}}{40 \text{ hr-cm}}$$

$$\omega = \left(\frac{100 \text{ mi}}{40 \text{ hr-cm}}\right)\left(\frac{5280 \text{ ft}}{1 \text{ mi}}\right)\left(\frac{12 \text{ in.}}{1 \text{ ft}}\right)\left(\frac{2.54 \text{ cm}}{1 \text{ in.}}\right) \times \left(\frac{1 \text{ hr}}{60 \text{ min}}\right)\left(\frac{1 \text{ min}}{60 \text{ s}}\right)$$

$$\omega = \frac{(100)(5280)(12)(2.54)}{(60)(60)(40)} \frac{\text{rad}}{\text{s}} = \mathbf{111.76} \, \frac{\textbf{rad}}{\textbf{s}}$$

6. (18) Impecunious $= i$

Destitute $= d$

$$\frac{i}{d} = \frac{14}{3}, \; i + d = 4420$$

$$\frac{14}{14 + 3} = \frac{i}{i + d}$$

$$\frac{14}{17} = \frac{i}{4420}$$

$$i = \mathbf{3640}$$

7. (38) x = average quality of third group

$$\text{Overall average quality} = \frac{\text{overall quality}}{\text{overall number}}$$

$$8 = \frac{\text{overall quality}}{4 + 10 + 16}$$

$$8 = \frac{4(6) + 10(4) + 16x}{30}$$

$$240 = 24 + 40 + 16x$$

$$16x = 176$$

$$x = \mathbf{11}$$

8. (16) $\frac{P_1V_1}{T_1} = \frac{P_2V_2}{T_2}$

$$P_2 = \frac{P_1V_1T_2}{T_1V_2}$$

$$P_2 = \frac{(4 \text{ atm})(5 \text{ L})(500 \text{ K})}{(400 \text{ K})(7 \text{ L})} = \frac{25}{7} = \mathbf{3.57 \text{ atm}}$$

9. (58) Equation of the perpendicular line:

$y = -2x + b$

$4 = -2(-2) + b$

$b = 0$

$y = -2x$

Point of intersection:

$\frac{1}{2}x - 1 = -2x$

$\frac{5}{2}x = 1$

$x = \frac{2}{5}$

$y = -2x = -2\left(\frac{2}{5}\right) = -\frac{4}{5}$

$\left(\frac{2}{5}, -\frac{4}{5}\right)$ and $(-2, 4)$

$$D = \sqrt{\left(\frac{2}{5} + 2\right)^2 + \left(-\frac{4}{5} - 4\right)^2}$$

$$= \sqrt{\left(\frac{12}{5}\right)^2 + \left(-\frac{24}{5}\right)^2} = \sqrt{\frac{720}{25}} = \mathbf{\frac{12\sqrt{5}}{5}}$$

10. (58) $y = -2x^2 - 8x - 4$

$y = -2(x^2 + 4x + \quad) - 4$

$y = -2(x^2 + 4x + 4) - 4 + 8$

$\mathbf{y = -2(x + 2)^2 + 4}$

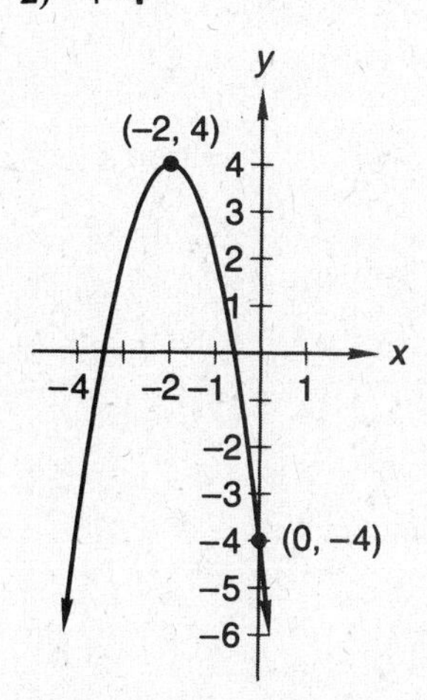

11. (57) Function = $\cos\theta$

Centerline = −3

Amplitude = 5

Period = 360°

Phase angle = 135°

$\mathbf{y = -3 + 5\cos(\theta - 135°)}$

12. (57) Function = $\cos\theta$

Centerline = 2

Amplitude = 6

Period = 180°

Phase angle = 0°

Coefficient = $\frac{360°}{180°} = 2$

$\mathbf{y = 2 + 6\cos 2\theta}$

13. (56)

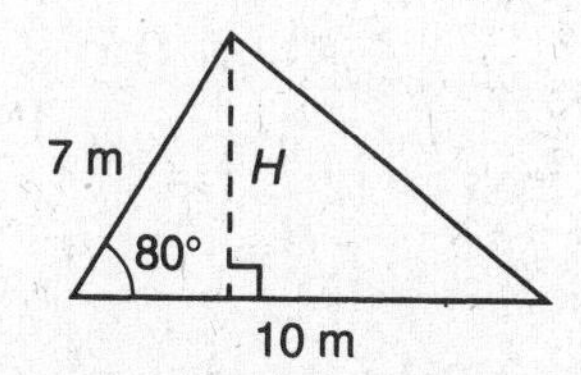

Note: Figure not drawn to scale

$\sin 80° = \frac{H}{7}$

$H = 7\sin 80° = 6.8937$ m

Area $= \frac{1}{2}BH = \frac{1}{2}(10)(6.8937) = \mathbf{34.47\ m^2}$

14. (56)

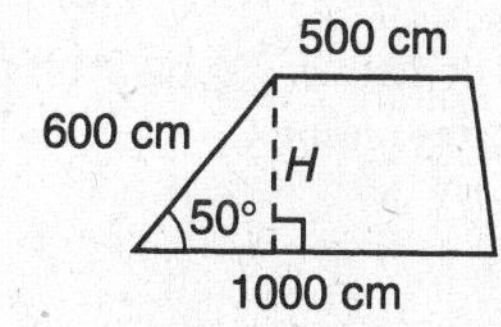

$\sin 50° = \frac{H}{600}$

$H = 600\sin 50° = 459.62667$ cm

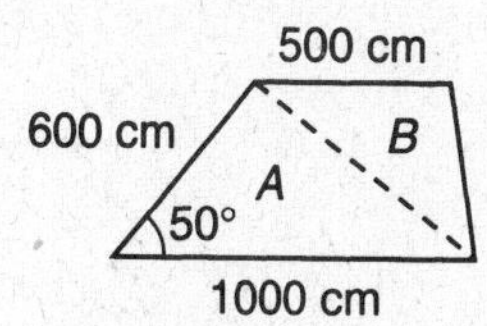

Total Area = Area A + Area B

$= \frac{1}{2}(1000)(459.62667)$

$+ \frac{1}{2}(500)(459.62667)$

$= \mathbf{344{,}720\ cm^2}$

15. (56)

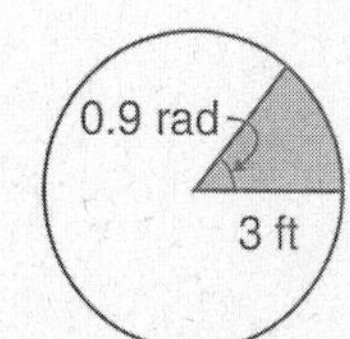

$A_{sector} = \pi r^2\left(\frac{0.9}{2\pi}\right) = \pi(3)^2\left(\frac{0.9}{2\pi}\right) = \mathbf{4.05\ ft^2}$

16. (56)

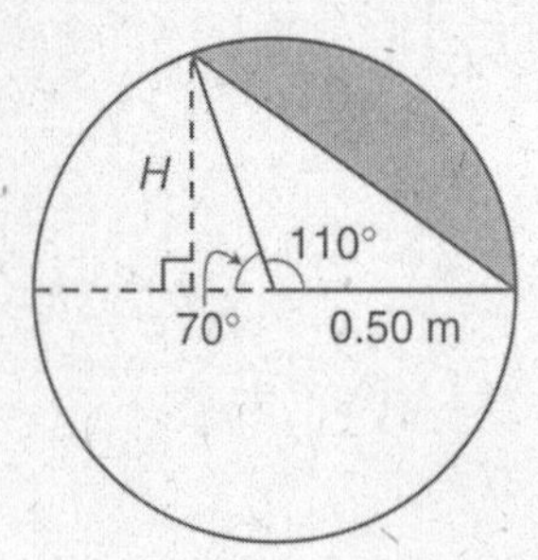

$H = 0.5 \sin 70° = 0.4698 \text{ m}$

$A_{\text{segment}} = A_{\text{sector}} - A_{\text{triangle}}$

$= \pi r^2\left(\frac{110°}{360°}\right) - \frac{1}{2}rH$

$= \pi(0.5)^2\left(\frac{110°}{360°}\right) - \frac{1}{2}(0.5)(0.4698)$

$= 0.24 - 0.12 = \mathbf{0.12\ m^2}$

17. (56) $\begin{cases} y \geq x^2 + 4x + 2 & \text{(parabola)} \\ y < -x + 1 & \text{(line)} \end{cases}$

The region must be on or above the parabola and below the line. This indicates region ***C***.

Test point: $(-1, 1)$

Parabola: $1 \geq 1 - 4 + 2$

$1 \geq -1$ True

line: $1 < -(-1) + 1$

$1 < 2$ True

18. (53) $\left(250\ \frac{\text{km}}{\text{s}}\right)\left(\frac{1000\text{ m}}{1\text{ km}}\right)\left(\frac{100\text{ cm}}{1\text{ m}}\right)\left(\frac{1\text{ in.}}{2.54\text{ cm}}\right)$

$\times \left(\frac{1\text{ ft}}{12\text{ in.}}\right)\left(\frac{1\text{ mi}}{5280\text{ ft}}\right)\left(\frac{60\text{ s}}{1\text{ min}}\right)\left(\frac{60\text{ min}}{1\text{ hr}}\right)$

$= \mathbf{\frac{(250)(1000)(100)(60)(60)}{(2.54)(12)(5280)}\ \frac{mi}{hr} = 559{,}234.07\ \frac{mi}{hr}}$

19. (50) $2 \cos \theta = \sqrt{3}$

$\cos \theta = \frac{\sqrt{3}}{2}$

$\theta = \mathbf{30°, 330°}$

20. (52) $2 \cos 5\theta - \sqrt{3} = 0$

$\cos 5\theta = \frac{\sqrt{3}}{2}$

$5\theta = 30°, 330°, 390°, 690°, 750°, 1050°, 1110°, 1410°, 1470°, 1770°$

$\theta = \mathbf{6°, 66°, 78°, 138°, 150°, 210°, 222°, 282°, 294°, 354°}$

21. (52) $\tan 3\theta = -1$

$3\theta = 135°, 315°, 495°, 675°, 855°, 1035°$

$\theta = \mathbf{45°, 105°, 165°, 225°, 285°, 345°}$

22. (37,39) $\sqrt{(x + 2)^2 + (y - 3)^2} = \sqrt{(x - 4)^2 + (y - 5)^2}$

$x^2 + 4x + 4 + y^2 - 6y + 9$

$= x^2 - 8x + 16 + y^2 - 10y + 25$

$12x + 4y = 28$

$\mathbf{3x + y - 7 = 0}$

23. (32)

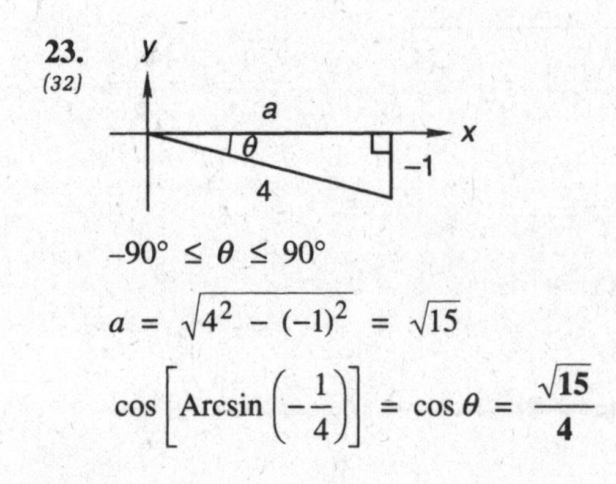

$-90° \leq \theta \leq 90°$

$a = \sqrt{4^2 - (-1)^2} = \sqrt{15}$

$\cos\left[\text{Arcsin}\left(-\frac{1}{4}\right)\right] = \cos \theta = \mathbf{\frac{\sqrt{15}}{4}}$

24. (48)

$\sec^2 \frac{3\pi}{4} - 1 = \left(-\sec \frac{\pi}{4}\right)^2 - 1$

$= (-\sqrt{2})^2 - 1 = 2 - 1 = \mathbf{1}$

25. (48) $\tan^2 \frac{3\pi}{4} + 2 \sec^2 \frac{5\pi}{4} + \cos \frac{5\pi}{2}$

$= \left(-\tan \frac{\pi}{4}\right)^2 + 2\left(-\sec \frac{\pi}{4}\right)^2 + \cos \frac{\pi}{2}$

$= (-1)^2 + 2(-\sqrt{2})^2 + 0$

$= 1 + 4 = \mathbf{5}$

26. (40,51) (a) $\frac{2}{3}\log_7 27 - \log_7(x + 2) = \log_7 3$

$$\log_7 27^{\frac{2}{3}} - \log_7(x + 2) = \log_7 3$$
$$\log_7 9 - \log_7 3 = \log_7(x + 2)$$
$$\log_7 \frac{9}{3} = \log_7(x + 2)$$
$$3 = x + 2$$
$$\mathbf{x = 1}$$

(b) $\frac{3}{4}\ln 16 - \ln x = \ln 4$

$$\ln 16^{\frac{3}{4}} - \ln 4 = \ln x$$
$$\ln 8 - \ln 4 = \ln x$$
$$\frac{8}{4} = x$$
$$\mathbf{x = 2}$$

27. (40) $f(g(x)) = \dfrac{1}{1 - \dfrac{x}{x - 1}} = \dfrac{1}{\dfrac{x - 1 - x}{x - 1}} = \dfrac{x - 1}{-1}$

$$g(f(x)) = \frac{\dfrac{1}{1 - x}}{\dfrac{1}{1 - x} - 1}$$
$$= \frac{\dfrac{1}{1 - x}}{\dfrac{1 - 1 + x}{1 - x}} = \frac{\dfrac{1}{1 - x}}{\dfrac{x}{1 - x}} = \frac{1}{x}$$

$f(g(x)) \neq x$

$g(f(x)) \neq x$

No, f and g are not inverse functions.

28. (21) (a) **Domain** $= \{x \in \mathbb{R} \mid \mathbf{-5 < x < 2}\}$

Range $= \{y \in \mathbb{R} \mid \mathbf{-3 \leq y < 6}\}$

(b) **Domain** $= \{x \in \mathbb{R} \mid \mathbf{-5 \leq x < 5}\}$

Range $= \{y \in \mathbb{R} \mid \mathbf{-4 \leq y \leq 4}\}$

29. (31) Replace x with $-x$ and y with $y + 2$.

$$y + 2 = \sqrt{-x}$$
$$\mathbf{y = \sqrt{-x} - 2}$$

30. (31) The graph of q is the graph of f translated four units to the left. Therefore the answer is **B.**

Problem Set 59

1. (55) $\dfrac{11!}{2!2!2!2!} = \mathbf{2{,}494{,}800}$

2. (45)

VOWELS			CONSONANTS				
3	2	1	5	4	3	2	1

$\rightarrow 3! \cdot 5! = \mathbf{720}$

3. (45)

Pres.	V.P.	Sec.
9	9	8

$\rightarrow 9 \cdot 9 \cdot 8 = \mathbf{648}$

4. (53) $v = r\omega$

$$v = (1 \text{ ft})\left(400\,\frac{\text{rad}}{\text{s}}\right)\left(\frac{1 \text{ mi}}{5280 \text{ ft}}\right)\left(\frac{60 \text{ s}}{1 \text{ min}}\right)\left(\frac{60 \text{ min}}{1 \text{ hr}}\right)$$
$$v = \mathbf{\frac{(400)(60)(60)}{5280}\,\frac{mi}{hr}} = \mathbf{272.73\,\frac{mi}{hr}}$$

5. (53) $\omega = \dfrac{v}{r} = \dfrac{40\,\dfrac{\text{km}}{\text{hr}}}{8 \text{ in.}} = \dfrac{40 \text{ km}}{8 \text{ in.-hr}}$

$$\omega = \left(\frac{40 \text{ km}}{8 \text{ in.-hr}}\right)\left(\frac{1000 \text{ m}}{1 \text{ km}}\right)\left(\frac{100 \text{ cm}}{1 \text{ m}}\right)$$
$$\times\left(\frac{1 \text{ in.}}{2.54 \text{ cm}}\right)\left(\frac{1 \text{ hr}}{60 \text{ min}}\right)$$
$$\omega = \frac{(40)(1000)(100)}{(8)(2.54)(60)}\,\frac{\text{rad}}{\text{min}}\left(\frac{1 \text{ rev}}{2\pi \text{ rad}}\right)$$
$$\omega = \mathbf{\frac{(40)(1000)(100)}{(2.54)(60)(8)(2\pi)}\,\frac{rev}{min}} = \mathbf{522.16\,\frac{rev}{min}}$$

6. (44) $\text{Rate} = \dfrac{4x + 4}{2y}\,\dfrac{\text{pencils}}{\text{dollars}}$

$$\text{Rate} \times \text{dollars} = \text{pencils}$$
$$\frac{4x + 4}{2y} \cdot 10 = \frac{20x + 20}{y} = \mathbf{\frac{20(x + 1)}{y}\text{ pencils}}$$

7. (25) $T_H = T_W + 3$

$$R_W T_W + R_H T_H = 8$$
$$\frac{2}{3}T_W + \frac{4}{7}(T_W + 3) = 8$$
$$\frac{2}{3}T_W + \frac{4}{7}T_W = 8 - \frac{12}{7}$$
$$\frac{26}{21}T_W = \frac{44}{7}$$
$$T_W = \mathbf{\frac{66}{13}\text{ days}}$$

8. (18)
$$P_N + D_N = 50$$
$$P_N = 50 - D_N$$
$$0.8P_N + 0.2D_N = 0.56(50)$$
$$0.8(50 - D_N) + 0.2D_N = 0.56(50)$$
$$40 - 0.8D_N + 0.2D_N = 28$$
$$-0.6D_N = -12$$
$$D_N = \textbf{20 gal of 20\%}$$

$$P_N = 50 - D_N$$
$$P_N = 50 - 20$$
$$P_N = \textbf{30 gal of 80\%}$$

9. (59)
$$2\log_5 x = \log_5 18 - \log_5 2$$
$$\log_5 x^2 = \log_5 \frac{18}{2}$$
$$\log_5 x^2 = \log_5 9$$
$$x^2 = 9$$
$$x = 3, -3 \quad (x \neq -3)$$
$$x = \mathbf{3}$$

10. (59)
$$\log_4 (x - 1) + \log_4 (x + 2) = 1$$
$$\log_4 [(x - 1)(x + 2)] = 1$$
$$(x - 1)(x + 2) = 4^1$$
$$x^2 + x - 2 - 4 = 0$$
$$(x + 3)(x - 2) = 0$$
$$x = -3, 2 \quad (x \neq -3)$$
$$x = \mathbf{2}$$

11. (59)
$$\log_4 x - \log_4 (2x - 1) = 1$$
$$\log_4 \frac{x}{2x - 1} = 1$$
$$\frac{x}{2x - 1} = 4^1$$
$$8x - 4 = x$$
$$7x = 4$$
$$x = \mathbf{\frac{4}{7}}$$

12. (59)
$$43^{\log_{43} 6} = \mathbf{6}$$

13. (59)
$$2^{\log_2 5 + \log_2 6} = 2^{\log_2 (5 \cdot 6)} = 2^{\log_2 30} = \mathbf{30}$$

14. (58) Equation of the perpendicular line:
$$y = -\frac{1}{2}x + b$$
$$1 = -\frac{1}{2}(2) + b$$
$$1 = -1 + b$$
$$b = 2$$
$$y = -\frac{1}{2}x + 2$$

Point of intersection:
$$-\frac{1}{2}x + 2 = 2x - 1$$
$$-\frac{5}{2}x = -3$$
$$x = \frac{6}{5}$$
$$y = -\frac{1}{2}x + 2$$
$$= -\frac{1}{2}\left(\frac{6}{5}\right) + 2 = -\frac{3}{5} + \frac{10}{5} = \frac{7}{5}$$

$\left(\frac{6}{5}, \frac{7}{5}\right)$ and $(2, 1)$

$$D = \sqrt{\left(\frac{6}{5} - 2\right)^2 + \left(\frac{7}{5} - 1\right)^2}$$
$$= \sqrt{\frac{16}{25} + \frac{4}{25}} = \mathbf{\frac{2\sqrt{5}}{5}}$$

15. (58)
$$y - 3x^2 + 6x - 5 = 0$$
$$y = 3x^2 - 6x + 5$$
$$y = 3(x^2 - 2x \qquad) + 5$$
$$y = 3(x^2 - 2x + 1) + 5 - 3$$
$$\mathbf{y = 3(x - 1)^2 + 2}$$

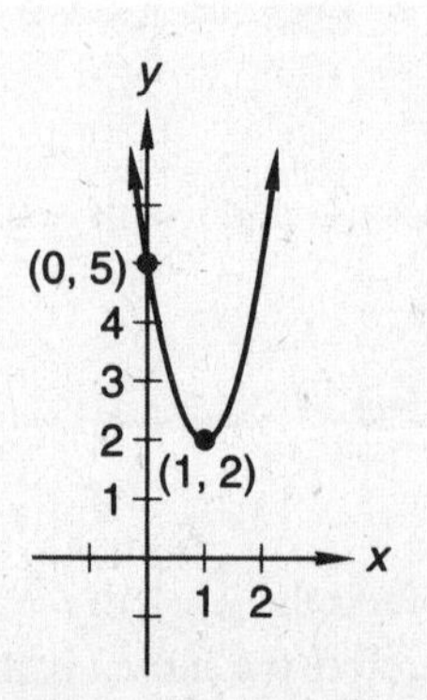

16. (57) Function = $\sin \theta$

Centerline = 3

Amplitude = 8

Period = 360°

Phase angle = 45°

$\mathbf{y = 3 + 8 \sin (\theta - 45°)}$

17. (57) Function = $\sin \theta$

Centerline = 4

Amplitude = 7

Period = 180°

Phase angle = 0°

$$\text{Coefficient} = \frac{360°}{180°} = 2$$

$\mathbf{y = 4 + 7 \sin 2\theta}$

18. (56)

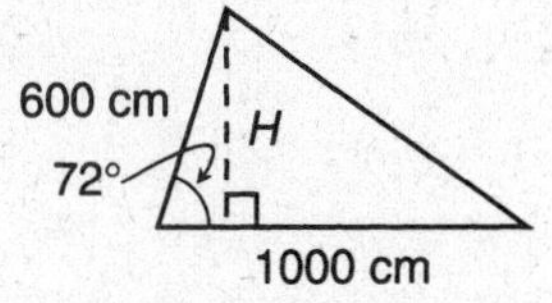

$$\sin 72° = \frac{H}{600}$$

$$H = 600 \sin 72° = 570.6339 \text{ cm}$$

$$\text{Area} = \frac{1}{2}BH$$

$$= \frac{1}{2}(1000)(570.6339) = \mathbf{285{,}316.95\ cm^2}$$

19. (56)

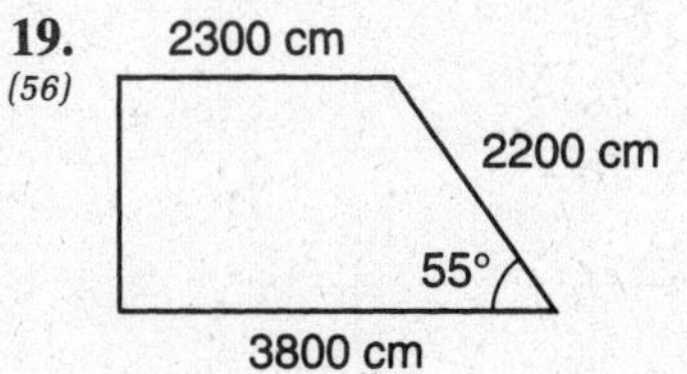

$$\sin 55° = \frac{H}{2200}$$

$$H = 2200 \sin 55° = 1802.1345 \text{ cm}$$

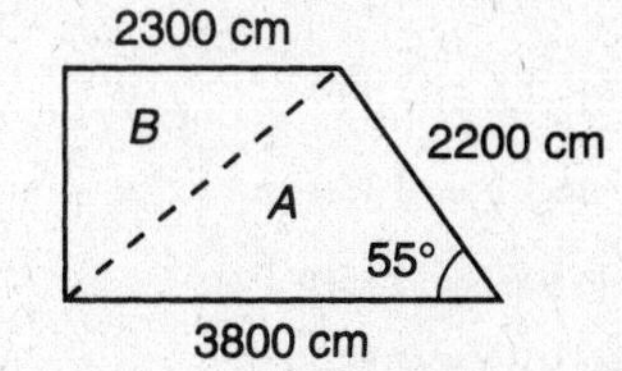

$$\text{Total Area} = \text{Area } A + \text{Area } B$$

$$= \frac{1}{2}(3800)(1802.1345) + \frac{1}{2}(2300)(1802.1345)$$

$$= \mathbf{5{,}496{,}510.22\ cm^2}$$

20. (56)

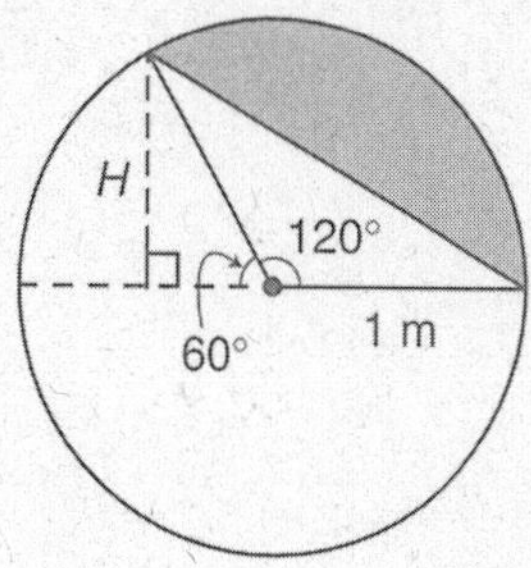

$$H = 1 \sin 60° = 0.866 \text{ m}$$

$$A_{\text{segment}} = A_{\text{sector}} - A_{\text{triangle}}$$

$$= \pi r^2\left(\frac{120°}{360°}\right) - \frac{1}{2}rH$$

$$= \pi(1)^2\left(\frac{120°}{360°}\right) - \frac{1}{2}(1)(0.866)$$

$$= 1.0472 - 0.433 = \mathbf{0.61\ m^2}$$

21. (56)
$$\begin{cases} y \le x^2 + 2x + 4 & \text{(parabola)} \\ y > 2x - 2 & \text{(line)} \end{cases}$$

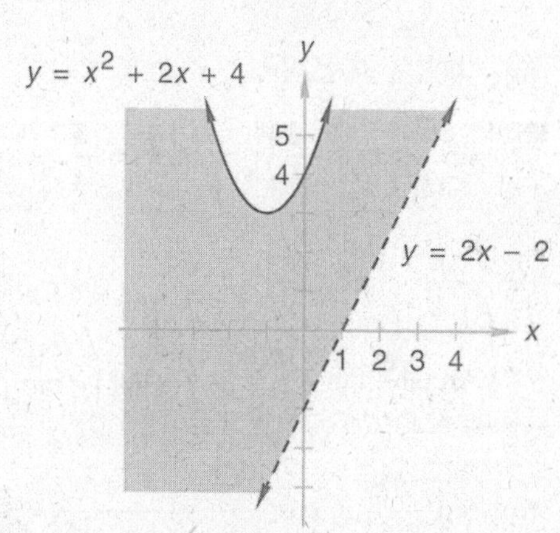

The region must be on or below the parabola and above the line.

Test point: (–1, 0)

Parabola: $0 \le (-1)^2 + 2(-1) + 4$

$0 \le 3$ True

Line: $0 > 2(-1) - 2$

$0 > -4$ True

22. (53)
$$\left(30 \frac{\text{cm}}{\text{s}}\right)\left(\frac{1 \text{ m}}{100 \text{ cm}}\right)\left(\frac{1 \text{ km}}{1000 \text{ m}}\right)\left(\frac{60 \text{ s}}{1 \text{min}}\right)\left(\frac{60 \text{ min}}{1 \text{ hr}}\right)$$

$$= \mathbf{\frac{(30)(60)(60)}{(100)(1000)} \frac{km}{hr} = 1.08 \frac{km}{hr}}$$

23. (50) $\sqrt{3}\tan\theta = 1$

$$\tan\theta = \frac{1}{\sqrt{3}}$$

$$\theta = \mathbf{30°, 210°}$$

24. (52) $\sqrt{3}\tan 3\theta - 1 = 0$

$$\sqrt{3}\tan 3\theta = 1$$

$$\tan 3\theta = \frac{1}{\sqrt{3}}$$

$$3\theta = 30°, 210°, 390°, 570°, 750°, 930°$$

$$\theta = \mathbf{10°, 70°, 130°, 190°, 250°, 310°}$$

25. (51) $\log_{10} 20{,}000 = 4.3010$

$$20{,}000 = \mathbf{10^{4.3010}}$$

26. (51) $\ln 20{,}000 = 9.9035$

$$20{,}000 = \mathbf{e^{9.9035}}$$

27. (47,48) (a) $-90° \le \theta \le 90°$

$$\sin\left[\text{Arcsin}\left(\frac{3}{4}\right)\right] = \sin\theta = \mathbf{\frac{3}{4}}$$

(b) $-\frac{\pi}{2} < \theta < \frac{\pi}{2}$

$$\text{Arctan}\left(\tan\frac{11\pi}{6}\right) = \text{Arctan}\left(-\frac{1}{\sqrt{3}}\right) = \mathbf{-\frac{\pi}{6}}$$

(c) $\sin^2\frac{7\pi}{6} + \cos^2\frac{7\pi}{6} - \frac{1}{\sec\frac{13\pi}{6}} - \cos\frac{13\pi}{6}$

$$= \left(-\sin\frac{\pi}{6}\right)^2 + \left(-\cos\frac{\pi}{6}\right)^2 - \cos\frac{\pi}{6} - \cos\frac{\pi}{6}$$

$$= \left(-\frac{1}{2}\right)^2 + \left(-\frac{\sqrt{3}}{2}\right)^2 - 2\left(\frac{\sqrt{3}}{2}\right) = \mathbf{1 - \sqrt{3}}$$

28. (42) $\mathbf{(x - 3)^2 + (y - 2)^2 = 10^2}$

29. (31) Replace x with $-x$ then replace x with $x + 3$.

$$\mathbf{y = \sqrt{-(x + 3)}}$$

30. (31) The graph of g is the graph of f translated one unit to the right. Therefore the answer is **A.**

Problem Set 60

1. (45)

3	2	1	2

$\rightarrow\ 3! \cdot 2 = \mathbf{12}$

2. (55) $\frac{11!}{3!} = \mathbf{6{,}652{,}800}$

3. (53) $v = r\omega$

$$v = (16 \text{ in.})\left(367\frac{\text{rad}}{\text{min}}\right)\left(\frac{1 \text{ ft}}{12 \text{ in.}}\right)\left(\frac{1 \text{ mi}}{5280 \text{ ft}}\right)\left(\frac{60 \text{ min}}{1 \text{ hr}}\right)$$

$$v = \mathbf{\frac{(16)(367)(60)}{(12)(5280)}\frac{mi}{hr} = 5.56\frac{mi}{hr}}$$

4. (53) $\omega = \frac{v}{r}$

$$\omega = \frac{5\frac{\text{km}}{\text{hr}}}{5 \text{ cm}} = \frac{5 \text{ km}}{5 \text{ cm-hr}}$$

$$\omega = \left(\frac{5 \text{ km}}{5 \text{ cm-hr}}\right)\left(\frac{1000 \text{ m}}{1 \text{ km}}\right)\left(\frac{100 \text{ cm}}{1 \text{ m}}\right)\left(\frac{1 \text{ hr}}{60 \text{ min}}\right)$$

$$\omega = \mathbf{\frac{(5)(1000)(100)}{(60)(5)}\frac{rad}{min} = 1666.67\frac{rad}{min}}$$

5. (36)

$$\begin{cases} \text{(a)}\ (J + W)T_D = D_D \\ \text{(b)}\ (J - W)T_U = D_U \end{cases}$$

$$T_D = T_U + 1,\ J = 8W$$

(b)
$$(J - W)T_U = D_U$$
$$(8W - W)T_U = 700$$
$$7WT_U = 700$$
$$WT_U = 100$$

(a)
$$(J + W)T_D = D_D$$
$$(8W + W)(T_U + 1) = 1350$$
$$9W(T_U + 1) = 1350$$
$$9WT_U + 9W = 1350$$
$$9(100) + 9W = 1350$$
$$9W = 450$$
$$\mathbf{W = 50 \text{ mph}}$$

$$J = 8W$$
$$J = 8(50)$$
$$\mathbf{J = 400 \text{ mph}}$$

6. (18)

$$\begin{cases} \text{(a)} & \dfrac{N_R}{N_W} = \dfrac{1}{2} \\ \text{(b)} & \dfrac{N_G}{N_W} = \dfrac{5}{4} \\ \text{(c)} & N_R + N_W + N_G = 22 \end{cases}$$

(a) $\dfrac{N_R}{N_W} = \dfrac{1}{2}$

$N_R = \dfrac{1}{2}N_W$

(b) $\dfrac{N_G}{N_W} = \dfrac{5}{4}$

$N_G = \dfrac{5}{4}N_W$

(c) $N_R + N_W + N_G = 22$

$$\left(\frac{1}{2}N_W\right) + N_W + \left(\frac{5}{4}N_W\right) = 22$$

$$\frac{11}{4}N_W = 22$$

$$\mathbf{N_W = 8}$$

$N_R = \dfrac{1}{2}N_W$ $\qquad N_G = \dfrac{5}{4}N_W$

$N_R = \dfrac{1}{2}(8)$ $\qquad N_G = \dfrac{5}{4}(8)$

$\mathbf{N_R = 4}$ $\qquad \mathbf{N_G = 10}$

7. (60) $\cos^2\theta - \cos\theta = 0$

$\cos\theta(\cos\theta - 1) = 0$

$\cos\theta = 0$ $\qquad \cos\theta = 1$

$\theta = 90°, 270°$ $\qquad \theta = 0°$

$\theta = \mathbf{0°, 90°, 270°}$

8. (60) $\sin^2\theta - 1 = 0$

$(\sin\theta - 1)(\sin\theta + 1) = 0$

$\sin\theta - 1 = 0$ $\qquad \sin\theta + 1 = 0$

$\sin\theta = 1$ $\qquad \sin\theta = -1$

$\theta = 90°$ $\qquad \theta = 270°$

$\theta = \mathbf{90°, 270°}$

9. (59) $3\log_6 x = \log_6 24 - \log_6 3$

$\log_6 x^3 = \log_6 \dfrac{24}{3}$

$x^3 = 8$

$x = \mathbf{2}$

10. (59) $\log x + \log(x - 3) = 1$

$\log[x(x - 3)] = 1$

$x(x - 3) = 10^1$

$x^2 - 3x - 10 = 0$

$(x - 5)(x + 2) = 0$

$x = 5, -2 \qquad (x \neq -2)$

$x = \mathbf{5}$

11. (59) $2\log_2 x - \log_2\left(x - \dfrac{1}{2}\right) = \log_3 3$

$\log_2 x^2 - \log_2\left(x - \dfrac{1}{2}\right) = 1$

$$\log_2\left[\frac{x^2}{\left(x - \frac{1}{2}\right)}\right] = 1$$

$$\frac{x^2}{\left(x - \frac{1}{2}\right)} = 2^1$$

$x^2 = 2x - 1$

$x^2 - 2x + 1 = 0$

$(x - 1)^2 = 0$

$x = \mathbf{1}$

12. (59) $32^{\log_{32} 7} = \mathbf{7}$

13. (59) $3^{\log_3 6 - \log_3 2} = 3^{\log_3 \frac{6}{2}} = 3^{\log_3 3} = \mathbf{3}$

14. (58) Equation of the perpendicular line:

$y = -x + b$

$6 = -(-4) + b$

$b = 2$

$y = -x + 2$

Point of intersection:

$x + 2 = -x + 2$

$2x = 0$

$x = 0$

$y = -x + 2 = 0 + 2 = 2$

$(0, 2)$ and $(-4, 6)$

$D = \sqrt{(0 + 4)^2 + (2 - 6)^2}$

$= \sqrt{16 + 16} = \mathbf{4\sqrt{2}}$

15. (58)
$y = 6x + 3x^2$

$y = 3(x^2 + 2x + \quad)$

$y = 3(x^2 + 2x + 1) - 3$

$\mathbf{y = 3(x + 1)^2 - 3}$

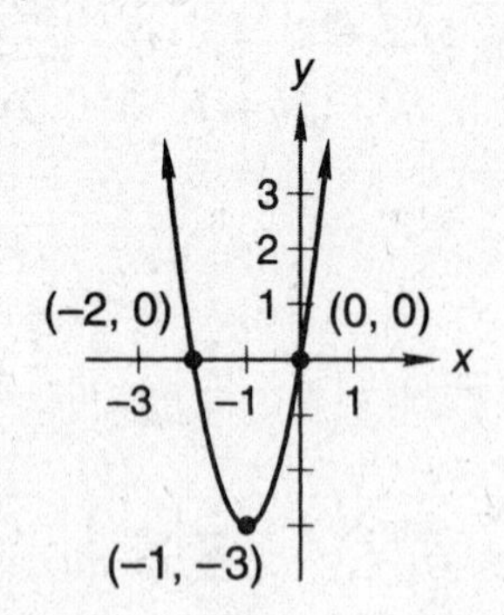

16. (57)
Function = $\cos\theta$

Centerline = -5

Amplitude = 15

Period = 360°

Phase angle = 45°

$\mathbf{y = -5 + 15\cos(\theta - 45°)}$

17. (57)
Function = $\cos x$

Centerline = -10

Amplitude = 20

Period = π

Phase angle = 0

Coefficient = $\dfrac{2\pi}{\pi} = 2$

$\mathbf{y = -10 + 20\cos 2x}$

18. (56)

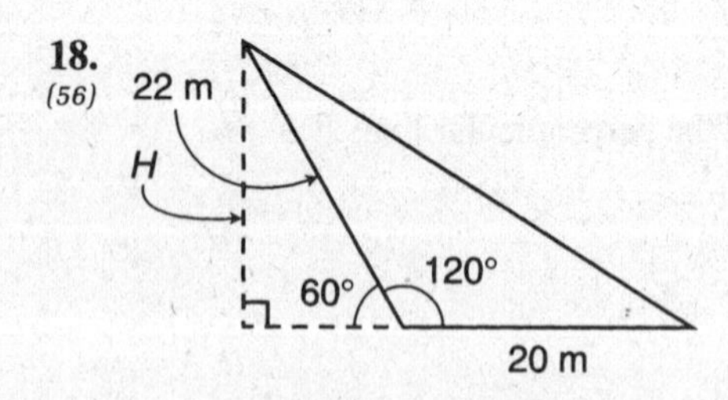

$H = 22\sin 60° = 19.053 \text{ m}$

Area $= \dfrac{1}{2}BH = \dfrac{1}{2}(20)(19.053) = \mathbf{190.53\ m^2}$

19. (56)

6 cm
9 cm
H
55°
15 cm

$H = 9\sin 55° = 7.3724 \text{ cm}$

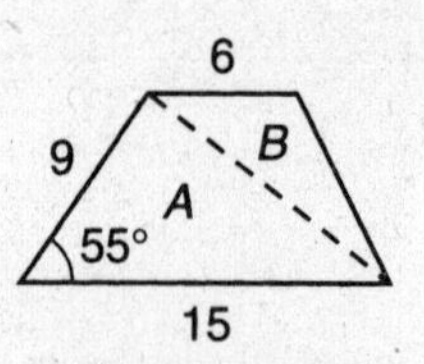

Total Area = Area A + Area B

$= \dfrac{1}{2}(15)(7.3724) + \dfrac{1}{2}(6)(7.3724)$

$= 55.30 + 22.11 = \mathbf{77.41\ cm^2}$

20. (56)

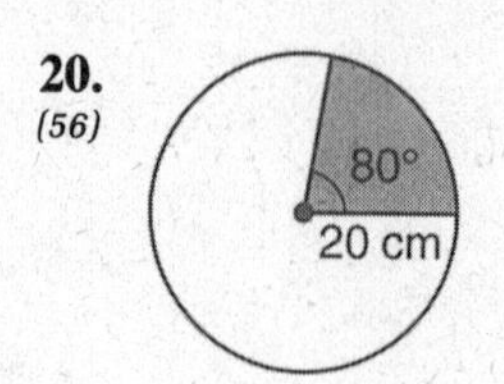

$A_{sector} = \pi r^2\left(\dfrac{80°}{360°}\right)$

$= \pi(20)^2\left(\dfrac{80°}{360°}\right) = \mathbf{279.25\ cm^2}$

21. (56)
$2.4 \text{ radians} \times \dfrac{180°}{\pi \text{ rad}} = 137.51°$

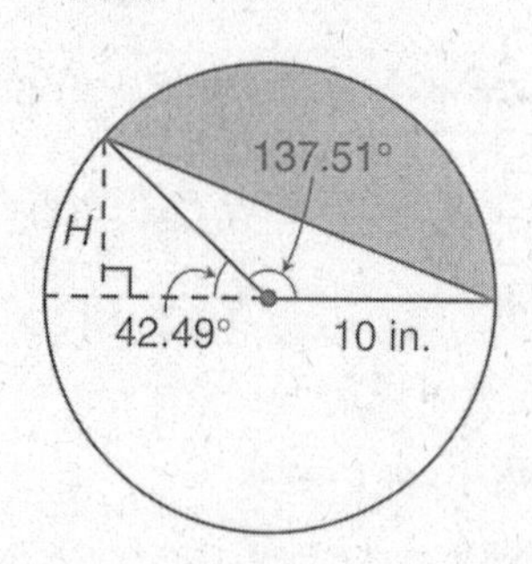

$H = 10\sin 42.49 = 6.7546 \text{ in.}$

$A_{segment} = A_{sector} - A_{triangle}$

$= \pi r^2\left(\dfrac{137.51°}{360°}\right) - \dfrac{1}{2}rH$

$= \pi(10)^2\left(\dfrac{137.51°}{360°}\right) - \dfrac{1}{2}(10)(6.7546)$

$= 120.0 - 33.77$

$= 86.23 \text{ in.}^2 \times \dfrac{1 \text{ ft}}{12 \text{ in.}} \times \dfrac{1 \text{ ft}}{12 \text{ in.}} = \mathbf{0.60\ ft^2}$

22. (56)
$\begin{cases} y \geq -x^2 + 4x - 1 \\ y \geq \dfrac{1}{2}x - 4 \end{cases}$

Test point: $(-1, 1)$

Parabola: $1 \geq -(-1)^2 + 4(-1) - 1$

$1 \geq -1 - 4 - 1$

$1 \geq -6$ True

Line: $1 \geq \frac{1}{2}(-1) - 4$

$1 \geq -\frac{9}{2}$ True

The system of inequalities designates the region above or on the parabola and above or on the line. Therefore the answer is **A.**

23. (52) $\sin 3\theta - 1 = 0$

$\sin 3\theta = 1$

$3\theta = 90°, 450°, 810°$

$\boldsymbol{\theta = 30°, 150°, 270°}$

24. (48) $\sin^2\left(\frac{3\pi}{4}\right) + \cos^2\left(\frac{3\pi}{4}\right)$

$= \left(\sin\frac{\pi}{4}\right)^2 + \left(-\cos\frac{\pi}{4}\right)^2$

$= \left(\frac{1}{\sqrt{2}}\right)^2 + \left(\frac{-1}{\sqrt{2}}\right)^2 = \frac{1}{2} + \frac{1}{2} = \mathbf{1}$

25. (32,48) (a) $-\frac{\pi}{2} \leq \theta \leq \frac{\pi}{2}$

$\text{Arcsin}\left(\sin\frac{\pi}{4}\right) = \boldsymbol{\frac{\pi}{4}}$

(b)

y

a

x

θ

−1

4

$-\frac{\pi}{2} \leq \theta \leq \frac{\pi}{2}$

$a = \sqrt{4^2 - (-1)^2} = \sqrt{15}$

$\cos^2\left[\text{Arcsin}\left(-\frac{1}{4}\right)\right] = \cos^2\theta = \left(\frac{\sqrt{15}}{4}\right)^2 = \boldsymbol{\frac{15}{16}}$

(c) $\sin^2\left(\frac{3\pi}{4}\right) - \cos^2\left(\frac{3\pi}{4}\right)$

$= \left(\sin\frac{\pi}{4}\right)^2 - \left(-\cos\frac{\pi}{4}\right)^2$

$= \left(\frac{1}{\sqrt{2}}\right)^2 - \left(\frac{-1}{\sqrt{2}}\right)^2$

$= \frac{1}{2} - \frac{1}{2} = \mathbf{0}$

26. (37,39) $\sqrt{(x+2)^2 + (y-4)^2} = \sqrt{(x-6)^2 + (y-2)^2}$

$x^2 + 4x + 4 + y^2 - 8y + 16$

$= x^2 - 12x + 36 + y^2 - 4y + 4$

$16x - 4y = 20$

$\frac{4}{5}x - \frac{1}{5}y = 1$

$\boldsymbol{\frac{x}{\frac{5}{4}} + \frac{y}{-5} = 1}$

27. (46) $\left[x - \left(-1 - \sqrt{3}i\right)\right]\left[x - \left(-1 + \sqrt{3}i\right)\right] = 0$

$\left(x + 1 + \sqrt{3}i\right)\left(x + 1 - \sqrt{3}i\right) = 0$

$x^2 + x - \sqrt{3}xi + x + 1 - \sqrt{3}i$

$+ \sqrt{3}xi + \sqrt{3}i - 3i^2 = 0$

$\mathbf{x^2 + 2x + 4 = 0}$

28. (40) $(f \circ g)(x) = \dfrac{5\left(\frac{7x+4}{2x-5}\right) + 4}{2\left(\frac{7x+4}{2x-5}\right) - 7}$

$= \dfrac{5(7x+4) + 4(2x-5)}{2(7x+4) - 7(2x-5)} = \dfrac{43x}{43} = x$

$(g \circ f)(x) = \dfrac{7\left(\frac{5x+4}{2x-7}\right) + 4}{2\left(\frac{5x+4}{2x-7}\right) - 5}$

$= \dfrac{7(5x+4) + 4(2x-7)}{2(5x+4) - 5(2x-7)} = \dfrac{43x}{43} = x$

$(f \circ g)(x) = (g \circ f)(x) = x$

Yes, f and g are inverse functions.

29. (31) Replace y with $-y$ and x with $x - 3$.

$-y = -|x - 3 + 3|$

$y = |x|$

$\mathbf{g(x) = |x|}$

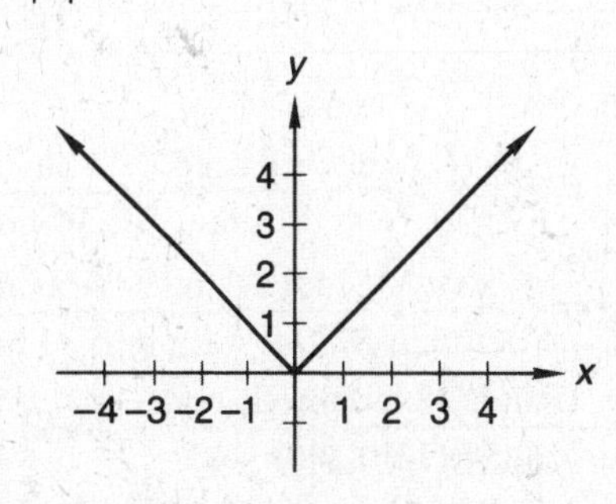

30. (21) $f(x) = x(x + 3)$

$$\frac{f(x+h) - f(x-h)}{2h}$$
$$= \frac{(x + h)(x + h + 3) - (x - h)(x - h + 3)}{2h}$$
$$= \frac{(x^2 + 2xh + 3x + 3h + h^2)}{2h} - \frac{(x^2 - 2xh + 3x - 3h + h^2)}{2h}$$
$$= \frac{4xh + 6h}{2h} = \mathbf{2x + 3}$$

Problem Set 61

1. (38)

12	11	10

$\rightarrow$ $12 \cdot 11 \cdot 10 =$ **1320**

2. (45)

T	2	S	1	$\rightarrow 2 \cdot 1 = 2$
T	2	1	S	$\rightarrow 2 \cdot 1 = 2$
2	T	1	S	$\rightarrow 2 \cdot 1 = 2$
S	2	T	1	$\rightarrow 2 \cdot 1 = 2$
S	2	1	T	$\rightarrow 2 \cdot 1 = 2$
2	S	1	T	$\rightarrow 2 \cdot 1 = 2$
				12

3. (53)

$$\omega = \frac{v}{r} = \frac{40\,\frac{\text{km}}{\text{hr}}}{14\text{ in.}} = \frac{40\text{ km}}{14\text{ in.-hr}}$$
$$\omega = \left(\frac{40\text{ km}}{14\text{ in.-hr}}\right)\left(\frac{1000\text{ m}}{1\text{ km}}\right)\left(\frac{100\text{ cm}}{1\text{ m}}\right)\left(\frac{1\text{ in.}}{2.54\text{ cm}}\right) \times \left(\frac{1\text{ hr}}{60\text{ min}}\right)\left(\frac{1\text{ min}}{60\text{ s}}\right)$$
$$\omega = \frac{(40)(1000)(100)}{(2.54)(60)(60)(14)}\,\frac{\text{rad}}{\text{s}} = \mathbf{31.25\,\frac{rad}{s}}$$

4. (53)

$$v = r\omega$$
$$v = (16\text{ cm})\left(40\,\frac{\text{rev}}{\text{s}}\right)\left(\frac{2\pi\text{ rad}}{1\text{ rev}}\right)\left(\frac{1\text{ in.}}{2.54\text{ cm}}\right) \times \left(\frac{1\text{ ft}}{12\text{ in.}}\right)\left(\frac{1\text{ mi}}{5280\text{ ft}}\right)\left(\frac{60\text{ s}}{1\text{ min}}\right)\left(\frac{60\text{ min}}{1\text{ hr}}\right)$$
$$v = \frac{(16)(40)(2\pi)(60)(60)}{(2.54)(12)(5280)}\,\frac{\text{mi}}{\text{hr}} = \mathbf{89.95\,\frac{mi}{hr}}$$

5. (38,44) Time downstream $= \frac{m}{d}$, time upstream $= \frac{m}{u}$

$$\text{Total time} = \frac{m}{u} + \frac{m}{d} = \frac{m(u + d)}{ud}$$
$$\text{Average speed} = \frac{\text{total distance}}{\text{total time}} = \frac{2m}{\frac{m(u+d)}{ud}} = \mathbf{\frac{2ud}{u + d}\text{ knots}}$$

6. (25)

$$\text{Rate} = \frac{200}{40} = 5\,\frac{\text{dollars}}{\text{hr}}$$
$$\text{New Rate} = 1.3 \times \text{Rate} = 1.3(5) = 6.5\,\frac{\text{dollars}}{\text{hr}}$$
$$\text{New Rate} \times T = \text{dollars}$$
$$6.5(38) = \mathbf{\$247}$$

7. (25)

$$\begin{cases} \text{(a)} & 5E_N = 4L_N + 38 \\ \text{(b)} & 2(E_N + 5) = 2(L_N + 5) + 8 \end{cases}$$

(b) $2(E_N + 5) = 2(L_N + 5) + 8$
$$2E_N + 10 = 2L_N + 10 + 8$$
$$2E_N = 2L_N + 8$$
$$E_N = L_N + 4$$

(a) $5E_N = 4L_N + 38$
$$5(L_N + 4) = 4L_N + 38$$
$$5L_N + 20 = 4L_N + 38$$
$$L_N = 18$$

$$E_N = L_N + 4 = (18) + 4 = 22$$

So in 13 years, **Early** = **35 yr** and **Lucy** = **31 yr.**

8. (18)

$$\begin{cases} \text{(a)} & \frac{T}{H} = \frac{2}{1} \\ \text{(b)} & 2T = 5U + 1 \\ \text{(c)} & H + T + U = 15 \end{cases}$$

(b) $2T = 5U + 1$
$$T = \frac{5}{2}U + \frac{1}{2}$$

(a) $\frac{T}{H} = \frac{2}{1}$
$$H = \frac{1}{2}T$$
$$H = \frac{1}{2}\left(\frac{5}{2}U + \frac{1}{2}\right)$$
$$H = \left(\frac{5}{4}U + \frac{1}{4}\right)$$

(c)
$$H + T + U = 15$$
$$\left(\frac{5}{4}U + \frac{1}{4}\right) + \left(\frac{5}{2}U + \frac{1}{2}\right) + U = 15$$
$$\frac{19}{4}U = \frac{57}{4}$$
$$U = 3$$

(b)
$$2T = 5U + 1$$
$$2T = 5(3) + 1$$
$$T = 8$$

(a)
$$H + T + U = 15$$
$$H + (8) + (3) = 15$$
$$H = 4$$

The number is **483.**

9. (61) Range $= 8 - 3 = \mathbf{5}$

Mean $= \dfrac{3 + 4 + 5 + 7 + 8 + 8}{6} = \mathbf{5.83}$

Median $= \dfrac{5 + 7}{2} = \mathbf{6}$

Mode $= \mathbf{8}$

$$\text{Variance} = \frac{1}{6}\left[(3 - 5.83)^2 + (4 - 5.83)^2 + (5 - 5.83)^2 + (7 - 5.83)^2 + (8 - 5.83)^2 + (8 - 5.83)^2\right]$$
$$= \mathbf{3.81}$$

Standard deviation $= \sqrt{3.81} = \mathbf{1.95}$

10. (61) $\mu = 65,\ \sigma = 3$

About 68% of the data lie within one standard deviation of the mean.

$\mu \pm \sigma = 65 \pm 3 = 62, 68$

68% of the data lie between 62 and 68.

About 95% of the data lie within two standard deviations of the mean.

$\mu \pm 2\sigma = 65 \pm 2(3) = 59, 71$

95% of the data lie between 59 and 71.

About 99% of the data lie within three standard deviations of the mean.

$\mu \pm 3\sigma = 65 \pm 3(3) = 56, 74$

99% of the data lie between 56 and 74.

11. (61)

STEM	LEAF
3	1, 0
4	0, 4, 2, 7
5	5, 7, 3, 2
6	2, 9

12. (61) We use a calculator to get

$\bar{x} = 57.5$ $Q_1 = 54$
$\sigma x = 11.06$ Med $= 60.5$
min $= 30$ $Q_3 = 64$
max $= 72$

*

30 Min | 39 | 54 Q_1 | 60.5 Med | 64 Q_3 | 72 Max

13. (52)
$$\cos 3\theta + \frac{\sqrt{2}}{2} = 0$$
$$\cos 3\theta = -\frac{\sqrt{2}}{2}$$
$$3\theta = 135°, 225°, 495°, 585°, 855°, 945°$$
$$\theta = \mathbf{45°, 75°, 165°, 195°, 285°, 315°}$$

14. (60)
$$\sin^2\theta - \frac{1}{4} = 0$$
$$\left(\sin\theta - \frac{1}{2}\right)\left(\sin\theta + \frac{1}{2}\right) = 0$$
$$\left(\sin\theta - \frac{1}{2}\right) = 0 \qquad \left(\sin\theta + \frac{1}{2}\right) = 0$$
$$\sin\theta = \frac{1}{2} \qquad \sin\theta = -\frac{1}{2}$$
$$\theta = 30°, 150° \qquad \theta = 210°, 330°$$

$\theta = \mathbf{30°, 150°, 210°, 330°}$

15. (60)
$$\cos^2\theta - \frac{1}{2}\cos\theta = 0$$
$$\cos\theta\left(\cos\theta - \frac{1}{2}\right) = 0$$
$$\cos\theta = 0 \qquad \cos\theta - \frac{1}{2} = 0$$
$$\theta = 90°, 270° \qquad \cos\theta = \frac{1}{2}$$
$$\theta = 60°, 300°$$

$\theta = \mathbf{60°, 90°, 270°, 300°}$

16. (59)

$$2 \log_5 x = \log_5 16 - \log_5 4$$
$$\log_5 x^2 = \log_5 \frac{16}{4}$$
$$x^2 = 4$$
$$x = \pm 2 \qquad (x \neq -2)$$
$$x = \mathbf{2}$$

17. (59)

$$\log x + \log (x - 9) = 1$$
$$\log (x)(x - 9) = 1$$
$$x^2 - 9x = 10^1$$
$$x^2 - 9x - 10 = 0$$
$$(x - 10)(x + 1) = 0$$
$$x = 10, -1 \qquad (x \neq -1)$$
$$x = \mathbf{10}$$

18. (59)

$$5^{\log_5 22 - \log_5 2} + 3 \log_2 2^4 = 5^{\log_5 \frac{22}{2}} + \log_2 (2^4)^3$$
$$= 5^{\log_5 11} + \log_2 2^{12} = 11 + 12 = \mathbf{23}$$

19. (58) Equation of the perpendicular line:

$$y = -x + b$$
$$4 = -(2) + b$$
$$b = 6$$
$$y = -x + 6$$

Point of intersection:

$$x - 1 = -x + 6$$
$$2x = 7$$
$$x = \frac{7}{2}$$

$$y = -x + 6 = -\left(\frac{7}{2}\right) + 6 = \frac{5}{2}$$

$$\left(\frac{7}{2}, \frac{5}{2}\right) \text{ and } (2, 4)$$

$$D = \sqrt{\left(\frac{7}{2} - 2\right)^2 + \left(\frac{5}{2} - 4\right)^2}$$
$$= \sqrt{\frac{9}{4} + \frac{9}{4}} = \mathbf{\frac{3\sqrt{2}}{2}}$$

20. (58)

$$y - 2x^2 + 4x = -7$$
$$y = 2x^2 - 4x - 7$$
$$y = 2(x^2 - 2x + \quad) - 7$$
$$y = 2(x^2 - 2x + 1) - 7 - 2$$
$$\mathbf{y = 2(x - 1)^2 - 9}$$

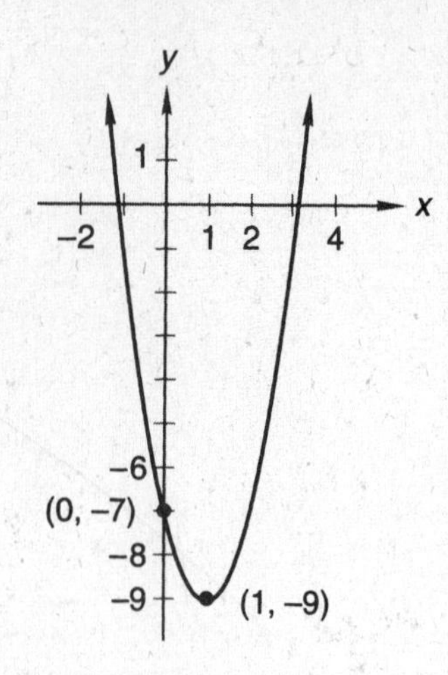

21. (57)

Function = $\sin x$

Centerline = -2

Amplitude = 6

Period = 2π

Phase shift = $\frac{\pi}{4}$

$$\mathbf{y = -2 + 6 \sin \left(x - \frac{\pi}{4}\right)}$$

22. (57)

Function = $\sin \theta$

Centerline = 2

Amplitude = 5

Period = $90°$

Phase angle = $0°$

Coefficient = $\frac{360°}{90°} = 4$

$$\mathbf{y = 2 + 5 \sin 4\theta}$$

23. (56)

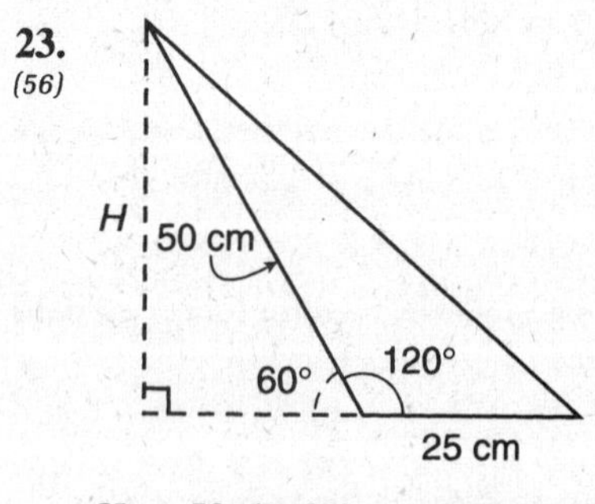

$$H = 50 \sin 60° = 43.3013 \text{ cm}$$

$$\text{Area} = \frac{1}{2}BH = \frac{1}{2}(25)(43.3013) = \mathbf{541.27 \text{ cm}^2}$$

24. (56)

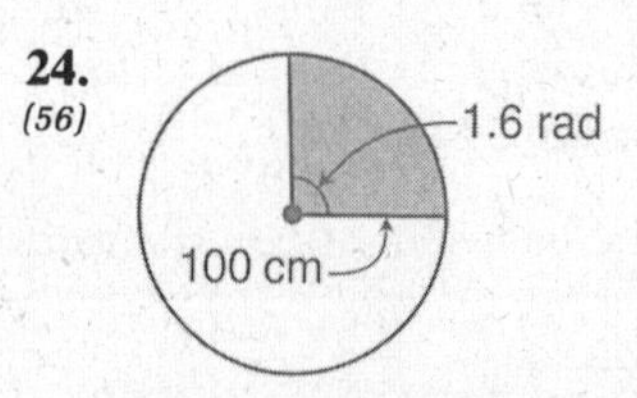

$$A_{\text{sector}} = \pi r^2 \left(\frac{1.6}{2\pi}\right) = \pi(100)^2 \left(\frac{1.6}{2\pi}\right) = \mathbf{8000 \text{ cm}^2}$$

25. (56)

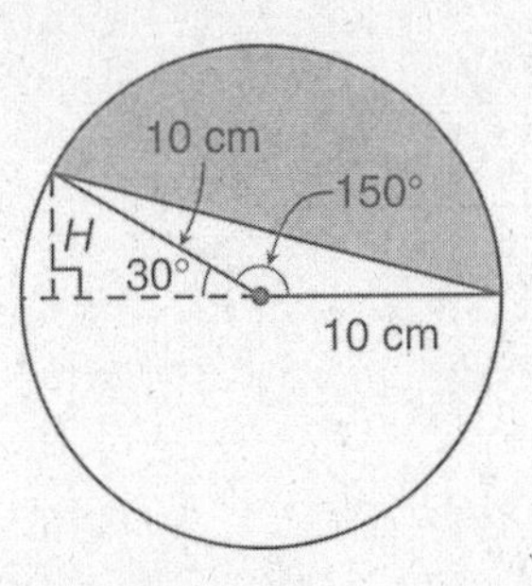

$H = 10 \sin 30° = 5 \text{ cm}$

$$A_{\text{segment}} = A_{\text{sector}} - A_{\text{triangle}}$$
$$= \pi(10)^2\left(\frac{150°}{360°}\right) - \frac{1}{2}(5)(10)$$
$$= \mathbf{105.90\ cm^2}$$

26. (56) $\begin{cases} y \le x^2 & \text{(parabola)} \\ x^2 + y^2 \le 4 & \text{(circle)} \end{cases}$

The region is within or on the circle and beneath or on the parabola.

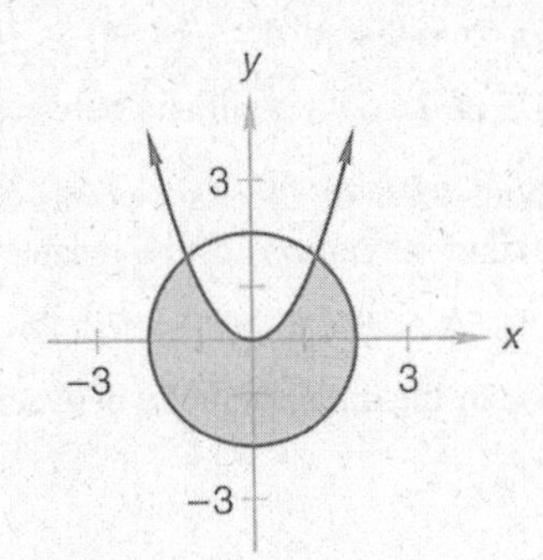

Test point: (0, –1)

Parabola: $-1 \le 0^2$

$-1 \le 0$ True

Line: $(0)^2 + (-1)^2 \le 4$

$1 \le 4$ True

27. (36,39) $2f\left(\frac{13\pi}{6}\right)g\left(\frac{13\pi}{6}\right) - f\left(\frac{13\pi}{3}\right)$

$$= 2 \sin \frac{13\pi}{6} \cos \frac{13\pi}{6} - \sin \frac{13\pi}{3}$$
$$= 2 \sin \frac{\pi}{6} \cos \frac{\pi}{6} - \sin \frac{\pi}{3}$$
$$= 2\left(\frac{1}{2}\right)\left(\frac{\sqrt{3}}{2}\right) - \frac{\sqrt{3}}{2} = \mathbf{0}$$

28. (36,41) $\cot(-450°) - \dfrac{\cos(-450°)}{\sin(-450°)}$

$$= \cot(-450°) - \cot(-450°) = \mathbf{0}$$

29. (32)

$-90° < \theta < 90°$

$$c = \sqrt{5^2 + 12^2} = 13$$

$$\sin\left[\text{Arctan}\left(-\frac{5}{12}\right)\right] = \sin\theta = \mathbf{-\frac{5}{13}}$$

30. (48) $m_0 = \dfrac{4 - 2}{4 - (-4)} = \dfrac{2}{8} = \dfrac{1}{4}$

$$(x_m, y_m) = \left(\frac{-4 + 4}{2}, \frac{2 + 4}{2}\right) = (0, 3)$$

The slope of a perpendicular line is the negative reciprocal, so $m = -4$.

$$y = -4x + b$$
$$3 = -4(0) + b$$
$$b = 3$$

$$\mathbf{y = -4x + 3}$$

Problem Set 62

1. (62) $C = mN + b$

$$\begin{cases} \text{(a)} & 2050 = m200 + b \\ \text{(b)} & 350 = m30 + b \end{cases}$$

$$\begin{array}{rrcl} \text{(a)} & 2050 & = & m200 + b \\ -1\text{(b)} & -350 & = & -m30 - b \\ \hline & 1700 & = & m170 \\ & m & = & 10 \end{array}$$

(b) $350 = m30 + b$

$$350 = (10)30 + b$$
$$b = 50$$

$$\mathbf{C = 10N + 50}$$

2. (38)

10	9	8	7	6

$\rightarrow\ 10 \cdot 9 \cdot 8 \cdot 7 \cdot 6 = \mathbf{30{,}240}$

3. (55) $\dfrac{(5 + 4 + 2)!}{5!4!2!} = \mathbf{6930}$

4. (45)

4	3	1	2	1

$\rightarrow\ 4 \cdot 3 \cdot 1 \cdot 2 \cdot 1 = \mathbf{24}$

5. $v = r\omega$
(53)

$$v = (20\text{ cm})\left(723\ \frac{\text{rad}}{\text{min}}\right)\left(\frac{1\text{ in.}}{2.54\text{ cm}}\right)\left(\frac{1\text{ ft}}{12\text{ in.}}\right)\left(\frac{1\text{ min}}{60\text{ s}}\right)$$

$$v = \frac{\mathbf{(20)(723)}}{\mathbf{(2.54)(12)(60)}}\ \frac{\mathbf{ft}}{\mathbf{s}} = \mathbf{7.91\ \frac{ft}{s}}$$

6. (36)

$$\begin{cases} \text{(a)}\ (D + W)T_D = D_D \\ \text{(b)}\ (D - W)T_U = D_U \end{cases}$$

$T_U = T_D + 1,\ D = 2W$

(a)
$$(D + W)T_D = D_D$$
$$(2W + W)T_D = 12$$
$$3WT_D = 12$$
$$WT_D = 4$$

(b)
$$(D - W)T_U = D_U$$
$$(2W - W)(T_D + 1) = 6$$
$$WT_D + W = 6$$
$$(4) + W = 6$$
$$\mathbf{W = 2\ mph}$$

$D = 2W$

$D = 2(2)$

$\mathbf{D = 4\ mph}$

7. $P = \dfrac{k}{V}$
(18)

$k = PV = (4\text{ atm})(500\text{ L}) = 2000\text{ atm}\cdot\text{L}$

$$P = \frac{2000\text{ atm}\cdot\text{L}}{V} = \frac{2000\text{ atm}\cdot\text{L}}{50\text{ L}} = \mathbf{40\ atm}$$

8. $R = \dfrac{kP}{W^3}$
(18)

$$k = \frac{RW^3}{P} = \frac{200(10^3)}{50} = 4000$$

$$R = \frac{4000P}{W^3} = \frac{4000(60)}{5^3} = \mathbf{1920}$$

9. (62)

$$\begin{cases} \text{(a)}\ mx + ny = c \\ \text{(b)}\ dx + ey = f \end{cases}$$

$$\begin{aligned} -d\text{(a)}\quad & -dmx - ndy = -cd \\ m\text{(b)}\quad & dmx + mey = mf \\ \hline & (me - nd)y = mf - cd \end{aligned}$$

$$\mathbf{y = \frac{mf - cd}{me - nd}}$$

10. Range $= 7 - 1 = \mathbf{6}$
(61)

$$\text{Mean} = \frac{1 + 2 + 3 + 4 + 7 + 7}{6} = \mathbf{4}$$

$$\text{Median} = \frac{3 + 4}{2} = \mathbf{3.5}$$

Mode $= \mathbf{7}$

$$\text{Variance} = \frac{(1 - 4)^2 + (2 - 4)^2 + (3 - 4)^2}{6} + \frac{(4 - 4)^2 + (7 - 4)^2 + (7 - 4)^2}{6}$$

$$= \mathbf{5.33}$$

Standard deviation $= \sqrt{5.33} = \mathbf{2.31}$

11. $\mu = 84,\ \sigma = 6$
(61)

(a) About 68% of the data points lie within one standard deviation of the mean.

$\mu \pm \sigma = 84 \pm 6 = 78, 90$

68% of the data points lie between **78 and 90.**

(b) About 95% of the data points lie within two standard deviations of the mean.

$\mu \pm 2\sigma = 84 \pm 2(6) = 72, 96$

95% of the data points lie between **72 and 96.**

12. We use a calculator to get
(61)

$\bar{x} = 557.11$ $\quad Q_1 = 479$

$\sigma x = 74.38$ $\quad$ Med $= 542$

min $= 465$ $\quad Q_3 = 622.5$

max $= 689$

Mean = 557.11

Standard deviation = 74.38

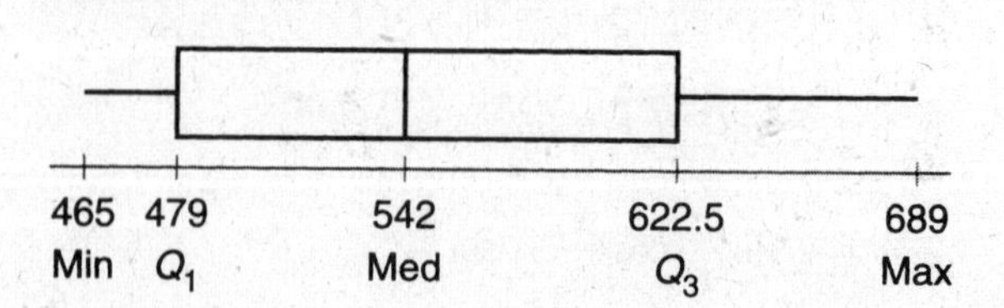

13. $\cos 4\theta - 1 = 0$
(52)

$$\cos 4\theta = 1$$
$$4\theta = 0°, 360°, 720°, 1080°$$
$$\boldsymbol{\theta = 0°, 90°, 180°, 270°}$$

14. (60)

$$\tan^2\theta = 1$$
$$\tan^2\theta - 1 = 0$$
$$(\tan\theta - 1)(\tan\theta + 1) = 0$$
$$\tan\theta - 1 = 0 \qquad \tan\theta + 1 = 0$$
$$\tan\theta = 1 \qquad \tan\theta = -1$$
$$\theta = 45°, 225° \qquad \theta = 135°, 315°$$

$\boldsymbol{\theta = 45°, 135°, 225°, 315°}$

15. (59) $3\log_6 x = \log_6 16 - \log_6 2$

$$\log_6 x^3 = \log_6 \frac{16}{2}$$
$$x^3 = 8$$
$$x = \mathbf{2}$$

16. (59) $\log_7(x + 9) - \log_7 x = \log_7 7$

$$\log_7 \frac{x + 9}{x} = \log_7 7$$
$$\frac{x + 9}{x} = 7$$
$$x + 9 = 7x$$
$$6x = 9$$
$$x = \mathbf{\frac{3}{2}}$$

17. (59) $6^{2\log_6 2} + 5\log_4 4 = 6^{\log_6 2^2} + \log_4 4^5$

$= 2^2 + 5 = \mathbf{9}$

18. (58) Equation of the perpendicular line:

$y = -x + b$
$1 = -(3) + b$
$b = 4$
$y = -x + 4$

Point of intersection:

$x - 4 = -x + 4$
$2x = 8$
$x = 4$
$y = -x + 4 = -(4) + 4 = 0$
$(4, 0)$ and $(3, 1)$

$D = \sqrt{(4 - 3)^2 + (0 - 1)^2} = \sqrt{1 + 1} = \boldsymbol{\sqrt{2}}$

19. (57) Function = $\cos\theta$

Centerline = -1
Amplitude = 4
Period = 360°
Phase angle = 135°
$\boldsymbol{y = -1 + 4\cos(\theta - 135°)}$

20. (57) Function = $\cos x$

Centerline = -3
Amplitude = 5
Period = 6π
Phase angle = 0

$$\text{Coefficient} = \frac{2\pi}{6\pi} = \frac{1}{3}$$

$$\boldsymbol{y = -3 + 5\cos\frac{1}{3}x}$$

21. (56)

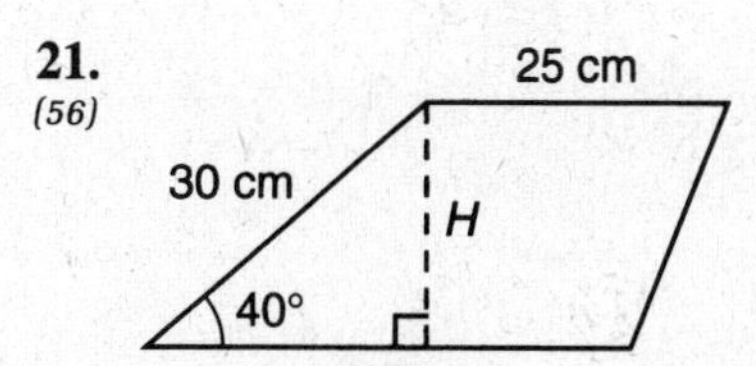

$H = 30\sin 40° = 19.2836$ cm

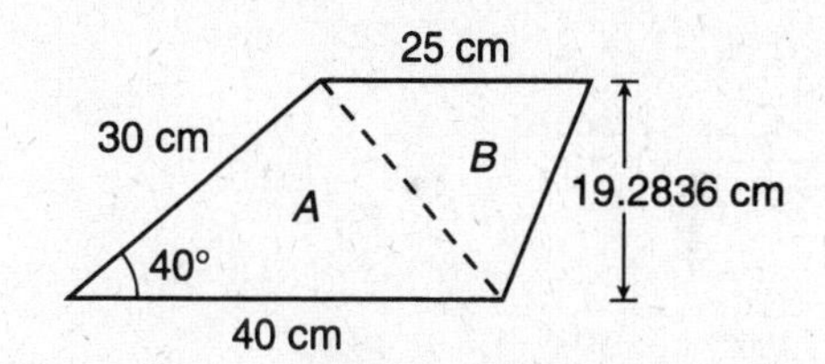

Total area = Area A + Area B

$$= \frac{1}{2}(40)(19.2836) + \frac{1}{2}(25)(19.2836)$$
$$= 385.67 + 241.05 = \mathbf{626.72\ cm^2}$$

22. (1)

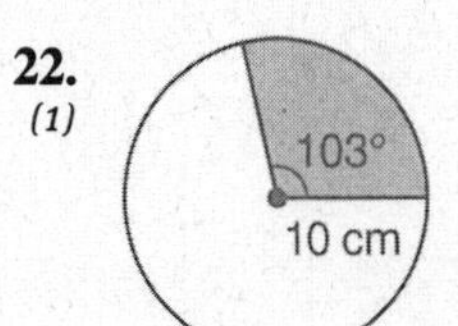

$$A_{sector} = \pi(10)^2\left(\frac{103°}{360°}\right) = \mathbf{89.88\ cm^2}$$

23. (56)

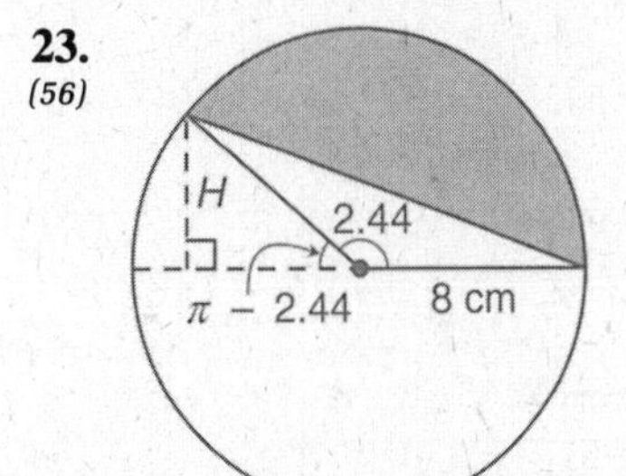

$H = 8\sin 0.7016 = 5.1635$ cm

$$A_{segment} = A_{sector} - A_{triangle}$$
$$= \pi(8)^2\left(\frac{2.44}{2\pi}\right) - \frac{1}{2}(8)(5.1635)$$
$$= 78.08 - 20.65 = \mathbf{57.43\ cm^2}$$

24. (56) $\begin{cases} y \le (x+3)^2 & \text{(parabola)} \\ y \le -x + 2 & \text{(line)} \end{cases}$

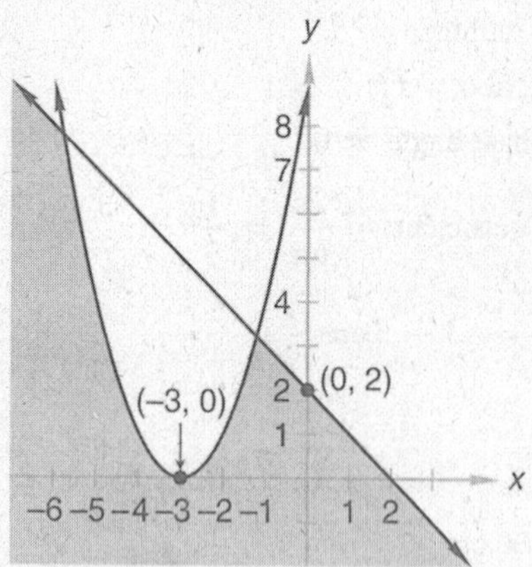

Test point: (1, 1)

Parabola: $1 \le (1 + 3)^2$

$1 \le 16$ true

Line: $1 \le -1 + 2$

$1 \le 1$ true

25. (36,39) $2 \sin \frac{19\pi}{6} \cos \frac{19\pi}{6} - \sin \frac{19\pi}{3}$

$= 2\left(-\sin \frac{\pi}{6}\right)\left(-\cos \frac{\pi}{6}\right) - \sin \frac{\pi}{3}$

$= 2\left(-\frac{1}{2}\right)\left(-\frac{\sqrt{3}}{2}\right) - \frac{\sqrt{3}}{2}$

$= \frac{\sqrt{3}}{2} - \frac{\sqrt{3}}{2} = \mathbf{0}$

26. (48) $\tan^2(-390°) + 1 = (-\tan 30°)^2 + 1$

$= \left(-\frac{1}{\sqrt{3}}\right)^2 + 1 = \mathbf{\frac{4}{3}}$

27. (32)

y
a
θ
x
−5
13

$-90° \le \theta \le 90°$

$a = \sqrt{13^2 - (-5)^2} = \sqrt{144} = 12$

$\tan\left[\text{Arcsin}\left(-\frac{5}{13}\right)\right] = \tan\theta = \mathbf{-\frac{5}{12}}$

28. (37,39) $\sqrt{(x+4)^2 + (y+4)^2} = \sqrt{(x-6)^2 + (y-2)^2}$

$x^2 + 8x + 16 + y^2 + 8y + 16$

$= x^2 - 12x + 36 + y^2 - 4y + 4$

$20x + 12y - 8 = 0$

$\mathbf{5x + 3y - 2 = 0}$

29. (31) The graph of g is the graph of f translated 7 units to the left and 5 units down. Therefore the answer is **D.**

30. (40) $f(g(x)) = \dfrac{1}{\dfrac{5x+1}{x} - 5} = \dfrac{1}{\dfrac{5x + 1 - 5x}{x}} = x$

$g(f(x)) = \dfrac{5\left(\dfrac{1}{x-5}\right) + 1}{\dfrac{1}{x-5}}$

$= \left(\dfrac{5 + x - 5}{x - 5}\right)\left(\dfrac{x-5}{1}\right) = x$

$f(g(x)) = g(f(x)) = x$

Yes, f and g are inverse functions.

Problem Set 63

1. (62) $N_R = mN_B + b$

(a) $\begin{cases} 11 = m2 + b \\ 35 = m8 + b \end{cases}$ (b)

(b) $35 = m8 + b$

-1(a) $-11 = -m2 - b$

$24 = m6$

$m = 4$

(a) $11 = m2 + b$

$11 = (4)2 + b$

$b = 3$

$N_R = 4N_B + 3$

$N_R = 4(6) + 3$

$\mathbf{N_R = 27}$

2. (55) $\frac{7!}{4!3!} = \mathbf{35}$

3. (45)

O FIRST

O	4	3	2	1

→ $4! = 24$

E FIRST

E	4	3	2	1

→ $4! = 24$

$\mathbf{48}$

4. (45)

B	G	B	G	B

3	2	2	1	1

$\rightarrow\ 3 \cdot 2 \cdot 2 \cdot 1 \cdot 1 = \mathbf{12}$

5. (53) $\omega = \dfrac{v}{r} = \dfrac{60\,\dfrac{\text{km}}{\text{hr}}}{12\text{ in.}} = \dfrac{60\text{ km}}{12\text{ in.-hr}}$

$$\omega = \left(\frac{60\text{ km}}{12\text{ in.-hr}}\right)\left(\frac{1000\text{ m}}{1\text{ km}}\right)\left(\frac{100\text{ cm}}{1\text{ m}}\right)\left(\frac{1\text{ in.}}{2.54\text{ cm}}\right) \times \left(\frac{1\text{ hr}}{60\text{ min}}\right)\left(\frac{1\text{ min}}{60\text{ s}}\right)$$

$$\omega = \mathbf{\frac{(60)(100)(1000)}{(2.54)(12)(60)(60)}\ \frac{rad}{s}} = \mathbf{54.68\ \frac{rad}{s}}$$

6. (44) $\text{Time} = \dfrac{\text{distance}}{\text{rate}}$

$T_M = \dfrac{m}{a},\ T_K = \dfrac{m}{b}$

$T_K - T_M = \dfrac{m}{b} - \dfrac{m}{a} = \mathbf{\dfrac{ma - mb}{ab}\ hr}$

7. (44) $\text{Rate} = \dfrac{a}{b}$

$\text{New rate} = \dfrac{a}{b} + c = \dfrac{a + bc}{b}$

$\text{New time} = \dfrac{\text{distance}}{\text{new rate}} = \dfrac{1000}{\dfrac{a + bc}{b}} = \mathbf{\dfrac{1000b}{a + bc}\ hr}$

8. (18) $\dfrac{\text{Onlookers}}{\text{Bystanders}} = \dfrac{17}{4}$

$\dfrac{\text{Onlookers}}{\text{Total}} = \dfrac{17}{17 + 4}$

$\dfrac{\text{Onlookers}}{1050} = \dfrac{17}{21}$

Onlookers = 850

Onlookers + Bystanders = Total

(850) + Bystanders = (1050)

Bystanders = 200

9. (63)

$$x^2 + y^2 + 8x - 6y - 11 = 0$$
$$(x^2 + 8x \quad) + (y^2 - 6y \quad) = 11$$
$$(x^2 + 8x + 16) + (y^2 - 6y + 9) = 36$$
$$\mathbf{(x + 4)^2 + (y - 3)^2 = 6^2}$$

Center = (–4, 3); radius = 6

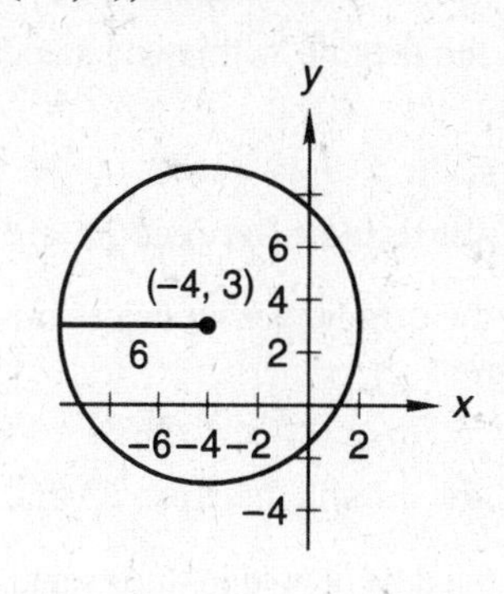

10. (63)

$$x^2 + y^2 - 8x + 2y + 13 = 0$$
$$(x^2 - 8x \quad) + (y^2 + 2y \quad) = -13$$
$$(x^2 - 8x + 16) + (y^2 + 2y + 1) = 4$$
$$\mathbf{(x - 4)^2 + (y + 1)^2 = 2^2}$$

Center = (4, –1); radius = 2

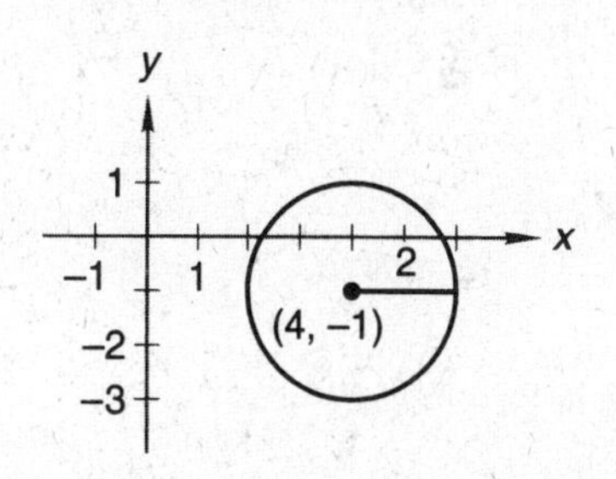

11. (62)

$$\begin{cases} \text{(a)}\ ax + by = c \\ \text{(b)}\ mx + ny = f \end{cases}$$

$$\begin{aligned} n(\text{a}) \qquad anx + bny &= cn \\ -b(\text{b}) \qquad -mbx - bny &= -bf \\ \hline x(an - mb) &= cn - bf \end{aligned}$$

$$\mathbf{x = \frac{cn - bf}{an - mb}}$$

12. (61) $24 + 16 + 18 + x + 28 = 100$

$x = 14$

Point = **14**

Range = **14**

Median = **18**

Standard deviation

$$= \sqrt{\frac{(-4)^2 + (4)^2 + (2)^2 + (6)^2 + (-8)^2}{5}} = \mathbf{5.22}$$

13. (61) $\mu = 76,\ \sigma = 4$

68% of the data lie within one standard deviation of the mean.

$\mu \pm \sigma = 76 \pm 4 = 72, 80$

68% of the data lie between 72 and 80.

95% of the data lie within two standard deviations of the mean.

$\mu \pm 2\sigma = 76 \pm 2(4) = 68, 84$

95% of the data lie between 68 and 84.

99% of the data lie within three standard deviations of the mean.

$\mu \pm 3\sigma = 76 \pm 3(4) = 64, 88$

99% of the data lie between 64 and 88.

14. (61)

STEM	LEAF
5	6
6	2, 7, 3, 5, 4
7	6, 0, 8, 2, 4
8	9, 3, 0, 1

15. (60)

$$\cos^2\theta - \frac{1}{4} = 0$$

$$\left(\cos\theta - \frac{1}{2}\right)\left(\cos\theta + \frac{1}{2}\right) = 0$$

$\cos\theta - \frac{1}{2} = 0$ $\qquad$ $\cos\theta + \frac{1}{2} = 0$

$\cos\theta = \frac{1}{2}$ $\qquad$ $\cos\theta = -\frac{1}{2}$

$\theta = 60°, 300°$ $\qquad$ $\theta = 120°, 240°$

$\theta =$ **60°, 120°, 240°, 300°**

16. (60) $\cos\theta\sin\theta - \sin\theta = 0$

$\sin\theta(\cos\theta - 1) = 0$

$\sin\theta = 0$ $\qquad$ $\cos\theta - 1 = 0$

$\theta = 0°, 180°$ $\qquad$ $\cos\theta = 1$

$\theta = 0°$

$\theta =$ **0°, 180°**

17. (59) $2\log_8 x = \log_8 6 + \log_8 2 - \log_8 1$

$$\log_8 x^2 = \log_8 \frac{(6 \cdot 2)}{1}$$

$x^2 = 12$

$x = \mathbf{2\sqrt{3}}$

18. (59) $\log_{\frac{1}{2}}(x + 3) - \log_{\frac{1}{2}}(x - 2) = \log_{\frac{1}{2}} 2$

$$\log_{\frac{1}{2}}\left(\frac{x + 3}{x - 2}\right) = \log_{\frac{1}{2}} 2$$

$$\frac{x + 3}{x - 2} = 2$$

$x + 3 = 2x - 4$

$x = \mathbf{7}$

19. (58) $y - x + 6 = 0$

$y = x - 6$

Equation of the perpendicular line:

$y = -x + b$

$4 = -(-2) + b$

$b = 2$

$y = -x + 2$

Point of intersection:

$x - 6 = -x + 2$

$2x = 8$

$x = 4$

$y = -x + 2 = -(4) + 2 = -2$

$(4, -2)$ and $(-2, 4)$

$D = \sqrt{[4 - (-2)]^2 + [(-2) - 4]^2}$

$= \sqrt{36 + 36} = \mathbf{6\sqrt{2}}$

20. (54,58) $y = 2x^2 - 12x + 9$

$y = 2(x^2 - 6x \quad\ ) + 9$

$y = 2(x^2 - 6x + 9) - 18 + 9$

$\mathbf{y = 2(x - 3)^2 - 9}$

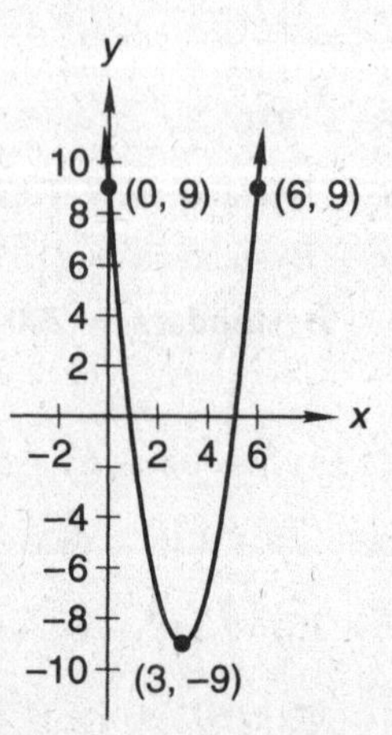

21. (47,57) Function = $\sin x$

Centerline = 4

Amplitude = 8

Period = 2π

Phase angle = $\dfrac{\pi}{4}$

$$\mathbf{y = 4 + 8\sin\left(x - \frac{\pi}{4}\right)}$$

22. (47,57) Function = $\sin x$

Centerline = –3

Amplitude = 5

Period = 4π

Phase angle = 0

Coefficient = $\dfrac{2\pi}{4\pi} = \dfrac{1}{2}$

$$\mathbf{y = -3 + 5\sin\frac{1}{2}x}$$

23. (56)

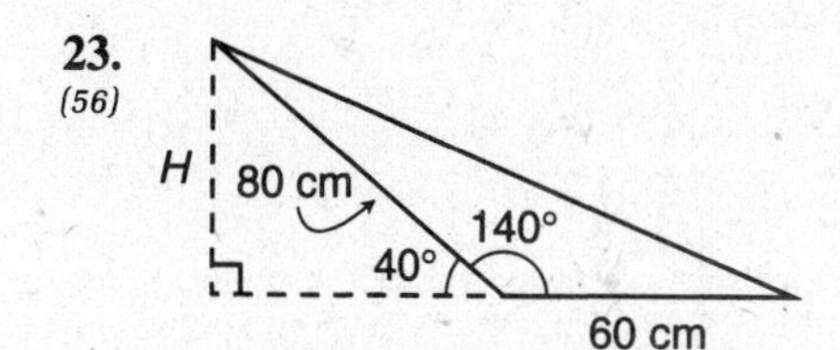

$H = 80 \sin 40° = 51.423$ cm

Area $= \dfrac{1}{2}BH = \dfrac{1}{2}(60)(51.423) = \mathbf{1542.69\ cm^2}$

24. (56)

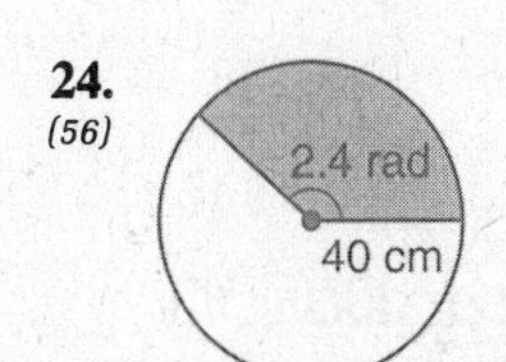

$$A_{sector} = \pi r^2\left(\frac{2.4}{2\pi}\right) = (40)^2 1.2 = \mathbf{1920\ cm^2}$$

25. (56)

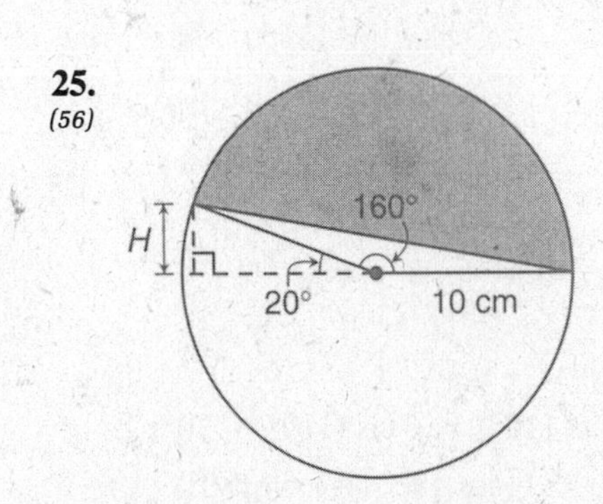

$H = 10 \sin 20° = 3.42$ cm

$$A_{segment} = A_{sector} - A_{triangle}$$
$$= \pi(10)^2\left(\frac{160°}{360°}\right) - \frac{1}{2}(10)(3.42)$$
$$= 139.63 - 17.10 = \mathbf{122.53\ cm^2}$$

26. (56)
$$\begin{cases} x^2 + y^2 \le 9 & \text{(circle)} \\ y \ge x & \text{(line)} \\ y \ge -x & \text{(line)} \end{cases}$$

The region must be on or inside the circle and on or above each line.

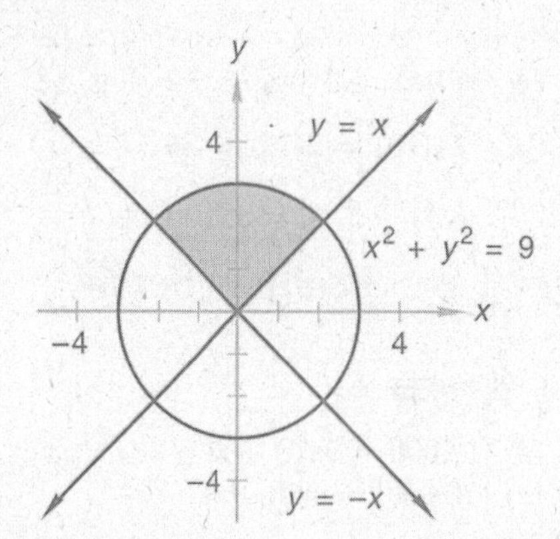

Test point = (0, 1)

Circle: $x^2 + y^2 \le 9$

$(0)^2 + (1)^2 \le 9$

$1 \le 9$ True

Line: $y \ge x$

$1 \ge 0$ True

Line: $y \ge -x$

$1 \ge 0$ True

27. (48) $\sin^2 \dfrac{3\pi}{2} + \cos^2 \dfrac{3\pi}{2} = \left(-\sin\dfrac{\pi}{2}\right)^2 + \left(\cos\dfrac{\pi}{2}\right)^2$

$= (-1)^2 + 0^2 = \mathbf{1}$

28. (48) $\cos^2(-405°) - \sin^2(-405°)$

$= \cos^2(-45°) - \sin^2(-45°)$

$= (\cos 45°)^2 - (-\sin 45°)^2$

$= \left(\dfrac{1}{\sqrt{2}}\right)^2 - \left(-\dfrac{1}{\sqrt{2}}\right)^2$

$= \dfrac{1}{2} - \dfrac{1}{2} = \mathbf{0}$

29. (32)

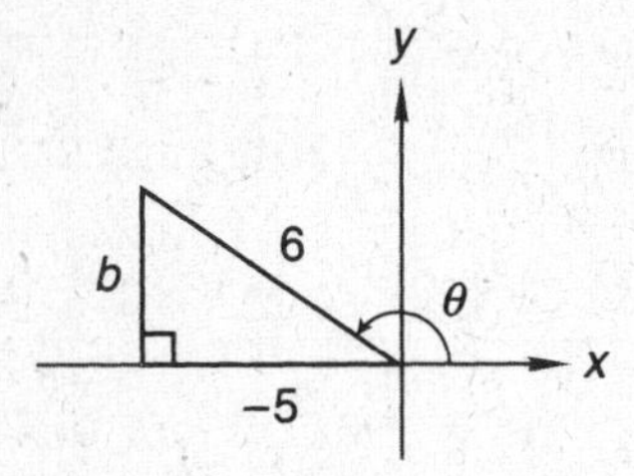

$0° \le \theta \le 180°$

$b = \sqrt{36 - 25} = \sqrt{11}$

$\sin\left[\text{Arccos}\left(-\frac{5}{6}\right)\right] = \sin\theta = \mathbf{\frac{\sqrt{11}}{6}}$

30. (24,40) $(g \circ f)(x) = 2\log_8(8^x) = \log_8(8^x)^2 = 2x$

$(g \circ f)(13) = 2(13) = \mathbf{26}$

Problem Set 64

1. (62) $C = mW + b$

(a) $15{,}000 = m10 + b$
(b) $20{,}000 = m20 + b$

$$\begin{aligned} \text{(b)}\quad 20{,}000 &= m20 + b \\ -2\text{(a)}\quad -30{,}000 &= -m20 - 2b \\ \hline -10{,}000 &= -b \\ b &= 10{,}000 \end{aligned}$$

(a) $15{,}000 = m10 + b$

$15{,}000 = m10 + (10{,}000)$

$5000 = m10$

$m = 500$

$C = 500W + 10{,}000$

$= 500(30) + 10{,}000 = \mathbf{\$25{,}000}$

2. (45) | 2 | 5 | 5 | $\rightarrow 2 \cdot 5 \cdot 5 = \mathbf{50}$

3. (55) $\frac{7!}{4!3!} = \mathbf{35}$

4. (53) $v = r\omega$

$v = (10\text{ in.})\left(75\,\frac{\text{rad}}{\text{min}}\right)\left(\frac{2.54\text{ cm}}{1\text{ in.}}\right)\left(\frac{1\text{ m}}{100\text{ cm}}\right)$

$\times\left(\frac{1\text{ km}}{1000\text{ m}}\right)\left(\frac{60\text{ min}}{1\text{ hr}}\right)$

$v = \mathbf{\frac{(10)(75)(2.54)(60)}{(100)(1000)}\,\frac{km}{hr} = 1.14\,\frac{km}{hr}}$

5. (38) Overall average rate $= \frac{\text{Total distance}}{\text{Total time}}$

$= \frac{(m + k)\text{ mi}}{(x + y)\text{ hr}}$

Time $= \frac{\text{distance}}{\text{overall average rate}}$

$= \frac{150}{\frac{(m + k)}{(x + y)}} = \mathbf{\frac{150(x + y)}{(m + k)}\text{ hr}}$

6. (44) $T_R = \frac{m}{r},\ T_B = \frac{m}{b}$

$T_B - T_R = \frac{m}{b} - \frac{m}{r} = \mathbf{\frac{mr - mb}{br}\text{ hr}}$

7. (44) Rate $= \frac{k^2t}{d}\,\frac{\text{pencils}}{\text{dollars}}$

Pencils = rate · dollars

$= \frac{k^2t}{d} \cdot 100 = \mathbf{\frac{100k^2t}{d}\text{ pencils}}$

8. (64) (a) $6 + 2i$

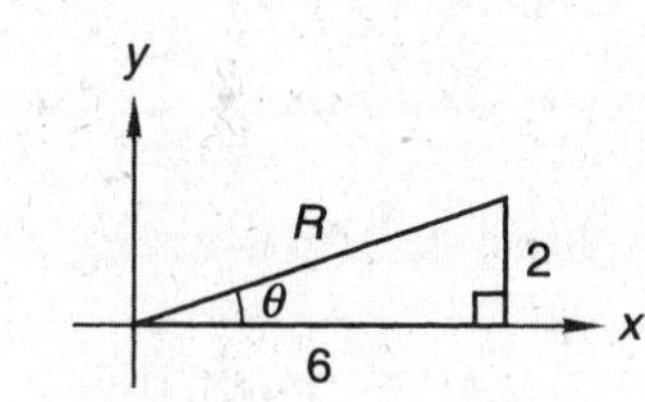

$R = \sqrt{6^2 + 2^2} = \sqrt{40} = 2\sqrt{10}$

$\tan\theta = \frac{2}{6}$

$\theta = 18.43°$

$2\sqrt{10}\text{ cis }18.43° = \mathbf{6.32\text{ cis }18.43°}$

(b) $5\text{ cis }30° = 5(\cos 30° + i\sin 30°)$

$= 5\left(\frac{\sqrt{3}}{2} + \frac{1}{2}i\right) = \mathbf{\frac{5\sqrt{3}}{2} + \frac{5}{2}i}$

9. (64) $(6\text{ cis }300°)(2\text{ cis }30°) = 12\text{ cis }(300° + 30°)$

$= 12\text{ cis }330° = 12(\cos 330° + i\sin 330°)$

$= 12\left(\frac{\sqrt{3}}{2} - \frac{1}{2}i\right) = \mathbf{6\sqrt{3} - 6i}$

10. (64) $(3\text{ cis }70°)(2\text{ cis }110°) = 6\text{ cis }(110° + 70°)$

$= 6\text{ cis }(180°) = 6(\cos 180° + i\sin 180°)$

$= 6(-1 + 0i) = \mathbf{-6}$

11. (63)

$$x^2 + y^2 - 4x = 0$$

$$(x^2 - 4x \quad) + y^2 = 0$$

$$(x^2 - 4x + 4) + y^2 = 4$$

$$\mathbf{(x - 2)^2 + y^2 = 2^2}$$

Center = (2, 0); radius = 2

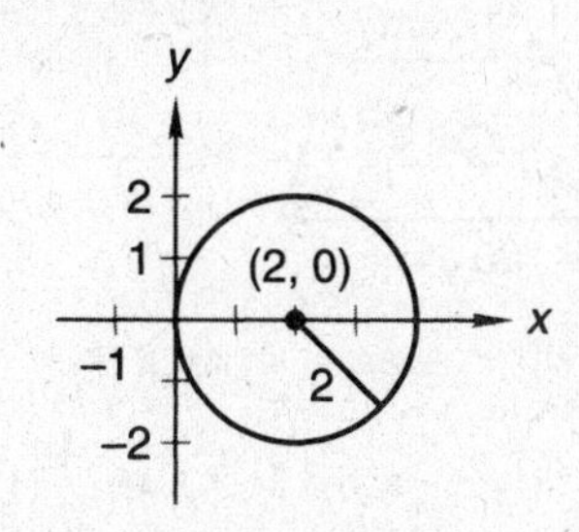

12. (63)

$$x^2 + y^2 + 6x - 4y + 9 = 0$$

$$(x^2 + 6x \quad) + (y^2 - 4y \quad) = -9$$

$$(x^2 + 6x + 9) + (y^2 - 4y + 4) = 4$$

$$\mathbf{(x + 3)^2 + (y - 2)^2 = 2^2}$$

Center = (–3, 2); radius = 2

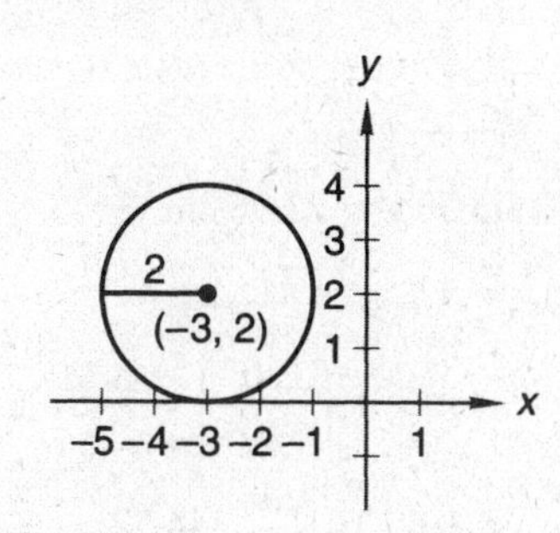

13. (62)

(a) $\begin{cases} cx + by = d \end{cases}$
(b) $\begin{cases} px + qy = f \end{cases}$

$$\begin{array}{rl} -p(\text{a}) & -cpx - bpy = -dp \\ c(\text{b}) & \underline{\;cpx + cqy = cf\;} \\ & y(cq - bp) = cf - dp \end{array}$$

$$\mathbf{y = \frac{cf - dp}{cq - bp}}$$

14. (61) 95% of the data lie within two standard deviations of the mean.

$\mu \pm 2\sigma = 17 \pm 2(5) = 7, 27$

95% of the data lie between **7 and 27.**

15. (61) We use a calculator to get:

$\bar{x} = 28.93$ $\quad Q_1 = 19$

$\sigma x = 9.67$ $\quad$ Med = 32

min = 12 $\quad Q_3 = 35$

max = 48

Mean = 28.93

Standard deviation = 9.67

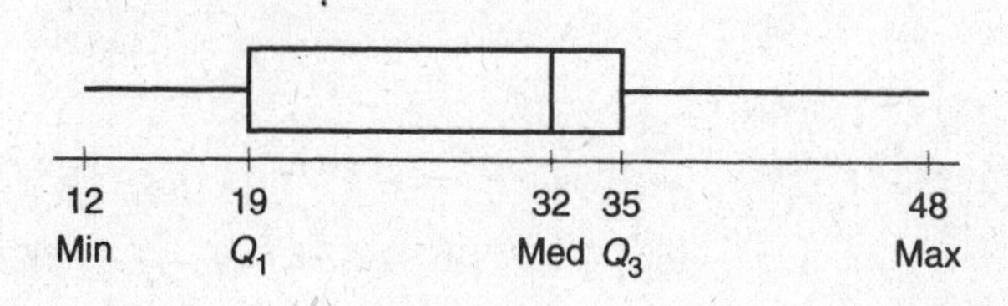

16. (50)

$$\sin\theta + 1 = 0$$

$$\sin\theta = -1$$

$$\theta = \mathbf{270°}$$

17. (60)

$$\tan^2\theta - 3 = 0$$

$$(\tan\theta - \sqrt{3})(\tan\theta + \sqrt{3}) = 0$$

$$\tan\theta - \sqrt{3} = 0 \qquad \tan\theta + \sqrt{3} = 0$$

$$\tan\theta = \sqrt{3} \qquad \tan\theta = -\sqrt{3}$$

$$\theta = 60°, 240° \qquad \theta = 120°, 300°$$

$$\theta = \mathbf{60°, 120°, 240°, 300°}$$

18. (60)

$$\tan\theta\sin\theta - \sin\theta = 0$$

$$\sin\theta(\tan\theta - 1) = 0$$

$$\sin\theta = 0 \qquad \tan\theta - 1 = 0$$

$$\theta = 0°, 180° \qquad \tan\theta = 1$$

$$\theta = 45°, 225°$$

$$\theta = \mathbf{0°, 45°, 180°, 225°}$$

19. (59)

$$2\log_8 x = \log_8(2x - 1)$$

$$\log_8 x^2 = \log_8(2x - 1)$$

$$x^2 = 2x - 1$$

$$x^2 - 2x + 1 = 0$$

$$(x - 1)^2 = 0$$

$$x = \mathbf{1}$$

20. (59)
$$\log_3 (x - 2) + \log_3 x = 1$$
$$\log_3 [(x - 2)x] = 1$$
$$(x - 2)x = 3^1$$
$$x^2 - 2x = 3$$
$$x^2 - 2x - 3 = 0$$
$$(x - 3)(x + 1) = 0$$
$$x = 3, -1 \qquad (x \neq -1)$$
$$x = \mathbf{3}$$

21. (59)
$$8^{2 \log_8 13} - 2 \log_7 7^3 = 8^{\log_8 13^2} - \log_7 (7^3)^2$$
$$= 13^2 - \log_7 7^6 = 13^2 - 6 = 169 - 6 = \mathbf{163}$$

22. (58)
$$y - x + 4 = 0$$
$$y = x - 4$$

Equation of the perpendicular line:

$y = -x + b$

$6 = -(-4) + b$

$b = 2$

$y = -x + 2$

Point of intersection:

$x - 4 = -x + 2$

$2x = 6$

$x = 3$

$y = x - 4 = (3) - 4 = -1$

$(3, -1)$ and $(-4, 6)$

$$D = \sqrt{[3 - (-4)]^2 + (-1 - 6)^2}$$
$$= \sqrt{7^2 + (-7)^2} = \mathbf{7\sqrt{2}}$$

23. (47,57) Function = $\cos x$

Centerline = -1

Amplitude = 5

Period = 360°

Phase angle = 135°

$\mathbf{y = -1 + 5 \cos (x - 135°)}$

24. (47,57) Function = $-\cos x$

Centerline = 6

Amplitude = 4

Period = 3π

$$\text{Coefficient} = \frac{2\pi}{3\pi} = \frac{2}{3}$$

$$\mathbf{y = 6 - 4 \cos \frac{2}{3}x}$$

25. (56)

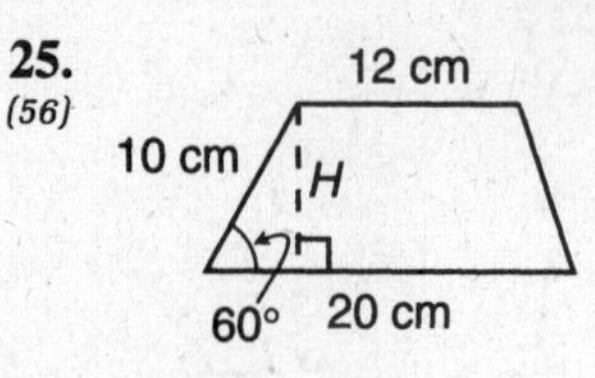

$$H = 10 \sin 60° = 10\frac{\sqrt{3}}{2} = 5\sqrt{3} \text{ cm}$$

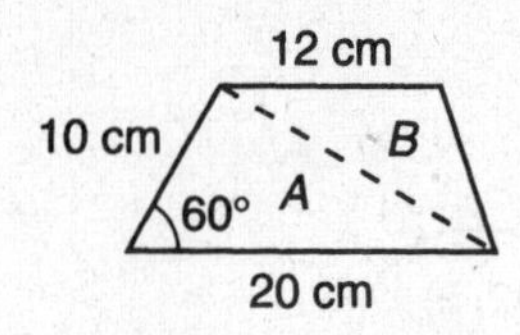

Total area = Area A + Area B

$$= \frac{1}{2}(20)(5\sqrt{3}) + \frac{1}{2}(12)(5\sqrt{3})$$
$$= 80\sqrt{3} \text{ cm}^2 = \mathbf{138.56 \text{ cm}^2}$$

26. (56)

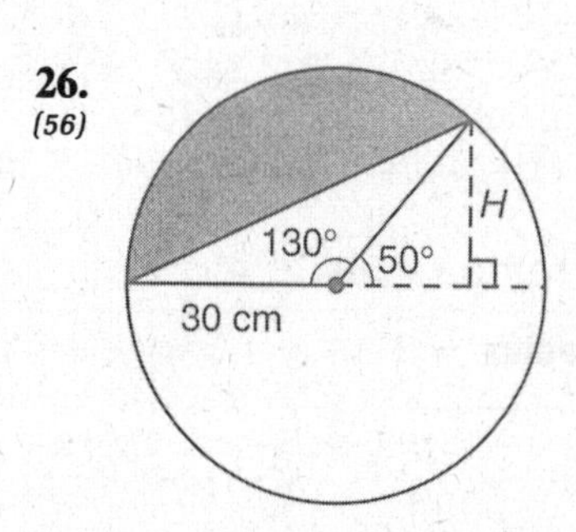

$$H = 30 \sin 50° = 22.98 \text{ cm}$$

$$A_{\text{segment}} = A_{\text{sector}} - A_{\text{triangle}}$$
$$= \pi(30)^2\left(\frac{130°}{360°}\right) - \frac{1}{2}(30)(22.98)$$
$$= 1021.018 - 344.72 = \mathbf{676.30 \text{ cm}^2}$$

27. (56)
$$\begin{cases} y \geq x^2 - 2 & \text{(parabola)} \\ y \leq 2 - x^2 & \text{(parabola)} \end{cases}$$

The region must be on or above the parabola $y = x^2 - 2$ and on or below the parabola $y = 2 - x^2$.

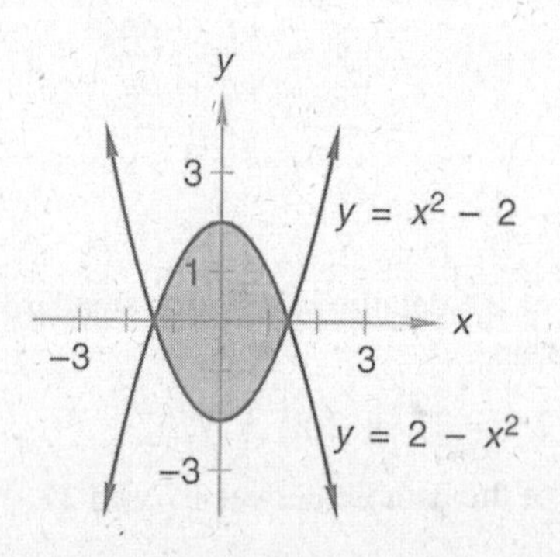

Test point = (0, 0)

$y \geq x^2 - 2$ | $y \leq 2 - x^2$
$0 \geq 0^2 - 2$ | $0 \leq 2 - 0^2$
$0 \geq -2$ True | $0 \leq 2$ True

28. (48) $\cos^2 \frac{5\pi}{4} - \sin^2 \frac{5\pi}{4} + \tan^2 \frac{5\pi}{4}$

$= \left(-\cos \frac{\pi}{4}\right)^2 - \left(-\sin \frac{\pi}{4}\right)^2 + \left(\tan \frac{\pi}{4}\right)^2$

$= \left(-\frac{1}{\sqrt{2}}\right)^2 - \left(-\frac{1}{\sqrt{2}}\right)^2 + (1)^2$

$= \frac{1}{2} - \frac{1}{2} + 1 = \mathbf{1}$

29. (48) $\tan^3 0° - \cot^2(-405°) + \csc^2(-315°)$

$= \tan^3 0° - (-\cot 45°)^2 + (\csc 45°)^2$

$= 0^3 - (-1)^2 + (\sqrt{2})^2 = 0 - 1 + 2 = \mathbf{1}$

30. (37,39) $\sqrt{(x+4)^2 + (y-6)^2} = \sqrt{(x-6)^2 + (y-8)^2}$

$(x+4)^2 + (y-6)^2 = (x-6)^2 + (y-8)^2$

$x^2 + 8x + 16 + y^2 - 12y + 36$

$= x^2 - 12x + 36 + y^2 - 16y + 64$

$20x + 4y = 48$

$\frac{5}{12}x + \frac{1}{12}y = 1$

$\mathbf{\frac{x}{\frac{12}{5}} + \frac{y}{12} = 1}$

Problem Set 65

1. (62) $C = mH + b$

(a) $1200 = m40 + b$
(b) $800 = m20 + b$

(a) $1200 = m40 + b$
-1(b) $-800 = -m20 - b$

$400 = m20$

$m = 20$

(a) $1200 = m40 + b$

$1200 = (20)40 + b$

$b = 400$

$C = 20H + 400$

$= 20(30) + 400 = \mathbf{\$1000}$

2. (45) | 1 | 5 | 5 | → $1 \cdot 5 \cdot 5 = 25$

This accounts for three-digit counting numbers less than 400 with all even digits. We also must include 400 itself, so add one.

$25 + 1 = \mathbf{26}$

3. (55) $\frac{11!}{4!4!2!} = \mathbf{34{,}650}$

4. (53) $\omega = \frac{v}{r} = \frac{45 \frac{\text{km}}{\text{hr}}}{30 \text{ cm}} = \frac{45 \text{ km}}{30 \text{ cm-hr}}$

$\omega = \left(\frac{45 \text{ km}}{30 \text{ cm-hr}}\right)\left(\frac{1000 \text{ m}}{1 \text{ km}}\right)\left(\frac{100 \text{ cm}}{1 \text{ m}}\right)\left(\frac{1 \text{ hr}}{60 \text{ min}}\right)$

$\omega = \mathbf{\frac{(45)(1000)(100)}{(30)(60)} \frac{rad}{min} = 2500 \frac{rad}{min}}$

5. (39) $42.6°\left(\frac{\pi}{180°}\right)(3960) = 937.2\pi = \mathbf{2944.30 \text{ miles}}$

6. (38) Average rate $= \frac{\text{total distance}}{\text{total time}}$

$= \frac{m + x}{h + h + 4} = \frac{m + x}{2h + 4}$

Time $= \frac{\text{distance}}{\text{average rate}}$

$= \frac{50}{\frac{m+x}{2h+4}} = \mathbf{\frac{50(2h+4)}{m+x} \text{ hr}}$

7. (44) Time $= \frac{\text{distance}}{\text{rate}}$

Kyle's time $= \frac{x}{p}$ hr

Keith's time $= \frac{x}{R}$ hr

Difference in time $= \frac{x}{p} - \frac{x}{R} = \mathbf{\frac{Rx - px}{Rp} \text{ hr}}$

8. (44) Rate $= \frac{k^2x + m}{d} \frac{\text{pencils}}{\text{dollar}}$

Pencils = rate × dollars

$= \frac{k^2x + m}{d} \cdot 500$

$= \mathbf{\frac{500(k^2x + m)}{d} \text{ pencils}}$

9. (65)

$$\cos x - \sqrt{1 - \cos^2 x} = 0$$
$$\cos x = \sqrt{1 - \cos^2 x}$$
$$\cos^2 x = 1 - \cos^2 x$$
$$2\cos^2 x = 1$$
$$\cos^2 x = \frac{1}{2}$$
$$\left(\cos x - \frac{1}{\sqrt{2}}\right)\left(\cos x + \frac{1}{\sqrt{2}}\right) = 0$$

$$\cos x - \frac{1}{\sqrt{2}} = 0 \qquad \cos x + \frac{1}{\sqrt{2}} = 0$$
$$\cos x = \frac{1}{\sqrt{2}} \qquad \cos x = -\frac{1}{\sqrt{2}}$$

Discard $-\frac{1}{\sqrt{2}}$ because when it is substituted for $\cos x$ in the original equation, $-\frac{2}{\sqrt{2}} \neq 0$.

$x =$ **45°, 315°**

10. (65)

$$\left(\sin x - \frac{\sqrt{3}}{2}\right)\left(\sin x + \frac{1}{2}\right) = 0$$

$$\sin x - \frac{\sqrt{3}}{2} = 0 \qquad \sin x + \frac{1}{2} = 0$$
$$\sin x = \frac{\sqrt{3}}{2} \qquad \sin x = -\frac{1}{2}$$
$$x = \frac{\pi}{3}, \frac{2\pi}{3} \qquad x = \frac{7\pi}{6}, \frac{11\pi}{6}$$

$$x = \mathbf{\frac{\pi}{3}, \frac{2\pi}{3}, \frac{7\pi}{6}, \frac{11\pi}{6}}$$

11. (65) $y = \log_2 x$

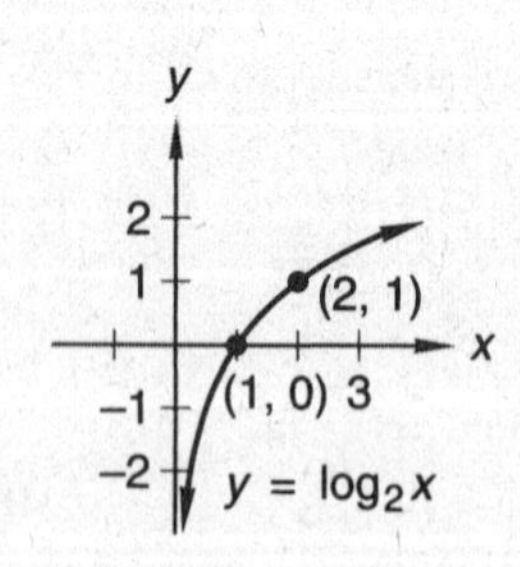

12. (64) (a) $3 + 5i$

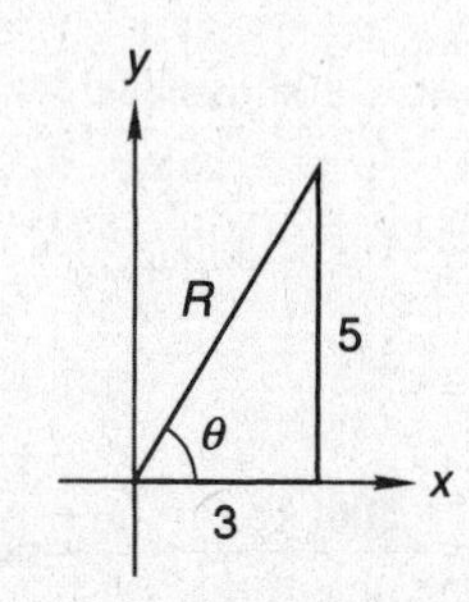

$$R = \sqrt{3^2 + 5^2} = \sqrt{34} = 5.83$$
$$\tan\theta = \frac{5}{3}$$
$$\theta = 59.04°$$

5.83 cis 59.04°

(b) $6 \text{ cis } 60° = 6(\cos 60° + i \sin 60°)$

$$= 6\left(\frac{1}{2} + \frac{\sqrt{3}}{2}i\right) = \mathbf{3 + 3\sqrt{3}i}$$

13. (64) $(5 \text{ cis } 20°)(6 \text{ cis } 70°) = 30 \text{ cis } 90°$

$$= 30(\cos 90° + i \sin 90°) = 30(0 + 1i) = \mathbf{30i}$$

14. (64) $[6 \text{ cis } (-30°)](3 \text{ cis } 90°) = 18 \text{ cis } 60°$

$$= 18(\cos 60° + i \sin 60°)$$
$$= 18\left(\frac{1}{2} + \frac{\sqrt{3}}{2}i\right) = \mathbf{9 + 9\sqrt{3}i}$$

15. (63)

$$x^2 + y^2 + 10x - 75 = 0$$
$$(x^2 + 10x + \quad) + y^2 = 75$$
$$(x^2 + 10x + 25) + y^2 = 100$$
$$\mathbf{(x + 5)^2 + y^2 = 10^2}$$

Center = (–5, 0); radius = 10

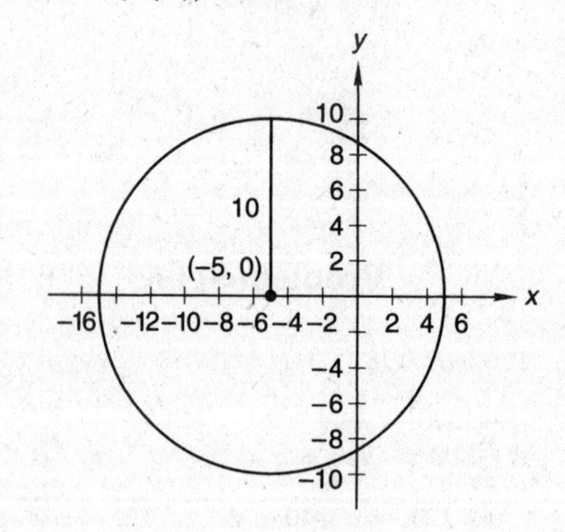

16. (62)

$$\begin{cases} \text{(a)} & ax + dy = g \\ \text{(b)} & cx + fy = h \end{cases}$$

$$\begin{array}{rl} f\text{(a)} & afx + dfy = gf \\ -d\text{(b)} & -cdx - dfy = -hd \\ \hline & x(af - cd) = gf - hd \end{array}$$

$$x = \mathbf{\frac{gf - hd}{af - cd}}$$

17. (61) Range $= 9 - 1 = \mathbf{8}$

Mean $= \dfrac{7 + 9 + 5 + 3 + 1 + 9}{6} = \mathbf{5.67}$

Median $= \dfrac{5 + 7}{2} = \mathbf{6}$

Mode $= \mathbf{9}$

$$\text{Variance} = \frac{1}{6}[(7 - 5.67)^2 + (9 - 5.67)^2 + (5 - 5.67)^2 + (3 - 5.67)^2 + (1 - 5.67)^2 + (9 - 5.67)^2] = \mathbf{8.89}$$

Standard deviation $= \sqrt{8.89} = \mathbf{2.98}$

18. (61) –1 and 1 is one standard deviation from the mean. Approximately **68%** of the data lies between –1 and 1.

19. (52) $\cos 3\theta + \dfrac{\sqrt{3}}{2} = 0$

$$\cos 3\theta = -\frac{\sqrt{3}}{2}$$

$$3\theta = 150°, 210°, 510°, 570, 870°, 930°$$

$$\theta = \mathbf{50°, 70°, 170°, 190°, 290°, 310°}$$

20. (59)

$$2 \ln x = \ln (6 - x)$$

$$\ln x^2 = \ln (6 - x)$$

$$x^2 = 6 - x$$

$$x^2 + x - 6 = 0$$

$$(x - 2)(x + 3) = 0$$

$$x = 2, -3 \qquad (x \neq -3)$$

$$x = \mathbf{2}$$

21. (59) $3 \ln x = \ln 8 + 3 \ln 2$

$$\ln x^3 = \ln 8 + \ln 2^3$$

$$\ln x^3 = \ln (8 \cdot 8)$$

$$x^3 = 64$$

$$x = \mathbf{4}$$

22. (59) $7^{2 \log_7 3} + 8^{\log_8 6 - \log_8 3} - 4 \log_9 9^{\frac{1}{2}}$

$$= 7^{\log_7 3^2} + 8^{\log_8 \frac{6}{3}} - \log_9 \left(9^{\frac{1}{2}}\right)^4$$

$$= 7^{\log_7 9} + 8^{\log_8 2} - \log_9 9^2 = 9 + 2 - 2 = \mathbf{9}$$

23. (47,57) Function $= \sin x$

Centerline $= 1$

Amplitude $= 5$

Period $= 2\pi$

Phase angle $= \dfrac{3\pi}{4}$

$$\mathbf{y = 1 + 5 \sin \left(x - \frac{3\pi}{4}\right)}$$

24. (47,57) Function $= \sin \theta$

Centerline $= 6$

Amplitude $= 10$

Period $= 7\pi$

Phase angle $= 0$

Coefficient $= \dfrac{2\pi}{7\pi} = \dfrac{2}{7}$

$$\mathbf{y = 6 + 10 \sin \frac{2}{7}\theta}$$

25. (56)

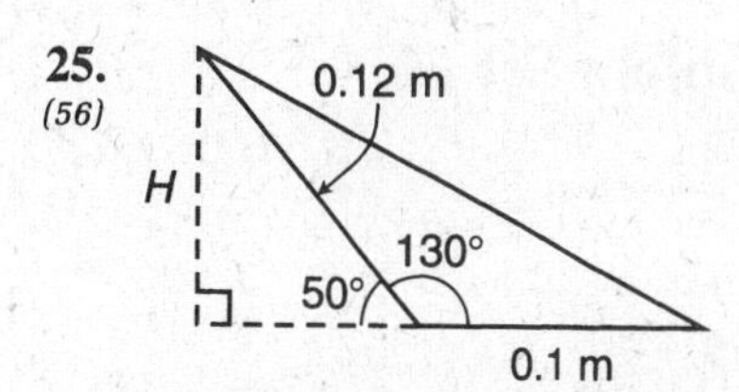

$H = 0.12 \sin 50° = 0.09193$ m

Area $= \dfrac{1}{2}BH = \dfrac{1}{2}(0.1)(0.09193) = \mathbf{0.0046\ m^2}$

26. (47)

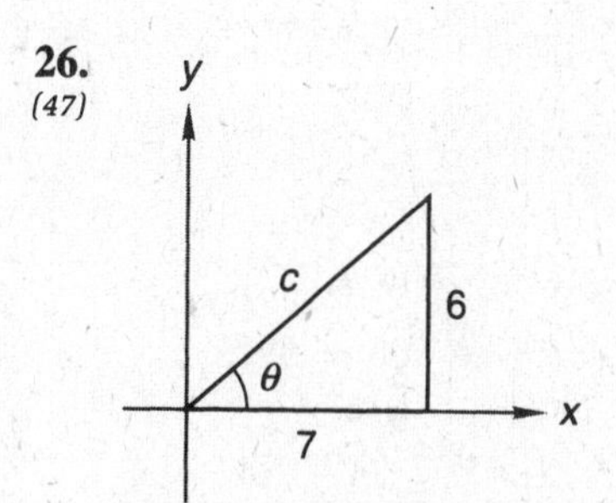

$-90° < \theta < 90°$

$c = \sqrt{6^2 + 7^2} = \sqrt{85}$

$$\cos\left(\text{Arctan}\,\frac{6}{7}\right) = \cos\theta = \frac{7}{\sqrt{85}} = \mathbf{\frac{7\sqrt{85}}{85}}$$

27. (48) $\sin^3 \dfrac{\pi}{2} - \cos^2 \dfrac{\pi}{6} + \tan^4\left(-\dfrac{\pi}{4}\right)$

$$= (1)^3 - \left(\frac{\sqrt{3}}{2}\right)^2 + (-1)^4 = 1 - \frac{3}{4} + 1 = \mathbf{\frac{5}{4}}$$

28. (53) $\left(600\,\frac{\text{km}}{\text{hr}}\right)\left(\frac{1000\text{ m}}{1\text{ km}}\right)\left(\frac{100\text{ cm}}{1\text{ m}}\right)\left(\frac{1\text{ in.}}{2.54\text{ cm}}\right)$

$\times\left(\frac{1\text{ ft}}{12\text{ in.}}\right)\left(\frac{1\text{ mi}}{5280\text{ ft}}\right)$

$= \frac{\mathbf{(600)(1000)(100)}}{\mathbf{(2.54)(12)(5280)}}\,\frac{\mathbf{mi}}{\mathbf{hr}} = \mathbf{372.82\,\frac{mi}{hr}}$

29. (46) $x^2 + 3x + 7 = 0$

$x = \frac{-3 \pm \sqrt{3^2 - 4(1)(7)}}{2(1)}$

$x = \frac{-3 \pm \sqrt{9 - 28}}{2} = \frac{-3 \pm \sqrt{19}i}{2}$

$\left(\mathbf{x + \frac{3}{2} - \frac{\sqrt{19}}{2}i}\right)\left(\mathbf{x + \frac{3}{2} + \frac{\sqrt{19}}{2}i}\right)$

30. (31) The graph of g is the graph of f translated ten units to the left and two units down. Therefore the answer is **D.**

Problem Set 66

1. (62) $C = mL + b$

$\begin{cases} \text{(a)}\ 1350 = m10 + b \\ \text{(b)}\ 2200 = m20 + b \end{cases}$

$$\begin{aligned} \text{(b)}\quad 2200 &= m20 + b \\ -2\text{(a)}\quad -2700 &= -m20 - 2b \\ \hline -500 &= -b \\ b &= 500 \end{aligned}$$

$$\begin{aligned} \text{(a)}\quad 1350 &= m10 + b \\ 1350 &= m10 + (500) \\ m10 &= 850 \\ m &= 85 \end{aligned}$$

$C = 85L + 500 = 85(5) + 500 = \mathbf{\$925}$

2. (55) $\frac{9!}{6!3!} = \mathbf{84}$

3. (55) $(N - 1)! = (6 - 1)! = 5! = \mathbf{120}$

4. (53) $\omega = \frac{v}{r} = \frac{12\,\frac{\text{mi}}{\text{hr}}}{13\text{ in.}} = \frac{12\text{ mi}}{13\text{ in.-hr}}$

$\omega = \left(\frac{12\text{ mi}}{13\text{ in.-hr}}\right)\left(\frac{1\text{ hr}}{60\text{ min}}\right)\left(\frac{5280\text{ ft}}{1\text{ mi}}\right)\left(\frac{12\text{ in.}}{1\text{ ft}}\right)$

$\omega = \frac{\mathbf{(12)(5280)(12)}}{\mathbf{(60)(13)}}\,\frac{\mathbf{rad}}{\mathbf{min}} = \mathbf{974.77\,\frac{rad}{min}}$

5. (38,44) Overall average rate $= \frac{\text{total distance}}{\text{total time}}$

$= \frac{k + z}{p + p + 6}$

$= \frac{k + z}{2p + 6}\,\frac{\text{mi}}{\text{hr}}$

Time $= \frac{\text{distance}}{\text{overall average rate}}$

$= \frac{740\text{ mi}}{\frac{k + z\text{ mi}}{2p + 6\text{ hr}}} = \mathbf{\frac{740(2p + 6)}{k + z}\ hr}$

6. (44) $T_M = \frac{m}{a}$, $T_J = \frac{m}{z}$

$T_M - T_J = \frac{m}{a} - \frac{m}{z} = \mathbf{\frac{mz - ma}{az}\ hr}$

7. (44) Rate $= \frac{p^2k}{m}\,\frac{\text{cars}}{\text{dollars}}$

Cars = rate × dollars

$= \frac{p^2k}{m}\cdot 10{,}000 = \mathbf{\frac{10{,}000p^2k}{m}\ cars}$

8. (66) Function $= -\cos x$

Centerline $= -1$

Amplitude $= 8$

Period $= 180°$

Phase angle $= 0°$

Coefficient $= \frac{360°}{180°} = 2$

$\mathbf{y = -1 - 8\cos 2x}$

9. (66) Function $= \sin x$

Centerline $= 3$

Amplitude $= 5$

Period $= 2\pi$

Phase angle $= \frac{3\pi}{4}$

$\mathbf{y = 3 + 5\sin\left(x - \frac{3\pi}{4}\right)}$

10. (65)

$$\tan x - \sqrt{1 - 2\tan^2 x} = 0$$
$$\tan x = \sqrt{1 - 2\tan^2 x}$$
$$\tan^2 x = 1 - 2\tan^2 x$$
$$3\tan^2 x = 1$$
$$\tan^2 x = \frac{1}{3}$$
$$\tan^2 x - \frac{1}{3} = 0$$
$$\left(\tan x + \frac{1}{\sqrt{3}}\right)\left(\tan x - \frac{1}{\sqrt{3}}\right) = 0$$

$$\tan x + \frac{1}{\sqrt{3}} = 0 \qquad \tan x - \frac{1}{\sqrt{3}} = 0$$
$$\tan x = -\frac{1}{\sqrt{3}} \qquad \tan x = \frac{1}{\sqrt{3}}$$

Discard $-\frac{1}{\sqrt{3}}$ because if we substitute it for $\tan x$ in the original equation, $\tan x - \sqrt{1 - 2\tan^2 x}$ does not equal 0.

$x = \mathbf{30°, 210°}$

11. (60) $\left(\cos x - \frac{1}{2}\right)\left(\sin x + \frac{1}{2}\right) = 0$

$$\cos x - \frac{1}{2} = 0 \qquad \sin x + \frac{1}{2} = 0$$
$$\cos x = \frac{1}{2} \qquad \sin x = -\frac{1}{2}$$
$$x = \frac{\pi}{3}, \frac{5\pi}{3} \qquad x = \frac{7\pi}{6}, \frac{11\pi}{6}$$

$$x = \boldsymbol{\frac{\pi}{3}, \frac{7\pi}{6}, \frac{5\pi}{3}, \frac{11\pi}{6}}$$

12. (65) $y = \log_3 x$

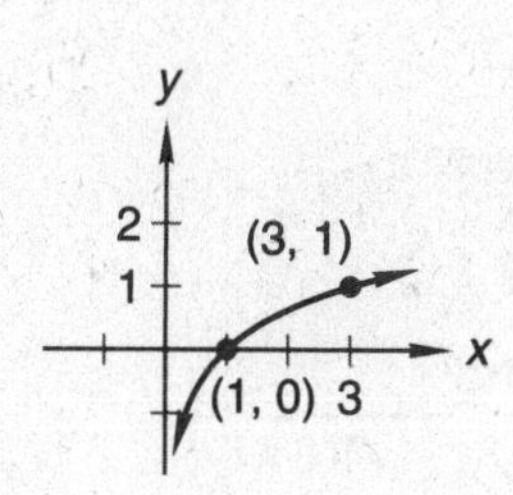

13. (64) (a) $4 + 3i$

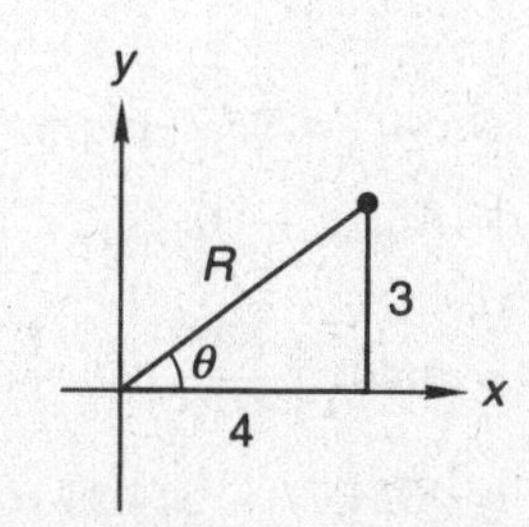

$$R = \sqrt{4^2 + 3^2} = \sqrt{25} = 5$$
$$\tan\theta = \frac{3}{4}$$
$$\theta = 36.87°$$

5 cis 36.87°

(b) $5 \text{ cis } 150° = 5(\cos 150° + i\sin 150°)$

$= 5(-\cos 30° + i\sin 30°)$

$= 5\left(-\frac{\sqrt{3}}{2} + \frac{1}{2}i\right) = \boldsymbol{-\frac{5}{2}\sqrt{3} + \frac{5}{2}i}$

14. (64) $(3 \text{ cis } 40°)[2 \text{ cis } (-50°)] = 6 \text{ cis } (-10°) = \mathbf{6\ cis\ 350°}$

15. (64) $[8 \text{ cis } (-450°)](2 \text{ cis } 60°)$

$= 16 \text{ cis } (-390°)$

$= 16[\cos(-390°) + i\sin(-390°)]$

$= 16(\cos 30° - i\sin 30°)$

$= 16\left(\frac{\sqrt{3}}{2} - \frac{1}{2}i\right) = \mathbf{8\sqrt{3} - 8i}$

16. (63)

$$x^2 - 8x + y^2 + 6y = -16$$
$$(x^2 - 8x \quad) + (y^2 + 6y \quad) = -16$$
$$(x^2 - 8x + 16) + (y^2 + 6y + 9) = 9$$
$$\mathbf{(x - 4)^2 + (y + 3)^2 = 3^2}$$

Center = (4, –3); radius = 3

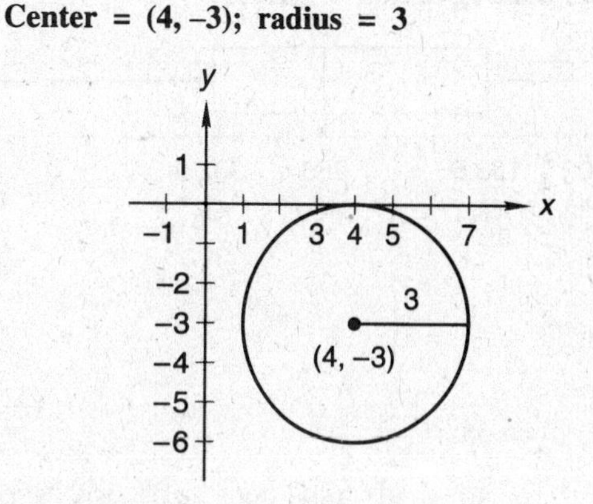

17. (62)

(a) $\begin{cases} ax + by = c \\ dx + fy = g \end{cases}$ (b)

$$f(\text{a}) \quad afx + bfy = fc$$
$$-b(\text{b}) \quad -bdx - bfy = -bg$$
$$\overline{x(af - bd) = cf - bg}$$
$$\mathbf{x = \frac{cf - bg}{af - bd}}$$

18. (66) $a = 3,\ b = 2,\ c = 6,\ d = 2,\ f = -4,\ g = 12$

$$x = \frac{cf - bg}{af - bd}$$

$$x = \frac{(6)(-4) - (2)(12)}{(3)(-4) - (2)(2)}$$

$$x = \frac{-24 - 24}{-12 - 4} = \frac{-48}{-16} = \mathbf{3}$$

19. (61) Range = **7**

$$\text{Mean} = \frac{2 + 7 + 6 + 1 + 0 + 2}{6} = \frac{18}{6} = \mathbf{3}$$

$$\text{Median} = \frac{2 + 2}{2} = \mathbf{2}$$

Mode = **2**

Variance

$$= \frac{(-1)^2 + 4^2 + 3^2 + (-2)^2 + (-3)^2 + (-1)^2}{6}$$

$$= \frac{40}{6} = \mathbf{6.67}$$

Standard deviation = $\sqrt{6.67}$ = **2.58**

20. (61) We use a calculator to get

```
 x̄ = 264.13        Q1 = 163.5
σx = 110.25       Med = 266
min = 102          Q3 = 337.5
max = 491
```

Mean = 264.13

Standard deviation = 110.25

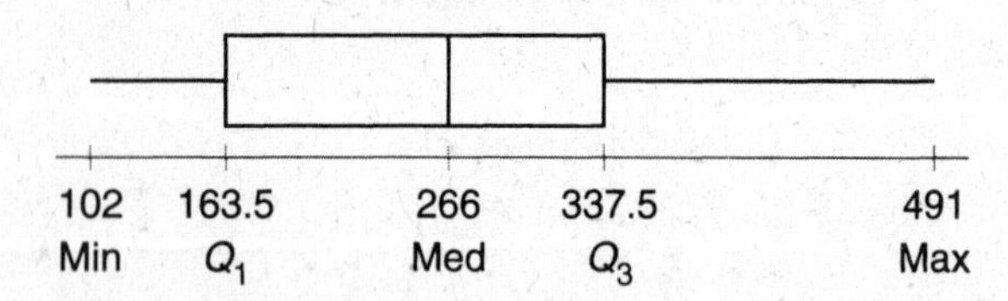

21. (52) $\sin 3\theta - \frac{1}{2} = 0$

$$\sin 3\theta = \frac{1}{2}$$

$$3\theta = 30°, 150°, 390°, 510°, 750°, 870°$$

$$\theta = \mathbf{10°, 50°, 130°, 170°, 250°, 290°}$$

22. (59)

$$2 \ln x = \ln (6x - 8)$$

$$\ln x^2 = \ln (6x - 8)$$

$$x^2 = 6x - 8$$

$$x^2 - 6x + 8 = 0$$

$$(x - 4)(x - 2) = 0$$

$$x = \mathbf{2, 4}$$

23. (59) $3 \log_{12} x = \frac{3}{4} \log_{12} 16 + \frac{3}{2} \log_{12} 4$

$$\log_{12} x^3 = \log_{12} 16^{\frac{3}{4}} + \log_{12} 4^{\frac{3}{2}}$$

$$\log_{12} x^3 = \log_{12} 8 + \log_{12} 8$$

$$\log_{12} x^3 = \log_{12} 64$$

$$x^3 = 64$$

$$x = \mathbf{4}$$

24. (59) $3^{2 \log_3 7 - 3 \log_3 2} + 4 \log_3 3^3$

$$= 3^{\log_3 7^2 - \log_3 2^3} + \log_3 (3^3)^4$$

$$= 3^{\log_3 \frac{49}{8}} + \log_3 3^{12}$$

$$= \frac{49}{8} + 12 = \frac{49 + 96}{8} = \mathbf{\frac{145}{8}}$$

25. (56)

4 m
2 m
80°
H
5 m

$H = 2 \sin 80° = 1.9696$ m

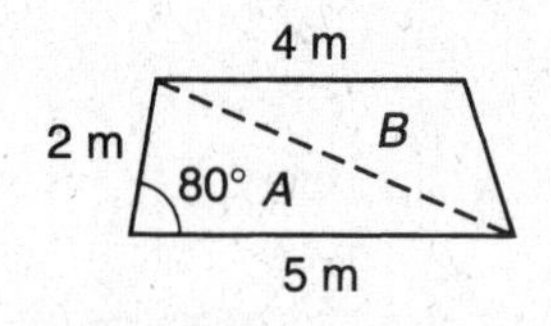

Total Area = Area A + Area B

$$= \frac{1}{2}(5)(1.9696) + \frac{1}{2}(4)(1.9696)$$

$$= 4.92 + 3.94 = \mathbf{8.86\ m^2}$$

26. (56)

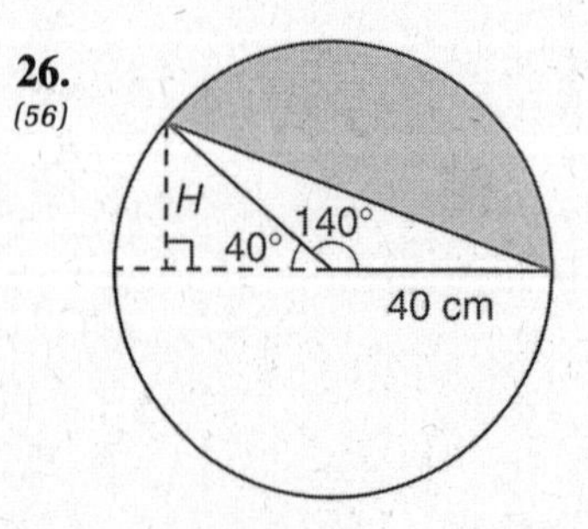

$H = 40 \sin 40° = 25.7115$ cm

$$A_{\text{segment}} = A_{\text{sector}} - A_{\text{triangle}}$$

$$= \pi(40)^2\left(\frac{140°}{360°}\right) - \frac{1}{2}(40)(25.7115)$$

$$= 1954.77 - 514.23 = \mathbf{1440.54\ cm^2}$$

27. (48) $\cot^2 \frac{4\pi}{3} - \tan^2 \frac{7\pi}{6} - \sec^3\left(-\frac{\pi}{3}\right)$

$= \cot^2 \frac{\pi}{3} - \tan^2 \frac{\pi}{6} - \sec^3 \frac{\pi}{3}$

$= \left(\frac{1}{\sqrt{3}}\right)^2 - \left(\frac{1}{\sqrt{3}}\right)^2 - \left(\frac{2}{1}\right)^3$

$= \frac{1}{3} - \frac{1}{3} - 8 = \mathbf{-8}$

28. (48) $\sin^2(-420°) + \sin^2 420° + \cos^2 270°$

$= \sin^2(-60°) + \sin^2(60°) + \cos^2 270°$

$= (-\sin 60°)^2 + (\sin 60°)^2 + \cos^2 270°$

$= \left(-\frac{\sqrt{3}}{2}\right)^2 + \left(\frac{\sqrt{3}}{2}\right)^2 + 0$

$= \frac{3}{4} + \frac{3}{4} = \mathbf{\frac{3}{2}}$

29. (46) $\left[x - (2 + \sqrt{5}i)\right]\left[x - (2 - \sqrt{5}i)\right] = 0$

$(x - 2 - \sqrt{5}i)(x - 2 + \sqrt{5}i) = 0$

$x^2 - 2x + \sqrt{5}xi - 2x + 4 - 2\sqrt{5}i$

$- \sqrt{5}xi + 2\sqrt{5}i - 5i^2 = 0$

$\mathbf{x^2 - 4x + 9 = 0}$

30. (21) $f(x) = \frac{1}{2x}$

$\dfrac{\frac{1}{2(x+h)} - \frac{1}{2(x-h)}}{2h} = \dfrac{\frac{(x-h)-(x+h)}{2(x+h)(x-h)}}{2h}$

$= \frac{-2h}{4h(x+h)(x-h)} = \mathbf{\frac{-1}{2(x+h)(x-h)}}$

Problem Set 67

1. (45) | 5 | 4 | 3 | $\rightarrow$ $5 \cdot 4 \cdot 3 = \mathbf{60}$

2. (36) (a) $\begin{cases}(B + W)T_D = D_D\\(B - W)T_U = D_U\end{cases}$ (b)

$B = W + 16,\ T_D = T_U$

(b) $(B - W)T = D_U$

$[(W + 16) - W]T = 32$

$16T = 32$

$T = 2$

(a) $(B + W)T = D_D$

$[(W + 16) + W](2) = 48$

$(2W + 16)(2) = 48$

$4W + 32 = 48$

$4W = 16$

$\mathbf{W = 4}$ **mph**

$B = W + 16$

$B = (4) + 16$

$\mathbf{B = 20}$ **mph**

3. (25) (a) $\begin{cases}T_F R_F = D_F\\T_H R_H = D_H\end{cases}$ (b)

$R_H = R_F + 200,\ T_F = 3T_H$

$R_H = R_F + 200$

$R_F = R_H - 200$

(a) $T_F R_F = D_F$

$(3T_H)(R_H - 200) = 1800$

$3T_H R_H - 600T_H = 1800$

$3(1200) - 600T_H = 1800$

$-600T_H = -1800$

$\mathbf{T_H = 3}$ **hr**

$T_F = 3T_H$

$T_F = 3(3)$

$\mathbf{T_F = 9}$ **hr**

(a) $T_F R_F = D_F$

$(9)R_F = 1800$

$\mathbf{R_F = 200}$ **mph**

$R_H = R_F + 200$

$R_H = (200) + 200$

$\mathbf{R_H = 400}$ **mph**

4. (18) (a) $\begin{cases}3e + 5f = \$5.50\\4e + 2f = \$5.00\end{cases}$ (b)

2(a) $6e + 10f = \$11.00$

−5(b) $-20e - 10f = -\$25.00$

$-14e = -\$14.00$

$\mathbf{e = \$1.00}$ **per dozen eggs**

(a) $3e + 5f = \$5.50$

$3(1) + 5f = \$5.50$

$5f = \$2.50$

$\mathbf{f = 50¢}$ **per lb of flour**

5. (18)
$$\begin{cases} \text{(a)}\ T = U + 1 \\ \text{(b)}\ 10U + T = 10T + U - T \end{cases}$$

(b)
$$\begin{aligned} 10U + T &= 10T + U - T \\ 9U &= 8T \\ 9U &= 8(U + 1) \\ 9U &= 8U + 8 \\ U &= 8 \end{aligned}$$

(a) $T = U + 1 = (8) + 1 = 9$

The original number was **98.**

6. (18)
$$\begin{cases} \text{(a)}\ U = 2T + 1 \\ \text{(b)}\ 10U + T + 10T + U = 77 \end{cases}$$

(b)
$$\begin{aligned} 10U + T + 10T + U &= 77 \\ 11U + 11T &= 77 \\ 11(2T + 1) + 11T &= 77 \\ 22T + 11 + 11T &= 77 \\ 33T &= 66 \\ T &= 2 \end{aligned}$$

(a) $U = 2T + 1 = 2(2) + 1 = 5$

The original number was **25.**

7. (67) (a) $\text{antilog}_2 5.31 = 2^{5.31} = \mathbf{39.67}$

(b) $\text{antilog}_{10} 5.31 = 10^{5.31} = \mathbf{204{,}173.79}$

8. (67) (a) $\text{antilog}_e 2 = e^2$

We estimate this as $3^2 = \mathbf{9}$.

(b) $\text{antilog}_e (-2) = \dfrac{1}{e^2}$

We estimate this as $3^{-2} = \dfrac{1}{3^2} = \mathbf{\dfrac{1}{9}}$.

9. (65) $y = \ln x$

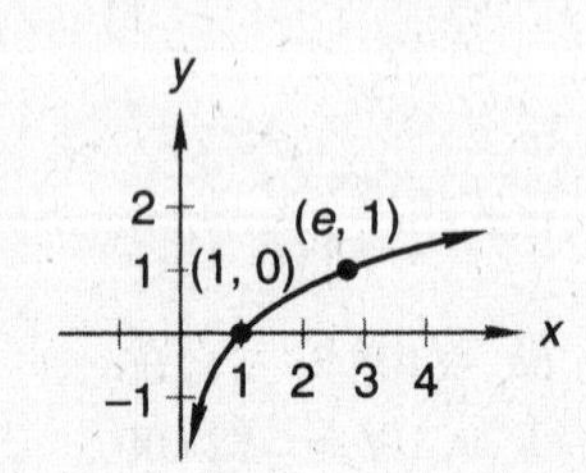

10. (66) Function = $\sin\theta$

Centerline = 4

Amplitude = 6

Period = 3π

Phase angle = 0

$$\text{Coefficient} = \frac{2\pi}{3\pi} = \frac{2}{3}$$

$$\mathbf{y = 4 + 6 \sin \frac{2}{3}\theta}$$

11. (66) Function = $\cos x$

Centerline = 3

Amplitude = 5

Period = 720°

Phase angle = 270°

$$\text{Coefficient} = \frac{360°}{720°} = \frac{1}{2}$$

$$\mathbf{y = 3 + 5 \cos \frac{1}{2}(x - 270°)}$$

12. (64) (a) $5 - 12i$

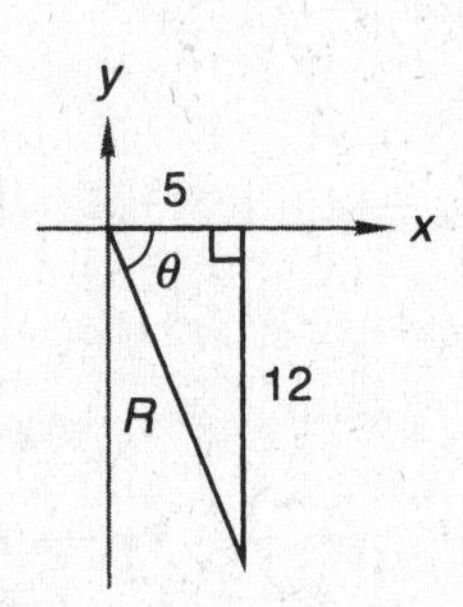

$$R = \sqrt{5^2 + 12^2} = \sqrt{169} = 13$$

$$\tan\theta = \frac{12}{5}$$

$$\theta = 67.38°$$

The polar angle is $360° - 67.38° = 292.62°$.

13 cis 292.62°

(b) $7 \text{ cis } 210° = 7(\cos 210° + i \sin 210°)$

$$= 7\left[-\frac{\sqrt{3}}{2} + i\left(-\frac{1}{2}\right)\right] = \mathbf{-\frac{7\sqrt{3}}{2} - \frac{7}{2}i}$$

13. (64) $(4 \text{ cis } 20°)(2 \text{ cis } 40°) = 8 \text{ cis } 60°$

$$= 8(\cos 60° + i \sin 60°)$$

$$= 8\left(\frac{1}{2} + \frac{\sqrt{3}}{2}i\right) = \mathbf{4 + 4\sqrt{3}i}$$

14. (64) $[3 \text{ cis } (-200°)](2 \text{ cis } 50°) = 6 \text{ cis } (-150°)$

$$= 6[\cos(-150°) + i \sin(-150°)]$$

$$= 6\left[-\frac{\sqrt{3}}{2} - i\left(\frac{1}{2}\right)\right] = \mathbf{-3\sqrt{3} - 3i}$$

15. (63)

$$x^2 + y^2 - 4x + 6y - 3 = 0$$

$$(x^2 - 4x + \quad) + (y^2 + 6y + \quad) = 3$$

$$(x^2 - 4x + 4) + (y^2 + 6y + 9) = 16$$

$$\mathbf{(x - 2)^2 + (y + 3)^2 = 4^2}$$

Center = (2, –3); radius = 4

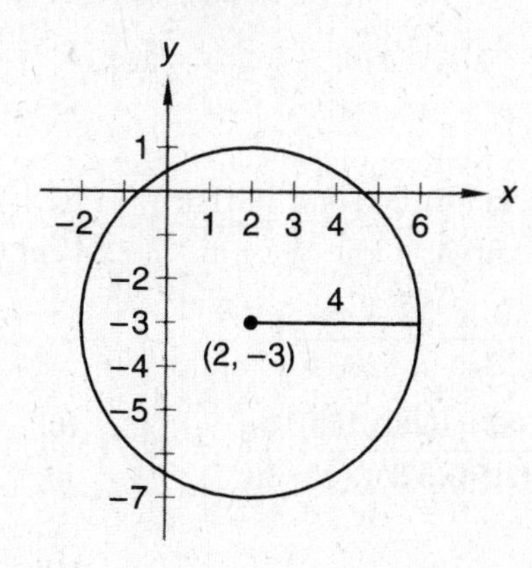

16. (62)

$$\begin{cases} \text{(a)} & cx + fy = a \\ \text{(b)} & dx + gy = b \end{cases}$$

$$\begin{aligned} -d\text{(a)} \quad & -cdx - dfy = -ad \\ c\text{(b)} \quad & \underline{cdx + cgy = bc} \\ & y(cg - df) = bc - ad \end{aligned}$$

$$\mathbf{y = \frac{bc - ad}{cg - df}}$$

17. (66) $a = 5,\ b = 7,\ c = 2,\ d = 3,\ f = -3,\ g = 2$

$$y = \frac{bc - ad}{cg - df}$$

$$= \frac{(7)(2) - (5)(3)}{(2)(2) - (3)(-3)}$$

$$= \frac{14 - 15}{4 + 9} = \mathbf{-\frac{1}{13}}$$

18. (61) 95% lie between –2 and 2, and the distribution is symmetric about 0. So **47.5%** lie between 0 and 2.

19. (61)

STEM	LEAF
1	7
2	
3	
4	6
5	6, 8
6	4, 6, 8
7	3, 5, 0, 1
8	3, 7, 5, 2
9	2

$\bar{x} = 68.31$ $\quad Q_1 = 61$
$\sigma x = 17.82$ $\quad$ Med = 70.5
min = 17 $\quad Q_3 = 82.5$
max = 92

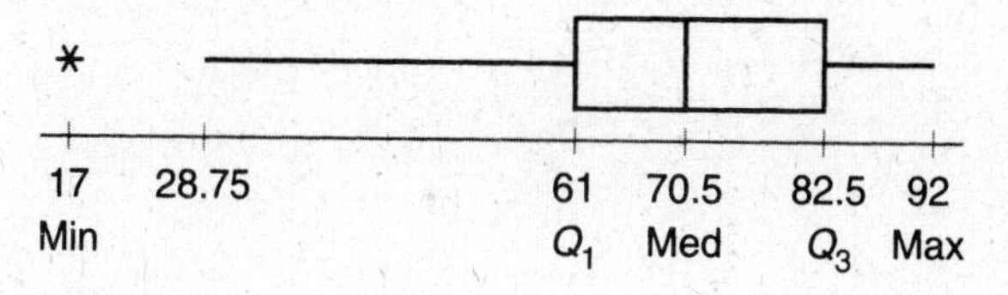

20. (60) $(\tan x + 1)\left(\cos x - \frac{1}{2}\right) = 0$

$$\tan x + 1 = 0 \qquad\qquad \cos x - \frac{1}{2} = 0$$

$$\tan x = -1 \qquad\qquad \cos x = \frac{1}{2}$$

$$x = \frac{3\pi}{4}, \frac{7\pi}{4} \qquad\qquad x = \frac{\pi}{3}, \frac{5\pi}{3}$$

$$x = \mathbf{\frac{\pi}{3}, \frac{3\pi}{4}, \frac{5\pi}{3}, \frac{7\pi}{4}}$$

21. (65) $\sqrt{\sin\theta} - \frac{\sqrt{2}}{2} = 0$

$$\sqrt{\sin\theta} = \frac{\sqrt{2}}{2}$$

$$\sin\theta = \frac{2}{4}$$

$$\theta = \frac{1}{2}$$

$$\theta = \mathbf{30°, 150°}$$

22. (52) $\cos 3\theta + \frac{1}{2} = 0$

$$\cos 3\theta = -\frac{1}{2}$$

$$3\theta = 120°, 240°, 480°, 600°, 840°, 960°$$

$$\theta = \mathbf{40°, 80°, 160°, 200°, 280°, 320°}$$

23. (48) $\tan^2(-405°) - \sec^2(-405°) + \sin^2 495°$

$$= (-\tan 45°)^2 - (\sec 45°)^2 + (\sin 45°)^2$$

$$= (-1)^2 - \left(\frac{\sqrt{2}}{1}\right)^2 + \left(\frac{1}{\sqrt{2}}\right)^2$$

$$= 1 - 2 + \frac{1}{2} = \mathbf{-\frac{1}{2}}$$

24. (59)

$$\ln x + \ln x = \ln(4x - 3)$$
$$\ln x^2 = \ln(4x - 3)$$
$$x^2 = 4x - 3$$
$$x^2 - 4x + 3 = 0$$
$$(x - 1)(x - 3) = 0$$
$$x = \mathbf{1, 3}$$

25. (59) $3 \log_7 x = \log_7 81 - \log_7 3$

$$\log_7 x^3 = \log_7 \left(\frac{81}{3}\right)$$
$$x^3 = 27$$
$$x = \mathbf{3}$$

26. (59) $2^{\log_2 3} - 3^{2 \log_3 2} + 5^{\log_5 4 + \log_5 6} - 4 \log 10^5$

$$= 3 - 3^{\log_3 2^2} + 5^{\log_5 4(6)} - \log 10^{20}$$
$$= 3 - 4 + 24 - 20 = \mathbf{3}$$

27. (56) $\begin{cases} y \geq (x - 3)^2 + 2 & \text{(parabola)} \\ y > x^2 & \text{(parabola)} \end{cases}$

The region must be on or above the parabola $y = (x - 3)^2 + 2$ and above the parabola $y = x^2$.

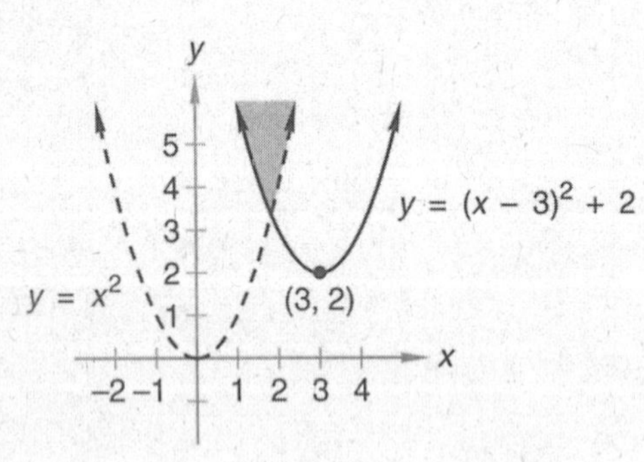

28. (56)

H
40 cm
135°
45°
28 cm

$$H = 40 \sin 45° = 28.2843 \text{ cm}$$

$$\text{Area} = \frac{1}{2}BH = \frac{1}{2}(28)(28.2843) = \mathbf{395.98\ cm^2}$$

29. (46) $x^2 - 4x + 6 = 0$

$$x = \frac{4 \pm \sqrt{(-4)^2 - 4(1)(6)}}{2(1)}$$

$$x = \frac{4 \pm \sqrt{16 - 24}}{2} = \frac{4 \pm 2\sqrt{2}i}{2} = 2 \pm \sqrt{2}i$$

$$\mathbf{(x - 2 - \sqrt{2}i)(x - 2 + \sqrt{2}i)}$$

30. (53) $\left(80 \frac{\text{km}}{\text{hr}}\right)\left(\frac{1000 \text{ m}}{1 \text{ km}}\right)\left(\frac{100 \text{ cm}}{1 \text{ m}}\right)\left(\frac{1 \text{ in.}}{2.54 \text{ cm}}\right)$

$$\times \left(\frac{1 \text{ ft}}{12 \text{ in.}}\right)\left(\frac{1 \text{ mi}}{5280 \text{ ft}}\right)$$

$$= \mathbf{\frac{(80)(1000)(100)}{(2.54)(5280)(12)} \frac{mi}{hr} = 49.71 \frac{mi}{hr}}$$

Problem Set 68

1. (62) $C = mN + b$

$$\begin{aligned} &\text{(a)} \quad 250 = m20 + b \\ &\text{(b)} \quad 325 = m30 + b \end{aligned}$$

$$\begin{aligned} 3\text{(a)} \quad & 750 = m60 + 3b \\ -2\text{(b)} \quad & -650 = -m60 - 2b \\ \hline & 100 = b \end{aligned}$$

$$\text{(a)} \quad 250 = m20 + b$$
$$250 = m20 + (100)$$
$$m20 = 150$$
$$m = 7.5$$

$$C = 7.5N + 100 = 7.5(50) + 100 = \mathbf{\$475}$$

2. (55) $\dfrac{9!}{5!4!} = \mathbf{126}$

3. (55) $\dfrac{6!}{3!2!} = \mathbf{60}$

4. (53) $r = \dfrac{v}{\omega}$

$$r = \frac{15 \frac{\text{mi}}{\text{hr}}}{400 \frac{\text{rad}}{\text{min}}} = \frac{15 \text{ mi-min}}{400 \text{ hr}}$$

$$r = \left(\frac{15 \text{ mi-min}}{400 \text{ hr}}\right)\left(\frac{1 \text{ hr}}{60 \text{ min}}\right)\left(\frac{5280 \text{ ft}}{1 \text{ mi}}\right)\left(\frac{12 \text{ in.}}{1 \text{ ft}}\right)$$

$$r = \mathbf{\frac{(15)(5280)(12)}{(60)(400)} \text{ in.} = 39.60 \text{ in.}}$$

5. (44) Average rate $= \dfrac{\text{total distance}}{\text{total time}}$

$$= \frac{m + s}{u + u + 2} = \frac{m + s}{2u + 2}$$

$$\text{Time} = \frac{100}{\dfrac{m + s}{2u + 2}} = \mathbf{\frac{100(2u + 2)}{m + s}\ hr}$$

6. (44) Rate $= \dfrac{w^3a^2}{g} \dfrac{\text{stereos}}{\text{dollars}}$

Rate × price = stereos

$$\left(\frac{w^3a^2}{g} \frac{\text{stereos}}{\text{dollars}}\right)(750 \text{ dollars}) = \mathbf{\frac{750w^3a^2}{g}\ stereos}$$

7. (7) $N, N + 1, N + 2, N + 3$

$$(N + 1)(N + 3) = 2(N)(N + 2) + 3$$
$$N^2 + 4N + 3 = 2N^2 + 4N + 3$$
$$N^2 = 0$$
$$N = 0$$

0, 1, 2, 3

8. (68) Directrix: $y = k - p = -3$

Focus: $(h, k + p) = (0, 3)$

$$k + p = 3$$
$$k - p = -3$$
$$2k = 0$$
$$k = 0$$

Vertex: $(h, k) = (0, 0)$

$$k + p = 3$$
$$(0) + p = 3$$
$$p = 3$$

$$y - k = \frac{1}{4p}(x - h)^2$$
$$y - (0) = \frac{1}{4(3)}[x - (0)]^2$$
$$y = \frac{1}{12}x^2$$

Parabola: $\mathbf{y = \frac{1}{12}x^2}$

Vertex: (0, 0)

9. (68) Vertex: $(h, k) = (0, 1)$

Focus: $(h, k + p) = (0, 3)$

$$k + p = 3$$
$$(1) + p = 3$$
$$p = 2$$

Directrix: $y = k - p = (1) - (2) = -1$

Axis of symmetry: $x = h = 0$

$$y - k = \frac{1}{4p}(x - h)^2$$
$$y - (1) = \frac{1}{4(2)}[x - (0)]^2$$
$$y - 1 = \frac{1}{8}x^2$$
$$y = \frac{1}{8}x^2 + 1$$

Parabola: $\mathbf{y = \frac{1}{8}x^2 + 1}$

Directrix: $\mathbf{y = -1}$

Axis of symmetry: $\mathbf{x = 0}$

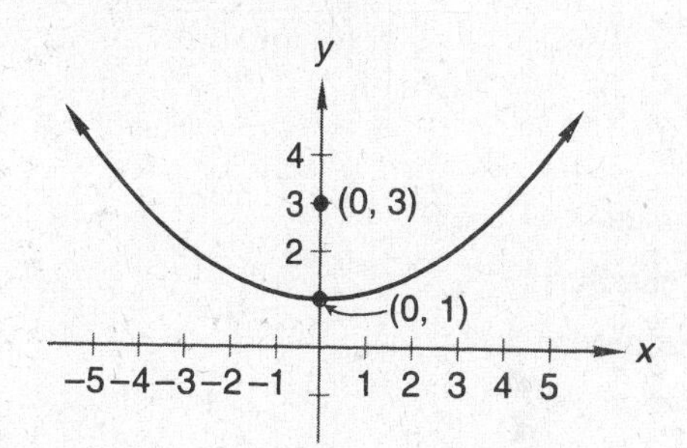

10. (68) $y = \dfrac{1}{100}x^2$

$$a = \frac{1}{4p}$$
$$\frac{1}{100} = \frac{1}{4p}$$
$$p = 25$$

The receiver should be placed **25 ft** above the vertex.

11. (61) (a) Approximately **68%** of the scores lie between 74 and 82, which is within one standard deviation of the mean.

(b) The scores are symmetrically distributed around the mean, so **34%** of the scores are between 78 and 82.

12. (61)

STEM	LEAF
2	53, 94, 11, 75, 92, 52
3	78, 12, 87, 10, 66, 96, 51
4	22, 08, 00

$\bar{x} = 331.69$ Q1 = 283.5
$\sigma x = 63.10$ Med = 331.5
min = 211   Q3 = 391.5
max = 422

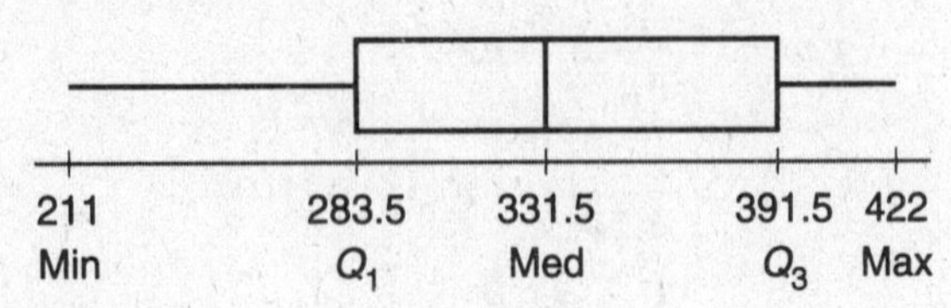

13. (65) $y = \log_{\frac{3}{2}} x$

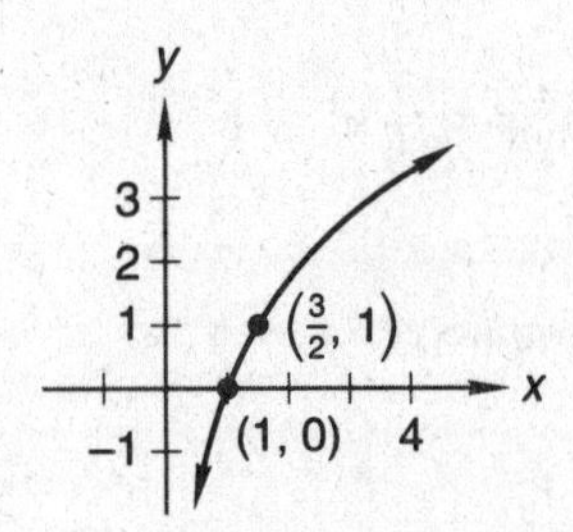

14. (66) Function = $\cos\theta$

Centerline = 3

Amplitude = 3

Period = 2

Phase angle = 0

Coefficient = $\frac{2\pi}{2} = \pi$

$\mathbf{y = 3 + 3\cos \pi\theta}$

15. (66) Function = $-\sin x$

Centerline = 1

Amplitude = 3

Period = 100°

Phase angle = 0°

Coefficient = $\frac{360°}{100°} = \frac{18}{5}$

$\mathbf{y = 1 - 3\sin \frac{18}{5}x}$

16. (64) $(3 \text{ cis } 80°)(5 \text{ cis } 310°) = 15 \text{ cis } 390°$

$= 15 \text{ cis } 30° = 15(\cos 30° + i \sin 30°)$

$= 15\left(\frac{\sqrt{3}}{2} + i\frac{1}{2}\right) = \mathbf{\frac{15\sqrt{3}}{2} + \frac{15}{2}i}$

17. (63)

$$x^2 + y^2 + 10x - 2y = 1$$
$$(x^2 + 10x + \quad) + (y^2 - 2y + \quad) = 1$$
$$(x^2 + 10x + 25) + (y^2 - 2y + 1) = 27$$
$$(x + 5)^2 + (y - 1)^2 = (\sqrt{27})^2$$

Center = (–5, 1); radius = $\sqrt{27}$

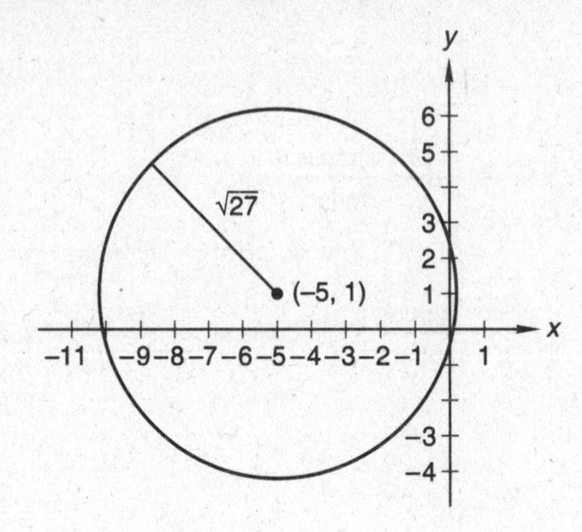

18. (62)

(a) $ax - dy = s$
(b) $bx + wy = k$

w(a) $\quad wax - wdy = ws$

d(b) $\quad bdx + wdy = kd$

$x(wa + bd) = ws + kd$

$$\mathbf{x = \frac{ws + kd}{wa + bd}}$$

19. (66)

$$\begin{cases} 3x - 5y = 10 \\ 4x + 2y = 22 \end{cases}$$

$a = 3,\ b = 4,\ d = 5,\ w = 2,\ s = 10,\ k = 22$

$$x = \frac{ws + kd}{wa + bd}$$
$$= \frac{(2)(10) + (22)(5)}{(2)(3) + (4)(5)} = \frac{20 + 110}{6 + 20} = \frac{130}{26} = \mathbf{5}$$

20. (60) $(\tan x)(2\sin x + 1) = 0$

$\tan x = 0 \qquad 2\sin x + 1 = 0$

$x = 0, \pi \qquad 2\sin x = -1$

$\sin x = -\frac{1}{2}$

$x = \frac{7\pi}{6}, \frac{11\pi}{6}$

$\mathbf{x = 0, \pi, \frac{7\pi}{6}, \frac{11\pi}{6}}$

21. (60)

$\cos\theta - \cos\theta\tan\theta = 0$

$\cos\theta(1 - \tan\theta) = 0$

$\cos\theta = 0 \qquad\qquad 1 - \tan\theta = 0$

$\theta = 90°, 270° \qquad\qquad \tan\theta = 1$

$\theta = 45°, 225°$

Since $\tan 90°$ and $\tan 270°$ are undefined, the expression does not make sense when $\theta = 90°$ or $\theta = 270°$.

$\theta = \mathbf{45°, 225°}$

22. (52)

$\sin 4\theta + 1 = 0$

$\sin 4\theta = -1$

$4\theta = 270°, 630°, 990°, 1350°$

$\theta = \mathbf{67.5°, 157.5°, 247.5°, 337.5°}$

23. (47)

$-90° < \theta < 90°$

$\text{Arctan}\,[\tan(-120°)] = \text{Arctan}\,(\tan 60°) = \mathbf{60°}$

24. (48)

$\csc^2\left(-\frac{3\pi}{2}\right) + \tan^3 4\pi - \sin^2\left(-\frac{3\pi}{2}\right)$

$= \left(\csc\frac{\pi}{2}\right)^2 + (\tan 0)^3 - \left(\sin\frac{\pi}{2}\right)^2$

$= (1)^2 + (0)^3 - (1)^2 = \mathbf{0}$

25. (24,48)

$(f - g)(510°) = \sec^2(510°) - \tan^2(510°)$

$= \sec^2 150° - \tan^2 150°$

$= (-\sec 30°)^2 - (-\tan 30°)^2$

$= \left(-\frac{2}{\sqrt{3}}\right)^2 - \left(-\frac{1}{\sqrt{3}}\right)^2$

$= \frac{4}{3} - \frac{1}{3} = \mathbf{1}$

26. (59)

$2\ln(x - 1) = \ln(3x - 5)$

$\ln(x - 1)^2 = \ln(3x - 5)$

$(x - 1)^2 = 3x - 5$

$x^2 - 2x + 1 = 3x - 5$

$x^2 - 5x + 6 = 0$

$(x - 2)(x - 3) = 0$

$x = \mathbf{2, 3}$

27. (59)

$3\log_2 5 - \log_2 x = \log_2 x$

$\log_2 5^3 = \log_2 x + \log_2 x$

$\log_2 125 = \log_2 x^2$

$125 = x^2$

$x = \mathbf{5\sqrt{5}}$

28. (59)

$e^{3\ln 5} - 10^{\log 3 + 2\log 2} - \ln e^{-3}$

$= e^{\ln 5^3} - 10^{\log 3 + \log 2^2} - (-3)$

$= e^{\ln 125} - 10^{\log(3)(4)} + 3$

$= 125 - 12 + 3 = \mathbf{116}$

29. (53,56)

6 in.

H

50°

5 in.

$H = 6\sin 50° = 4.5963$ in.

$\text{Area} = \frac{1}{2}(5)(4.5963) = 11.491 \text{ in.}^2$

$11.491 \text{ in.}^2 \times \frac{1 \text{ ft}^2}{144 \text{ in.}^2} = \mathbf{0.080 \textbf{ ft}^2}$

30. (67)

(a) $\text{antilog}_5 3 = 5^3 = \mathbf{125}$

(b) $\text{antilog}_3 4 = 3^4 = \mathbf{81}$

Problem Set 69

1. (38)

$\begin{cases} \text{(a)}\ R_G T_G = D_G \\ \text{(b)}\ R_B T_B = D_B \end{cases}$

$T_G = 2 + T_B,\ T_G + T_B = 6$

$T_G + T_B = 6$

$(2 + T_B) + T_B = 6$

$2T_B = 4$

$T_B = 2$

$T_G = 2 + T_B = 2 + (2) = 4$

(a) $R_G T_G = D_G$

$R_G(4) = 160$

$\mathbf{R_G = 40 \textbf{ mph}}$

(b) $R_B T_B = D_B$

$R_B(2) = 160$

$\mathbf{R_B = 80 \textbf{ mph}}$

2. (18) $p = \dfrac{\sqrt{m}y^2}{x^2}$

$$\frac{\sqrt{4m}(2y)^2}{\left(\frac{x}{2}\right)^2} = \frac{2\sqrt{m}4y^2}{\frac{x^2}{4}} = 32\frac{\sqrt{m}y^2}{x^2} = 32p$$

Therefore, **p is multiplied by 32.**

3. (1) $4(90 - A) + 40 = 2(180 - A)$

$$360 - 4A + 40 = 360 - 2A$$
$$2A = 40$$
$$A = \mathbf{20°}$$

4. (28) Distance $= x$

Time $= k$

New time $= k - 20$

New rate $= \dfrac{\text{distance}}{\text{new time}} = \mathbf{\dfrac{x}{k - 20}\ \dfrac{mi}{min}}$

5. (18) $s + 1.2s + (1.2)[(1.2)s] = 29{,}120$

$$3.64s = 29{,}120$$
$$s = \mathbf{\$8000}$$

6. (69) $\begin{vmatrix} -4 & 6 \\ 5 & 2 \end{vmatrix}$

$(-4)(2) - (5)(6) = -8 - 30 = \mathbf{-38}$

7. (69) $\begin{vmatrix} x & 2 \\ 3 & x - 1 \end{vmatrix} = 4$

$$x(x - 1) - (3)(2) = 4$$
$$x^2 - x - 10 = 0$$
$$x = \frac{-(-1) \pm \sqrt{(-1)^2 - 4(1)(-10)}}{2(1)} = \mathbf{\frac{1 \pm \sqrt{41}}{2}}$$

8. (68) Vertex: $(h, k) = (0, 0)$

Focus: $(h, k + p) = \left(0, -\dfrac{3}{10}\right)$

$$k + p = -\frac{3}{10}$$
$$0 + p = -\frac{3}{10}$$
$$p = -\frac{3}{10}$$

Directrix: $y = k - p$

$$y = 0 - \left(-\frac{3}{10}\right)$$
$$y = \frac{3}{10}$$

Axis of symmetry: $x = h = 0$

$$y - k = \frac{1}{4p}(x - h)^2$$
$$y - 0 = \frac{1}{4\left(-\frac{3}{10}\right)}(x - 0)^2$$
$$y = -\frac{5}{6}x^2$$

Parabola: $\mathbf{y = -\frac{5}{6}x^2}$

Directrix: $\mathbf{y = \frac{3}{10}}$

Axis of symmetry: $\mathbf{x = 0}$

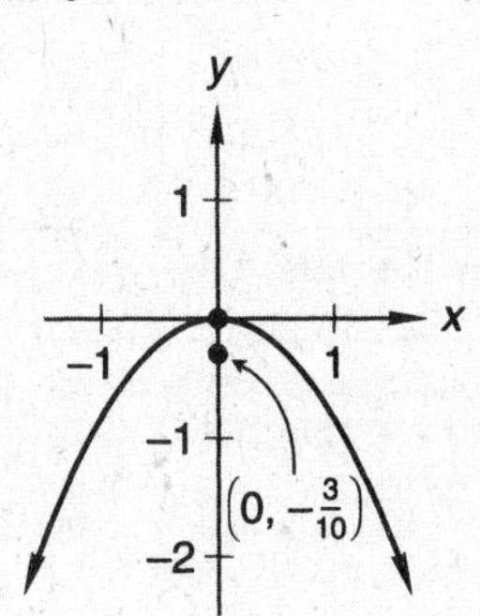

9. (68) Vertex: $(h, k) = (-3, 2)$

Focus: $(h, k + p) = (-3, -1)$

$$k + p = -1$$
$$(2) + p = -1$$
$$p = -3$$

Directrix: $y = k - p = (2) - (-3) = 5$

Axis of symmetry: $x = h = -3$

$$y - k = \frac{1}{4p}(x - h)^2$$
$$y - 2 = \frac{1}{4(-3)}[x - (-3)]^2$$
$$y = -\frac{1}{12}(x + 3)^2 + 2$$

Directrix: $\mathbf{y = 5}$

Axis of symmetry: $\mathbf{x = -3}$

Parabola: $\mathbf{y = -\frac{1}{12}(x + 3)^2 + 2}$

10. (61) Mean $= \dfrac{-2 + 3 - 4 + 5 - 1 + 0}{6} = \mathbf{0.17}$

Median $= \dfrac{-1 + 0}{2} = \mathbf{-0.5}$

Mode: All the numbers appear only once. Therefore the **mode does not exist.**

Variance

$$= \frac{1}{6}\left[\left(-2 - \frac{1}{6}\right)^2 + \left(3 - \frac{1}{6}\right)^2 + \left(-4 - \frac{1}{6}\right)^2 + \left(5 - \frac{1}{6}\right)^2 + \left(-1 - \frac{1}{6}\right)^2 + \left(0 - \frac{1}{6}\right)^2\right]$$

$= \mathbf{9.14}$

Standard deviation $= \sqrt{9.14} = \mathbf{3.02}$

11. (61) Mean $= 0$, $\sigma = 1$

In a normal distribution, 68% of the data would lie between –1 and 1, or within σ of the mean. Since the data is evenly distributed, 34% lie between –1 and 0. Likewise, 95% of the data would lie between –2 and 2, or within 2σ of the mean. So, 47.5% lie between 0 and 2. Therefore, the percentage between –1 and 2, is 34% + 47.5% = **81.5%.**

12. (65) $y = \log_{\frac{5}{3}} x$

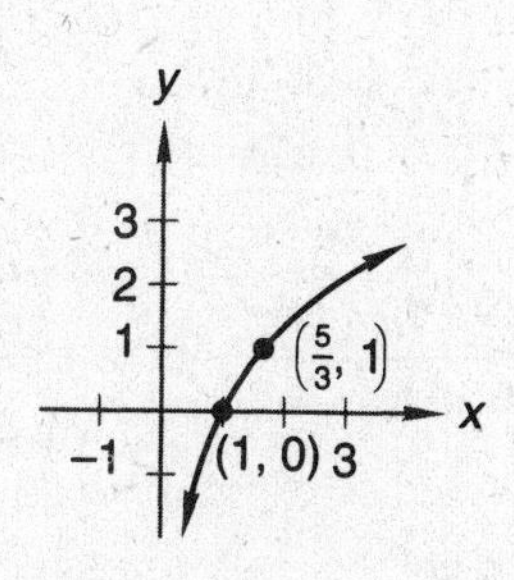

13. (66) Function = $\cos x$

Centerline = 5

Amplitude = 4

Period = π

Phase angle $= -\dfrac{3\pi}{8}$

Coefficient $= \dfrac{2\pi}{\pi} = 2$

$$\mathbf{y = 5 + 4\cos 2\left(x + \frac{3\pi}{8}\right)}$$

14. (66) Function = $\sin\theta$

Centerline = –3

Amplitude = 10

Period = 120°

Phase angle = 20°

Coefficient $= \dfrac{360°}{120°} = 3$

$\mathbf{y = -3 + 10\sin 3(\theta - 20°)}$

15. (64) $(6 \text{ cis } 215°)(2 \text{ cis } 205°) = 12 \text{ cis } 420°$

$= 12 \text{ cis } 60° = 12(\cos 60° + i \sin 60°)$

$= 12\left(\dfrac{1}{2} + i\dfrac{\sqrt{3}}{2}\right) = \mathbf{6 + 6\sqrt{3}i}$

16. (64) (a) $6 - 2i$

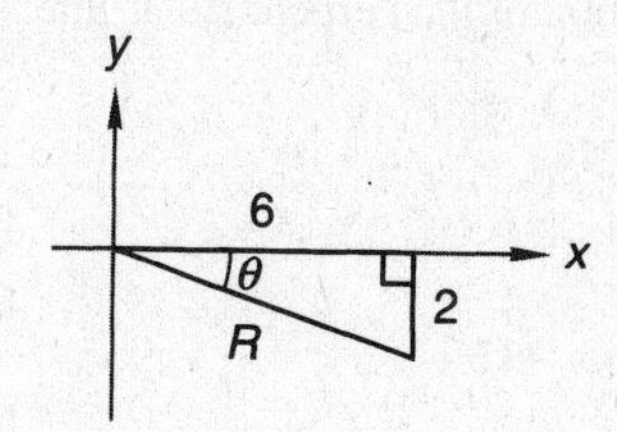

$R = \sqrt{6^2 + (-2)^2} = 2\sqrt{10} = 6.32$

$\tan\theta = \dfrac{2}{6}$

$\theta = 18.43°$

The polar angle is 360° – 18.43° = 341.57°.

6.32 cis 341.57°

(b) $5 \text{ cis}\left(-\dfrac{13\pi}{4}\right) = 5 \text{ cis}\left(-\dfrac{5\pi}{4}\right)$

$= 5\left(-\cos\dfrac{\pi}{4} + i\sin\dfrac{\pi}{4}\right)$

$= 5\left[\left(-\dfrac{\sqrt{2}}{2}\right) + i\left(\dfrac{\sqrt{2}}{2}\right)\right] = \mathbf{-\dfrac{5\sqrt{2}}{2} + \dfrac{5\sqrt{2}}{2}i}$

17. (62)
$$\begin{cases} \text{(a)} & ax + by = c \\ \text{(b)} & px + qy = d \end{cases}$$

$$\begin{array}{rl} -p\text{(a)} & -pax - pby = -pc \\ a\text{(b)} & pax + qay = ad \\ \hline & y(aq - pb) = ad - pc \end{array}$$

$$\mathbf{y = \frac{ad - pc}{aq - pb}}$$

18. (56)
$$\begin{cases} y \geq (x-3)^2 + 1 & \text{(parabola)} \\ y < x & \text{(line)} \end{cases}$$

The area is beneath the line and on or above the parabola.

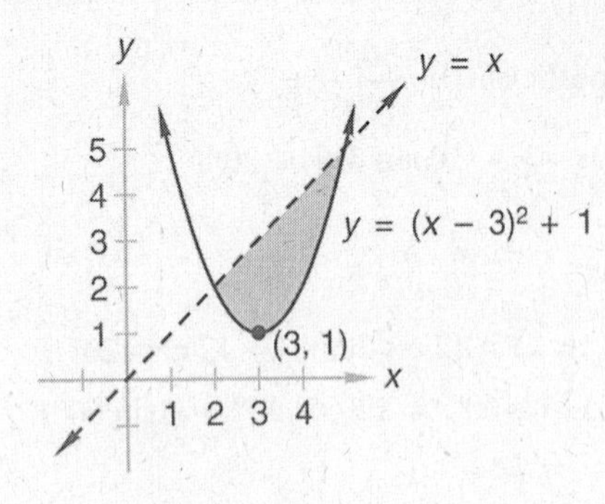

19. (58) $x - y + 3 = 0$

$$y = x + 3$$

Equation of the perpendicular line:

$$y = -x + b$$
$$(8) = -(-3) + b$$
$$b = 5$$
$$y = -x + 5$$

Point of intersection:

$$-x + 5 = x + 3$$
$$2x = 2$$
$$x = 1$$

$$y = x + 3 = (1) + 3 = 4$$

(1, 4) and (−3, 8)

$$D = \sqrt{[1 - (-3)]^2 + (4 - 8)^2} = \sqrt{32} = \mathbf{4\sqrt{2}}$$

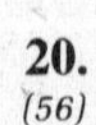

20. (56)

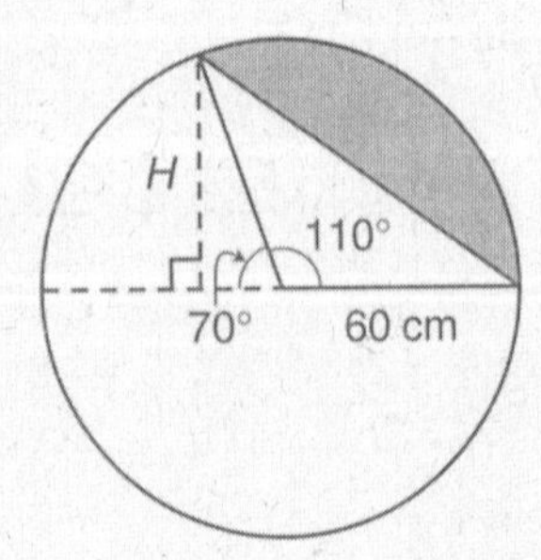

$$H = 60 \sin 70° = 56.3816 \text{ cm}$$

$$A_{\text{segment}} = A_{\text{sector}} - A_{\text{triangle}}$$
$$= \left(\frac{110°}{360°}\right)(\pi)(60)^2 - \left(\frac{1}{2}\right)(60)(56.3816)$$
$$= 3455.76 - 1691.45 = \mathbf{1764.31\ cm^2}$$

21. (60) $(\tan\theta - \sqrt{3})(\tan\theta + \sqrt{3}) = 0$

$$\tan\theta - \sqrt{3} = 0 \qquad \tan\theta + \sqrt{3} = 0$$
$$\tan\theta = \sqrt{3} \qquad \tan\theta = -\sqrt{3}$$
$$\theta = \frac{\pi}{3}, \frac{4\pi}{3} \qquad \theta = \frac{2\pi}{3}, \frac{5\pi}{3}$$

$$\theta = \mathbf{\frac{\pi}{3}, \frac{2\pi}{3}, \frac{4\pi}{3}, \frac{5\pi}{3}}$$

22. (60) $\sin x + 2\sin^2 x = 0$

$$\sin x(1 + 2\sin x) = 0$$
$$\sin x = 0 \qquad 1 + 2\sin x = 0$$
$$x = 0, \pi \qquad 2\sin x = -1$$
$$\sin x = -\frac{1}{2}$$
$$x = \frac{7\pi}{6}, \frac{11\pi}{6}$$

$$x = \mathbf{0, \pi, \frac{7\pi}{6}, \frac{11\pi}{6}}$$

23. (60) $\tan^2 x - \frac{1}{3} = 0$

$$\left(\tan x - \frac{1}{\sqrt{3}}\right)\left(\tan x + \frac{1}{\sqrt{3}}\right) = 0$$
$$\tan x - \frac{1}{\sqrt{3}} = 0 \qquad \tan x + \frac{1}{\sqrt{3}} = 0$$
$$\tan x = \frac{1}{\sqrt{3}} \qquad \tan x = -\frac{1}{\sqrt{3}}$$
$$x = \frac{\pi}{6}, \frac{7\pi}{6} \qquad x = \frac{5\pi}{6}, \frac{11\pi}{6}$$

$$x = \mathbf{\frac{\pi}{6}, \frac{5\pi}{6}, \frac{7\pi}{6}, \frac{11\pi}{6}}$$

24. (32) $0° \leq \theta \leq 180°$

$$\text{Arccos}(\cos 210°) = \text{Arccos}\left(-\frac{\sqrt{3}}{2}\right) = \mathbf{150°}$$

25. (48) $\cot^2(-510°) - \csc^3(-510°) - \tan^2(-510°)$

$$= \cot^2(210°) - \csc^3(210°) - \tan^2(210°)$$
$$= (\cot 30°)^2 - (-\csc 30°)^3 - (\tan 30°)^2$$
$$= (\sqrt{3})^2 - (-2)^3 - \left(\frac{1}{\sqrt{3}}\right)^2$$
$$= 3 - (-8) - \frac{1}{3} = 10\frac{2}{3} = \mathbf{\frac{32}{3}}$$

26. (46)

$$[x - (\sqrt{2} + i)][x - (\sqrt{2} - i)] = 0$$
$$(x - \sqrt{2} - i)(x - \sqrt{2} + i) = 0$$
$$x^2 - \sqrt{2}x + ix - \sqrt{2}x + 2 - \sqrt{2}i$$
$$- ix + \sqrt{2}i - i^2 = 0$$
$$\mathbf{x^2 - 2\sqrt{2}x + 3 = 0}$$

27. (56)

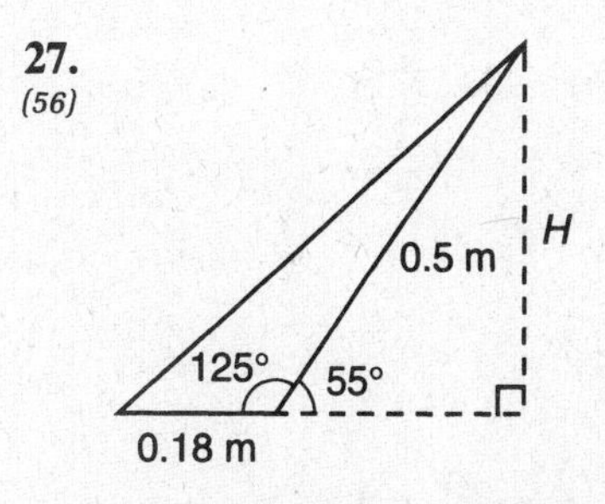

$H = 0.5 \sin 55° = 0.4096$ m

$\text{Area} = \frac{1}{2}BH = \frac{1}{2}(0.18)(0.4096) = \mathbf{0.037\ m^2}$

28. (59)

$$\frac{2}{3}\log_4 8 - 3\log_4 x = \log_4 4$$
$$\log_4 8^{\frac{2}{3}} - \log_4 x^3 = \log_4 4$$
$$\log_4 4 - \log_4 x^3 = \log_4 4$$
$$\log_4 4 - \log_4 4 = \log_4 x^3$$
$$\log_4 \frac{4}{4} = \log_4 x^3$$
$$x^3 = 1$$
$$x = \mathbf{1}$$

29. (67)

(a) $\text{antilog}_7 2.78 = 7^{2.78} = \mathbf{223.55}$

(b) $\text{antilog}_{13} 2.78 = 13^{2.78} = \mathbf{1249.58}$

30. (34)

$$(f \circ g)(x) = \ln(\log x)$$
$$f(g(x)) = \ln(\log x)$$
$$\mathbf{f(x) = \ln x}$$
$$\mathbf{g(x) = \log x}$$

Problem Set 70

1. (62) $S = mA + b$

(a) $\begin{cases} 20 = m1000 + b \\ 30 = m3000 + b \end{cases}$ (b)

$$\begin{aligned} \text{(a)}\quad 20 &= m1000 + b \\ -\text{(b)}\quad -30 &= -m3000 - b \\ \hline -10 &= -m2000 \\ m &= 0.005 \end{aligned}$$

$$\begin{aligned} \text{(a) } 20 &= m1000 + b \\ 20 &= (0.005)1000 + b \\ 20 &= 5 + b \\ b &= 15 \end{aligned}$$

$$S = 0.005A + 15$$
$$= 0.005(6000) + 15 = \mathbf{45\ squirrels}$$

2. (25)

$$R_I T_I + R_M T_M = \text{jobs}$$
$$\left(\frac{5}{4}\frac{\text{jobs}}{\text{hr}}\right)(6\text{ hr}) + R_M(6\text{ hr}) = 9\text{ jobs}$$
$$\frac{30}{4}\text{ jobs} + R_M(6\text{ hr}) = 9\text{ jobs}$$
$$R_M(6\text{ hr}) = \frac{3}{2}\text{ jobs}$$
$$R_M = \frac{1}{4}\frac{\text{jobs}}{\text{hr}}$$

Mike would take **4 hr** to complete one job.

3. (36)

(a) $\begin{cases} (B - W)T_U = D_U \\ (B + W)T_D = D_D \end{cases}$ (b)

$B = 10W,\ T_U = T_D - 2$

$$\begin{aligned} \text{(b)}\quad (B + W)T_D &= D_D \\ (10W + W)T_D &= 264 \\ 11WT_D &= 264 \\ WT_D &= 24 \end{aligned}$$

$$\begin{aligned} \text{(a)}\quad (B - W)T_U &= D_U \\ (10W - W)(T_D - 2) &= 180 \\ 9W(T_D - 2) &= 180 \\ 9WT_D - 18W &= 180 \\ 9(24) - 18W &= 180 \\ 18W &= 36 \\ \mathbf{W} &= \mathbf{2\ mph} \end{aligned}$$

$B = 10W$

$B = 10(2)$

$\mathbf{B = 20\ mph}$

4. (18) $J = \dfrac{H^2 A^{\frac{1}{3}}}{S}$

$$\frac{(2H)^2(8A)^{\frac{1}{3}}}{4S} = \frac{(4)H^2(2)A^{\frac{1}{3}}}{4S} = 2\,\frac{H^2A^{\frac{1}{3}}}{S} = 2J$$

Therefore, ***J* is doubled.**

5. (55) $\dfrac{11!}{3!3!} = \mathbf{1{,}108{,}800}$

6. (70) For $5\frac{1}{2}$ inches:

$$z \text{ score} = \frac{x - \mu}{\sigma} = \frac{5\frac{1}{2} - 5}{\frac{1}{2}} = 1$$

Percentile = 0.8413

For 4 inches:

$$z \text{ score} = \frac{x - \mu}{\sigma} = \frac{4 - 5}{\frac{1}{2}} = -2$$

Percentile = 0.0228

0.8413 − 0.0228 = 0.8185

81.85%

7. (70) $z \text{ score} = \dfrac{x - \mu}{\sigma} = \dfrac{87 - 75}{8} = \dfrac{12}{8} = 1.5$

Percentile = 0.9332

93.32%

8. (69) $\begin{vmatrix} 5 & 1 \\ 7 & 2 \end{vmatrix}$

$(5)(2) - (7)(1) = 10 - 7 = \mathbf{3}$

9. (69) $\begin{vmatrix} x - 2 & 4 \\ 1 & x + 1 \end{vmatrix} = 0$

$(x - 2)(x + 1) - 1(4) = 0$

$x^2 - 2x + x - 2 - 4 = 0$

$x^2 - x - 6 = 0$

$(x - 3)(x + 2) = 0$

$x = \mathbf{3, -2}$

10. (68) Vertex: $(h, k) = (2, -1)$

Focus: $(h, k + p) = (2, 4)$

$k + p = 4$

$(-1) + p = 4$

$p = 5$

Directrix: $y = k - p = (-1) - (5) = -6$

Axis of symmetry: $x = h = 2$

$$y - k = \frac{1}{4p}(x - h)^2$$

$$y + 1 = \frac{1}{(4)(5)}(x - 2)^2$$

$$y = \frac{1}{20}(x - 2)^2 - 1$$

Parabola: $\mathbf{y = \frac{1}{20}(x - 2)^2 - 1}$

Directrix: $\mathbf{y = -6}$

Axis of symmetry: $\mathbf{x = 2}$

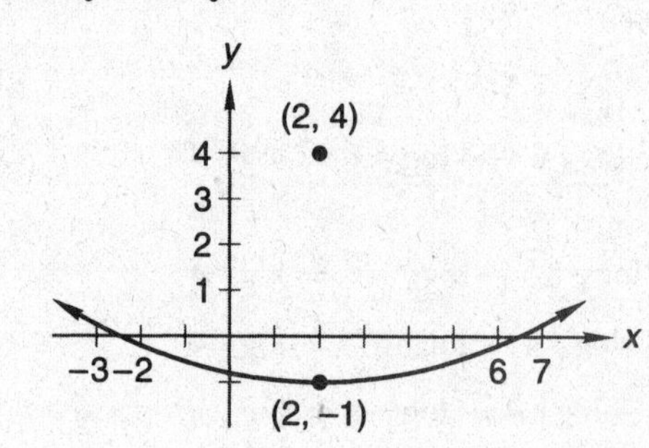

11. (65) $y = \log_9 x$

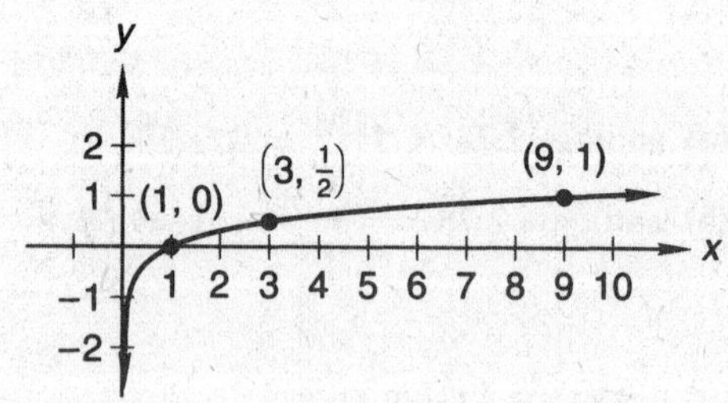

12. (63)

$$3x^2 - 12x + 3y^2 + 18y - 24 = 0$$

$$x^2 - 4x + y^2 + 6y - 8 = 0$$

$$(x^2 - 4x \quad) + (y^2 + 6y \quad) = 8$$

$$(x^2 - 4x + 4) + (y^2 + 6y + 9) = 21$$

$$(x - 2)^2 + (y + 3)^2 = 21$$

$$\mathbf{(x - 2)^2 + (y + 3)^2 = (\sqrt{21})^2}$$

Center = (2, −3); radius = $\mathbf{\sqrt{21}}$

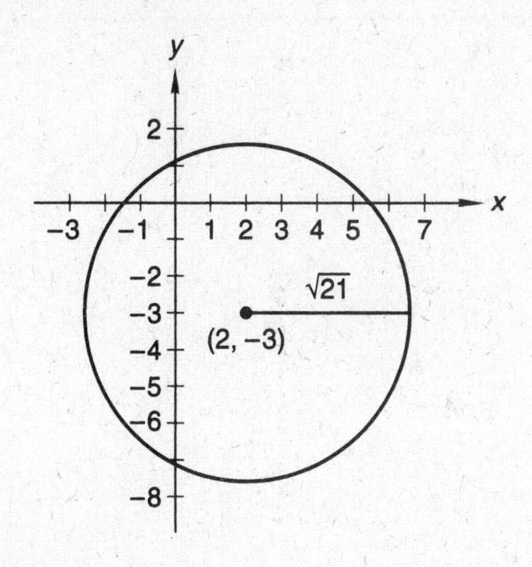

13. (66) Function = sin x

Centerline = −9

Amplitude = 31

Period = 240°

Phase angle = −10°

$$\text{Coefficient} = \frac{360°}{240°} = \frac{3}{2}$$

$$\mathbf{y = -9 + 31 \sin \frac{3}{2}(x + 10°)}$$

14. (66) Function = cos x

Centerline = 12

Amplitude = 2

Period = 3π

$$\text{Phase angle} = \frac{\pi}{2}$$

$$\text{Coefficient} = \frac{2\pi}{3\pi} = \frac{2}{3}$$

$$\mathbf{y = 12 + 2 \cos \frac{2}{3}\left(x - \frac{\pi}{2}\right)}$$

15. (62)

$$\begin{cases} \text{(a)}\ dx - wy = s \\ \text{(b)}\ kx + jy = a \end{cases}$$

$$\begin{aligned} j(a)\quad & jdx - jwy = js \\ w(b)\quad & kwx + jwy = aw \\ \hline & (jd + kw)x = js + aw \end{aligned}$$

$$\mathbf{x = \frac{js + aw}{jd + kw}}$$

16. (64) $(5 \text{ cis } 72°)[-3 \text{ cis } (-312°)] = -15 \text{ cis } (-240°)$

$= -15[\cos(-240°) + i \sin(-240°)]$

$= -15(-\cos 60° + i \sin 60°)$

$$= -15\left(-\frac{1}{2} + \frac{\sqrt{3}}{2}i\right) = \mathbf{\frac{15}{2} - \frac{15\sqrt{3}}{2}i}$$

17. (64) $\left(6 \text{ cis } \frac{4\pi}{3}\right)\left(2 \text{ cis } \frac{\pi}{3}\right) = 12 \text{ cis } \frac{5\pi}{3}$

$$= 12\left(\cos \frac{5\pi}{3} + i \sin \frac{5\pi}{3}\right) = 12\left(\cos \frac{\pi}{3} - i \sin \frac{\pi}{3}\right)$$

$$= 12\left(\frac{1}{2} - i\frac{\sqrt{3}}{2}\right) = \mathbf{6 - 6\sqrt{3}i}$$

18. (60) $(\sqrt{2} \sin x - 1)(\sqrt{3} - 2 \cos x) = 0$

$$\begin{aligned} \sqrt{2} \sin x - 1 &= 0 & \sqrt{3} - 2 \cos x &= 0 \\ \sqrt{2} \sin x &= 1 & 2 \cos x &= \sqrt{3} \\ \sin x &= \frac{1}{\sqrt{2}} & \cos x &= \frac{\sqrt{3}}{2} \\ x &= \frac{\pi}{4}, \frac{3\pi}{4} & x &= \frac{\pi}{6}, \frac{11\pi}{6} \end{aligned}$$

$$\mathbf{x = \frac{\pi}{6}, \frac{\pi}{4}, \frac{3\pi}{4}, \frac{11\pi}{6}}$$

19. (60) $2 \cos^2 x + \cos x - 1 = 0$

$(2 \cos x - 1)(\cos x + 1) = 0$

$$\begin{aligned} 2 \cos x - 1 &= 0 & \cos x + 1 &= 0 \\ 2 \cos x &= 1 & \cos x &= -1 \\ \cos x &= \frac{1}{2} & x &= \pi \\ x &= \frac{\pi}{3}, \frac{5\pi}{3} \end{aligned}$$

$$\mathbf{x = \frac{\pi}{3}, \pi, \frac{5\pi}{3}}$$

20. (47) $-90° < \theta < 90°$

$\text{Arctan}[\tan(-210°)] = \text{Arctan}[\tan(-30°)] = \mathbf{-30°}$

21. (48) $\sec^2 \frac{5\pi}{4} - \cos^2 \frac{7\pi}{4} = \left(-\sec \frac{\pi}{4}\right)^2 - \left(\cos \frac{\pi}{4}\right)^2$

$$= (-\sqrt{2})^2 - \left(\frac{1}{\sqrt{2}}\right)^2 = 2 - \frac{1}{2} = \mathbf{\frac{3}{2}}$$

22. (61) $\text{Mean} = \frac{1 + 1 + 3 + 4 + 7 + 8}{6} = \frac{24}{6} = \mathbf{4}$

$\text{Median} = \frac{3 + 4}{2} = \mathbf{3.5}$

Mode = **1**

$\text{Variance} = \frac{3^2 + 3^2 + 1^2 + 0^2 + 3^2 + 4^2}{6}$

$= \mathbf{7.33}$

Standard deviation = $\sqrt{7.33}$ = **2.71**

23. (56) $\begin{cases} x^2 + y^2 \geq 4 & \text{(circle)} \\ y \geq 2x^2 & \text{(parabola)} \end{cases}$

The region must be on or outside the circle and on or above the parabola.

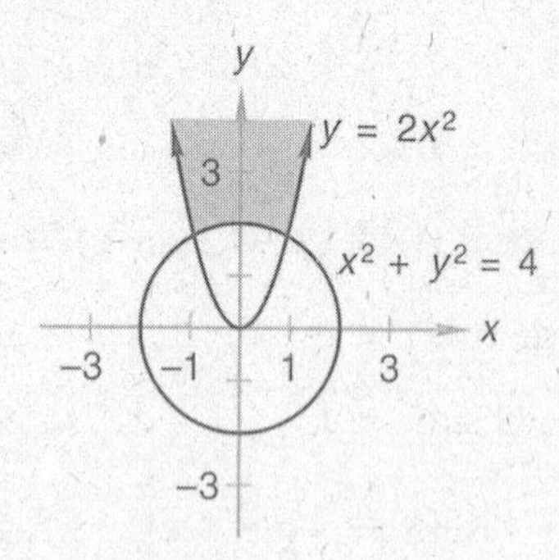

24. (56)

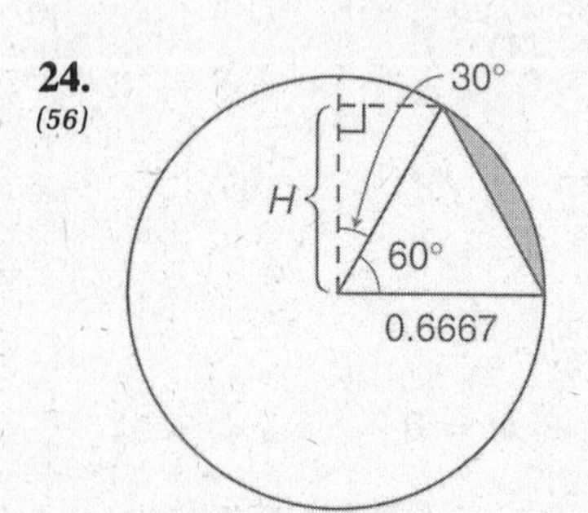

$$H = 0.6667 \cos 30° = 0.5774 \text{ ft}$$

$$A_{\text{segment}} = A_{\text{sector}} - A_{\text{triangle}}$$

$$= \pi(0.6667)^2\left(\frac{60°}{360°}\right) - \frac{1}{2}(0.6667)(0.5774)$$

$$= 0.2327 - 0.1925 = \mathbf{0.040\ ft^2}$$

25. (46) $\left[x - (2\sqrt{3} - 2i)\right]\left[x - (2\sqrt{3} + 2i)\right] = 0$

$$(x - 2\sqrt{3} + 2i)(x - 2\sqrt{3} - 2i) = 0$$

$$x^2 - 2\sqrt{3}x - 2ix - 2\sqrt{3}x + 4(3)$$

$$+ 4\sqrt{3}i + 2ix - 4\sqrt{3}i - 4i^2 = 0$$

$$\mathbf{x^2 - 4\sqrt{3}x + 16 = 0}$$

26. (58) $3x - 4y - 2 = 0$

$$4y = 3x - 2$$

$$y = \frac{3}{4}x - \frac{1}{2}$$

Equation of the perpendicular line:

$$y = -\frac{4}{3}x + b$$

$$2 = -\frac{4}{3}(5) + b$$

$$b = \frac{26}{3}$$

$$y = -\frac{4}{3}x + \frac{26}{3}$$

Point of intersection:

$$-\frac{4}{3}x + \frac{26}{3} = \frac{3}{4}x - \frac{1}{2}$$

$$-16x + 104 = 9x - 6$$

$$110 = 25x$$

$$x = \frac{22}{5}$$

$$y = \frac{3}{4}x - \frac{1}{2} = \frac{3}{4}\left(\frac{22}{5}\right) - \frac{1}{2} = \frac{14}{5}$$

$\left(\frac{22}{5}, \frac{14}{5}\right)$ and $(5, 2)$

$$D = \sqrt{\left(\frac{22}{5} - 5\right)^2 + \left(\frac{14}{5} - 2\right)^2}$$

$$= \sqrt{\left(-\frac{3}{5}\right)^2 + \left(\frac{4}{5}\right)^2} = \sqrt{\frac{9}{25} + \frac{16}{25}} = \mathbf{1}$$

27. (56)

5 in.
H
70°
6 in.

$$H = 5 \sin 70° = 4.6985 \text{ in.}$$

$$\text{Area} = \frac{1}{2}BH = \frac{1}{2}(6)(4.6985) = \mathbf{14.10\ in.^2}$$

28. (59) $\frac{3}{4}\log_7 625 + 2\log_7 x = 2$

$$\log_7 625^{\frac{3}{4}} + \log_7 x^2 = 2$$

$$\log_7 125x^2 = 2$$

$$125x^2 = 7^2$$

$$x^2 = \frac{49}{125}$$

$$x = \frac{7}{5\sqrt{5}}$$

$$x = \mathbf{\frac{7\sqrt{5}}{25}}$$

29. (59) $2^{\log_2 3 + \log_2 5 - \log_2 7} - \log_5 5$

$$= 2^{\log_2[3(5)] - \log_2 7} - \log_5 5$$

$$= 2^{\log_2 \frac{15}{7}} - \log_5 5 = \frac{15}{7} - 1 = \mathbf{\frac{8}{7}}$$

30. $f(x) = \sqrt{1 - x^2}$
(21)

$$y = \sqrt{1 - x^2}$$
$$y^2 = 1 - x^2$$
$$x^2 + y^2 = 1$$

This is the equation of a circle centered at the origin and having a radius of 1. However, in the original equation the square root of $1 - x^2$ must be positive. Therefore the answer is **D.**

Problem Set 71

1. (36)

$$\begin{cases} \text{(a)}\ (B + W)T_D = D_D \\ \text{(b)}\ (B - W)T_U = D_U \end{cases}$$

$B = 5W,\ T_D = T_U + 2$

(b) $(B - W)T_U = D_U$

$$(5W - W)T_U = 64$$
$$4WT_U = 64$$
$$WT_U = 16$$

(a) $(B + W)T_D = D_D$

$$(5W + W)(T_U + 2) = 144$$
$$6W(T_U + 2) = 144$$
$$6WT_U + 12W = 144$$
$$6(16) + 12W = 144$$
$$12W = 48$$
$$\mathbf{W = 4\ mph}$$

$$B = 5W$$
$$B = 5(4)$$
$$\mathbf{B = 20\ mph}$$

2. (18)

$$\begin{cases} \text{(a)}\ 4(N_R + N_B) = 3N_W + 3 \\ \text{(b)}\ 5(N_B + N_W) = 8N_R + 13 \\ \text{(c)}\ N_W = N_B + 5 \end{cases}$$

(a) $4(N_R + N_B) = 3N_W + 3$

$$4N_R + 4N_B = 3(N_B + 5) + 3$$
$$4N_R + 4N_B = 3N_B + 15 + 3$$

(a′) $4N_R + N_B = 18$

(b) $5(N_B + N_W) = 8N_R + 13$

$$5N_B + 5N_W = 8N_R + 13$$
$$5N_B + 5(N_B + 5) = 8N_R + 13$$
$$10N_B + 25 = 8N_R + 13$$

(b′) $-8N_R + 10N_B = -12$

(b′) $-8N_R + 10N_B = -12$

2(a′) $8N_R + 2N_B = 36$

$$12N_B = 24$$
$$\mathbf{N_B = 2}$$

(c) $N_W = N_B + 5$

$$N_W = (2) + 5$$
$$\mathbf{N_W = 7}$$

(a′) $4N_R + N_B = 18$

$$4N_R + (2) = 18$$
$$4N_R = 16$$
$$\mathbf{N_R = 4}$$

3. (45)

M	3	O	2	1	→ $3 \cdot 2 \cdot 1 = 6$
M	3	2	O	1	→ $3 \cdot 2 \cdot 1 = 6$
M	3	2	1	O	→ $3 \cdot 2 \cdot 1 = 6$
3	M	2	O	1	→ $3 \cdot 2 \cdot 1 = 6$
3	M	2	1	O	→ $3 \cdot 2 \cdot 1 = 6$
O	3	M	2	1	→ $3 \cdot 2 \cdot 1 = 6$
3	2	M	1	O	→ $3 \cdot 2 \cdot 1 = 6$
O	3	2	M	1	→ $3 \cdot 2 \cdot 1 = 6$
3	O	2	M	1	→ $3 \cdot 2 \cdot 1 = 6$
O	3	2	1	M	→ $3 \cdot 2 \cdot 1 = 6$
3	O	2	1	M	→ $3 \cdot 2 \cdot 1 = 6$
3	2	O	1	M	→ $3 \cdot 2 \cdot 1 = 6$
					72

4. (25) $R_T = \dfrac{12\text{ jobs}}{15\text{ hr}} = \dfrac{4}{5}\dfrac{\text{jobs}}{\text{hr}}$

$$R_T T_T + R_I T_I = \text{jobs}$$
$$\frac{4}{5}(20) + R_I(20) = 25$$
$$20R_I = 9$$
$$R_I = \frac{9}{20}\frac{\text{jobs}}{\text{hr}}$$

$$TR_I = \text{jobs}$$
$$T\left(\frac{9}{20}\right) = 1$$
$$T = \mathbf{\frac{20}{9}\ hr}$$

5. (18) $x = \dfrac{ky^2}{z\sqrt{w}}$

$$\frac{k(3y)^2}{2z\sqrt{4w}} = \frac{9ky^2}{4z\sqrt{w}} = \frac{9}{4}(x)$$

Therefore, x **is multiplied by** $\mathbf{\dfrac{9}{4}}$.

6. (18)
$$\begin{cases} \text{(a)}\ L + W + H = 150 \\ \text{(b)}\ L = 2(W + H) \\ \text{(c)}\ 2W = 1 + H \end{cases}$$

(a) $L + W + H = 150$

$2(W + H) + W + H = 150$

(a′) $3W + 3H = 150$

(c) $2W = 1 + H$

$H = 2W - 1$

(a′) $3W + 3H = 150$

$3W + 3(2W - 1) = 150$

$9W - 3 = 150$

$W = \dfrac{153}{9}$

$W = 17$

(c) $2W = 1 + H$

$H = 2(17) - 1$

$H = 33$

(b) $L = 2(W + H)$

$L = 2(17 + 33)$

$L = 100$

Length = 100 cm

Width = 17 cm

Height = 33 cm

7. (71) $\dfrac{x^2}{16} + \dfrac{y^2}{9} = 1$

Let $x = 0$ | Let $y = 0$

$\dfrac{0}{16} + \dfrac{y^2}{9} = 1$ | $\dfrac{x^2}{16} + \dfrac{0}{9} = 1$

$y = \pm 3$ | $x = \pm 4$

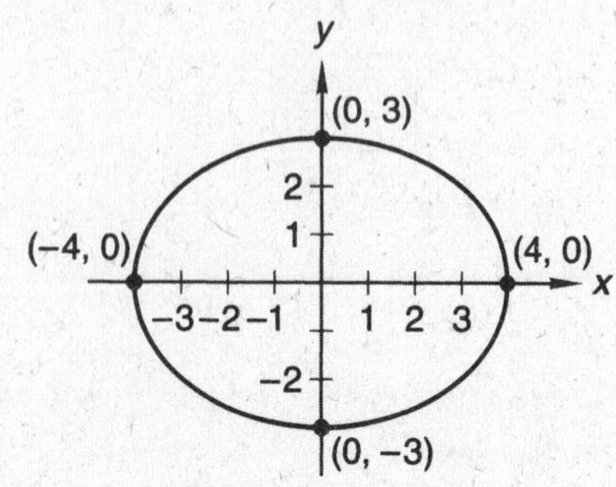

8. (71) $9x^2 + 4y^2 = 36$

$$\mathbf{\frac{x^2}{4} + \frac{y^2}{9} = 1}$$

Let $x = 0$ | Let $y = 0$

$\dfrac{0}{4} + \dfrac{y^2}{9} = 1$ | $\dfrac{x^2}{4} + \dfrac{0}{9} = 1$

$y = \pm 3$ | $x = \pm 2$

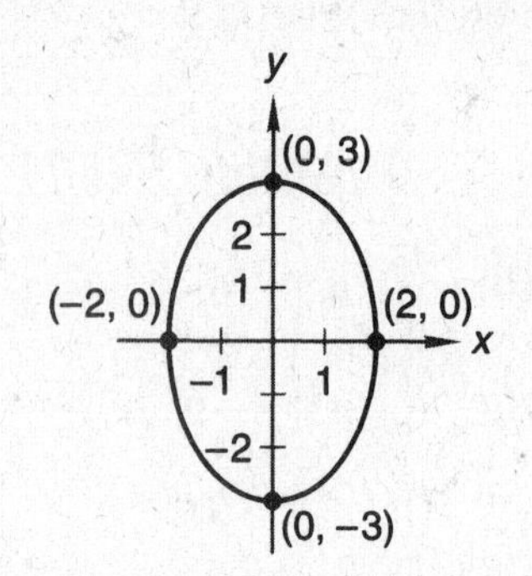

9. (70) $z \text{ score} = \dfrac{135 - 120}{10} = 1.5$

Percentile $= 0.9332$

$1 - 0.9332 = 0.0668$

6.68%

10. (69)
$$\begin{vmatrix} x + 1 & 1 \\ 3 & x - 1 \end{vmatrix} = 0$$

$(x + 1)(x - 1) - (3)(1) = 0$

$x^2 - 4 = 0$

$x^2 = 4$

$\mathbf{x = \pm 2}$

11. (68) Vertex: $(h, k) = (0, 0)$

Focus: $(h, k + p) = (0, -2)$

$k + p = -2$

$(0) + p = -2$

$p = -2$

Directrix: $y = k - p = (0) - (-2) = 2$

Axis of symmetry: $x = h = 0$

$y - k = \dfrac{1}{4p}(x - h)^2$

$y - 0 = \dfrac{1}{4(-2)}(x - 0)^2$

$y = -\dfrac{1}{8}x^2$

Parabola: $\mathbf{y = -\dfrac{1}{8}x^2}$

Directrix: $\mathbf{y = 2}$

Axis of symmetry: $\mathbf{x = 0}$

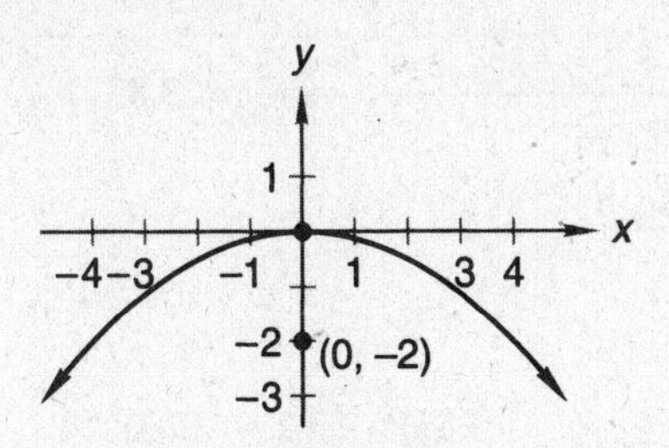

12. (68) Vertex: $(h, k) = (1, 2)$

Focus: $(h, k + p) = (1, 4)$

$$k + p = 4$$
$$(2) + p = 4$$
$$p = 2$$

Directrix: $y = k - p = (2) - (2) = 0$

Axis of symmetry: $x = h = 1$

$$y - k = \frac{1}{4p}(x - h)^2$$
$$y - 2 = \frac{1}{4(2)}(x - 1)^2$$
$$y = \frac{1}{8}(x - 1)^2 + 2$$

Parabola: $\mathbf{y = \frac{1}{8}(x - 1)^2 + 2}$

Directrix: $\mathbf{y = 0}$

Axis of symmetry: $\mathbf{x = 1}$

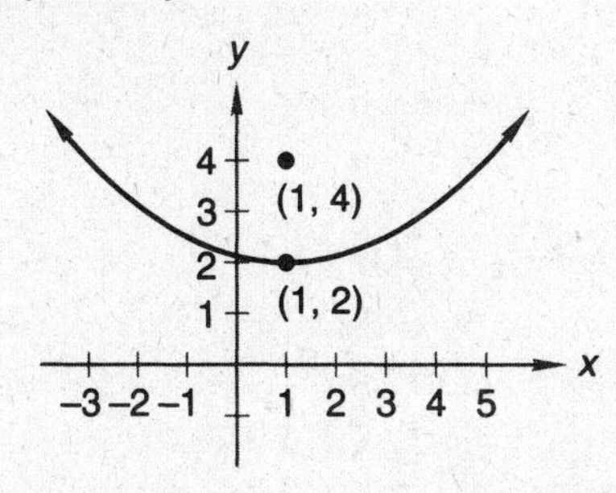

13. (65) $y = \log_{2.2} x$

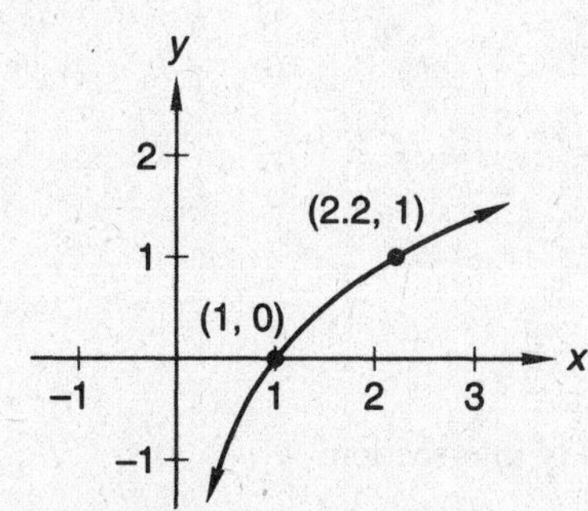

14. (63)

$$x^2 + y^2 + 8x - 12y + 43 = 0$$
$$(x^2 + 8x \quad) + (y^2 - 12y \quad) = -43$$
$$(x^2 + 8x + 16) + (y^2 - 12y + 36) = 9$$
$$(x + 4)^2 + (y - 6)^2 = 3^2$$

Center = (−4, 6); radius = 3

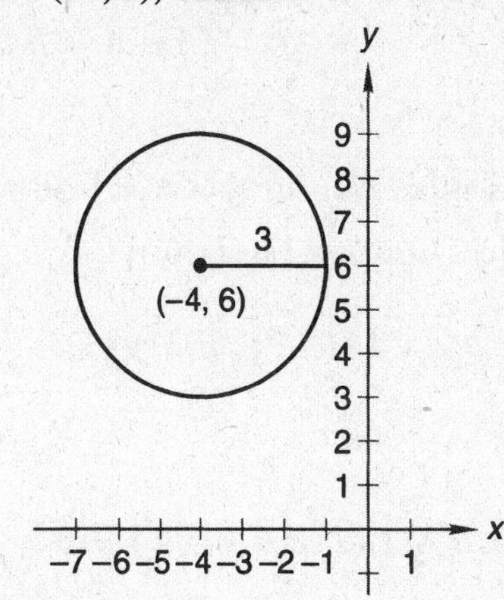

15. (66) Function = cos x

Centerline = −3

Amplitude = 8

Period = 1440°

Phase angle = 180°

$$\text{Coefficient} = \frac{360°}{1440°} = \frac{1}{4}$$

$$\mathbf{y = -3 + 8\cos\frac{1}{4}(x - 180°)}$$

16. (66) Function = cos x

Centerline = 1

Amplitude = 2

Period = 4π

$$\text{Phase angle} = -\frac{\pi}{2}$$

$$\text{Coefficient} = \frac{2\pi}{4\pi} = \frac{1}{2}$$

$$\mathbf{y = 1 + 2\cos\frac{1}{2}\left(x + \frac{\pi}{2}\right)}$$

17. (64) (a) $-7 - 8i$

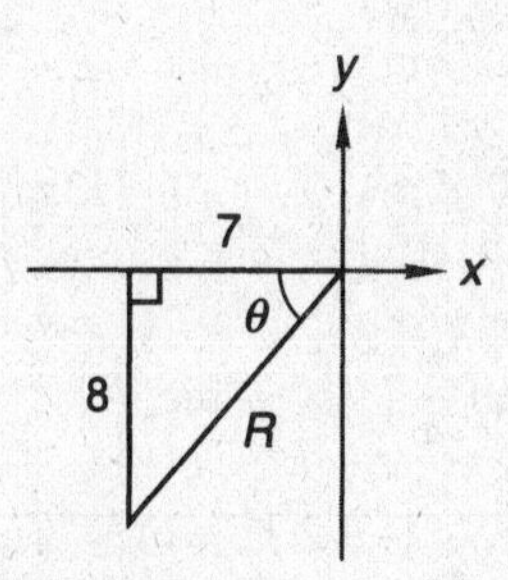

$$R = \sqrt{7^2 + 8^2} = 10.63$$

$$\tan\theta = \frac{8}{7}$$

$$\theta = 48.81°$$

The polar angle is 180° + 48.81° = 228.81°.

10.63 cis 228.81°

(b) $7 \text{ cis } 270° = 7(\cos 270° + i \sin 270°)$
$= 7(0 - 1i) = \mathbf{0 - 7i}$

18. (64) $(3 \text{ cis } 40°)[2 \text{ cis } (-70°)] = 6 \text{ cis } (-30°)$
$= 6[\cos(-30°) + i \sin(-30°)]$
$= 6\left(\frac{\sqrt{3}}{2} - \frac{1}{2}i\right) = \mathbf{3\sqrt{3} - 3i}$

19. (60) $(2\cos\theta + 1)(2\sin\theta + \sqrt{2}) = 0$

$2\cos\theta = -1$ $\quad$ $2\sin\theta = -\sqrt{2}$

$\cos\theta = -\frac{1}{2}$ $\quad$ $\sin\theta = -\frac{\sqrt{2}}{2}$

$\theta = 120°, 240°$ $\quad$ $\theta = 225°, 315°$

$\theta = \mathbf{120°, 225°, 240°, 315°}$

20. (60)
$\cot^2 x = 3$
$\cot^2 x - 3 = 0$
$(\cot x - \sqrt{3})(\cot x + \sqrt{3}) = 0$

$\cot x = \sqrt{3}$ $\quad$ $\cot x = -\sqrt{3}$

$x = 30°, 210°$ $\quad$ $x = 150°, 330°$

$x = \mathbf{30°, 150°, 210°, 330°}$

21. (60) $\cot^2 x + \cot x = 0$
$\cot x(\cot x + 1) = 0$

$\cot x = 0$ $\quad$ $\cot x = -1$

$x = 90°, 270°$ $\quad$ $x = 135°, 315°$

$x = \mathbf{90°, 135°, 270°, 315°}$

22. (48) $\tan^2\left(-\frac{13\pi}{4}\right) - \sec^2\left(-\frac{13\pi}{4}\right)$
$= \left(-\tan\frac{\pi}{4}\right)^2 - \left(-\sec\frac{\pi}{4}\right)^2$
$= (-1)^2 - (-\sqrt{2})^2 = 1 - 2 = \mathbf{-1}$

23. (48) $\sin^2(-480°) + \cos^2(-480°)$
$= (-\sin 60°)^2 + (-\cos 60°)^2$
$= \left(-\frac{\sqrt{3}}{2}\right)^2 + \left(-\frac{1}{2}\right)^2 = \frac{3}{4} + \frac{1}{4} = \mathbf{1}$

24. (61) Mean $= \frac{5 + 1 + 7 + 2 + 5 + 3}{6} = \mathbf{3.83}$

Median $= \frac{5 + 3}{2} = \mathbf{4}$

Mode $= \mathbf{5}$

Variance $= \frac{1}{6}[(1 - 3.83)^2 + (2 - 3.83)^2$
$+ (3 - 3.83)^2 + (5 - 3.83)^2$
$+ (5 - 3.83)^2 + (7 - 3.83)^2]$
$= \mathbf{4.14}$

Standard deviation $= \sqrt{4.14} = \mathbf{2.03}$

25. (56) $\begin{cases} x^2 + y^2 \le 3^2 & \text{(circle)} \\ y \le x^2 & \text{(parabola)} \end{cases}$

This system of inequalities designates a region within and on the circle and below and on the parabola.

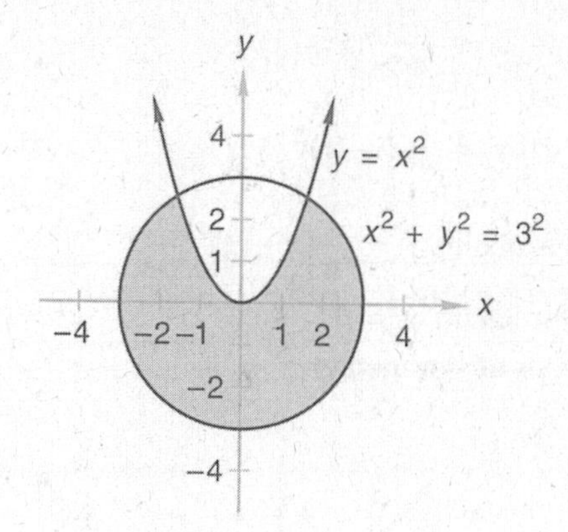

26. (58) $2x - 5y + 4 = 0$
$5y = 2x + 4$
$y = \frac{2}{5}x + \frac{4}{5}$

Equation of the perpendicular line:

$y = -\frac{5}{2}x + b$

$-2 = -\frac{5}{2}(0) + b$

$b = -2$

$y = -\frac{5}{2}x - 2$

Point of intersection:

$\frac{2}{5}x + \frac{4}{5} = -\frac{5}{2}x - 2$

$\frac{29}{10}x = -\frac{14}{5}$

$x = -\frac{28}{29}$

$$y = -\frac{5}{2}x - 2 = -\frac{5}{2}\left(-\frac{28}{29}\right) - 2 = \frac{12}{29}$$

$\left(-\frac{28}{29}, \frac{12}{29}\right)$ and $(0, -2)$

$$D = \sqrt{\left[\left(-\frac{28}{29}\right) - 0\right]^2 + \left[\frac{12}{29} - (-2)\right]^2}$$

$$= \sqrt{\left(-\frac{28}{29}\right)^2 + \left(\frac{70}{29}\right)^2}$$

$$= \frac{\sqrt{5684}}{29} = \mathbf{\frac{14\sqrt{29}}{29}}$$

27. (56)

8 cm, H, 28°, 12 cm

$H = 8 \sin 28° = 3.76$ cm

$$\text{Area} = \frac{1}{2}(12)(3.76)$$

$$= 22.56 \text{ cm}^2 \times \frac{1 \text{ in.}^2}{(2.54 \text{ cm})^2} \times \frac{1 \text{ ft}^2}{(12 \text{ in.})^2}$$

$$= \mathbf{0.024 \text{ ft}^2}$$

28. (59)

$$\frac{3}{4}\log_8 16 + 2\log_8 x = \log_8 x + 1$$

$$\log_8 16^{\frac{3}{4}} + \log_8 x^2 - \log_8 x = 1$$

$$\log_8 \frac{8x^2}{x} = 1$$

$$\log_8 8x = 1$$

$$8x = 8^1$$

$$x = \mathbf{1}$$

29. (59)

(a) $\ln e - \log 10 = 1 - 1 = \mathbf{0}$

(b) $\ln e^3 + \ln e^{\frac{1}{2}} = 3 + \frac{1}{2} = \mathbf{\frac{7}{2}}$

(c) $\log_2 4 + 4^{\log_2 15 - \log_2 5}$

$$= \log_2 2^2 + 2^{2 \log_2 \frac{15}{5}}$$

$$= \log_2 2^2 + 2^{2 \log_2 3}$$

$$= 2 + 3^2 = \mathbf{11}$$

30. (46) $x^2 + 7x + 15 = 0$

$$x = \frac{-7 \pm \sqrt{7^2 - 4(1)(15)}}{2(1)}$$

$$x = \frac{-7 \pm \sqrt{11}i}{2} = -\frac{7}{2} \pm \frac{\sqrt{11}}{2}i$$

$$\mathbf{\left(x + \frac{7}{2} - \frac{\sqrt{11}}{2}i\right)\left(x + \frac{7}{2} + \frac{\sqrt{11}}{2}i\right)}$$

Problem Set 72

1. (55) $\frac{9!}{2!2!} = \mathbf{90{,}720}$

2. (28) $r = \frac{d}{t} = \frac{y}{m - 2}\ \mathbf{\frac{yd}{min}}$

3. (44) $RWT = \text{marks}$

$Rmw = z$

$$R = \frac{z}{mw}\ \frac{\text{marks}}{\text{men-weeks}}$$

$$RWT = 1000$$

$$\left(\frac{z}{mw}\right)W(1) = 1000$$

$$W = \mathbf{\frac{1000wm}{z}\ \text{men}}$$

4. (18)

(a) $\frac{N_B}{N_Q} = \frac{14}{1}$

(b) $N_B + N_Q = 2550$

(b) $N_B + N_Q = 2550$

$$N_Q = 2550 - N_B$$

(a) $\frac{N_B}{N_Q} = \frac{14}{1}$

$$N_B = 14N_Q$$

$$N_B = 14(2550 - N_B)$$

$$N_B = 35700 - 14N_B$$

$$15N_B = 35700$$

$$N_B = \mathbf{2380 \text{ beauties}}$$

5. (18)

$$\begin{cases} T + U = 7 & \text{(a)} \\ H + T = 5 & \text{(b)} \\ T = 4H & \text{(c)} \end{cases}$$

(b) $H + T = 5$

$H + (4H) = 5$

$5H = 5$

$H = 1$

(c) $T = 4H$

$T = 4(1)$

$T = 4$

(a) $T + U = 7$

$(4) + U = 7$

$U = 3$

The number is **143.**

6. (1) $10(90 - A) = 3(180 - A) + 318$

$900 - 10A = 540 - 3A + 318$

$7A = 42$

$A = \mathbf{6°}$

7. (72) $A = 180° - 40° - 30°$

$\mathbf{A = 110°}$

$$\frac{a}{\sin 110°} = \frac{8}{\sin 30°}$$

$$a = \frac{8 \sin 110°}{\sin 30°}$$

$\mathbf{a = 15.04}$

$$\frac{m}{\sin 40°} = \frac{8}{\sin 30°}$$

$$m = \frac{8 \sin 40°}{\sin 30°}$$

$\mathbf{m = 10.28}$

8. (72) $B = 180° - 130° - 22°$

$\mathbf{B = 28°}$

$$\frac{c}{\sin 22°} = \frac{12}{\sin 130°}$$

$$c = \frac{12 \sin 22°}{\sin 130°}$$

$\mathbf{c = 5.87}$

$$\frac{b}{\sin 28°} = \frac{12}{\sin 130°}$$

$$b = \frac{12 \sin 28°}{\sin 130°}$$

$\mathbf{b = 7.35}$

9. (71) $\dfrac{x^2}{16} + \dfrac{y^2}{25} = 1$

Let $x = 0$

$$\frac{0}{16} + \frac{y^2}{25} = 1$$

$y = \pm 5$

Let $y = 0$

$$\frac{x^2}{16} + \frac{0}{25} = 1$$

$x = \pm 4$

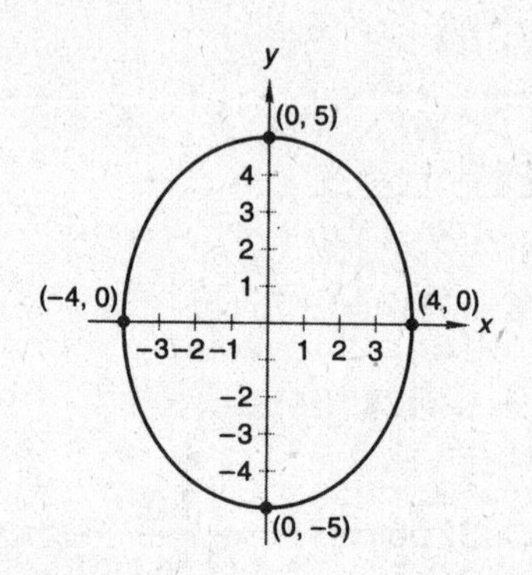

10. (71) $4x^2 + 16y^2 = 64$

$$\mathbf{\frac{x^2}{16} + \frac{y^2}{4} = 1}$$

Let $x = 0$

$$\frac{0}{16} + \frac{y^2}{4} = 1$$

$y = \pm 2$

Let $y = 0$

$$\frac{x^2}{16} + \frac{0}{4} = 1$$

$x = \pm 4$

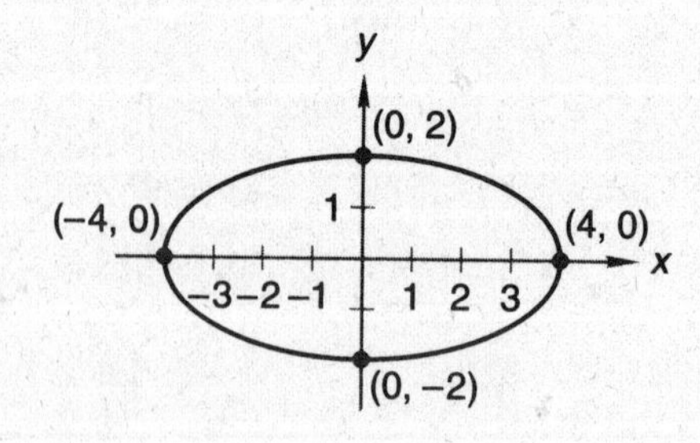

11. (70) $z_1 = \dfrac{x_1 - \mu}{\sigma} = \dfrac{7.6 - 8}{0.25} = \dfrac{-0.4}{0.25} = -1.6$

Percentile $= 0.0548$

$z_2 = \dfrac{x_2 - \mu}{\sigma} = \dfrac{8.2 - 8}{0.25} = \dfrac{0.2}{0.25} = 0.8$

Percentile $= 0.7881$

$0.7881 - 0.0548 = 0.7333$

73.33%

12. (69)

$$\begin{vmatrix} a-1 & 1 \\ 2 & a \end{vmatrix} = 4$$

$$(a-1)a - 2(1) = 4$$

$$a^2 - a - 6 = 0$$

$$(a+2)(a-3) = 0$$

$$a = \mathbf{-2, 3}$$

13. (68)

$$y = -\frac{1}{8}x^2$$

$$\frac{1}{4p}x^2 = -\frac{1}{8}x^2$$

$$4p = -8$$

$$p = -2$$

Focus: $(0, p) = (0, -2)$

Directrix: $y = -p = -(-2) = 2$

Focus = (0, –2)

Vertex = (0, 0)

Directrix: $y = 2$

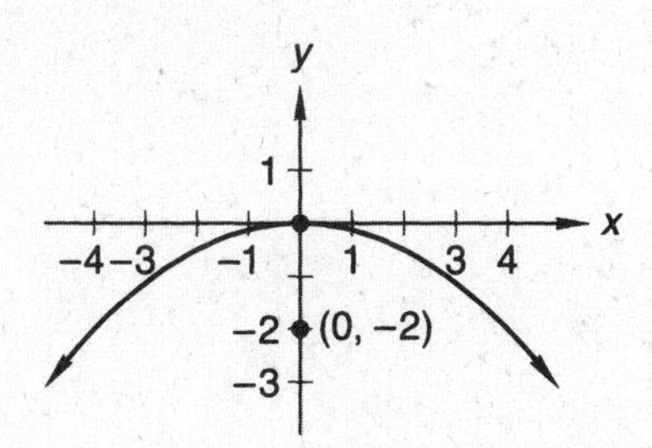

14. (68)

$$y = \frac{1}{50}x^2$$

$$\frac{1}{4p}x^2 = \frac{1}{50}x^2$$

$$4p = 50$$

$$p = \frac{50}{4} = \mathbf{12.5\ ft}$$

15. (65) $y = \log_{3.5} x$

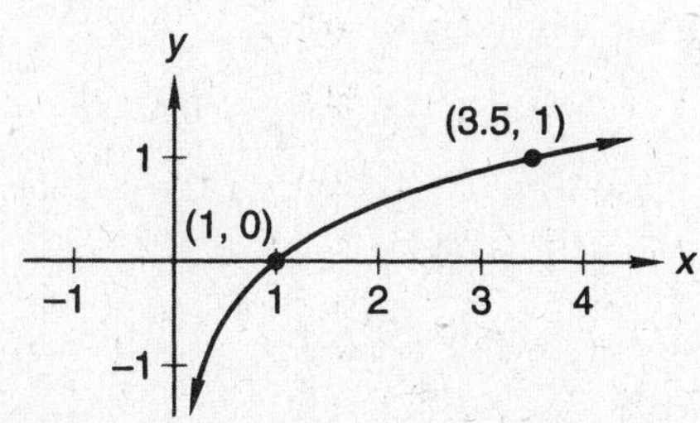

16. (66) Function = $\sin x$

Centerline = –3

Amplitude = 5

Period = 4π

Phase angle = $\frac{\pi}{2}$

Coefficient = $\frac{2\pi}{4\pi} = \frac{1}{2}$

$$\mathbf{y = -3 + 5 \sin \frac{1}{2}\left(x - \frac{\pi}{2}\right)}$$

17. (66) Coefficient = $\frac{2\pi}{\pi} = 2$

$$\mathbf{y = -10 + 20 \sin 2x}$$

18. (64) (a) $-8 + 6i$

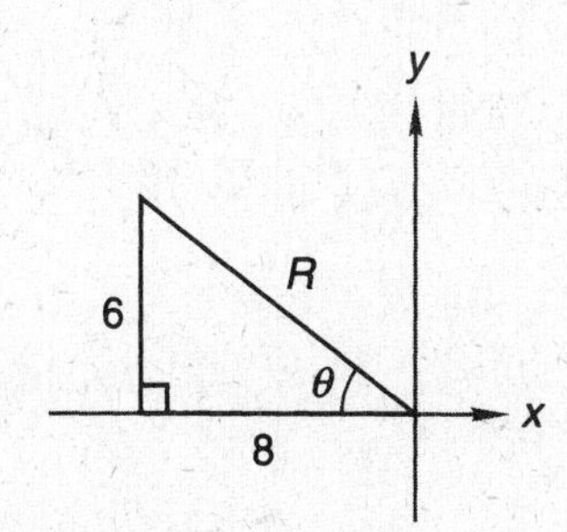

$$R = \sqrt{8^2 + 6^2} = 10$$

$$\tan \theta = \frac{6}{8}$$

$$\theta = 36.87°$$

The polar angle is $180° - 36.87° = 143.13°$.

10 cis 143.13°

(b) $3 \text{ cis } (-420°) = 3 \text{ cis } (-60°)$

$$= 3[\cos(-60°) + i \sin(-60°)]$$

$$= \mathbf{\frac{3}{2} - \frac{3\sqrt{3}}{2}i}$$

19. (64)

$$\left(6 \text{ cis } \frac{3\pi}{5}\right)\left(3 \text{ cis } \frac{2\pi}{5}\right)$$

$$= 18 \text{ cis } (\pi)$$

$$= 18(\cos \pi + i \sin \pi)$$

$$= 18(-1 + 0i) = \mathbf{-18 + 0i}$$

20. (60)
$$\sin^2 x \cos x = \cos x$$
$$\sin^2 x \cos x - \cos x = 0$$
$$\cos x(\sin^2 x - 1) = 0$$
$$\cos x(\sin x - 1)(\sin x + 1) = 0$$

$\cos x = 0$ $\quad$ $\sin x = 1$ $\quad$ $\sin x = -1$

$x = \frac{\pi}{2}, \frac{3\pi}{2}$ $\quad$ $x = \frac{\pi}{2}$ $\quad$ $x = \frac{3\pi}{2}$

$$x = \mathbf{\frac{\pi}{2}, \frac{3\pi}{2}}$$

21. (60)
$$2 \sin^2 t + \sin t - 1 = 0$$
$$(2 \sin t - 1)(\sin t + 1) = 0$$

$\sin t = \frac{1}{2}$ $\quad$ $\sin t = -1$

$t = \frac{\pi}{6}, \frac{5\pi}{6}$ $\quad$ $t = \frac{3\pi}{2}$

$$t = \mathbf{\frac{\pi}{6}, \frac{5\pi}{6}, \frac{3\pi}{2}}$$

22. (52)
$$\tan 2\theta - \sqrt{3} = 0$$
$$\tan 2\theta = \sqrt{3}$$
$$2\theta = \frac{\pi}{3}, \frac{4\pi}{3}, \frac{7\pi}{3}, \frac{10\pi}{3}$$
$$\theta = \mathbf{\frac{\pi}{6}, \frac{2\pi}{3}, \frac{7\pi}{6}, \frac{5\pi}{3}}$$

23. (48)
$$\sin^2\left(-\frac{2\pi}{3}\right) + \cos^2\left(-\frac{2\pi}{3}\right)$$
$$= \left(-\sin\frac{\pi}{3}\right)^2 + \left(-\cos\frac{\pi}{3}\right)^2$$
$$= \left(-\frac{\sqrt{3}}{2}\right)^2 + \left(-\frac{1}{2}\right)^2 = \frac{3}{4} + \frac{1}{4} = \mathbf{1}$$

24. (48)
$$\cot^3 90° - \sec^2(-225°) + \csc^2 150°$$
$$= \cot^3 90° - (-\sec 45°)^2 + \csc^2 30°$$
$$= 0 - (-\sqrt{2})^2 + (2)^2 = -2 + 4 = \mathbf{2}$$

25. (61)
$$\text{Mean} = \frac{2 + 6 + 7 + 7 + 8 + 3}{6} = \frac{11}{2} = \mathbf{5.5}$$

Median = **6.5**

Mode = **7**

$$\text{Variance} = \frac{1}{6}\big[(2 - 5.5)^2 + (6 - 5.5)^2 + (7 - 5.5)^2 + (7 - 5.5)^2 + (8 - 5.5)^2 + (3 - 5.5)^2\big]$$
$$= \mathbf{4.92}$$

$$\text{Standard deviation} = \sqrt{4.92} = \mathbf{2.22}$$

26. (56)
$$\begin{cases} x^2 + (y + 2)^2 \le 9 & \text{(big circle)} \\ x^2 + (y + 2)^2 \ge 1 & \text{(little circle)} \end{cases}$$

The region must be on or inside the circle $x^2 + (y + 2)^2 \le 9$ and on or outside the circle $x^2 + (y + 2)^2 \ge 1$.

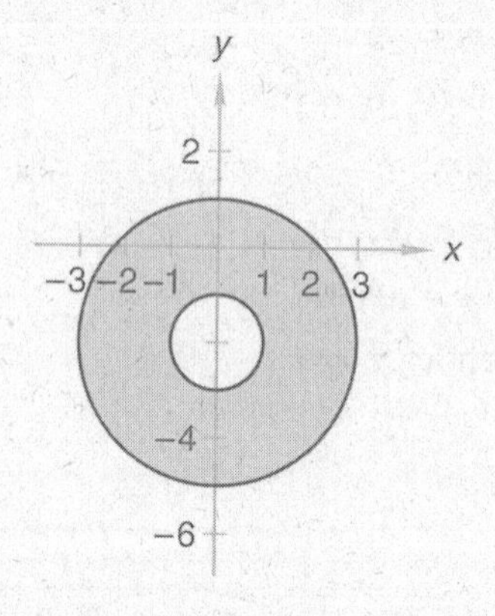

27. (59)
$$2 \log_7 x + \log_7 49 = 5^{\log_5 2}$$
$$\log_7 x^2 + \log_7 7^2 = 2$$
$$\log_7 7^2x^2 = 2$$
$$7^2x^2 = 7^2$$
$$x^2 = 1$$
$$x = \pm 1 \quad (x \ne -1)$$
$$x = \mathbf{1}$$

28. (59)
$$\frac{1}{2} \log_8 16 + \log_8 x = 2 \log_8 (x + 1)$$
$$\log_8 16^{\frac{1}{2}} + \log_8 x = \log_8 (x + 1)^2$$
$$\log_8 4x = \log_8 (x + 1)^2$$
$$4x = (x + 1)^2$$
$$4x = x^2 + 2x + 1$$
$$x^2 - 2x + 1 = 0$$
$$(x - 1)^2 = 0$$
$$x = \mathbf{1}$$

29. $x^2 + 3x + 10 = 0$
(46)

$$x = \frac{-3 \pm \sqrt{3^2 - 4(1)(10)}}{2(1)}$$

$$x = \frac{-3 \pm \sqrt{-31}}{2} = -\frac{3}{2} \pm \frac{\sqrt{31}}{2}i$$

$$\left(\boldsymbol{x + \frac{3}{2} - \frac{\sqrt{31}}{2}i}\right)\left(\boldsymbol{x + \frac{3}{2} + \frac{\sqrt{31}}{2}i}\right)$$

30. $(f \circ g)(x) = \log\left\{3\left[\frac{1}{3}\left(10^x - 1\right)\right] + 1\right\}$
(40)

$$= \log\left[\left(10^x - 1\right) + 1\right]$$

$$= \log\left(10^x\right) = x$$

$$(g \circ f)(x) = \frac{1}{3}\left[10^{\log(3x+1)} - 1\right]$$

$$= \frac{1}{3}\left[(3x + 1) - 1\right] = \frac{1}{3}(3x) = x$$

$(f \circ g)(x) = (g \circ f)(x) = x$

Yes, f and g are inverse functions.

Problem Set 73

1. $\frac{7!}{3!2!2!} = \mathbf{210}$
(55)

2. $R_{\text{increase}} = R_{\text{after}} - R_{\text{before}}$
(44)

$$= \frac{1000}{M - 5} - \frac{1000}{M}$$

$$= \frac{1000M - 1000(M - 5)}{M(M - 5)}$$

$$= \frac{\mathbf{5000}}{\boldsymbol{M(M - 5)}} \textbf{ dollars}$$

3. $\frac{\text{Part that remains}}{\text{Whole homework}} = \frac{\boldsymbol{h - w}}{\boldsymbol{h}}$
(44)

4. $W = 3S$
(18)

$2W = 2(3S) = 6S$

Therefore, her father is **6 times** older than her son.

5. (a) $\begin{cases} 3L = 2G \\ L + G = 70 \end{cases}$ (b)
(18)

(b) $L + G = 70$

$L = 70 - G$

(a)
$$3L = 2G$$
$$3(70 - G) = 2G$$
$$210 - 3G = 2G$$
$$5G = 210$$
$$G = 42$$

$L = 70 - G = 70 - (42) = 28$

The numbers are **28** and **42.**

6. $\boldsymbol{\frac{x + 2}{x + 1}}$
(18)

7.
(73)

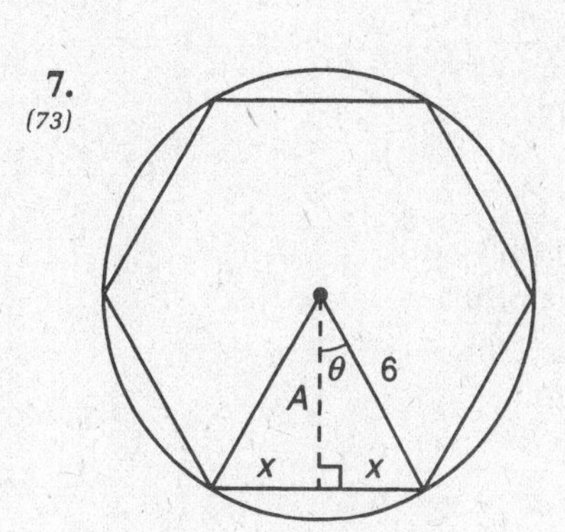

$$\theta = \frac{1}{2}\left(\frac{360°}{6}\right) = 30°$$

$x = 6 \sin 30° = 3$ in.

$A = 6 \cos 30° = 5.1962$ in.

$\text{Area}_{\Delta} = \frac{1}{2}(3)(5.1962) = 7.7943 \text{ in.}^2$

$\text{Area} = (12)(7.7943) = \mathbf{93.53 \text{ in.}^2}$

8.
(73)

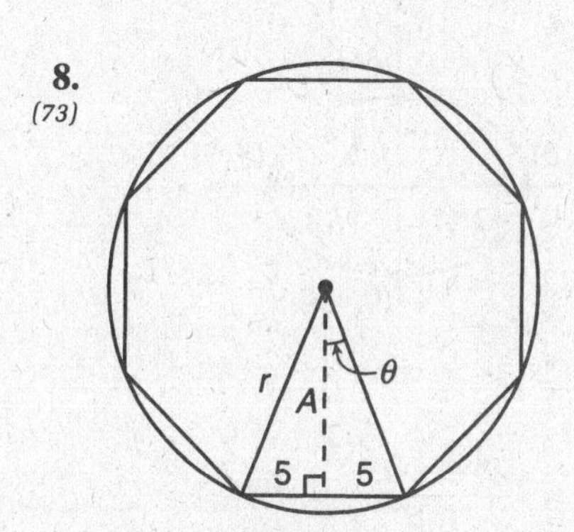

$$\theta = \frac{1}{2}\left(\frac{360°}{8}\right) = 22.5°$$

$$A = \frac{5}{\tan 22.5°} = 12.071 \text{ cm}$$

$\text{Area}_{\Delta} = \frac{1}{2}(5)(12.071) = 30.1775 \text{ cm}^2$

$\text{Area} = (16)(30.1775) = \mathbf{482.84 \text{ cm}^2}$

9. (5) $S_L = 9S_S$

$$\frac{S_S}{S_L} = \frac{1}{9}$$

$$\frac{A_S}{A_L} = \frac{(S_S)^2}{(S_L)^2} = \left(\frac{S_S}{S_L}\right)^2 = \left(\frac{1}{9}\right)^2 = \mathbf{\frac{1}{81}}$$

10. (72) $A = 180° - 130° - 30°$

$\mathbf{A = 20°}$

$$\frac{b}{\sin 30°} = \frac{15}{\sin 130°}$$

$$b = \frac{15 \sin 30°}{\sin 130°}$$

$\mathbf{b = 9.79}$

$$\frac{a}{\sin 20°} = \frac{15}{\sin 130°}$$

$$a = \frac{15 \sin 20°}{\sin 130°}$$

$\mathbf{a = 6.70}$

11. (71) $4x^2 + 9y^2 = 36$

$$\mathbf{\frac{x^2}{9} + \frac{y^2}{4} = 1}$$

Let $x = 0$

$$\frac{0}{9} + \frac{y^2}{4} = 1$$

$y = \pm 2$

Let $y = 0$

$$\frac{x^2}{9} + \frac{0}{4} = 1$$

$x = \pm 3$

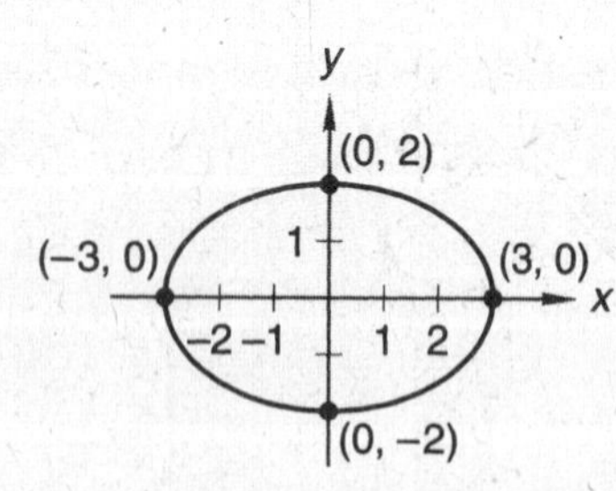

12. (70) $z_+ = \dfrac{x - \mu}{\sigma} = \dfrac{4.14 - 4.0}{0.1} = 1.4$

Percentile $= 0.9192$

$z_- = \dfrac{x - \mu}{\sigma} = \dfrac{3.94 - 4.0}{0.1} = -0.6$

Percentile $= 0.2743$

$0.9192 - 0.2743 = 0.6449$

64.49%

13. (69)

$$\begin{vmatrix} b + 2 & 1 \\ -12 & b \end{vmatrix} = 12$$

$(b + 2)b - (-12)(1) = 12$

$b(b + 2) = 0$

$\mathbf{b = -2, 0}$

14. (68) Vertex: $(h, k) = (3, -2)$

Focus: $(h, k + p) = (3, 5)$

$k + p = 5$

$(-2) + p = 5$

$p = 7$

Directrix: $y = k - p = (-2) - (7) = -9$

Axis of symmetry: $x = h = 3$

$$y - k = \frac{1}{4p}(x - h)^2$$

$$y - (-2) = \frac{1}{4(7)}(x - 3)^2$$

$$y = \frac{1}{28}(x - 3)^2 - 2$$

Parabola: $\mathbf{y = \frac{1}{28}(x - 3)^2 - 2}$

Directrix: $\mathbf{y = -9}$

Axis of symmetry: $\mathbf{x = 3}$

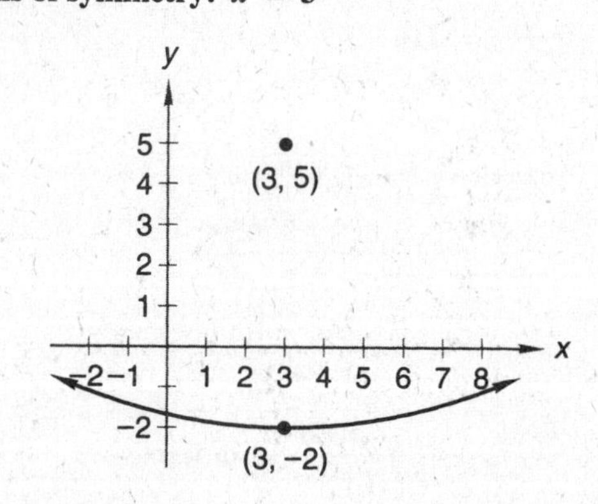

15. (68) $y = \dfrac{1}{9}x^2$

$$\frac{1}{4p}x^2 = \frac{1}{9}x^2$$

$$p = \frac{9}{4}$$

Focus: $(0, p) = \left(0, \dfrac{9}{4}\right)$

Directrix: $y = -p = -\dfrac{9}{4}$

Axis of symmetry: $x = h = 0$

Focus $= \left(0, \frac{9}{4}\right)$

Directrix: $y = -\frac{9}{4}$

Vertex = (0, 0)

Axis of symmetry: $x = 0$

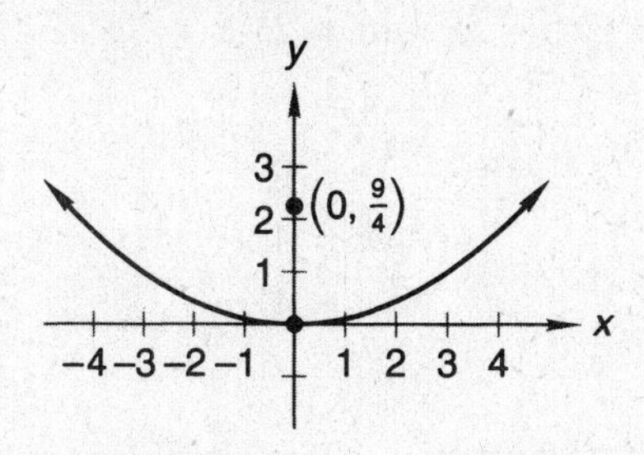

16. $y = \log_{\frac{7}{4}} x$
(65)

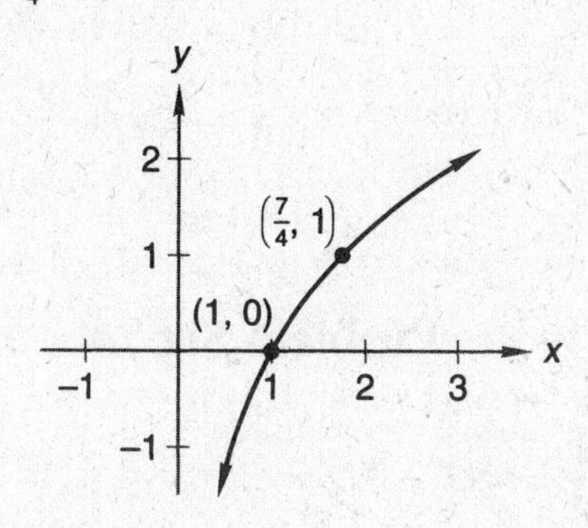

17. Function = $\cos x$
(66)
Centerline = −3

Amplitude = 2

Period = 216°

Phase angle = 81°

Coefficient $= \frac{360°}{216°} = \frac{5}{3}$

$$\mathbf{y = -3 + 2\cos\frac{5}{3}(x - 81°)}$$

18. $\mathbf{y = -2 + 6\sin\left(x - \frac{\pi}{4}\right)}$
(66)

19. $(3 \text{ cis } 420°)(4 \text{ cis } 330°) = 12 \text{ cis } 750°$
(64)
$= 12 \text{ cis } 30° = 12(\cos 30° + i \sin 30°)$

$= 12\left(\frac{\sqrt{3}}{2} + \frac{1}{2}i\right) = \mathbf{6\sqrt{3} + 6i}$

20.
(62)

$$\begin{cases} \text{(a)} & ax + by = c \\ \text{(b)} & dx + fy = g \end{cases}$$

$$\begin{aligned} -d\text{(a)} \quad & -dax - bdy = -cd \\ a\text{(b)} \quad & dax + afy = ag \\ \hline & (af - bd)y = ag - cd \\ & \mathbf{y = \frac{ag - cd}{af - bd}} \end{aligned}$$

21. $\begin{cases} 2x + 6y = 14 \\ x - 4y = 8 \end{cases}$
(66)

$a = 2,\ b = 6,\ c = 14,\ d = 1,\ f = -4,\ g = 8$

$$y = \frac{ag - cd}{af - bd}$$

$$y = \frac{2(8) - 14(1)}{2(-4) - 6(1)}$$

$$y = \frac{2}{-14}$$

$$\mathbf{y = -\frac{1}{7}}$$

22. $\sin\theta - 2\sin^2\theta = 0$
(60)

$\sin\theta\,(1 - 2\sin\theta) = 0$

$\sin\theta = 0$
$\theta = 0°, 180°$

$2\sin\theta = 1$
$\sin\theta = \frac{1}{2}$
$\theta = 30°, 150°$

$\mathbf{\theta = 0°, 30°, 150°, 180°}$

23. $\sec^2\theta - 4 = 0$
(60)

$(\sec\theta - 2)(\sec\theta + 2) = 0$

$\sec\theta = 2$
$\theta = 60°, 300°$

$\sec\theta = -2$
$\theta = 120°, 240°$

$\mathbf{\theta = 60°, 120°, 240°, 300°}$

24. $2\cos^2 x - 3\cos x - 2 = 0$
(60)

$(2\cos x + 1)(\cos x - 2) = 0$

$2\cos x = -1$

$\cos x = -\frac{1}{2}$

$x = \mathbf{120°, 240°}$

$\cos x = 2$
no answer

25. (48) $\tan^2(-690°) - \sec^2(-690°) - \cot^2 690°$

$= \tan^2(30°) - \sec^2(30°) - \cot^2(-30°)$

$= \left(\frac{1}{\sqrt{3}}\right)^2 - \left(\frac{2}{\sqrt{3}}\right)^2 - (-\sqrt{3})^2$

$= \frac{1}{3} - \frac{4}{3} - 3 = \mathbf{-4}$

26. (39,48) $\cos^2\left(\frac{\pi}{4}\right) + \sin^2\left(\frac{3\pi}{4}\right)\cos^2\left(\frac{5\pi}{4}\right)$

$= \left(\frac{1}{\sqrt{2}}\right)^2 + \left(\frac{1}{\sqrt{2}}\right)^2\left(-\frac{1}{\sqrt{2}}\right)^2$

$= \frac{1}{2} + \left(\frac{1}{2}\right)\left(\frac{1}{2}\right) = \frac{1}{2} + \frac{1}{4} = \mathbf{\frac{3}{4}}$

27. (56) $\begin{cases} 16 \geq x^2 + y^2 & \text{(circle)} \\ y > x - 2 & \text{(line)} \end{cases}$

The region must be on or inside the circle and above the line.

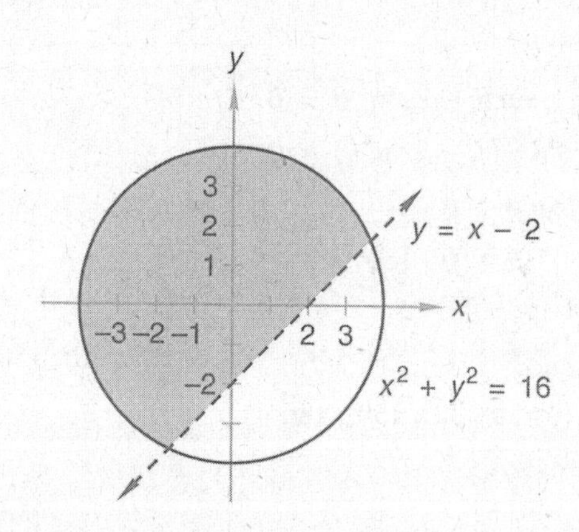

28. (59) $\frac{1}{3}\ln 27 + 2\ln x = \ln(2 - x)$

$\ln 27^{\frac{1}{3}} + \ln x^2 = \ln(2 - x)$

$\ln(3x^2) = \ln(2 - x)$

$3x^2 = 2 - x$

$3x^2 + x - 2 = 0$

$(3x - 2)(x + 1) = 0$

$x = \frac{2}{3}, -1 \quad (x \neq -1)$

$x = \mathbf{\frac{2}{3}}$

29. (59) $\log_8(x - 2) + \log_8(x - 3) = \frac{1}{3}$

$\log_8(x - 2)(x - 3) = \frac{1}{3}$

$(x - 2)(x - 3) = 8^{\frac{1}{3}}$

$x^2 - 5x + 6 = 2$

$x^2 - 5x + 4 = 0$

$(x - 4)(x - 1) = 0$

$x = 4, 1 \quad (x \neq 1)$

$x = \mathbf{4}$

30. (37) $\sqrt{(x + 6)^2 + (y - 4)^2} = \sqrt{(x - 4)^2 + (y - 9)^2}$

$x^2 + 12x + 36 + y^2 - 8y + 16$

$= x^2 - 8x + 16 + y^2 - 18y + 81$

$10y = -20x + 45$

$\mathbf{y = -2x + \frac{9}{2}}$

Problem Set 74

1. (55) $\frac{12!}{8!4!} = \mathbf{495}$

2. (18) $\begin{cases} \text{(a)}\ L_1 = 3L_2 \\ \text{(b)}\ L_2 = 3L_3 \\ \text{(c)}\ L_1 + L_2 + L_3 = 1 \end{cases}$

(c) $L_1 + L_2 + L_3 = 1$

$(3L_2) + L_2 + L_3 = 1$

$4L_2 + L_3 = 1$

$4(3L_3) + L_3 = 1$

$13L_3 = 1$

$L_3 = \mathbf{\frac{1}{13}}$

3. (53) **(36y − 12f − n) in.**

4. (25) $\text{Rate}_{\text{Sat}} = \frac{40.25}{2.5} = 16.10\ \frac{\text{dollars}}{\text{hr}}$

$2.5 \times \text{Rate} = \text{Rate}_{\text{Sat}}$

$\text{Rate} = \frac{\text{Rate}_{\text{Sat}}}{2.5} = \frac{16.10}{2.5} =$ **\$6.44 per hour**

5. (R) $\dfrac{\text{Part saved}}{\text{Money earned}} = \dfrac{E - P}{E}$

6. (44) Item price − cost of item = profit per item

$$\frac{D}{N} - x = \frac{P}{N}$$

$$x = \frac{D}{N} - \frac{P}{N}$$

$$x = \mathbf{\frac{D - P}{N}\ \frac{dollars}{item}}$$

7. (38,44) $D_1 = m,\ D_2 = m,\ T_1 = \dfrac{m}{p},\ T_2 = \dfrac{m}{k}$

Overall average speed $= \dfrac{\text{total distance}}{\text{total time}}$

$$= \frac{m + m}{\dfrac{m}{p} + \dfrac{m}{k}} = \frac{2m}{\dfrac{m(k + p)}{pk}} = \mathbf{\frac{2pk}{k + p}\ mph}$$

8. (74) $x = \dfrac{\begin{vmatrix} -4 & -3 \\ 8 & 2 \end{vmatrix}}{\begin{vmatrix} 2 & -3 \\ 4 & 2 \end{vmatrix}} = \dfrac{-8 + 24}{4 + 12} = \dfrac{16}{16} = 1$

$$y = \frac{\begin{vmatrix} 2 & -4 \\ 4 & 8 \end{vmatrix}}{\begin{vmatrix} 2 & -3 \\ 4 & 2 \end{vmatrix}} = \frac{16 + 16}{4 + 12} = \frac{32}{16} = 2$$

$\mathbf{x = 1;\ y = 2}$

9. (74) $x = \dfrac{\begin{vmatrix} -2 & 1 \\ -3 & -2 \end{vmatrix}}{\begin{vmatrix} 3 & 1 \\ 1 & -2 \end{vmatrix}} = \dfrac{4 + 3}{-6 - 1} = \dfrac{7}{-7} = -1$

$$y = \frac{\begin{vmatrix} 3 & -2 \\ 1 & -3 \end{vmatrix}}{\begin{vmatrix} 3 & 1 \\ 1 & -2 \end{vmatrix}} = \frac{-9 + 2}{-6 - 1} = \frac{-7}{-7} = 1$$

$\mathbf{x = -1;\ y = 1}$

10. (73)

θ
10 cm
A
x

$$\theta = \frac{1}{2}\left(\frac{360°}{12}\right) = 15°$$

$x = 10 \sin 15° = 2.5882$ cm

$A = 10 \cos 15° = 9.66$ cm

$\text{Area}_\Delta = \dfrac{1}{2}(2.5882)(9.66) = 12.5\ \text{cm}^2$

Area $= 24(12.5) = \mathbf{300\ cm^2}$

11. (73)

θ
A
r
2 cm

$$\theta = \frac{1}{2}\left(\frac{360°}{6}\right) = 30°$$

$$r = \frac{2}{\sin 30°} = \mathbf{4\ cm}$$

12. (5) $S_L = 1.5S_S$

$2S_L = 3S_S$

$$\frac{S_L}{S_S} = \frac{3}{2}$$

$$\frac{A_L}{A_S} = \frac{(S_L)^2}{(S_S)^2} = \left(\frac{S_L}{S_S}\right)^2 = \left(\frac{3}{2}\right)^2 = \mathbf{\frac{9}{4}}$$

13. (72) $A = 180° - 15° - 30°$

$\mathbf{A = 135°}$

$$\frac{a}{\sin 135°} = \frac{20}{\sin 30°}$$

$$a = \frac{20 \sin 135°}{\sin 30°}$$

$\mathbf{a = 28.28}$

$$\frac{b}{\sin 15°} = \frac{20}{\sin 30°}$$

$$b = \frac{20 \sin 15°}{\sin 30°}$$

$\mathbf{b = 10.35}$

14. $25x^2 + 16y^2 = 400$
(71)

$$\frac{\mathbf{x^2}}{\mathbf{16}} + \frac{\mathbf{y^2}}{\mathbf{25}} = \mathbf{1}$$

Let $x = 0$ Let $y = 0$

$\frac{0}{16} + \frac{y^2}{25} = 1$ $\frac{x^2}{16} + \frac{0}{25} = 1$

$y = \pm 5$ $x = \pm 4$

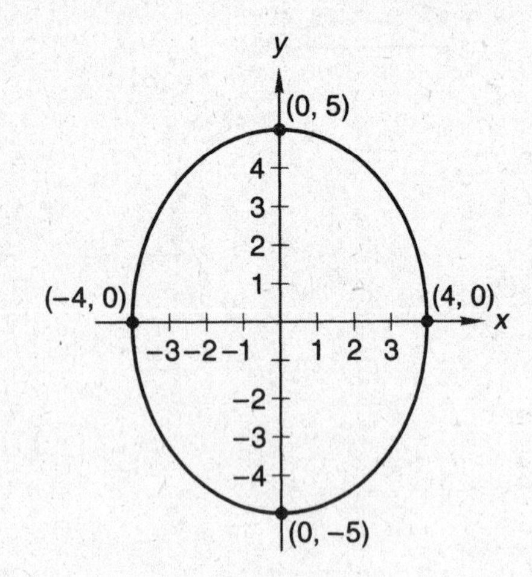

15. $z = \frac{x - \mu}{\sigma} = \frac{120 - 100}{8} = 2.5$
(70)

Percentile = 0.9938

$1 - 0.9938 = 0.0062$

0.62%

16. $\begin{vmatrix} x - 1 & 5 \\ 3 & x \end{vmatrix} = 5$
(69)

$(x - 1)x - (3)5 = 5$

$x^2 - x - 20 = 0$

$(x - 5)(x + 4) = 0$

$x = \mathbf{-4, 5}$

17. Vertex: $(h, k) = (-2, 1)$
(68)

Focus: $(h, k + p) = (-2, -2)$

$k + p = -2$

$(1) + p = -2$

$p = -3$

Directrix: $y = k - p = (1) - (-3) = 4$

Axis of symmetry: $x = h = -2$

$$y - k = \frac{1}{4p}(x - h)^2$$

$$y - (1) = \frac{1}{4(-3)}[x - (-2)]^2$$

$$y = -\frac{1}{12}(x + 2)^2 + 1$$

Parabola: $\mathbf{y = -\frac{1}{12}(x + 2)^2 + 1}$

Directrix: $\mathbf{y = 4}$

Axis of symmetry: $\mathbf{x = -2}$

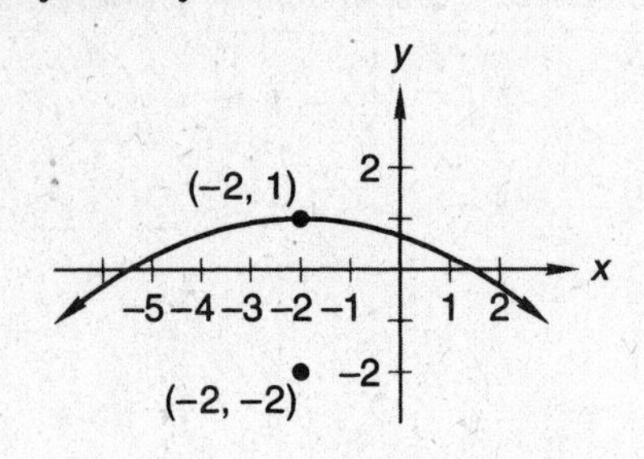

18. $y = \log_{\frac{8}{3}} x$
(65)

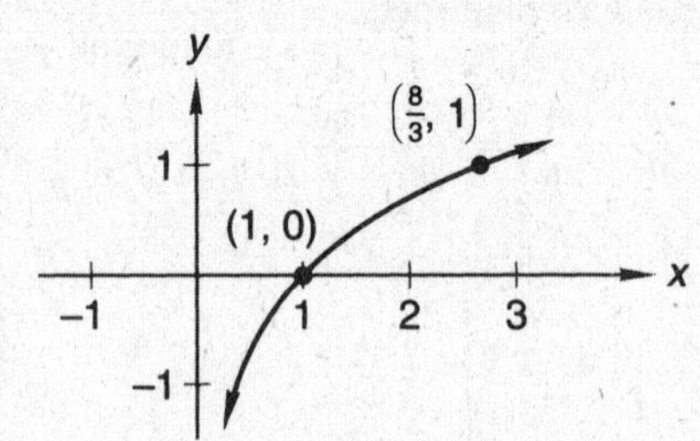

19.
(63)

$$x^2 + 10x + y^2 - 4y + 20 = 0$$

$$(x^2 + 10x \quad) + (y^2 - 4y \quad) = -20$$

$$(x^2 + 10x + 25) + (y^2 - 4y + 4) = 9$$

$$(x + 5)^2 + (y - 2)^2 = 3^2$$

Center = (−5, 2); radius = 3

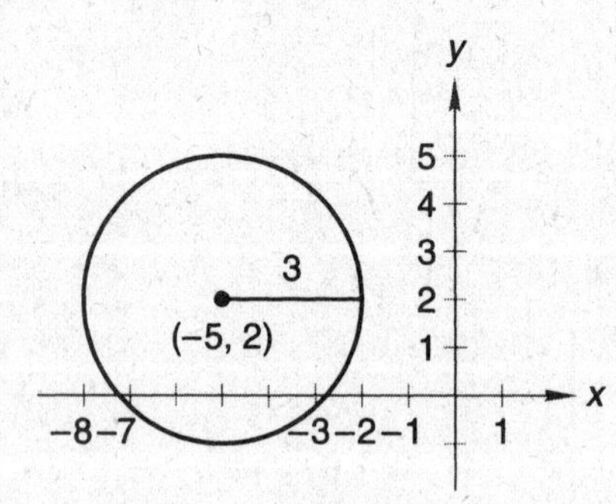

20. Function = $\sin x$
(66)

Centerline = 3

Amplitude = 11

Period = 240°

Phase angle = 40°

Coefficient = $\frac{360°}{240°} = \frac{3}{2}$

$\mathbf{y = 3 + 11 \sin \frac{3}{2}(x - 40°)}$

21. $y = -1 + 14 \sin (x + 45°)$
(66)

22. $\left[4 \text{ cis}\left(-\frac{\pi}{3}\right)\right]\left[6 \text{ cis}\left(\frac{4\pi}{3}\right)\right]$
(64)

$= 24 \text{ cis } \pi = 24 \cos \pi + 24i \sin \pi$

$= 24(-1 + 0i) = \mathbf{-24 + 0i}$

23. (a) $2 - 3i$
(64)

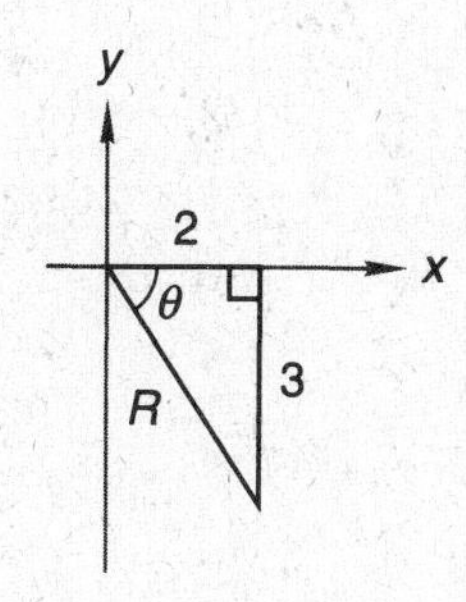

$R = \sqrt{2^2 + 3^2} = \sqrt{13} = 3.61$

$\tan \theta = \frac{3}{2}$

$\theta = 56.31°$

The polar angle is $360° - 56.31° = 303.69°$.

3.61 cis 303.69°

(b) $13 \text{ cis } \frac{5\pi}{6}$

$= 13\left(\cos \frac{5\pi}{6} + i \sin \frac{5\pi}{6}\right)$

$= 13\left(-\frac{\sqrt{3}}{2} + \frac{1}{2}i\right)$

$= \mathbf{-\frac{13\sqrt{3}}{2} + \frac{13}{2}i}$

24. $4 \cos^2 t - 3 = 0$
(60)

$\cos^2 t - \frac{3}{4} = 0$

$\left(\cos t - \frac{\sqrt{3}}{2}\right)\left(\cos t + \frac{\sqrt{3}}{2}\right) = 0$

$\cos t = \frac{\sqrt{3}}{2}$ $\qquad$ $\cos t = -\frac{\sqrt{3}}{2}$

$t = \frac{\pi}{6}, \frac{11\pi}{6}$ $\qquad$ $t = \frac{5\pi}{6}, \frac{7\pi}{6}$

$t = \mathbf{\frac{\pi}{6}, \frac{5\pi}{6}, \frac{7\pi}{6}, \frac{11\pi}{6}}$

25. $2 \cos 2\theta + 1 = 0$
(52)

$\cos 2\theta = -\frac{1}{2}$

$2\theta = \frac{2\pi}{3}, \frac{4\pi}{3}, \frac{8\pi}{3}, \frac{10\pi}{3}$

$\theta = \mathbf{\frac{\pi}{3}, \frac{2\pi}{3}, \frac{4\pi}{3}, \frac{5\pi}{3}}$

26. $\sec x - \sqrt{2 \sec x - 1} = 0$
(65)

$\sec^2 x = 2 \sec x - 1$

$\sec^2 x - 2 \sec x + 1 = 0$

$(\sec x - 1)^2 = 0$

$\sec x = 1$

$x = \mathbf{0}$

27. $\csc^2\left(-\frac{3\pi}{4}\right) - \cot^2\left(-\frac{3\pi}{4}\right)$
(48)

$+ \sec^2\left(-\frac{3\pi}{4}\right) - \tan^2\left(-\frac{3\pi}{4}\right)$

$= \left(-\csc \frac{\pi}{4}\right)^2 - \left(\cot \frac{\pi}{4}\right)^2$

$+ \left(-\sec \frac{\pi}{4}\right)^2 - \left(\tan \frac{\pi}{4}\right)^2$

$= (-\sqrt{2})^2 - (1)^2 + (-\sqrt{2})^2 - (-1)^2$

$= 2 - 1 + 2 - 1 = \mathbf{2}$

28. $\frac{1}{4} \ln 16 + 2 \ln x = \ln (6 - x)$
(59)

$\ln 16^{\frac{1}{4}} + \ln x^2 = \ln (6 - x)$

$\ln 2x^2 = \ln (6 - x)$

$2x^2 = 6 - x$

$2x^2 + x - 6 = 0$

$(2x - 3)(x + 2) = 0$

$x = \frac{3}{2}, -2$ $\qquad (x \neq -2)$

$x = \mathbf{\frac{3}{2}}$

29. (a) $b^{\log_b 24 - \log_b 6 + 2 \log_b 3}$
(59)

$= b^{\log_b \frac{(24)(3^2)}{6}} = \frac{24(9)}{6} = \mathbf{36}$

(b) $2 \ln e - 2 \log 10^2 = 2(1) - 2(2) = \mathbf{-2}$

(c) $2 \ln e^4 + 4 \ln e^{\frac{1}{2}} - \log 10^5$

$= 2(4) + 4\left(\frac{1}{2}\right) - 5 = \mathbf{5}$

30. (46)
$$x^2 + 20 = 0$$
$$x^2 = -20$$
$$x = \pm\sqrt{-20} = \pm 2\sqrt{5}\,i$$
$$\mathbf{(x - 2\sqrt{5}i)(x + 2\sqrt{5}i)}$$

Problem Set 75

1. (44)
$$c = 4.40 + 1.10(m - 3)$$
$$= 4.40 + 1.10m - 3.30$$
$$= \mathbf{\$1.10 + \$1.10m}$$

2. (44)
$$R_{\text{after}} - R_{\text{before}} = \frac{D}{p - 20} - \frac{D}{p}$$
$$= \frac{Dp}{(p - 20)p} - \frac{D(p - 20)}{p(p - 20)}$$
$$= \mathbf{\frac{20D}{p^2 - 20p}\ \frac{\text{dollars}}{\text{student}}}$$

3. (18)
$$p = \frac{km^2}{\sqrt{N}}$$
$$\frac{k(4m)^2}{\sqrt{9N}} = \frac{16}{3}\frac{km^2}{\sqrt{N}} = \frac{16}{3}p$$

Therefore, **p is multiplied by $\frac{16}{3}$.**

4. (25)
$$\text{Distance} = \left(\frac{\text{miles}}{\text{gallon}}\right)(\text{gallons of gas})$$
$$D_1 = \left(\frac{7 \text{ miles}}{\text{gallon}}\right)(1 \text{ gallon}) = 7 \text{ miles}$$
$$D_2 = \left(\frac{7 \text{ miles}}{\frac{1}{3} \text{ gallon}}\right)(1 \text{ gallon}) = 21 \text{ miles}$$
$$D_2 - D_1 = \mathbf{14 \text{ miles}}$$

5. (75)
$${}_8C_3 = \frac{8!}{(8 - 3)!3!} = \mathbf{56 \text{ triangles}}$$

6. (75)
$${}_8C_4 = \frac{8!}{(8 - 4)!4!} = \mathbf{70}$$

7. (74)
$$x = \frac{\begin{vmatrix} -3 & -4 \\ -4 & 3 \end{vmatrix}}{\begin{vmatrix} 3 & -4 \\ 2 & 3 \end{vmatrix}} = \frac{-9 - 16}{9 - (-8)} = -\frac{25}{17}$$
$$y = \frac{\begin{vmatrix} 3 & -3 \\ 2 & -4 \end{vmatrix}}{\begin{vmatrix} 3 & -4 \\ 2 & 3 \end{vmatrix}} = \frac{-12 - (-6)}{9 - (-8)} = -\frac{6}{17}$$
$$\mathbf{x = -\frac{25}{17};\ y = -\frac{6}{17}}$$

8. (73)

θ
10 cm
A
x

$$\theta = \frac{1}{2}\left(\frac{360°}{10}\right) = 18°$$
$$A = 10 \cos 18° = 9.511 \text{ cm}$$
$$x = 10 \sin 18° = 3.09 \text{ cm}$$
$$\text{Area}_\Delta = \frac{1}{2}(3.09)(9.511) = 14.6945 \text{ cm}^2$$
$$\text{Area} = (20)(14.6945) = \mathbf{293.89 \text{ cm}^2}$$

9. (73)

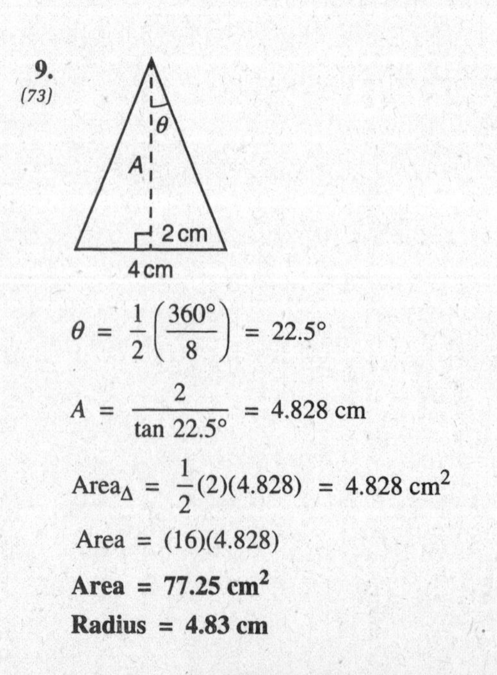

$$\theta = \frac{1}{2}\left(\frac{360°}{8}\right) = 22.5°$$
$$A = \frac{2}{\tan 22.5°} = 4.828 \text{ cm}$$
$$\text{Area}_\Delta = \frac{1}{2}(2)(4.828) = 4.828 \text{ cm}^2$$
$$\text{Area} = (16)(4.828)$$
$$\mathbf{Area = 77.25 \text{ cm}^2}$$
$$\mathbf{Radius = 4.83 \text{ cm}}$$

10. (5)

$$S_L = 2.5S_S$$
$$2S_L = 5S_S$$
$$\frac{S_L}{S_S} = \frac{5}{2}$$
$$\frac{A_L}{A_S} = \frac{(S_L)^2}{(S_S)^2} = \left(\frac{S_L}{S_S}\right)^2 = \left(\frac{5}{2}\right)^2 = \mathbf{\frac{25}{4}}$$

11. (72)

$$A = 180° - 70° - 60°$$
$$\mathbf{A = 50°}$$
$$\frac{a}{\sin 50°} = \frac{8}{\sin 60°}$$
$$a = \frac{8 \sin 50°}{\sin 60°}$$
$$\mathbf{a = 7.08}$$
$$\frac{b}{\sin 70°} = \frac{8}{\sin 60°}$$
$$b = \frac{8 \sin 70°}{\sin 60°}$$
$$\mathbf{b = 8.68}$$

12. (71)

$$5x^2 + 8y^2 = 40$$
$$\mathbf{\frac{x^2}{8} + \frac{y^2}{5} = 1}$$

Let $x = 0$

$$\frac{0}{8} + \frac{y^2}{5} = 0$$
$$y = \pm\sqrt{5}$$

Let $y = 0$

$$\frac{x^2}{8} + \frac{0}{5} = 1$$
$$x = \pm\sqrt{8}$$

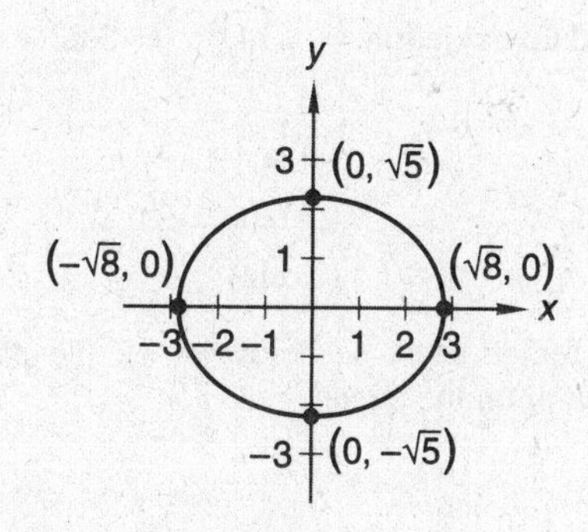

13. (69)

$$\begin{vmatrix} 3 & k \\ k+1 & k \end{vmatrix} = 1$$
$$3(k) - (k + 1)k = 1$$
$$3k - k^2 - k = 1$$
$$k^2 - 2k + 1 = 0$$
$$(k - 1)^2 = 0$$
$$\mathbf{k = 1}$$

14. (68)

$$y = -\frac{1}{16}x^2$$
$$\frac{1}{4p}x^2 = -\frac{1}{16}x^2$$
$$p = -4$$

Focus: $(0, p) = (0, -4)$

Directrix: $y = -p = -(-4) = 4$

Focus = (0, −4)

Vertex = (0, 0)

Directrix: $y = 4$

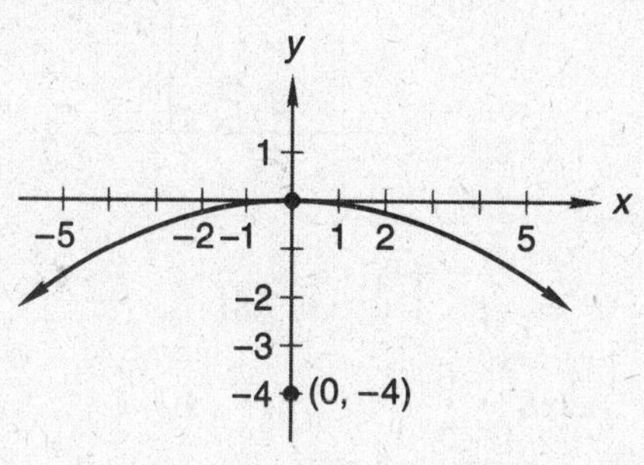

15. (65)

$$y = 1 + \log_2 x$$

16. (66)

Function = $\cos x$

Centerline = 3

Amplitude = 4

Period = 540°

Phase angle = −90°

$$\text{Coefficient} = \frac{360°}{540°} = \frac{2}{3}$$
$$\mathbf{y = 3 + 4\cos\frac{2}{3}(x + 90°)}$$

17. (66)

$$\text{Coefficient} = \frac{2\pi}{4\pi} = \frac{1}{2}$$
$$\mathbf{y = 10 + 6\sin\frac{1}{2}\left(x - \frac{\pi}{2}\right)}$$

18. (64) $[5 \text{ cis } (-120°)](2 \text{ cis } 660°) = 10 \text{ cis } 540°$

$= 10 \text{ cis } 180° = 10 (\cos 180° + i \sin 180°)$

$= 10(-1 + 0i) = \mathbf{-10 + 0i}$

19. (64) (a) $5 + 6i$

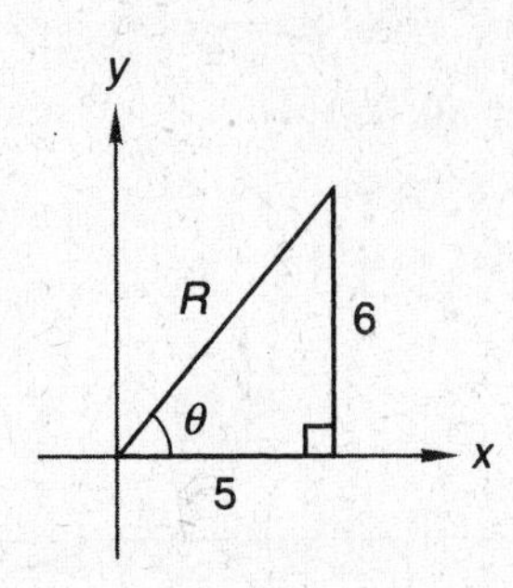

$R = \sqrt{5^2 + 6^2} = 7.81$

$\tan \theta = \dfrac{6}{5}$

$\theta = 50.19°$

7.81 cis 50.19°

(b) $-7 \text{ cis } (-585°) = -7 \text{ cis } (-225°)$

$= -7(-\cos 45° + i \sin 45°)$

$= -7\left(-\dfrac{\sqrt{2}}{2} + \dfrac{\sqrt{2}}{2}i\right)$

$= \mathbf{\dfrac{7\sqrt{2}}{2} - \dfrac{7\sqrt{2}}{2}i}$

20. (60)

$$4 \sin^2 \theta - 3 = 0$$

$$\left(\sin \theta - \frac{\sqrt{3}}{2}\right)\left(\sin \theta + \frac{\sqrt{3}}{2}\right) = 0$$

$\sin \theta = \dfrac{\sqrt{3}}{2}$ $\qquad$ $\sin \theta = -\dfrac{\sqrt{3}}{2}$

$\theta = 60°, 120°$ $\qquad$ $\theta = 240°, 300°$

$\theta = \mathbf{60°, 120°, 240°, 300°}$

21. (52) $2 \sin \dfrac{\theta}{3} - 1 = 0$

$\sin \dfrac{\theta}{3} = \dfrac{1}{2}$

$\dfrac{\theta}{3} = 30°$

$\theta = \mathbf{90°}$

22. (60) $\sin^2 \theta - 4 \sin \theta + 3 = 0$

$(\sin \theta - 3)(\sin \theta - 1) = 0$

$\sin \theta = 3$ $\qquad$ $\sin \theta = 1$

no answer $\qquad$ $\theta = \mathbf{90°}$

23. (48) $\cot^2 \dfrac{14\pi}{3} - \csc^2 \dfrac{14\pi}{3} + \sin^2 \dfrac{14\pi}{3} + \cos^2 \dfrac{14\pi}{3}$

$= \cot^2 \dfrac{2\pi}{3} - \csc^2 \dfrac{2\pi}{3} + \sin^2 \dfrac{2\pi}{3} + \cos^2 \dfrac{2\pi}{3}$

$= \left(-\cot \dfrac{\pi}{3}\right)^2 - \left(\csc \dfrac{\pi}{3}\right)^2$

$+ \left(\sin \dfrac{\pi}{3}\right)^2 + \left(-\cos \dfrac{\pi}{3}\right)^2$

$= \left(-\dfrac{1}{\sqrt{3}}\right)^2 - \left(\dfrac{2}{\sqrt{3}}\right)^2 + \left(\dfrac{\sqrt{3}}{2}\right)^2 + \left(-\dfrac{1}{2}\right)^2$

$= \dfrac{1}{3} - \dfrac{4}{3} + \dfrac{3}{4} + \dfrac{1}{4} = \mathbf{0}$

24. (61) Mean $= \dfrac{3 + 6 + 9 - 2 + 0 + 0}{6} = \mathbf{2.67}$

Median $= \mathbf{1.5}$

Mode $= \mathbf{0}$

Variance $= \dfrac{1}{6}\left[(3 - 2.67)^2 + (6 - 2.67)^2\right.$

$+ (9 - 2.67)^2 + (-2 - 2.67)^2$

$\left.+ (0 - 2.67)^2 + (0 - 2.67)^2\right]$

$= \mathbf{14.56}$

Standard deviation $= \sqrt{14.56} = \mathbf{3.82}$

25. (56) $\begin{cases} y \le x^2 - 4 & \text{(parabola)} \\ x^2 + y^2 \le 9 & \text{(circle)} \end{cases}$

The region must be below or on the parabola and inside or on the circle.

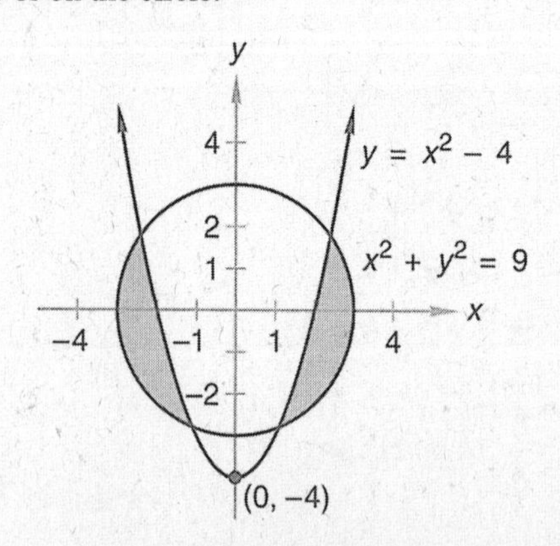

26. (58) Equation of the perpendicular line, using the point-slope formula:

$$y - y_1 = m(x - x_1)$$

$$y - 1 = -\frac{1}{3}(x - 1)$$

$$y = -\frac{1}{3}x + \frac{4}{3}$$

Point of intersection:

$$-\frac{1}{3}x + \frac{4}{3} = 3x + 2$$

$$\frac{10}{3}x = -\frac{2}{3}$$

$$x = -\frac{1}{5}$$

$$y = 3x + 2 = 3\left(-\frac{1}{5}\right) + 2 = \frac{7}{5}$$

$\left(-\frac{1}{5}, \frac{7}{5}\right)$ and $(1, 1)$

$$D = \sqrt{\left(1 + \frac{1}{5}\right)^2 + \left(1 - \frac{7}{5}\right)^2}$$

$$= \sqrt{\frac{36}{25} + \frac{4}{25}} = \sqrt{\frac{40}{25}} = \mathbf{\frac{2\sqrt{10}}{5}}$$

27. (59) $\frac{1}{3}\ln 8 + 2\ln x = \ln(-3x + 2)$

$$\ln 8^{\frac{1}{3}} + \ln x^2 = \ln(-3x + 2)$$

$$\ln 2x^2 = \ln(-3x + 2)$$

$$2x^2 + 3x - 2 = 0$$

$$(2x - 1)(x + 2) = 0$$

$$x = \frac{1}{2}, -2 \qquad (x \neq -2)$$

$$x = \mathbf{\frac{1}{2}}$$

28. (59) (a) $h^{3\log_h 2 - \log_h 2 - \log_h 6}$

$$= h^{\log_h 2^3 - \log_h 2 - \log_h 6}$$

$$= h^{\log_h \frac{8}{(2)(6)}} = h^{\log_h \frac{2}{3}} = \mathbf{\frac{2}{3}}$$

(b) $5\log 10^3 + 2\log 10^{\frac{1}{2}} - \ln e^2$

$$= \log 10^{15} + \log 10 - \ln e^2$$

$$= 15 + 1 - 2 = \mathbf{14}$$

(c) $\log 10^{\frac{1}{2}} + \ln e^{\frac{3}{2}} - 6\ln e^{\frac{1}{3}}$

$$= \log 10^{\frac{1}{2}} + \ln e^{\frac{3}{2}} - \ln e^2$$

$$= \frac{1}{2} + \frac{3}{2} - 2 = \mathbf{0}$$

29. (67) (a) $\text{antilog}_6(-2) = 6^{-2} = \mathbf{\frac{1}{36}}$

(b) $\text{antilog}_2(-6) = 2^{-6} = \mathbf{\frac{1}{64}}$

30. (46) $2\left(x^2 - \frac{5}{2}x + 3\right) = 0$

$$x = \frac{\frac{5}{2} \pm \sqrt{\frac{25}{4} - 4(1)(3)}}{2} = \frac{5}{4} \pm \frac{\sqrt{23}i}{4}$$

$$\mathbf{2\left(x - \frac{5}{4} - \frac{\sqrt{23}}{4}i\right)\left(x - \frac{5}{4} + \frac{\sqrt{23}}{4}i\right)}$$

Problem Set 76

1. (75) $_{10}C_8 = \frac{10!}{8!2!} = \textbf{45 committees}$

2. (75) $_{10}C_6 = \frac{10!}{6!4!} = \textbf{210 groups}$

3. (R)

$$1.2P = 480$$

$$P = 400$$

$$1.15P = 1.15(400) = \mathbf{\$460}$$

$$76(460) = \mathbf{\$34{,}960}$$

4. (25)

$$\begin{cases} \text{(a)}\ M_N = R_N + 1 \\ \text{(b)}\ 3(M_N - 8) = 2(R_N - 8) + 6 \end{cases}$$

(b)

$$3(M_N - 8) = 2(R_N - 8) + 6$$

$$3M_N - 24 = 2R_N - 16 + 6$$

$$3M_N - 2R_N = 14$$

$$3(R_N + 1) - 2R_N = 14$$

$$R_N = 11$$

(a) $M_N = R_N + 1 = (11) + 1 = 12$

$M_N + 17 = \boldsymbol{M} = \textbf{29 yr}$

$R_N + 17 = \boldsymbol{R} = \textbf{28 yr}$

5. (28) Rate $= \frac{\text{distance}}{\text{time}} = \frac{\boldsymbol{y} \textbf{ yards}}{\boldsymbol{m} \textbf{ - 15 minutes}}$

6. (76) $\cos x \tan x = \cos x \cdot \dfrac{\sin x}{\cos x} = \sin x = \dfrac{1}{\dfrac{1}{\sin x}} = \dfrac{1}{\csc x}$

7. (76) $\dfrac{\cot x}{\csc x} = \dfrac{\dfrac{\cos x}{\sin x}}{\dfrac{1}{\sin x}} = \dfrac{\cos x}{\sin x} \cdot \dfrac{\sin x}{1} = \cos x$

8. (76) $-\sin(-\theta)\cos(90° - \theta) = -(-\sin\theta)\sin\theta = \sin^2\theta$

9. (74) $x = \dfrac{\begin{vmatrix} 5 & -4 \\ 6 & 2 \end{vmatrix}}{\begin{vmatrix} 6 & -4 \\ 3 & 2 \end{vmatrix}} = \dfrac{10 + 24}{12 + 12} = \dfrac{34}{24} = \dfrac{17}{12}$

$y = \dfrac{\begin{vmatrix} 6 & 5 \\ 3 & 6 \end{vmatrix}}{\begin{vmatrix} 6 & -4 \\ 3 & 2 \end{vmatrix}} = \dfrac{36 - 15}{12 + 12} = \dfrac{21}{24} = \dfrac{7}{8}$

$\mathbf{x = \dfrac{17}{12}; \; y = \dfrac{7}{8}}$

10. (73)

θ 12 in.
A
x

$\theta = \dfrac{1}{2}\left(\dfrac{360°}{5}\right) = 36°$

$x = 12 \sin 36° = 7.0534$ in.

$A = 12 \cos 36° = 9.7082$ in.

Radius = 9.71 in.

$\text{Area}_\Delta = \dfrac{1}{2}(7.0534)(9.7082) = 34.2379 \text{ in.}^2$

Area = (10)(34.2379)

Area = 342.38 in.²

11. (73)

θ r
2.5 in.

$\theta = \dfrac{1}{2}\left(\dfrac{360°}{8}\right) = 22.5°$

$\sin 22.5° = \dfrac{2.5}{r}$

$r = \dfrac{2.5}{\sin 22.5°} = $ **6.53 in.**

12. (72) $B = 180° - 110° - 40°$

$\mathbf{B = 30°}$

$\dfrac{a}{\sin 110°} = \dfrac{8}{\sin 40°}$

$a = \dfrac{8 \sin 110°}{\sin 40°}$

$\mathbf{a = 11.70}$ **ft**

$\dfrac{b}{\sin 30°} = \dfrac{8}{\sin 40°}$

$b = \dfrac{8 \sin 30°}{\sin 40°}$

$\mathbf{b = 6.22}$ **ft**

13. (71) $6x^2 + 3y = 36$

$\mathbf{\dfrac{x^2}{6} + \dfrac{y^2}{12} = 1}$

Let $x = 0$

$\dfrac{0}{6} + \dfrac{y^2}{12} = 1$

$y = \pm\sqrt{12}$

Let $y = 0$

$\dfrac{x^2}{6} + \dfrac{0}{12} = 1$

$x = \pm\sqrt{6}$

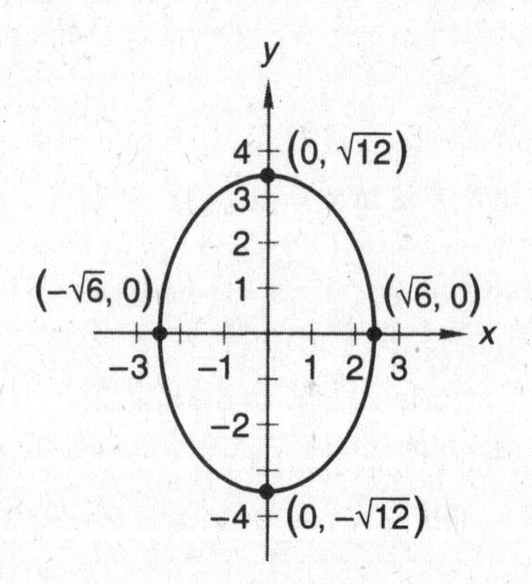

14. (61,70) $\mu = 750$

$\sigma^2 = 2500$

$\sigma = 50$

$z_1 = \dfrac{x_1 - \mu}{\sigma} = \dfrac{780 - 750}{50} = 0.6$

Percentile = 0.7257

$z_2 = \dfrac{x_2 - \mu}{\sigma} = \dfrac{740 - 750}{50} = -0.2$

Percentile = 0.4207

$0.7257 - 0.4207 = 0.305$

30.5%

15. (69)

$$\begin{vmatrix} 1-a & 3 \\ 2 & 2-a \end{vmatrix} = 0$$

$$(1 - a)(2 - a) - 2(3) = 0$$

$$2 - 3a + a^2 - 6 = 0$$

$$a^2 - 3a - 4 = 0$$

$$(a - 4)(a + 1) = 0$$

$$a = \mathbf{-1, 4}$$

16. (68) Vertex: $(h, k) = (-3, -2)$

Focus: $(h, k + p) = (-3, 5)$

$$k + p = 5$$

$$(-2) + p = 5$$

$$p = 7$$

Directrix: $y = k - p = (-2) - 7 = -9$

Axis of symmetry: $x = h = -3$

$$y - k = \frac{1}{4p}(x - h)^2$$

$$y + 2 = \frac{1}{4(7)}(x + 3)^2$$

$$y = \frac{1}{28}(x + 3)^2 - 2$$

Parabola: $\boldsymbol{y = \frac{1}{28}(x + 3)^2 - 2}$

Directrix: $\boldsymbol{y = -9}$

Axis of symmetry: $\boldsymbol{x = -3}$

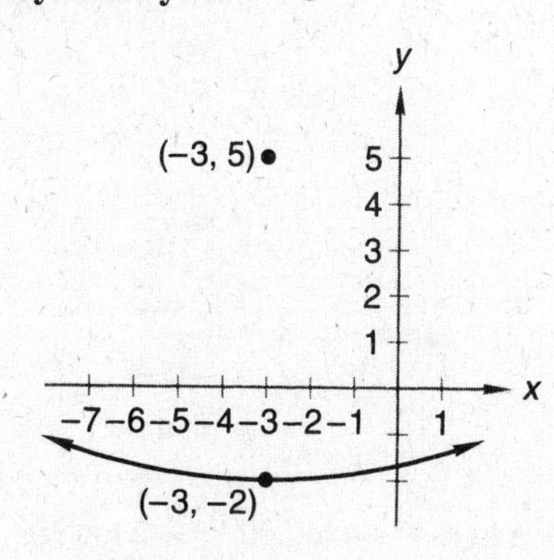

17. (65) $y = -\log_2 x$

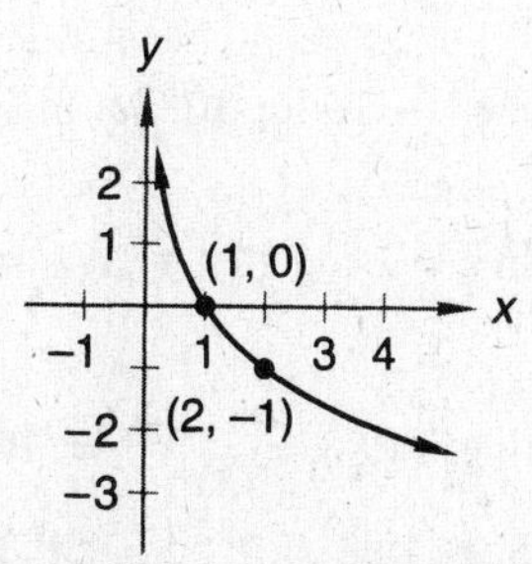

18. (63)

$$2x^2 + 2y^2 - 4x + 4y - 4 = 0$$

$$x^2 + y^2 - 2x + 2y = 2$$

$$(x^2 - 2x \quad) + (y^2 + 2y \quad) = 2$$

$$(x^2 - 2x + 1) + (y^2 + 2y + 1) = 4$$

$$(x - 1)^2 + (y + 1)^2 = 2^2$$

Center = (1, –1); radius = 2

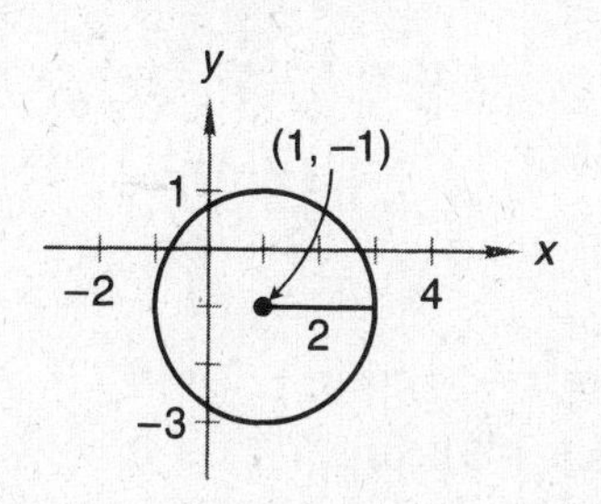

19. (66) Period = 540°

$$\text{Coefficient} = \frac{360°}{540°} = \frac{2}{3}$$

$$\boldsymbol{y = 3 + 6 \cos \frac{2}{3}(\theta - 135°)}$$

20. (66)

$$\text{Coefficient} = \frac{360°}{90°} = 4$$

$$\boldsymbol{y = -3 + \frac{3}{2} \sin 4(x + 30°)}$$

21. (64)

$$\left[6 \text{ cis} \left(-\frac{7\pi}{8}\right)\right]\left(3 \text{ cis} \frac{13\pi}{8}\right)$$

$$= 18 \text{ cis} \frac{6\pi}{8} = 18 \text{ cis} \frac{3\pi}{4}$$

$$= 18\left(\cos \frac{3\pi}{4} + i \sin \frac{3\pi}{4}\right)$$

$$= 18\left(-\frac{\sqrt{2}}{2} + \frac{\sqrt{2}}{2}i\right) = \boldsymbol{-9\sqrt{2} + 9\sqrt{2}i}$$

22. (60)

$$\csc^2 x - 1 = 0$$

$$(\csc x - 1)(\csc x + 1) = 0$$

$$\csc x = 1 \qquad \csc x = -1$$

$$x = \frac{\pi}{2} \qquad x = \frac{3\pi}{2}$$

$$x = \boldsymbol{\frac{\pi}{2}, \frac{3\pi}{2}}$$

23. $2\cos 4\theta + 1 = 0$
(52)

$$\cos 4\theta = -\frac{1}{2}$$

$$4\theta = \frac{2\pi}{3}, \frac{4\pi}{3}, \frac{8\pi}{3}, \frac{10\pi}{3}, \frac{14\pi}{3}, \frac{16\pi}{3}, \frac{20\pi}{3}, \frac{22\pi}{3}$$

$$\theta = \mathbf{\frac{\pi}{6}, \frac{\pi}{3}, \frac{2\pi}{3}, \frac{5\pi}{6}, \frac{7\pi}{6}, \frac{4\pi}{3}, \frac{5\pi}{3}, \frac{11\pi}{6}}$$

24. $\cos^2 x + 4\cos x + 3 = 0$
(60)

$$(\cos x + 3)(\cos x + 1) = 0$$

$\cos x = -3$ $\qquad$ $\cos x = -1$

no answer $\qquad$ $x = \pi$

25. $\tan^2(-510°) - \sec^2(-510°) + \sin^2(-510°) + \cos^2(-510°)$
(48)

$$= \tan^2(210°) - \sec^2(210°) + \sin^2(210°) + \cos^2(210°)$$

$$= \tan^2 30° - (-\sec 30°)^2 + (-\sin 30°)^2 + (-\cos 30°)^2$$

$$= \left(\frac{1}{\sqrt{3}}\right)^2 - \left(-\frac{2}{\sqrt{3}}\right)^2 + \left(-\frac{1}{2}\right)^2 + \left(-\frac{\sqrt{3}}{2}\right)^2$$

$$= \frac{1}{3} - \frac{4}{3} + \frac{1}{4} + \frac{3}{4} = (-1) + (1) = \mathbf{0}$$

26. $3x + 2y = 10$
(58)

$$y = -\frac{3}{2}x + 5$$

Equation of the perpendicular line, using the point-slope formula:

$$y - y_1 = m(x - x_1)$$

$$y - 3 = \frac{2}{3}(x - 2)$$

$$y = \frac{2}{3}x + \frac{5}{3}$$

Point of intersection:

$$\frac{2}{3}x + \frac{5}{3} = -\frac{3}{2}x + 5$$

$$\frac{13}{6}x = \frac{10}{3}$$

$$x = \frac{20}{13}$$

$$y = \frac{2}{3}x + \frac{5}{3} = \frac{2}{3}\left(\frac{20}{13}\right) + \frac{5}{3} = \frac{40}{39} + \frac{65}{39}$$

$$= \frac{105}{39} = \frac{35}{13}$$

$\left(\frac{20}{13}, \frac{35}{13}\right)$ and (2, 3)

$$D = \sqrt{\left(\frac{20}{13} - 2\right)^2 + \left(\frac{35}{13} - 3\right)^2}$$

$$= \sqrt{\left(-\frac{6}{13}\right)^2 + \left(-\frac{4}{13}\right)^2} = \sqrt{\frac{52}{169}} = \mathbf{\frac{2\sqrt{13}}{13}}$$

27. $\log_7(x + 1) + \log_7(2x - 3) = \log_7 4x$
(59)

$$\log_7(x + 1)(2x - 3) = \log_7 4x$$

$$(x + 1)(2x - 3) = 4x$$

$$2x^2 - x - 3 = 4x$$

$$2x^2 - 5x - 3 = 0$$

$$(2x + 1)(x - 3) = 0$$

$$x = -\frac{1}{2}, 3 \quad \left(x \neq -\frac{1}{2}\right)$$

$$x = \mathbf{3}$$

28. $\frac{1}{4}\log_{\frac{1}{2}} 16 - 2\log_{\frac{1}{2}} x = 3$
(59)

$$\log_{\frac{1}{2}} 16^{\frac{1}{4}} - \log_{\frac{1}{2}} x^2 = 3$$

$$\log_{\frac{1}{2}} \frac{2}{x^2} = 3$$

$$\frac{2}{x^2} = \frac{1}{8}$$

$$x^2 = 16$$

$$x = \pm 4 \quad (x = -4)$$

$$x = \mathbf{4}$$

29. (a) $5 \ln e^{-2} - 4 \log 10^{-3} + 2 \log 10^5$
(59)

$$= \ln e^{-10} - \log 10^{-12} + \log 10^{10}$$

$$= -10 - (-12) + 10 = \mathbf{12}$$

(b) $5 \ln e^{\frac{1}{3}} - 10 \log 10^{\frac{1}{6}} + \ln 1$

$$= \ln e^{\frac{5}{3}} - \log 10^{\frac{5}{3}} + \ln 1 = \frac{5}{3} - \frac{5}{3} + 0 = \mathbf{0}$$

30. (a) $\text{antilog}_{17} 1.88 = 17^{1.88} = \mathbf{205.70}$
(67)

(b) $\text{antilog}_7 1.88 = 7^{1.88} = \mathbf{38.80}$

Problem Set 77

1. (75) $_9C_5 = \frac{9!}{5!4!} = $ **126 teams**

2. (38) $T_1 = \frac{200 \text{ mi}}{50 \frac{\text{mi}}{\text{hr}}} = 4 \text{ hr}$

$T_2 = \frac{400 \text{ mi}}{40 \frac{\text{mi}}{\text{hr}}} = 10 \text{ hr}$

$$\text{Average rate} = \frac{\text{total distance}}{\text{total time}} = \frac{600 \text{ mi}}{(4 + 10) \text{ hr}} = \mathbf{42.86 \text{ mph}}$$

3. (R) $T = \$200(1 - 0.12 - 0.06) = \$164$

$4T = 4(\$164) = $ **\$656**

4. (R) $\text{Taxes} = (\$6{,}400{,}000)\left(\frac{\$0.80}{\$100}\right) = \$51{,}200$

$\text{Still due} = (0.02)(\$51{,}200) = $ **\$1024**

5. (7) $N, N + 1, N + 2$

$$N(N + 2) = (N + 1) + 19$$
$$N(N + 2) - (N + 1) = 19$$
$$N^2 + 2N - N - 1 = 19$$
$$N^2 + N - 20 = 0$$
$$(N + 5)(N - 4) = 0$$
$$N = -5, 4$$

N must be positive, so the numbers are **4, 5, 6.**

6. (R) $\mathbf{\dfrac{x + 2}{x + 1}}$

7. (77) $(x + y)^4$

Term	①	②	③	④	⑤
For x	4	3	2	1	0
For y	0	1	2	3	4
Coefficient	1	4	6	4	1

$\mathbf{x^4 + 4x^3y + 6x^2y^2 + 4xy^3 + y^4}$

8. (77) $(x + y)^7$

Term	①	②	③	④	⑤...⑧
For x	7	6	5	4	3... 0
For y	0	1	2	3	4... 7
Coefficient	1	7	21	35	35... 1

$\mathbf{21x^5y^2}$

9. (76) $\sin x \sec x = \sin x \frac{1}{\cos x} = \frac{\sin x}{\cos x} = \tan x$

10. (76) $\sec x \cot x = \frac{1}{\cos x} \cdot \frac{\cos x}{\sin x} = \frac{1}{\sin x} = \csc x$

11. (76) $\sin(-\theta) \tan(90° - \theta) = (-\sin\theta) \cot\theta$

$= (-\sin\theta)\left(\frac{\cos\theta}{\sin\theta}\right) = -\cos\theta$

12. (74) $x = \frac{\begin{vmatrix} 6 & -2 \\ 7 & -1 \end{vmatrix}}{\begin{vmatrix} 4 & -2 \\ 3 & -1 \end{vmatrix}} = \frac{-6 + 14}{-4 + 6} = 4$

$y = \frac{\begin{vmatrix} 4 & 6 \\ 3 & 7 \end{vmatrix}}{\begin{vmatrix} 4 & -2 \\ 3 & -1 \end{vmatrix}} = \frac{28 - 18}{-4 + 6} = 5$

$\mathbf{x = 4; y = 5}$

13. (73)

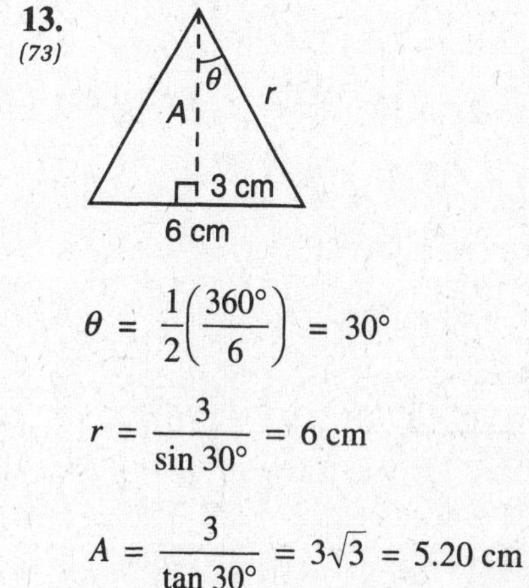

$\theta = \frac{1}{2}\left(\frac{360°}{6}\right) = 30°$

$r = \frac{3}{\sin 30°} = 6 \text{ cm}$

$A = \frac{3}{\tan 30°} = 3\sqrt{3} = 5.20 \text{ cm}$

Radius of circumscribed circle = 6 cm

Radius of inscribed circle = 5.20 cm

14. (72) $A = 180° - 120° - 30°$

$\mathbf{A = 30°}$

$\frac{a}{\sin 30°} = \frac{10}{\sin 30°}$

$\mathbf{a = 10 \text{ cm}}$

15. (71)
$$\frac{8x^2}{16} + \frac{2y^2}{16} = \frac{16}{16}$$
$$\mathbf{\frac{x^2}{2} + \frac{y^2}{8} = 1}$$

Let $x = 0$ | Let $y = 0$

$$\frac{0}{2} + \frac{y^2}{8} = 1 \qquad \frac{x^2}{2} + \frac{0}{8} = 1$$
$$y = \pm\sqrt{8} \qquad x = \pm\sqrt{2}$$

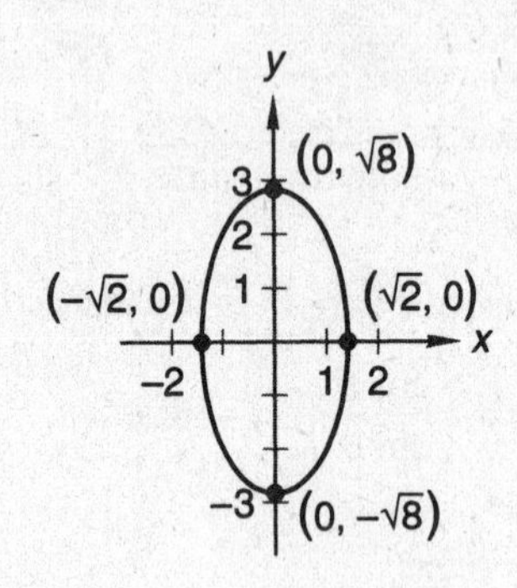

16. (70) $z = \frac{x - \mu}{\sigma} = \frac{82 - 75}{5} = 1.4$

Percentile $= 0.9192$

$1 - 0.9192 = 0.0808$

8.08%

17. (69)
$$\begin{vmatrix} 1 + s & -2 \\ s & 2s - 1 \end{vmatrix} = 0$$
$$(1 + s)(2s - 1) - (s)(-2) = 0$$
$$2s + 2s^2 - 1 - s + 2s = 0$$
$$2s^2 + 3s - 1 = 0$$
$$s = \frac{-3 \pm \sqrt{3^2 - 4(2)(-1)}}{2(2)} = \mathbf{\frac{-3 \pm \sqrt{17}}{4}}$$

18. (68) Vertex: $(h, k) = (4, 3)$

Focus: $(h, k + p) = (4, 0)$

$$k + p = 0$$
$$(3) + p = 0$$
$$p = -3$$

Directrix: $y = k - p = 6$

Axis of symmetry: $x = h = 4$

$$y - k = \frac{1}{4p}(x - h)^2$$
$$y - 3 = \frac{1}{4(-3)}(x - 4)^2$$
$$y = -\frac{1}{12}(x - 4)^2 + 3$$

Parabola: $\mathbf{y = -\frac{1}{12}(x - 4)^2 + 3}$

Directrix: $\mathbf{y = 6}$

Axis of symmetry: $\mathbf{x = 4}$

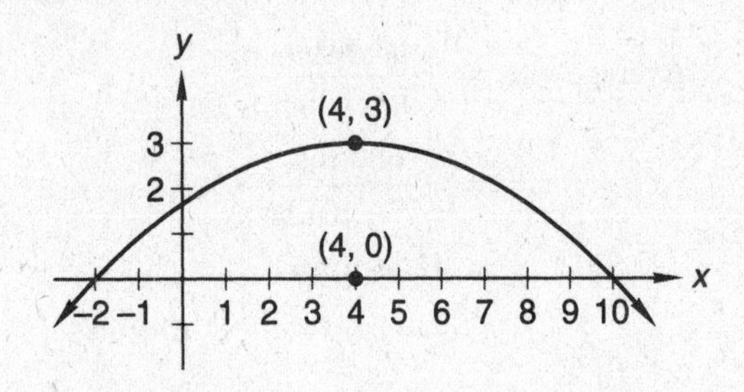

19. (66) Period $= \frac{8\pi}{3}$

Coefficient $= \frac{2\pi}{\frac{8\pi}{3}} = \frac{3}{4}$

$$\mathbf{y = 11 + \cos\frac{3}{4}\left(x + \frac{\pi}{3}\right)}$$

20. (66) Coefficient $= \frac{360°}{540°} = \frac{2}{3}$

$$\mathbf{y = -3 + 7\sin\frac{2}{3}(x - 60°)}$$

21. (64) $[4 \text{ cis } (-30°)][2 \text{ cis } (-270°)] = 8 \text{ cis } (-300°)$

$= 8 \text{ cis } 60° = 8(\cos 60° + i \sin 60°)$

$= 8\left(\frac{1}{2} + \frac{\sqrt{3}}{2}i\right) = \mathbf{4 + 4\sqrt{3}i}$

22. (60)
$$3\cot^2 x - 1 = 0$$
$$\cot^2 x - \frac{1}{3} = 0$$
$$\left(\cot x - \frac{1}{\sqrt{3}}\right)\left(\cot x + \frac{1}{\sqrt{3}}\right) = 0$$
$$\cot x = \frac{1}{\sqrt{3}} \qquad \cot x = -\frac{1}{\sqrt{3}}$$
$$x = 60°, 240° \qquad x = 120°, 300°$$
$$x = \mathbf{60°, 120°, 240°, 300°}$$

23. (52) $\sqrt{3} \tan 4\theta = 1$

$$\tan 4\theta = \frac{1}{\sqrt{3}}$$

$$4\theta = 30°, 210°, 390°, 570°, 750°, 930°, 1110°, 1290°$$

$$\theta = \mathbf{7.5°, 52.5°, 97.5°, 142.5°, 187.5°, 232.5°, 277.5°, 322.5°}$$

24. (48) $f\left(\frac{19\pi}{3}\right) - g\left(\frac{19\pi}{3}\right) = \csc^2\left(\frac{19\pi}{3}\right) - \cot^2\left(\frac{19\pi}{3}\right)$

$$= \left(\csc\frac{\pi}{3}\right)^2 - \left(\cot\frac{\pi}{3}\right)^2 = \left(\frac{2}{\sqrt{3}}\right)^2 - \left(\frac{1}{\sqrt{3}}\right)^2$$

$$= \frac{4}{3} - \frac{1}{3} = \mathbf{1}$$

25. (58) $x - y + 2 = 0$

$$y = x + 2$$

Equation of the perpendicular line, using the point-slope formula:

$$y - y_1 = m(x - x_1)$$
$$y + 4 = -(x - 1)$$
$$y = -x - 3$$

Point of intersection:

$$-x - 3 = x + 2$$
$$2x = -5$$
$$x = -\frac{5}{2}$$

$$y = -x - 3 = -\left(-\frac{5}{2}\right) - 3 = -\frac{1}{2}$$

$\left(-\frac{5}{2}, -\frac{1}{2}\right)$ and $(1, -4)$

$$D = \sqrt{\left(1 + \frac{5}{2}\right)^2 + \left(-4 + \frac{1}{2}\right)^2}$$

$$= \sqrt{\left(\frac{7}{2}\right)^2 + \left(-\frac{7}{2}\right)^2} = \sqrt{\frac{49}{4}(2)} = \mathbf{\frac{7\sqrt{2}}{2}}$$

26. (56)

$$x^2 + y^2 - 6x + 4y + 12 \le 0$$
$$(x^2 - 6x + 9) + (y^2 + 4y + 4) \le -12 + 9 + 4$$
$$(x - 3)^2 + (y + 2)^2 \le 1$$

$$2x^2 + 2y^2 + 8y \le 10$$
$$x^2 + (y^2 + 4y \quad) \le 5$$
$$x^2 + (y^2 + 4y + 4) \le 5 + 4$$
$$x^2 + (y + 2)^2 \le 3^2$$

The inequality indicates the region within or on both circles.

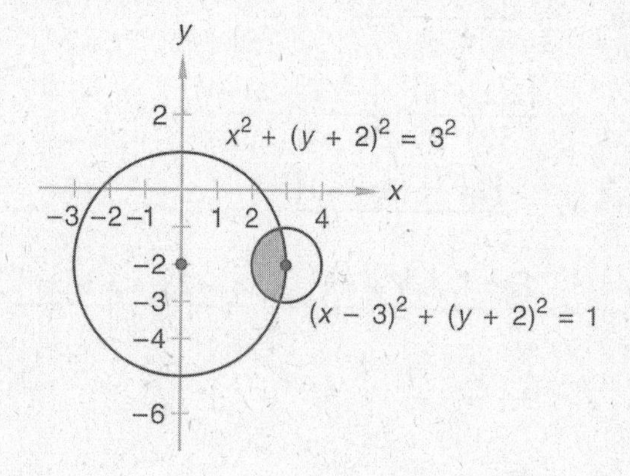

27. (3) $h = \sqrt{3^2 + 4^2} = 5$

$$\frac{x}{2} = \frac{5}{4}$$
$$x = \frac{10}{4}$$
$$x = \mathbf{\frac{5}{2}}$$

28. (59) $\log_8 (x - 1) - \log_8 (x - 2) = 2 \log_8 3$

$$\log_8 \frac{(x - 1)}{(x - 2)} = \log_8 3^2$$
$$\frac{(x - 1)}{(x - 2)} = 9$$
$$x - 1 = 9x - 18$$
$$8x = 17$$
$$x = \mathbf{\frac{17}{8}}$$

29. (59) (a) $3 \log_7 7 + 7^{2 \log_7 3 - \log_7 3}$

$$= \log_7 7^3 + 7^{\log_7 \frac{3^2}{3}}$$
$$= 3 + 3 = \mathbf{6}$$

(b) $4 \log 10^2 - 2 \log 10 + \ln e^{-2}$

$$= \log 10^8 - \log 10^2 + \ln e^{-2}$$
$$= 8 - 2 + (-2) = \mathbf{4}$$

(c) $2 \ln e^{\frac{2}{3}} - \log 10^{\frac{1}{2}} - 4 \log_3 3^{\frac{1}{12}}$

$= \ln e^{\frac{4}{3}} - \log 10^{\frac{1}{2}} - \log_3 3^{\frac{4}{12}}$

$= \frac{4}{3} - \frac{1}{2} - \frac{1}{3} = \mathbf{\frac{1}{2}}$

30. (21) $\frac{f(x + h) - f(x)}{h}$

$= \frac{[2(x + h)^2 - 3(x + h) + 1]}{h} - \frac{[2x^2 - 3x + 1]}{h}$

$= \frac{2x^2 + 4xh + 2h^2 - 3x - 3h + 1 - 2x^2}{h} + \frac{3x - 1}{h}$

$= \frac{4xh - 3h + 2h^2}{h} = \mathbf{4x - 3 + 2h}$

Problem Set 78

1. (75) ${}_9C_5 = \frac{9!}{5!4!} =$ **126 committees**

2. (38) $\text{Avg}_4 = \frac{\text{Sum}_4}{4}$

$\text{Sum}_4 = 4 \cdot \text{Avg}_4$

$\text{Sum}_4 = 4(72) = 288$

$\text{Avg}_5 = \frac{\text{Sum}_4 + T_5}{5}$

$\frac{288 + T_5}{5} = 70$

$288 + T_5 = 5(70)$

$T_5 = 350 - 288$

$T_5 = \mathbf{62}$

3. (38) $\text{Avg}_5 = \frac{\text{Sum}_5}{5}$

$\text{Sum}_5 = 5 \cdot \text{Avg}_5$

$\text{Sum}_5 = 5(80°) = 400°$

$\text{Avg}_6 = \frac{\text{Sum}_5 + T_6}{6}$

$\frac{400 + T_6}{6} = 83°$

$400 + T_6 = 6(83°)$

$T_6 = 498° - 400°$

$T_6 = \mathbf{98°}$

4. (44) Rate $\times N =$ price

$a \times K = aK$

Change = cents proffered − price in cents

= **(25*Q* − *Ka*) cents**

5. (25) $T_{SL} = T_{SP} + 2$

$R_{SL}T_{SL} + R_{SP}T_{SP} = \text{jobs}$

$\left(\frac{2}{3}\right)(T_{SP} + 2) + \left(\frac{5}{3}\right)T_{SP} = 6$

$\frac{2}{3}T_{SP} + \frac{4}{3} + \frac{5}{3}T_{SP} = 6$

$\frac{7}{3}T_{SP} = \frac{14}{3}$

$T_{SP} = \mathbf{2\ hr}$

6. (R) $\begin{cases} \text{(a)}\ 0.06A + 0.08B = 220 \\ \text{(b)}\ A + B = 3000 \end{cases}$

(b) $A + B = 3000$

$A = 3000 - B$

100(a) $6A + 8B = 22{,}000$

$6(3000 - B) + 8B = 22{,}000$

$2B + 18{,}000 = 22{,}000$

$2B = 4000$

$B = 2000$

$A = 3000 - B = 3000 - 2000 = 1000$

***A*: $1000**

***B*: $2000**

7. (78) $\frac{x^2}{16} - \frac{y^2}{4} = 1$

Let $y = 0$ | Let $x = 0$

$\frac{x^2}{16} - \frac{0}{4} = 1$ | $\frac{0}{16} - \frac{y^2}{4} = 1$

$x = \pm 4$ | $y = \pm 2i$

From the values of x and y, we see that $m_1 = \frac{2}{4} = \frac{1}{2}$ and $m_2 = -\frac{2}{4} = -\frac{1}{2}$ and we can determine the vertices and the equations of the asymptotes.

Vertices = (4, 0), (−4, 0)

Asymptotes: $\mathbf{y = \frac{1}{2}x;\ y = -\frac{1}{2}x}$

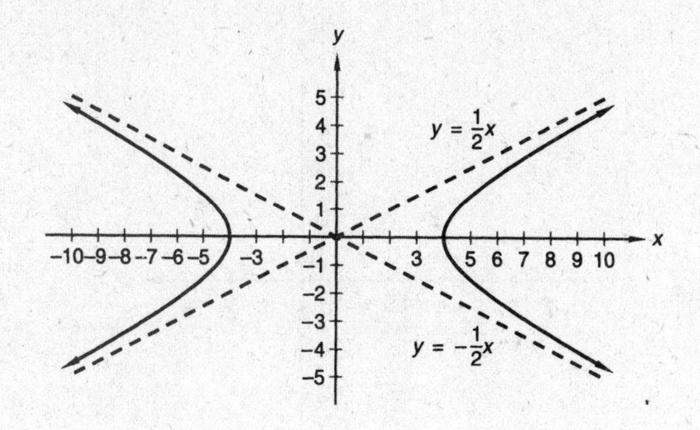

8. $4y^2 - 9x^2 = 36$
(78)

$$\frac{y^2}{9} - \frac{x^2}{4} = 1$$

Let $x = 0$ — Let $y = 0$

$$\frac{y^2}{9} - \frac{0}{4} = 1 \qquad \frac{0}{9} - \frac{x^2}{4} = 1$$

$$y = \pm 3 \qquad x = \pm 2i$$

From the values of x and y, we see that $m_1 = \frac{3}{2}$ and $m_2 = -\frac{3}{2}$ and we can determine the vertices and the equations of the asymptotes.

Vertices = (0, 3), (0, –3)

Asymptotes: $y = \frac{3}{2}x;\ y = -\frac{3}{2}x$

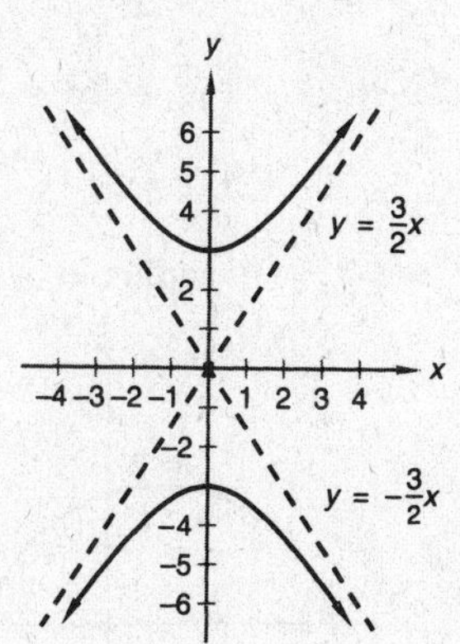

9. $(a + b)^5$
(77)

$$= \mathbf{a^5 + 5a^4b + 10a^3b^2 + 10a^2b^3 + 5ab^4 + b^5}$$

10. $\mathbf{15a^2b^4}$
(77)

11. $\sin\theta \csc(90° - \theta) = \sin\theta \sec\theta$
(76)

$$= \sin\theta \frac{1}{\cos\theta} = \tan\theta$$

12. $\sec(90° - \theta)\tan\theta = \csc\theta\tan\theta$
(76)

$$= \frac{1}{\sin\theta} \cdot \frac{\sin\theta}{\cos\theta} = \frac{1}{\cos\theta} = \sec\theta$$

13. (74)

$$x = \frac{\begin{vmatrix} 5 & -3 \\ -3 & -4 \end{vmatrix}}{\begin{vmatrix} 4 & -3 \\ 2 & -4 \end{vmatrix}} = \frac{-20 - 9}{-16 + 6} = \frac{-29}{-10} = \frac{29}{10}$$

$$y = \frac{\begin{vmatrix} 4 & 5 \\ 2 & -3 \end{vmatrix}}{\begin{vmatrix} 4 & -3 \\ 2 & -4 \end{vmatrix}} = \frac{-12 - 10}{-16 + 6} = \frac{-22}{-10} = \frac{11}{5}$$

$$\mathbf{x = \frac{29}{10};\ y = \frac{11}{5}}$$

14. Side $= \frac{50 \text{ cm}}{10}$
(73)

Side = 5 cm

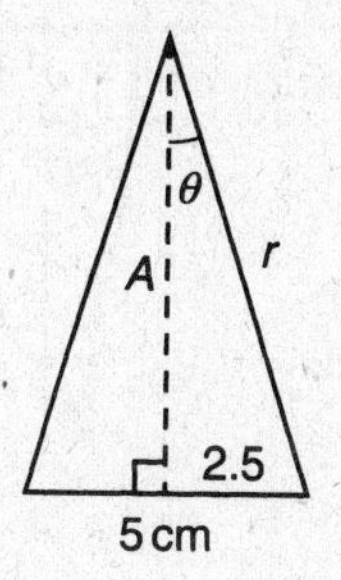

$$\theta = \frac{1}{2}\left(\frac{360°}{10}\right) = 18°$$

$$r = \frac{2.5}{\sin 18°}$$

Radius = 8.09 cm

15. $\frac{S_L}{S_S} = \frac{5}{3}$
(5)

$$\frac{A_S}{A_L} = \left(\frac{S_S}{S_L}\right)^2 = \left(\frac{3}{5}\right)^2 = \mathbf{\frac{9}{25}}$$

16. $A = 180° - 130° - 20°$
(72)

$\mathbf{A = 30°}$

$$\frac{a}{\sin 30°} = \frac{6}{\sin 20°}$$

$$a = \frac{6 \sin 30°}{\sin 20°}$$

$\mathbf{a = 8.77\ m}$

$$\frac{b}{\sin 130°} = \frac{6}{\sin 20°}$$

$$b = \frac{6 \sin 130°}{\sin 20°}$$

$\mathbf{b = 13.44\ m}$

17. $49x^2 + 4y^2 = 196$
(71)

$$\frac{\mathbf{x^2}}{\mathbf{4}} + \frac{\mathbf{y^2}}{\mathbf{49}} = \mathbf{1}$$

Let $x = 0$ Let $y = 0$

$$\frac{0}{4} + \frac{y^2}{49} = 1 \qquad \frac{x^2}{4} + \frac{0}{49} = 1$$

$$y = \pm 7 \qquad x = \pm 2$$

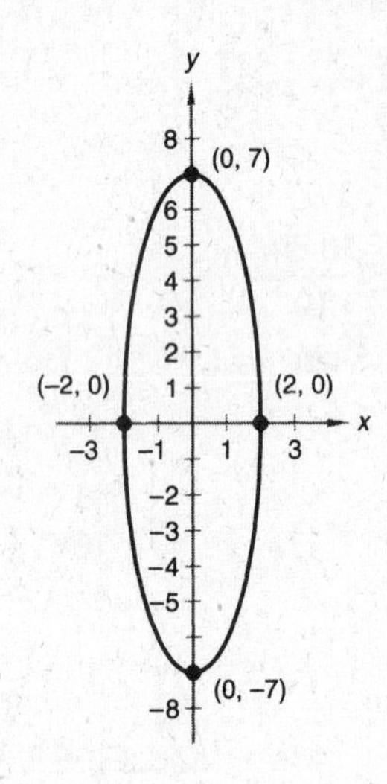

18. *(69)*

$$\begin{vmatrix} d+2 & 3d-1 \\ d-4 & d+2 \end{vmatrix} = 0$$

$$(d + 2)^2 - (d - 4)(3d - 1) = 0$$

$$d^2 + 4d + 4 - (3d^2 - 13d + 4) = 0$$

$$-2d^2 + 17d = 0$$

$$-d(2d - 17) = 0$$

$$d = \mathbf{0}, \frac{\mathbf{17}}{\mathbf{2}}$$

19. Vertex: $(h, k) = (-2, -6)$
(68) Focus: $(h, k + p) = (-2, 4)$

$$k + p = 4$$

$$(-6) + p = 4$$

$$p = 10$$

Directrix: $y = k - p = -16$
Axis of symmetry: $x = h = -2$

$$y - k = \frac{1}{4p}(x - h)^2$$

$$y + 6 = \frac{1}{4(10)}(x + 2)^2$$

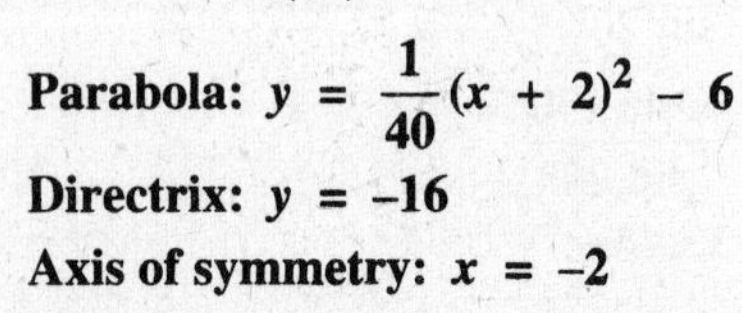

Parabola: $\mathbf{y = \frac{1}{40}(x + 2)^2 - 6}$
Directrix: $\mathbf{y = -16}$
Axis of symmetry: $\mathbf{x = -2}$

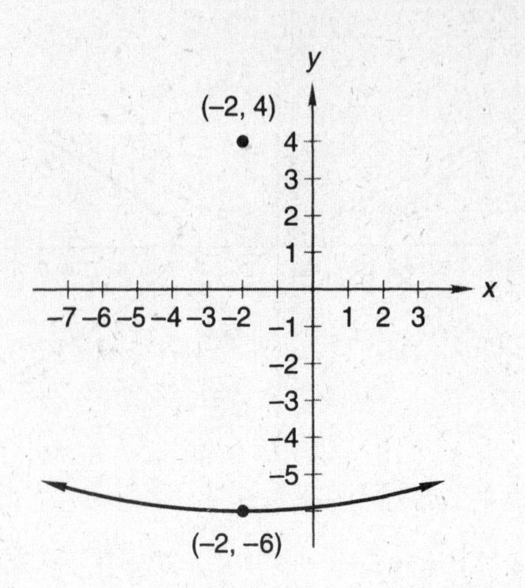

20. $y = -\log_{10} x$
(65)

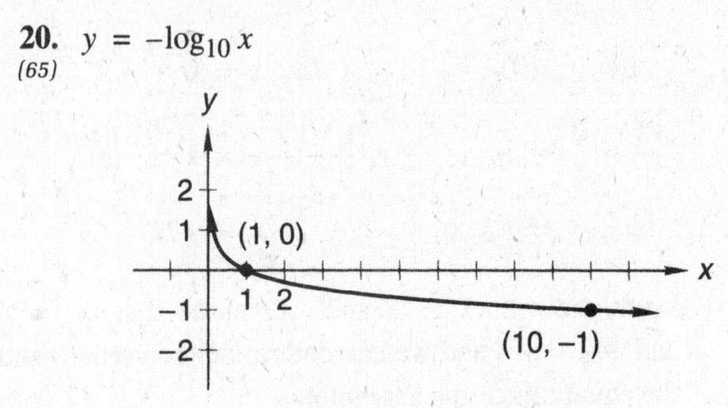

21. *(63)*

$$3x^2 + 6x + 6y + 3y^2 - 33 = 0$$

$$x^2 + 2x + y^2 + 2y - 11 = 0$$

$$(x^2 + 2x \quad) + (y^2 + 2y \quad) = 11$$

$$(x^2 + 2x + 1) + (y^2 + 2y + 1) = 13$$

$$(x + 1)^2 + (y + 1)^2 = (\sqrt{13})^2$$

Center = (−1, −1); radius = $\mathbf{\sqrt{13}}$

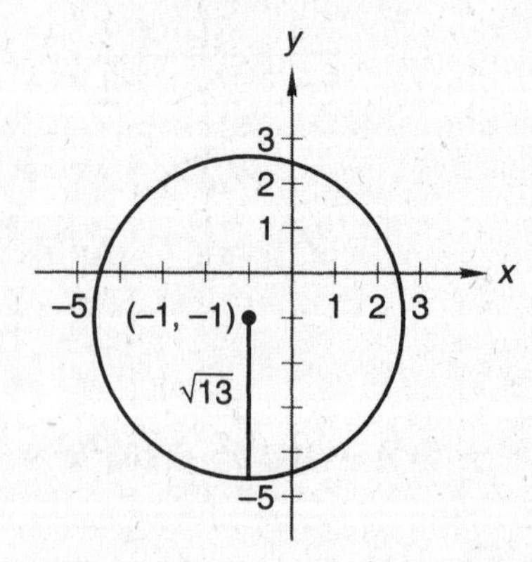

22. $\mathbf{y = 5 + 4 \sin \frac{2}{3}\left(x - \frac{3\pi}{2}\right)}$
(66)

23. $\left(3 \text{ cis } \frac{10\pi}{9}\right)\left(2 \text{ cis } \frac{5\pi}{9}\right) = 6 \text{ cis } \frac{5\pi}{3}$
(64)

$$= 6\left(\cos \frac{5\pi}{3} + i \sin \frac{5\pi}{3}\right) = 6\left(\frac{1}{2} - \frac{\sqrt{3}}{2}i\right)$$

$$= \mathbf{3 - 3\sqrt{3}i}$$

24. (52)

$$\cos 3\theta + 1 = 0$$
$$\cos 3\theta = -1$$
$$3\theta = \pi, 3\pi, 5\pi$$
$$\theta = \frac{\pi}{3}, \pi, \frac{5\pi}{3}$$

25. (60)

$$\cos 2\theta(\csc\theta - 1) = 0$$

$$\cos 2\theta = 0 \qquad \csc\theta = 1$$
$$2\theta = \frac{\pi}{2}, \frac{3\pi}{2}, \frac{5\pi}{2}, \frac{7\pi}{2} \qquad \theta = \frac{\pi}{2}$$
$$\theta = \frac{\pi}{4}, \frac{3\pi}{4}, \frac{5\pi}{4}, \frac{7\pi}{4}$$

$$\theta = \frac{\pi}{4}, \frac{\pi}{2}, \frac{3\pi}{4}, \frac{5\pi}{4}, \frac{7\pi}{4}$$

26. (32)

(a) $0° \le \theta \le 180°$

$$\text{Arccos}(\cos 240°) = \text{Arccos}\left(-\frac{1}{2}\right) = \mathbf{120°}$$

(b) $-\frac{\pi}{2} \le \theta \le \frac{\pi}{2}$

$$\text{Arcsin}\left(\sin\frac{5\pi}{3}\right) = \text{Arcsin}\left(-\frac{\sqrt{3}}{2}\right) = -\frac{\pi}{3}$$

27. (61)

$$\text{Mean} = \frac{5 + 2 + 7 + 1 + 5 + 3}{6} = \mathbf{3.83}$$

Median = **4**

Mode = **5**

$$\text{Variance} = \frac{1}{6}[(5 - 3.83)^2 + (2 - 3.83)^2 + (7 - 3.83)^2 + (1 - 3.83)^2 + (5 - 3.83)^2 + (3 - 3.83)^2] = \mathbf{4.14}$$

$$\text{Standard Deviation} = \sqrt{4.14} = \mathbf{2.03}$$

28. (58)

$$\frac{x}{4} + y = 1$$
$$y = -\frac{x}{4} + 1$$

Equation of the perpendicular line:

$$y - y_1 = m(x - x_1)$$
$$y + 2 = 4(x + 1)$$
$$y = 4x + 2$$

Point of intersection:

$$4x + 2 = -\frac{x}{4} + 1$$
$$\frac{17x}{4} = -1$$
$$x = -\frac{4}{17}$$

$$y = 4x + 2 = 4\left(-\frac{4}{17}\right) + 2 = -\frac{16}{17} + \frac{34}{17} = \frac{18}{17}$$

$\left(-\frac{4}{17}, \frac{18}{17}\right)$ and $(-1, -2)$

$$D = \sqrt{\left(-\frac{4}{17} + 1\right)^2 + \left(\frac{18}{17} + 2\right)^2}$$
$$= \sqrt{\left(\frac{13}{17}\right)^2 + \left(\frac{52}{17}\right)^2} = \sqrt{\frac{2873}{289}} = \frac{13\sqrt{17}}{17}$$

29. (59)

$$2\ln x - \ln\left(x - \frac{1}{4}\right) = \ln 4$$
$$\ln x^2 - \ln\left(x - \frac{1}{4}\right) = \ln 4$$
$$\ln\frac{x^2}{\left(x - \frac{1}{4}\right)} = \ln 4$$
$$\frac{x^2}{\left(x - \frac{1}{4}\right)} = 4$$
$$x^2 = 4x - 1$$
$$x^2 - 4x + 1 = 0$$
$$x = \frac{4 \pm \sqrt{4^2 - 4(1)(1)}}{2(1)}$$
$$x = \frac{4 \pm 2\sqrt{3}}{2} = \mathbf{2 \pm \sqrt{3}}$$

30. (59)

$$2\log_2 x + \log_2 5 = 2\log_2 3$$
$$\log_2 x^2 + \log_2 5 = \log_2 3^2$$
$$\log_2 5x^2 = \log_2 9$$
$$5x^2 = 9$$
$$x^2 = \frac{9}{5}$$
$$x = \pm\frac{3}{\sqrt{5}} \qquad \left(x \ne -\frac{3\sqrt{5}}{5}\right)$$
$$x = \frac{3\sqrt{5}}{5}$$

Problem Set 79

1. (75) $_{12}C_9 = \frac{12!}{9!3!} = \textbf{220 committees}$

2. (55) $\frac{8!}{2!} = \textbf{20,160}$

3. (62) $C = mN + b$

$$\begin{cases} \text{(a)} & 5100 = m10 + b \\ \text{(b)} & 2600 = m5 + b \end{cases}$$

$$\begin{array}{rrcl} \text{(a)} & 5100 & = & m10 + b \\ -1\text{(b)} & -2600 & = & -m5 - b \\ \hline & 2500 & = & m5 \\ & m & = & 500 \end{array}$$

(a) $5100 = m10 + b$

$5100 = (500)10 + b$

$b = 100$

$C = 500N + 100 = 500(2) + 100 = \textbf{\$1100}$

4. (36)

$$\begin{cases} \text{(a)} & (B - W)T_U = D_U \\ \text{(b)} & (B + W)T_D = D_D \end{cases}$$

$B = 3W$

$T_D = T_U - 2$

(a) $(B - W)T_U = D_U$

$2WT_U = 32$

$WT_U = 16$

(b) $(B + W)T_D = D_D$

$4W(T_U - 2) = 16$

$4WT_U - 8W = 16$

$4(16) - 8W = 16$

$8W = 48$

$\boldsymbol{W = 6}$ **mph**

$B = 3W$

$B = 3(6)$

$\boldsymbol{B = 18}$ **mph**

5. (25) $RWT = \text{jobs}$

$R(81)(24) = 1$

$R = \frac{1}{1944} \; \frac{\text{jobs}}{\text{worker-days}}$

$RWT = \text{jobs}$

$T = \frac{\text{jobs}}{RW} = \frac{1}{\left(\frac{1}{1944}\right)\left(\frac{4}{3}\right)(81)} = 18 \text{ days}$

Days saved $= 24 - 18 =$ **6 days**

6. (79) $(3 \text{ cis } 35°)^3 = 3^3 \text{ cis } (3 \cdot 35°) = \textbf{27 cis 105°}$

7. (79) $(1 - \sqrt{3}i)^5 = (2 \text{ cis } 300°)^5$

$= 2^5 \text{ cis } (5 \cdot 300°) = 32 \text{ cis } 1500°$

$= 32 \text{ cis } 60° = 32 (\cos 60° + i \sin 60°)$

$= 32\left(\frac{1}{2} + \frac{\sqrt{3}}{2}i\right) = \mathbf{16 + 16\sqrt{3}i}$

8. (79) $(8i)^{\frac{1}{3}} = (8 \text{ cis } 90°)^{\frac{1}{3}} = 8^{\frac{1}{3}} \text{ cis} \left(\frac{90°}{3} + n\frac{360°}{3}\right)$

$8^{\frac{1}{3}} \text{ cis} \left(\frac{90°}{3}\right) = \textbf{2 cis 30°}$

$8^{\frac{1}{3}} \text{ cis} \left(\frac{90°}{3} + 120°\right) = \textbf{2 cis 150°}$

$8^{\frac{1}{3}} \text{ cis} \left(\frac{90°}{3} + 240°\right) = \textbf{2 cis 270°}$

9. (79) $(-1)^{\frac{1}{2}} = (\text{cis } 180°)^{\frac{1}{2}} = 1^{\frac{1}{2}} \text{ cis} \left(\frac{180°}{2} + n\frac{360°}{2}\right)$

$1^{\frac{1}{2}} \text{ cis} \left(\frac{180°}{2}\right) = 1 \text{ cis } 90°$

$= 1(\cos 90° + i \sin 90°) = 1(0 + i) = \boldsymbol{i}$

$1^{\frac{1}{2}} \text{ cis} \left(\frac{180°}{2} + 180°\right) = 1 \text{ cis } 270°$

$= 1(\cos 270° + i \sin 270°) = 1(0 - i) = \boldsymbol{-i}$

10. (78) $\frac{x^2}{16} - \frac{y^2}{9} = 1$

Let $y = 0$

$\frac{x^2}{16} - \frac{0}{9} = 1$

$x = \pm 4$

Let $x = 0$

$\frac{0}{16} - \frac{y^2}{9} = 1$

$y = \pm 3i$

From the values of x and y, we see that $m_1 = \frac{3}{4}$ and $m_2 = -\frac{3}{4}$ and we can determine the vertices and the equations of the asymptotes.

Vertices = (4, 0), (–4, 0)

Asymptotes: $y = \frac{3}{4}x$; $y = -\frac{3}{4}x$

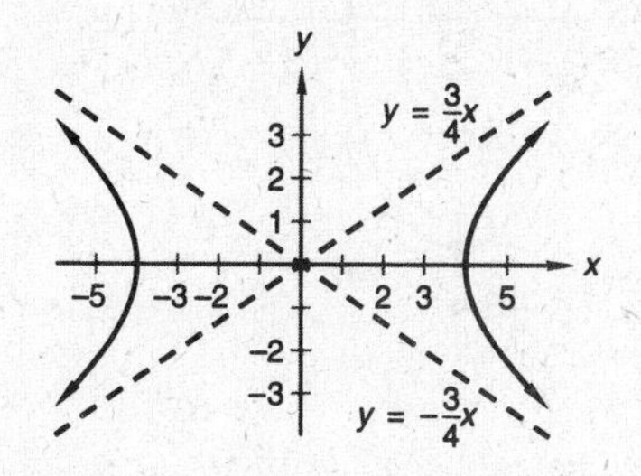

11. $9x^2 - 4y^2 = 36$
(78)

$$\frac{x^2}{4} - \frac{y^2}{9} = 1$$

Let $y = 0$ Let $x = 0$

$$\frac{x^2}{4} - \frac{0}{9} = 1 \qquad \frac{0}{4} - \frac{y^2}{9} = 1$$

$$x = \pm 2 \qquad y = \pm 3i$$

From the values of x and y, we see that $m_1 = \frac{3}{2}$ and $m_2 = -\frac{3}{2}$ and we can determine the vertices and the equations of the asymptotes.

Vertices = (2, 0), (–2, 0)

Asymptotes: $y = \frac{3}{2}x$; $y = -\frac{3}{2}x$

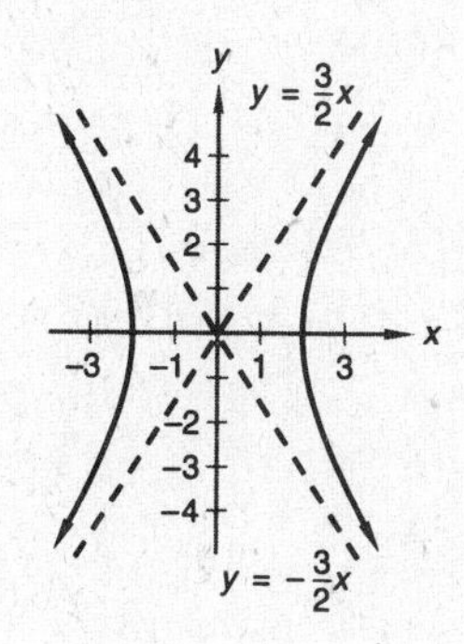

12. $(x + y)^6$
(77)

Term	①	②	③	④	⑤	⑥	⑦
For x	6	5	4	3	2	1	0
For y	0	1	2	3	4	5	6
Coefficient	1	6	15	20	15	6	1

$\mathbf{x^6 + 6x^5y + 15x^4y^2 + 20x^3y^3 + 15x^2y^4}$
$\mathbf{+ 6xy^5 + y^6}$

13. $(a + b)^9$
(77)

Term	①...⑤	⑥	⑦	⑧	⑨	⑩
For a	9... 5	4	3	2	1	0
For b	0... 4	5	6	7	8	9
Coefficient	1...126	126	64	36	9	1

$\mathbf{126a^4b^5}$

14. (76)

$$\frac{\tan\theta}{\sec\theta} = \frac{\frac{\sin\theta}{\cos\theta}}{\frac{1}{\cos\theta}} = \frac{\sin\theta}{\cos\theta}\cdot\frac{\cos\theta}{1} = \sin\theta$$

15. $\sin(90° - \theta)\sec(90° - \theta) = \cos\theta\csc\theta$
(76)

$$= \cos\theta \cdot \frac{1}{\sin\theta} = \frac{\cos\theta}{\sin\theta} = \cot\theta$$

16. (74)

$$x = \frac{\begin{vmatrix} 8 & -3 \\ 5 & 2 \end{vmatrix}}{\begin{vmatrix} 5 & -3 \\ 4 & 2 \end{vmatrix}} = \frac{16 + 15}{10 + 12} = \frac{31}{22}$$

$$y = \frac{\begin{vmatrix} 5 & 8 \\ 4 & 5 \end{vmatrix}}{\begin{vmatrix} 5 & -3 \\ 4 & 2 \end{vmatrix}} = \frac{25 - 32}{10 + 12} = -\frac{7}{22}$$

$\mathbf{x = \frac{31}{22}; y = -\frac{7}{22}}$

17. Side $= \frac{49 \text{ ft}}{7} = 7$ ft
(73)

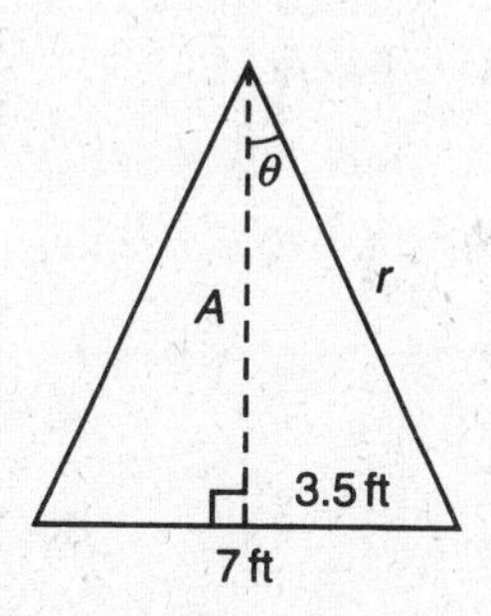

$$\theta = \frac{1}{2}\left(\frac{360°}{7}\right) = 25.71°$$

$$r = \frac{3.5}{\sin 25.71°} = \mathbf{8.07 \text{ ft}}$$

18. Side $= \frac{96\text{ ft}}{12}$
(73)
Side = 8 ft

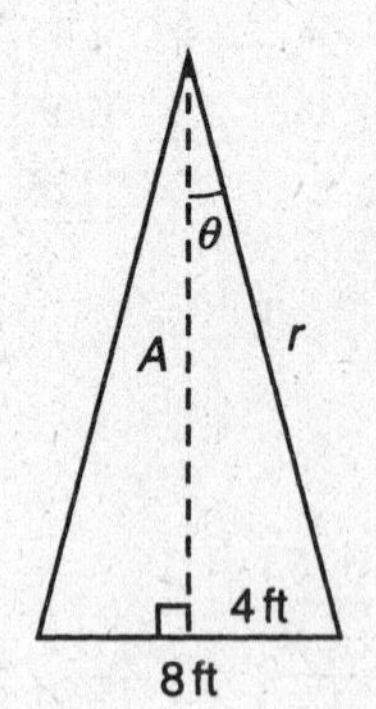

$$\theta = \frac{1}{2}\left(\frac{360°}{12}\right) = 15°$$

$$A = \frac{4}{\tan 15°} = 14.9282 \text{ ft}$$

$$\text{Area}_\Delta = \frac{1}{2}(4)(14.9282) = 29.8564 \text{ ft}^2$$

Area = (24)(29.8564)

Area = 716.55 ft²

19. $B = 180° - 120° - 20°$
(72)
$\boldsymbol{B = 40°}$

$$\frac{a}{\sin 120°} = \frac{5}{\sin 20°}$$

$$a = \frac{5 \sin 120°}{\sin 20°}$$

$\boldsymbol{a = 12.66}$ **cm**

$$\frac{b}{\sin 40°} = \frac{5}{\sin 20°}$$

$$b = \frac{5 \sin 40°}{\sin 20°}$$

$\boldsymbol{b = 9.40}$ **cm**

20. $16x^2 + 4y^2 = 64$
(71)

$$\boldsymbol{\frac{x^2}{4} + \frac{y^2}{16} = 1}$$

Let $x = 0$

$$\frac{0}{4} + \frac{y^2}{16} = 1$$

$$y = \pm 4$$

Let $y = 0$

$$\frac{x^2}{4} + \frac{0}{16} = 1$$

$$x = \pm 2$$

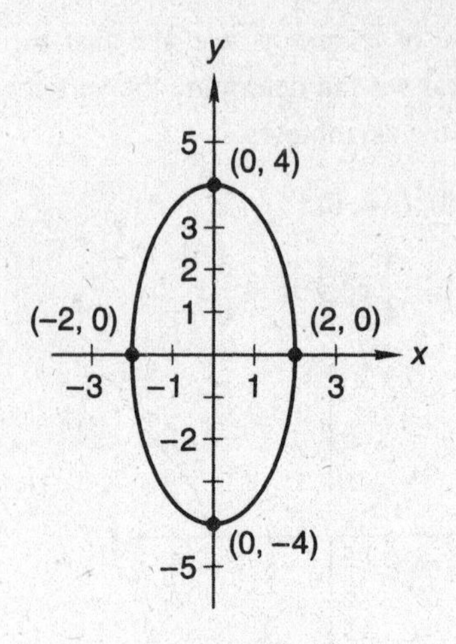

21. $z = \frac{x - \mu}{\sigma} = \frac{6.9 - 6.8}{0.2} = 0.5$
(70)
Percentile = 0.6915
1 − 0.6915 = 0.3085
30.85%

22.
(69)

$$\begin{vmatrix} x + 2 & 2x \\ x - 1 & x - 3 \end{vmatrix} + 8 = 0$$

$$(x + 2)(x - 3) - (x - 1)2x + 8 = 0$$

$$x^2 - x - 6 - 2x^2 + 2x + 8 = 0$$

$$-x^2 + x + 2 = 0$$

$$x^2 - x - 2 = 0$$

$$(x - 2)(x + 1) = 0$$

$$\boldsymbol{x = -1, 2}$$

23. Vertex: $(h, k) = (-4, 2)$
(68)
Focus: $(h, k + p) = (-4, 6)$

$k + p = 6$

$(2) + p = 6$

$p = 4$

Directrix: $y = k - p = -2$

Axis of symmetry: $x = h = -4$

$$y - k = \frac{1}{4p}(x - h)$$

$$y - 2 = \frac{1}{4(4)}(x + 4)^2$$

Parabola: $\boldsymbol{y = \frac{1}{16}(x + 4)^2 + 2}$

Directrix: $\boldsymbol{y = -2}$

Axis of symmetry: $\boldsymbol{x = -4}$

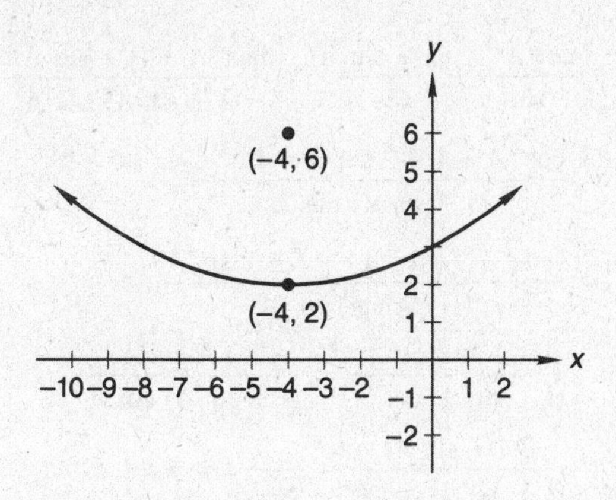

24. (66) Period = 120°

$$\text{Coefficient} = \frac{360°}{120°} = 3$$

$$\mathbf{y = -1 + 5\cos 3(x - 50°)}$$

25. (64)

$$[4 \text{ cis } (-300°)](2 \text{ cis } 30°)$$
$$= (4)(2) \text{ cis } (-300° + 30°) = 8 \text{ cis } (-270°)$$
$$= 8 \text{ cis } 90° = 8(\cos 90° + i \sin 90°)$$
$$= 8(0 + i) = \mathbf{8i}$$

26. (60)

$$2\sqrt{2}\sin^2\theta - 12\sin\theta + 5\sqrt{2} = 0$$
$$(\sqrt{2}\sin\theta - 5)(2\sin\theta - \sqrt{2}) = 0$$

$\sin\theta = \frac{5}{\sqrt{2}}$ $\qquad$ $\sin\theta = \frac{\sqrt{2}}{2}$

no answer $\qquad$ $\theta = \mathbf{45°, 135°}$

27. (52)

$$2\cos 4\theta + 1 = 0$$
$$2\cos 4\theta = -1$$
$$\cos 4\theta = -\frac{1}{2}$$
$$4\theta = 120°, 240°, 480°, 600°, 840°, 960°, 1200°, 1320°$$
$$\theta = \mathbf{30°, 60°, 120°, 150°, 210°, 240°, 300°, 330°}$$

28. (58)

$$x - 3y + 5 = 0$$
$$y = \frac{1}{3}x + \frac{5}{3}$$

Equation of the perpendicular line:

$$y - y_1 = m(x - x_1)$$
$$y - 3 = -3(x - 1)$$
$$y = -3x + 6$$

Point of intersection:

$$-3x + 6 = \frac{x}{3} + \frac{5}{3}$$
$$-9x + 18 = x + 5$$
$$10x = 13$$
$$x = \frac{13}{10}$$

$$y = -3x + 6 = -3\left(\frac{13}{10}\right) + 6 = \frac{-39}{10} + \frac{60}{10} = \frac{21}{10}$$

$\left(\frac{13}{10}, \frac{21}{10}\right)$ and $(1, 3)$

$$D = \sqrt{\left(1 - \frac{13}{10}\right)^2 + \left(3 - \frac{21}{10}\right)^2}$$
$$= \sqrt{\left(-\frac{3}{10}\right)^2 + \left(\frac{9}{10}\right)^2} = \sqrt{\frac{90}{100}} = \mathbf{\frac{3\sqrt{10}}{10}}$$

29. (59)

$$\ln(x + 2) - \ln(3x - 4) = \ln 3$$
$$\ln\frac{(x + 2)}{(3x - 4)} = \ln 3$$
$$(x + 2) = 3(3x - 4)$$
$$x + 2 = 9x - 12$$
$$8x = 14$$
$$x = \mathbf{\frac{7}{4}}$$

30. (48)

$$m_0 = \frac{-8 - (-2)}{4 - (-6)} = \frac{-6}{10} = -\frac{3}{5}$$

$$(x_m, y_m) = \left(\frac{-6 + 4}{2}, \frac{-2 - 8}{2}\right) = (-1, -5)$$

The slope of the perpendicular is the negative reciprocal so, $m = \frac{5}{3}$.

$$y - y_m = m(x - x_m)$$
$$y + 5 = \frac{5}{3}(x + 1)$$
$$\mathbf{y = \frac{5}{3}x - \frac{10}{3}}$$

Problem Set 80

1. (45)

4	3	1	2	1

= **24**

2. (28) Distance traveled = rate × time = mh

Distance left = 300 − distance traveled
= **(300 − *hm*) mi**

3. (25) $\frac{3}{4} - \frac{1}{2} = \frac{1}{4}$ cup white sugar needed

$$1 \text{ cup brown sugar} = \frac{3}{4} \text{ cup white sugar}$$

$$\frac{1}{3} \text{ cup brown sugar} = \frac{1}{4} \text{ cup white sugar}$$

$\mathbf{\frac{1}{3}}$ **cup**

4. (25) $R = \frac{\frac{4}{7}\text{ part}}{2\text{ hr}} = \frac{2}{7}\frac{\text{part}}{\text{hr}}$

$$1 - \frac{4}{7} = \frac{3}{7} \text{ to be filled}$$

$$\frac{3}{7} = RT$$

$$\frac{3}{7} = \left(\frac{2}{7}\right)T$$

$$T = \mathbf{\frac{3}{2}\ hr}$$

5. (44) $RWT = \text{jobs}$

$$R = \frac{\text{jobs}}{WT} = \frac{50}{M(1)} = \frac{50}{M}\frac{\text{jobs}}{\text{worker-day}}$$

$$RWT = \text{jobs}$$

$$T = \frac{\text{jobs}}{RW} = \frac{20}{\left(\frac{50}{M}\right)(10)} = \frac{20}{\frac{500}{M}} = \mathbf{\frac{M}{25}\ days}$$

6. (18)
$$\begin{cases} \text{(a)}\ H + T + U = 17 \\ \text{(b)}\ T = H - 5 \\ \text{(c)}\ 2U = 4T + 8 \end{cases}$$

(c) $2U = 4T + 8$

$$U = 2T + 4$$
$$U = 2(H - 5) + 4$$
$$U = 2H - 6$$

(a)
$$H + T + U = 17$$
$$H + (H - 5) + (2H - 6) = 17$$
$$4H - 11 = 17$$
$$4H = 28$$
$$H = 7$$

(b) $T = H - 5 = (7) - 5 = 2$

$U = 2H - 6 = 2(7) - 6 = 8$

The number was **728.**

7. (80) $\frac{\sec^2 x - \tan^2 x}{1 + \cot^2 x} = \frac{1}{\csc^2 x} = \sin^2 x$

8. (80) $\frac{\cos A}{1 + \sin A} + \frac{1 + \sin A}{\cos A} = \frac{\cos^2 A + (1 + \sin A)^2}{(1 + \sin A)\cos A}$

$$= \frac{\cos^2 A + 1 + 2\sin A + \sin^2 A}{(1 + \sin A)\cos A}$$

$$= \frac{(\sin^2 A + \cos^2 A) + 1 + 2\sin A}{(1 + \sin A)\cos A}$$

$$= \frac{1 + 1 + 2\sin A}{(1 + \sin A)\cos A} = \frac{2 + 2\sin A}{(1 + \sin A)\cos A}$$

$$= \frac{2(1 + \sin A)}{(1 + \sin A)\cos A} = \frac{2}{\cos A} = 2\sec A$$

9. (80) $\frac{1}{\tan A} + \tan A = \frac{1 + \tan^2 A}{\tan A} = \frac{\sec^2 A}{\tan A} = \frac{\frac{1}{\cos^2 A}}{\frac{\sin A}{\cos A}}$

$$= \frac{1}{\cos^2 A} \cdot \frac{\cos A}{\sin A} = \frac{1}{\cos A} \cdot \frac{1}{\sin A} = \sec A \csc A$$

10. (79) $(2 \text{ cis } 300°)^6 = 2^6 \text{ cis } (6 \cdot 300°) = 64 \text{ cis } (1800°)$

$$= \mathbf{64\ cis\ 0°}$$

11. (79) $(27 \text{ cis } 36°)^{\frac{1}{3}} = 27^{\frac{1}{3}} \text{ cis}\left(\frac{36°}{3} + n\frac{360°}{3}\right)$

$$27^{\frac{1}{3}} \text{ cis } \frac{36°}{3} = \mathbf{3\ cis\ 12°}$$

$$27^{\frac{1}{3}} \text{ cis}\left(\frac{36°}{3} + 120°\right) = \mathbf{3\ cis\ 132°}$$

$$27^{\frac{1}{3}} \text{ cis}\left(\frac{36°}{3} + 240°\right) = \mathbf{3\ cis\ 252°}$$

12. (79) $(-16)^{\frac{1}{4}} = (16 \text{ cis } 180°)^{\frac{1}{4}}$

$$= 16^{\frac{1}{4}} \text{ cis}\left(\frac{180°}{4} + n\frac{360°}{4}\right)$$

$$16^{\frac{1}{4}} \text{ cis } \frac{180°}{4} = 2 \text{ cis } 45°$$
$$= 2(\cos 45° + i \sin 45°)$$
$$= \mathbf{\sqrt{2} + \sqrt{2}i}$$

$$16^{\frac{1}{4}} \text{ cis}\left(\frac{180°}{4} + 90°\right) = 2 \text{ cis } 135°$$
$$= 2(\cos 135° + i \sin 135°)$$
$$= \mathbf{-\sqrt{2} + \sqrt{2}i}$$

$$16^{\frac{1}{4}} \text{ cis}\left(\frac{180°}{4} + 180°\right) = 2 \text{ cis } 225°$$
$$= 2(\cos 225° + i \sin 225°)$$
$$= -\sqrt{2} - \sqrt{2}i$$

$$16^{\frac{1}{4}} \text{ cis}\left(\frac{180°}{4} + 270°\right) = 2 \text{ cis } 315°$$
$$= 2(\cos 315° + i \sin 315°)$$
$$= \sqrt{2} - \sqrt{2}i$$

13. (78) $16x^2 - 25y^2 = 400$

$$\mathbf{\frac{x^2}{25} - \frac{y^2}{16} = 1}$$

Let $y = 0$

$$\frac{x^2}{25} - \frac{0}{16} = 1$$
$$x = \pm 5$$

Let $x = 0$

$$\frac{0}{25} - \frac{y^2}{16} = 1$$
$$y = \pm 4i$$

From the values of x and y, we see that $m_1 = \frac{4}{5}$ and $m_2 = -\frac{4}{5}$ and we can determine the vertices and the equations of the asymptotes.

Vertices = (5, 0), (–5, 0)

Asymptotes: $\mathbf{y = \frac{4}{5}x;\ y = -\frac{4}{5}x}$

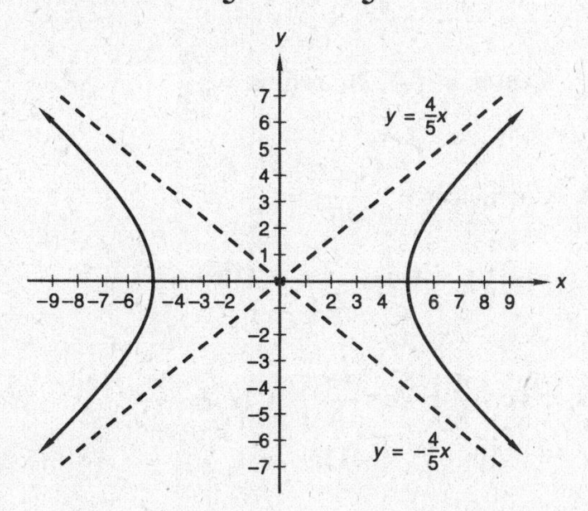

14. (77) $(a + b)^3 = \mathbf{a^3 + 3a^2b + 3ab^2 + b^3}$

15. (77) $\mathbf{15x^4y^2}$

16. (74) $$x = \frac{\begin{vmatrix} 7 & -2 \\ 9 & 4 \end{vmatrix}}{\begin{vmatrix} 4 & -2 \\ 3 & 4 \end{vmatrix}} = \frac{28 + 18}{16 + 6} = \frac{46}{22} = \frac{23}{11}$$

$$y = \frac{\begin{vmatrix} 4 & 7 \\ 3 & 9 \end{vmatrix}}{\begin{vmatrix} 4 & -2 \\ 3 & 4 \end{vmatrix}} = \frac{36 - 21}{16 + 6} = \frac{15}{22}$$

$$\mathbf{x = \frac{23}{11};\ y = \frac{15}{22}}$$

17. (73) $\text{Side} = \dfrac{39 \text{ in.}}{13}$

Side = 3 in.

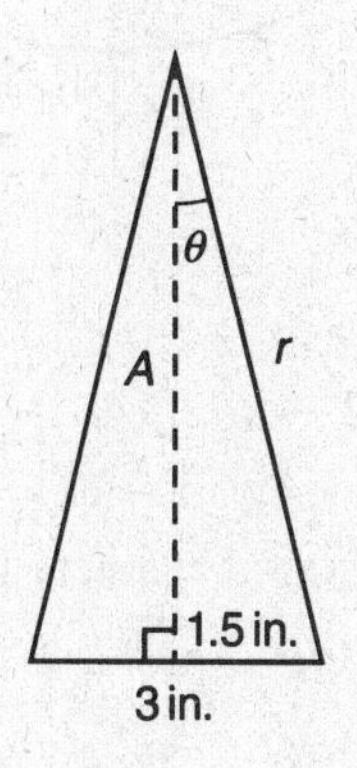

$$\theta = \frac{1}{2}\left(\frac{360°}{13}\right) = 13.8462°$$

$$A = \frac{1.5}{\tan 13.8462°} = 6.0857$$

$$\text{Area}_\Delta = \frac{1}{2}(1.5)(6.0857) = 4.5643 \text{ in.}^2$$

$$\text{Area} = (26)(4.5643) \text{ in.}^2$$

Area = 118.67 in.²

$$r = \frac{1.5}{\sin 13.8462°}$$

Radius = 6.27 in.

18. (72) $C = 180° - 145° - 20°$

$\mathbf{C = 15°}$

$$\frac{b}{\sin 145°} = \frac{12}{\sin 20°}$$
$$b = \frac{12 \sin 145°}{\sin 20°}$$

$\mathbf{b = 20.12}$ **in.**

$$\frac{c}{\sin 15°} = \frac{12}{\sin 20°}$$
$$c = \frac{12 \sin 15°}{\sin 20°}$$

$\mathbf{c = 9.08}$ **in.**

19. (71) $9x^2 + 25y^2 = 225$

$$\frac{x^2}{25} + \frac{y^2}{9} = 1$$

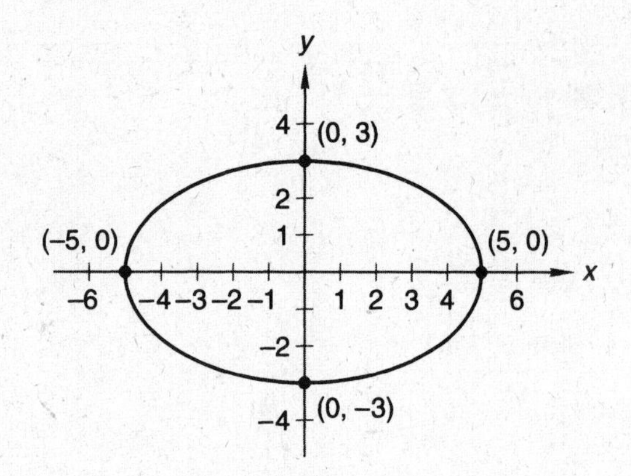

20. (69)

$$\begin{vmatrix} 2-k & -1 \\ -4 & 1-3k \end{vmatrix} = 0$$

$$(2 - k)(1 - 3k) - (-4)(-1) = 0$$

$$2 - 7k + 3k^2 - 4 = 0$$

$$3k^2 - 7k - 2 = 0$$

$$k = \frac{7 \pm \sqrt{7^2 - 4(3)(-2)}}{2(3)} = \mathbf{\frac{7}{6} \pm \frac{\sqrt{73}}{6}}$$

21. (68) Vertex: $(h, k) = (0, 0)$

Focus: $(h, k + p) = \left(0, \frac{7}{5}\right)$

$$k + p = \frac{7}{5}$$

$$0 + p = \frac{7}{5}$$

$$p = \frac{7}{5}$$

Directrix: $y = k - p = -\frac{7}{5}$

Axis of symmetry: $x = h = 0$

$$y - k = \frac{1}{4p}(x - h)^2$$

$$y - 0 = \frac{1}{4\left(\frac{7}{5}\right)}(x - 0)^2$$

Parabola: $\mathbf{y = \frac{5}{28}x^2}$

Directrix: $\mathbf{y = -\frac{7}{5}}$

Axis of symmetry: $\mathbf{x = 0}$

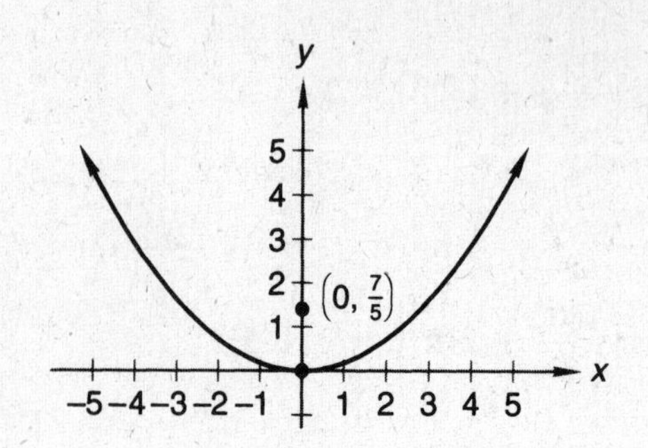

22. (65) $y = -\log_6 x$

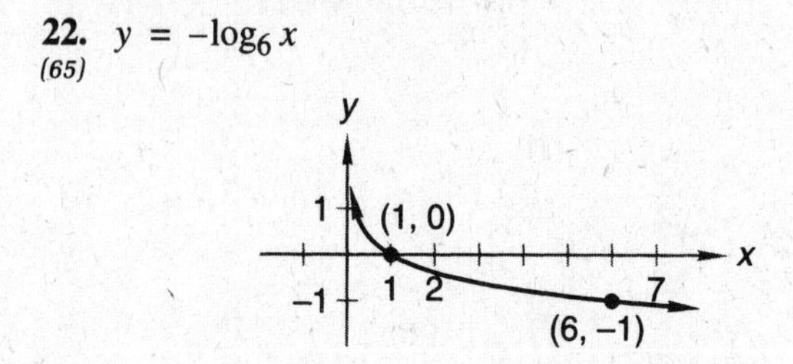

23. (63)

$$3x^2 + 3y^2 + 6x - 12y - \frac{4}{3} = 0$$

$$x^2 + 2x + y^2 - 4y = \frac{4}{9}$$

$$(x^2 + 2x + 1) + (y^2 - 4y + 4) = \frac{4}{9} + 1 + 4$$

$$(x + 1)^2 + (y - 2)^2 = \frac{49}{9}$$

$$(x + 1)^2 + (y - 2)^2 = \left(\frac{7}{3}\right)^2$$

Center = (−1, 2); radius = $\mathbf{\frac{7}{3}}$

24. (66) Coefficient $= \frac{360°}{240°} = \frac{3}{2}$

$$\mathbf{y = -4 + 6\cos\frac{3}{2}(x + 110°)}$$

25. (64) $\left(3 \text{ cis } \frac{\pi}{6}\right)\left(4 \text{ cis } \frac{4\pi}{3}\right) = 12 \text{ cis } \frac{3\pi}{2}$

$= 12(0 - i) = \mathbf{-12i}$

26. (65)

$$\sqrt{3}\cos x - \sqrt{2 - \cos^2 x} = 0$$

$$\sqrt{3}\cos x = \sqrt{2 - \cos^2 x}$$

$$3\cos^2 x = 2 - \cos^2 x$$

$$4\cos^2 x - 2 = 0$$

$$\cos^2 x - \frac{1}{2} = 0$$

$$\left(\cos x - \frac{1}{\sqrt{2}}\right)\left(\cos x + \frac{1}{\sqrt{2}}\right) = 0$$

$$\cos x = \pm\frac{1}{\sqrt{2}}$$

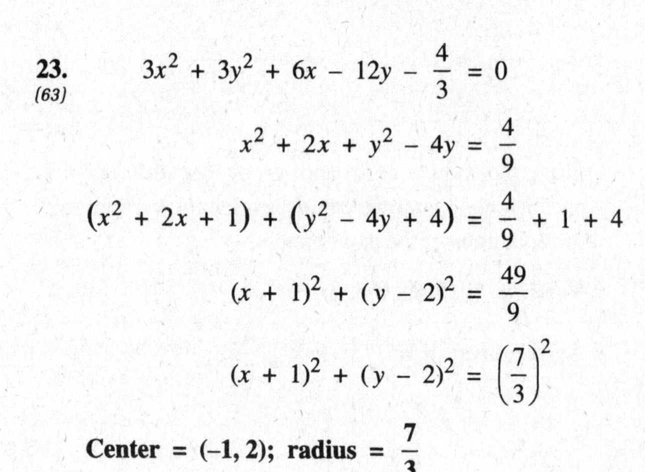

$$\sqrt{3}\cos x = \sqrt{2 - \cos^2 x}$$

$$\sqrt{3}\left(-\frac{1}{\sqrt{2}}\right) \stackrel{?}{=} \sqrt{2 - \left(-\frac{1}{\sqrt{2}}\right)^2}$$

$$-\frac{\sqrt{3}}{\sqrt{2}} \stackrel{?}{=} \sqrt{2 - \frac{1}{2}}$$

$$-\frac{\sqrt{3}}{\sqrt{2}} \neq \frac{\sqrt{3}}{\sqrt{2}}$$

so $-\frac{1}{\sqrt{2}}$ is invalid

$x = \mathbf{\frac{\pi}{4}, \frac{7\pi}{4}}$

27. (52)

$$\sqrt{3}\tan\frac{\theta}{4} - 1 = 0$$

$$\sqrt{3}\tan\frac{\theta}{4} = 1$$

$$\tan\frac{\theta}{4} = \frac{1}{\sqrt{3}}$$

$$\frac{\theta}{4} = \frac{\pi}{6}$$

$$\theta = \mathbf{\frac{2\pi}{3}}$$

28. (32)

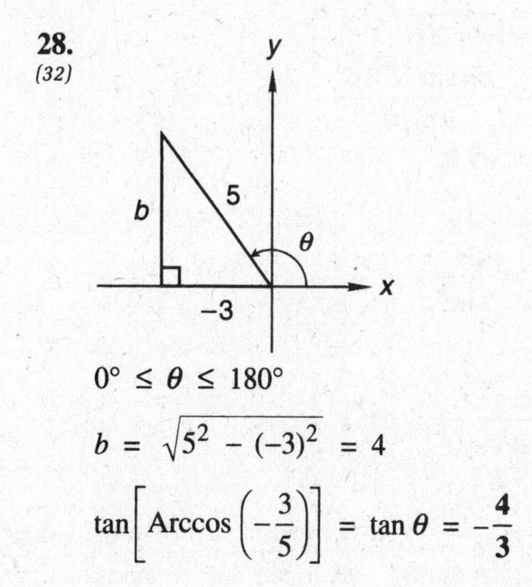

$0° \leq \theta \leq 180°$

$$b = \sqrt{5^2 - (-3)^2} = 4$$

$$\tan\left[\text{Arccos}\left(-\frac{3}{5}\right)\right] = \tan\theta = \mathbf{-\frac{4}{3}}$$

29. (59)

$$\log_7(2x - 1) - \log_7(3x - 3) = \log_7 5$$

$$\log_7\frac{(2x - 1)}{(3x - 3)} = \log_7 5$$

$$2x - 1 = 5(3x - 3)$$

$$2x - 1 = 15x - 15$$

$$13x = 14$$

$$x = \mathbf{\frac{14}{13}}$$

30. (59)

$$2\log_6 6 + 6^{2\log_6 3 - \log_6 3}$$

$$= \log_6 6^2 + 6^{\log_6 3^2 - \log_6 3}$$

$$= 2 + 6^{\log_6 \frac{9}{3}} = 2 + 3 = \mathbf{5}$$

Problem Set 81

1. (75)

$${}_{10}C_6 = \frac{10!}{6!4!} = \textbf{210 teams}$$

2. (38)

$$\text{Avg}_4 = \frac{\text{sum}_4}{4}$$

$$(70) = \frac{\text{sum}_4}{4}$$

$$\text{sum}_4 = (70)(4) = 280$$

$$\text{Avg}_5 = \frac{\text{sum}_4 + x}{5}$$

$$60 = \frac{(280) + x}{5}$$

$$300 = 280 + x$$

$$x = \mathbf{20}$$

3. (18)

$$\begin{cases} \text{(a)}\ x + y = 150 \\ \text{(b)}\ \frac{x}{27} + \frac{y}{23} = 6 \end{cases}$$

(b) $\frac{x}{27} + \frac{y}{23} = 6$

$$x + \frac{27y}{23} = 162$$

$$x = 162 - \frac{27}{23}y$$

(a) $x + y = 150$

$$\left(162 - \frac{27}{23}y\right) + y = 150$$

$$162 - \frac{4}{23}y = 150$$

$$\frac{4}{23}y = 12$$

$$y = \left(\frac{23}{4}\right)12$$

$$y = 69$$

(a) $x + y = 150$

$$x = 150 - (69)$$

$$x = 81$$

The numbers are **81** and **69.**

4. (18)
$$\begin{cases} (a)\ A + B + C = 180° \\ (b)\ B - A = 10° \\ (c)\ C - B = 25° \end{cases}$$

$$\begin{array}{lrl} \text{(a)} & A + B + C &= 180° \\ \text{(b)} & -A + B &= 10° \\ \hline \text{(d)} & 2B + C &= 190° \\ -1\text{(c)} & B - C &= -25° \\ \hline & 3B &= 165° \\ & B &= 55° \end{array}$$

(c) $C - B = 25°$

$C - (55°) = 25°$

$C = 80°$

(b) $B - A = 10°$

$(55°) - A = 10°$

$A = 45°$

45°; 55°; 80°

5. (25) $RWT = \text{jobs}$

$R(5)(2) = 3$

$$R = \frac{3}{10}\,\frac{\text{jobs}}{\text{worker-days}}$$

$RWT = \text{jobs}$

$$W = \frac{\text{jobs}}{RT}$$

$$W = \frac{18}{\left(\frac{3}{10}\right)(10)} = \textbf{6 workers}$$

6. (81) $a^2 = b^2 + c^2 - 2bc\cos A$

$$a = \sqrt{8^2 + 6^2 - 2(8)(6)\cos 40°}$$

$\mathbf{a = 5.14}$ **in.**

$$\frac{6}{\sin C} = \frac{a}{\sin 40°}$$

$$\sin C = \frac{6\sin 40°}{5.1439}$$

$\mathbf{C = 48.57°}$

$B = 180° - 40° - 48.57°$

$\mathbf{B = 91.43°}$

7. (81) $b^2 = a^2 + c^2 - 2ac\cos B$

$(10)^2 = (7)^2 + (5)^2 - 2(7)(5)\cos B$

$26 = -70\cos B$

$$\cos B = -\frac{26}{70}$$

$\mathbf{B = 111.80°}$

$$\frac{7}{\sin A} = \frac{10}{\sin B}$$

$$\sin A = \frac{7\sin 111.80°}{10}$$

$\mathbf{A = 40.54°}$

$C = 180° - 111.80° - 40.54°$

$\mathbf{C = 27.66°}$

8. (81)

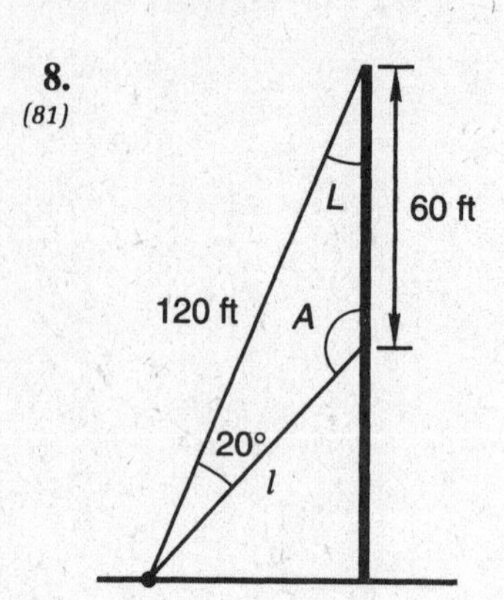

$$\frac{120}{\sin A} = \frac{60}{\sin 20°}$$

$$\sin A = \frac{120\sin 20°}{60}$$

$A = 136.84°$

$L = 180° - 20° - 136.84° = 23.16°$

$$\frac{l}{\sin 23.16°} = \frac{60}{\sin 20°}$$

$$l = \frac{60\sin 23.16°}{\sin 20°}$$

$l =$ **69 ft**

9. (80)
$$\frac{1 - \cos^2 x}{\sec^2 x - 1} = \frac{\sin^2 x}{\tan^2 x} = \frac{\sin^2 x}{\frac{\sin^2 x}{\cos^2 x}}$$

$$= \frac{\sin^2 x}{1}\cdot\frac{\cos^2 x}{\sin^2 x} = \cos^2 x$$

10. (80)
$$\frac{1}{1 + \sin A} + \frac{1}{1 - \sin A} = \frac{1 - \sin A + 1 + \sin A}{(1 + \sin A)(1 - \sin A)}$$

$$= \frac{2}{1 - \sin^2 A} = \frac{2}{\cos^2 A} = 2\sec^2 A$$

11. (79)
$$(-1 - i)^6 = \left(\sqrt{2}\operatorname{cis} 225°\right)^6 = \left(\sqrt{2}\right)^6 \operatorname{cis}(6\cdot 225°)$$

$$= 8\operatorname{cis} 1350° = 8\operatorname{cis} 270° = 8(0 - i) = \mathbf{-8i}$$

12. (79)
$$(8\operatorname{cis} 45°)^{\frac{1}{3}} = 8^{\frac{1}{3}}\operatorname{cis}\left(\frac{45°}{3} + n\frac{360°}{3}\right)$$

$= \mathbf{2\ cis\ 15°, 2\ cis\ 135°, 2\ cis\ 255°}$

13. (79) $(-81)^{\frac{1}{4}} = (81 \text{ cis } 180°)^{\frac{1}{4}}$

$= 81^{\frac{1}{4}} \text{ cis}\left(\frac{180°}{4} + n\frac{360°}{4}\right)$

$= 3 \text{ cis } 45°, 3 \text{ cis } 135°, 3 \text{ cis } 225°, 3 \text{ cis } 315°$

$= \mathbf{\frac{3\sqrt{2}}{2} + \frac{3\sqrt{2}}{2}i, -\frac{3\sqrt{2}}{2} + \frac{3\sqrt{2}}{2}i,}$

$\mathbf{-\frac{3\sqrt{2}}{2} - \frac{3\sqrt{2}}{2}i, \frac{3\sqrt{2}}{2} - \frac{3\sqrt{2}}{2}i}$

14. (78) $4x^2 - 25y^2 = 100$

$\mathbf{\frac{x^2}{25} - \frac{y^2}{4} = 1}$

Let $y = 0$ Let $x = 0$

$\frac{x^2}{25} = 1$ $-\frac{y^2}{4} = 1$

$x = \pm 5$ $y = \pm 2i$

Vertices = (5, 0), (–5, 0)

Asymptotes: $\mathbf{y = \frac{2}{5}x;\ y = -\frac{2}{5}x}$

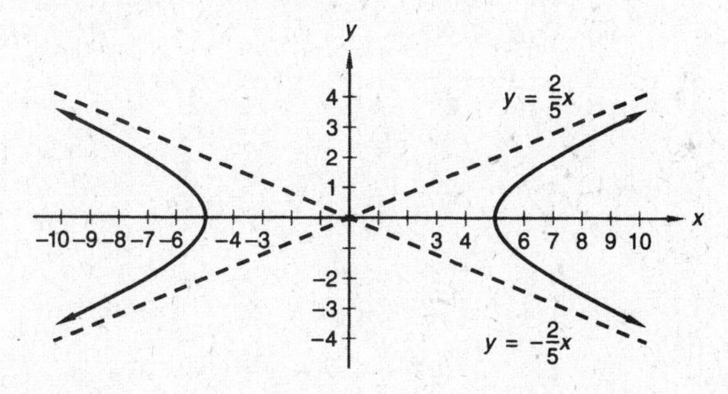

15. (77) $(m + n)^4$

Term	①	②	③	④	⑤
For m	4	3	2	1	0
For n	0	1	2	3	4
Coefficient	1	4	6	4	1

$\mathbf{m^4 + 4m^3n + 6m^2n^2 + 4mn^3 + n^4}$

16. (77) $(x + y)^7$

Term	①	②	③	④	⑤...⑧
For x	7	6	5	4	3... 0
For y	0	1	2	3	4... 7
Coefficient	1	7	21	35	35... 1

$\mathbf{35x^4y^3}$

17. (73) Side $= \frac{36 \text{ in.}}{12} = 3$ in.

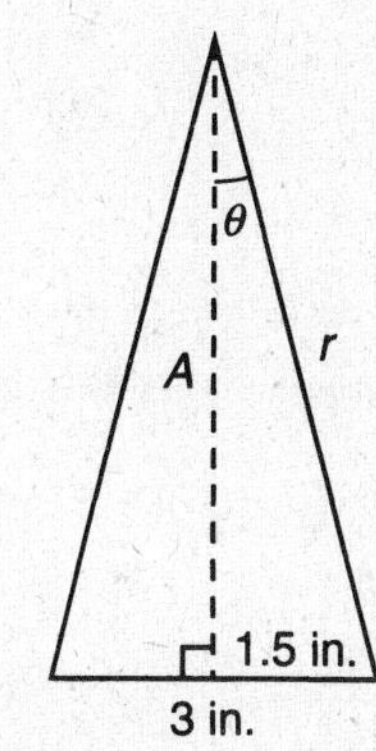

$\theta = \frac{1}{2}\left(\frac{360°}{12}\right) = 15°$

$A = \frac{1.5}{\tan 15°} = 5.5981$ in.

$\text{Area}_\Delta = \frac{1}{2}(1.5)(5.5981) = 4.1986 \text{ in.}^2$

Area = (24)(4.1986)

Area = 100.77 in.²

$r = \frac{1.5}{\sin 15°}$

Radius = 5.80 in.

18. (71) $4x^2 + 25y^2 = 100$

$\mathbf{\frac{x^2}{25} + \frac{y^2}{4} = 1}$

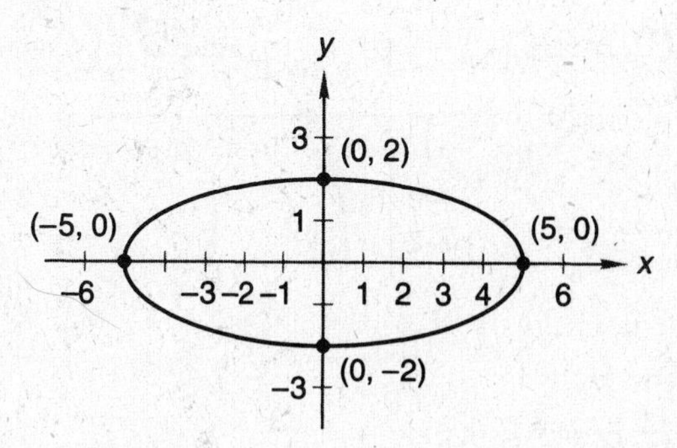

19. (70) $z_1 = \frac{x_1 - \mu}{\sigma} = \frac{60 - 62}{2.5} = -0.8$

Percentile 0.2119

$z_2 = \frac{x_2 - \mu}{\sigma} = \frac{63 - 62}{2.5} = 0.4$

Percentile 0.6554

$0.6554 - 0.2119 = 0.4435$

44.35%

20. (68) Vertex: $(h, k) = (4, 2)$

Focus: $(h, k + p) = (4, -2)$

$k + p = -2$

$(2) + p = -2$

$p = -4$

Directrix: $y = k - p = 6$

Axis of symmetry: $x = h = 4$

$$y - k = \frac{1}{4p}(x - h)^2$$

$$y - 2 = \frac{1}{4(-4)}(x - 4)^2$$

Parabola: $\mathbf{y = -\frac{1}{16}(x - 4)^2 + 2}$

Directrix: $\mathbf{y = 6}$

Axis of symmetry: $\mathbf{x = 4}$

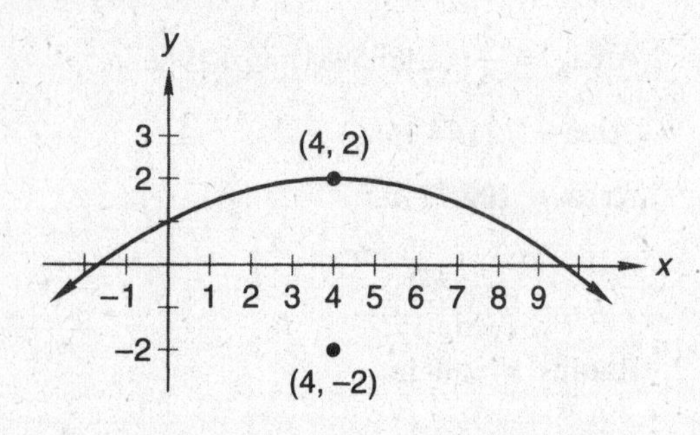

21. (65) $y = \log_2(x - 1)$

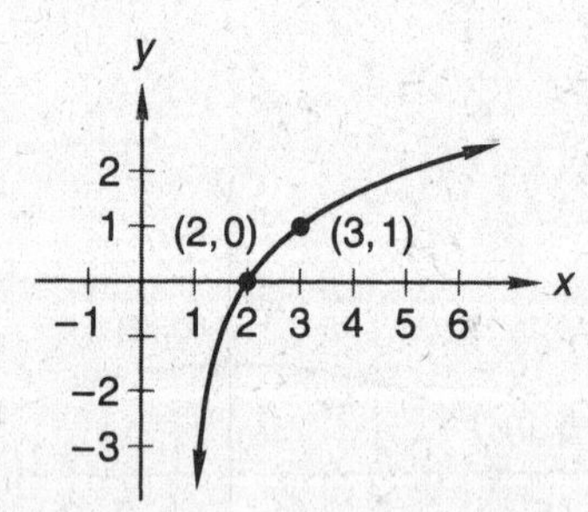

22. (66) Period = 320°

$$\text{Coefficient} = \frac{360°}{320°} = \frac{9}{8}$$

$$\mathbf{y = 10 + 2\sin\frac{9}{8}(\theta - 90°)}$$

23. (64) $(2 \text{ cis } 30°)[4 \text{ cis }(-90°)] = 8 \text{ cis }(-60°)$

$$= 8\left(\frac{1}{2} - \frac{\sqrt{3}}{2}i\right) = \mathbf{4 - 4\sqrt{3}i}$$

24. (52) $\sqrt{3}\tan 3\theta + 1 = 0$

$$\tan 3\theta = -\frac{1}{\sqrt{3}}$$

$3\theta = 150°, 330°, 510°, 690°, 870°, 1050°$

$\theta = \mathbf{50°, 110°, 170°, 230°, 290°, 350°}$

25. (60) $\cos\theta + 2\cos\theta\csc\theta = 0$

$\cos\theta(1 + 2\csc\theta) = 0$

$\cos\theta = 0$

$\theta = \mathbf{90°, 270°}$

$1 + 2\csc\theta = 0$

$\csc\theta = -\frac{1}{2}$

no answer

26. (32)

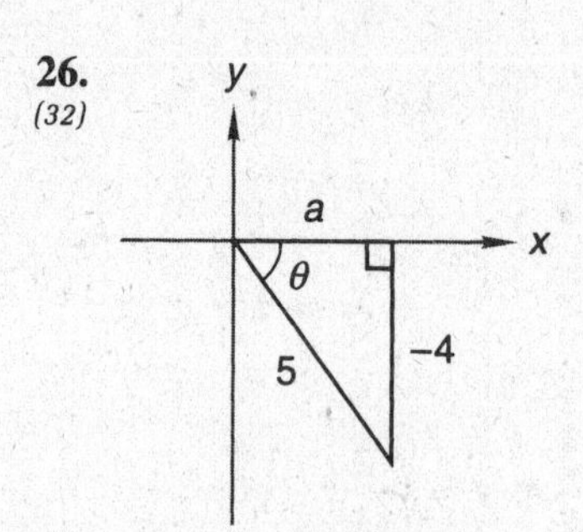

$-90° \le \theta \le 90°$

$$a = \sqrt{5^2 - 4^2} = 3$$

$$\cos\left[\text{Arcsin}\left(-\frac{4}{5}\right)\right] = \cos\theta = \mathbf{\frac{3}{5}}$$

27. (56)

$$2x^2 + 2y^2 - 4x + 8y - 8 \ge 0$$

$$x^2 - 2x + y^2 + 4y \ge 4$$

$$(x^2 - 2x + 1) + (y^2 + 4y + 4) \ge 4 + 1 + 4$$

$$(x - 1)^2 + (y + 2)^2 \ge 9$$

$$\begin{cases} (x - 1)^2 + (y + 2)^2 \ge 3^2 & \text{(circle)} \\ y < -(x - 1)^2 - 2 & \text{(parabola)} \end{cases}$$

The inequality indicates the region outside or on the circle and below the parabola.

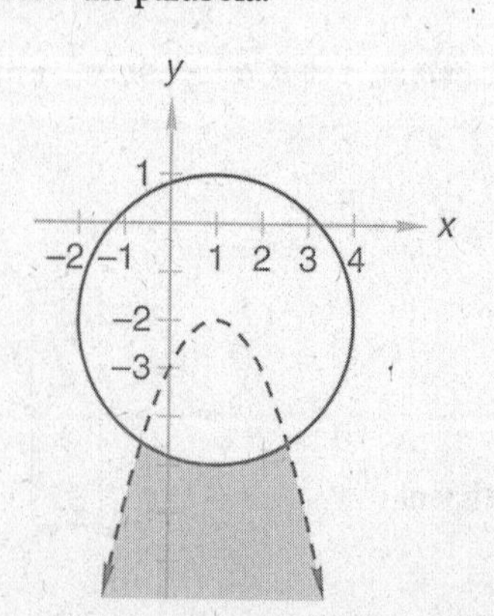

28. (59)
$$\log_{\frac{1}{2}}(3x - 2) - \log_{\frac{1}{2}}(x + 2) = -\log_{\frac{1}{2}} 4$$
$$\log_{\frac{1}{2}}\left(\frac{3x - 2}{x + 2}\right) = \log_{\frac{1}{2}} 4^{-1}$$
$$\frac{(3x - 2)}{(x + 2)} = \frac{1}{4}$$
$$4(3x - 2) = x + 2$$
$$12x - 8 = x + 2$$
$$11x = 10$$
$$x = \mathbf{\frac{10}{11}}$$

29. (59)
$$\log 10(x - 1) = 0$$
$$10(x - 1) = 10^0$$
$$10x - 10 = 1$$
$$10x = 11$$
$$x = \mathbf{\frac{11}{10}}$$

30. (40)
$$\frac{2 \log_9 6 - \log_9 12}{4 \log_9 3} = \frac{\log_9 \frac{6^2}{12}}{\log_9 3^4} = \frac{\log_9 3}{\log_9 81}$$
$$= \frac{\log_9 9^{\frac{1}{2}}}{\log_9 9^2} = \frac{\frac{1}{2}}{2} = \mathbf{\frac{1}{4}}$$

Problem Set 82

1. (55)
$$\frac{20!}{10!6!4!} = \textbf{38,798,760 patterns}$$

2. (7)
$$11N,\ 11(N + 1),\ 11(N + 2)$$
$$4[11N + 11(N + 2)] + 66 = 10[11(N + 1)]$$
$$4(22N + 22) + 66 = 10(11N + 11)$$
$$88N + 88 + 66 = 110N + 110$$
$$22N = 44$$
$$N = 2$$

The numbers are **22, 33, 44.**

3. (18)
$$\begin{cases} \text{(a)}\ h + l + w = 18 \\ \text{(b)}\ h = \frac{4}{5}(l + w) \\ \text{(c)}\ h = 4(l - w) \end{cases}$$

(b)
$$h = \frac{4}{5}(l + w)$$
$$5h = 4l + 4w$$
$$5h - 4l - 4w = 0$$

$$5h - 4l - 4w = 0$$
$$4\text{(a)}\quad 4h + 4l + 4w = 72$$
$$9h = 72$$
$$h = 8$$

Height = 8 ft

(b) $h = \frac{4}{5}(l + w)$
$$5h = 4l + 4w$$

(c) $h = 4(l - w)$
$$h = 4l - 4w$$

$$5h = 4l + 4w$$
$$h = 4l - 4w$$
$$6h = 8l$$
$$6(8) = 8l$$
$$l = 6$$

Length = 6 ft

(a)
$$h + l + w = 18$$
$$(8) + (6) + w = 18$$
$$w = 4$$

Width = 4 ft

4. (28) Distance traveled = RT = rh miles

Distance to go = t − distance traveled
= $\mathbf{(t - rh)}$ **miles**

5. (25)
$$RWT = \text{jobs}$$
$$R(7)(5) = 3$$
$$R = \frac{3}{35}\ \frac{\text{jobs}}{\text{women-days}}$$
$$RWT = \text{jobs}$$
$$T = \frac{\text{jobs}}{RW} = \frac{27}{\left(\frac{3}{35}\right)(9)} = \textbf{35 days}$$

6. (44)
$$R_1T_1 + R_2T_2 = \text{jobs}$$
$$\frac{a}{3}T + \frac{3}{b}T = 13$$
$$T\left(\frac{a}{3} + \frac{3}{b}\right) = 13$$
$$T = \frac{13}{\frac{ab + 9}{3b}}$$
$$T = \mathbf{\frac{39b}{ab + 9}}\ \textbf{days}$$

7. (82)

$$
\begin{aligned}
6^{3x+2} &= 4^{2x-1} \\
\log 6^{3x+2} &= \log 4^{2x-1} \\
(3x + 2)\log 6 &= (2x - 1)\log 4 \\
(3x + 2)(0.7782) &= (2x - 1)(0.6021) \\
2.3346x + 1.5564 &= 1.2042x - 0.6021 \\
1.1304x &= -2.1585 \\
x &= \mathbf{-1.91}
\end{aligned}
$$

8. (82)

$$
\begin{aligned}
10^{-3x-4} &= 5^{2x-1} \\
\log 10^{-3x-4} &= \log 5^{2x-1} \\
(-3x - 4)\log 10 &= (2x - 1)\log 5 \\
-3x - 4 &= (2x - 1)(0.6990) \\
-3x - 4 &= 1.398x - 0.6990 \\
-4.398x &= 3.301 \\
x &= \mathbf{-0.75}
\end{aligned}
$$

9. (81) $a^2 = b^2 + c^2 - 2bc \cos A$

$$a = \sqrt{10^2 + 9^2 - 2(10)(9)\cos 45°}$$

$\mathbf{a = 7.33\ cm}$

$$\frac{10}{\sin B} = \frac{7.3294}{\sin 45°}$$

$$\sin B = \frac{10 \sin 45°}{7.3294}$$

$\mathbf{B = 74.74°}$

$C = 180° - 45° - 74.74°$

$\mathbf{C = 60.26°}$

10. (81)

$$
\begin{aligned}
b^2 &= a^2 + c^2 - 2ac \cos B \\
12^2 &= 10^2 + 4^2 - 2(10)(4)\cos B \\
28 &= -80 \cos B \\
\cos B &= -\frac{28}{80} \\
B &= \mathbf{110.49°}
\end{aligned}
$$

$$\frac{10}{\sin A} = \frac{12}{\sin 110.49°}$$

$$\sin A = \frac{10 \sin 110.49°}{12}$$

$\mathbf{A = 51.32°}$

$C = 180° - 51.32° - 110.49°$

$\mathbf{C = 18.19°}$

11. (80)

$$\frac{\sin A}{1 + \cos A} + \frac{1 + \cos A}{\sin A}$$

$$= \frac{\sin^2 A + (1 + \cos A)^2}{(1 + \cos A)\sin A}$$

$$= \frac{\sin^2 A + 1 + 2\cos A + \cos^2 A}{(1 + \cos A)\sin A}$$

$$= \frac{1 + 1 + 2\cos A}{(1 + \cos A)\sin A} = \frac{2 + 2\cos A}{(1 + \cos A)\sin A}$$

$$= \frac{2(1 + \cos A)}{(1 + \cos A)\sin A} = \frac{2}{\sin A} = 2 \csc A$$

12. (80)

$$\frac{\csc^2 \theta - \cot^2 \theta}{1 + \cot^2 \theta} = \frac{1}{\csc^2 \theta} = \sin^2 \theta$$

13. (80)

$$\frac{\sin x}{\csc x} + \frac{\cos x}{\sec x} = \frac{\sin x}{\frac{1}{\sin x}} + \frac{\cos x}{\frac{1}{\cos x}}$$

$$= \sin^2 x + \cos^2 x = 1$$

14. (79)

$$
\begin{aligned}
(1 \text{ cis } 12°)^{30} &= 1^{30} \text{ cis } (30 \cdot 12°) \\
&= 1 \text{ cis } 360° \\
&= \mathbf{1\ cis\ 0°}
\end{aligned}
$$

15. (79)

$$(16 \text{ cis } 60°)^{\frac{1}{4}} = 16^{\frac{1}{4}} \text{ cis}\left(\frac{60°}{4} + n\frac{360°}{4}\right)$$

$= \mathbf{2\ cis\ 15°, 2\ cis\ 105°, 2\ cis\ 195°, 2\ cis\ 285°}$

16. (79)

$$(-1)^{\frac{1}{3}} = (1 \text{ cis } 180°)^{\frac{1}{3}} = 1^{\frac{1}{3}} \text{ cis}\left(\frac{180°}{3} + n\frac{360°}{3}\right)$$

$= 1 \text{ cis } 60°, 1 \text{ cis } 180°, 1 \text{ cis } 300°$

$$= \mathbf{\frac{1}{2} + \frac{\sqrt{3}}{2}i, -1, \frac{1}{2} - \frac{\sqrt{3}}{2}i}$$

17. (78) $9x^2 - 36y^2 = 324$

$$\mathbf{\frac{x^2}{36} - \frac{y^2}{9} = 1}$$

Let $y = 0$

$$\frac{x^2}{36} = 1$$

$x = \pm 6$

Let $x = 0$

$$-\frac{y^2}{9} = 1$$

$y = \pm 3i$

Vertices = (6, 0), (−6, 0)

Asymptotes: $\mathbf{y = \frac{1}{2}x;\ y = -\frac{1}{2}x}$

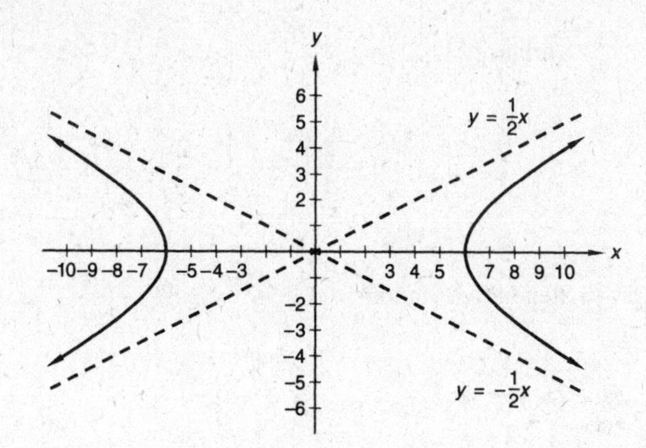

18. $(p + q)^5$
(77)

Term	①	②	③	④	⑤	⑥
For p	5	4	3	2	1	0
For q	0	1	2	3	4	5
Coefficient	1	5	10	10	5	1

$\boldsymbol{p^5 + 5p^4q + 10p^3q^2 + 10p^2q^3 + 5pq^4 + q^5}$

19. $(x + 1)^{10}$
(77)

Term	①	②	③	④	⑤	⑥	⑦ . . . ⑪
For x	10	9	8	7	6	5	4 . . . 0
For 1	0	1	2	3	4	5	6 . . . 10
Coeff.	1	10	45	120	210	252	210 . . . 1

$252x^5 1^5 = \mathbf{252x^5}$

20. Side $= \dfrac{42}{14} = 3$
(73)

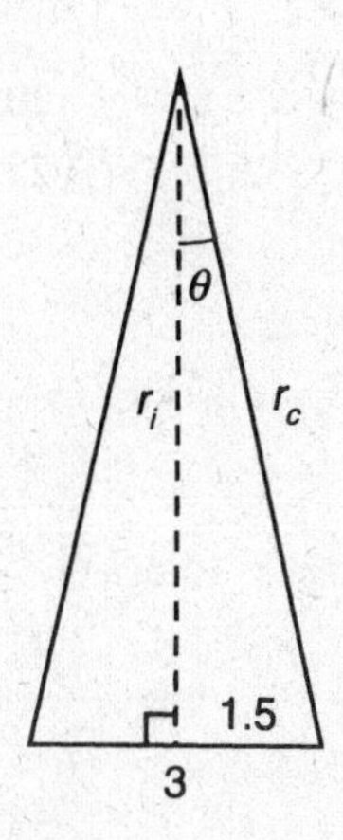

$$\theta = \frac{1}{2}\left(\frac{360°}{14}\right) = 12.86°$$

$$r_c = \frac{1.5}{\sin 12.86°} = 6.74$$

Radius of circumscribed circle = 6.74

$$r_i = \frac{1.5}{\tan 12.86°} = 6.57$$

Radius of inscribed circle = 6.57

21. $9x^2 + 36y^2 = 324$
(71)

$$\boldsymbol{\frac{x^2}{36} + \frac{y^2}{9} = 1}$$

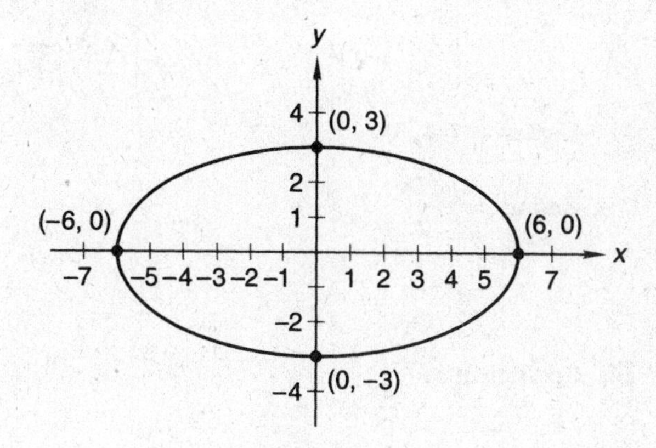

22. Vertex: $(h, k) = (0, 0)$
(68)

Focus: $(h, k + p) = (0, 6)$

$k + p = 6$

$0 + p = 6$

$p = 6$

Directrix: $y = k - p = -6$

Axis of symmetry: $x = h = 0$

$$y - k = \frac{1}{4p}(x - h)^2$$

$$y - 0 = \frac{1}{4(6)}(x - 0)^2$$

Parabola: $\boldsymbol{y = \frac{1}{24}x^2}$

Directrix: $\boldsymbol{y = -6}$

Axis of symmetry: $\boldsymbol{x = 0}$

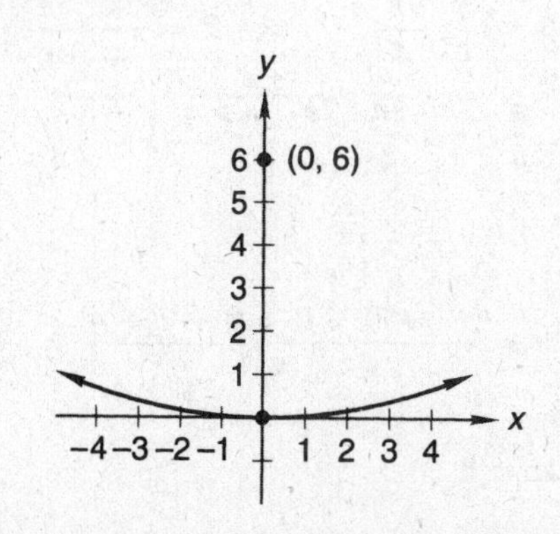

23. (63)

$$2x^2 + 2y^2 - 12y + 4x + 1 = 0$$
$$2x^2 + 4x + 2y^2 - 12y = -1$$
$$x^2 + 2x + y^2 - 6y = -\frac{1}{2}$$
$$(x^2 + 2x + 1) + (y^2 - 6y + 9) = -\frac{1}{2} + 1 + 9$$
$$(x + 1)^2 + (y - 3)^2 = \frac{19}{2}$$
$$(x + 1)^2 + (y - 3)^2 = \left(\frac{\sqrt{38}}{2}\right)^2$$

Center $= \mathbf{(-1, 3)}$

Radius $= \mathbf{\frac{\sqrt{38}}{2}}$

24. (66) Coefficient $= \frac{2\pi}{\frac{\pi}{2}} = 4$

$$\mathbf{y = 2 + 9 \sin 4\left(x - \frac{\pi}{8}\right)}$$

25. (60)

$$\csc^2 \theta - 4 = 0$$
$$(\csc \theta - 2)(\csc \theta + 2) = 0$$

$\csc \theta = 2$ $\qquad$ $\csc \theta = -2$

$\sin \theta = \frac{1}{2}$ $\qquad$ $\sin \theta = -\frac{1}{2}$

$\theta = \frac{\pi}{6}, \frac{5\pi}{6}$ $\qquad$ $\theta = \frac{7\pi}{6}, \frac{11\pi}{6}$

$$\theta = \mathbf{\frac{\pi}{6}, \frac{5\pi}{6}, \frac{7\pi}{6}, \frac{11\pi}{6}}$$

26. (60) $(\sqrt{2} \cos \theta - 1)(\sqrt{2} \cos \theta + 1) = 0$

$\cos \theta = \frac{1}{\sqrt{2}}$ $\qquad$ $\cos \theta = -\frac{1}{\sqrt{2}}$

$\theta = \frac{\pi}{4}, \frac{7\pi}{4}$ $\qquad$ $\theta = \frac{3\pi}{4}, \frac{5\pi}{4}$

$$\theta = \mathbf{\frac{\pi}{4}, \frac{3\pi}{4}, \frac{5\pi}{4}, \frac{7\pi}{4}}$$

27. (61) Mean $= \frac{-2 + 0 + 0 + 0 + 2 + 4}{6} = \mathbf{\frac{2}{3}}$

Median $= \mathbf{0}$

Mode $= \mathbf{0}$

$$\text{Variance} = \frac{1}{6}\left[\left(-2 - \frac{2}{3}\right)^2 + \left(0 - \frac{2}{3}\right)^2 + \left(0 - \frac{2}{3}\right)^2 + \left(0 - \frac{2}{3}\right)^2 + \left(2 - \frac{2}{3}\right)^2 + \left(4 - \frac{2}{3}\right)^2\right]$$
$$= \mathbf{3.56}$$

Standard deviation $= \sqrt{3.56} = \mathbf{1.89}$

28. (58)

$$2x + y = 3$$
$$y = -2x + 3$$

Equation of the perpendicular line:

$$y - y_1 = m(x - x_1)$$
$$y - 1 = \frac{1}{2}(x - 2)$$
$$y = \frac{1}{2}x$$

Point of intersection:

$$\frac{1}{2}x = -2x + 3$$
$$\frac{5}{2}x = 3$$
$$x = \frac{6}{5}$$
$$y = \frac{1}{2}x = \frac{1}{2}\left(\frac{6}{5}\right) = \frac{3}{5}$$

$\left(\frac{6}{5}, \frac{3}{5}\right)$ and $(2, 1)$

$$D = \sqrt{\left(2 - \frac{6}{5}\right)^2 + \left(1 - \frac{3}{5}\right)^2}$$
$$= \sqrt{\left(\frac{4}{5}\right)^2 + \left(\frac{2}{5}\right)^2} = \sqrt{\frac{20}{25}} = \mathbf{\frac{2\sqrt{5}}{5}}$$

29. (59) $\frac{2}{3} \log_7 8 + \log_7 x - \log_7 (x - 2) = \log_7 9$

$$\log_7 8^{\frac{2}{3}} + \log_7 x - \log_7 (x - 2) = \log_7 9$$
$$\log_7 \frac{4x}{(x - 2)} = \log_7 9$$
$$\frac{4x}{(x - 2)} = 9$$
$$4x = 9x - 18$$
$$5x = 18$$
$$x = \mathbf{\frac{18}{5}}$$

30. (70) $z_1 = \dfrac{x_1 - \mu}{\sigma} = \dfrac{38 - 40}{2.5} = -0.8$

Percentile 0.2119

$z_2 = \dfrac{x_2 - \mu}{\sigma} = \dfrac{41.5 - 40}{2.5} = 0.6$

Percentile 0.7257

$0.7257 - 0.2119 = 0.5138$

51.38%

Problem Set 83

1. (81)

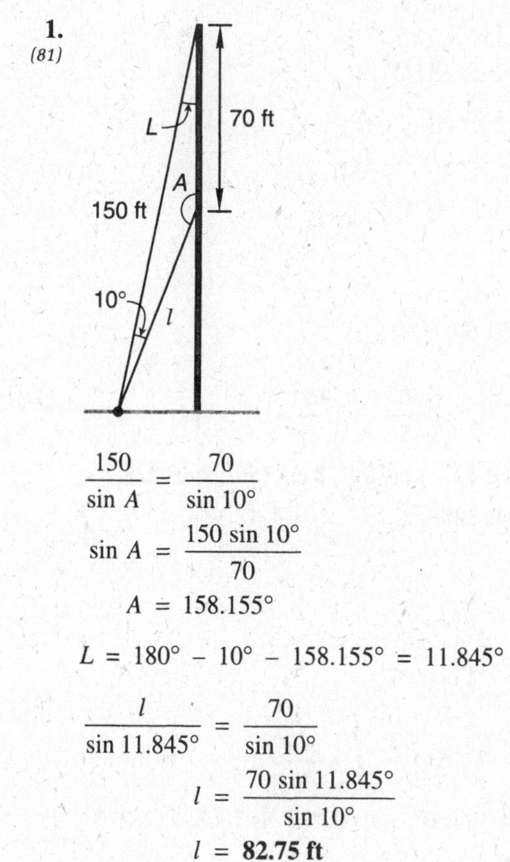

$$\frac{150}{\sin A} = \frac{70}{\sin 10°}$$

$$\sin A = \frac{150 \sin 10°}{70}$$

$$A = 158.155°$$

$$L = 180° - 10° - 158.155° = 11.845°$$

$$\frac{l}{\sin 11.845°} = \frac{70}{\sin 10°}$$

$$l = \frac{70 \sin 11.845°}{\sin 10°}$$

$$l = \mathbf{82.75\ ft}$$

2. (62) $C = mN + b$

(a) $850 = m20 + b$
(b) $450 = m10 + b$

(a) $850 = m20 + b$
-1(b) $-450 = -m10 - b$

$400 = m10$

$m = 40$

(a) $850 = m20 + b$

$850 = (40)20 + b$

$b = 50$

$C = 40N + 50 = 40(2) + 50 = 80 + 50 = \mathbf{\$130}$

3. (44) $RWT = \text{jobs}$

$$R = \frac{\text{jobs}}{WT} = \frac{J}{M(1)} = \frac{J}{M}\ \frac{\text{jobs}}{\text{man-day}}$$

$RWT = \text{jobs}$

$$T = \frac{\text{jobs}}{RW}$$

$$= \frac{k}{\left(\frac{J}{M}\right)(M - 5)} = \mathbf{\frac{Mk}{J(M - 5)}\ days}$$

4. (83)

	1	2	3	4	5	6
1	2	3	4	5	6	7
2	3	4	5	6	7	8
3	4	5	6	7	8	9
4	5	6	7	8	9	(10)
5	6	7	8	9	(10)	(11)
6	7	8	9	(10)	(11)	(12)

$$P(\text{sum} > 9) = \frac{6}{36} = \mathbf{\frac{1}{6}}$$

5. (83) $\dfrac{5}{17} \cdot \dfrac{12}{17} = \mathbf{\dfrac{60}{289}}$

6. (83) $\dfrac{5}{17} \cdot \dfrac{12}{16} = \dfrac{60}{272} = \mathbf{\dfrac{15}{68}}$

7. (82)

$$7^{2x-4} = 5^{3x+2}$$
$$\log 7^{2x-4} = \log 5^{3x+2}$$
$$(2x - 4) \log 7 = (3x + 2) \log 5$$
$$(2x - 4)(0.8451) = (3x + 2)(0.6990)$$
$$1.6902x - 3.3804 = 2.097x + 1.398$$
$$-4.7784 = 0.4068x$$
$$x = \mathbf{-11.75}$$

8. (82)

$$10^{3x-1} = 5^{4x-2}$$
$$\log 10^{3x-1} = \log 5^{4x-2}$$
$$(3x - 1) \log 10 = (4x - 2) \log 5$$
$$3x - 1 = (4x - 2)(0.6990)$$
$$3x - 1 = 2.796x - 1.398$$
$$0.204x = -0.398$$
$$x = \mathbf{-1.95}$$

9. (81) $a^2 = b^2 + c^2 - 2bc \cos A$

$$a = \sqrt{12^2 + 8^2 - 2(12)(8)\cos 60°}$$

$$\boldsymbol{a = 10.58 \text{ in.}}$$

$$\frac{12}{\sin B} = \frac{10.583}{\sin 60°}$$

$$\sin B = \frac{12 \sin 60°}{10.583}$$

$$\boldsymbol{B = 79.11°}$$

$$C = 180° - 60° - 79.11°$$

$$\boldsymbol{C = 40.89°}$$

10. (81) $b^2 = a^2 + c^2 - 2ab \cos B$

$$14^2 = 10^2 + 8^2 - 2(10)(8)\cos B$$

$$196 = 100 + 64 - 160 \cos B$$

$$32 = -160 \cos B$$

$$\cos B = -\frac{32}{160}$$

$$\boldsymbol{B = 101.54°}$$

$$\frac{14}{\sin 101.54°} = \frac{8}{\sin C}$$

$$\sin C = \frac{8 \sin 101.54°}{14}$$

$$\boldsymbol{C = 34.05°}$$

$$A = 180° - 101.54° - 34.05°$$

$$\boldsymbol{A = 44.41°}$$

11. (80) $\dfrac{\sin A}{1 - \cos A} + \dfrac{1 - \cos A}{\sin A}$

$$= \frac{\sin^2 A + (1 - \cos A)^2}{(1 - \cos A)\sin A}$$

$$= \frac{\sin^2 A + 1 - 2\cos A + \cos^2 A}{(1 - \cos A)\sin A}$$

$$= \frac{2 - 2\cos A}{(1 - \cos A)\sin A}$$

$$= \frac{2(1 - \cos A)}{(1 - \cos A)\sin A}$$

$$= \frac{2}{\sin A} = 2 \csc A$$

12. (80) $\dfrac{\sec^2\theta - \tan^2\theta}{\tan^2\theta + 1} = \dfrac{1}{\sec^2\theta} = \cos^2\theta$

13. (80) $\dfrac{1}{\tan(-x)} + \tan(-x) = \dfrac{1}{-\tan x} - \tan x$

$$= \frac{-1 - \tan^2 x}{\tan x} = \frac{-(1 + \tan^2 x)}{\tan x} = \frac{-\sec^2 x}{\tan x}$$

$$= \frac{-\dfrac{1}{\cos^2 x}}{\dfrac{\sin x}{\cos x}} = -\frac{1}{\cos^2 x} \cdot \frac{\cos x}{\sin x}$$

$$= -\frac{1}{\cos x} \cdot \frac{1}{\sin x} = -\sec x \csc x$$

14. (79) $(-1 + i)^{10}$

$$= \left(\sqrt{2} \text{ cis } 135°\right)^{10}$$

$$= \left(\sqrt{2}\right)^{10} \text{ cis } (1350°)$$

$$= 32 \text{ cis } 270° = \boldsymbol{-32i}$$

15. (79) $(32 \text{ cis } 60°)^{\frac{1}{5}}$

$$= 32^{\frac{1}{5}} \text{ cis}\left(\frac{60°}{5} + n\frac{360°}{5}\right)$$

$$= \textbf{2 cis 12°, 2 cis 84°, 2 cis 156°, 2 cis 228°, 2 cis 300°}$$

16. (79) $1^{\frac{1}{4}} = (1 \text{ cis } 0°)^{\frac{1}{4}}$

$$= 1^{\frac{1}{4}} \text{ cis}\left(\frac{0°}{4} + n\frac{360°}{4}\right)$$

$$= 1 \text{ cis } 0°, 1 \text{ cis } 90°, 1 \text{ cis } 180°, 1 \text{ cis } 270°$$

$$= \boldsymbol{1, i, -1, -i}$$

17. (78) $9y^2 - 16x^2 = 144$

$$\boldsymbol{\frac{y^2}{16} - \frac{x^2}{9} = 1}$$

Let $x = 0$: $\dfrac{y^2}{16} = 1$, $y = \pm 4$

Let $y = 0$: $-\dfrac{x^2}{9} = 1$, $x = \pm 3i$

Vertices = (0, 4), (0, –4)

Asymptotes: $\boldsymbol{y = \frac{4}{3}x;\ y = -\frac{4}{3}x}$

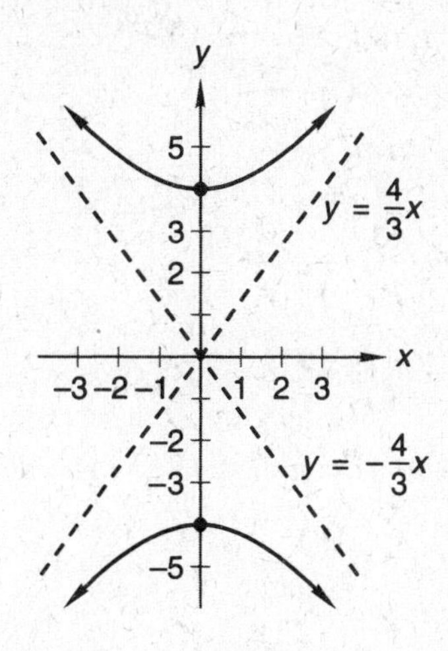

18. $(x + z)^6$
(77)

Term	①	②	③	④	⑤	⑥	⑦
For x	6	5	4	3	2	1	0
For z	0	1	2	3	4	5	6
Coefficient	1	6	15	20	15	6	1

$\mathbf{20x^3z^3}$

19. (74)

$$x = \frac{\begin{vmatrix} -1 & 3 \\ 7 & 5 \end{vmatrix}}{\begin{vmatrix} 5 & 3 \\ 4 & 5 \end{vmatrix}} = \frac{-5 - 21}{25 - 12} = \frac{-26}{13} = -2$$

$$y = \frac{\begin{vmatrix} 5 & -1 \\ 4 & 7 \end{vmatrix}}{\begin{vmatrix} 5 & 3 \\ 4 & 5 \end{vmatrix}} = \frac{35 + 4}{25 - 12} = \frac{39}{13} = 3$$

$\mathbf{x = -2;\ y = 3}$

20. (73)

$$\text{Side} = \frac{96 \text{ cm}}{12} = 8 \text{ cm}$$

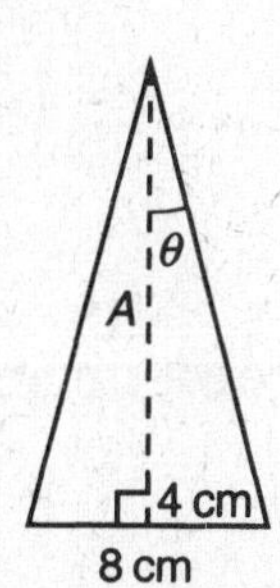

$$\theta = \frac{1}{2}\left(\frac{360°}{12}\right) = 15°$$

$$A = \frac{4}{\tan 15°} = 14.9282$$

$$\text{Area}_\Delta = \frac{1}{2}(4)(14.9282) = 29.8564 \text{ cm}^2$$

$$\text{Area} = (24)(29.8564)$$

$$\textbf{Area} = \mathbf{716.55 \text{ cm}^2}$$

$$\textbf{Radius} = \mathbf{14.93 \text{ cm}}$$

21. $16x^2 + 9y^2 = 144$
(71)

$$\mathbf{\frac{x^2}{9} + \frac{y^2}{16} = 1}$$

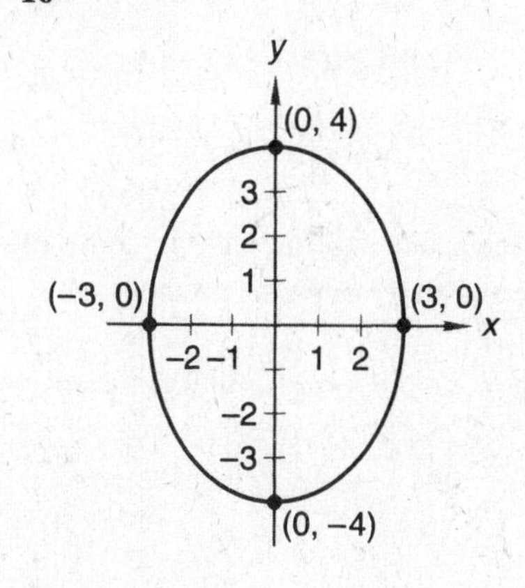

22. Period = 80°
(66)

$$\text{Coefficient} = \frac{360°}{80°} = \frac{9}{2}$$

$$\mathbf{y = -8 + 6 \cos \frac{9}{2}(x - 30°)}$$

23. (60)

$$\sec^2 \theta + \sec \theta - 2 = 0$$
$$(\sec \theta - 1)(\sec \theta + 2) = 0$$

$\sec \theta = 1$ $\qquad$ $\sec \theta = -2$

$\theta = 0°$ $\qquad$ $\theta = 120°, 240°$

$\mathbf{\theta = 0°, 120°, 240°}$

24. $\csc 2\theta - 2 = 0$
(52)

$$\csc 2\theta = 2$$
$$2\theta = 30°, 150°, 390°, 510°$$
$$\mathbf{\theta = 15°, 75°, 195°, 255°}$$

25. (32)

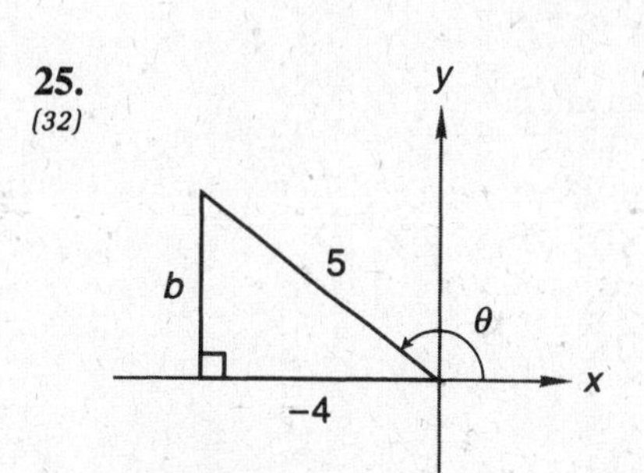

$$0° \le \theta \le 180°$$

$$b = \sqrt{5^2 - (-4)^2} = 3$$

$$\tan\left[\text{Arccos}\left(-\frac{4}{5}\right)\right] = \tan \theta = \mathbf{-\frac{3}{4}}$$

26. (56)

$$y^2 + x^2 + 2x - 6y \ge 6$$
$$(x^2 + 2x + 1) + (y^2 - 6y + 9) \ge 6 + 1 + 9$$
$$(x + 1)^2 + (y - 3)^2 \ge 4^2 \qquad \text{(circle)}$$

$$8y - x^2 - 2x \ge -63$$
$$8y \ge x^2 + 2x - 63$$
$$8y \ge (x^2 + 2x + 1) - 63 - 1$$
$$8y \ge (x + 1)^2 - 64$$
$$y \ge \frac{1}{8}(x + 1)^2 - 8 \qquad \text{(parabola)}$$

The inequality indicates the region outside or on the circle and above or on the parabola.

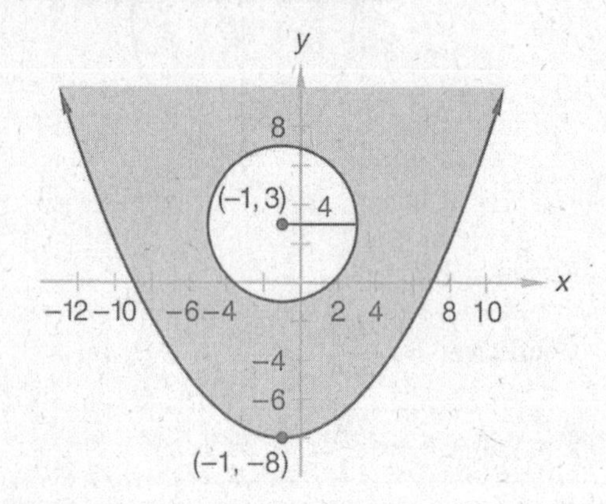

27. (58) $3x - 2y = 1$

$$2y = 3x - 1$$
$$y = \frac{3}{2}x - \frac{1}{2}$$

Equation of the perpendicular line:

$$y - y_1 = m(x - x_1)$$
$$y - 2 = -\frac{2}{3}(x - 4)$$
$$y = -\frac{2}{3}x + \frac{14}{3}$$

Point of intersection:

$$-\frac{2}{3}x + \frac{14}{3} = \frac{3}{2}x - \frac{1}{2}$$
$$-4x + 28 = 9x - 3$$
$$13x = 31$$
$$x = \frac{31}{13}$$

$$y = \frac{3}{2}x - \frac{1}{2} = \frac{3}{2}\left(\frac{31}{13}\right) - \frac{1}{2} = \frac{40}{13}$$

$\left(\frac{31}{13}, \frac{40}{13}\right)$ and $(4, 2)$

$$D = \sqrt{\left(4 - \frac{31}{13}\right)^2 + \left(2 - \frac{40}{13}\right)^2}$$
$$= \sqrt{\left(\frac{21}{13}\right)^2 + \left(-\frac{14}{13}\right)^2} = \sqrt{\frac{637}{169}} = \mathbf{\frac{7\sqrt{13}}{13}}$$

28. (59)

$$x = \log_8 16 - \log_8 4$$
$$x = \log_8 \frac{16}{4}$$
$$x = \log_8 4$$
$$8^x = 4$$
$$(2^3)^x = 2^2$$
$$2^{3x} = 2^2$$
$$3x = 2$$
$$x = \mathbf{\frac{2}{3}}$$

29. (59) $\log_5 10 - \log_5 2 + 5^{3 \log_5 2 - \log_5 4}$

$$= \log_5 \left(\frac{10}{2}\right) + 5^{\log_5 8 - \log_5 4}$$
$$= \log_5 5 + 5^{\log_5 \frac{8}{4}}$$
$$= 1 + 2 = \mathbf{3}$$

30. (46) $x^2 + 0x + 27 = 0$

$$x = \frac{0 \pm \sqrt{0 - 4(1)(27)}}{2(1)}$$
$$x = \frac{\pm\sqrt{-108}}{2}$$
$$x = \frac{\pm 6\sqrt{3}i}{2} = \pm 3\sqrt{3}i$$

$$\mathbf{(x - 3\sqrt{3}i)(x + 3\sqrt{3}i)}$$

Problem Set 84

1. (83) $\frac{1}{2} \cdot \frac{1}{2} \cdot \frac{1}{2} = \mathbf{\frac{1}{8}}$

2. (83)
(a) $\frac{4}{7} \cdot \frac{4}{7} = \mathbf{\frac{16}{49}}$

(b) $\frac{4}{7} \cdot \frac{3}{6} = \frac{12}{42} = \mathbf{\frac{2}{7}}$

3. (38)
$$\text{Avg}_5 = \frac{\text{sum}_5}{5}$$
$$100° = \frac{\text{sum}_5}{5}$$
$$\text{sum}_5 = 500°$$
$$\text{Avg}_6 = \frac{\text{sum}_5 + T}{6}$$
$$90° = \frac{500° + T}{6}$$
$$540° = 500° + T$$
$$T = \mathbf{40°}$$

4. (28) $$\text{New rate} = \frac{\text{distance}}{\text{new time}} = \mathbf{\frac{k}{p + m}\ \frac{yd}{min}}$$

5. (44)
$$RWT = \text{jobs}$$
$$R(10)(5) = P$$
$$R = \frac{P}{50}\ \frac{\text{jobs}}{\text{worker-day}}$$
$$RWT = \text{jobs}$$
$$T = \frac{\text{jobs}}{RW}$$
$$= \frac{k}{\left(\frac{P}{50}\right)(10 + 5)} = \frac{50k}{p(15)} = \mathbf{\frac{10k}{3p}\ days}$$

6. (25)
$$R_DT + R_CT = J$$
$$\frac{4}{3}(6) + R_C(6) = 11$$
$$6R_C = 3$$
$$R_C = \frac{1}{2}\ \frac{\text{job}}{\text{days}}$$
$$J = R_CT_C = \left(\frac{1}{2}\ \frac{\text{job}}{\text{days}}\right)(4\text{ days}) = \mathbf{2\ jobs}$$

7. (84) $y = -2 + 4\cos\frac{1}{2}(x + 30°)$

$$\text{Period} = \frac{360°}{\frac{1}{2}} = 720°$$

Phase angle $= -30°$

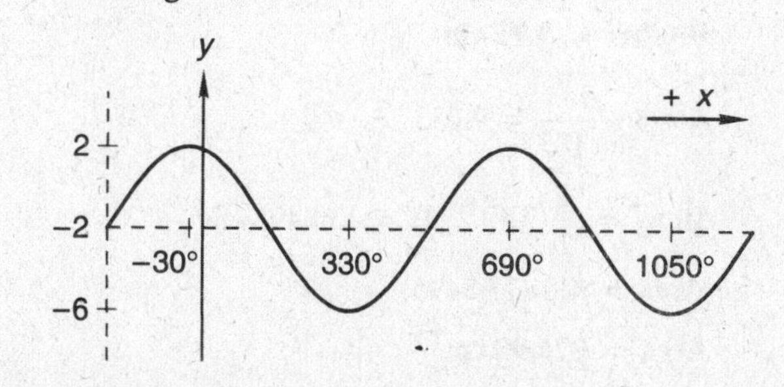

8. (84) $y = -1 + 8\sin 2(x - 45°)$

$$\text{Period} = \frac{360°}{2} = 180°$$

Phase angle $= 45°$

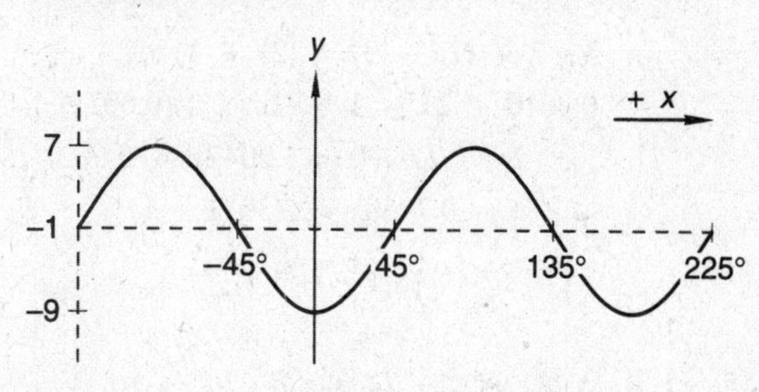

9. (80,84)
$$\frac{\csc^4 x - \cot^4 x}{\csc^2 x + \cot^2 x} + \cot^2 x$$
$$= \frac{(\csc^2 x + \cot^2 x)(\csc^2 x - \cot^2 x)}{\csc^2 x + \cot^2 x} + \cot^2 x$$
$$= \csc^2 x - \cot^2 x + \cot^2 x = \csc^2 x$$

10. (80,84)
$$\cos x - \cos x \sin^2 x = \cos x(1 - \sin^2 x)$$
$$= \cos x \cos^2 x = \cos^3 x$$

11. (80,84)
$$\frac{\sec^2\theta - 1}{\cot\theta} = \frac{\tan^2\theta}{\cot\theta} = \frac{\tan^2\theta}{\frac{1}{\tan\theta}} = \tan^3\theta$$

12. (80,84)
$$\cos(-\theta)\csc(-\theta) = \cos\theta(-\csc\theta)$$
$$= \cos\theta\left(-\frac{1}{\sin\theta}\right) = -\frac{\cos\theta}{\sin\theta}$$
$$= -\cot\theta = \cot(-\theta)$$

13. (82)
$$8^{3x-1} = 5^{2x+1}$$
$$\log 8^{3x-1} = \log 5^{2x+1}$$
$$(3x - 1)\log 8 = (2x + 1)\log 5$$
$$(3x - 1)(0.9031) = (2x + 1)(0.6990)$$
$$2.7093x - 0.9031 = 1.3980x + 0.6990$$
$$1.3113x = 1.6021$$
$$x = \mathbf{1.22}$$

14. (82)
$$2 \cdot 10^{2x-3} = 4^{2x+1}$$
$$\log\left(2 \cdot 10^{2x-3}\right) = \log 4^{2x+1}$$
$$\log 2 + \log 10^{2x-3} = \log 4^{2x+1}$$
$$\log 2 + (2x - 3)\log 10 = (2x + 1)\log 4$$
$$\log 2 + (2x - 3) = (2x + 1)\log 4$$
$$0.3010 + 2x - 3 = (2x + 1)(0.6021)$$
$$2x - 2.6990 = 1.2042x + 0.6021$$
$$0.7958x = 3.3011$$
$$x = \mathbf{4.15}$$

15. (81)
$$b^2 = a^2 + c^2 - 2ac\cos B$$
$$10^2 = 7^2 + 8^2 - 2(7)(8)\cos B$$
$$-13 = -112\cos B$$
$$\cos B = \frac{13}{112}$$
$$\mathbf{B = 83.33°}$$

$$\frac{7}{\sin A} = \frac{10}{\sin 83.33°}$$
$$\sin A = \frac{7\sin 83.33°}{10}$$
$$\mathbf{A = 44.05°}$$

$$C = 180° - 44.05° - 83.33°$$
$$\mathbf{C = 52.62°}$$

16. (81)
$$q^2 = 8^2 + 10^2 - 2(8)(10)\cos 130°$$
$$q = \sqrt{8^2 + 10^2 - 2(8)(10)\cos 130°}$$
$$= \sqrt{64 + 100 - 160(-0.6428)}$$
$$= \sqrt{266.85}$$
$$\mathbf{q = 16.34\text{ cm}}$$

17. (79) $\left(\frac{1}{2}\operatorname{cis} 100°\right)^3 = \left(\frac{1}{2}\right)^3 \operatorname{cis}(3 \cdot 100°) = \mathbf{\frac{1}{8}\,cis\,300°}$

18. (79) $(16\operatorname{cis} 120°)^{\frac{1}{4}} = 2\operatorname{cis}\left(30° + n\frac{360°}{4}\right)$

$= 2\operatorname{cis} 30°, 2\operatorname{cis} 120°, 2\operatorname{cis} 210°, 2\operatorname{cis} 300°$

$= \mathbf{\sqrt{3} + i,\ -1 + \sqrt{3}i,\ -\sqrt{3} - i,\ 1 - \sqrt{3}i}$

19. (79) $8^{\frac{1}{3}} = (8\operatorname{cis} 360°)^{\frac{1}{3}} = 2\operatorname{cis}\left(0° + n\frac{360°}{3}\right)$

$= 2\operatorname{cis} 0°, 2\operatorname{cis} 120°, 2\operatorname{cis} 240°$

$= \mathbf{2,\ -1 + \sqrt{3}i,\ -1 - \sqrt{3}i}$

20. (78) $4x^2 - 25y^2 = 400$

$$\mathbf{\frac{x^2}{100} - \frac{y^2}{16} = 1}$$

Let $y = 0$: $\frac{x^2}{100} = 1$, $x = \pm 10$

Let $x = 0$: $-\frac{y^2}{16} = 1$, $y = \pm 4i$

Vertices = (10, 0), (–10, 0)

Asymptotes: $\mathbf{y = \frac{2}{5}x;\ y = -\frac{2}{5}x}$

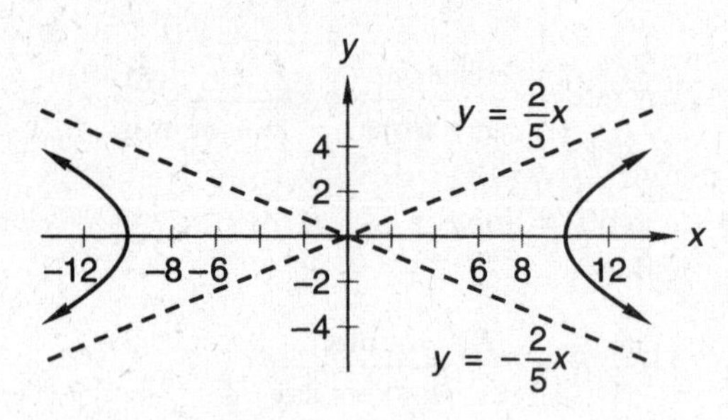

21. (77)

Term	①	②	③	④	⑤	⑥	⑦	⑧
For a	7	6	5	4	3	2	1	0
For c	0	1	2	3	4	5	6	7
Coeff.	1	7	21	35	35	21	7	1

$\mathbf{a^7 + 7a^6c + 21a^5c^2 + 35a^4c^3 + 35a^3c^4}$
$\mathbf{+ 21a^2c^5 + 7ac^6 + c^7}$

22. (73)

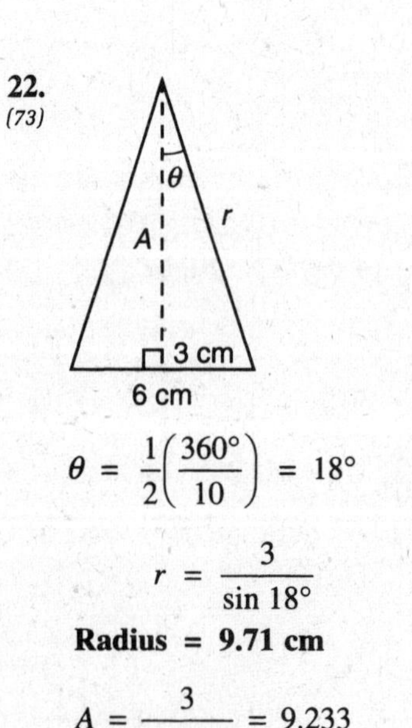

$$\theta = \frac{1}{2}\left(\frac{360°}{10}\right) = 18°$$
$$r = \frac{3}{\sin 18°}$$
Radius = 9.71 cm

$$A = \frac{3}{\tan 18°} = 9.233$$

$$\text{Area}_\Delta = \frac{1}{2}(3)(9.233) = 13.8495\text{ cm}^2$$
$$\text{Area} = (20)(13.8495)$$
Area = 276.99 cm²

23. (71)
$$2x^2 + 25y^2 = 50$$
$$\frac{x^2}{25} + \frac{y^2}{2} = 1$$

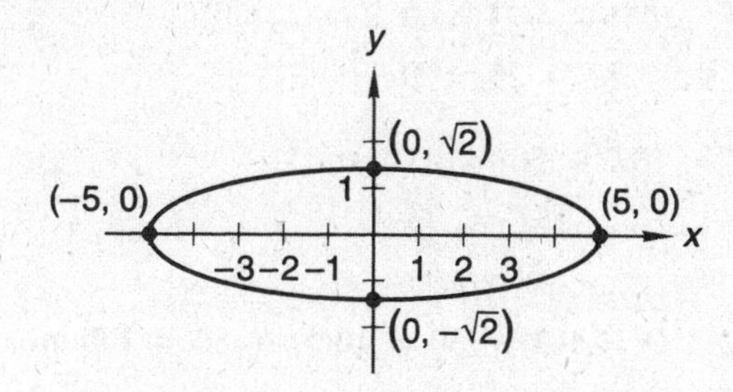

24. (63)
$$x^2 + y^2 + 6x - 4y - 15 = 0$$
$$x^2 + 6x + y^2 - 4y = 15$$
$$(x^2 + 6x + 9) + (y^2 - 4y + 4) = 15 + 9 + 4$$
$$(x + 3)^2 + (y - 2)^2 = (\sqrt{28})^2$$

Center = (−3, 2)

Radius = $\mathbf{2\sqrt{7}}$

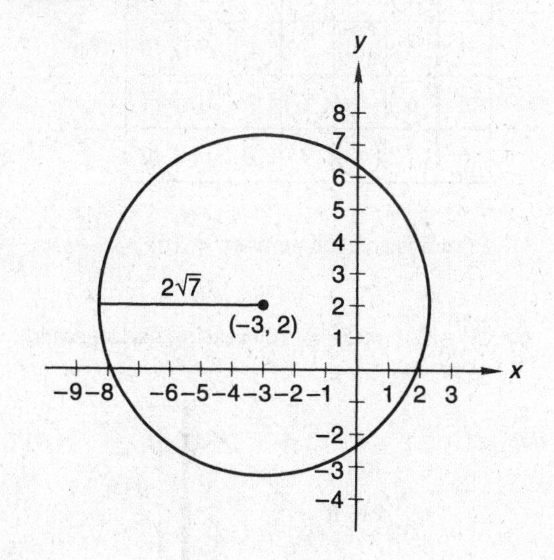

25. (64) (5 cis 205°)(2 cis 155°)

= 10 cis 360° = 10(1 + 0i) = **10**

26. (52) $2\cos\frac{\theta}{3} + \sqrt{3} = 0$

$$\cos\frac{\theta}{3} = -\frac{\sqrt{3}}{2}$$
$$\frac{\theta}{3} = \frac{5\pi}{6}, \frac{7\pi}{6}$$
$$\theta = \frac{5\pi}{2}, \frac{7\pi}{2}$$

Since the domain of θ is $0 \le \theta < 2\pi$ and $\theta > 2\pi$, there is **no solution.**

27. (60)
$$3\sec^2 x - 4 = 0$$
$$\sec^2 x - \frac{4}{3} = 0$$
$$\left(\sec x - \frac{2}{\sqrt{3}}\right)\left(\sec x + \frac{2}{\sqrt{3}}\right) = 0$$

$\sec x = \frac{2}{\sqrt{3}}$ $\qquad$ $\sec x = -\frac{2}{\sqrt{3}}$

$x = \frac{\pi}{6}, \frac{11\pi}{6}$ $\qquad$ $x = \frac{5\pi}{6}, \frac{7\pi}{6}$

$$x = \boldsymbol{\frac{\pi}{6}, \frac{5\pi}{6}, \frac{7\pi}{6}, \frac{11\pi}{6}}$$

28. (58) Equation of the perpendicular line:

$$y = -\frac{1}{3}x$$

Point of intersection:

$$3x + 1 = -\frac{1}{3}x$$
$$\frac{10}{3}x = -1$$
$$x = -\frac{3}{10}$$
$$y = -\frac{1}{3}x = -\frac{1}{3}\left(-\frac{3}{10}\right) = \frac{1}{10}$$

$\left(-\frac{3}{10}, \frac{1}{10}\right)$ and (0, 0)

$$D = \sqrt{\left(0 + \frac{3}{10}\right)^2 + \left(0 - \frac{1}{10}\right)^2}$$
$$= \sqrt{\frac{9}{100} + \frac{1}{100}} = \boldsymbol{\frac{\sqrt{10}}{10}}$$

29. (59) $\frac{3}{2}\log_8 4 + 2\log_8 x = \log_8 16$

$$\log_8 4^{\frac{3}{2}} + \log_8 x^2 = \log_8 16$$
$$\log_8 8x^2 = \log_8 16$$
$$8x^2 = 16$$
$$x^2 = 2$$
$$x = \pm\sqrt{2} \qquad (x \ne -\sqrt{2})$$
$$x = \boldsymbol{\sqrt{2}}$$

30. (70) $z = \frac{x - \mu}{\sigma} = \frac{728 - 700}{40} = 0.7$

Percentile 0.7580

1 − 0.7580 = 0.2420

24.20%

Problem Set 85

1. (85) $R_B = 1,\ R_L = \frac{1}{12},\ T_B = T_L$

Big Hand:

$$R_BT_B = S + 30$$
$$(1)T = S + 30$$
$$S = T - 30$$

Little Hand:

$$R_LT_L = S$$
$$\left(\frac{1}{12}\right)T = S$$
$$S = \frac{T}{12}$$

$$\frac{T}{12} = T - 30$$
$$\frac{11}{12}T = 30$$
$$T = \frac{360}{11} = \mathbf{32\frac{8}{11}\ min}$$

2. (85) $R_B = 1,\ R_L = \frac{1}{12},\ T_L = T_B$

Big Hand:

$$R_BT_B = S + 45$$
$$(1)T = S + 45$$
$$S = T - 45$$

Little Hand:

$$R_LT_L = S$$
$$\left(\frac{1}{12}\right)T = S$$
$$S = \frac{T}{12}$$

$$\frac{T}{12} = T - 45$$
$$\frac{11}{12}T = 45$$
$$T = \frac{540}{11} = \mathbf{49\frac{1}{11}\ min}$$

3. (25) First trip:

$$T_1 = T + 5$$
$$R_1T_1 = D$$
$$(10)(T + 5) = D$$
$$10T + 50 = D$$

Second trip:

$$T_2 = T - 1$$
$$R_2T_2 = D$$
$$(25)(T - 1) = D$$
$$25T - 25 = D$$

$$10T + 50 = 25T - 25$$
$$15T = 75$$
$$T = 5$$

$$D = 10T + 50 = 10(5) + 50 = \mathbf{100\ mi}$$

4. (83) The first two draws do not affect the outcome of the third draw. Therefore the probability remains $\frac{4}{7}$.

5. (83)

	1	2	3	4	5	6
1	(2)	3	(4)	5	(6)	7
2	3	(4)	5	(6)	7	(8)
3	(4)	5	(6)	7	(8)	9
4	5	(6)	7	(8)	9	10
5	(6)	7	(8)	9	10	11
6	7	(8)	9	10	11	12

$$P(\text{sum is an even number} < 10) = \frac{14}{36} = \mathbf{\frac{7}{18}}$$

6. (55) $(N - 1)! = 5! =$ **120 seating arrangements**

7. (81)

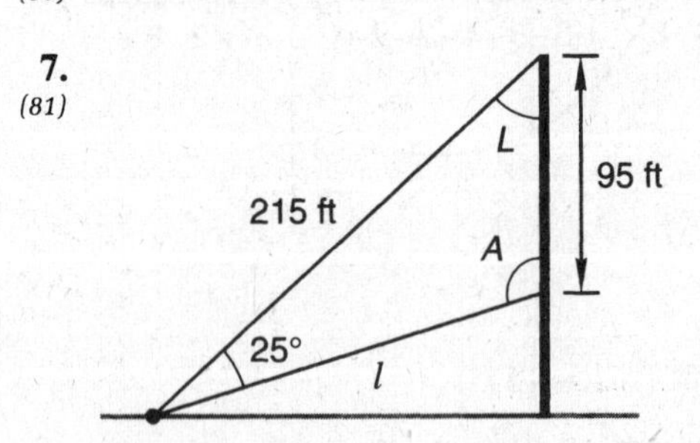

$$\frac{215}{\sin A} = \frac{95}{\sin 25°}$$
$$\sin A = \frac{215 \sin 25°}{95}$$
$$A = 106.97°$$

$$L = 180° - 25° - 106.97° = 48.03°$$

$$\frac{l}{\sin 48.03°} = \frac{95}{\sin 25°}$$
$$l = \frac{95 \sin 48.03°}{\sin 25°}$$
$$l = \mathbf{167.13\ ft}$$

8. (7)

$$N(N + 2) = 3(N + 4)$$
$$N^2 + 2N = 3N + 12$$
$$N^2 - N - 12 = 0$$
$$(N - 4)(N + 3) = 0$$
$$N = 4, -3$$

-3 is odd so it is invalid.

The numbers are **4, 6, 8.**

9. (85)

$$3 \tan^2 \theta = 7 \sec \theta - 5$$
$$3(\sec^2\theta - 1) - 7 \sec \theta + 5 = 0$$
$$3 \sec^2 \theta - 7 \sec \theta + 2 = 0$$
$$(3 \sec \theta - 1)(\sec \theta - 2) = 0$$

$3 \sec \theta = 1$

$\sec \theta = \frac{1}{3}$

no answer

$\sec \theta = 2$

$\theta =$ **60°, 300°**

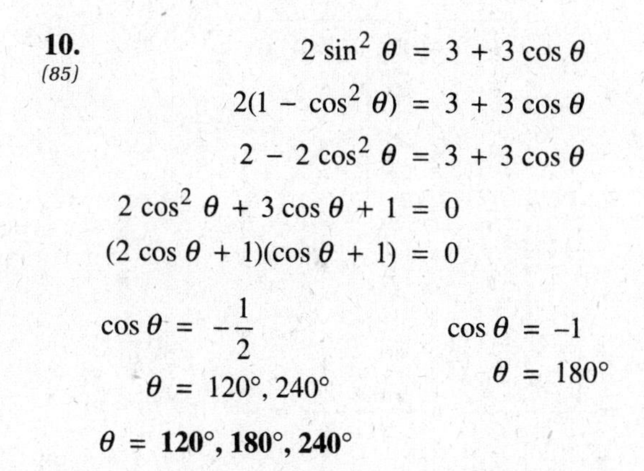

10. (85)

$$2 \sin^2 \theta = 3 + 3 \cos \theta$$
$$2(1 - \cos^2 \theta) = 3 + 3 \cos \theta$$
$$2 - 2 \cos^2 \theta = 3 + 3 \cos \theta$$
$$2 \cos^2 \theta + 3 \cos \theta + 1 = 0$$
$$(2 \cos \theta + 1)(\cos \theta + 1) = 0$$

$\cos \theta = -\frac{1}{2}$

$\theta = 120°, 240°$

$\cos \theta = -1$

$\theta = 180°$

$\theta =$ **120°, 180°, 240°**

11. (85)

$$2\tan^2 \theta = \sec \theta - 1$$
$$2(\sec^2 \theta - 1) = \sec \theta - 1$$
$$2\sec^2 \theta - \sec \theta - 1 = 0$$
$$(2\sec \theta + 1)(\sec \theta - 1) = 0$$

$2 \sec \theta = -1$

$\sec \theta = -\frac{1}{2}$

no answer

$\sec \theta = 1$

$\theta =$ **0°**

12. (84) $y = -2 + 5 \sin 3\left(x - \frac{\pi}{3}\right)$

Period $= \frac{2\pi}{3}$

Phase angle $= \frac{\pi}{3}$

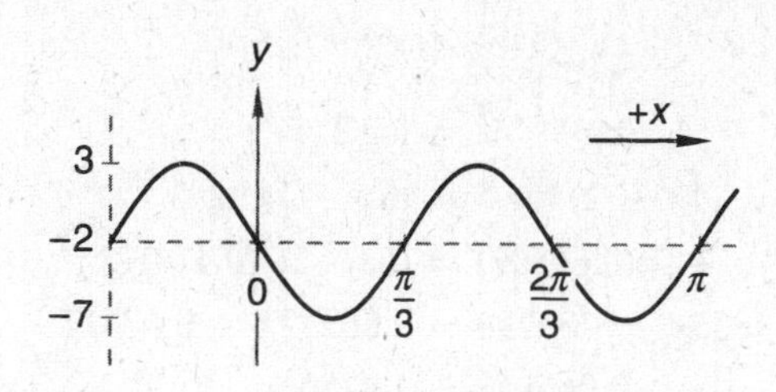

13. (84) $y = 3 + 5 \cos \frac{1}{2}(x - 90°)$

Period $= \dfrac{360°}{\frac{1}{2}} = 720°$

Phase angle $= 90°$

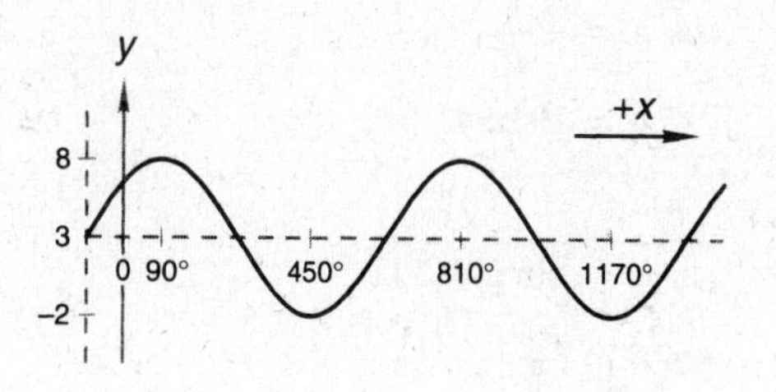

14. (80,84) $\dfrac{\cos^4 x - \sin^4 x}{\cos^2 x - \sin^2 x}$

$$= \frac{(\cos^2 x - \sin^2 x)(\cos^2 x + \sin^2 x)}{\cos^2 x - \sin^2 x}$$
$$= \cos^2 x + \sin^2 x = 1$$

15. (80,84) $\tan x + \cot x = \dfrac{\sin x}{\cos x} + \dfrac{\cos x}{\sin x}$

$$= \frac{\sin^2 x + \cos^2 x}{\cos x \sin x} = \frac{1}{\cos x \sin x} = \sec x \csc x$$

16. (80,84) $\dfrac{1}{1 + \cos x} + \dfrac{1}{1 - \cos x}$

$$= \frac{1 - \cos x + 1 + \cos x}{1 - \cos^2 x}$$
$$= \frac{2}{\sin^2 x} = 2 \csc^2 x$$

17. (82)

$$10^{3x-2} = 7^{2x+3}$$
$$\log 10^{3x-2} = \log 7^{2x+3}$$
$$3x - 2 = (2x + 3) \log 7$$
$$3x - 2 = (2x + 3)(0.8451)$$
$$3x - 2 = 1.6902x + 2.5353$$
$$1.3098x = 4.5353$$
$$x = \mathbf{3.46}$$

18. (82)

$$5^{4x+2} = 3^{6x-1}$$
$$\log 5^{4x+2} = \log 3^{6x-1}$$
$$(4x + 2)\log 5 = (6x - 1)\log 3$$
$$(4x + 2)(0.69897) = (6x - 1)(0.47712)$$
$$2.79855x + 1.39794 = 2.86272x - 0.47712$$
$$0.06684x = 1.87506$$
$$x = \mathbf{28.05}$$

19. (81)

$$7^2 = 5^2 + 4^2 - 2(5)(4)\cos A$$
$$49 = 25 + 16 - 40\cos A$$
$$8 = -40\cos A$$
$$\cos A = -\frac{8}{40}$$
$$\mathbf{A = 101.54°}$$

$$\frac{5}{\sin B} = \frac{7}{\sin 101.54°}$$
$$\sin B = \frac{5 \sin 101.54°}{7}$$
$$\sin B = 0.6999$$
$$\mathbf{B = 44.42°}$$

$$C = 180° - 101.54° - 44.42°$$
$$\mathbf{C = 34.04°}$$

20. (81)

$$p^2 = 8^2 + 10^2 - 2(8)(10)\cos 140°$$
$$p^2 = 286.57$$
$$\mathbf{p = 16.93\ cm}$$

$$H = 8 \sin 40° = 5.142$$
$$\text{Area} = \frac{1}{2}BH$$
$$\text{Area} = \frac{1}{2}(10)(5.142)$$
$$\textbf{Area} = \mathbf{25.71\ cm^2}$$

21. (79)

$$(1 - \sqrt{3}i)^4 = [2 \text{ cis } (-60°)]^4$$
$$= 16 \text{ cis } (-240°) = 16 \text{ cis } 120°$$
$$= 16\left(-\frac{1}{2} + \frac{\sqrt{3}}{2}i\right) = \mathbf{-8 + 8\sqrt{3}i}$$

22. (79)

$$(16 \text{ cis } 240°)^{\frac{1}{4}} = 2 \text{ cis } \left(60° + n\frac{360°}{4}\right)$$
$$= 2 \text{ cis } 60°, 2 \text{ cis } 150°, 2 \text{ cis } 240°, 2 \text{ cis } 330°$$
$$= \mathbf{1 + \sqrt{3}i,\ -\sqrt{3} + i,\ -1 - \sqrt{3}i,\ \sqrt{3} - i}$$

23. (78)

$$4y^2 - 9x^2 = 144$$
$$\frac{y^2}{36} - \frac{x^2}{16} = 1$$

Vertices = (0, 6), (0, –6)

Asymptotes: $y = \frac{3}{2}x;\ y = -\frac{3}{2}x$

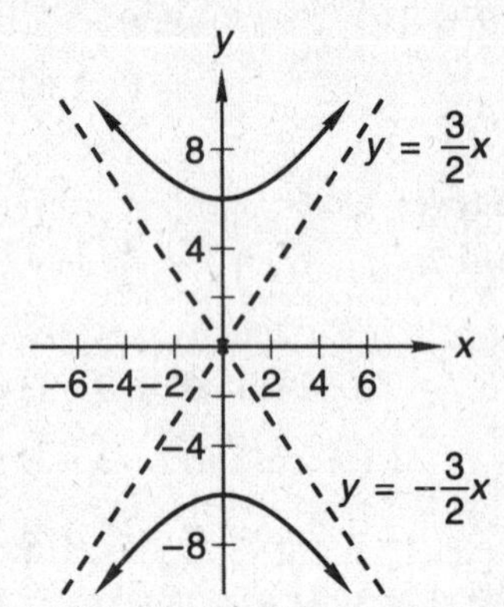

24. (74)

$$x = \frac{\begin{vmatrix} -4 & -3 \\ 8 & 2 \end{vmatrix}}{\begin{vmatrix} 2 & -3 \\ 1 & 2 \end{vmatrix}} = \frac{-8 + 24}{4 + 3} = \frac{16}{7}$$

$$y = \frac{\begin{vmatrix} 2 & -4 \\ 1 & 8 \end{vmatrix}}{\begin{vmatrix} 2 & -3 \\ 1 & 2 \end{vmatrix}} = \frac{16 + 4}{4 + 3} = \frac{20}{7}$$

$$\mathbf{x = \frac{16}{7};\ y = \frac{20}{7}}$$

25. (73)

$$\text{Side} = \frac{32 \text{ cm}}{8} = 4 \text{ cm}$$

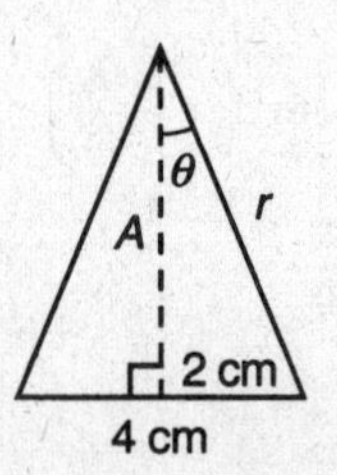

$$\theta = \frac{1}{2}\left(\frac{360°}{8}\right) = 22.5°$$

$$r = \frac{2}{\sin 22.5°}$$

Radius = 5.23 cm

$$A = \frac{2}{\tan 22.5°} = 4.828 \text{ cm}$$

$$\text{Area}_\Delta = \frac{1}{2}(2)(4.828) = 4.828 \text{ cm}^2$$

$$\text{Area} = (16)(4.828)$$

$$\textbf{Area} = \mathbf{77.25\ cm^2}$$

26. (71) $9x^2 + 4y^2 = 72$

$$\boldsymbol{\frac{x^2}{8} + \frac{y^2}{18} = 1}$$

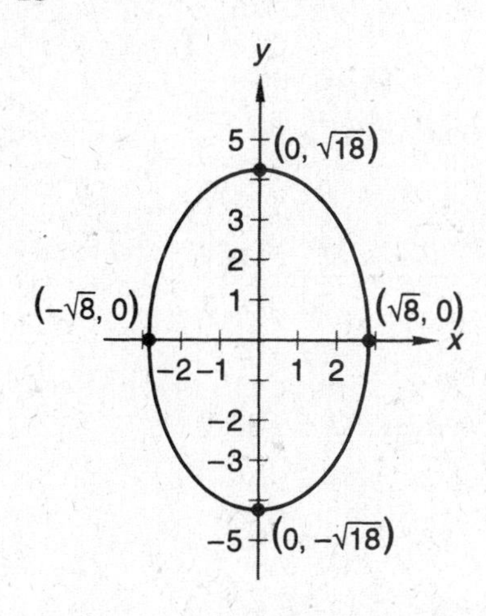

27. (64) $\left(6 \text{ cis } \frac{2\pi}{3}\right)\left(3 \text{ cis } \frac{5\pi}{6}\right) = 18 \text{ cis } \frac{3\pi}{2}$

$= 18(0 - i) = \mathbf{-18i}$

28. (61) $\text{Mean} = \frac{-2 + (-2) + 1 + 3 + 0}{5} = \mathbf{0}$

Median = **0**

Mode = **−2**

$$\text{Variance} = \frac{(-2-0)^2 + (-2-0)^2 + (1-0)^2}{5} + \frac{(3-0)^2 + (0-0)^2}{5}$$

$= \mathbf{3.60}$

Standard deviation $= \sqrt{3.60} = \mathbf{1.90}$

29. (59) $\log_3(x + 2) - \log_3(3x + 2) = \log_3 4$

$$\log_3 \frac{x+2}{3x+2} = \log_3 4$$

$$\frac{x+2}{3x+2} = 4$$

$$x + 2 = 12x + 8$$

$$11x = -6$$

$$x = \mathbf{-\frac{6}{11}}$$

30. (56)

10 cm
H
50°
10 cm

$H = 10 \sin 50° = 7.66$ cm

$$\text{Area}_{\text{segment}} = \text{Area}_{\text{sector}} - \text{Area}_{\text{triangle}}$$

$$= \pi(10)^2\left(\frac{50°}{360°}\right) - \frac{1}{2}(10)(7.66)$$

$$= 43.63 - 38.30 = \mathbf{5.33\ cm^2}$$

Problem Set 86

1. (85) $R_B = 1,\ R_L = \frac{1}{12},\ T_B = T_L$

Big hand:

$$R_B T_B = S + 15$$

$$(1)T = S + 15$$

$$S = T - 15$$

Little hand:

$$R_L T_L = S$$

$$\left(\frac{1}{12}\right)T = S$$

$$S = \frac{1}{12}T$$

$$\frac{T}{12} = T - 15$$

$$\frac{11}{12}T = 15$$

$$T = \frac{180}{11}$$

$$T = \mathbf{16\frac{4}{11}\ minutes}$$

2. (83) $\frac{13}{52} \cdot \frac{39}{51} = \frac{1}{4} \cdot \frac{13}{17} = \mathbf{\frac{13}{68}}$

3. (83) $\frac{10}{36} = \mathbf{\frac{5}{18}}$

4. (62) $O = mI + b$

$$\begin{cases} (a) & 45 = m10 + b \\ (b) & 37 = m8 + b \end{cases}$$

$$\begin{array}{rrcl} (a) & 45 &=& m10 + b \\ -1(b) & -37 &=& -m8 - b \\ \hline & 8 &=& m2 \\ & m &=& 4 \end{array}$$

(a) $45 = m10 + b$

$45 = (4)10 + b$

$b = 5$

$\mathbf{O = 4I + 5}$

5. (25) Distance $= d$; time $= t$

1st trip: $d = 30(t + 2)$

2nd trip: $d = 64t$

$30(t + 2) = 64t$

$30t + 60 = 64t$

$34t = 60$

$t = \frac{30}{17}$

$d = 64\left(\frac{30}{17}\right) = \frac{1920}{17}$ km $= \mathbf{112\frac{16}{17}}$ **km**

6. (7) $N, N + 2, N + 4$

$N(N + 4) = -(N + 4) + 18$

$N^2 + 4N = -N - 4 + 18$

$N^2 + 5N - 14 = 0$

$(N - 2)(N + 7) = 0$

$N = 2, -7$ (-7 is invalid)

$N = 2$

The numbers are **2, 4, 6.**

7. (86) $a_1 = -10$

$d = 6$

$a_1, a_1 + d, a_1 + 2d, a_1 + 3d, a_1 + 4d$

−10, −4, 2, 8, 14

8. (86) $a_1 = 5$, $d = -4$, $n = 30$

$a_n = a_1 + (n - 1)d$

$a_{30} = a_1 + 29d = (5) + 29(-4) = \mathbf{-111}$

9. (86) $a_1 = 3$, $a_5 = -13$

$a_5 = a_1 + 4d$

$-13 = 3 + 4d$

$d = -4$

$a_1 + d, a_1 + 2d, a_1 + 3d$

−1, −5, −9

10. (86) $a_{10} = a_1 + 9d = -30$

$a_{20} = a_1 + 19d = 40$

$$\begin{cases} (a) & a_1 + 9d = -30 \\ (b) & a_1 + 19d = 40 \end{cases}$$

$$\begin{array}{rrcl} (a) & a_1 + 9d &=& -30 \\ -(b) & -a_1 - 19d &=& -40 \\ \hline & -10d &=& -70 \\ & d &=& 7 \end{array}$$

(a) $a_1 + 9d = -30$

$a_1 + 9(7) = -30$

$a_1 = -93$

−93, −86, −79, −72, −65

11. (86) $a_1 = 2x + 3y$

$a_6 = a_1 + 5d = 7x + 8y$

$$\begin{cases} (a) & a_1 = 2x + 3y \\ (b) & a_1 + 5d = 7x + 8y \end{cases}$$

(b) $a_1 + 5d = 7x + 8y$

$(2x + 3y) + 5d = 7x + 8y$

$5d = 5x + 5y$

$d = x + y$

$a_n = a_1 + (n - 1)d$

$a_8 = a_1 + 7d$

$= (2x + 3y) + 7(x + y)$

$= 2x + 3y + 7x + 7y$

$= \mathbf{9x + 10y}$

12. (60)
$$2\sin^2 x + 3\sin x + 1 = 0$$
$$(2\sin x + 1)(\sin x + 1) = 0$$
$$2\sin x = -1 \qquad \sin x = -1$$
$$\sin x = -\frac{1}{2} \qquad x = \frac{3\pi}{2}$$
$$x = \frac{7\pi}{6}, \frac{11\pi}{6}$$
$$x = \mathbf{\frac{7\pi}{6}, \frac{3\pi}{2}, \frac{11\pi}{6}}$$

13. (85)
$$2\tan^2\theta = 3\sec\theta - 3$$
$$2(\sec^2\theta - 1) = 3\sec\theta - 3$$
$$2\sec^2\theta - 3\sec\theta + 1 = 0$$
$$(2\sec\theta - 1)(\sec\theta - 1) = 0$$
$$2\sec\theta = 1 \qquad \sec\theta = 1$$
$$\sec\theta = \frac{1}{2} \qquad \theta = \mathbf{0}$$
no answer

14. (52)
$$-1 - \sqrt{3}\tan\frac{\theta}{2} = 0$$
$$-\sqrt{3}\tan\frac{\theta}{2} = 1$$
$$\tan\frac{\theta}{2} = -\frac{1}{\sqrt{3}}$$
$$\frac{\theta}{2} = \frac{5\pi}{6}, \frac{11\pi}{6}$$
$$\theta = \frac{5\pi}{3}, \frac{11\pi}{3}$$
Since $\frac{11\pi}{3} > 2\pi$, $\theta = \mathbf{\frac{5\pi}{3}}$ only.

15. (84) $y = 4 + 7\sin 2\left(x + \frac{\pi}{6}\right)$

Period $= \frac{2\pi}{2} = \pi$

Phase angle $= -\frac{\pi}{6}$

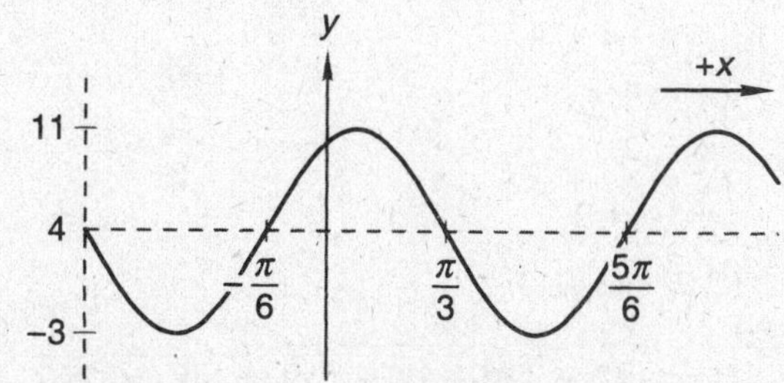

16. (84) $y = 1 + 5\cos\frac{1}{3}(x - 105°)$

Period $= \frac{360°}{\frac{1}{3}} = 1080°$

Phase angle $= 105°$

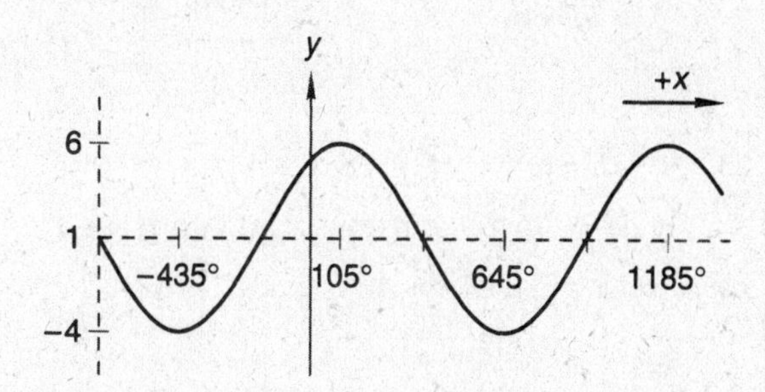

17. (80,84)
$$\frac{\sec^2 x}{\sec^2 x - 1} = \frac{\sec^2 x}{\tan^2 x} = \frac{\frac{1}{\cos^2 x}}{\frac{\sin^2 x}{\cos^2 x}}$$
$$= \frac{1}{\cos^2 x}\cdot\frac{\cos^2 x}{\sin^2 x} = \frac{1}{\sin^2 x} = \csc^2 x$$

18. (80,84)
$$\frac{\cos^2\theta}{\sin\theta} + \sin\theta = \frac{\cos^2\theta + \sin^2\theta}{\sin\theta}$$
$$= \frac{1}{\sin\theta} = \csc\theta$$

19. (80,84)
$$\frac{\cos(-x)}{\sin(-x)\cot(-x)} = \frac{\cos x}{(-\sin x)(-\cot x)}$$
$$= \frac{\cos x}{\sin x\left(\frac{\cos x}{\sin x}\right)} = 1$$

20. (82)
$$10^{3x-2} = 5^{2x-1}\cdot 10$$
$$\log 10^{3x-2} = \log\left(5^{2x-1}\cdot 10\right)$$
$$(3x - 2)\log 10 = (2x - 1)\log 5 + \log 10$$
$$3x - 2 = (2x - 1)(0.6990) + 1$$
$$3x - 2 = 1.398x - 0.6990 + 1$$
$$1.602x = 2.301$$
$$x = \mathbf{1.44}$$

21. (82)
$$3^{5x-3} = 9^{2-x}$$
$$3^{5x-3} = (3^2)^{2-x}$$
$$3^{5x-3} = 3^{4-2x}$$
$$5x - 3 = 4 - 2x$$
$$7x = 7$$
$$x = \mathbf{1}$$

22. $c^2 = a^2 + b^2 - 2ab\cos C$
(81)

$c = \sqrt{5^2 + 7^2 - 2(5)(7)\cos 50°}$

$c = \sqrt{25 + 49 - 70(0.6428)}$

$c = \sqrt{29.004}$

$\mathbf{c = 5.39\ m}$

23. $a^2 = b^2 + c^2 - 2bc\cos A$
(81)

$(1.0)^2 = (0.8)^2 + (0.7)^2 - 2(0.8)(0.7)\cos A$

$-0.13 = -1.12\cos A$

$\cos A = \dfrac{0.13}{1.12}$

$\mathbf{A = 83.33°}$

$\dfrac{0.8}{\sin B} = \dfrac{1.0}{\sin 83.33°}$

$\sin B = \dfrac{0.8\sin 83.33°}{1.0}$

$\mathbf{B = 52.62°}$

$C = 180° - 83.33° - 52.62°$

$\mathbf{C = 44.05°}$

24. $(8\text{ cis }270°)^{\frac{1}{3}} = 2\text{ cis}\left(90° + n\dfrac{360°}{3}\right)$
(79)

$= 2\text{ cis }90°, 2\text{ cis }210°, 2\text{ cis }330°$

$= \mathbf{2i,\ -\sqrt{3} - i,\ \sqrt{3} - i}$

25. $16x^2 - 9y^2 = 144$
(78)

$\mathbf{\dfrac{x^2}{9} - \dfrac{y^2}{16} = 1}$

Vertices = (3, 0), (–3, 0)

Asymptotes: $\mathbf{y = \dfrac{4}{3}x;\ y = -\dfrac{4}{3}x}$

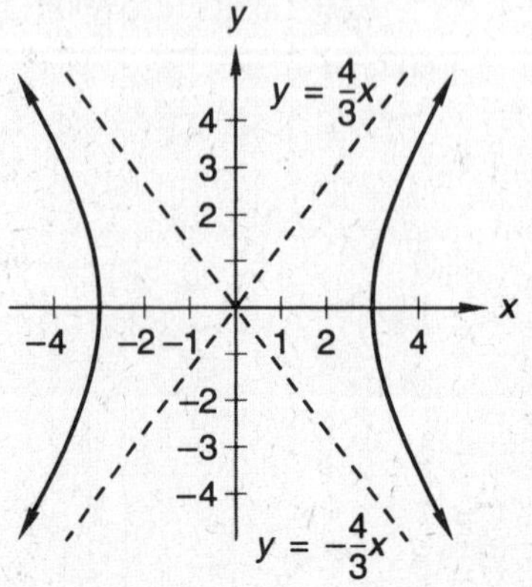

26. $(r + s)^3$
(77)

Term	①	②	③	④
For r	3	2	1	0
For s	0	1	2	3
Coefficient	1	3	3	1

$\mathbf{r^3 + 3r^2s + 3rs^2 + s^3}$

27. $z = \dfrac{x - \mu}{\sigma} = \dfrac{435 - 300}{90} = 1.5$
(70)

Percentile = 0.9332

Yes, percentile = 0.9332.

28.
(56)

$$\begin{cases} \dfrac{x^2}{16} + \dfrac{y^2}{4} \le 1 & \text{(ellipse)} \\ \dfrac{x^2}{4} + \dfrac{y^2}{16} \le 1 & \text{(ellipse)} \end{cases}$$

The inequality indicates the region within or on both ellipses.

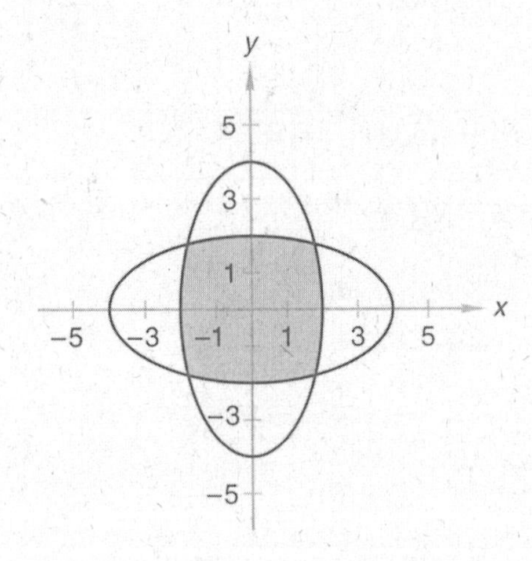

29. $x + y = 4$
(58)

$y = -x + 4$

Equation of the perpendicular line:

$y - y_1 = m(x - x_1)$

$y - 0 = (1)(x - 0)$

$y = x$

Point of intersection:

$x = -x + 4$

$2x = 4$

$x = 2$

$y = x = 2$

(2, 2) and (0, 0)

$D = \sqrt{(2 - 0)^2 + (2 - 0)^2} = \sqrt{8} = \mathbf{2\sqrt{2}}$

30. (59) $\frac{2}{3}\log_5 8 - \log_5 (x - 4) = 1$

$$\log_5 8^{\frac{2}{3}} - \log_5 (x - 4) = 1$$

$$\log_5 \left(\frac{4}{x - 4}\right) = 1$$

$$\frac{4}{x - 4} = 5^1$$

$$4 = 5x - 20$$

$$5x = 24$$

$$x = \mathbf{\frac{24}{5}}$$

Problem Set 87

1. (85) $R_B = 1,\ R_L = \frac{1}{12},\ T_B = T_L$

Big hand:

$$R_B T_B = S + 25$$

$$(1)T = S + 25$$

$$S = T - 25$$

Little hand:

$$R_L T_L = S$$

$$\left(\frac{1}{12}\right)T = S$$

$$S = \frac{T}{12}$$

$$\frac{T}{12} = T - 25$$

$$\frac{11}{12}T = 25$$

$$T = \frac{300}{11}\text{ min} = \mathbf{27\frac{3}{11}\text{ min}}$$

2. (83)

	1	2	3	4	5	6
1	(2)	(3)	(4)	(5)	6	7
2	(3)	(4)	(5)	6	7	8
3	(4)	(5)	6	7	8	9
4	(5)	6	7	8	9	10
5	6	7	8	9	10	(11)
6	7	8	9	10	(11)	(12)

$$P(\text{sum} > 10 \text{ or sum} < 6) = \mathbf{\frac{13}{36}}$$

3. (83) (a) $\left(\frac{2}{9}\right)\left(\frac{3}{9}\right) = \mathbf{\frac{2}{27}}$

(b) $\left(\frac{2}{9}\right)\left(\frac{3}{8}\right) = \mathbf{\frac{1}{12}}$

4. (25) $\begin{cases} \text{(a)}\ 60(T + 1) = D \\ \text{(b)}\ (T - 1) = D \end{cases}$

$$60(T + 1) = 90(T - 1)$$

$$60T + 60 = 90T - 90$$

$$30T = 150$$

$$T = 5$$

(a) $D = 60(T + 1) = 60(5 + 1) = \mathbf{360\text{ km}}$

5. (28) New rate $= \dfrac{\text{distance}}{\text{new time}} = \mathbf{\dfrac{x}{y - k}\ \dfrac{\text{yd}}{\text{min}}}$

6. (7) $N,\ N + 3,\ N + 6$

$$3[N + (N + 6)] = 4(N + 3) + 42$$

$$3N + 3N + 18 = 4N + 12 + 42$$

$$2N = 36$$

$$N = 18$$

The numbers are **18, 21, 24.**

7. (87) $\sin 15°$

$= \sin(60° - 45°)$

$= \sin 60° \cos 45° - \cos 60° \sin 45°$

$= \left(\frac{\sqrt{3}}{2}\right)\left(\frac{\sqrt{2}}{2}\right) - \left(\frac{1}{2}\right)\left(\frac{\sqrt{2}}{2}\right) = \mathbf{\frac{\sqrt{6} - \sqrt{2}}{4}}$

8. (87) $\cos\left(\theta - \frac{\pi}{4}\right)$

$= \cos\theta \cos\frac{\pi}{4} + \sin\theta \sin\frac{\pi}{4}$

$= \cos\theta\left(\frac{\sqrt{2}}{2}\right) + \sin\theta\left(\frac{\sqrt{2}}{2}\right)$

$= \mathbf{\frac{\sqrt{2}}{2}(\cos\theta + \sin\theta)}$

9. (87)
$$\tan(A + B) = \frac{\sin(A + B)}{\cos(A + B)} = \frac{\sin A \cos B + \cos A \sin B}{\cos A \cos B - \sin A \sin B} = \frac{\frac{\sin A \cos B}{\cos A \cos B} + \frac{\cos A \sin B}{\cos A \cos B}}{\frac{\cos A \cos B}{\cos A \cos B} - \frac{\sin A \sin B}{\cos A \cos B}} = \frac{\tan A + \tan B}{1 - \tan A \tan B}$$

10. (87)
$$\tan 75° = \tan(30° + 45°) = \frac{\tan 30° + \tan 45°}{1 - \tan 30° \tan 45°}$$
$$= \frac{\frac{1}{\sqrt{3}} + 1}{1 - \frac{1}{\sqrt{3}}(1)} = \frac{\frac{1 + \sqrt{3}}{\sqrt{3}}}{\frac{\sqrt{3} - 1}{\sqrt{3}}} = \frac{1 + \sqrt{3}}{\sqrt{3} - 1} \cdot \frac{\sqrt{3} + 1}{\sqrt{3} + 1}$$
$$= \frac{\sqrt{3} + 1 + 3 + \sqrt{3}}{3 - 1} = \frac{4 + 2\sqrt{3}}{2} = \mathbf{2 + \sqrt{3}}$$

11. (86) $a_1 = 6,\ d = -3,\ n = 10$

$a_n = a_1 + (n - 1)d$

$a_{10} = a_1 + 9d = 6 + 9(-3) = \mathbf{-21}$

12. (86) $a_1 = 6,\ a_6 = 106$

$a_n = a_1 + (n - 1)d$

$a_6 = a_1 + 5d$

$106 = 6 + 5d$

$5d = 100$

$d = 20$

$a_1 + d,\ a_1 + 2d,\ a_1 + 3d,\ a_1 + 4d$

26, 46, 66, 86

13. (86)
$$\begin{cases} \text{(a)}\ a_1 + 9d = 39 \\ \text{(b)}\ a_1 + 3d = 15 \end{cases}$$

(a) $a_1 + 9d = 39$

−(b) $-a_1 - 3d = -15$

$6d = 24$

$d = 4$

(b) $a_1 + 3d = 15$

$a_1 + 3(4) = 15$

$a_1 = 3$

$a_1,\ a_1 + d,\ a_1 + 2d,\ a_1 + 3d$

3, 7, 11, 15

14. (85)
$$2\cos^2 x + \sin x + 1 = 0$$
$$2(1 - \sin^2 x) + \sin x + 1 = 0$$
$$2 - 2\sin^2 x + \sin x + 1 = 0$$
$$2\sin^2 x - \sin x - 3 = 0$$
$$(2\sin x - 3)(\sin x + 1) = 0$$

$2\sin x = 3$ $\sin x = -1$

$\sin x = \frac{3}{2}$ $x = \mathbf{270°}$

no answer

15. (85)
$$\tan^2 x - 3\sec x + 3 = 0$$
$$(\sec^2 x - 1) - 3\sec x + 3 = 0$$
$$\sec^2 x - 3\sec x + 2 = 0$$
$$(\sec x - 1)(\sec x - 2) = 0$$

$\sec x = 1$ $\sec x = 2$

$x = 0°$ $x = 60°, 300°$

$x = \mathbf{0°, 60°, 300°}$

16. (52) $1 + 2\cos 2\theta = 0$

$\cos 2\theta = -\frac{1}{2}$

$2\theta = 120°, 240°, 480°, 600°$

$\theta = \mathbf{60°, 120°, 240°, 300°}$

17. (84) $y = -4 - 2\cos(x - 45°)$

Period = 360°

Phase angle = 45°

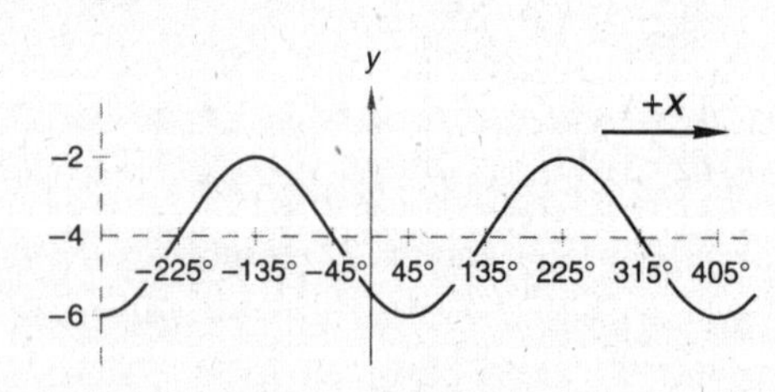

18. (84) $y = -2 + 3\sin\frac{2}{3}(x + 90°)$

$$\text{Period} = \frac{360°}{\frac{2}{3}} = 540°$$

Phase angle = −90°

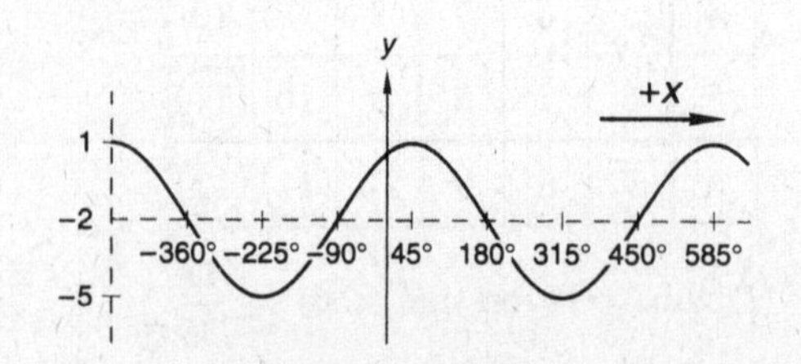

19. (80,84) $\sec x + \sec x \tan^2 x = \sec x(1 + \tan^2 x)$

$= \sec x \sec^2 x = \sec^3 x$

20. (80,84) $\dfrac{\sin^2 x}{\cos x} + \cos x = \dfrac{\sin^2 x + \cos^2 x}{\cos x} = \dfrac{1}{\cos x} = \sec x$

21. (80,84) $\dfrac{\tan(-\theta)\cos(-\theta)}{\sin(-\theta)} = \dfrac{-\tan\theta\cos\theta}{-\sin\theta}$

$= \dfrac{\frac{\sin\theta}{\cos\theta}\cdot\cos\theta}{\sin\theta} = \dfrac{\sin\theta}{\sin\theta} = 1$

22. (82)

$$3^{3x+2} = 5^{6x-1}$$
$$\log 3^{3x+2} = \log 5^{6x-1}$$
$$(3x + 2)\log 3 = (6x - 1)\log 5$$
$$(3x + 2)(0.47712) = (6x - 1)(0.69897)$$
$$1.4314x + 0.9542 = 4.1938x - 0.69897$$
$$2.7624x = 1.6532$$
$$x = \mathbf{0.60}$$

23. (82)

$$\frac{6^{2x-4}}{10^{3x+1}} = 1$$
$$\log\frac{6^{2x-4}}{10^{3x+1}} = \log 1$$
$$\log 6^{2x-4} - \log 10^{3x+1} = 0$$
$$(2x - 4)\log 6 - (3x + 1) = 0$$
$$(2x - 4)(0.77815) - 3x - 1 = 0$$
$$1.5563x - 3.1126 - 3x - 1 = 0$$
$$-1.4437x = 4.1126$$
$$x = \mathbf{-2.85}$$

24. (81) $a = \sqrt{10^2 + 11^2 - 2(10)(11)\cos 20°}$

$= 3.7773$

$\boldsymbol{a = 3.78}$ **cm**

$$\frac{a}{\sin 20°} = \frac{10}{\sin B}$$
$$\sin B = \frac{10\sin 20°}{3.7773}$$
$$\boldsymbol{B = 64.89°}$$

$C = 180° - 20° - 64.89°$

$\boldsymbol{C = 95.11°}$

25. (79) $(16 \text{ cis } 40°)^{\frac{1}{4}} = 2 \text{ cis}\left(10° + n\dfrac{360°}{4}\right)$

$=$ **2 cis 10°, 2 cis 100°, 2 cis 190°, 2 cis 280°**

26. (78) $25y^2 - 4x^2 = 100$

$$\frac{y^2}{4} - \frac{x^2}{25} = 1$$

Vertices = (0, 2), (0, –2)

Asymptotes: $\boldsymbol{y = \frac{2}{5}x;\ y = -\frac{2}{5}x}$

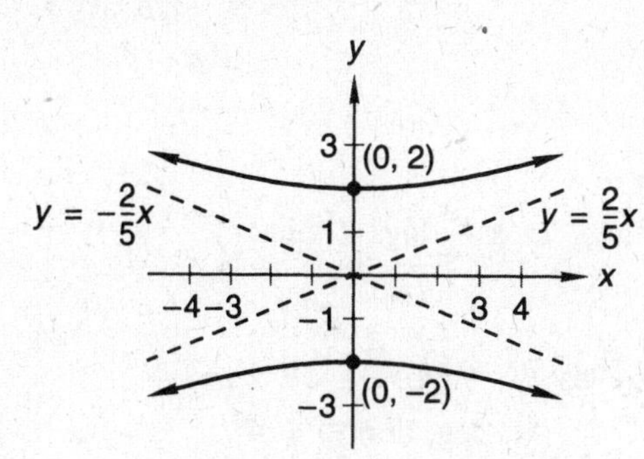

27. (74) $x = \dfrac{\begin{vmatrix} 3 & -5 \\ -10 & 2 \end{vmatrix}}{\begin{vmatrix} 4 & -5 \\ 3 & 2 \end{vmatrix}} = \dfrac{6 - 50}{8 + 15} = -\dfrac{44}{23}$

$y = \dfrac{\begin{vmatrix} 4 & 3 \\ 3 & -10 \end{vmatrix}}{\begin{vmatrix} 4 & -5 \\ 3 & 2 \end{vmatrix}} = \dfrac{-40 - 9}{8 + 15} = -\dfrac{49}{23}$

$\boldsymbol{x = -\frac{44}{23};\ y = -\frac{49}{23}}$

28. (73) Side $= \dfrac{30 \text{ in.}}{6} = 5$ in.

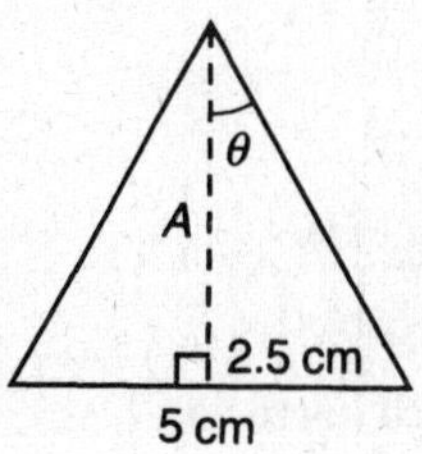

$$\theta = \frac{1}{2}\left(\frac{360°}{6}\right) = 30°$$

$$A = \frac{2.5}{\tan 30°} = 4.33$$

Radius = 4.33 in.

$\text{Area}_\Delta = \frac{1}{2}(2.5)(4.33) = 5.4125 \text{ in.}^2$

Area = (12)(5.4125)

Area = 64.95 in.²

29. (59)
$$\log(x+1) + \log(x-2) = 1$$
$$\log(x+1)(x-2) = 1$$
$$(x+1)(x-2) = 10^1$$
$$x^2 - x - 12 = 0$$
$$(x+3)(x-4) = 0$$
$$x = -3, 4 \quad (x \neq -3)$$
$$\mathbf{x = 4}$$

30. (59)
$$\log_3 7x + \frac{2}{3}\log_3 27 = 4$$
$$\log_3 7x + \log_3 27^{\frac{2}{3}} = 4$$
$$\log_3 (7x)(9) = 4$$
$$63x = 3^4$$
$$x = \frac{81}{63}$$
$$x = \mathbf{\frac{9}{7}}$$

Problem Set 88

1. (88) $A_0 = 400,\ A_{10} = 4000$
$$A_t = A_0 e^{kt}$$
$$4000 = 400e^{k10}$$
$$e^{k10} = 10$$
$$10k = \ln 10$$
$$k = \frac{\ln 10}{10}$$
$$k = 0.23$$
$$\mathbf{A_t = 400e^{0.23t}}$$
$$A_{12} = 400e^{0.23(12)}$$
$$A_{12} = \mathbf{6320\ bacteria}$$

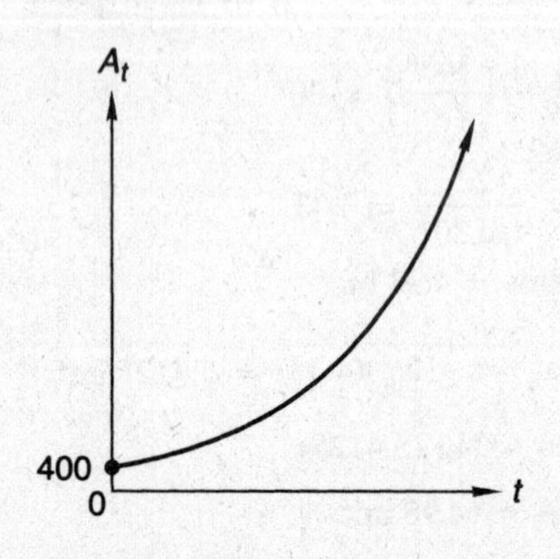

2. (88) $A_0 = 40,\ A_{50} = 20$
$$A_t = A_0 e^{kt}$$
$$20 = 40e^{k50}$$
$$\frac{20}{40} = e^{k50}$$
$$\ln\frac{1}{2} = 50k$$
$$k = \frac{\ln\frac{1}{2}}{50}$$
$$k = -0.014$$
$$\mathbf{A_t = 40e^{-0.014t}}$$
$$A_{100} = 40e^{-0.014(100)}$$
$$A_{100} = \mathbf{9.86\ grams}$$

3. (88) $A_0 = 100,\ A_4 = 20$
$$A_t = A_0 e^{kt}$$
$$20 = 100e^{4k}$$
$$\frac{1}{5} = e^{4k}$$
$$\ln\frac{1}{5} = 4k$$
$$k = \frac{\ln\frac{1}{5}}{4}$$
$$k = -0.40$$
$$\mathbf{A_t = 100e^{-0.40t}}$$
$$\left(\frac{1}{2}\right)(100) = 100e^{-0.40t}$$
$$e^{-0.40t} = \frac{1}{2}$$
$$-0.40t = \ln\frac{1}{2}$$
$$t = \frac{\ln\frac{1}{2}}{-0.40} = 1.73$$

Half-life = 1.73 yr

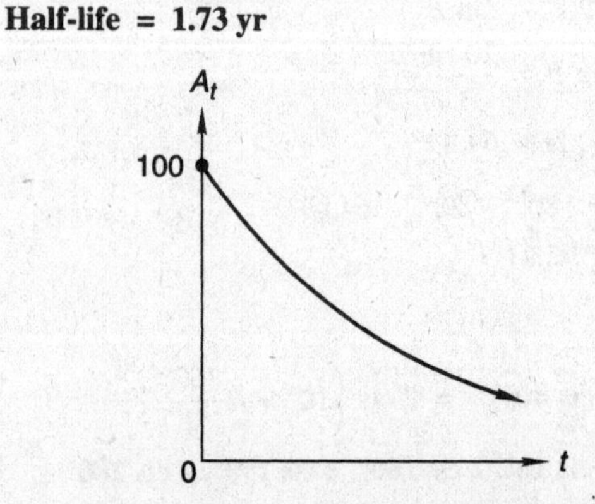

4. $R_B = 1,\ R_L = \dfrac{1}{12},\ T_B = T_L$
(85)

Big hand:

$$R_B T_B = S + 15$$
$$(1)T = S + 15$$
$$S = T - 15$$

Little hand:

$$R_L T_L = S$$
$$S = \frac{1}{12}T$$
$$\frac{1}{12}T = T - 15$$
$$\frac{11}{12}T = 15$$
$$T = \frac{180}{11} = 16\frac{4}{11}\text{ min}$$

3:16:22

5. $\left(\dfrac{26}{52}\right)\left(\dfrac{25}{51}\right) = \left(\dfrac{1}{2}\right)\left(\dfrac{25}{51}\right) = \mathbf{\dfrac{25}{102}}$
(83)

6.
(83)

	1	2	3	4	5	6
1	(2)	(3)	(4)	5	6	7
2	(3)	(4)	5	6	7	8
3	(4)	5	6	7	8	9
4	5	6	7	8	9	10
5	6	7	8	9	10	11
6	7	8	9	10	11	12

$$P(\text{sum} < 5) \times P(\text{sum} < 5) = \left(\frac{6}{36}\right)\left(\frac{6}{36}\right) = \mathbf{\frac{1}{36}}$$

7. $N,\ N + 1,\ N + 2,\ N + 3$
(7)

$$2[(N)(N + 3)] = 8 + (N + 1)(N + 2)$$
$$2N^2 + 6N = 8 + N^2 + 3N + 2$$
$$N^2 + 3N - 10 = 0$$
$$(N + 5)(N - 2) = 0$$
$$N = -5, 2 \qquad (N \neq -5)$$
$$N = 2$$

The numbers are **2, 3, 4, 5.**

8. $\cos 75° = \cos(45° + 30°)$
(87)

$$= \cos 45° \cos 30° - \sin 45° \sin 30°$$
$$= \left(\frac{1}{\sqrt{2}} \cdot \frac{\sqrt{3}}{2}\right) - \left(\frac{1}{\sqrt{2}} \cdot \frac{1}{2}\right)$$
$$= \frac{\sqrt{3}\sqrt{2}}{4} - \frac{\sqrt{2}}{4} = \mathbf{\frac{\sqrt{6} - \sqrt{2}}{4}}$$

9. $\sin\left(\theta - \dfrac{\pi}{6}\right) = \sin\theta \cos\dfrac{\pi}{6} - \cos\theta \sin\dfrac{\pi}{6}$
(87)

$$= \mathbf{\frac{\sqrt{3}}{2}\sin\theta - \frac{1}{2}\cos\theta}$$

10. $\tan(A - B) = \dfrac{\sin(A - B)}{\cos(A - B)}$
(87)

$$= \frac{\sin A \cos B - \cos A \sin B}{\cos A \cos B + \sin A \sin B}$$
$$= \frac{\dfrac{\sin A \cos B}{\cos A \cos B} - \dfrac{\cos A \sin B}{\cos A \cos B}}{\dfrac{\cos A \cos B}{\cos A \cos B} + \dfrac{\sin A \sin B}{\cos A \cos B}}$$
$$= \frac{\tan A - \tan B}{1 + \tan A \tan B}$$

11. $a_1 = 2,\ d = -4,\ n = 20$
(86)

$$a_n = a_1 + (n - 1)d$$
$$a_{20} = 2 + 19(-4) = \mathbf{-74}$$

12. (a) $\begin{cases} a_1 + 7d = -8 \\ a_1 + 9d = -12 \end{cases}$ (b)
(86)

$$-(a)\quad -a_1 - 7d = 8$$
$$(b)\quad a_1 + 9d = -12$$
$$2d = -4$$
$$d = -2$$

$$(a)\quad a_1 + 7d = -8$$
$$a_1 + 7(-2) = -8$$
$$a_1 = 6$$

6, 4, 2

13. $\sqrt{3}\tan^2 x + 2\tan x - \sqrt{3} = 0$
(60)

$$\left(\sqrt{3}\tan x - 1\right)\left(\tan x + \sqrt{3}\right) = 0$$

$$\tan x = \frac{1}{\sqrt{3}} \qquad\qquad \tan x = -\sqrt{3}$$
$$x = \frac{\pi}{6}, \frac{7\pi}{6} \qquad\qquad x = \frac{2\pi}{3}, \frac{5\pi}{3}$$

$$x = \mathbf{\frac{\pi}{6}, \frac{2\pi}{3}, \frac{7\pi}{6}, \frac{5\pi}{3}}$$

14. (60)
$$\tan^2\theta - \tan\theta = 0$$
$$\tan\theta(\tan\theta - 1) = 0$$

$$\tan\theta = 0 \qquad\qquad \tan\theta = 1$$
$$\theta = 0, \pi \qquad\qquad \theta = \frac{\pi}{4}, \frac{5\pi}{4}$$

$$\theta = \mathbf{0}, \frac{\boldsymbol{\pi}}{\mathbf{4}}, \boldsymbol{\pi}, \frac{\mathbf{5}\boldsymbol{\pi}}{\mathbf{4}}$$

15. (52) $\sec 4\theta = 1$
$$4\theta = 0, 2\pi, 4\pi, 6\pi$$
$$\theta = \mathbf{0}, \frac{\boldsymbol{\pi}}{\mathbf{2}}, \boldsymbol{\pi}, \frac{\mathbf{3}\boldsymbol{\pi}}{\mathbf{2}}$$

16. (84) $y = 2 + 3\cos\frac{1}{3}(x + \pi)$

$$\text{Period} = \frac{2\pi}{\frac{1}{3}} = 6\pi$$

Phase angle $= -\pi$

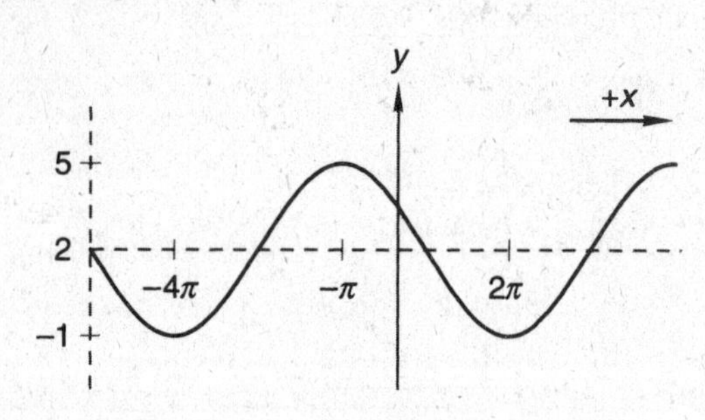

17. (84) $y = -7 + 10\sin\frac{9}{5}x$

$$\text{Period} = \frac{360°}{\frac{9}{5}} = 200°$$

Phase angle $= 0°$

y
+x
3
−7
−17
100°
200°

18. (80,84)
$$\frac{\cos\theta}{1 - \sin\theta} - \frac{\cos\theta}{1 + \sin\theta}$$
$$= \frac{\cos\theta(1 + \sin\theta) - \cos\theta(1 - \sin\theta)}{1 - \sin^2\theta}$$
$$= \frac{\cos\theta + \sin\theta\cos\theta - \cos\theta + \cos\theta\sin\theta}{\cos^2\theta}$$
$$= \frac{2\cos\theta\sin\theta}{\cos^2\theta} = \frac{2\sin\theta}{\cos\theta} = 2\tan\theta$$

19. (80,84) $(1 - \sin\theta)(1 + \sin\theta) = 1 - \sin^2\theta = \cos^2\theta$

20. (80,84)
$$\frac{\cos^3 x + \sin^3 x}{\cos x + \sin x}$$
$$= \frac{(\cos x + \sin x)(\cos^2 x - \sin x\cos x + \sin^2 x)}{\cos x + \sin x}$$
$$= \cos^2 x - \sin x\cos x + \sin^2 x = 1 - \sin x\cos x$$

21. (82)
$$4^{3x-1} = 8^{x+1}$$
$$(2^2)^{(3x-1)} = (2^3)^{(x+1)}$$
$$2^{6x-2} = 2^{3x+3}$$
$$6x - 2 = 3x + 3$$
$$3x = 5$$
$$x = \frac{\mathbf{5}}{\mathbf{3}}$$

22. (82)
$$\frac{9^{8x-3}}{5^{2x-3}} = 1$$
$$\log 9^{8x-3} - \log 5^{2x-3} = \log 1$$
$$(8x - 3)\log 9 - (2x - 3)\log 5 = 0$$
$$(8x - 3)(0.9542) - (2x - 3)(0.69897) = 0$$
$$7.6339x - 2.8626 - 1.3979x + 2.09691 = 0$$
$$6.236x = 0.76569$$
$$x = \mathbf{0.12}$$

23. (81)
$$a^2 = 8^2 + 9^2 - 2(8)(9)\cos 120°$$
$$a^2 = 64 + 81 - 144(-0.5)$$
$$a^2 = 217$$
$$\boldsymbol{a} = \mathbf{14.73\ m}$$

$$h = 8\sin 60° = 6.928\text{ m}$$

$$\text{Area} = \frac{1}{2}(9)(6.928)$$
$$\textbf{Area} = \mathbf{31.18\ m^2}$$

24. (79) $(27 \text{ cis } 30°)^{\frac{1}{3}} = 3 \text{ cis}\left(10° + n\frac{360°}{3}\right)$

$= \mathbf{3 \text{ cis } 10°, 3 \text{ cis } 130°, 3 \text{ cis } 250°}$

25. (74) $x = \dfrac{\begin{vmatrix} 2 & -4 \\ 4 & 3 \end{vmatrix}}{\begin{vmatrix} 3 & -4 \\ 2 & 3 \end{vmatrix}} = \dfrac{6 + 16}{9 + 8} = \dfrac{22}{17}$

$y = \dfrac{\begin{vmatrix} 3 & 2 \\ 2 & 4 \end{vmatrix}}{\begin{vmatrix} 3 & -4 \\ 2 & 3 \end{vmatrix}} = \dfrac{12 - 4}{9 + 8} = \dfrac{8}{17}$

$\mathbf{x = \frac{22}{17}; \; y = \frac{8}{17}}$

26. (73) Side $= \dfrac{30 \text{ cm}}{5} = 6 \text{ cm}$

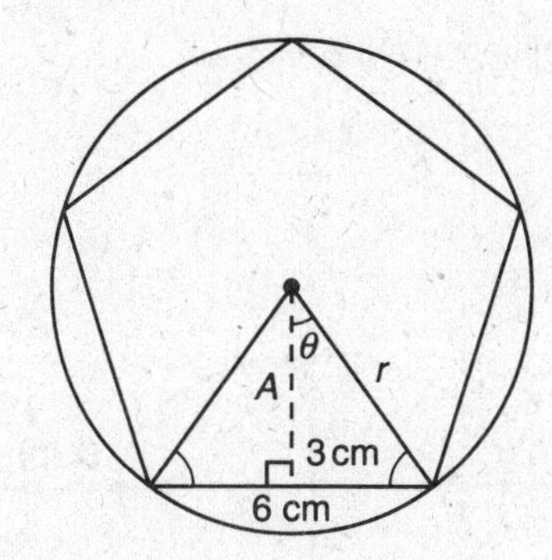

$\theta = \dfrac{1}{2}\left(\dfrac{360°}{5}\right) = 36°$

$r = \dfrac{3}{\sin 36°} = 5.104 \text{ cm}$

$A = \dfrac{3}{\tan 36°} = 4.129$

$\text{Area}_\Delta = \dfrac{1}{2}(3)(4.129) = 6.1935 \text{ cm}^2$

$\text{Area} = (10)(6.1935)$

Area of pentagon = $\mathbf{61.94 \text{ cm}^2}$

$A_{\text{circle}} = \pi r^2$

$= \pi(5.104)^2$

Area of circle = $\mathbf{81.84 \text{ cm}^2}$

27. (70) $z = \dfrac{x - \mu}{\sigma} = \dfrac{590 - 400}{100} = 1.9$

Percentile = 0.9713

97.13% > 95%

Yes, percentile = 0.9713.

28. (59) $\dfrac{2}{3}\log_2 27 + \log_2(3x + 2) = 2$

$\log_2\left(27^{\frac{2}{3}}(3x + 2)\right) = 2$

$9(3x + 2) = 2^2$

$27x + 18 = 4$

$27x = -14$

$x = \mathbf{-\frac{14}{27}}$

29. (59) $2\log_3 2 + \log_3(x - 1) = \log_3(x + 2)$

$\log_3 2^2(x - 1) = \log_3(x + 2)$

$4x - 4 = x + 2$

$3x = 6$

$x = \mathbf{2}$

30. (59) $4^{2\log_2 3} - 3\log_2 4 = \left(2^2\right)^{\log_2 3^2} - 3\log_2\left(2^2\right)$

$= 2^{\log_2 9^2} - 3\log_2 2^2 = 81 - 3(2) = \mathbf{75}$

Problem Set 89

1. (88) $A_0 = 40,\ A_6 = 2000$

$A_t = A_0 e^{kt}$

$2000 = 40e^{6k}$

$50 = e^{6k}$

$\ln 50 = 6k$

$k = \dfrac{\ln 50}{6}$

$k = 0.65$

$\mathbf{A_t = 40e^{0.65t}}$

$A_8 = 40e^{0.65(8)}$

$A_8 = $ **7251 rabbits**

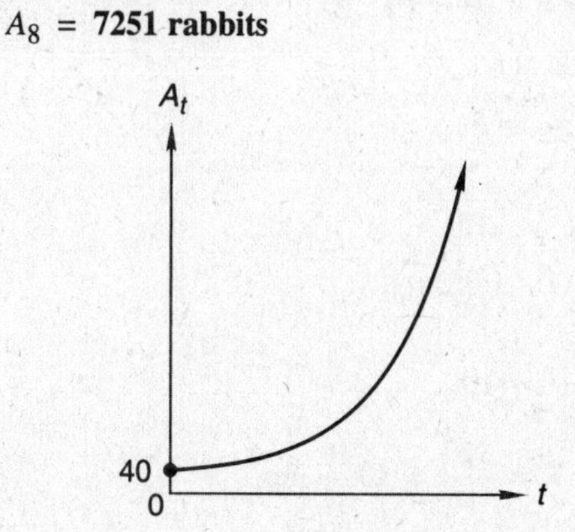

2. (88) $A_0 = 40{,}000,\ A_{10} = 30{,}000$

$$A_t = A_0e^{kt}$$
$$30{,}000 = 40{,}000e^{10k}$$
$$e^{10k} = \frac{3}{4}$$
$$10k = \ln\frac{3}{4}$$
$$k = \frac{\ln\frac{3}{4}}{10}$$
$$k = -0.029$$
$$\mathbf{A_t = 40{,}000e^{-0.029t}}$$
$$\frac{1}{2}(40{,}000) = 40{,}000e^{-0.029t}$$
$$0.5 = e^{-0.029t}$$
$$\ln 0.5 = -0.029t$$
$$t = \frac{\ln 0.5}{-0.029}$$
$$t = \mathbf{23.90\ hr}$$

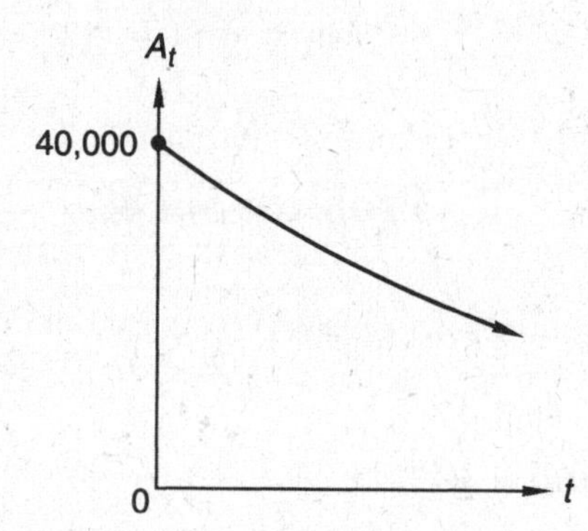

3. (85) $R_B = 1,\ R_L = \frac{1}{12},\ T_B = T_L$

Big hand:

$$R_BT_B = S + 10$$
$$(1)T = S + 10$$
$$S = T - 10$$

Little hand:

$$R_LT_L = S$$
$$S = \frac{1}{12}T$$
$$\frac{1}{12}T = T - 10$$
$$\frac{11}{12}T = 10$$
$$T = \frac{120}{11} = \mathbf{10\frac{10}{11}\ min}$$

4. (83)

$$P(BBR) = \frac{26}{52}\cdot\frac{25}{51}\cdot\frac{26}{50} = \frac{13}{102}$$
$$P(BRB) = \frac{26}{52}\cdot\frac{26}{51}\cdot\frac{25}{50} = \frac{13}{102}$$
$$P(RBB) = \frac{26}{52}\cdot\frac{26}{51}\cdot\frac{25}{50} = \frac{13}{102}$$
$$\frac{39}{102} = \mathbf{\frac{13}{34}}$$

5. (83) $\frac{3}{7}\cdot\frac{3}{7}\cdot\frac{2}{6} = \mathbf{\frac{3}{49}}$

6. (7) $N,\ N + 2,\ N + 4$

$$N + (N + 4) = (N + 2) + 6$$
$$2N + 4 = N + 8$$
$$N = 4$$

The numbers are **4, 6, 8.**

7. (89)

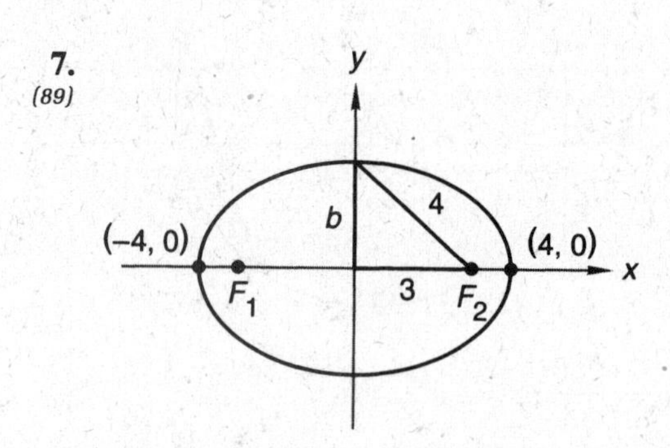

$$c^2 + b^2 = a^2$$
$$3^2 + b^2 = 4^2$$
$$b^2 = 7$$
$$b = \sqrt{7}$$
$$\frac{x^2}{a^2} + \frac{y^2}{b^2} = 1$$
$$\mathbf{\frac{x^2}{16} + \frac{y^2}{7} = 1}$$

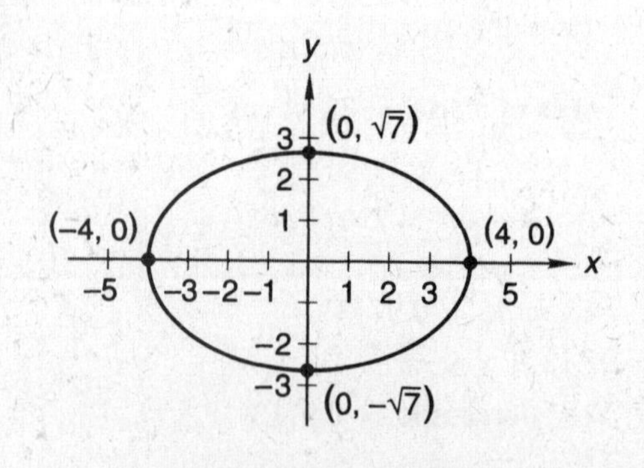

8. (89)

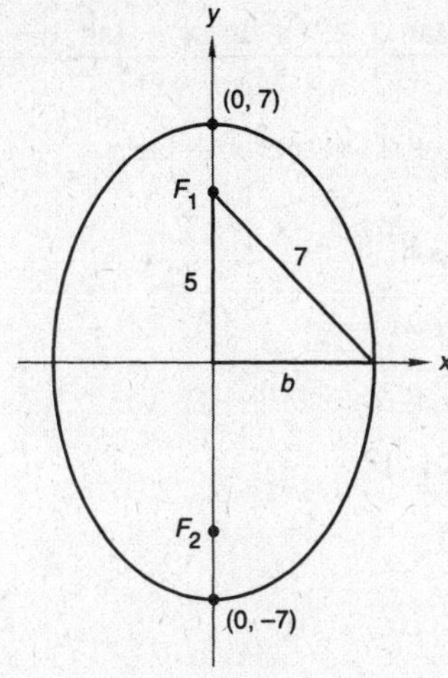

$$c^2 + b^2 = a^2$$
$$5^2 + b^2 = 7^2$$
$$b^2 = 24$$
$$b = \sqrt{24}$$
$$\frac{x^2}{b^2} + \frac{y^2}{a^2} = 1$$
$$\mathbf{\frac{x^2}{24} + \frac{y^2}{49} = 1}$$

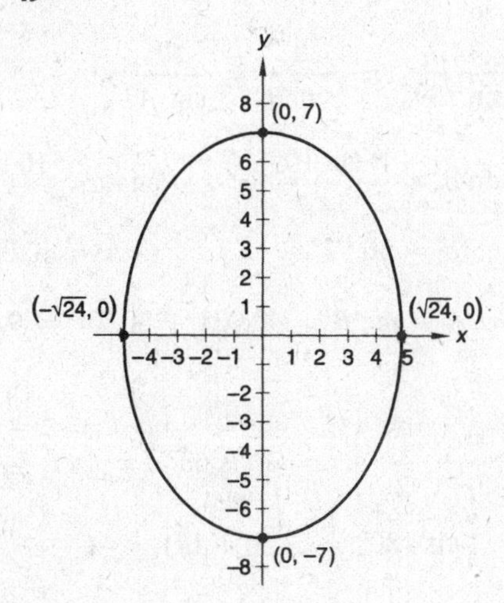

9. (89) $a = \frac{1}{2}(10) = 5$

$$b = \frac{1}{2}(4) = 2$$
$$\frac{x^2}{b^2} + \frac{y^2}{a^2} = 1$$
$$\mathbf{\frac{x^2}{4} + \frac{y^2}{25} = 1}$$

10. (87) $\sin\left(x - \frac{\pi}{2}\right) = \sin x \cos \frac{\pi}{2} - \cos x \sin \frac{\pi}{2}$

$= \sin x(0) - \cos x(1) = \mathbf{-\cos x}$

11. (87) $\tan 75° = \tan(45° + 30°)$

$$= \frac{\tan 45° + \tan 30°}{1 - \tan 45° \tan 30°}$$
$$= \frac{1 + \frac{1}{\sqrt{3}}}{1 - (1)\left(\frac{1}{\sqrt{3}}\right)} = \frac{\frac{\sqrt{3}+1}{\sqrt{3}}}{\frac{\sqrt{3}-1}{\sqrt{3}}}$$
$$= \frac{\sqrt{3}+1}{\sqrt{3}-1} \cdot \frac{\sqrt{3}+1}{\sqrt{3}+1} = \frac{3 + 2\sqrt{3} + 1}{3 - 1}$$
$$= \frac{4 + 2\sqrt{3}}{2} = \mathbf{2 + \sqrt{3}}$$

12. (87) $\tan 15° = \tan(60° - 45°)$

$$= \frac{\tan 60° - \tan 45°}{1 + \tan 60° \tan 45°}$$
$$= \frac{\sqrt{3} - 1}{1 + (\sqrt{3})(1)} \cdot \frac{1 - \sqrt{3}}{1 - \sqrt{3}}$$
$$= \frac{\sqrt{3} - 3 - 1 + \sqrt{3}}{1 - 3} = \frac{-4 + 2\sqrt{3}}{-2} = \mathbf{2 - \sqrt{3}}$$

13. (86) $a_n = a_1 + (n - 1)d$

$a_{16} = a_1 + 15d = 3 + (15)(-3) = \mathbf{-42}$

14. (86)
$$\begin{cases} \text{(a)}\ a_1 + 8d = -46 \\ \text{(b)}\ a_1 + 3d = -16 \end{cases}$$

$$\begin{array}{rl} \text{(a)} & a_1 + 8d = -46 \\ -\text{(b)} & -a_1 - 3d = 16 \\ \hline & 5d = -30 \\ & d = -6 \end{array}$$

(a) $a_1 + 8d = -46$

$a_1 + 8(-b) = -46$

$a_1 = 2$

2, −4, −10, −16

15. (60) $\sqrt{3}\cot^2 x + 2\cot x - \sqrt{3} = 0$

$(\sqrt{3}\cot x - 1)(\cot x + \sqrt{3}) = 0$

$\cot x = \frac{1}{\sqrt{3}}$ $\cot x = -\sqrt{3}$

$x = 60°, 240°$ $x = 150°, 330°$

$x = \mathbf{60°, 150°, 240°, 330°}$

16. (60)

$$3\sec^2 x + 5\sec x - 2 = 0$$
$$(3\sec x - 1)(\sec x + 2) = 0$$

$\sec x = \dfrac{1}{3}$ no answer

$\sec x = -2$

$x = \mathbf{120°, 240°}$

17. (52)

$$\csc 4x + 2 = 0$$
$$\csc 4x = -2$$
$$4x = 210°, 330°, 570°, 690°, 930°, 1050°, 1290°, 1410°$$
$$x = \mathbf{52.5°, 82.5°, 142.5°, 172.5°, 232.5°, 262.5°, 322.5°, 352.5°}$$

18. (84) $y = 2\cos\left(x - \dfrac{\pi}{2}\right)$

Period $= 2\pi$

Phase angle $= \dfrac{\pi}{2}$

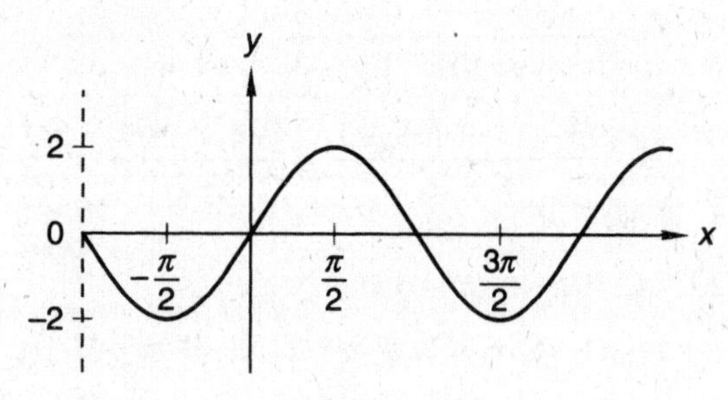

19. (84) $y = -3 + 10\sin 3(x - 20°)$

Period $= \dfrac{360°}{3} = 120°$

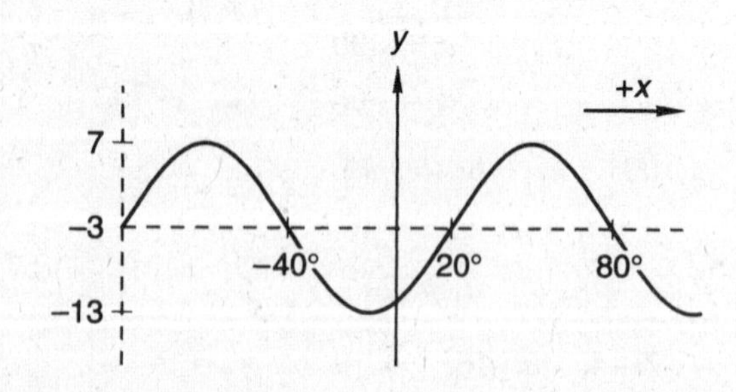

20. (80,84)

$$\frac{\sin\theta}{1 - \cos\theta} - \frac{\sin\theta}{1 + \cos\theta}$$
$$= \frac{\sin\theta(1 + \cos\theta) - \sin\theta(1 - \cos\theta)}{1 - \cos^2\theta}$$
$$= \frac{2\sin\theta\cos\theta}{\sin^2\theta} = \frac{2\cos\theta}{\sin\theta} = 2\cot\theta$$

21. (80,84)

$$\frac{\tan^3 x + 1}{\tan x + 1} = \frac{(\tan x + 1)(\tan^2 x - \tan x + 1)}{\tan x + 1}$$
$$= \tan^2 x - \tan x + 1 = \sec^2 x - \tan x$$

22. (82)

$$9^{3-2x} = 27^{x+4}$$
$$3^{2(3-2x)} = 3^{3(x+4)}$$
$$2(3 - 2x) = 3(x + 4)$$
$$6 - 4x = 3x + 12$$
$$7x = -6$$
$$x = \mathbf{-\frac{6}{7}}$$

23. (82)

$$\frac{8^{3x-1}}{4^{2x-1}} = \frac{1}{2}$$
$$\frac{2^{3(3x-1)}}{2^{2(2x-1)}} = 2^{-1}$$
$$2^{(9x-3)-(4x-2)} = 2^{-1}$$
$$5x - 1 = -1$$
$$x = \mathbf{0}$$

24. (72)

$$\frac{14.7}{\sin 109°} = \frac{8}{\sin B} = \frac{10}{\sin A}$$

$\sin B = \dfrac{8\sin 109°}{14.7}$ $\qquad$ $\sin A = \dfrac{10\sin 109°}{14.7}$

$B = 30.97°$ $\qquad$ $A = 40.03°$

$$m\angle A - m\angle B = 40.03° - 30.97° = \mathbf{9.06°}$$

25. (79)

$$\left(\frac{1}{2} + \frac{\sqrt{3}}{2}i\right)^9 = (1\text{ cis }60°)^9 = 1\text{ cis }540°$$
$$= 1\text{ cis }180° = 1(-1 + 0i) = \mathbf{-1}$$

26. (79)

$$(-16i)^{\frac{1}{2}} = (16\text{ cis }270°)^{\frac{1}{2}} = 4\text{ cis}\left(135° + n\frac{360°}{2}\right)$$
$$= 4\text{ cis }135°, 4\text{ cis }315°$$
$$= \mathbf{-2\sqrt{2} + 2\sqrt{2}i,\ 2\sqrt{2} - 2\sqrt{2}i}$$

27. (73)

θ r_c r_i 2.5 cm 5 cm

$$\theta = \frac{1}{2}\left(\frac{360°}{5}\right) = 36°$$

$$r_c = \frac{2.5}{\sin 36°} = 4.25 \text{ cm}$$

Radius of circumscribed circle = 4.25 cm

$$r_i = \frac{2.5}{\tan 36°} = 3.44 \text{ cm}$$

Radius of inscribed circle = 3.44 cm

28. (59) $\frac{3}{4} \ln 16 + 2 \ln x = \ln (2x + 1)$

$$\ln (16)^{\frac{3}{4}} x^2 = \ln (2x + 1)$$
$$8x^2 = 2x + 1$$
$$8x^2 - 2x - 1 = 0$$
$$(4x + 1)(2x - 1) = 0$$
$$x = -\frac{1}{4}, \frac{1}{2} \quad \left(x \neq -\frac{1}{4}\right)$$
$$x = \mathbf{\frac{1}{2}}$$

29. (59) $-\frac{1}{2} \log_2 (x - 1) = 1 + \log_2 5$

$$-\frac{1}{2} \log_2 (x - 1) - \log_2 5 = 1$$
$$\log_2 \frac{1}{5(x - 1)^{\frac{1}{2}}} = 1$$
$$\frac{1}{5(x - 1)^{\frac{1}{2}}} = 2^1$$
$$5(x - 1)^{\frac{1}{2}} = \frac{1}{2}$$
$$(x - 1)^{\frac{1}{2}} = \frac{1}{10}$$
$$x - 1 = \frac{1}{100}$$
$$x = \mathbf{\frac{101}{100}}$$

30. (24) $(f \circ g)(x) = \left[2(x^2 - 1) + 3\right]^2 + 1$

$$= (2x^2 - 2 + 3)^2 + 1$$
$$= (2x^2 + 1)^2 + 1$$
$$= 4x^4 + 4x^2 + 1 + 1$$
$$= \mathbf{4x^4 + 4x^2 + 2}$$

Problem Set 90

1. (88) $\mathbf{A_0 = 80,}$ $A_2 = 400$

$$A_t = A_0 e^{kt}$$
$$400 = 80e^{2k}$$
$$5 = e^{2k}$$
$$\ln 5 = 2k$$
$$k = \frac{\ln 5}{2}$$
$$\mathbf{k = 0.80}$$

$$\mathbf{A_t = 80e^{0.80t}}$$
$$2000 = 80e^{0.80t}$$
$$25 = e^{0.80t}$$
$$\ln 25 = 0.80t$$
$$t = \frac{\ln 25}{0.80}$$
$$\mathbf{t = 4.02 \text{ yr}}$$

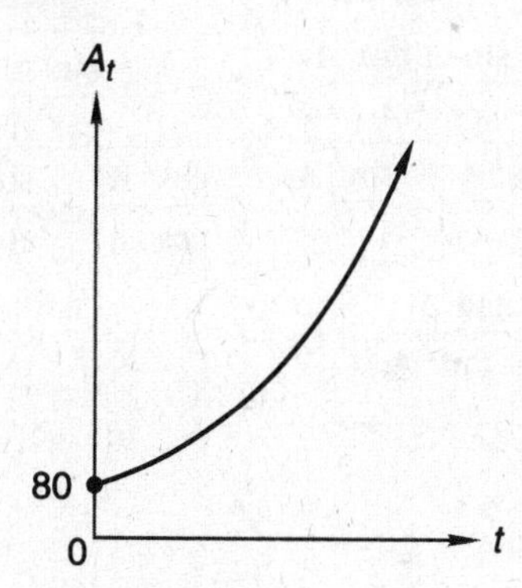

2. (88) $A_0 = 2000,\ A_4 = 500$

$$A_t = A_0 e^{kt}$$
$$500 = 2000e^{4k}$$
$$0.25 = e^{4k}$$
$$\ln 0.25 = 4k$$
$$k = \frac{\ln 0.25}{4}$$
$$k = -0.35$$
$$\mathbf{A_t = 2000e^{-0.35t}}$$

3. (83) $\frac{13}{52} \cdot \frac{13}{52} = \frac{1}{4} \cdot \frac{1}{4} = \mathbf{\frac{1}{16}}$

4. (83) $\frac{1}{6} \cdot \frac{1}{6} \cdot \frac{3}{6} = \frac{1}{6} \cdot \frac{1}{6} \cdot \frac{1}{2} = \mathbf{\frac{1}{72}}$

5. (83) The results of the first two flips do not affect the results of the third flip. Therefore the answer is $\frac{1}{2}$.

6. (25) $R_m = R_w = 2R_c$

$$(10R_m + 20R_w + 30R_c)5 = 38{,}250$$
$$[10(2R_c) + 20(2R_c) + 30R_c]5 = 38{,}250$$
$$(20R_c + 40R_c + 30R_c)5 = 38{,}250$$
$$450R_c = 38{,}250$$
$$R_c = \$85$$

$R_m = 2R_c = 2(\$85) = \mathbf{\$170}$

7. (90) $\cos 2A = \cos(A + A) = \cos A \cos A - \sin A \sin A$

$= \cos^2 A - \sin^2 A$

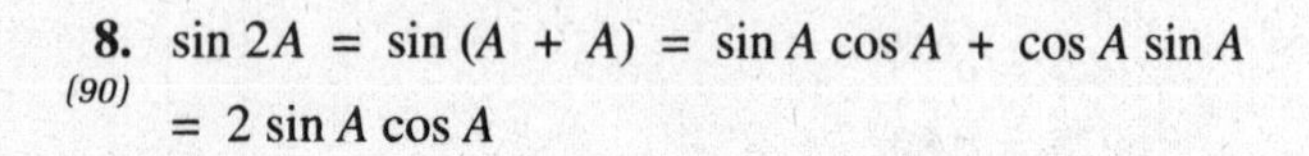

8. (90) $\sin 2A = \sin(A + A) = \sin A \cos A + \cos A \sin A$

$= 2 \sin A \cos A$

9. (90) $\tan 2A = \dfrac{\sin 2A}{\cos 2A} = \dfrac{2 \sin A \cos A}{\cos^2 A - \sin^2 A}$

$$= \frac{\dfrac{2 \sin A \cos A}{\cos^2 A}}{\dfrac{\cos^2 A - \sin^2 A}{\cos^2 A}} = \frac{2\dfrac{\sin A}{\cos A}}{\dfrac{\cos^2 A}{\cos^2 A} - \dfrac{\sin^2 A}{\cos^2 A}}$$

$$= \frac{2 \tan A}{1 - \tan^2 A}$$

10. (89)

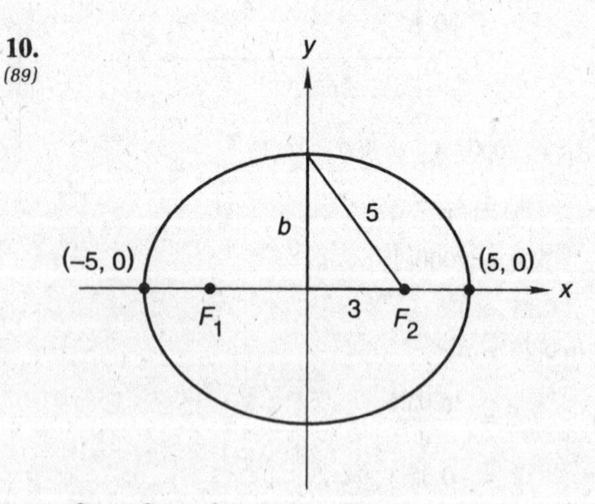

$$c^2 + b^2 = a^2$$
$$3^2 + b^2 = 5^2$$
$$b^2 = 16$$
$$b = 4$$

$$\frac{x^2}{a^2} + \frac{y^2}{b^2} = 1$$

$$\mathbf{\frac{x^2}{25} + \frac{y^2}{16} = 1}$$

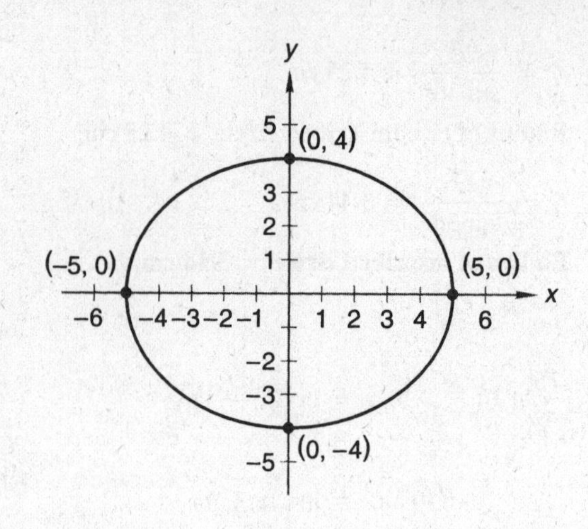

11. (89)

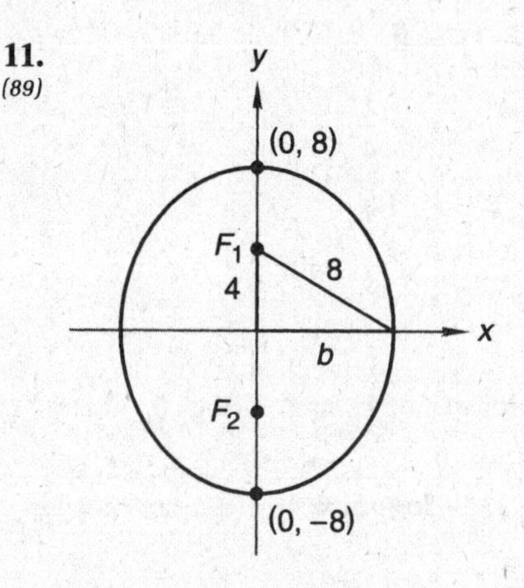

$$c^2 + b^2 = a^2$$
$$4^2 + b^2 = 8^2$$
$$b^2 = 48$$
$$b = \sqrt{48}$$

$$\frac{x^2}{b^2} + \frac{y^2}{a^2} = 1$$

$$\mathbf{\frac{x^2}{48} + \frac{y^2}{64} = 1}$$

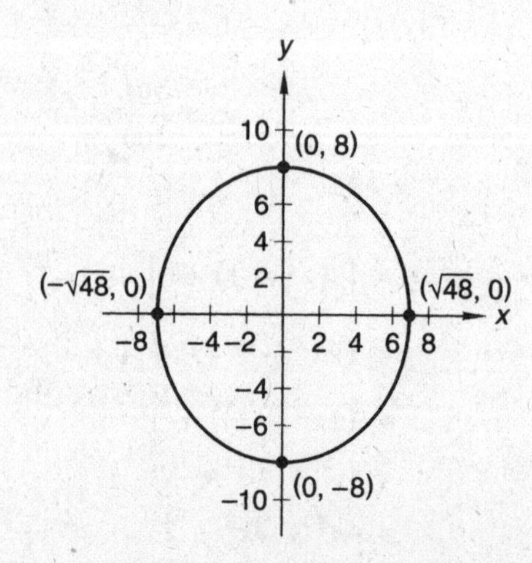

12. (89) $a = \frac{1}{2}(12) = 6$

$b = \frac{1}{2}(4) = 2$

$\frac{x^2}{a^2} + \frac{y^2}{b^2} = 1$

$\mathbf{\frac{x^2}{36} + \frac{y^2}{4} = 1}$

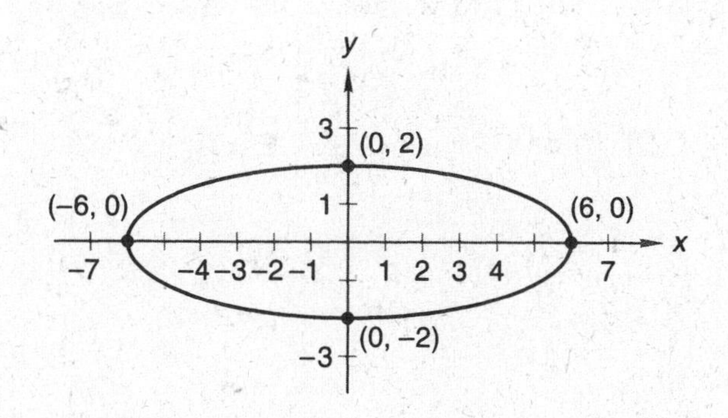

13. (87) $\sin 105° = \sin(60° + 45°)$

$= \sin 60° \cos 45° + \cos 60° \sin 45°$

$= \left(\frac{\sqrt{3}}{2} \cdot \frac{\sqrt{2}}{2}\right) + \left(\frac{1}{2} \cdot \frac{\sqrt{2}}{2}\right)$

$= \frac{\sqrt{6}}{4} + \frac{\sqrt{2}}{4} = \mathbf{\frac{\sqrt{6} + \sqrt{2}}{4}}$

14. (86) $a_1 = 4,\ d = 2,\ n = 18$

$a_n = a_1 + (n - 1)d$

$a_{18} = a_1 + 17d = (4) + 17(2) = \mathbf{38}$

15. (86) (a) $a_1 + 9d = -22$
(b) $a_1 + 2d = -8$

(a) $a_1 + 9d = -22$
−(b) $-a_1 - 2d = 8$

$7d = -14$

$d = -2$

(b) $a_1 + 2d = -8$

$a_1 + 2(-2) = -8$

$a_1 = -4$

−4, −6, −8

16. (85) $2\cos^2 y - 9\sin y + 3 = 0$

$2(1 - \sin^2 y) - 9\sin y + 3 = 0$

$2\sin^2 y + 9\sin y - 5 = 0$

$(2\sin y - 1)(\sin y + 5) = 0$

$\sin y = \frac{1}{2}$ $\quad\quad$ $\sin y = -5$
no answer

$y = \mathbf{\frac{\pi}{6}, \frac{5\pi}{6}}$

17. (52) $\sec 2x + 2 = 0$

$\sec 2x = -2$

$2x = \frac{2\pi}{3}, \frac{4\pi}{3}, \frac{8\pi}{3}, \frac{10\pi}{3}$

$x = \mathbf{\frac{\pi}{3}, \frac{2\pi}{3}, \frac{4\pi}{3}, \frac{5\pi}{3}}$

18. (84) $y = -3\sin\left(x - \frac{\pi}{4}\right)$

Period $= 2\pi$

Phase angle $= \frac{\pi}{4}$

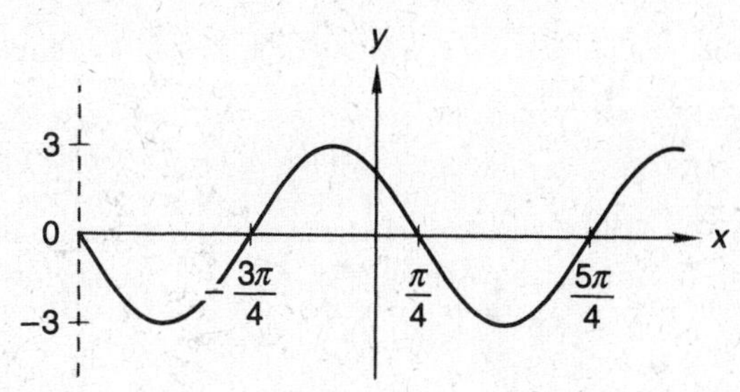

19. (84) $y = -2 + 5\cos 4(\theta - 40°)$

Period $= \frac{360°}{4} = 90°$

Phase angle $= 40°$

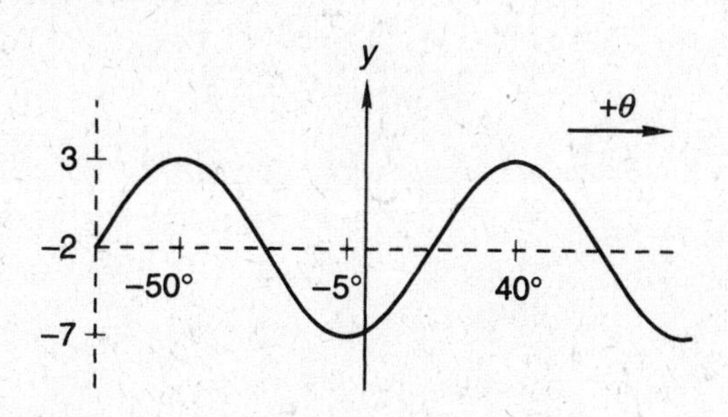

20. (80,84) $\sec^2 x \sin^2 x + \sin^2 x \csc^2 x$

$= \frac{1}{\cos^2 x}\sin^2 x + \sin^2 x \frac{1}{\sin^2 x}$

$= \tan^2 x + 1 = \sec^2 x$

21. (80,84) $\sin^2\theta + \tan^2\theta + \cos^2\theta = 1 + \tan^2\theta = \sec^2\theta$

22. (81)
$$12^2 = 10^2 + 8^2 - 2(10)(8)\cos B$$
$$144 = 164 - 160\cos B$$
$$\cos B = \frac{20}{160}$$
$$B = \mathbf{82.82°}$$

23. (74)
$$x = \frac{\begin{vmatrix} -6 & 2 \\ 2 & -4 \end{vmatrix}}{\begin{vmatrix} 5 & 2 \\ 3 & -4 \end{vmatrix}} = \frac{24 - 4}{-20 - 6} = \frac{20}{-26} = -\frac{10}{13}$$
$$y = \frac{\begin{vmatrix} 5 & -6 \\ 3 & 2 \end{vmatrix}}{\begin{vmatrix} 5 & 2 \\ 3 & -4 \end{vmatrix}} = \frac{10 + 18}{-20 - 6} = \frac{28}{-26} = -\frac{14}{13}$$
$$\mathbf{x = -\frac{10}{13};\ y = -\frac{14}{13}}$$

24. (56)
$$x^2 + 2x + y^2 + 4y \le 11$$
$$(x^2 + 2x + 1) + (y^2 + 4y + 4) \le 11 + 1 + 4$$
$$(x + 1)^2 + (y + 2)^2 \le 4^2 \text{ (circle)}$$

$x - y < 3$; $y > x - 3$ (line)

$x - y > 0$; $y < x$ (line)

The inequality indicates the region inside or on the circle, above the line $y = x - 3$ and below the line $y = x$.

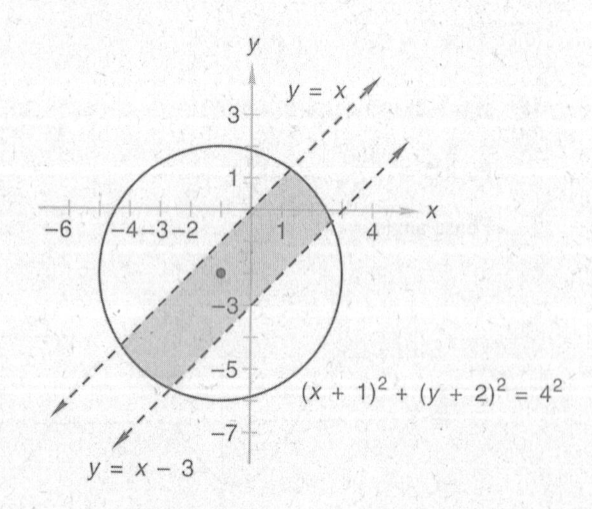

25. (82)
$$27^{2-3x} = 81^{x-5}$$
$$3^{3(2-3x)} = 3^{4(x-5)}$$
$$6 - 9x = 4x - 20$$
$$13x = 26$$
$$x = \mathbf{2}$$

26. (82)
$$7^{2x-1} = 8^{3x+2}$$
$$(2x - 1)\log 7 = (3x + 2)\log 8$$
$$(2x - 1)(0.8451) = (3x + 2)(0.9031)$$
$$1.6902x - 0.8451 = 2.7093x + 1.8062$$
$$-1.0191x = 2.6513$$
$$x = \mathbf{-2.60}$$

27. (59)
$$\ln x + \ln x = 2\ln 2$$
$$2\ln x = 2\ln 2$$
$$x = \mathbf{2}$$

28. (59)
$$2^{\log_2 9 - 2\log_2 6 + 4\log_2 2} + 3\log_2 2$$
$$= 2^{\log_2 \frac{(9)(2^4)}{6^2}} + \log_2 2^3$$
$$= \frac{9 \cdot 16}{36} + 3 = 4 + 3 = \mathbf{7}$$

29. (67)

(a) $\text{antilog}_4(-1) = 4^{-1} = \mathbf{\frac{1}{4}}$

(b) $\text{antilog}_{\frac{1}{3}}(-1) = \left(\frac{1}{3}\right)^{-1} = \mathbf{3}$

30. (70)
$$z = \frac{x - \mu}{\sigma} = \frac{0.300 - 0.270}{0.015} = 2$$

Percentile = 0.9772

$1 - 0.9772 = 0.0228$

2.28%

Problem Set 91

1. (85) $R_L = \frac{1}{12}$, $R_B = 1$, $T_B = T_L$

Big hand:
$$R_B T_B = S + 40$$
$$(1)T = S + 40$$
$$S = T - 40$$

Little hand:
$$R_L T_L = S$$
$$S = \frac{1}{12}T$$

$$\frac{1}{12}T = T - 40$$
$$\frac{11}{12}T = 40$$
$$T = \frac{480}{11} = \mathbf{43\frac{7}{11}\ min}$$

2. (18) $x - \frac{1}{8}x - 5000 - \frac{1}{2}\left(x - \frac{1}{8}x - 5000\right) = 4500$

$\frac{7}{8}x - 5000 - \frac{1}{2}x + \frac{1}{16}x + 2500 = 4500$

$\frac{7}{16}x = 7000$

$x = \frac{16(7000)}{7}$

$x = $ **16,000 marbles**

3. (83) $\frac{1}{6} \cdot \frac{1}{6} \cdot \frac{4}{6} = \frac{4}{216} = \mathbf{\frac{1}{54}}$

4. (88) $\mathbf{A_0 = 30,}\ A_3 = 900$

$A_t = A_0e^{kt}$

$900 = 30e^{3k}$

$30 = e^{3k}$

$\ln 30 = 3k$

$k = \frac{\ln 30}{3}$

$\mathbf{k = 1.13}$

$\mathbf{A_t = 30e^{1.13t}}$

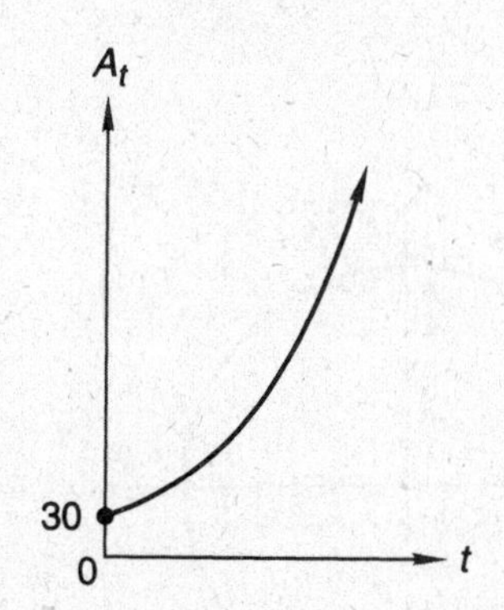

5. (88) $A_0 = 42,\ A_{10} = 7,\ A_t = 30$

$A_t = A_0e^{kt}$

$7 = 42e^{10k}$

$\frac{1}{6} = e^{10k}$

$\ln \frac{1}{6} = 10k$

$k = \frac{\ln \frac{1}{6}}{10}$

$k = -0.18$

$\mathbf{A_t = 42e^{-0.18t}}$

$30 = 42e^{-0.18t}$

$\frac{5}{7} = e^{-0.18t}$

$\ln \frac{5}{7} = -0.18t$

$t = \frac{\ln \frac{5}{7}}{-0.18}$

$\mathbf{t = 1.87\ yr}$

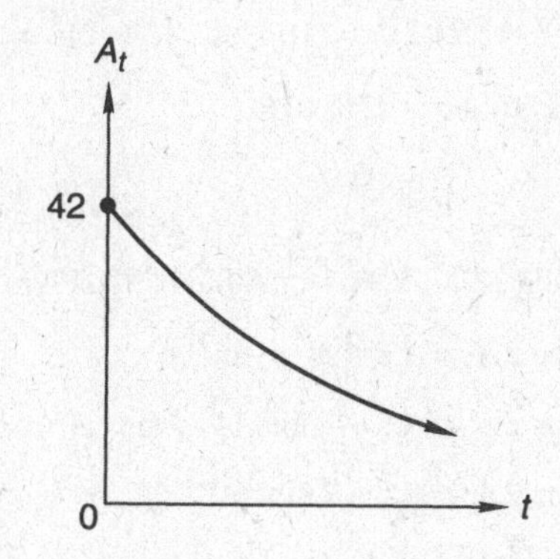

6. (25) $R_1T_1 + R_2T_2 = \text{cisterns}$

$\frac{1}{20}T + \frac{1}{30}T = 1$

$\frac{5}{60}T = 1$

$T = $ **12 hr**

7. (91) $a_1 = -4,\ r = -2$

$a_6 = a_1r^5 = (-4)(-2)^5 = \mathbf{128}$

8. (91)

①	②	③	④
4,	___,	___,	108
a_1,	a_1r,	a_1r^2,	a_1r^3

$a_1r^3 = 108$

$(4)r^3 = 108$

$r^3 = 27$

$r = 3$

$a_1r = (4)(3) = 12$

$a_1r^2 = (4)(3)^2 = 36$

12, 36

9. (91)

①	②	③	④	⑤
2,	___,	___,	___,	32
a_1,	a_1r,	a_1r^2,	a_1r^3,	a_1r^4

$$a_1r^4 = 32$$
$$(2)r^4 = 32$$
$$r^4 = 16$$
$$r = \pm 2$$

$a_1r = (2)(2) = 4$ or $(2)(-2) = -4$

$a_1r^2 = (2)(2)^2 = 8$ or $(2)(-2)^2 = 8$

$a_1r^3 = (2)(2)^3 = 16$ or $(2)(-2)^3 = -16$

4, 8, 16 or −4, 8, −16

10. (90) Using $\cos^2 A = 1 - \sin^2 A$, we find:

$$\cos 2A = \cos^2 A - \sin^2 A$$
$$\cos 2A = (1 - \sin^2 A) - \sin^2 A$$
$$\cos 2A = 1 - 2\sin^2 A$$
$$2\sin^2 A = 1 - \cos 2A$$
$$\sin^2 A = \frac{1 - \cos 2A}{2}$$
$$\sin A = \pm\sqrt{\frac{1 - \cos 2A}{2}}$$
$$\sin \frac{A}{2} = \pm\sqrt{\frac{1 - \cos A}{2}}$$

11. (90) $\sin 15° = \pm\sqrt{\frac{1 - \cos 30°}{2}} = \pm\sqrt{\frac{1 - \frac{\sqrt{3}}{2}}{2}}$

$= \pm\sqrt{\frac{2 - \sqrt{3}}{4}} = \mathbf{\frac{\sqrt{2 - \sqrt{3}}}{2}}$

12. (90) Using $\sin^2 A = 1 - \cos^2 A$, we find:

$$\cos 2A = \cos^2 A - \sin^2 A$$
$$\cos 2A = \cos^2 A - (1 - \cos^2 A)$$
$$\cos 2A = 2\cos^2 A - 1$$
$$2\cos^2 A = 1 + \cos 2A$$
$$\cos^2 A = \frac{1 + \cos 2A}{2}$$
$$\cos A = \pm\sqrt{\frac{1 + \cos 2A}{2}}$$
$$\cos \frac{A}{2} = \pm\sqrt{\frac{1 + \cos A}{2}}$$

13. (90) $\cos 15° = \sqrt{\frac{1 + \cos 30°}{2}} = \sqrt{\frac{1 + \frac{\sqrt{3}}{2}}{2}}$

$= \sqrt{\frac{2 + \sqrt{3}}{4}} = \mathbf{\frac{\sqrt{2 + \sqrt{3}}}{2}}$

14. (89)

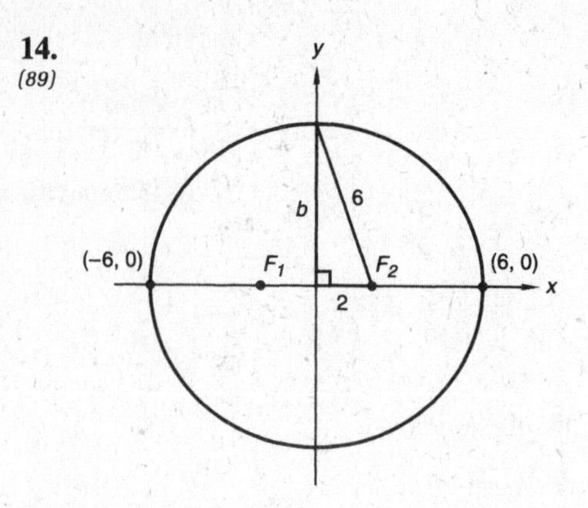

$$a^2 = c^2 + b^2$$
$$6^2 = 2^2 + b^2$$
$$b^2 = 32$$
$$b = \sqrt{32}$$
$$\frac{x^2}{a^2} + \frac{y^2}{b^2} = 1$$
$$\mathbf{\frac{x^2}{36} + \frac{y^2}{32} = 1}$$

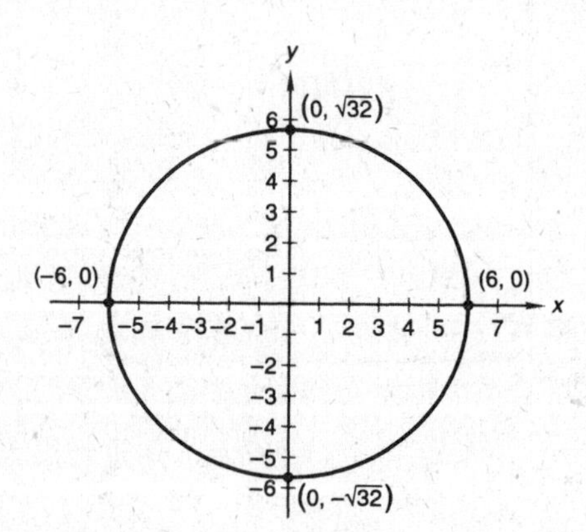

15. (89) $a = \frac{14}{2} = 7$

$b = \frac{10}{2} = 5$

$$\frac{x^2}{a^2} + \frac{y^2}{b^2} = 1$$
$$\mathbf{\frac{x^2}{25} + \frac{y^2}{49} = 1}$$

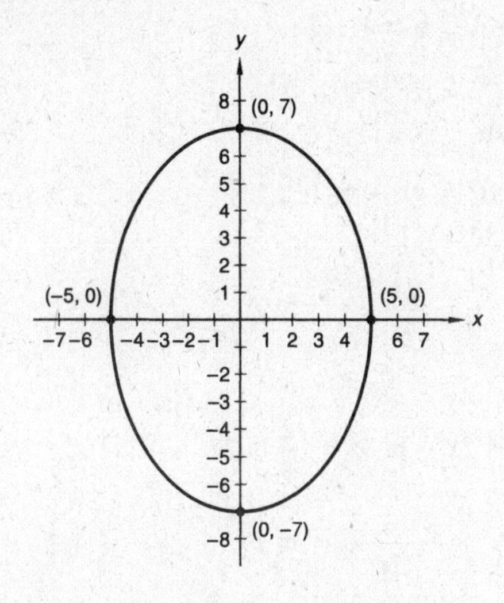

16. (87) $\cos 105° = \cos(45° + 60°)$

$= \cos 45° \cos 60° - \sin 45° \sin 60°$

$= \left(\frac{\sqrt{2}}{2} \cdot \frac{1}{2}\right) - \left(\frac{\sqrt{2}}{2} \cdot \frac{\sqrt{3}}{2}\right)$

$= \frac{\sqrt{2}}{4} - \frac{\sqrt{6}}{4} = \mathbf{\frac{\sqrt{2} - \sqrt{6}}{4}}$

17. (86) $a_1 = 6$

$a_{10} = a_1 + 9d$

$-30 = 6 + 9d$

$9d = -36$

$d = -4$

$a_1,\ a_1 + d,\ a_1 + 2d,\ a_1 + 3d$

6, 2, –2, –6

18. (85) $2 \sin^2 y - 9 \cos y + 3 = 0$

$2(1 - \cos^2 y) - 9 \cos y + 3 = 0$

$2 \cos^2 y + 9 \cos y - 5 = 0$

$(2 \cos y - 1)(\cos y + 5) = 0$

$\cos y = \frac{1}{2}$ $\qquad$ $\cos y = -5$

$y = \mathbf{60°, 300°}$ $\qquad$ no answer

19. (52) $\sqrt{3} + 2 \cos 3\theta = 0$

$\cos 3\theta = -\frac{\sqrt{3}}{2}$

$3\theta = 150°, 210°, 510°, 570°, 870°, 930°$

$\theta = \mathbf{50°, 70°, 170°, 190°, 290°, 310°}$

20. (84) $y = 2 - 3 \cos (x + 40°)$

Period = 360°

Phase angle = –40°

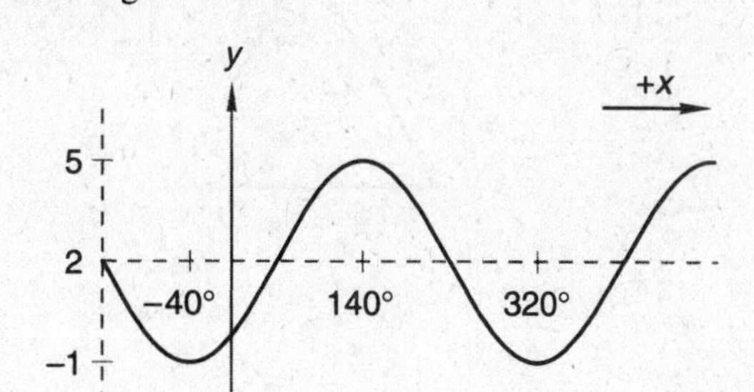

21. (84) $y = \frac{3}{2} + \frac{3}{2} \sin 2\left(x + \frac{\pi}{6}\right)$

Period $= \frac{2\pi}{2} = \pi$

Phase angle $= -\frac{\pi}{6}$

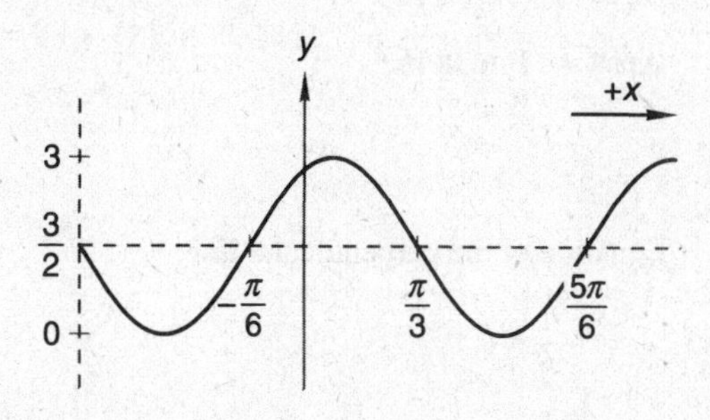

22. (80,84) $\frac{1 - \cos^2 \theta}{1 + \tan^2 \theta} = \frac{\sin^2 \theta}{\sec^2 \theta} = \sin^2 \theta \cos^2 \theta$

23. (80,84) $\csc^2 y \sec^2 y - \sec^2 y = \sec^2 y(\csc^2 y - 1)$

$= \sec^2 y \cot^2 y = \frac{1}{\cos^2 y} \cdot \frac{\cos^2 y}{\sin^2 y}$

$= \frac{1}{\sin^2 y} = \csc^2 y$

24. (72) $\frac{19.04}{\sin 130°} = \frac{10}{\sin B}$

$\sin B = \frac{10 \sin 130°}{19.04}$

$\mathbf{B = 23.73°}$

$\frac{19.04}{\sin 130°} = \frac{11}{\sin A}$

$\sin A = \frac{11 \sin 130°}{19.04}$

$\mathbf{A = 26.27°}$

25. (73) Side $= \dfrac{48 \text{ in.}}{6} = 8$ in.

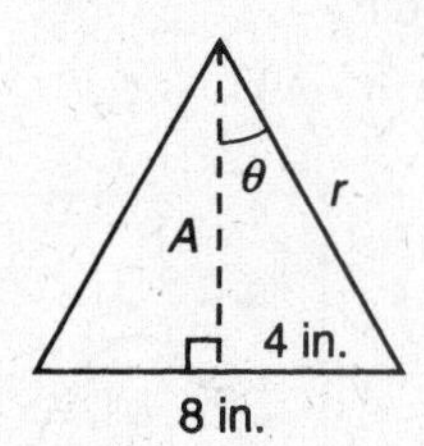

$$\theta = \frac{1}{2}\left(\frac{360°}{6}\right) = 30°$$

$$r = \frac{4}{\sin 30°}$$

Radius = 8 in.

$$A = 8\cos 30° = 8\left(\frac{\sqrt{3}}{2}\right) = 4\sqrt{3} \text{ in.}$$

$$\text{Area}_\Delta = \frac{1}{2}(4)(4\sqrt{3}) = 8\sqrt{3} \text{ in.}^2$$

$$\text{Area} = (12)(8\sqrt{3})$$

Area = 166.28 in.²

26. (58) $y = 3x - 1$

Equation of the perpendicular line:

$$y = -\frac{1}{3}x + b$$

$$2 = -\frac{1}{3}(2) + b$$

$$b = \frac{8}{3}$$

$$y = -\frac{1}{3}x + \frac{8}{3}$$

Point of intersection:

$$3x - 1 = -\frac{1}{3}x + \frac{8}{3}$$

$$\frac{10}{3}x = \frac{11}{3}$$

$$x = \frac{11}{10}$$

$$y = 3x - 1 = 3\left(\frac{11}{10}\right) - 1 = \frac{23}{10}$$

$\left(\frac{11}{10}, \frac{23}{10}\right)$ and (2, 2)

$$D = \sqrt{\left(2 - \frac{11}{10}\right)^2 + \left(2 - \frac{23}{10}\right)^2}$$

$$= \sqrt{\left(\frac{9}{10}\right)^2 + \left(-\frac{3}{10}\right)^2} = \sqrt{\frac{90}{100}} = \mathbf{\frac{3\sqrt{10}}{10}}$$

27. (82)

$$16^{3x-4} = 4^{2x+1} \cdot 8$$

$$2^{4(3x-4)} = 2^{2(2x+1)} \cdot 2^3$$

$$2^{12x-16} = 2^{4x+2+3}$$

$$12x - 16 = 4x + 5$$

$$8x = 21$$

$$x = \mathbf{\frac{21}{8}}$$

28. (59) $\ln(2x + 3) - \ln 3x = \frac{2}{3}\ln 27$

$$\ln\frac{2x + 3}{3x} = \ln 27^{\frac{2}{3}}$$

$$\frac{2x + 3}{3x} = 9$$

$$2x + 3 = 27x$$

$$25x = 3$$

$$x = \mathbf{\frac{3}{25}}$$

29. (59) $\log(2x + 1) = 3\log 2 - \log 4$

$$\log(2x + 1) = \log\frac{2^3}{4}$$

$$2x + 1 = 2$$

$$2x = 1$$

$$x = \mathbf{\frac{1}{2}}$$

30. (59) $3^{\log_3 4 - 2\log_3 8 + 4\log_3 2} - \frac{1}{2}\log_3 3^2$

$$= 3^{\log_3 \frac{4 \cdot 2^4}{8^2}} - \log_3(3^2)^{\frac{1}{2}}$$

$$= 3^{\log_3 1} - \log_3 3 = 1 - 1 = \mathbf{0}$$

Problem Set 92

1. (85) $R_L = \frac{1}{12}$, $R_B = 1$, $T_B = T_L$

Big hand:

$$R_BT_B = S + 15$$

$$(1)T = S + 15$$

$$S = T - 15$$

Little hand:

$$R_LT_L = S$$

$$S = \frac{1}{12}T$$

$$\frac{1}{12}T = T - 15$$

$$\frac{11}{12}T = 15$$

$$T = \frac{15 \cdot 12}{11}$$

$$T = \mathbf{16\frac{4}{11}\ min}$$

2. $A_0 = 10{,}000,\ A_{40} = 9000$
(88)

$$A_t = A_0e^{kt}$$

$$9000 = 10{,}000e^{40k}$$

$$0.9 = e^{40k}$$

$$\ln 0.9 = 40k$$

$$k = \frac{\ln 0.9}{40}$$

$$k = -0.0026$$

$$\mathbf{A_t = 10{,}000e^{-0.0026t}}$$

$$5000 = 10{,}000e^{-0.0026t}$$

$$0.5 = e^{-0.0026t}$$

$$\ln 0.5 = -0.0026t$$

$$t = \frac{\ln 0.5}{-0.0026}$$

$$t = \mathbf{266.60\ min}$$

3. $A_0 = 100,\ A_6 = 1000$
(88)

$$A_t = A_0e^{kt}$$

$$1000 = 100e^{6k}$$

$$10 = e^{6k}$$

$$\ln 10 = 6k$$

$$k = \frac{\ln 10}{6}$$

$$k = 0.38$$

$$\mathbf{A_t = 100e^{0.38t}}$$

$$2000 = 100e^{0.38t}$$

$$20 = e^{0.38t}$$

$$\ln 20 = 0.38t$$

$$t = \mathbf{7.88\ mo}$$

4. $S = mT + b$
(62)

$$\begin{cases} \text{(a)}\ 8 = m10 + b \\ \text{(b)}\ 13 = m20 + b \end{cases}$$

$$\begin{array}{rrcl} \text{(a)} & 8 & = & m10 + b \\ -\text{(b)} & -13 & = & -m20 - b \\ \hline & -5 & = & -m10 \end{array}$$

$$m = \frac{1}{2}$$

(a) $8 = m10 + b$

$$8 = \frac{1}{2}(10) + b$$

$$b = 3$$

$$S = \frac{1}{2}T + 3 = \frac{1}{2}(8) + 3 = \mathbf{7}$$

5. $N,\ N + 2,\ N + 4$
(7)

$$N(N + 4) = 5(N + 2) + 20$$

$$N^2 + 4N = 5N + 30$$

$$N^2 - N - 30 = 0$$

$$(N + 5)(N - 6) = 0$$

$$N = 6, -5 \qquad (N \neq -5)$$

6, 8, 10

6. $P(K \cup B) = P(K) + P(B) - P(K \cap B)$
(92)

$$= \frac{4}{52} + \frac{26}{52} - \frac{2}{52} = \frac{28}{52} = \mathbf{\frac{7}{13}}$$

7. $_nP_r = \dfrac{n!}{(n - r)!}$
(92)

$$_8P_5 = \frac{8!}{(8 - 5)!}$$

$$\mathbf{_8P_5 = \frac{8!}{3!} = 6720}$$

$$_nC_r = \frac{n!}{r!(n - r)!}$$

$$_8C_5 = \frac{8!}{5!(8 - 5)!}$$

$$\mathbf{_8C_5 = \frac{8!}{5!3!} = 56}$$

8. $a_1 = 2,\ r = -\frac{1}{2},\ n = 4$
(91)

$$a_n = a_1r^{n-1}$$

$$a_4 = 2\left(-\frac{1}{2}\right)^3 = \mathbf{-\frac{1}{4}}$$

9. (91)

①	②	③	④
2,	___,	___,	−16
a_1,	a_1r,	a_1r^2,	a_1r^3

$$a_1r^3 = -16$$
$$2r^3 = -16$$
$$r^3 = -8$$
$$r = -2$$
$$a_1r = 2 \cdot (-2) = -4$$
$$a_1r^2 = 2 \cdot (-2)^2 = 8$$

−4, 8

10. (90) Using $\cos^2 A = 1 - \sin^2 A$, we find:

$$\cos(A + B) = \cos A \cos B - \sin A \sin B$$
$$\cos(A + A) = \cos A \cos A - \sin A \sin A$$
$$\cos 2A = \cos^2 A - \sin^2 A$$
$$\cos 2A = (1 - \sin^2 A) - \sin^2 A$$
$$\cos 2A = 1 - 2\sin^2 A$$
$$2\sin^2 A = 1 - \cos 2A$$
$$\sin^2 A = \frac{1 - \cos 2A}{2}$$
$$\sin A = \pm\sqrt{\frac{1 - \cos 2A}{2}}$$
$$\sin\frac{A}{2} = \pm\sqrt{\frac{1 - \cos A}{2}}$$

11. (90) $\sin 165° = \sqrt{\frac{1 - \cos 330°}{2}} = \sqrt{\frac{1 - \cos 30°}{2}}$

$$= \sqrt{\frac{1 - \frac{\sqrt{3}}{2}}{2}} = \sqrt{\frac{2 - \sqrt{3}}{4}} = \boldsymbol{\frac{\sqrt{2 - \sqrt{3}}}{2}}$$

12. (87) $\cos\left(x + \frac{\pi}{2}\right) = \cos x \cos\frac{\pi}{2} - \sin x \sin\frac{\pi}{2}$

$$= (\cos x)(0) - (\sin x)(1) = \boldsymbol{-\sin x}$$

13. (89) $a = \frac{10}{2} = 5,\ b = \frac{8}{2} = 4$

$$\frac{x^2}{b^2} + \frac{y^2}{a^2} = 1$$
$$\boldsymbol{\frac{x^2}{16} + \frac{y^2}{25} = 1}$$

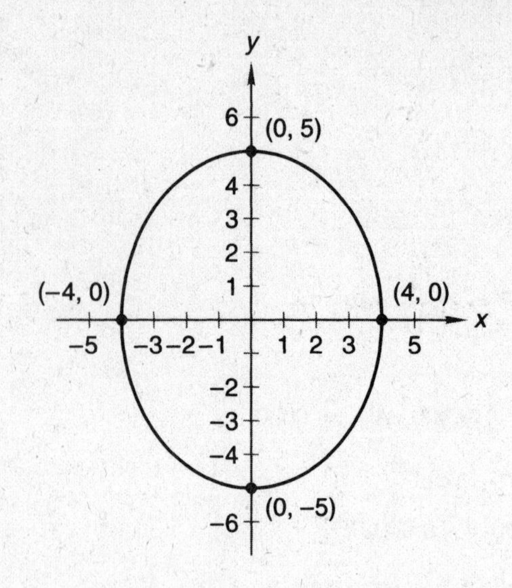

14. (86) $a_4 = -2,\ a_{10} = 10$

(a) $a_1 + 3d = -2$
(b) $a_1 + 9d = 10$

(a) $a_1 + 3d = -2$
−(b) $-a_1 - 9d = -10$

$$-6d = -12$$
$$d = 2$$

(a) $a_1 + 3(2) = -2$

$$a_1 = -8$$

−8, −6, −4, −2

15. (85)

$$\tan^2\theta - \sec\theta - 1 = 0$$
$$(\sec^2\theta - 1) - \sec\theta - 1 = 0$$
$$\sec^2\theta - \sec\theta - 2 = 0$$
$$(\sec\theta - 2)(\sec\theta + 1) = 0$$

$\sec\theta = 2$ $\qquad$ $\sec\theta = -1$

$\theta = \frac{\pi}{3}, \frac{5\pi}{3}$ $\qquad$ $\theta = \pi$

$$\boldsymbol{\theta = \frac{\pi}{3}, \pi, \frac{5\pi}{3}}$$

16. (52) $\sin 3\theta = 0$

$$3\theta = 0, \pi, 2\pi, 3\pi, 4\pi, 5\pi$$
$$\boldsymbol{\theta = 0, \frac{\pi}{3}, \frac{2\pi}{3}, \pi, \frac{4\pi}{3}, \frac{5\pi}{3}}$$

17. (84) $y = -3 + 5\cos\frac{1}{2}\left(x - \frac{3}{2}\pi\right)$

$$\text{Period} = \frac{2\pi}{\frac{1}{2}} = 4\pi$$

$$\text{Phase angle} = \frac{3\pi}{2}$$

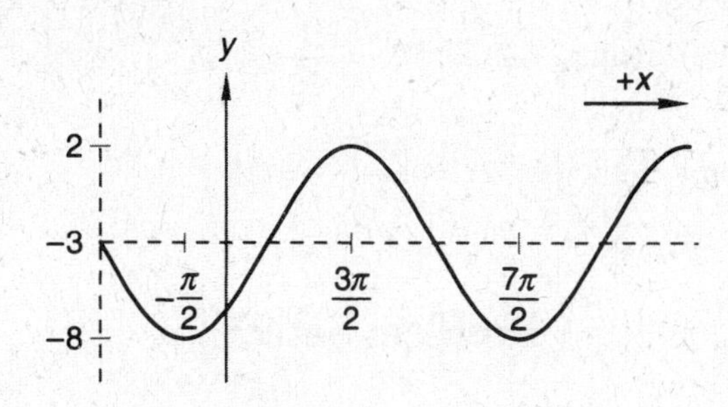

18. (80,84) $\frac{\sin^4 \theta - \cos^4 \theta}{\sin^2 \theta - \cos^2 \theta}$

$= \frac{(\sin^2 \theta + \cos^2 \theta)(\sin^2 \theta - \cos^2 \theta)}{\sin^2 \theta - \cos^2 \theta}$

$= \sin^2 \theta + \cos^2 \theta = 1$

19. (80,84) $\frac{\sec^2 x}{\sec^2 x - 1} = \frac{\sec^2 x}{\tan^2 x} = \frac{\frac{1}{\cos^2 x}}{\frac{\sin^2 x}{\cos^2 x}}$

$= \frac{1}{\cos^2 x} \cdot \frac{\cos^2 x}{\sin^2 x} = \frac{1}{\sin^2 x} = \csc^2 x$

20. (81) $8^2 = 5^2 + 6^2 - 2(5)(6) \cos B$

$3 = -60 \cos B$

$B = 92.87°$

$\frac{6}{\sin A} = \frac{8}{\sin 92.87°}$

$\sin A = \frac{6 \sin 92.87°}{8}$

$A = 48.51°$

$m\angle B - m\angle A = 92.87° - 48.51° = \mathbf{44.36°}$

21. (3) $\frac{c}{8} = \frac{\frac{4}{3}a}{a}$

$c = \left(\frac{4}{3}\right)(8) = \mathbf{\frac{32}{3}}$

22. (79) $(16 \text{ cis } 120°)^{\frac{1}{4}} = 16^{\frac{1}{4}} \text{ cis}\left(30° + n\frac{360°}{4}\right)$

$= 2 \text{ cis } 30°,\ 2 \text{ cis } 120°,\ 2 \text{ cis } 210°,\ 2 \text{ cis } 300°$

$= \mathbf{\sqrt{3} + i,\ -1 + \sqrt{3}i,\ -\sqrt{3} - i,\ 1 - \sqrt{3}i}$

23. (78) $16x^2 - 25y^2 = 800$

$\mathbf{\frac{x^2}{50} - \frac{y^2}{32} = 1}$

Let $y = 0$ | Let $x = 0$

$\frac{x^2}{50} = 1$ | $\frac{-y^2}{32} = 1$

$x^2 = 50$ | $y^2 = -32$

$x = \pm 5\sqrt{2}$ | $y = \pm 4\sqrt{2}i$

Vertices = $(5\sqrt{2}, 0), (-5\sqrt{2}, 0)$

Asymptotes: $y = \frac{4}{5}x$; $y = -\frac{4}{5}x$

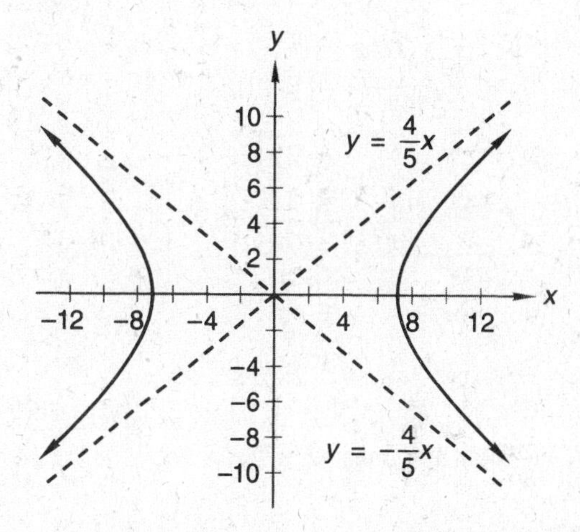

24. (73) Side $= \frac{40 \text{ cm}}{10} = 4$ cm

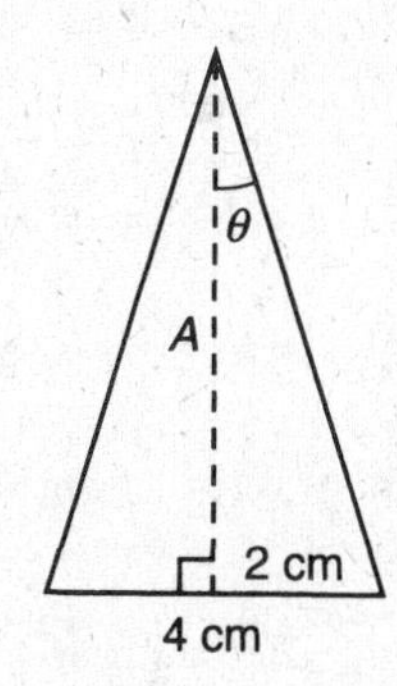

$\theta = \frac{1}{2}\left(\frac{360°}{10}\right) = 18°$

$A = \frac{2}{\tan 18°} = 6.1554$ cm

Radius of inscribed circle $= r_i = 6.1554$ cm

Perimeter $= 2\pi r_i$

$= 2\pi(6.1554)$

Perimeter = 38.68 cm

$\text{Area}_\Delta = \frac{1}{2}(2)(6.1554) = 6.1554 \text{ cm}^2$

Area = (20)(6.1554)

Area = 123.11 cm²

25. (70) $z = \frac{x - \mu}{\sigma} = \frac{3 - 7}{3} = -1.33$

$-1.3 > z > -1.4$

Percentile for $-1.3 = 0.0968$

Percentile for $-1.4 = 0.0808$

$0.0968 >$ percentile for $-1.33 > 0.0808 > 5\%$

No, Smith is not in the top 5%.

26. (82)

$$\frac{16^{3x-2}}{2^{4x-1}} = 4$$
$$\frac{(2^4)^{3x-2}}{2^{4x-1}} = 2^2$$
$$2^{4(3x-2)-(4x-1)} = 2^2$$
$$8x - 7 = 2$$
$$8x = 9$$
$$x = \mathbf{\frac{9}{8}}$$

27. (59) $\frac{2}{3}\log_7 8 - \log_7 (x - 1) = 1$

$$\log_7 \frac{8^{\frac{2}{3}}}{(x - 1)} = 1$$
$$\frac{4}{x - 1} = 7^1$$
$$4 = 7x - 7$$
$$7x = 11$$
$$x = \mathbf{\frac{11}{7}}$$

28. (59) $\log x + \log (x + 2) = \frac{1}{2}\log 9$

$$\log [(x)(x + 2)] = \log 9^{\frac{1}{2}}$$
$$x^2 + 2x = 3$$
$$x^2 + 2x - 3 = 0$$
$$(x + 3)(x - 1) = 0$$
$$x = -3, 1 \qquad (x \neq -3)$$
$$x = \mathbf{1}$$

29. (59) $8^{2\log_8 3 - \log_8 4} + \frac{9}{4}\log_8 8^{\frac{1}{3}}$

$$= 8^{\log_8 \frac{3^2}{4}} + \log_8 8^{\frac{1}{3}\cdot\frac{9}{4}}$$
$$= \frac{9}{4} + \frac{3}{4} = \frac{12}{4} = \mathbf{3}$$

30. (67) (a) $\text{antilog}_5 (-2) = 5^{-2} = \mathbf{\frac{1}{25}}$

(b) $\text{antilog}_{\frac{1}{5}} (-2) = \left(\frac{1}{5}\right)^{-2} = \mathbf{25}$

Problem Set 93

1. (92) $P(G \cup R) = P(G) + P(R) - P(G \cap R)$

$$= \frac{4}{9} + \frac{5}{9} - \frac{2}{9} = \mathbf{\frac{7}{9}}$$

2. (88) $A_0 = 400,\ A_{30} = 300$

$$A_t = A_0 e^{kt}$$
$$300 = 400e^{30k}$$
$$0.75 = e^{30k}$$
$$\ln 0.75 = 30k$$
$$k = \frac{\ln 0.75}{30}$$
$$k = -0.0096$$
$$\mathbf{A_t = 400e^{-0.0096t}}$$
$$200 = 400e^{-0.0096t}$$
$$0.5 = e^{-0.0096t}$$
$$\ln 0.5 = -0.0096t$$
$$t = \mathbf{72.20\ hr}$$

3. (88) $A_0 = 10,\ A_{40} = 80$

$$A_t = A_0 e^{kt}$$
$$80 = 10e^{40k}$$
$$8 = e^{40k}$$
$$\ln 8 = 40k$$
$$k = 0.051986$$
$$A_t = 10e^{0.051986t}$$
$$A_{120} = 10e^{0.051986(120)}$$
$$A_{120} = 10(512)$$
$$A_{120} = \mathbf{5120}$$

4. (85) $R_L = \frac{1}{12},\ R_B = 1,\ T_L = T_B$

Big hand:

$$R_B T_B = S + 25$$
$$(1)T = S + 25$$
$$S = T - 25$$

Little hand:

$$R_L T_L = S$$

$$S = \frac{1}{12}T$$

$$\frac{1}{12}T = T - 25$$

$$\frac{11}{12}T = 25$$

$$T = \frac{300}{11} = 27\frac{3}{11} \text{ min}$$

$$\mathbf{4{:}27\frac{3}{11}}$$

5. (92)

$$_nP_r = \frac{n!}{(n-r)!}$$

$$_9P_6 = \frac{9!}{(9-6)!}$$

$$\mathbf{_9P_6 = \frac{9!}{3!} = 60{,}480}$$

$$_nC_r = \frac{n!}{r!(n-r)!}$$

$$_9C_6 = \frac{9!}{6!(9-6)!}$$

$$\mathbf{_9C_6 = \frac{9!}{6!3!} = 84}$$

6. (93)

$$\frac{1 + \sin B}{\cos B} \cdot \frac{(1 - \sin B)}{(1 - \sin B)} = \frac{1 - \sin^2 B}{\cos B(1 - \sin B)}$$

$$= \frac{\cos^2 B}{\cos B(1 - \sin B)} = \frac{\cos B}{1 - \sin B}$$

7. (93)

$$\frac{\tan B + 1}{\tan B - 1} = \frac{\frac{\sin B}{\cos B} + 1}{\frac{\sin B}{\cos B} - 1} \cdot \frac{\frac{1}{\sin B}}{\frac{1}{\sin B}}$$

$$= \frac{\frac{1}{\cos B} + \frac{1}{\sin B}}{\frac{1}{\cos B} - \frac{1}{\sin B}} = \frac{\sec B + \csc B}{\sec B - \csc B}$$

8. (93)

$$(y \sin\theta + x\cos\theta)^2 + (y\cos\theta - x\sin\theta)^2$$

$$= y^2\sin^2\theta + 2yx\sin\theta\cos\theta + x^2\cos^2\theta$$
$$+ y^2\cos^2\theta - 2yx\cos\theta\sin\theta + x^2\sin^2\theta$$
$$= y^2\sin^2\theta + y^2\cos^2\theta + x^2\cos^2\theta + x^2\sin^2\theta$$
$$= y^2(\sin^2\theta + \cos^2\theta) + x^2(\cos^2\theta + \sin^2\theta)$$
$$= x^2 + y^2$$

9. (91)

①	②	③	④	⑤
2,	___,	___,	___,	162
a_1,	a_1r,	a_1r^2,	a_1r^3,	a_1r^4

$$a_1r^4 = 162$$
$$2r^4 = 162$$
$$r^4 = 81$$
$$r = \pm 3$$

$a_1r = 2(3) = 6$ or $2(-3) = -6$

$a_1r^2 = 2(3)^2 = 18$ or $2(-3)^2 = 18$

$a_1r^3 = 2(3)^3 = 54$ or $2(-3)^3 = -54$

6, 18, 54 or −6, 18, −54

10. (86) $a_5 = 12$, $a_{13} = -4$

$$\begin{cases} \text{(a)}\ a_1 + 4d = 12 \\ \text{(b)}\ a_1 + 12d = -4 \end{cases}$$

$$\begin{array}{rrcrcr} \text{(a)} & a_1 & + & 4d & = & 12 \\ -\text{(b)} & -a_1 & - & 12d & = & 4 \\ \hline & & & -8d & = & 16 \\ & & & d & = & -2 \end{array}$$

(a) $a_1 + 4(-2) = 12$

$$a_1 = 20$$

20, 18, 16, 14

11. (90) Using $\sin^2 A = 1 - \cos^2 A$, we find:

$$\cos(A + B) = \cos A\cos B - \sin A\sin B$$
$$\cos(A + A) = \cos A\cos A - \sin A\sin A$$
$$\cos 2A = \cos^2 A - \sin^2 A$$
$$\cos 2A = \cos^2 A - (1 - \cos^2 A)$$
$$\cos 2A = 2\cos^2 A - 1$$
$$2\cos^2 A = \cos 2A + 1$$
$$\cos^2 A = \frac{\cos 2A + 1}{2}$$
$$\cos A = \pm\sqrt{\frac{\cos 2A + 1}{2}}$$
$$\cos\frac{A}{2} = \pm\sqrt{\frac{\cos A + 1}{2}}$$

12. (87)

$$\sin\left(x - \frac{\pi}{4}\right) = \sin x\cos\frac{\pi}{4} - \cos x\sin\frac{\pi}{4}$$
$$= (\sin x)\frac{\sqrt{2}}{2} - (\cos x)\frac{\sqrt{2}}{2}$$
$$= \mathbf{\frac{\sqrt{2}}{2}(\sin x - \cos x)}$$

13. (87)
$$\sin 285° = \sin(240° + 45°)$$
$$= \sin 240° \cos 45° + \cos 240° \sin 45°$$
$$= \left(-\frac{\sqrt{3}}{2}\right)\left(\frac{\sqrt{2}}{2}\right) + \left(-\frac{1}{2}\right)\left(\frac{\sqrt{2}}{2}\right)$$
$$= -\frac{\sqrt{6}}{4} - \frac{\sqrt{2}}{4} = \mathbf{\frac{-\sqrt{6} - \sqrt{2}}{4}}$$

14. (89)
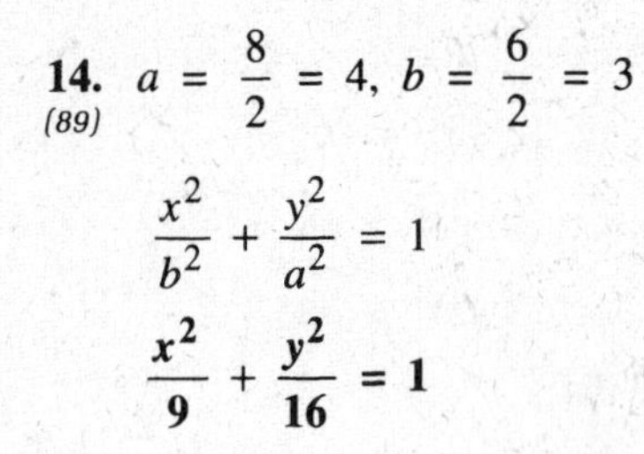
$$a = \frac{8}{2} = 4,\ b = \frac{6}{2} = 3$$
$$\frac{x^2}{b^2} + \frac{y^2}{a^2} = 1$$
$$\mathbf{\frac{x^2}{9} + \frac{y^2}{16} = 1}$$

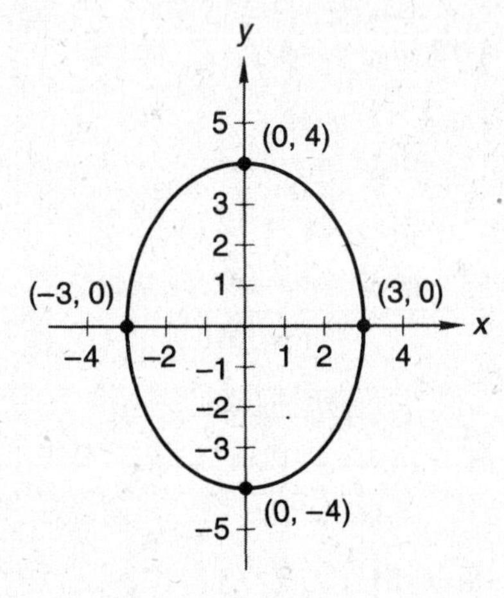

15. (85)
$$2\sin^2 x + 7\cos x + 2 = 0$$
$$2(1 - \cos^2 x) + 7\cos x + 2 = 0$$
$$2\cos^2 x - 7\cos x - 4 = 0$$
$$(2\cos x + 1)(\cos x - 4) = 0$$

$\cos x = -\frac{1}{2}$ $\qquad$ $\cos x = 4$

$x = \mathbf{120°, 240°}$ $\qquad$ no answer

16. (52)
$$\sqrt{3}\cot 3x + 1 = 0$$
$$\cot 3x = -\frac{1}{\sqrt{3}}$$
$$3x = 120°, 300°, 480°, 660°, 840°, 1020°$$
$$x = \mathbf{40°, 100°, 160°, 220°, 280°, 340°}$$

17. (84)
$$y = 2 + 5\sin\frac{5}{4}(x + 38°)$$
$$\text{Period} = \frac{360°}{\frac{5}{4}} = 288°$$
Phase angle = −38°

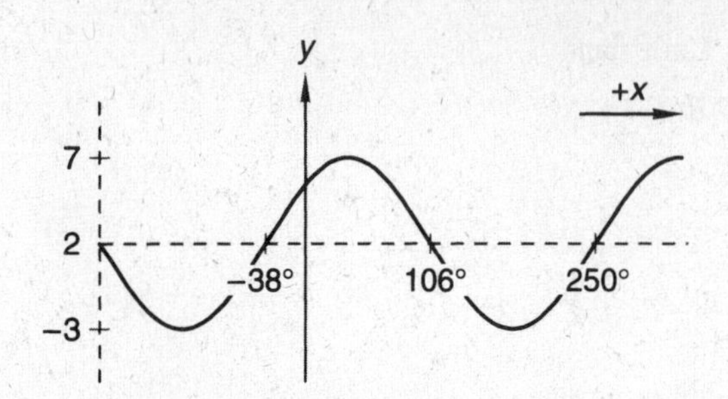

18. (84)
$$y = 4 + 6\cos\frac{1}{2}\left(x - \frac{3\pi}{2}\right)$$
$$\text{Period} = \frac{2\pi}{\frac{1}{2}} = 4\pi$$
$$\text{Phase angle} = \frac{3\pi}{2}$$

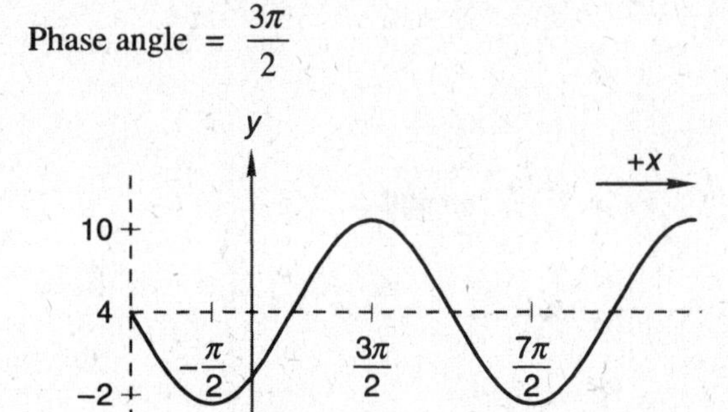

19. (81)
$$a^2 = 8^2 + 9^2 - 2(8)(9)\cos 30°$$
$$a^2 = 20.2923$$
$$a = \mathbf{4.50\ cm}$$

20. (79)
$$(32 \text{ cis } 20°)^{\frac{1}{5}} = 2 \text{ cis}\left(4° + n\frac{360°}{5}\right)$$
$$= \mathbf{2\ cis\ 4°, 2\ cis\ 76°, 2\ cis\ 148°, 2\ cis\ 220°, 2\ cis\ 292°}$$

21. (78)
$$9x^2 - 4y^2 = 72$$
$$\mathbf{\frac{x^2}{8} - \frac{y^2}{18} = 1}$$

Let $y = 0$ $\qquad$ Let $x = 0$

$\frac{x^2}{8} = 1$ $\qquad$ $\frac{y^2}{18} = 1$

$x^2 = 8$ $\qquad$ $y^2 = -18$

$x = \pm 2\sqrt{2}$ $\qquad$ $y = \pm 3\sqrt{2}i$

Vertices = $\mathbf{(2\sqrt{2}, 0), (-2\sqrt{2}, 0)}$

Asymptotes: $\mathbf{y = \frac{3}{2}x;\ y = -\frac{3}{2}x}$

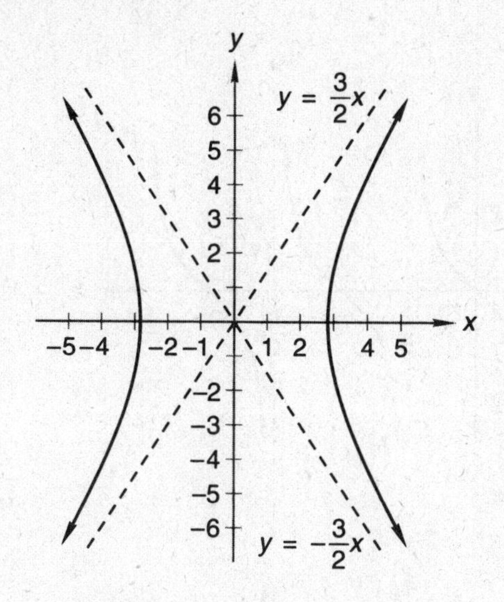

22. (74)

$$x = \frac{\begin{vmatrix} 9 & 3 \\ -4 & -2 \end{vmatrix}}{\begin{vmatrix} 1 & 3 \\ 7 & -2 \end{vmatrix}} = \frac{-18 + 12}{-2 - 21} = \frac{-6}{-23}$$

$$y = \frac{\begin{vmatrix} 1 & 9 \\ 7 & -4 \end{vmatrix}}{\begin{vmatrix} 1 & 3 \\ 7 & -2 \end{vmatrix}} = \frac{-4 - 63}{-23} = \frac{-67}{-23}$$

$$\mathbf{x = \frac{6}{23}; \ y = \frac{67}{23}}$$

23. (58) $x = -1$

Equation of the perpendicular line: $y = -7$

Point of intersection: $(-1, -7)$

$(-1, -7)$ and $(3, -7)$

$$D = \sqrt{[3 - (-1)]^2 + [-7 - (-7)]^2}$$
$$= \sqrt{4^2 + 0} = \mathbf{4}$$

24. (59)

$$\frac{3}{4}\log_8 16 + \log_8 (3x - 2) = 2$$
$$\log_8 \left(16^{\frac{3}{4}}\right)(3x - 2) = 2$$
$$8(3x - 2) = 8^2$$
$$24x - 16 = 64$$
$$24x = 80$$
$$x = \mathbf{\frac{10}{3}}$$

25. (82)

$$27^{2x+3} = 3^{x-4}$$
$$3^{3(2x+3)} = 3^{x-4}$$
$$6x + 9 = x - 4$$
$$5x = -13$$
$$x = \mathbf{-\frac{13}{5}}$$

26. (59)

$$\frac{1}{4}\log_3 81 - \log_3 (4x - 1) = \log_3 27$$
$$\log_3 \frac{81^{\frac{1}{4}}}{(4x - 1)} = \log_3 27$$
$$\frac{3}{4x - 1} = 27$$
$$3 = 108x - 27$$
$$108x = 30$$
$$x = \frac{30}{108} = \mathbf{\frac{5}{18}}$$

27. (59)

$$e^{-\ln 3 + \frac{1}{2}\ln 9 - \frac{1}{3}\ln 8} - \ln e$$
$$= e^{\ln \frac{(3^{-1})(9)^{\frac{1}{2}}}{8^{\frac{1}{3}}}} - \ln e$$
$$= e^{\ln \frac{\frac{1}{3}(3)}{2}} - \ln e$$
$$= \frac{1}{2} - 1 = \mathbf{-\frac{1}{2}}$$

28. (67)

(a) $\text{antilog}_6 2 = 6^2 = \mathbf{36}$

(b) $\text{antilog}_6 (-2) = 6^{-2} = \mathbf{\frac{1}{36}}$

29. (46) $x^2 + 9x + 35 = 0$

$$x = \frac{-9 \pm \sqrt{9^2 - 4(1)(35)}}{2(1)}$$
$$x = \frac{-9 \pm \sqrt{-59}}{2}$$
$$x = -\frac{9}{2} \pm \frac{\sqrt{59}}{2}i$$

$$\mathbf{\left(x + \frac{9}{2} + \frac{\sqrt{59}}{2}i\right)\left(x + \frac{9}{2} - \frac{\sqrt{59}}{2}i\right)}$$

30. (70) $z = \dfrac{x - \mu}{\sigma} = \dfrac{92 - 75}{10} = \dfrac{17}{10} = 1.7$

Percentile = 0.9554

$1 - 0.9554 = 0.0446$

4.46%

Problem Set 94

1. (92) $P(G \cup W) = P(G) + P(W) - P(G \cap W)$

$= \dfrac{4}{10} + \dfrac{7}{10} - \dfrac{3}{10} = \dfrac{8}{10} = \mathbf{\dfrac{4}{5}}$

2. (88) $A_0 = 4,\ A_{20} = 200$

$A_t = A_0 e^{kt}$

$200 = 4e^{20k}$

$50 = e^{20k}$

$\ln 50 = 20k$

$k = 0.20$

$\mathbf{A_t = 4e^{0.20t}}$

3. (88) $A_0 = 1,\ A_{100} = 0.95817$

$A_t = A_0 e^{kt}$

$0.95817 = 1e^{100k}$

$\ln 0.95817 = 100k$

$k = -0.00042730$

$A_t = e^{-0.00042730t}$

$0.5 = e^{-0.00042730t}$

$\ln 0.5 = -0.00042730t$

$t = \mathbf{1622.15\ yr}$

4. (55) $\dfrac{10!}{2!2!} = \mathbf{907{,}200}$

5. (44) $RWT = \text{jobs}$

$R(X)(9) = P$

$R = \dfrac{P}{9X}\ \dfrac{\text{jobs}}{\text{man-day}}$

$RWT = \text{jobs}$

$\left(\dfrac{P}{9X}\right)(Y)T = 20$

$T = \dfrac{(20)(9X)}{PY} = \mathbf{\dfrac{180X}{PY}\ days}$

6. (94) (a) $y = \tan x$

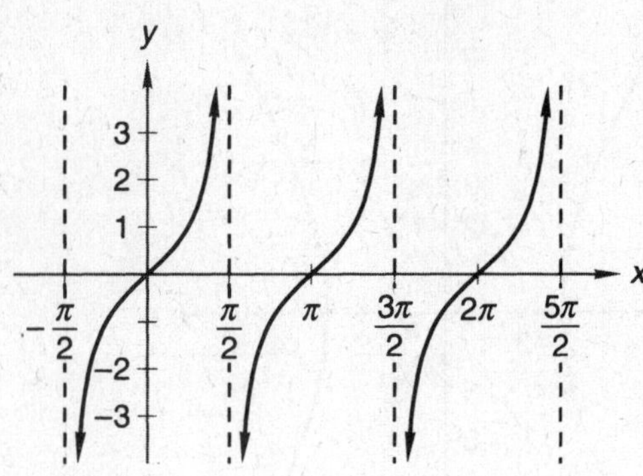

(b) $y = \cot x$

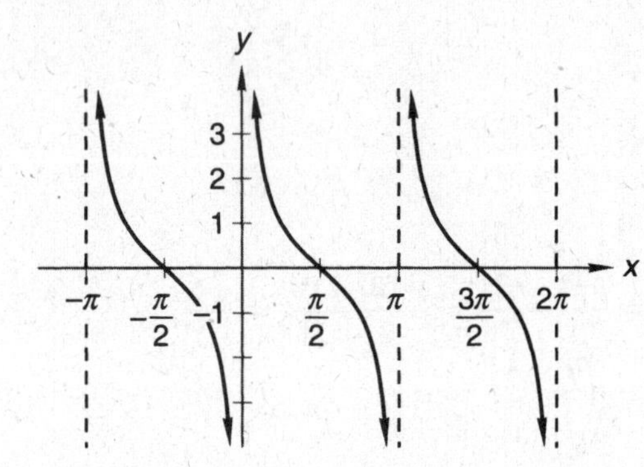

7. (94) (a) Centerline = 4

Amplitude = 4

From the graph shown, we see that the function is sec.

$\mathbf{y = 4 + 4 \sec x}$

(b) Centerline = −2

Amplitude = 4

From the graph shown, we see that the function is csc.

$\mathbf{y = -2 + 4 \csc x}$

8. (93) $\dfrac{1 + \cos x}{\sin x} \cdot \dfrac{(1 - \cos x)}{(1 - \cos x)} = \dfrac{1 - \cos^2 x}{\sin x\,(1 - \cos x)}$

$= \dfrac{\sin^2 x}{\sin x\,(1 - \cos x)} = \dfrac{\sin x}{1 - \cos x}$

9. (93) $\dfrac{\tan B + 1}{\tan B - 1} = \dfrac{\dfrac{\sin B}{\cos B} + 1}{\dfrac{\sin B}{\cos B} - 1} \cdot \dfrac{\dfrac{1}{\sin B}}{\dfrac{1}{\sin B}}$

$= \dfrac{\dfrac{1}{\cos B} + \dfrac{1}{\sin B}}{\dfrac{1}{\cos B} - \dfrac{1}{\sin B}} = \dfrac{\sec B + \csc B}{\sec B - \csc B}$

10. (93) $(1 + \sin x)^2 + (1 - \sin x)^2$

$= 1 + 2\sin x + \sin^2 x + 1 - 2\sin x + \sin^2 x$

$= 2 + 2\sin^2 x = 2(1 + \sin^2 x)$

$= 2[1 + (1 - \cos^2 x)] = 2(2 - \cos^2 x)$

$= 4 - 2\cos^2 x$

11. (90) $\tan 2A = \dfrac{\sin(A + A)}{\cos(A + A)}$

$= \dfrac{\sin A \cos A + \cos A \sin A}{\cos A \cos A - \sin A \sin A}$

$= \dfrac{\dfrac{\sin A \cos A}{\cos A \cos A} + \dfrac{\cos A \sin A}{\cos A \cos A}}{\dfrac{\cos A \cos A}{\cos A \cos A} - \dfrac{\sin A \sin A}{\cos A \cos A}}$

$= \dfrac{\tan A + \tan A}{1 - (\tan A)(\tan A)} = \dfrac{2\tan A}{1 - \tan^2 A}$

12. (90) $\sin 75° = \sin \dfrac{150°}{2} = \sqrt{\dfrac{1 - \cos 150°}{2}}$

$= \sqrt{\dfrac{1 - \left(-\dfrac{\sqrt{3}}{2}\right)}{2}} = \sqrt{\dfrac{2 + \sqrt{3}}{4}} = \mathbf{\dfrac{\sqrt{2 + \sqrt{3}}}{2}}$

13. (87) $\cos(-285°) = \cos(285°) = \cos(240° + 45°)$

$= \cos 240° \cos 45° - \sin 240° \sin 45°$

$= \left(-\dfrac{1}{2}\right)\left(\dfrac{\sqrt{2}}{2}\right) - \left(-\dfrac{\sqrt{3}}{2}\right)\left(\dfrac{\sqrt{2}}{2}\right) = \mathbf{\dfrac{\sqrt{6} - \sqrt{2}}{4}}$

14. (87) $\cos(\pi - x) = \cos\pi \cos x + \sin\pi \sin x$

$= (-1)\cos x + (0)\sin x = \mathbf{-\cos x}$

15. (85)

$2\cos^2 x + 7\sin x + 2 = 0$

$2(1 - \sin^2 x) + 7\sin x + 2 = 0$

$2\sin^2 x - 7\sin x - 4 = 0$

$(2\sin x + 1)(\sin x - 4) = 0$

$\sin x = -\dfrac{1}{2}$ $\qquad \sin x = 4$

$x = \mathbf{\dfrac{7\pi}{6}, \dfrac{11\pi}{6}}$ $\qquad$ no answer

16. (52) $\sqrt{3}\cot\dfrac{x}{2} - 1 = 0$

$\cot\dfrac{x}{2} = \dfrac{1}{\sqrt{3}}$

$\dfrac{x}{2} = \dfrac{\pi}{3}$

$x = \mathbf{\dfrac{2\pi}{3}}$

17. (84) $y = -3 + 2\sin\dfrac{5}{3}(x - 27°)$

$\text{Period} = \dfrac{360°}{\dfrac{5}{3}} = 216°$

Phase angle $= 27°$

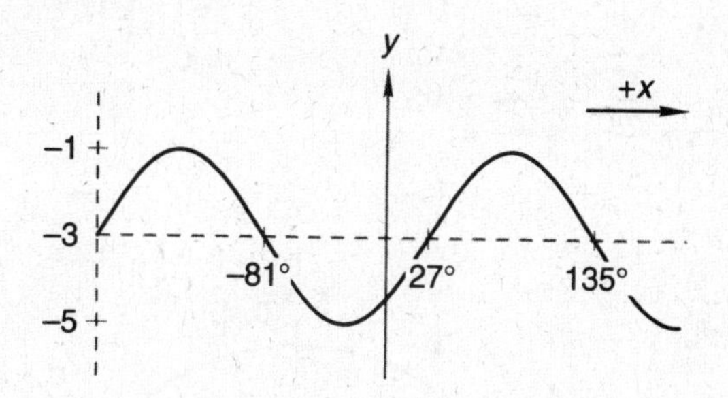

18. (81) $c = \sqrt{6^2 + 8^2 - 2(6)(8)\cos 60°} = \mathbf{7.21\ ft}$

19. (79) $(81 \text{ cis } 24°)^{\frac{1}{4}} = 3 \text{ cis}\left(6° + n\dfrac{360°}{4}\right)$

$= \mathbf{3\ cis\ 6°, 3\ cis\ 96°, 3\ cis\ 186°, 3\ cis\ 276°}$

20. (91)

①	②	③	④
–2,	___,	___,	16
a_1,	a_1r,	a_1r^2,	a_1r^3

$a_1r^3 = 16$

$-2r^3 = 16$

$r^3 = -8$

$r = -2$

$a_1r = -2 \cdot (-2) = 4$

$a_1r^2 = -2 \cdot (-2)^2 = -8$

4, –8

21. (86) $a_8 = 19,\ a_{10} = 25$

(a) $a_1 + 7d = 19$
(b) $a_1 + 9d = 25$

(a) $\quad a_1 + 7d = 19$
–(b) $\quad -a_1 - 9d = -25$

$-2d = -6$

$d = 3$

(a) $a_1 + 7d = 19$

$a_1 + 7(3) = 19$

$a_1 = -2$

–2, 1, 4

22. (89) $a = \frac{10}{2} = 5$

$b = \frac{8}{2} = 4$

$\frac{x^2}{a^2} + \frac{y^2}{b^2} = 1$

$\mathbf{\frac{x^2}{25} + \frac{y^2}{16} = 1}$

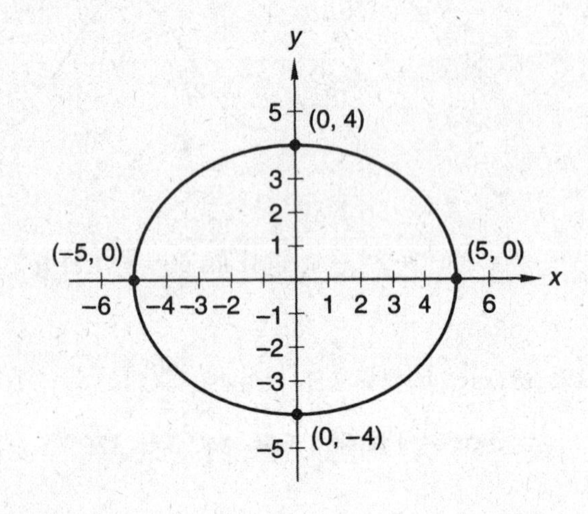

23. (73)

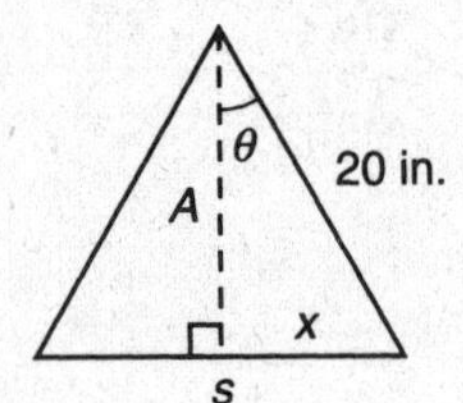

$\theta = \frac{1}{2}\left(\frac{360°}{6}\right) = 30°$

$x = 20 \sin 30° = 10 \text{ in.}$

$s = 2x = 2(10) = 20 \text{ in.}$

Perimeter = 6(20)

Perimeter = 120 in.

$A = 20 \cos 30° = 20\left(\frac{\sqrt{3}}{2}\right) = 10\sqrt{3}$

$\text{Area}_\Delta = \frac{1}{2}(10)(10\sqrt{3}) = 50\sqrt{3} \text{ in.}^2$

$\text{Area} = (12)(50\sqrt{3})$

Area = 1039.23 in.²

24. (58) $3x + 4y = 2$

$y = -\frac{3}{4}x + \frac{1}{2}$

Equation of the perpendicular line:

$y = \frac{4}{3}x + b$

$2 = \frac{4}{3}(-1) + b$

$b = \frac{10}{3}$

$y = \frac{4}{3}x + \frac{10}{3}$

Point of intersection:

$-\frac{3}{4}x + \frac{1}{2} = \frac{4}{3}x + \frac{10}{3}$

$\frac{25}{12}x = -\frac{17}{6}$

$x = -\frac{34}{25}$

$y = -\frac{3}{4}x + \frac{1}{2} = -\frac{3}{4}\left(-\frac{34}{25}\right) + \frac{1}{2} = \frac{76}{50} = \frac{38}{25}$

$\left(-\frac{34}{25}, \frac{38}{25}\right)$ and $(-1, 2)$

$D = \sqrt{\left[-1 - \left(-\frac{34}{25}\right)\right]^2 + \left(2 - \frac{38}{25}\right)^2}$

$= \sqrt{\left(\frac{9}{25}\right)^2 + \left(\frac{12}{25}\right)^2} = \sqrt{\frac{225}{625}} = \sqrt{\frac{9}{25}} = \mathbf{\frac{3}{5}}$

25. (82)

$12^{2x-2} = 4^{3x+1}$

$(2x - 2) \log 12 = (3x + 1) \log 4$

$(2x - 2)(1.0792) = (3x + 1)(0.60206)$

$2.1584x - 2.1584 = 1.80618x + 0.60206$

$0.35222x = 2.76046$

$x = \mathbf{7.84}$

26. (59) $\frac{2}{3} \log_3 27 - \log_3 (2x - 1) = 1$

$\log_3 \frac{27^{\frac{2}{3}}}{(2x - 1)} = 1$

$\frac{9}{2x - 1} = 3^1$

$6x - 3 = 9$

$6x = 12$

$x = \mathbf{2}$

27. (59) $\frac{1}{3}\log_5 8 - 2\log_5 x = 0$

$$\log_5 \frac{8^{\frac{1}{3}}}{x^2} = 0$$

$$\frac{2}{x^2} = 5^0$$

$$x^2 = 2$$

$$x = \sqrt{2}$$

28. (67) (a) $\text{antilog}_2(-3) = 2^{-3} = \mathbf{\frac{1}{8}}$

(b) $\text{antilog}_3(-2) = 3^{-2} = \mathbf{\frac{1}{9}}$

29. (46) $x^2 + 10x + 30 = 0$

$$x = \frac{-10 \pm \sqrt{10^2 - 4(1)(30)}}{2(1)}$$

$$x = \frac{-10 \pm 2\sqrt{5}i}{2}$$

$$x = -5 \pm \sqrt{5}i$$

$$\mathbf{(x + 5 - \sqrt{5}i)(x + 5 + \sqrt{5}i)}$$

30. (21,59) $g(2\ln 3 - \ln 2) = e^{2\ln 3 - \ln 2} = e^{\ln\frac{9}{2}} = \mathbf{\frac{9}{2}}$

Problem Set 95

1. (92) $P(G \cup R) = P(G) + P(R) - P(G \cap R)$

$$\frac{2}{3} = \frac{1}{2} + \frac{1}{3} - P(G \cap R)$$

$$P(G \cap R) = \frac{1}{2} + \frac{1}{3} - \frac{2}{3} = \mathbf{\frac{1}{6}}$$

2. (83) $\frac{3}{10} \cdot \frac{3}{10} \cdot \frac{2}{9} = \frac{18}{900} = \mathbf{\frac{1}{50}}$

3. (88) $A_0 = 600,\ A_{60} = 1000$

$$A_t = A_0 e^{kt}$$

$$1000 = 600e^{60k}$$

$$1.6667 = e^{60k}$$

$$\ln 1.6667 = 60k$$

$$k = 0.0085138$$

$$A_t = 600e^{0.0085138t}$$

$$5000 = 600e^{0.0085138t}$$

$$8.3333 = e^{0.0085138t}$$

$$\ln 8.3333 = 0.0085138t$$

$$t = \mathbf{249.04\ min}$$

4. (88) $A_0 = 600,\ A_{60} = 580$

$$A_t = A_0 e^{kt}$$

$$580 = 600e^{60k}$$

$$0.96667 = e^{60k}$$

$$\ln 0.96667 = 60k$$

$$k = -0.00056503$$

$$A_t = 600e^{-0.00056503t}$$

$$300 = 600e^{-0.00056503t}$$

$$0.5 = e^{-0.00056503t}$$

$$\ln 0.5 = -0.00056503t$$

$$t = \mathbf{1226.75\ min}$$

5. (18) Alcohol_1 + alcohol added = alcohol final

$$0.2(140) + A = 0.44(140 + A)$$

$$28 + A = 61.6 + 0.44A$$

$$0.56A = 33.6$$

$$A = \mathbf{60\ liters}$$

6. (95) $3 + 4i$

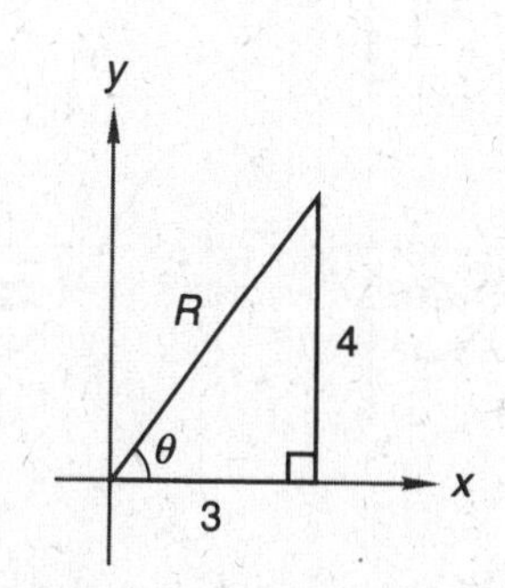

$$R = \sqrt{3^2 + 4^2} = 5$$

$$\tan\theta = \frac{4}{3}$$

$$\theta = 53.13°$$

$$3 + 4i = 5 \text{ cis } 53.13°$$

$$(5 \text{ cis } 53.13°)^{\frac{1}{4}} = 5^{\frac{1}{4}} \text{ cis}\left(\frac{53.13°}{4} + n\frac{360°}{4}\right)$$

= 1.50 cis 13.28°, 1.50 cis 103.28°, 1.50 cis 193.28°, 1.50 cis 283.28°

7. $2 + 3i$
(95)

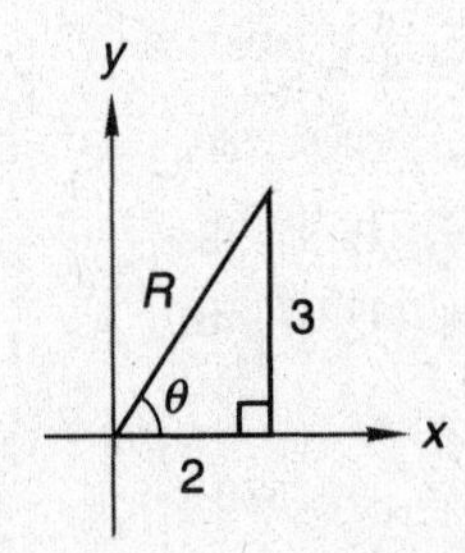

$$R = \sqrt{2^2 + 3^2} = \sqrt{13}$$

$$\tan \theta = \frac{3}{2}$$

$$\theta = 56.31°$$

$$2 + 3i = \sqrt{13} \text{ cis } 56.31°$$

$$\left(\sqrt{13} \text{ cis } 56.31°\right)^{\frac{1}{3}} = \left(13^{\frac{1}{2}}\right)^{\frac{1}{3}} \text{ cis}\left(\frac{56.31°}{3} + n\frac{360°}{3}\right)$$

$= \textbf{1.53 cis 18.77°, 1.53 cis 138.77°, 1.53 cis 258.77°}$

8. $y = 3 + 11 \cos \frac{3}{2}(x - 100°)$
(84)

$$\text{Period} = \frac{360°}{\frac{3}{2}} = 240°$$

Phase angle = 100°

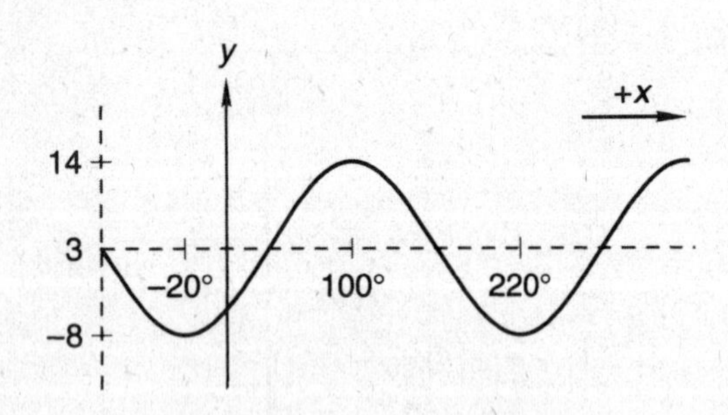

9. (a) $y = \sec x$
(94)

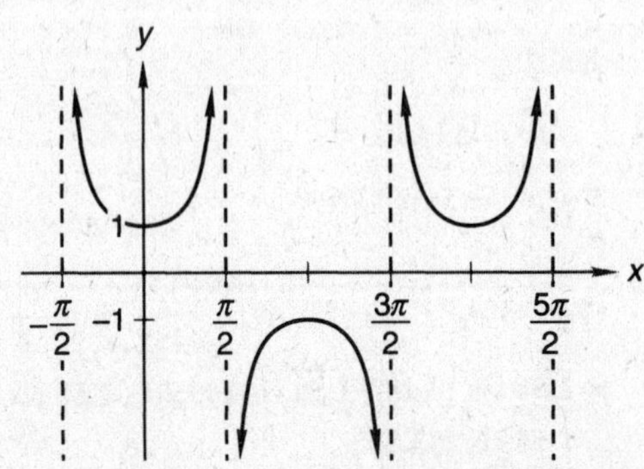

(b) $y = \csc x$

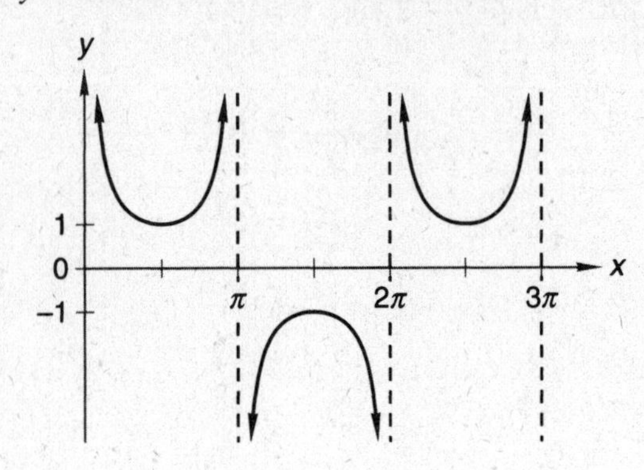

10. (a) $\boldsymbol{y = \tan \theta}$
(94)

(b) $\boldsymbol{y = \cot \theta}$

11.
(93)

$$\frac{\sin x}{1 + \cos x} + \frac{1 + \cos x}{\sin x} = \frac{\sin^2 x + (1 + \cos x)^2}{(1 + \cos x)\sin x}$$

$$= \frac{\sin^2 x + 1 + 2\cos x + \cos^2 x}{(1 + \cos x)\sin x}$$

$$= \frac{2 + 2\cos x}{(1 + \cos x)\sin x} = \frac{2(1 + \cos x)}{(1 + \cos x)\sin x} = 2 \csc x$$

12. $(1 + \tan x)^2 = 1 + 2\tan x + \tan^2 x$
(93)

$$= \sec^2 x + 2\tan x$$

13.
(93)

$$\frac{\sin^4 x - \cos^4 x}{2\sin^2 x - 1}$$

$$= \frac{(\sin^2 x + \cos^2 x)(\sin^2 x - \cos^2 x)}{2\sin^2 x - 1}$$

$$= \frac{\sin^2 x - (1 - \sin^2 x)}{2\sin^2 x - 1} = \frac{2\sin^2 x - 1}{2\sin^2 x - 1} = 1$$

14. Using $\sin^2 A = 1 - \cos^2 A$ and $A = \frac{x}{2}$, we find:
(90)

$$\cos(A + B) = \cos A \cos B - \sin A \sin B$$

$$\cos(A + A) = \cos^2 A - \sin^2 A$$

$$\cos 2A = \cos^2 A - (1 - \cos^2 A)$$

$$\cos 2A = 2\cos^2 A - 1$$

$$\cos^2 A = \frac{\cos 2A + 1}{2}$$

$$\cos A = \pm\sqrt{\frac{\cos 2A + 1}{2}}$$

$$\cos \frac{x}{2} = \pm\sqrt{\frac{\cos x + 1}{2}}$$

15. (87) $\cos 285° = \cos(240° + 45°)$

$= \cos 240° \cos 45° - \sin 240° \sin 45°$

$= \left(-\frac{1}{2}\right)\left(\frac{\sqrt{2}}{2}\right) - \left(-\frac{\sqrt{3}}{2}\right)\left(\frac{\sqrt{2}}{2}\right)$

$= -\frac{\sqrt{2}}{4} + \frac{\sqrt{6}}{4} = \mathbf{\frac{\sqrt{6} - \sqrt{2}}{4}}$

16. (87) $\sin\left(x + \frac{\pi}{4}\right) = \sin x \cos\frac{\pi}{4} + \cos x \sin\frac{\pi}{4}$

$= \sin x\left(\frac{\sqrt{2}}{2}\right) + \cos x\left(\frac{\sqrt{2}}{2}\right) = \mathbf{\frac{\sqrt{2}}{2}(\sin x + \cos x)}$

17. (85) $3\tan^2 x + 5\sec x + 1 = 0$

$3(\sec^2 x - 1) + 5\sec x + 1 = 0$

$3\sec^2 x + 5\sec x - 2 = 0$

$(3\sec x - 1)(\sec x + 2) = 0$

$\sec x = \frac{1}{3}$ $\quad$ $\sec x = -2$

no answer $\quad$ $x = \mathbf{120°, 240°}$

18. (52) $-1 + \tan 4x = 0$

$\tan 4x = 1$

$4x = 45°, 225°, 405°, 585°, 765°, 945°, 1125°, 1305°$

$x = \mathbf{11.25°, 56.25°, 101.25°, 146.25°, 191.25°, 236.25°, 281.25°, 326.25°}$

19. (81) $4^2 = 6^2 + 8^2 - 2(6)(8)\cos A$

$-84 = -96\cos A$

$\cos A = \frac{84}{96}$

$A = \mathbf{28.96°}$

20. (92) ${}_nP_r = \frac{n!}{(n - r)!}$

${}_7P_2 = \frac{7!}{(7 - 2)!} = \frac{7!}{5!}$

$\mathbf{{}_7P_2 = 42}$

${}_nC_r = \frac{n!}{r!(n - r)!}$

${}_7C_2 = \frac{7!}{2!(7 - 2)!} = \frac{7!}{2!5!}$

$\mathbf{{}_7C_2 = 21}$

21. (91) ① ② ③ ④

3, __, __, −24

a_1, a_1r, a_1r^2, a_1r^3

$a_1r^3 = -24$

$3r^3 = -24$

$r^3 = -8$

$r = -2$

$a_1r = 3 \cdot (-2) = -6$

$a_1r^2 = 3 \cdot (-2)^2 = 12$

−6, 12

22. (86) $a_4 = 4$, $a_{13} = 28$

$$\begin{cases} \text{(a)} & a_1 + 3d = 4 \\ \text{(b)} & a_1 + 12d = 28 \end{cases}$$

$$\begin{array}{rr} \text{(a)} & a_1 + 3d = 4 \\ -\text{(b)} & -a_1 - 12d = -28 \\ \hline & -9d = -24 \end{array}$$

$d = \frac{24}{9}$

$d = \frac{8}{3}$

(a) $a_1 + 3\left(\frac{8}{3}\right) = 4$

$a_1 = 4 - 8$

$a_1 = -4$

$\mathbf{-4, -\frac{4}{3}, \frac{4}{3}, 4, \frac{20}{3}}$

23. (89) $a = \frac{10}{2} = 5$, $b = \frac{4}{2} = 2$

$\frac{x^2}{a^2} + \frac{y^2}{b^2} = 1$

$\mathbf{\frac{x^2}{25} + \frac{y^2}{4} = 1}$

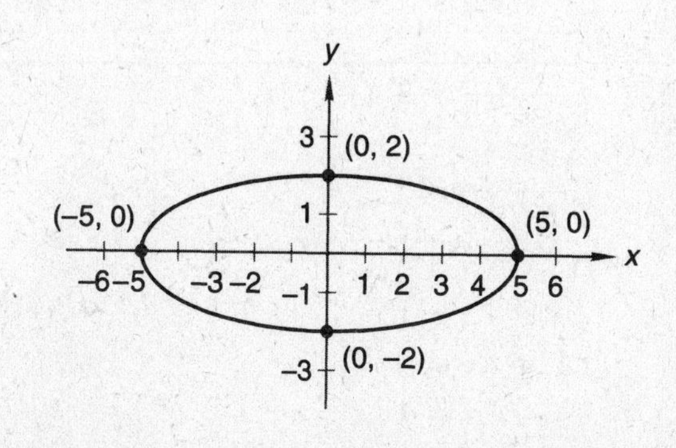

24. (78) $32x^2 - 18y^2 = 288$

$$\frac{x^2}{9} - \frac{y^2}{16} = 1$$

Let $y = 0$ Let $x = 0$

$x = \pm 3$ $y = \pm 4i$

Vertices = (3, 0), (–3, 0)

Asymptotes: $y = \frac{4}{3}x$; $y = -\frac{4}{3}x$

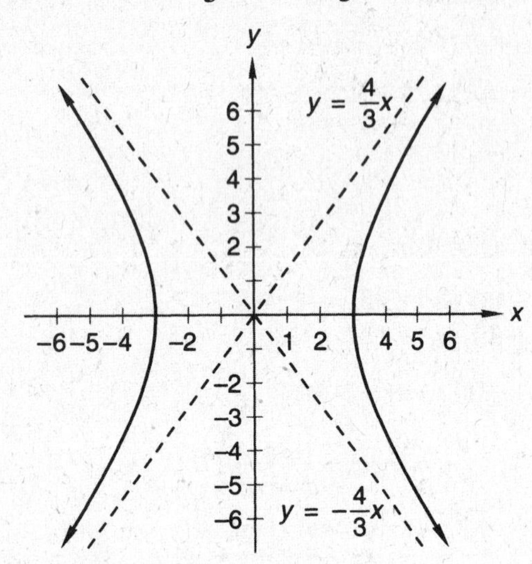

25. (73) Side $= \frac{35 \text{ in.}}{7} = 5$ in.

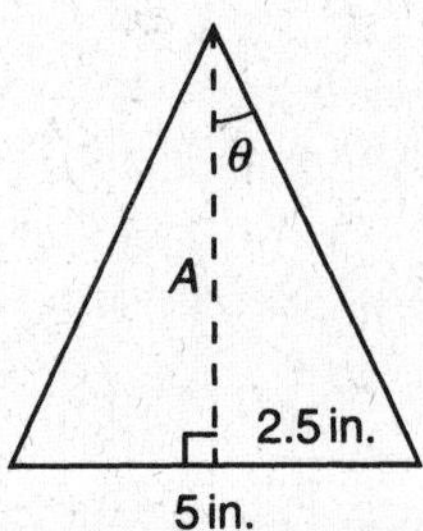

$$\theta = \frac{1}{2}\left(\frac{360°}{7}\right) = 25.71°$$

$$A = \frac{2.5}{\tan 25.71°} = 5.1913 \text{ in.}$$

$$\text{Area}_\Delta = \frac{1}{2}(2.5)(5.1913) = 6.48913 \text{ in.}^2$$

$$\text{Area} = (14)(6.48913) = \mathbf{90.85 \text{ in.}^2}$$

26. (82)

$$\frac{9^{3x-2}}{3^{2x-1}} = 3$$

$$\frac{3^{2(3x-2)}}{3^{2x-1}} = 3$$

$$3^{6x-4-2x+1} = 3^1$$

$$4x - 3 = 1$$

$$4x = 4$$

$$x = \mathbf{1}$$

27. (59) $\frac{1}{3}\log_2 27 - \log_2 (2x - 1) = 2$

$$\log_2 \frac{27^{\frac{1}{3}}}{2x - 1} = 2$$

$$\frac{3}{2x - 1} = 2^2$$

$$3 = 4(2x - 1)$$

$$3 = 8x - 4$$

$$8x = 7$$

$$x = \mathbf{\frac{7}{8}}$$

28. (59)

$$x = \log_{\frac{1}{3}} 18 - \log_{\frac{1}{3}} 6$$

$$x = \log_{\frac{1}{3}} \frac{18}{6}$$

$$x = \log_{\frac{1}{3}} 3$$

$$\left(\frac{1}{3}\right)^x = 3$$

$$x = \mathbf{-1}$$

29. (59) $5^{\log_5 7 - \log_5 3} - \log_5 5^2 = 5^{\log_5 \frac{7}{3}} - \log_5 5^2$

$$= \frac{7}{3} - 2 = \mathbf{\frac{1}{3}}$$

30. (24) $(g \circ f)(x) = \mathbf{\sqrt[3]{x} - 1}$

Problem Set 96

1. (88) $A_0 = 500$

$$A_t = A_0 e^{kt}$$

$$4000 = 500e^{0.005t}$$

$$8 = e^{0.005t}$$

$$\ln 8 = 0.005t$$

$$t = \mathbf{415.89 \text{ units of time}}$$

2. (88) $A_0 = 60{,}000$, $A_5 = 50{,}000$

$$A_t = A_0 e^{kt}$$

$$50{,}000 = 60{,}000e^{5k}$$

$$\ln \frac{5}{6} = 5k$$

$$k = \frac{\ln \frac{5}{6}}{5}$$

$$k = -0.036464$$

$$A_t = A_0 e^{-0.036464t}$$
$$10{,}000 = 60{,}000 e^{-0.036464t}$$
$$\ln \frac{1}{6} = -0.036464t$$
$$t = \mathbf{49.14\ min}$$

3. (85) $R_L = \dfrac{1}{12}$, $R_B = 1$, $T_B = T_L$

Big hand:

$R_B T_B = S + 5$

Little hand:

$R_L T_L = S$

$$R_B T_B = S + 5$$
$$R_B T_B = R_L T_L + 5$$
$$T = \frac{1}{12}T + 5$$
$$\frac{11}{12}T = 5$$
$$T = \frac{60}{11} = \mathbf{5\frac{5}{11}\ min}$$

4. (36) (a) $(B + 4)T_D = D_D$
(b) $(B - 4)T_U = D_U$

$T_D = T_U$

Upstream:

(b) $(B - 4)T_U = D_U$

$$T_U = \frac{3}{B - 4}$$

Downstream:

(a) $(B + 4)T_D = D_D$

$$(B + 4)\left(\frac{3}{B - 4}\right) = 27$$
$$(B + 4)3 = 27(B - 4)$$
$$3B + 12 = 27B - 108$$
$$24B = 120$$
$$B = \mathbf{5\ mph}$$

5. (96)
$$\frac{\cos^4 x - \sin^4 x}{\cos 2x}$$
$$= \frac{(\cos^2 x + \sin^2 x)(\cos^2 x - \sin^2 x)}{(\cos^2 x - \sin^2 x)}$$
$$= \cos^2 x + \sin^2 x = 1$$

6. (96) $(\sin x + \cos x)^2$
$$= \sin^2 x + 2 \sin x \cos x + \cos^2 x$$
$$= 1 + 2 \sin x \cos x = 1 + \sin 2x$$

7. (96)
$$\frac{1 - \cos x}{\sin x} \cdot \frac{(1 + \cos x)}{(1 + \cos x)}$$
$$= \frac{(1 + \cos x)(1 - \cos x)}{(1 + \cos x)\sin x}$$
$$= \frac{1 - \cos^2 x}{(1 + \cos x)\sin x}$$
$$= \frac{\sin^2 x}{(1 + \cos x)\sin x}$$
$$= \frac{\sin x}{1 + \cos x}$$

8. (96)

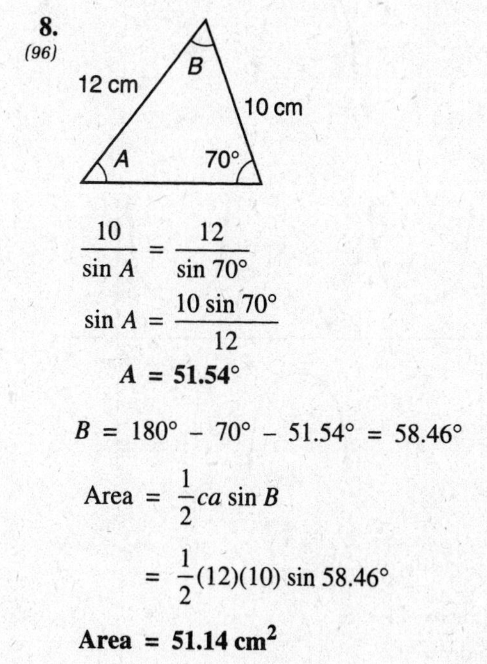

$$\frac{10}{\sin A} = \frac{12}{\sin 70°}$$
$$\sin A = \frac{10 \sin 70°}{12}$$
$$A = \mathbf{51.54°}$$

$$B = 180° - 70° - 51.54° = 58.46°$$

$$\text{Area} = \frac{1}{2}ca \sin B$$
$$= \frac{1}{2}(12)(10) \sin 58.46°$$
$$\textbf{Area} = \mathbf{51.14\ cm^2}$$

9. (94) (a) $y = \tan x$

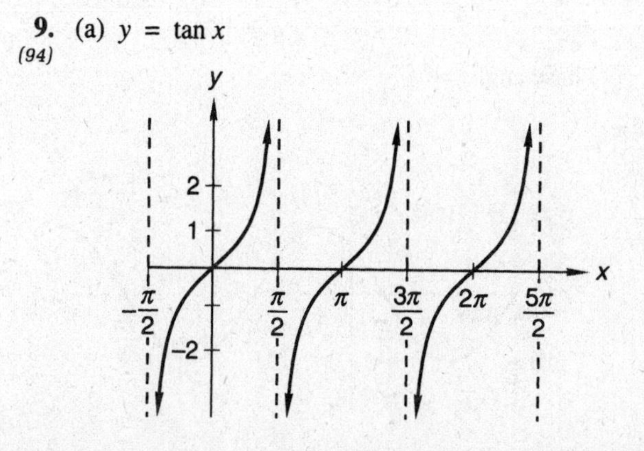

(b) $y = \cot x$

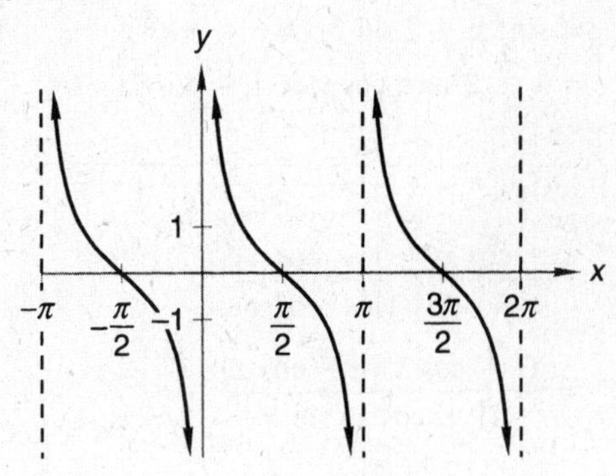

10. (a) $y = \sec x$
(94)

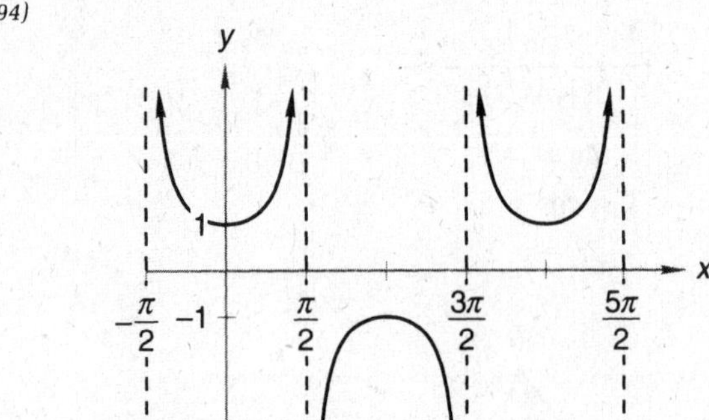

(b) $y = \csc x$

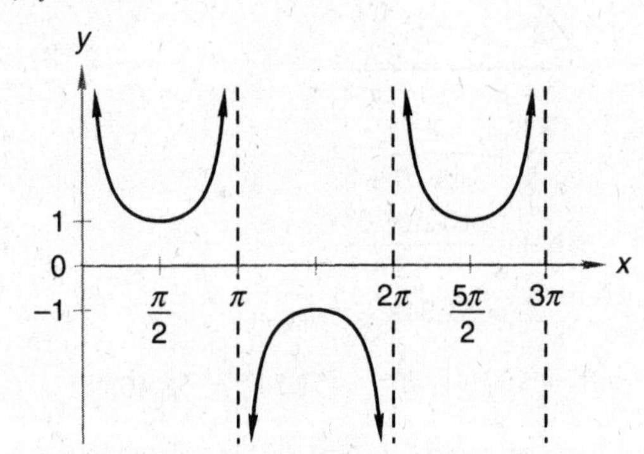

11. $y = 11 + \sin \frac{3}{4}\left(x + \frac{\pi}{3}\right)$
(84)

Centerline $= 11$

Amplitude $= 1$

$$\text{Period} = \frac{2\pi}{\frac{3}{4}} = \frac{8\pi}{3}$$

$$\text{Phase angle} = -\frac{\pi}{3}$$

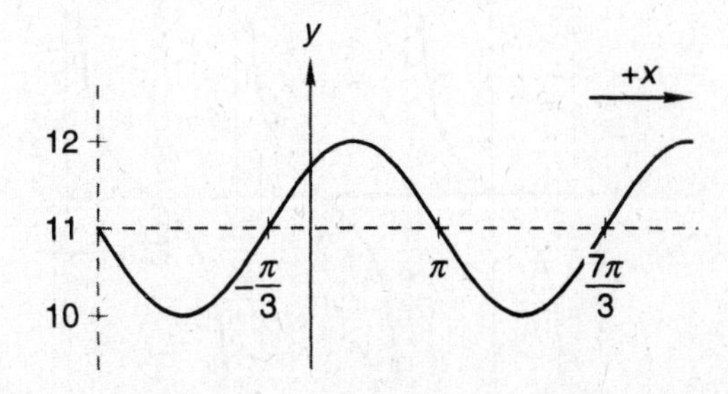

12. $\sin(-15°) = \sin(30° - 45°)$
(87)

$= \sin 30° \cos 45° - \cos 30° \sin 45°$

$$= \left(\frac{1}{2}\right)\left(\frac{\sqrt{2}}{2}\right) - \left(\frac{\sqrt{3}}{2}\right)\left(\frac{\sqrt{2}}{2}\right) = \mathbf{\frac{\sqrt{2} - \sqrt{6}}{4}}$$

13. $\cos\left(x - \frac{3\pi}{2}\right) = \cos x \cos \frac{3\pi}{2} + \sin x \sin \frac{3\pi}{2}$
(87)

$= (\cos x)(0) + \sin x(-1) = \mathbf{-\sin x}$

14.
(85)

$$2 \tan^2 x + 3 \sec x = 0$$
$$2(\sec^2 x - 1) + 3 \sec x = 0$$
$$2 \sec^2 x + 3 \sec x - 2 = 0$$
$$(2 \sec x - 1)(\sec x + 2) = 0$$

$\sec x = \frac{1}{2}$ $\qquad$ $\sec x = -2$

no answer $\qquad$ $x = \mathbf{\frac{2\pi}{3}, \frac{4\pi}{3}}$

15. $\sqrt{3} - 2 \sin 2\theta = 0$
(52)

$$\sin 2\theta = \frac{\sqrt{3}}{2}$$
$$2\theta = \frac{\pi}{3}, \frac{2\pi}{3}, \frac{7\pi}{3}, \frac{8\pi}{3}$$
$$\theta = \mathbf{\frac{\pi}{6}, \frac{\pi}{3}, \frac{7\pi}{6}, \frac{4\pi}{3}}$$

16. $3 + 4i$
(95)

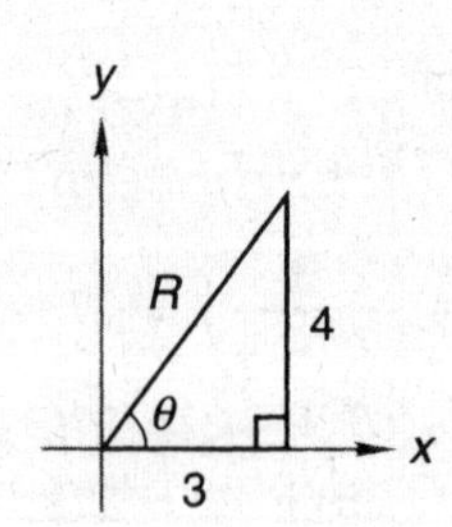

$R = \sqrt{3^2 + 4^2} = 5$

$\tan \theta = \frac{4}{3}$

$\theta = 53.13°$

$3 + 4i = 5 \operatorname{cis} 53.13°$

$$(5 \operatorname{cis} 53.13°)^{\frac{1}{2}} = 5^{\frac{1}{2}} \operatorname{cis}\left(\frac{53.13°}{2} + n\frac{360°}{2}\right)$$

$= \mathbf{2.24 \operatorname{cis} 26.57°, 2.24 \operatorname{cis} 206.57°}$

17. $1 - i$
(95)

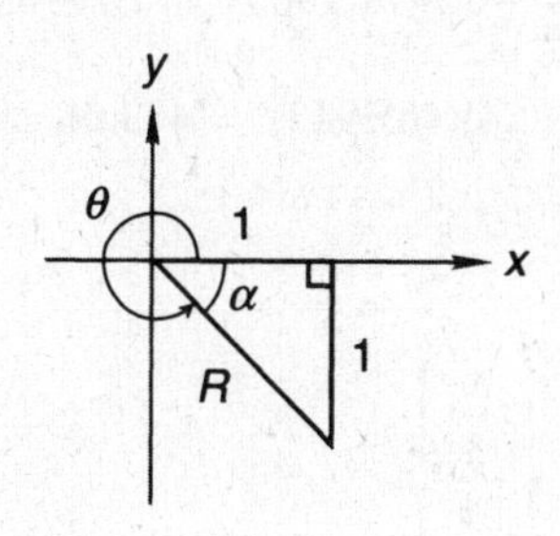

$$R = \sqrt{1^2 + 1^2} = \sqrt{2}$$

$$\tan \alpha = \frac{1}{1}$$

$$\alpha = 45°$$

$$\theta = 360° - 45° = 315°$$

$$1 - i = \sqrt{2} \text{ cis } 315°$$

$$(\sqrt{2} \text{ cis } 315°)^{\frac{1}{3}} = (\sqrt{2})^{\frac{1}{3}} \text{ cis}\left(\frac{315°}{3} + n\frac{360°}{3}\right)$$

$= \mathbf{1.12 \text{ cis } 105°, 1.12 \text{ cis } 225°, 1.12 \text{ cis } 345°}$

18. $_8P_3 = \dfrac{8!}{(8-3)!} = \dfrac{8!}{5!}$
(92)

$\mathbf{_8P_3 = 336}$

$$_8C_3 = \frac{8!}{(8-3)!3!} = \frac{8!}{5!3!}$$

$\mathbf{_8C_3 = 56}$

19. ① ② ③ ④
(91)

①	②	③	④
−5,	___,	___,	625
a_1,	a_1r,	a_1r^2,	a_1r^3

$$a_1r^3 = 625$$

$$(-5)r^3 = 625$$

$$r^3 = -125$$

$$r = -5$$

25, −125

20. $a_3 = -8,\ a_5 = 0$
(86)

$$\begin{cases} \text{(a)} & a_1 + 2d = -8 \\ \text{(b)} & a_1 + 4d = 0 \end{cases}$$

$$\begin{aligned} -\text{(a)} \quad -a_1 - 2d &= 8 \\ \text{(b)} \quad a_1 + 4d &= 0 \\ \hline 2d &= 8 \\ d &= 4 \end{aligned}$$

$$\begin{aligned} \text{(a)} \quad a_1 + 2d &= -8 \\ a_1 + 2(4) &= -8 \\ a_1 &= -16 \end{aligned}$$

−16, −12, −8

21.
(89)

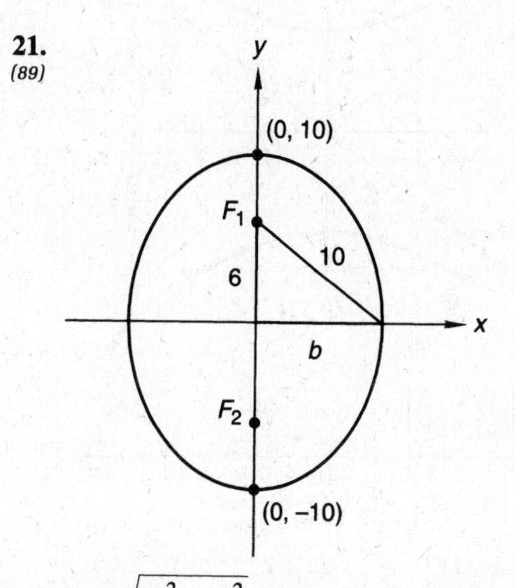

$$b = \sqrt{10^2 - 6^2} = 8$$

$$a = 10$$

$$\frac{x^2}{b^2} + \frac{y^2}{a^2} = 1$$

$$\mathbf{\frac{x^2}{64} + \frac{y^2}{100} = 1}$$

Length of major axis = 20

Length of minor axis = 16

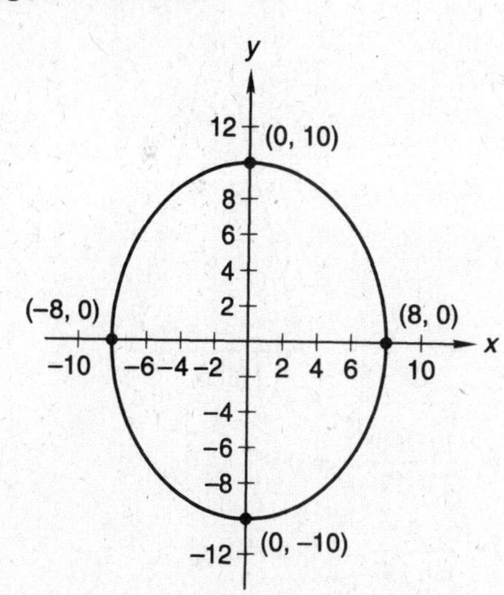

22. (78) $16y^2 - 9x^2 = 144$

$$\frac{y^2}{9} - \frac{x^2}{16} = 1$$

Let $x = 0$ Let $y = 0$

$y = \pm 3$ $x = \pm 4i$

Vertices = (0, 3), (0, –3)

Asymptotes: $y = \frac{3}{4}x$; $y = -\frac{3}{4}x$

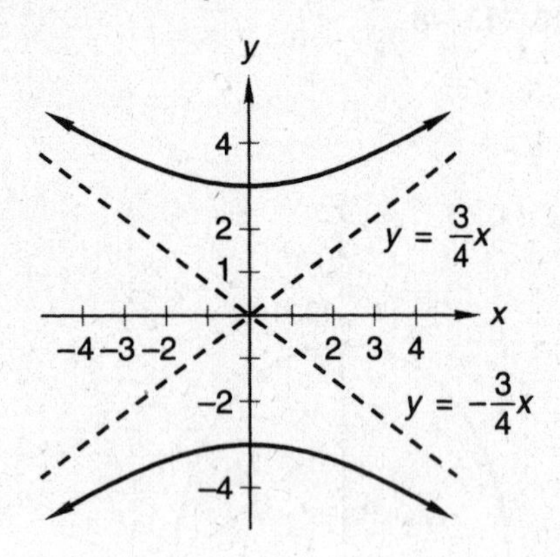

23. (74) $x = \dfrac{\begin{vmatrix} 11 & 3 \\ -7 & -2 \end{vmatrix}}{\begin{vmatrix} 5 & 3 \\ 3 & -2 \end{vmatrix}} = \dfrac{-22 + 21}{-10 - 9} = \dfrac{-1}{-19} = \dfrac{1}{19}$

$$y = \frac{\begin{vmatrix} 5 & 11 \\ 3 & -7 \end{vmatrix}}{\begin{vmatrix} 5 & 3 \\ 3 & -2 \end{vmatrix}} = \frac{-35 - 33}{-10 - 9} = \frac{-68}{-19} = \frac{68}{19}$$

$$\mathbf{x = \frac{1}{19};\ y = \frac{68}{19}}$$

24. (73) $\text{Side} = \dfrac{72 \text{ cm}}{12} = 6 \text{ cm}$

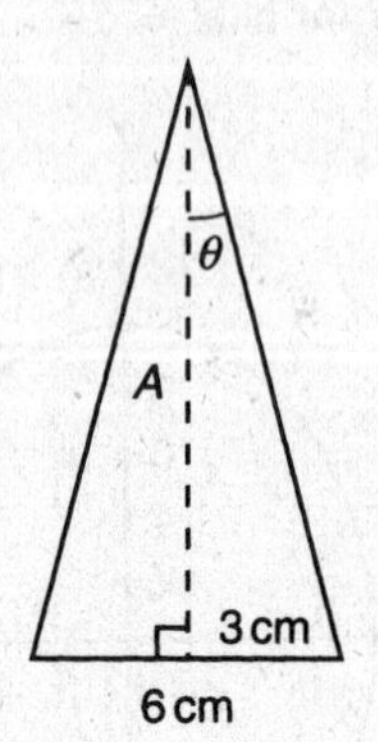

$$\theta = \frac{1}{2}\left(\frac{360°}{12}\right) = 15°$$

$$A = \frac{3}{\tan 15°} = 11.1962 \text{ cm}$$

$$\text{Area}_\Delta = \frac{1}{2}(3)(11.1962) = 16.7943 \text{ cm}^2$$

$$\text{Area} = (24)(16.7943) = \mathbf{403.06\ cm^2}$$

25. (56) $r = 6$

$$\text{Area} = \left(\frac{\frac{\pi}{3}}{2\pi}\right)\pi r^2$$

$$= \left(\frac{1}{6}\right)\pi(6)^2 = 6\pi \text{ in.}^2 = \mathbf{18.85\ in.^2}$$

26. (82)

$$12^{4x+2} = 8^{2x-3}$$
$$(4x + 2) \log 12 = (2x - 3) \log 8$$
$$(4x + 2)(1.0792) = (2x - 3)(0.9031)$$
$$4.3168x + 2.1584 = 1.8062x - 2.7093$$
$$2.5106x = -4.8677$$
$$x = \mathbf{-1.94}$$

27. (59)

$$\log_3 (x + 2) = 1 - \log_3 x$$
$$\log_3 (x + 2) + \log_3 x = 1$$
$$\log_3 (x + 2)(x) = 1$$
$$x^2 + 2x = 3^1$$
$$x^2 + 2x - 3 = 0$$
$$(x + 3)(x - 1) = 0$$
$$x = -3, 1 \quad (x \neq -3)$$
$$x = \mathbf{1}$$

28. (59) $\frac{2}{3} \log_{12} 8 - \log_{12} (3x - 2) = \frac{1}{2} \log_{12} 10{,}000$

$$\log_{12} \frac{8^{\frac{2}{3}}}{3x - 2} = \log_{12} 10{,}000^{\frac{1}{2}}$$
$$\frac{4}{3x - 2} = 100$$
$$4 = 100(3x - 2)$$
$$4 = 300x - 200$$
$$300x = 204$$
$$x = \frac{204}{300}$$
$$x = \mathbf{\frac{17}{25}}$$

29. (59) $2e^{\ln 2} + e^{2\ln 8 - \ln 2} = 2 \cdot 2 + e^{\ln \frac{8^2}{2}}$

$$= 4 + \frac{8^2}{2} = \mathbf{36}$$

30. (59)
$$7^{\log_7(x-y)+\log_7(x+y)} = 7^{\log_7(x-y)(x+y)}$$
$$= (x - y)(x + y) = \mathbf{x^2 - y^2}$$

Problem Set 97

1. (92)
$$P(10 \cup S) = P(10) + P(S) - P(10 \cap S)$$
$$= \frac{4}{52} + \frac{13}{52} - \frac{1}{52} = \frac{16}{52} = \mathbf{\frac{4}{13}}$$

2. (83)
$$\frac{4}{7} \cdot \frac{7}{10} = \mathbf{\frac{2}{5}}$$

3. (88)
$$A_0 = 50,\ A_{30} = 300$$
$$A_t = A_0 e^{kt}$$
$$300 = 50e^{30k}$$
$$6 = e^{30k}$$
$$\ln 6 = 30k$$
$$k = 0.059725$$

$$1500 = 50e^{0.059725t}$$
$$30 = e^{0.059725t}$$
$$\ln 30 = 0.059725t$$
$$t = \mathbf{56.95\ min}$$

4. (18)
$$\begin{cases} \text{(a)}\ H + S + C = 128 \\ \text{(b)}\ \frac{1}{2}S + 12 = C \\ \text{(c)}\ \frac{1}{2}C + 12 = H \end{cases}$$

$$2\text{(b)}\quad S + 24 = 2C$$
$$S = 2C - 24$$

$$\text{(a)}\quad H + S + C = 128$$
$$\left(\frac{1}{2}C + 12\right) + (2C - 24) + C = 128$$
$$\frac{7}{2}C = 140$$
$$\mathbf{C = 40}$$

$$S = 2C - 24$$
$$S = 2(40) - 24$$
$$\mathbf{S = 56}$$

$$\text{(c)}\quad \frac{1}{2}C + 12 = H$$
$$\frac{1}{2}(40) + 12 = H$$
$$\mathbf{H = 32}$$

5. (25)
$$\text{Rate} = \frac{500}{Y}\ \frac{\text{dollars}}{\text{yd}}$$
$$\text{New rate} = \frac{500}{Y} + 5 = \frac{500 + 5Y}{Y}\ \frac{\text{dollars}}{\text{yd}}$$
$$\text{New rate} \times Y = 750$$
$$\left(\frac{500 + 5Y}{Y}\right)Y = 750$$
$$500 + 5Y = 750$$
$$5Y = 250$$
$$Y = \mathbf{50\ yd}$$

6. (97)

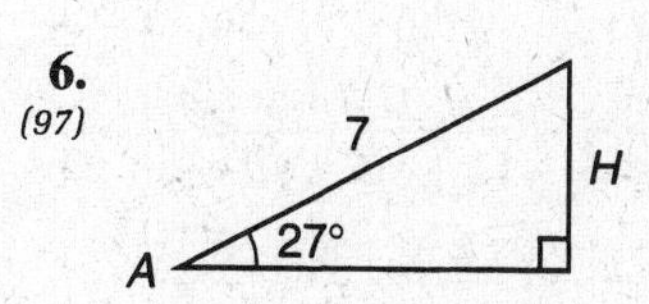

$H = 7 \sin 27° = 3.18$

$H < 5 < 7$, therefore two triangles are possible.

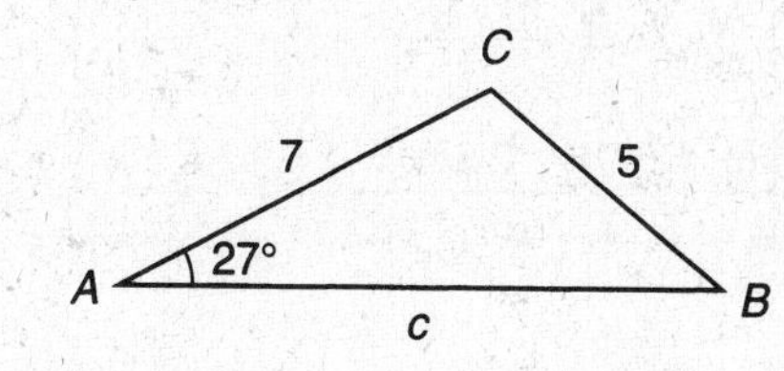

$$\frac{5}{\sin 27°} = \frac{7}{\sin B}$$
$$\sin B = \frac{7 \sin 27°}{5}$$
$$B = 39.46° \text{ or } 140.54°$$

If $B = 39.46°$,

$C = 180° - 27° - 39.46° = 113.54°$.

$$\frac{5}{\sin 27} = \frac{c}{\sin 113.54°}$$
$$c = \frac{5 \sin 113.54°}{\sin 27°}$$
$$c = \mathbf{10.10}$$

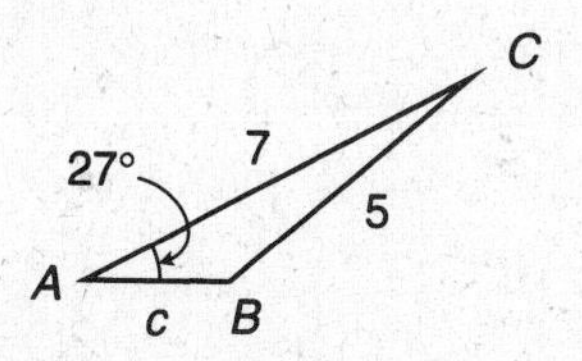

If $B = 140.54°$,

$C = 180° - 27° - 140.54° = 12.46°$.

$$\frac{5}{\sin 27°} = \frac{c}{\sin 12.46°}$$
$$c = \frac{5 \sin 12.46°}{\sin 27°}$$
$$c = \mathbf{2.38}$$

7. (97)

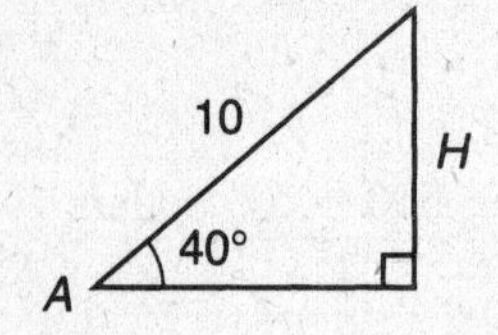

$H = 10 \sin 40° = 6.43$

$H > 5$

No such triangle exists.

8. (96) $\cos 2x + 2\sin^2 x = (1 - 2\sin^2 x) + 2\sin^2 x = 1$

9. (93) $\dfrac{\cos x}{1 - \sin x} - \dfrac{\cos x}{1 + \sin x}$

$= \dfrac{\cos x(1 + \sin x)}{(1 + \sin x)(1 - \sin x)} - \dfrac{\cos x(1 - \sin x)}{(1 + \sin x)(1 - \sin x)}$

$= \dfrac{\cos x + \cos x \sin x - \cos x + \cos x \sin x}{1 - \sin^2 x}$

$= \dfrac{2\cos x \sin x}{\cos^2 x} = \dfrac{2 \sin x}{\cos x} = 2\tan x$

10. (96) $\dfrac{2\sin x}{\sin 2x} = \dfrac{2 \sin x}{2\sin x \cos x} = \dfrac{1}{\cos x} = \sec x$

11. (96)

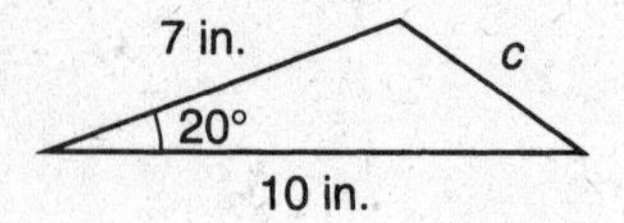

$c^2 = 7^2 + 10^2 - 2(7)(10)\cos 20°$

$c^2 = 17.44$

$\mathbf{c = 4.18}$ **in.**

$\text{Area} = \frac{1}{2}ab \sin C = \frac{1}{2}(10)(7)\sin 20° = 11.97 \text{ in.}^2$

$11.97 \text{ in.}^2 \times \dfrac{1 \text{ ft}}{12 \text{ in.}} \times \dfrac{1 \text{ ft}}{12 \text{ in.}} = 0.083 \text{ ft}^2$

Area = 0.083 ft²

12. (94) $y = 1 + \tan x$

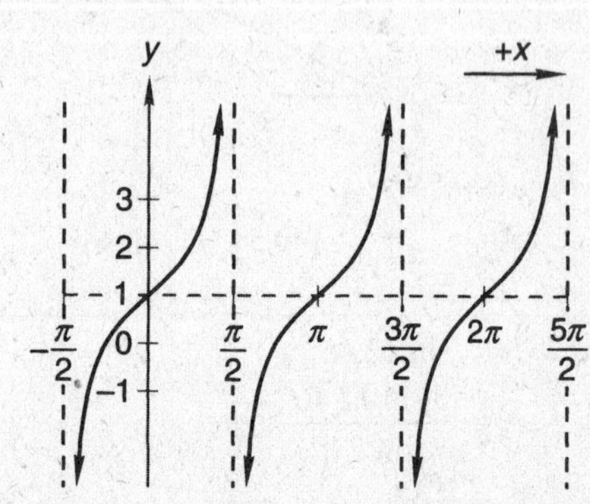

13. (94) $y = -1 + \cot x$

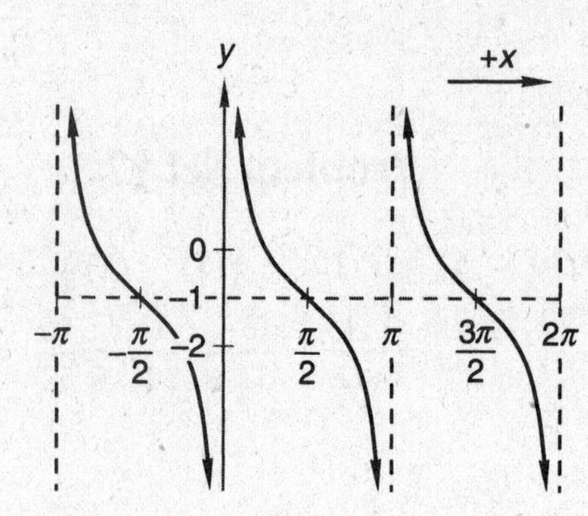

14. (84) $y = 3 + 4\cos \frac{2}{3}(x + 90°)$

Centerline = 3

Amplitude = 4

$\text{Period} = \dfrac{360°}{\frac{2}{3}} = 540°$

Phase angle = –90°

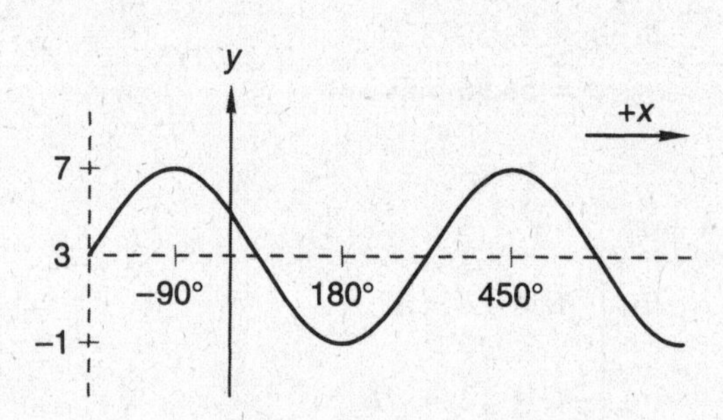

15. (90) $2 \sin 15° \cos 15° = \sin[(2)15°]$

$= \sin 30°$

$= \mathbf{\frac{1}{2}}$

16. (85)

$2\cos^2 x = 1 - \sin x$

$2(1 - \sin^2 x) - 1 + \sin x = 0$

$2\sin^2 x - \sin x - 1 = 0$

$(2\sin x + 1)(\sin x - 1) = 0$

$\sin x = -\frac{1}{2}$ $\qquad$ $\sin x = 1$

$x = 210°, 330°$ $\qquad$ $x = 90°$

$\mathbf{x = 90°, 210°, 330°}$

17. (52)
$$-2 + \sec 3\theta = 0$$
$$\sec 3\theta = 2$$
$$3\theta = 60°, 300°, 420°, 660°, 780°, 1020°$$
$$\theta = \mathbf{20°, 100°, 140°, 220°, 260°, 340°}$$

18. (95) $-5 + 8i$

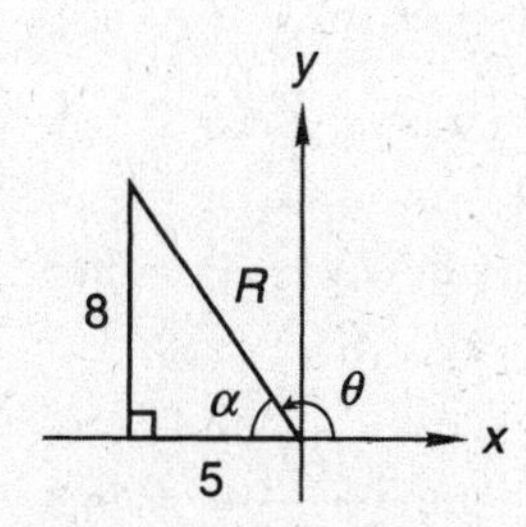

$$R = \sqrt{5^2 + 8^2} = \sqrt{89} = 9.43$$
$$\tan \alpha = \frac{8}{5}$$
$$\alpha = 57.99°$$
$$\theta = 180° - 57.99° = 122.01°$$
$$-5 + 8i = 9.43 \text{ cis } 122.01°$$
$$(9.43 \text{ cis } 122.01°)^{\frac{1}{3}}$$
$$= 9.43^{\frac{1}{3}} \text{ cis}\left(\frac{122.01°}{3} + n\frac{360°}{3}\right)$$
$$= \mathbf{2.11 \text{ cis } 40.67°, 2.11 \text{ cis } 160.67°, 2.11 \text{ cis } 280.67°}$$

19. (91)

①	②	③	④	⑤
2,	___,	___,	___,	32
a_1,	a_1r,	a_1r^2,	a_1r^3,	a_1r^4

$$a_1r^4 = 32$$
$$2r^4 = 32$$
$$r^4 = 16$$
$$r = \pm 2$$

4, 8, 16 or −4, 8, −16

20. (86) $a_5 = 0,\ a_8 = 9$

$$\begin{cases} \text{(a)}\ a_1 + 4d = 0 \\ \text{(b)}\ a_1 + 7d = 9 \end{cases}$$

$$-\text{(a)}\quad -a_1 - 4d = 0$$
$$\text{(b)}\quad a_1 + 7d = 9$$
$$3d = 9$$
$$d = 3$$

$$\text{(a)}\quad a_1 + 4d = 0$$
$$a_1 + 4(3) = 0$$
$$a_1 = -12$$

−12, −9, −6

21. (89)

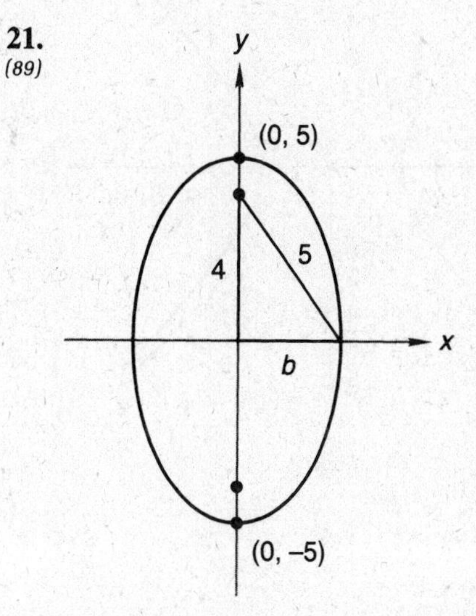

$$b = \sqrt{5^2 - 4^2} = 3$$
$$a = 5$$
$$\frac{x^2}{b^2} + \frac{y^2}{a^2} = 1$$
$$\mathbf{\frac{x^2}{9} + \frac{y^2}{25} = 1}$$

Length of major axis = 10

Length of minor axis = 6

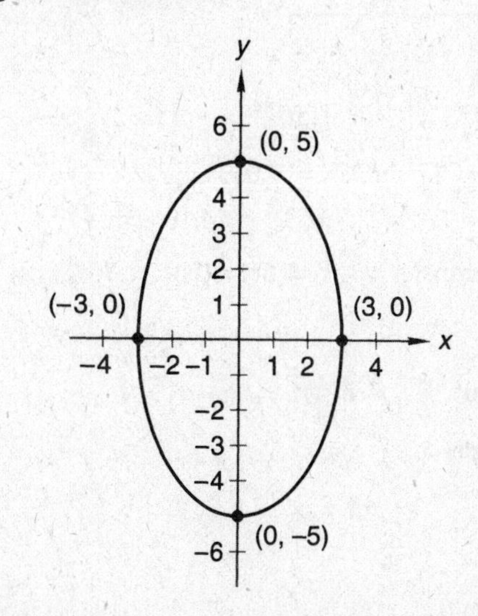

22. (78)
$$x^2 - y^2 = 4$$
$$\mathbf{\frac{x^2}{4} - \frac{y^2}{4} = 1}$$

Let $y = 0$ Let $x = 0$

$x = \pm 2$ $y = \pm 2i$

Vertices = (2, 0), (–2, 0)

Asymptotes: $y = x$; $y = -x$

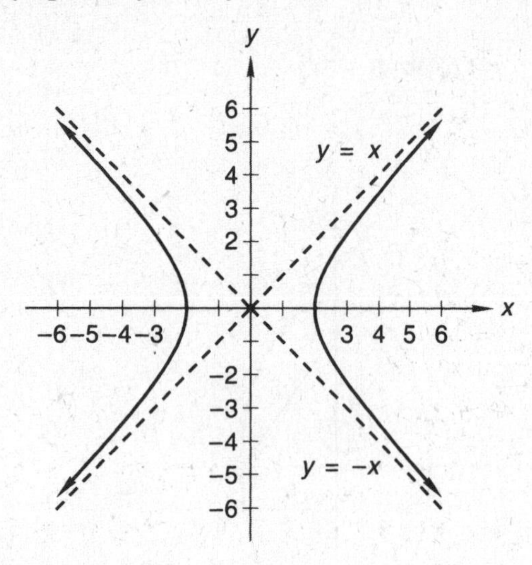

23. (1)
$5 \text{ m} = 500 \text{ cm}$

$$\text{Area} = \left(\frac{30}{360}\right)\pi r^2$$
$$= \left(\frac{30}{360}\right)\pi(500)^2$$
$$= \mathbf{65{,}449.85 \text{ cm}^2}$$

24. (73)

θ, 12, x, s

$$\theta = \frac{1}{2}\left(\frac{360^\circ}{5}\right) = 36^\circ$$
$$x = 12 \sin 36^\circ = 7.053 \text{ cm}$$
$$s = 2x = 2(7.053) = 14.106 \text{ cm}$$
$$\text{Perimeter} = 5s = 5(14.106) = \mathbf{70.53 \text{ cm}}$$

25. (59)
$$10^{2 \log 5} - e^{3 \ln 2} = \log_2 x$$
$$10^{\log 5^2} - e^{\ln 2^3} = \log_2 x$$
$$25 - 8 = \log_2 x$$
$$x = \mathbf{2^{17}}$$

26. (59)
$$2 \log x - \log (x - 1) = \log 4$$
$$\log \frac{x^2}{(x - 1)} = \log 4$$
$$x^2 = 4x - 4$$
$$x^2 - 4x + 4 = 0$$
$$(x - 2)^2 = 0$$
$$x = \mathbf{2}$$

27. (59)
$$\frac{3}{4} \ln 16 - \ln (2x - 4) = \ln 3$$
$$\ln \frac{16^{\frac{3}{4}}}{(2x - 4)} = \ln 3$$
$$\frac{8}{2x - 4} = 3$$
$$3(2x - 4) = 8$$
$$6x = 20$$
$$x = \mathbf{\frac{10}{3}}$$

28. (82)
$$10^{2x-4} = 100^{3x}$$
$$10^{2x-4} = 10^{2(3x)}$$
$$2x - 4 = 6x$$
$$4x = -4$$
$$x = \mathbf{-1}$$

29. (67)
(a) $\text{antilog}_6 (-3) = 6^{-3} = \mathbf{\frac{1}{216}}$

(b) $\text{antilog}_{\frac{1}{6}} (-3) = \left(\frac{1}{6}\right)^{-3} = \mathbf{216}$

30. (46)
$$x^2 + 5x + 17 = 0$$
$$x = \frac{-5 \pm \sqrt{25 - (4)(1)(17)}}{2(1)}$$
$$x = \frac{-5 \pm \sqrt{-43}}{2}$$
$$\mathbf{\left(x + \frac{5}{2} - \frac{\sqrt{43}}{2}i\right)\left(x + \frac{5}{2} + \frac{\sqrt{43}}{2}i\right)}$$

Problem Set 98

1. (83)
$$\frac{52-13}{52}\cdot\frac{52-13}{52}=\frac{3}{4}\cdot\frac{3}{4}=\mathbf{\frac{9}{16}}$$

2. (88) $A_0 = 400,\ A_{60} = 300$
$$A_t = A_0e^{kt}$$
$$300 = 400e^{60k}$$
$$\frac{3}{4} = e^{60k}$$
$$\ln\frac{3}{4} = 60k$$
$$k = \frac{\ln\frac{3}{4}}{60}$$
$$k = -0.0047947$$
$$A_{180} = 400e^{(-0.0047947)(180)} = \mathbf{168.75}$$

3. (41)
$${}_{10}P_5 = \frac{10!}{(10-5)!} = \mathbf{30{,}240}$$

4. (44) $R_1T_1 + R_2T_2 = \text{jobs}$
$$\frac{1}{m}T + \frac{1}{f}T = 1$$
$$\left(\frac{f+m}{mf}\right)T = 1$$
$$T = \mathbf{\frac{mf}{f+m}\ hr}$$

5. (28) Distance $= m$, rate $= \frac{m}{h}$, time $= h$

New distance $= m$

New time $= h + 3$
$$\text{New rate} = \frac{\text{new distance}}{\text{new time}} = \mathbf{\frac{m}{h+3}\ mph}$$

6. (18)
$$\text{Rate} = \frac{1350}{L}\ \frac{\text{dollars}}{\text{lap}}$$
$$\text{New rate} = \frac{1350}{L} + 2 = \frac{1350+2L}{L}\ \frac{\text{dollars}}{\text{lap}}$$
$$\text{New rate} \times L = 2250$$
$$\left(\frac{1350+2L}{L}\right)L = 2250$$
$$1350 + 2L = 2250$$
$$2L = 900$$
$$L = \mathbf{450\ laps}$$

7. (92)
$${}_7P_6 = \frac{7!}{(7-6)!}$$
$${}_7P_6 = \frac{7!}{1!} = \mathbf{5040}$$
$${}_7C_6 = \frac{7!}{6!(7-6)!}$$
$${}_7C_6 = \frac{7!}{6!1!} = \mathbf{7}$$

8. (98) $\log_6 81 = x$
$$6^x = 81$$
$$x\log 6 = \log 81$$
$$x = \frac{\log 81}{\log 6}$$
$$x = \mathbf{2.45}$$

9. (98)
$$x^{\sqrt{\log x}} = 10^8$$
$$\sqrt{\log x}\,\log x = 8$$
$$(\log x)^{\frac{3}{2}} = 8$$
$$\log x = 8^{\frac{2}{3}}$$
$$\log x = 4$$
$$x = \mathbf{10^4}$$

10. (98) $\log_2(\log_2 x) = 3$
$$\log_2 x = 2^3$$
$$\log_2 x = 8$$
$$x = \mathbf{2^8}$$

11. (98)
$$\log_3\sqrt{x} = \sqrt{\log_3 x}$$
$$\log_3 x^{\frac{1}{2}} = \sqrt{\log_3 x}$$
$$\frac{1}{2}\log_3 x = \sqrt{\log_3 x}$$
$$\frac{1}{4}(\log_3 x)^2 = \log_3 x$$
$$\frac{1}{4}(\log_3 x)^2 - \log_3 x = 0$$
$$\log_3 x\left(\frac{1}{4}\log_3 x - 1\right) = 0$$

$\log_3 x = 0$, $x = 3^0$, $x = 1$

$\frac{1}{4}\log_3 x = 1$, $\log_3 x = 4$, $x = 3^4$

1, 81

12. (97)

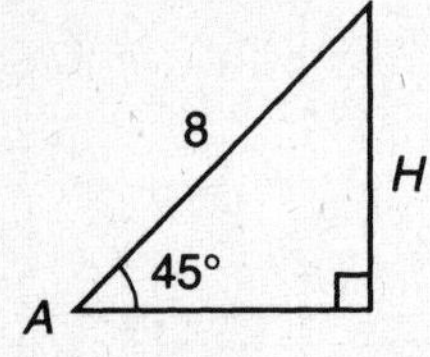

$H = 8 \sin 45° = 5.66$

$5.66 > 4$

No such triangle exists.

13. (97)

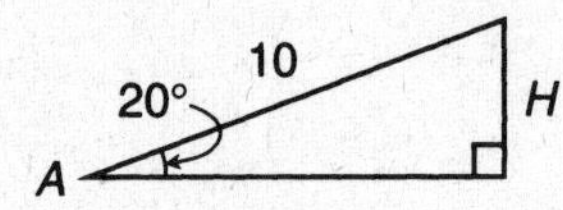

$H = 10 \sin 20° = 3.42$

$H < 6 < 10$, therefore two triangles are possible.

$$\frac{6}{\sin 20°} = \frac{10}{\sin B}$$

$$\sin B = \frac{10 \sin 20°}{6}$$

$$B = 34.75° \text{ or } 145.25°$$

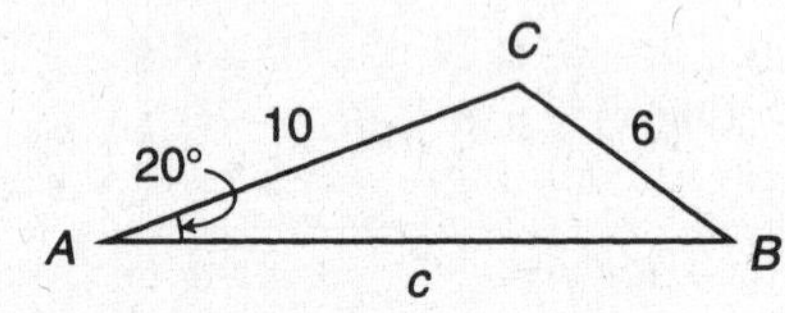

If $B = 34.75°$,

$C = 180° - 20° - 34.75° = 125.25°$.

$$\frac{c}{\sin 125.25°} = \frac{6}{\sin 20°}$$

$$c = \frac{6 \sin 125.25°}{\sin 20°}$$

$$c = \mathbf{14.33}$$

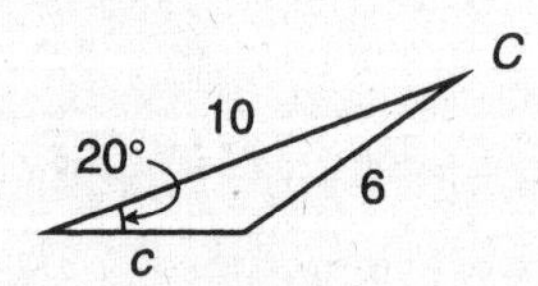

If $B = 145.25°$,

$C = 180° - 20° - 145.25° = 14.75°$.

$$\frac{c}{\sin 14.75°} = \frac{6}{\sin 20°}$$

$$c = \frac{6 \sin 14.75°}{\sin 20°}$$

$$c = \mathbf{4.47}$$

14. (96)

$$\frac{2 \cos 2x}{\sin 2x} = \frac{2(\cos^2 x - \sin^2 x)}{2 \sin x \cos x}$$

$$= \frac{\cos^2 x}{\sin x \cos x} - \frac{\sin^2 x}{\sin x \cos x}$$

$$= \frac{\cos x}{\sin x} - \frac{\sin x}{\cos x} = \cot x - \tan x$$

15. (96)

$$2 \csc 2x \cos x = \frac{2 \cos x}{\sin 2x}$$

$$= \frac{2 \cos x}{2 \sin x \cos x} = \frac{1}{\sin x} = \csc x$$

16. (96)

$$\frac{2 \tan x}{1 + \tan^2 x} = \frac{\frac{2 \sin x}{\cos x}}{\sec^2 x} = \frac{\frac{2 \sin x}{\cos x}}{\frac{1}{\cos^2 x}}$$

$$= \frac{2 \sin x}{\cos x} \cdot \frac{\cos^2 x}{1} = 2 \sin x \cos x = \sin 2x$$

17. (81)

$$5^2 = 8^2 + 8^2 - 2(8)(8) \cos A$$

$$\cos A = \frac{8^2 + 8^2 - 5^2}{2(8)(8)}$$

$$\cos A = 0.8047$$

$$A = \mathbf{36.42°}$$

$$B = C = \frac{180° - A}{2}$$

$$\mathbf{B = 71.79°}$$

$$\mathbf{C = 71.79°}$$

18. (84) $y = -1 + 5 \cos 3(x - 50°)$

Centerline $= -1$

Amplitude $= 5$

Period $= \frac{360°}{3} = 120°$

Phase angle $= 50°$

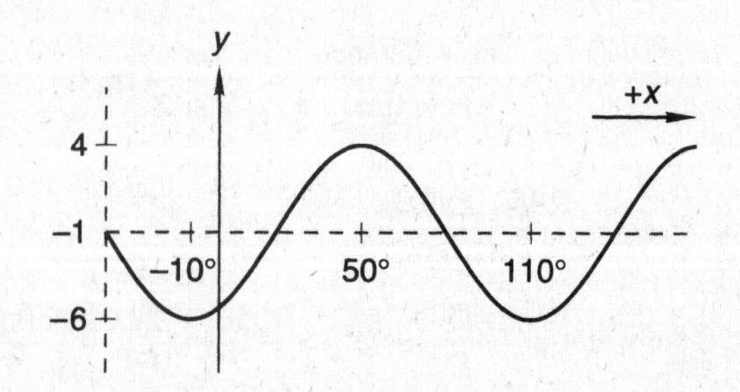

19. (94) $y = 2 + \sec x$

Centerline $= 2$

Amplitude $= 1$

Period $= 2\pi$

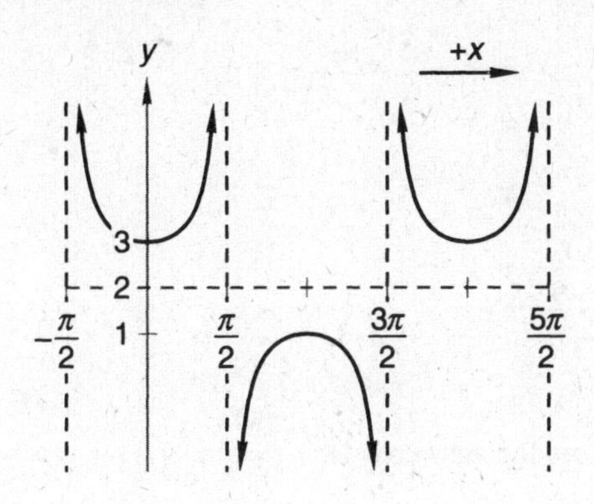

20. $y = -3 + \csc x$
(94)

Centerline = −3

Amplitude = 1

Period = 2π

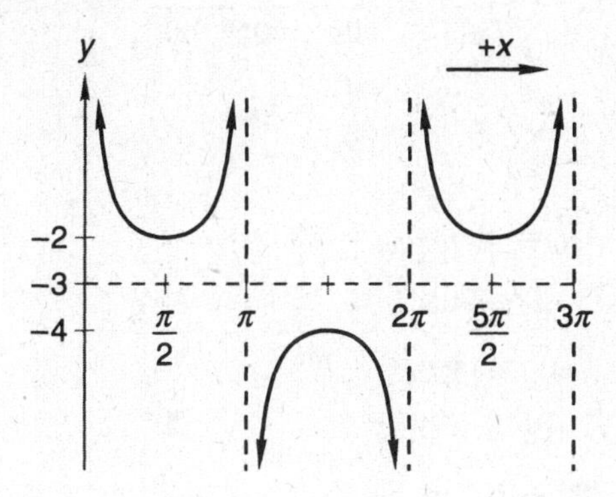

21. $\cos^2 15° - \sin^2 15° = \cos (2)(15°)$
(90,96)

$= \cos 30° = \mathbf{\frac{\sqrt{3}}{2}}$

22. $2\cos^2 x - \sqrt{3}\cos x = 0$
(60)

$\cos x(2\cos x - \sqrt{3}) = 0$

$\cos x = 0 \qquad x = \frac{\pi}{2}, \frac{3\pi}{2}$

$2\cos x = \sqrt{3} \qquad \cos x = \frac{\sqrt{3}}{2} \qquad x = \frac{\pi}{6}, \frac{11\pi}{6}$

$x = \mathbf{\frac{\pi}{6}, \frac{\pi}{2}, \frac{3\pi}{2}, \frac{11\pi}{6}}$

23. $2 - \sqrt{3}\sec 2\theta = 0$
(52)

$\sec 2\theta = \frac{2}{\sqrt{3}}$

$2\theta = \frac{\pi}{6}, \frac{11\pi}{6}, \frac{13\pi}{6}, \frac{23\pi}{6}$

$\theta = \mathbf{\frac{\pi}{12}, \frac{11\pi}{12}, \frac{13\pi}{12}, \frac{23\pi}{12}}$

24. $6 - 2i$
(95)

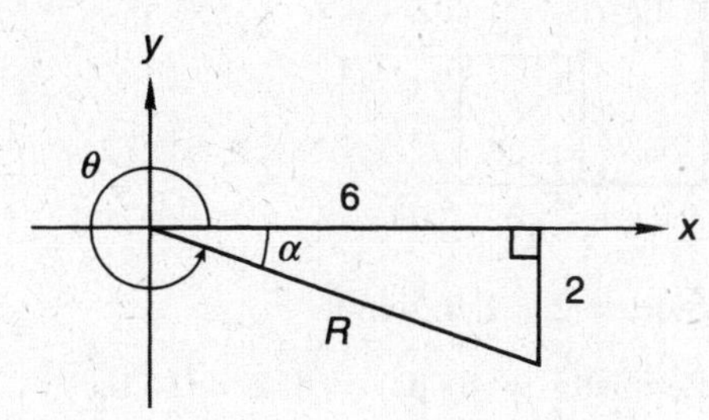

$R = \sqrt{6^2 + 2^2} = 6.32$

$\tan \alpha = \frac{2}{6}$

$\alpha = 18.43°$

$\theta = 360° - 18.43° = 341.57°$

$6 - 2i = 6.32 \text{ cis } 341.57°$

$(6.32 \text{ cis } 341.57°)^{\frac{1}{3}}$

$= 6.32^{\frac{1}{3}} \text{ cis} \left(\frac{341.57°}{3} + n\frac{360°}{3}\right)$

$=$ **1.85 cis 113.86°, 1.85 cis 233.86°, 1.85 cis 353.86°**

25. $a_4 = a_1 r^3 = 2(4)^3 = \mathbf{128}$
(91)

26. $25x^2 + 36y^2 = 1800$
(71,89)

$\mathbf{\frac{x^2}{72} + \frac{y^2}{50} = 1}$

$a = \sqrt{72} = 6\sqrt{2}$

$b = \sqrt{50} = 5\sqrt{2}$

Length of major axis = $12\sqrt{2}$

Length of minor axis = $10\sqrt{2}$

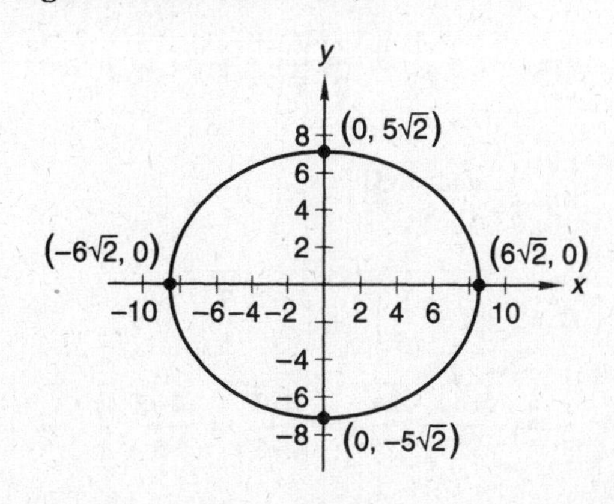

27. (73)

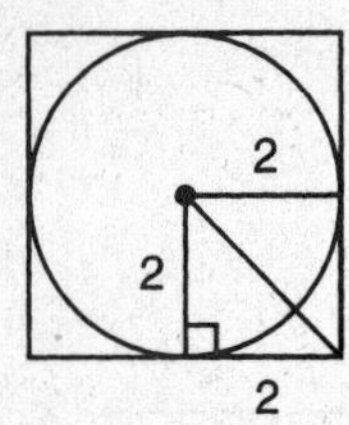

Side = 2 · 2 = 4 in.

Perimeter = 4(side) = 4(4) = **16 in.**

28. (5) $SF = \dfrac{\frac{3}{4}}{\frac{2}{3}} = \dfrac{9}{8}$

$\dfrac{\text{Area}_1}{\text{Area}_2} = \left(\dfrac{9}{8}\right)^2 = \mathbf{\dfrac{81}{64}}$

29. (58) $2x - 4y + 2 = 0$

$y = \dfrac{1}{2}x + \dfrac{1}{2}$

Equation of the perpendicular line:

$y = -2x + b$

$3 = -2(1) + b$

$b = 5$

$y = -2x + 5$

Point of intersection:

$\dfrac{1}{2}x + \dfrac{1}{2} = -2x + 5$

$\dfrac{5}{2}x = \dfrac{9}{2}$

$x = \dfrac{9}{5}$

$y = \dfrac{1}{2}x + \dfrac{1}{2} = \dfrac{1}{2}\left(\dfrac{9}{5}\right) + \dfrac{1}{2} = \dfrac{7}{5}$

$\left(\dfrac{9}{5}, \dfrac{7}{5}\right)$ and (1, 3)

$D = \sqrt{\left(1 - \dfrac{9}{5}\right)^2 + \left(3 - \dfrac{7}{5}\right)^2}$

$= \sqrt{\dfrac{16}{25} + \dfrac{64}{25}} = \sqrt{\dfrac{80}{25}} = \mathbf{\dfrac{4\sqrt{5}}{5}}$

30. (21)

$y = -\sqrt{x^2 - 9}$

$y^2 = x^2 - 9$

$9 = x^2 - y^2$

$\dfrac{x^2}{9} - \dfrac{y^2}{9} = 1$

$|x| \geq 3$

$f(x) \leq 0$

Therefore the answer is **D.**

Problem Set 99

1. (92) $P(B \cup G) = P(B) + P(G) - P(B \cap G)$

$= \dfrac{20}{60} + \dfrac{50}{60} - \dfrac{10}{60}$

$= \dfrac{60}{60} = \mathbf{1}$

2. (88) $A_{10} = 500,\ A_{20} = 400$

$A_t = A_0 e^{kt}$

$400 = 500e^{k(20-10)}$

$0.8 = e^{k10}$

$\ln 0.8 = 10k$

$k = -0.022314$

$\dfrac{1}{2}A_0 = A_0 e^{-0.022314t}$

$\dfrac{1}{2} = e^{-0.022314t}$

$\ln \dfrac{1}{2} = -0.022314t$

$t = \mathbf{31.06\ yr}$

3. (45)

S_1	S_2	3	2	1	→	3 × 2 × 1 = 6
S_2	S_1	3	2	1	→	3 × 2 × 1 = 6
3	S_1	S_2	2	1	→	3 × 2 × 1 = 6
3	S_2	S_1	2	1	→	3 × 2 × 1 = 6
3	2	S_1	S_2	1	→	3 × 2 × 1 = 6
3	2	S_2	S_1	1	→	3 × 2 × 1 = 6
3	2	1	S_1	S_2	→	3 × 2 × 1 = 6
3	2	1	S_2	S_1	→	3 × 2 × 1 = 6
						48

4. (44) Distance = F, rate = r

New distance = $F + 6$

New rate = $r + 4$

New time = $\dfrac{\text{new distance}}{\text{new rate}} = \mathbf{\dfrac{F + 6}{r + 4}}$ **hr**

5. (44) $R_B = \dfrac{m}{4}\dfrac{\text{jobs}}{\text{hr}}$

$R_BT_B + R_ST_S = \text{jobs}$

$\dfrac{m}{4}t + R_St = 2$

$R_St = 2 - \dfrac{m}{4}t$

$R_St = \dfrac{8 - mt}{4}$

$R_S = \mathbf{\dfrac{8 - mt}{4t}\dfrac{jobs}{hr}}$

6. (18) $V = kS^2\sqrt{E}$

$k(3S)^2\sqrt{4E} = k(9)S^2 2\sqrt{E} = 18(V)$

Therefore, **the value was multiplied by 18.**

7. (99)

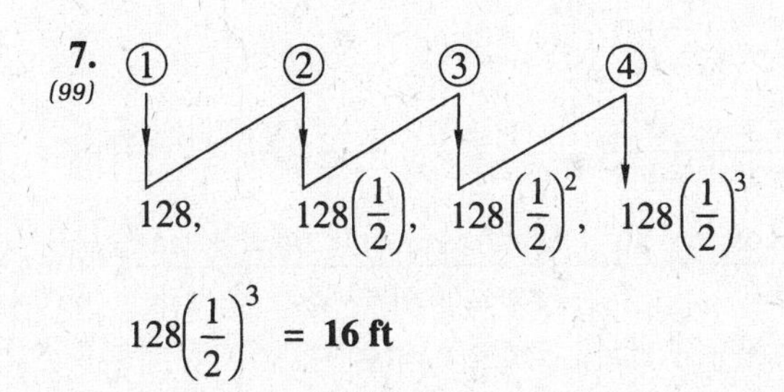

$128\left(\dfrac{1}{2}\right)^3 = \mathbf{16\ ft}$

8. (99) (a) $\begin{cases} L - S = 24 \\ \sqrt{LS} = 9 \end{cases}$ (b)

(a) $L - S = 24$

$S = L - 24$

(b) $\sqrt{LS} = 9$

$LS = 81$

$L(L - 24) = 81$

$L^2 - 24L - 81 = 0$

$(L + 3)(L - 27) = 0$

$L = -3, 27$

If $L = -3$,

$S = L - 24 = (-3) - 24 = -27$

If $L = 27$,

$S = L - 24 = (27) - 24 = 3$

$(L, S) = \mathbf{(-3, -27)\ and\ (27, 3)}$

9. (99) $1 + \sqrt{2}, 3 + 2\sqrt{2}, \ldots$

$r = \dfrac{3 + 2\sqrt{2}}{1 + \sqrt{2}} = \dfrac{3 + 2\sqrt{2}}{1 + \sqrt{2}} \cdot \dfrac{1 - \sqrt{2}}{1 - \sqrt{2}}$

$= \dfrac{3 + 2\sqrt{2} - 3\sqrt{2} - 4}{1 - 2} = 1 + \sqrt{2}$

$a_n = a_1r^{n-1}$

$a_4 = (1 + \sqrt{2})(1 + \sqrt{2})^3$

$= (1 + \sqrt{2})\left[1 + 3\sqrt{2} + 3(2) + (\sqrt{2})^3\right]$

$= (1 + \sqrt{2})(7 + 5\sqrt{2})$

$= 7 + 5\sqrt{2} + 7\sqrt{2} + 5(2)$

$= \mathbf{17 + 12\sqrt{2}}$

10. (99) Arithmetic mean $= \dfrac{x + y}{2} = \dfrac{8 + 22}{2} = \dfrac{30}{2} = \mathbf{15}$

Geometric mean $= \pm\sqrt{xy} = \pm\sqrt{8 \cdot 22} = \mathbf{\pm 4\sqrt{11}}$

11. (98) $\log_5 60 = x$

$5^x = 60$

$\log 5^x = \log 60$

$x \log 5 = \log 60$

$x = \dfrac{\log 60}{\log 5}$

$x = \mathbf{2.54}$

12. (98) $\log_6 50 = x$

$6^x = 50$

$\log 6^x = \log 50$

$x \log 6 = \log 50$

$x = \dfrac{\log 50}{\log 6}$

$x = \mathbf{2.18}$

13. (98) $\log_2 (\log_2 x) = 2$

$\log_2 x = 2^2$

$\log_2 x = 4$

$x = 2^4$

$x = \mathbf{16}$

14. (98)

$$\log_2 x^2 = 2(\log_2 x)^2$$
$$2\log_2 x = 2(\log_2 x)^2$$
$$(\log_2 x)^2 - \log_2 x = 0$$
$$\log_2 x(\log_2 x - 1) = 0$$

$\log_2 x = 0$ $\qquad$ $\log_2 x = 1$

$x = 2^0$ $\qquad$ $x = 2^1$

$x = 1$ $\qquad$ $x = 2$

$x =$ **1, 2**

15. (98)

$$\log_3 x^{\frac{1}{3}} = (\log_3 x)^{\frac{1}{2}}$$
$$\frac{1}{3}\log_3 x = (\log_3 x)^{\frac{1}{2}}$$
$$\left(\frac{1}{3}\log_3 x\right)^2 = \left[(\log_3 x)^{\frac{1}{2}}\right]^2$$
$$\frac{1}{9}(\log_3 x)^2 = \log_3 x$$
$$\frac{1}{9}(\log_3 x)^2 - \log_3 x = 0$$
$$\log_3 x\left(\frac{1}{9}\log_3 x - 1\right) = 0$$

$\log_3 x = 0$ $\qquad$ $\log_3 x = 9$

$x = 3^0$ $\qquad$ $x = 3^9$

$x = 1$

$x =$ **$1, 3^9$**

16. (97)

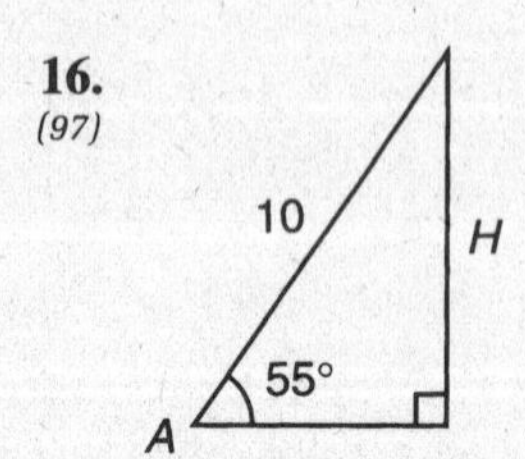

$H = 10 \sin 55° = 8.19$

$H < 9 < 10$, therefore two triangles are possible.

$$\frac{9}{\sin 55°} = \frac{10}{\sin B}$$
$$\sin B = \frac{10 \sin 55°}{9}$$
$$B = 65.53° \text{ or } 114.47°$$

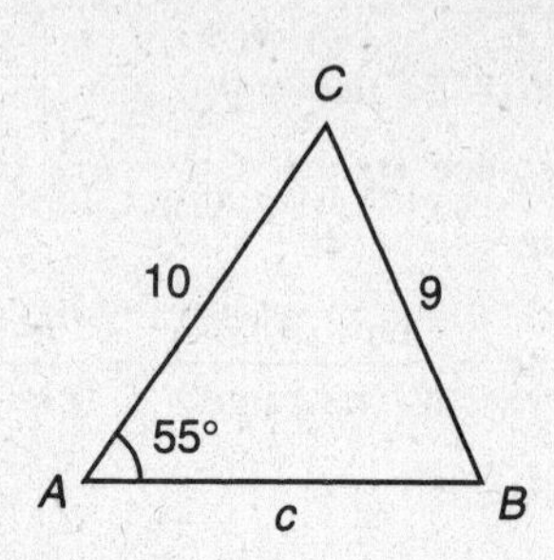

If $B = 65.53°$,

$C = 180° - 55° - 65.53° = 59.47°$

$$\frac{c}{\sin 59.47°} = \frac{9}{\sin 55°}$$
$$c = \frac{9 \sin 59.47°}{\sin 55°}$$
$$c = \mathbf{9.46}$$

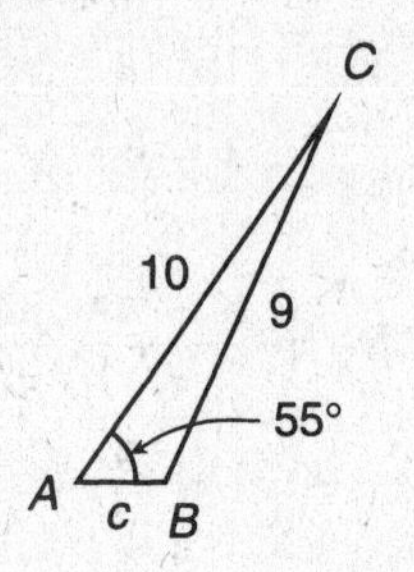

If $B = 114.47°$,

$C = 180° - 55° - 114.47° = 10.53°$

$$\frac{c}{\sin 10.53°} = \frac{9}{\sin 55°}$$
$$c = \frac{9 \sin 10.53°}{\sin 55°}$$
$$c = \mathbf{2.01}$$

17. (96)

$$\frac{\cos^3 x - \sin^3 x}{\cos x - \sin x}$$
$$= \frac{(\cos x - \sin x)(\cos^2 x + \cos x \sin x + \sin^2 x)}{\cos x - \sin x}$$
$$= \cos^2 x + \sin^2 x + \cos x \sin x$$
$$= 1 + \cos x \sin x = 1 + \frac{1}{2}\sin 2x$$

18. (96)

$$\frac{\cos 2x}{\cos^2 x} = \frac{\cos^2 x - \sin^2 x}{\cos^2 x}$$
$$= \frac{\cos^2 x}{\cos^2 x} - \frac{\sin^2 x}{\cos^2 x} = 1 - \tan^2 x$$

19. (96)

$$\frac{2}{\cot x - \tan x} = \frac{2}{\frac{1}{\tan x} - \tan x}$$
$$= \frac{2}{\frac{1 - \tan^2 x}{\tan x}} = \frac{2 \tan x}{1 - \tan^2 x} = \tan 2x$$

20. (81,96)

$$a^2 = b^2 + c^2 - 2bc\cos A$$

$$a^2 = 500^2 + 700^2 - 2(500)(700)\cos 60°$$

$$a^2 = 250{,}000 + 490{,}000 - 350{,}000$$

$$a^2 = 390{,}000$$

$$\mathbf{a = 624.50\text{ cm}}$$

$$\text{Area} = \frac{1}{2}bc\sin A$$

$$\text{Area} = \frac{1}{2}(500)(700)\sin 60°$$

$$\textbf{Area = 151,554.45 cm}^{\mathbf{2}}$$

21. (94) $y = -\tan x = -\dfrac{\sin x}{\cos x}$

Note: The graph has asymptotes where $\cos x = 0$ and crosses the x axis where $\sin x = 0$. The negative sign turns the graph of $y = \tan x$ about the x axis.

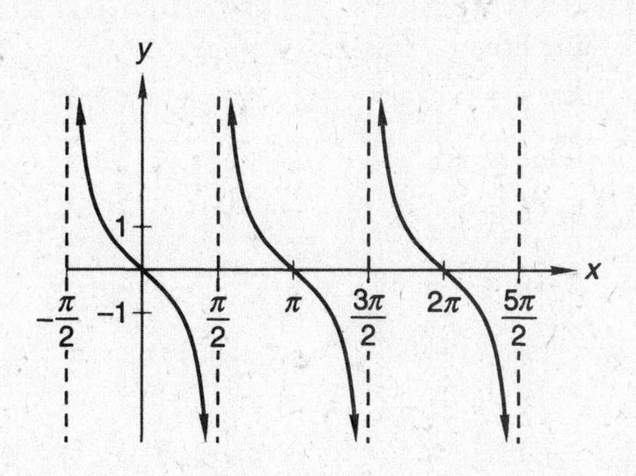

22. (94) $y = -\cot x = -\dfrac{\cos x}{\sin x}$

Note: The graph has asymptotes where $\sin x = 0$ and crosses the x axis where $\cos x = 0$. The negative sign turns the graph of $y = \cot x$ about the x axis.

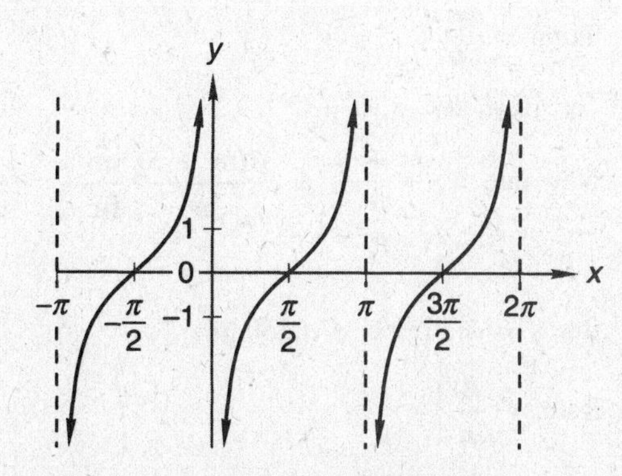

23. (94) (a) $\mathbf{y = \cot\theta}$

(b) $\mathbf{y = -3 + \sec x}$

24. (90,96) Use $2\cos^2 A - 1 = \cos 2A$.

$$2\cos^2 15° - 1 = \cos(2 \cdot 15°) = \cos 30° = \mathbf{\frac{\sqrt{3}}{2}}$$

25. (85)

$$\sin x = \cos x$$

$$\frac{\sin x}{\cos x} = 1$$

$$\tan x = 1$$

$$x = \mathbf{45°, 225°}$$

26. (52)

$$2 + \sqrt{3}\sec 4\theta = 0$$

$$\sec 4\theta = -\frac{2}{\sqrt{3}}$$

$$\cos 4\theta = -\frac{\sqrt{3}}{2}$$

$$4\theta = 150°, 210°, 510°, 570°, 870°, 930°, 1230°, 1290°$$

$$\theta = \mathbf{37.5°, 52.5°, 127.5°, 142.5°, 217.5°, 232.5°, 307.5°, 322.5°}$$

27. (95) $-3 - 2i$

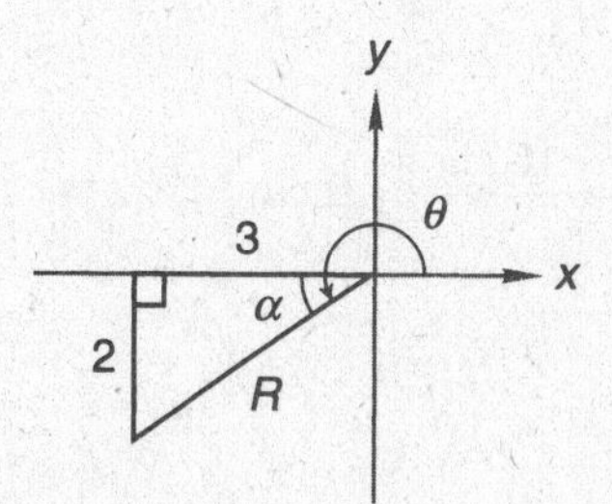

$$R = \sqrt{3^2 + 2^2} = 3.61$$

$$\tan\alpha = \frac{2}{3}$$

$$\alpha = 33.69°$$

$$\theta = 180° + 33.69° = 213.69°$$

$$-3 - 2i = 3.61\text{ cis }213.69°$$

$$(3.61\text{ cis }213.69)^{\frac{1}{4}} = 3.61^{\frac{1}{4}}\text{ cis}\left(\frac{213.69°}{4} + n\frac{360°}{4}\right)$$

$$= \mathbf{1.38\text{ cis }53.42°, 1.38\text{ cis }143.42°, 1.38\text{ cis }233.42°, 1.38\text{ cis }323.42°}$$

28. (89)

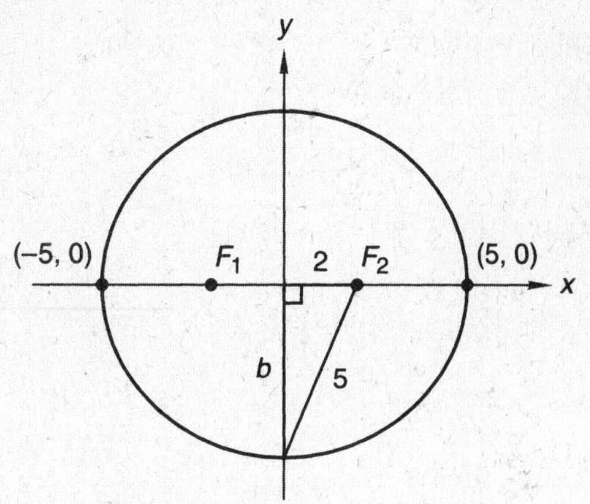

$$b = \sqrt{5^2 - 2^2} = \sqrt{21}$$

$$a = 5$$

$$\frac{x^2}{a^2} + \frac{y^2}{b^2} = 1$$

$$\mathbf{\frac{x^2}{25} + \frac{y^2}{21} = 1}$$

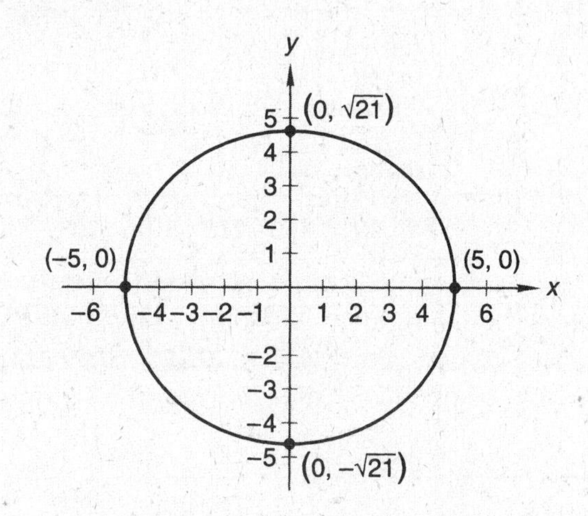

29. (73) Side $= \dfrac{16 \text{ cm}}{8} = 2 \text{ cm}$

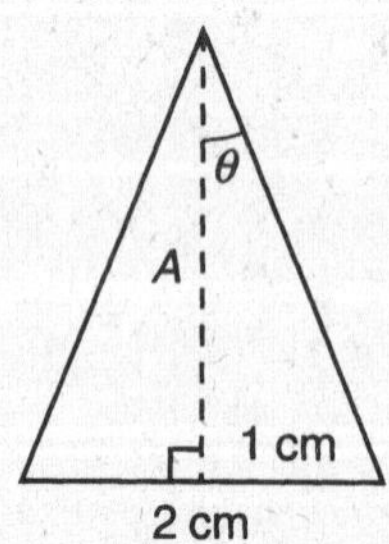

$$\theta = \frac{1}{2}\left(\frac{360°}{8}\right) = 22.5°$$

$$A = \frac{1}{\tan 22.5°} = 2.414 \text{ cm}$$

$$\text{Area}_{\Delta} = \frac{1}{2}(1)(2.414) = 1.207 \text{ cm}^2$$

$$\text{Area} = (16)\left(1.207 \text{ cm}^2\right) = \mathbf{19.31 \text{ cm}^2}$$

30. (46) $x^2 + x + 1 = 0$

$$x = \frac{-1 \pm \sqrt{1^2 - 4(1)(1)}}{2(1)}$$

$$x = \frac{-1 \pm \sqrt{-3}}{2} = -\frac{1}{2} \pm \frac{\sqrt{3}}{2}i$$

$$\mathbf{\left(x + \frac{1}{2} - \frac{\sqrt{3}}{2}i\right)\left(x + \frac{1}{2} + \frac{\sqrt{3}}{2}i\right)}$$

Problem Set 100

1. (99)

① ② ③ ④

$$81, \quad 81\left(\frac{2}{3}\right), \quad 81\left(\frac{2}{3}\right)^2, \quad 81\left(\frac{2}{3}\right)^3$$

$$81\left(\frac{2}{3}\right)^3 = \mathbf{24 \text{ ft}}$$

2. (85) $R_L = \dfrac{1}{12}$, $R_B = 1$, $T_B = T_L$

Big hand:

$$R_B T_B = S + 15$$

Little hand:

$$R_L T_L = S$$

$$R_B T_B = S + 15$$

$$R_B T_B = R_L T_L + 15$$

$$T = \frac{1}{12}T + 15$$

$$\frac{11}{12}T = 15$$

$$T = \frac{180}{11}$$

$$T = \mathbf{16\frac{4}{11} \text{ min}}$$

3. (28) Distance $= m$, rate $= R$

$$\text{Time} = \frac{m}{R}$$

New distance $= m + 5$

$$\text{New rate} = \frac{m + 5}{\frac{m}{R}} = \mathbf{\frac{R(m + 5)}{m} \frac{\text{mi}}{\text{hr}}}$$

4. (44) Items $= m$, price $= d$ dollars

$$\text{Rate} = \frac{d}{m}$$

$$\text{New rate} = \frac{d}{m} + k = \frac{d + km}{m}$$

New rate $\times N$ = price

$$\frac{d + km}{m}N = 14$$

$$N = \frac{\mathbf{14m}}{\mathbf{d + km}} \textbf{ items}$$

5. (18) $N_G = k\dfrac{\sqrt{N_R}}{(N_W)^2}$

$$k\frac{\sqrt{16N_R}}{(2N_W)^2} = k\frac{4\sqrt{N_R}}{4(N_W)^2} = k\frac{\sqrt{N_R}}{(N_W)^2} = N_G$$

Therefore, there is **no change in the number of greens.**

6. (92) $_{11}P_4 = \dfrac{11!}{(11 - 4)!} = \dfrac{11!}{7!}$

$\mathbf{_{11}P_4 = 7920}$

$_{11}C_4 = \dfrac{11!}{(11 - 4)!4!} = \dfrac{11!}{7!4!}$

$\mathbf{_{11}C_4 = 330}$

7. (100)

$$\begin{aligned} \cos(A + B) &= \cos A \cos B - \sin A \sin B \\ \cos(A - B) &= \cos A \cos B + \sin A \sin B \\ \hline \cos(A + B) + \cos(A - B) &= 2 \cos A \cos B \end{aligned}$$

$$\cos A \cos B = \frac{1}{2}[\cos(A + B) + \cos(A - B)]$$

8. (100)

$$\begin{aligned} \cos(A - B) &= \cos A \cos B + \sin A \sin B \\ -\cos(A + B) &= -\cos A \cos B + \sin A \sin B \\ \hline \cos(A - B) - \cos(A + B) &= 2 \sin A \sin B \end{aligned}$$

$$\sin A \sin B = \frac{1}{2}[\cos(A - B) - \cos(A + B)]$$

9. (100)

$$\begin{aligned} A + B &= x \\ A - B &= y \\ \hline 2A &= x + y \end{aligned} \qquad \begin{aligned} A + B &= x \\ -(A - B &= y) \\ \hline 2B &= x - y \end{aligned}$$

$$A = \frac{x + y}{2} \qquad B = \frac{x - y}{2}$$

$$\begin{aligned} \sin(A + B) &= \sin A \cos B + \cos A \sin B \\ -\sin(A - B) &= -\sin A \cos B + \cos A \sin B \\ \hline \sin(A + B) - \sin(A - B) &= 2 \cos A \sin B \end{aligned}$$

$$\sin x - \sin y = 2 \cos \frac{x + y}{2} \sin \frac{x - y}{2}$$

10. (100)

$$\begin{aligned} A + B &= x \\ A - B &= y \\ \hline 2A &= x + y \end{aligned} \qquad \begin{aligned} A + B &= x \\ -(A - B &= y) \\ \hline 2B &= x - y \end{aligned}$$

$$A = \frac{x + y}{2} \qquad B = \frac{x - y}{2}$$

$$\begin{aligned} \sin(A + B) &= \sin A \cos B + \cos A \sin B \\ \sin(A - B) &= \sin A \cos B - \cos A \sin B \\ \hline \sin(A + B) + \sin(A - B) &= 2 \sin A \cos B \end{aligned}$$

$$\sin x + \sin y = 2 \sin \frac{x + y}{2} \cos \frac{x - y}{2}$$

11. (99) Arithmetic mean $= \dfrac{x + y}{2} = \dfrac{9 + 21}{2} = \dfrac{30}{2} = \mathbf{15}$

Geometric mean $= \pm\sqrt{xy} = \pm\sqrt{9 \cdot 21} = \mathbf{\pm 3\sqrt{21}}$

12. (99) $\begin{cases} \text{(a) } L - S = 12 \\ \text{(b) } \sqrt{LS} = 8 \end{cases}$

(b) $\sqrt{LS} = 8$

$LS = 64$

$S = \dfrac{64}{L}$

(a) $L - S = 12$

$$L - \frac{64}{L} = 12$$

$$L^2 - 64 = 12L$$

$$L^2 - 12L - 64 = 0$$

$$(L + 4)(L - 16) = 0$$

$$L = 16, -4$$

If $L = 16$

$$S = \frac{64}{L} = \frac{64}{(16)} = 4$$

If $L = -4$

$$S = \frac{64}{L} = \frac{64}{(-4)} = -16$$

$(L, S) =$ **(16, 4) and (–4, –16)**

13. (98) $\log_8 50 = x$

$$8^x = 50$$

$$x \log 8 = \log 50$$

$$x = \frac{\log 50}{\log 8}$$

$$x = \mathbf{1.88}$$

14. $\log_5 70 = x$
(98)

$$5^x = 70$$
$$x \log 5 = \log 70$$
$$x = \frac{\log 70}{\log 5}$$
$$x = \mathbf{2.64}$$

15. $\log_3(\log_3 x) = 2$
(98)

$$\log_3 x = 3^2$$
$$x = \mathbf{3^9}$$

16. $\log_3 x^2 = \log_3 x$
(98)

$$2\log_3 x - \log_3 x = 0$$
$$\log_3 x = 0$$
$$x = 3^0$$
$$x = \mathbf{1}$$

17. $\frac{2}{3}\ln 8 + \ln x - \ln(3x - 2) = \ln 3$
(59,98)

$$\ln \frac{8^{\frac{2}{3}} x}{3x - 2} = \ln 3$$
$$\frac{4x}{3x - 2} = 3$$
$$4x = 9x - 6$$
$$5x = 6$$
$$x = \mathbf{\frac{6}{5}}$$

18.
(97)

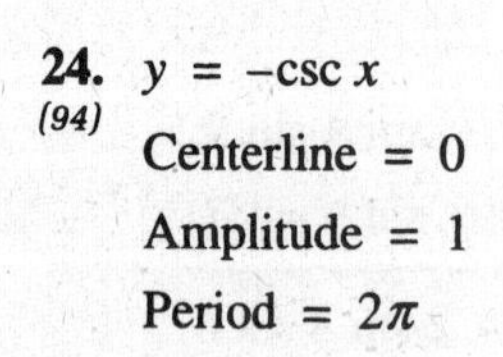

$H = 8 \sin 35° = 4.59$

$H > 4$

No such triangle exists.

19. $\cot x + \tan x = \frac{\cos x}{\sin x} + \frac{\sin x}{\cos x}$
(96)

$$= \frac{\cos^2 x + \sin^2 x}{\sin x \cos x} = \frac{1}{\sin x \cos x} \cdot \frac{2}{2}$$
$$= \frac{2}{2 \sin x \cos x} = \frac{2}{\sin 2x} = 2 \csc 2x$$

20. $\frac{\sin 2x}{\tan x} = \frac{2 \sin x \cos x}{\frac{\sin x}{\cos x}} = 2 \sin x \cos x \cdot \frac{\cos x}{\sin x}$
(96)

$$= 2\cos^2 x$$

21. $(\tan^2 x)(1 + \cot^2 x) = \tan^2 x \csc^2 x$
(93)

$$= \frac{\sin^2 x}{\cos^2 x} \cdot \frac{1}{\sin^2 x} = \frac{1}{\cos^2 x} = \frac{1}{1 - \sin^2 x}$$

22. $\frac{18}{\sin 100°} = \frac{10}{\sin A}$
(72)

$$\sin A = \frac{10 \sin 100°}{18}$$
$$A = \mathbf{33.17°}$$

$B = 180° - 100° - 33.17°$

$\mathbf{B = 46.83°}$

23. $y = 3 + 2\sin\frac{1}{2}(x - \pi)$
(84)

Centerline = 3

Amplitude = 2

Period $= \frac{2\pi}{\frac{1}{2}} = 4\pi$

Phase angle = π

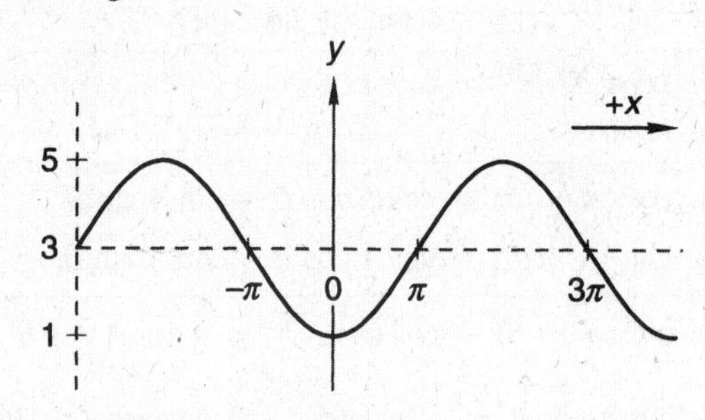

24. $y = -\csc x$
(94)

Centerline = 0

Amplitude = 1

Period = 2π

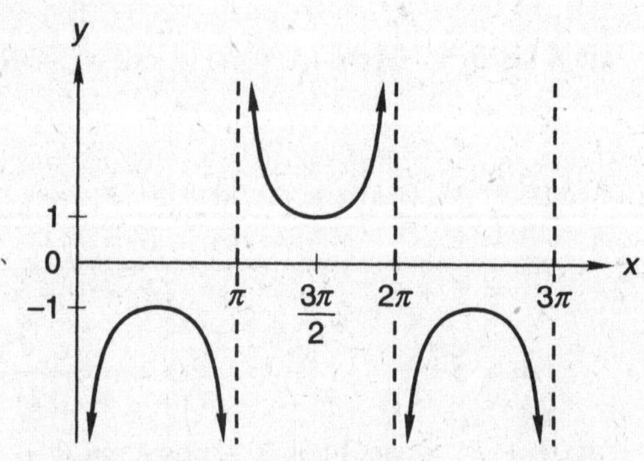

25. $\cos 110° \cos 50° + \sin 110° \sin 50°$
(87)

$$= \cos(110° - 50°) = \cos(60°) = \mathbf{\frac{1}{2}}$$

26. (85)

$$2\sin^2 x + 15\cos x - 9 = 0$$
$$2(1 - \cos^2 x) + 15\cos x - 9 = 0$$
$$2\cos^2 x - 15\cos x + 7 = 0$$
$$(2\cos x - 1)(\cos x - 7) = 0$$

$\cos x = \frac{1}{2}$ $\qquad$ $\cos x = 7$

no answer

$x = \mathbf{\frac{\pi}{3}, \frac{5\pi}{3}}$

27. (52) $\sqrt{3} - \cot 2\theta = 0$

$$\cot 2\theta = \sqrt{3}$$
$$2\theta = \frac{\pi}{6}, \frac{7\pi}{6}, \frac{13\pi}{6}, \frac{19\pi}{6}$$
$$\theta = \mathbf{\frac{\pi}{12}, \frac{7\pi}{12}, \frac{13\pi}{12}, \frac{19\pi}{12}}$$

28. (95) $4 - 3i = 5 \text{ cis } 323.13°$

$$(5 \text{ cis } 323.13°)^{\frac{1}{4}} = 5^{\frac{1}{4}} \text{ cis}\left(\frac{323.13°}{4} + n\frac{360°}{4}\right)$$

$=$ **1.50 cis 80.78°, 1.50 cis 170.78°, 1.50 cis 260.78°, 1.50 cis 350.78°**

29. (74)

$$x = \frac{\begin{vmatrix} 10 & -2 \\ -4 & 3 \end{vmatrix}}{\begin{vmatrix} 4 & -2 \\ -2 & 3 \end{vmatrix}} = \frac{22}{8} = \frac{11}{4}$$

$$y = \frac{\begin{vmatrix} 4 & 10 \\ -2 & -4 \end{vmatrix}}{\begin{vmatrix} 4 & -2 \\ -2 & 3 \end{vmatrix}} = \frac{4}{8} = \frac{1}{2}$$

$\mathbf{x = \frac{11}{4}; y = \frac{1}{2}}$

30. (59) $x3^{\log_3(x+1) - \log_3(x^2+x)} = x3^{\log_3 \frac{x+1}{x^2+x}}$

$$= x3^{\log_3 \frac{x+1}{x(x+1)}} = x3^{\log_3 \frac{1}{x}} = x \cdot \frac{1}{x} = \mathbf{1}$$

Problem Set 101

1. (91)

① ② ③ ④ ⑤

$$256, \quad 256\left(\frac{3}{4}\right), \quad 256\left(\frac{3}{4}\right)^2, \quad 256\left(\frac{3}{4}\right)^3, \quad 256\left(\frac{3}{4}\right)^4$$

$$D = 256\left(\frac{3}{4}\right)^4 = \mathbf{81 \text{ ft}}$$

2. (83) $\left(\frac{1}{3}\right)\left(\frac{4}{11}\right) = \mathbf{\frac{4}{33}}$

3. (88) $A_0 = 800,\ A_{20} = 760$

$$A_t = A_0 e^{kt}$$
$$760 = 800e^{20k}$$
$$0.95 = e^{20k}$$
$$\ln 0.95 = 20k$$
$$k = -0.0025647$$

$$400 = 800e^{-0.0025647t}$$
$$0.5 = e^{-0.0025647t}$$
$$\ln 0.5 = -0.0025647t$$
$$t = \mathbf{270.27 \text{ min}}$$

4. (36)
(a) $(B + W)T_D = D_D$
(b) $(B - W)T_U = D_U$

$B = 3W,\ T_D = T_U + 2$

(b) $(B - W)T_U = D_U$
$$(2W)T_U = 48$$
$$WT_U = 24$$

(a) $(B + W)T_D = D_D$
$$(4W)(T_U + 2) = 120$$
$$4WT_U + 8W = 120$$
$$4(24) + 8W = 120$$
$$8W = 24$$
$$\mathbf{W = 3 \text{ mph}}$$

$B = 3W$
$B = 3(3)$
$\mathbf{B = 9 \text{ mph}}$

5. (25)
(a) $R_O T_O = 360$
(b) $R_B T_B = 360$

$R_B = 2R_O,\ T_B = 9 - T_O$

(b) $R_B T_B = 360$
$$(2R_O)(9 - T_O) = 360$$
$$18R_O - 2R_O T_O = 360$$
$$18R_O - 2(360) = 360$$
$$18R_O = 1080$$
$$R_O = \mathbf{60 \text{ mph}}$$

6. (R)
$$\begin{cases} \dfrac{R}{G} = \dfrac{4}{3} \\ R + G = 63 \end{cases}$$

$$\frac{R}{R + G} = \frac{4}{4 + 3}$$
$$\frac{R}{63} = \frac{4}{7}$$
$$7R = 252$$
$$R = \mathbf{36}$$

7. (101)
$$\begin{vmatrix} 2 & 0 & -1 \\ 0 & 3 & -2 \\ 3 & 4 & -1 \end{vmatrix}$$

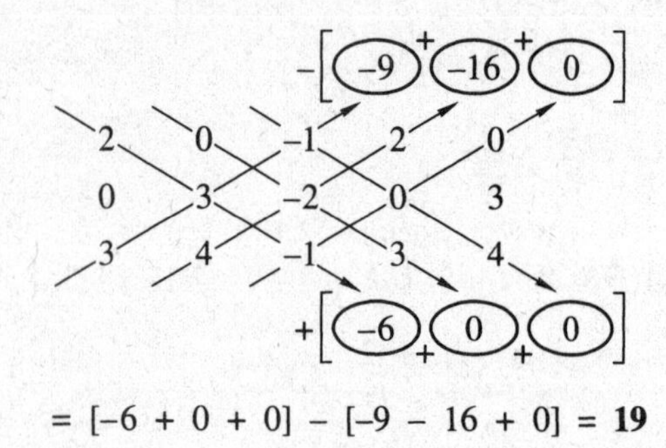

$$= [-6 + 0 + 0] - [-9 - 16 + 0] = \mathbf{19}$$

8. (101)
$$z = \frac{\begin{vmatrix} 2 & 3 & 2 \\ 0 & 1 & 5 \\ 1 & 2 & -1 \end{vmatrix}}{\begin{vmatrix} 2 & 3 & 1 \\ 0 & 1 & 2 \\ 1 & 2 & 0 \end{vmatrix}}$$

$$z = \frac{\begin{array}{c} -[(2)\ (20)\ (0)] \\ \begin{array}{ccccc} 2 & 3 & 2 & 2 & 3 \\ 0 & 1 & 5 & 0 & 1 \\ 1 & 2 & -1 & 1 & 2 \end{array} \\ +[(-2)\ (15)\ (0)] \end{array}}{\begin{array}{c} -[(1)\ (8)\ (0)] \\ \begin{array}{ccccc} 2 & 3 & 1 & 2 & 3 \\ 0 & 1 & 2 & 0 & 1 \\ 1 & 2 & 0 & 1 & 2 \end{array} \\ +[(0)\ (6)\ (0)] \end{array}}$$

$$z = \frac{[-2 + 15 + 0] - [2 + 20 + 0]}{[0 + 6 + 0] - [1 + 8 + 0]} = \frac{-9}{-3} = \mathbf{3}$$

9. (100)
$$\sin(A + B) = \sin A \cos B + \cos A \sin B$$
$$-\sin(A - B) = -\sin A \cos B + \cos A \sin B$$
$$\sin(A + B) - \sin(A - B) = 2 \cos A \sin B$$
$$\cos A \sin B = \frac{1}{2}[\sin(A + B) - \sin(A - B)]$$

10. (100)
$$\sin(A + B) = \sin A \cos B + \cos A \sin B$$
$$\sin(A - B) = \sin A \cos B - \cos A \sin B$$
$$\sin(A + B) + \sin(A - B) = 2 \sin A \cos B$$
$$\sin A \cos B = \frac{1}{2}[\sin(A + B) + \sin(A - B)]$$

11. (99)
$$\begin{cases} \text{(a)} & \sqrt{xy} = 4 \\ \text{(b)} & \dfrac{x + y}{2} = 5 \end{cases}$$

(b) $\dfrac{x + y}{2} = 5$
$$x + y = 10$$
$$x = 10 - y$$

(a) $\sqrt{xy} = 4$
$$xy = 16$$
$$(10 - y)y = 16$$
$$10y - y^2 = 16$$
$$y^2 - 10y + 16 = 0$$
$$(y - 2)(y - 8) = 0$$
$$y = 2, 8$$

If $y = 2$,
$x = 10 - y = 10 - (2) = 8.$

If $y = 8$,
$x = 10 - y = 10 - (8) = 2.$

2 and 8

12. (99)
$$\begin{cases} \text{(a)} & -\sqrt{LS} = -10 \\ \text{(b)} & L - S = 15 \end{cases}$$

(b) $L - S = 15$
$$S = L - 15$$

(a)
$$-\sqrt{LS} = -10$$
$$LS = 100$$
$$L(L - 15) = 100$$
$$L^2 - 15L - 100 = 0$$
$$(L - 20)(L + 5) = 0$$
$$L = 20, -5$$

If $L = 20$,
$S = L - 15 = (20) - 15 = 5.$

If $L = -5$,
$S = L - 15 = (-5) - 15 = -20.$

$(L, S) =$ **(20, 5) and (–5, –20)**

13. (98) (a) $\log_5 35 = x$
$$5^x = 35$$
$$x \log 5 = \log 35$$
$$x = \frac{\log 35}{\log 5}$$
$$x = \mathbf{2.21}$$

(b) $\log_6 40 = x$
$$6^x = 40$$
$$x \log 6 = \log 40$$
$$x = \frac{\log 40}{\log 6}$$
$$x = \mathbf{2.06}$$

14. (98)
$$x^{\ln x} = e^4$$
$$\ln\left(x^{\ln x}\right) = 4$$
$$(\ln x)(\ln x) = 4$$
$$(\ln x)^2 = 4$$
$$\ln x = \pm 2$$
$$x = \boldsymbol{e^2, e^{-2}}$$

15. (59) $\log_4 (x + 3) + \log_4 (x - 3) = 2$
$$\log_4 [(x + 3)(x - 3)] = 2$$
$$x^2 - 9 = 4^2$$
$$x^2 - 25 = 0$$
$$(x + 5)(x - 5) = 0$$
$$x = -5, 5 \qquad (x \neq -5)$$
$$x = \mathbf{5}$$

16. (98)
$$\log_4 \sqrt[4]{x} = \sqrt{\log_4 x}$$
$$\frac{1}{4} \log_4 x = \left(\log_4 x\right)^{\frac{1}{2}}$$
$$\left(\frac{1}{4} \log_4 x\right)^2 = \left[\left(\log_4 x\right)^{\frac{1}{2}}\right]^2$$
$$\frac{1}{16} (\log_4 x)^2 = \log_4 x$$
$$\frac{1}{16} (\log_4 x)^2 - \log_4 x = 0$$
$$\log_4 x\left(\frac{1}{16} \log_4 x - 1\right) = 0$$

$\log_4 x = 0$ $\qquad$ $\log_4 x = 16$

$x = 4^0$ $\qquad$ $x = 4^{16}$

$x = 1$

$x = \mathbf{1, 4^{16}}$

17. (97)

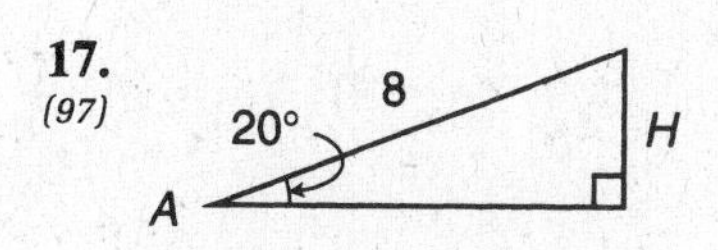

$H = 8 \sin 20° = 2.74$

$H < 4 < 8$, therefore two triangles are possible.
$$\frac{4}{\sin 20°} = \frac{8}{\sin B}$$
$$\sin B = \frac{8 \sin 20°}{4}$$
$$B = 43.16° \text{ or } 136.84°$$

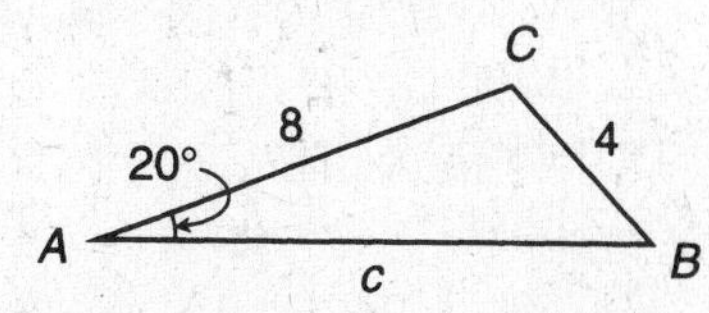

If $B = 43.16°$, $C = 180° - 20° - 43.16° = 116.84°$
$$\frac{c}{\sin 116.84°} = \frac{4}{\sin 20°}$$
$$c = \frac{4 \sin 116.84°}{\sin 20°}$$
$$c = \mathbf{10.44}$$

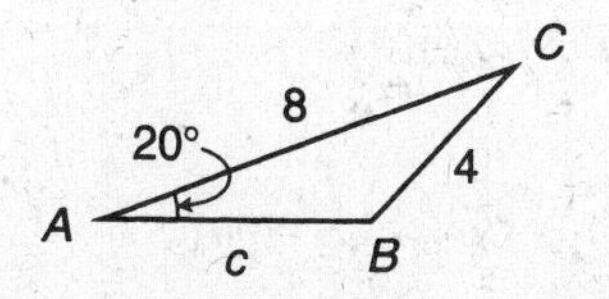

If $B = 136.84°$, $C = 180° - 20° - 136.84° = 23.16°$
$$\frac{c}{\sin 23.16°} = \frac{4}{\sin 20°}$$
$$c = \frac{4 \sin 23.16°}{\sin 20°}$$
$$c = \mathbf{4.60}$$

18. (96) $\dfrac{\sec 2x - 1}{2 \sec 2x} = \dfrac{\dfrac{1}{\cos 2x} - 1}{\dfrac{2}{\cos 2x}}$

$= \dfrac{\dfrac{1 - \cos 2x}{\cos 2x}}{\dfrac{2}{\cos 2x}} = \dfrac{1 - \cos 2x}{2}$

$= \dfrac{1 - (1 - 2\sin^2 x)}{2} = \dfrac{2 \sin^2 x}{2} = \sin^2 x$

19. (84) $(\cot^2 x)(1 + \tan^2 x) = \cot^2 x \sec^2 x$

$= \left(\dfrac{\cos^2 x}{\sin^2 x}\right)\left(\dfrac{1}{\cos^2 x}\right) = \dfrac{1}{\sin^2 x} = \dfrac{1}{1 - \cos^2 x}$

20. (93,96) $\dfrac{1}{2} \cot x \sec^2 x = \dfrac{1}{2} \dfrac{\cos x}{\sin x} \cdot \dfrac{1}{\cos^2 x}$

$= \dfrac{1}{2 \sin x \cos x} = \dfrac{1}{\sin 2x} = \csc 2x$

21. (96)

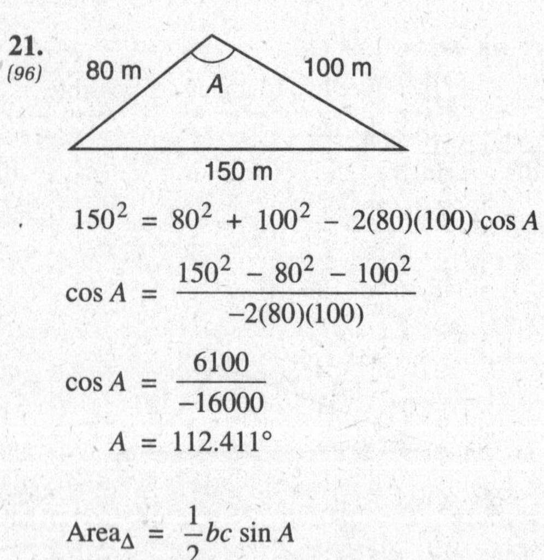

$150^2 = 80^2 + 100^2 - 2(80)(100) \cos A$

$\cos A = \dfrac{150^2 - 80^2 - 100^2}{-2(80)(100)}$

$\cos A = \dfrac{6100}{-16000}$

$A = 112.411°$

$\text{Area}_\Delta = \dfrac{1}{2} bc \sin A$

$= \dfrac{1}{2}(80)(100) \sin 112.411° = \mathbf{3697.89\ m^2}$

22. (84) $y = 10 + 2 \sin 4\left(x + \dfrac{\pi}{6}\right)$

$\text{Period} = \dfrac{2\pi}{4} = \dfrac{\pi}{2}$

y
+x
12
10
8
$-\dfrac{\pi}{6}$ $\dfrac{\pi}{12}$ $\dfrac{\pi}{3}$

23. (94) $y = -\sec x$

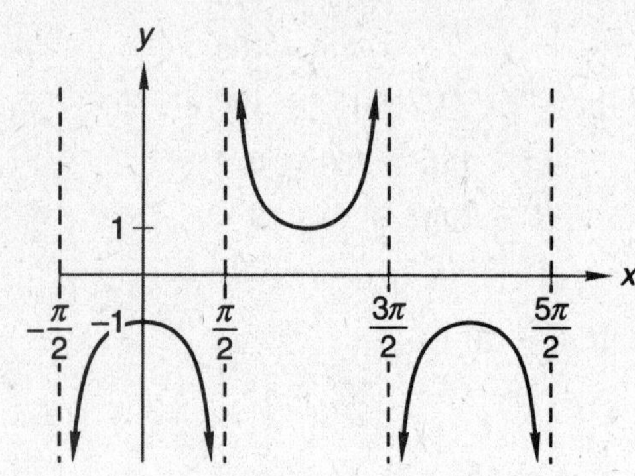

24. (90) $\cos 75° = \cos\left(\dfrac{150°}{2}\right) = \sqrt{\dfrac{1 + \cos 150°}{2}}$

$= \sqrt{\dfrac{1 - \dfrac{\sqrt{3}}{2}}{2}} = \mathbf{\dfrac{\sqrt{2 - \sqrt{3}}}{2}}$

25. (52) $2 \csc 3x + 4 = 0$

$\csc 3x = -2$

$3x = 210°, 330°, 570°, 690°, 930°, 1050°$

$x = \mathbf{70°, 110°, 190°, 230°, 310°, 350°}$

26. (85) $2\cos^2 x + 15 \sin x - 9 = 0$

$2(1 - \sin^2 x) + 15 \sin x - 9 = 0$

$2 \sin^2 x - 15 \sin x + 7 = 0$

$(2 \sin x - 1)(\sin x - 7) = 0$

$2 \sin x = 1$ $\qquad$ $\sin x = 7$

$\sin x = \dfrac{1}{2}$ $\qquad$ no answer

$x = \mathbf{30°, 150°}$

27. (95) $-4 + 2i = 4.47 \text{ cis } 153.43°$

$(4.47 \text{ cis } 153.43°)^{\frac{1}{3}} = 4.47^{\frac{1}{3}} \text{ cis}\left(\dfrac{153.43°}{3} + n\dfrac{360°}{3}\right)$

$= \mathbf{1.65 \text{ cis } 51.14°, 1.65 \text{ cis } 171.14°, 1.65 \text{ cis } 291.14°}$

28. (58) $x - y = 5$

$y = x - 5$

Equation of the perpendicular line:

$y = -x + b$

$1 = -(0) + b$

$b = 1$

$y = -x + 1$

Point of intersection:

$$x - 5 = -x + 1$$
$$2x = 6$$
$$x = 3$$

$$y = -x + 1 = -(3) + 1 = -2$$

(3, –2) and (0, 1)

$$D = \sqrt{(0 - 3)^2 + (1 - (-2))^2} = \sqrt{18} = \mathbf{3\sqrt{2}}$$

29. (67) (a) $\text{antilog}_4(-3) = 4^{-3} = \mathbf{\frac{1}{64}}$

(b) $\text{antilog}_{\frac{1}{4}}(-3) = \left(\frac{1}{4}\right)^{-3} = \mathbf{64}$

30. (70) $z = \frac{x - \mu}{\sigma} = \frac{32 - 36}{2} = -2$

Percentile = 0.0228

2.28%

Problem Set 102

1. (91) $243\left(\frac{1}{3}\right)^3 = \frac{243}{27} = \mathbf{9\text{ ft}}$

2. (92) $P(2 \cup R) = P(2) + P(R) - P(2 \cap R)$

$$= \frac{4}{52} + \frac{26}{52} - \frac{2}{52} = \frac{28}{52} = \mathbf{\frac{7}{13}}$$

3. (88) $A_0 = 0.04,\ A_{20} = 2.6$

$$A_t = A_0 e^{kt}$$
$$2.6 = 0.04e^{20k}$$
$$65 = e^{20k}$$
$$\ln 65 = 20k$$
$$k = \frac{\ln 65}{20}$$
$$k = 0.20872$$

$$16 = 0.04e^{0.20872t}$$
$$400 = e^{0.20872t}$$
$$\ln 400 = 0.20872t$$
$$t = \mathbf{28.71\text{ s}}$$

4. (36) $\begin{cases} \text{(a)}\ (B + W)T_D = D_D \\ \text{(b)}\ (B - W)T_U = D_U \end{cases}$

$B = 2W,\ T_U = T_D + 1$

(b) $(B - W)T_U = D_U$

$$(2W - W)T_U = 20$$
$$WT_U = 20$$

(a) $(B + W)T_D = D_D$

$$3W(T_U - 1) = 33$$
$$3WT_U - 3W = 33$$
$$3(20) - 3W = 33$$
$$3W = 27$$
$$\mathbf{W = 9\text{ mph}}$$

$$B = 2W$$
$$B = 2(9)$$
$$\mathbf{B = 18\text{ mph}}$$

5. (25) $\begin{cases} \text{(a)}\ W_N = 5P_N \\ \text{(b)}\ W_N - 5 = 6(P_N - 5) + 19 \end{cases}$

(b) $W_N - 5 = 6(P_N - 5) + 19$

$$(5P_N) - 5 = 6P_N - 30 + 19$$
$$P_N = 6$$

(a) $W_N = 5P_N = 5(6) = 30$

17 years from now:

$\mathbf{P = 23\text{ yr};\ W = 47\text{ yr}}$

6. (1,18) $\begin{cases} \text{(a)}\ 4(90^\circ - A_1) = (180^\circ - A_2) - 40^\circ \\ \text{(b)}\ A_1 + A_2 = 80^\circ \end{cases}$

(b) $A_1 + A_2 = 80^\circ$

$$A_2 = 80^\circ - A_1$$

(a) $4(90^\circ - A_1) = (180^\circ - A_2) - 40^\circ$

$$360^\circ - 4A_1 = 140^\circ - A_2$$
$$360^\circ - 4A_1 = 140^\circ - (80^\circ - A_1)$$
$$5A_1 = 300^\circ$$
$$\mathbf{A_1 = 60^\circ}$$

$$A_2 = 80^\circ - A_1$$
$$A_2 = 80^\circ - (60^\circ)$$
$$\mathbf{A_2 = 20^\circ}$$

7. (102) $(x - y)^7$

$F = x,\ S = -y$

Term	①	②	③	④	⑤...⑧
Exp. of F	7	6	5	4	3... 0
Exp. of S	0	1	2	3	4... 7
Coefficient	1	7	21	35	35... 1

$35F^4S^3 = 35x^4(-y)^3 = \mathbf{-35x^4y^3}$

8. (102) $(2a^2 - b^3)^3$

$F = 2a^2,\ S = -b^3$

Term	①	②	③	④
Exponents of F	3	2	1	0
Exponents of S	0	1	2	3
Coefficient	1	3	3	1

$(F + S)^3$

$= F^3 + 3F^2S + 3FS^2 + S^3$

$= (2a^2)^3 + 3(2a^2)^2(-b^3) + 3(2a^2)(-b^3)^2 + (-b^3)^3$

$= \mathbf{8a^6 - 12a^4b^3 + 6a^2b^6 - b^9}$

9. (101) $\begin{vmatrix} 0 & 1 & 2 \\ 3 & 0 & 1 \\ 1 & 5 & 6 \end{vmatrix}$

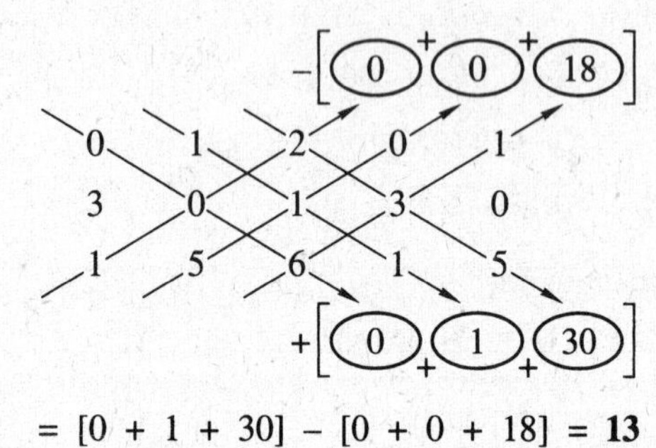

$= [0 + 1 + 30] - [0 + 0 + 18] = \mathbf{13}$

10. (101) $x = \dfrac{\begin{vmatrix} 0 & -1 & 1 \\ 4 & 3 & 2 \\ 5 & 2 & 0 \end{vmatrix}}{\begin{vmatrix} 3 & -1 & 1 \\ 0 & 3 & 2 \\ 1 & 2 & 0 \end{vmatrix}}$

Numerator: $-[15 + 0 + 0]$, $+[0 + (-10) + 8]$

Denominator: $-[3 + 12 + 0]$, $+[0 + (-2) + 0]$

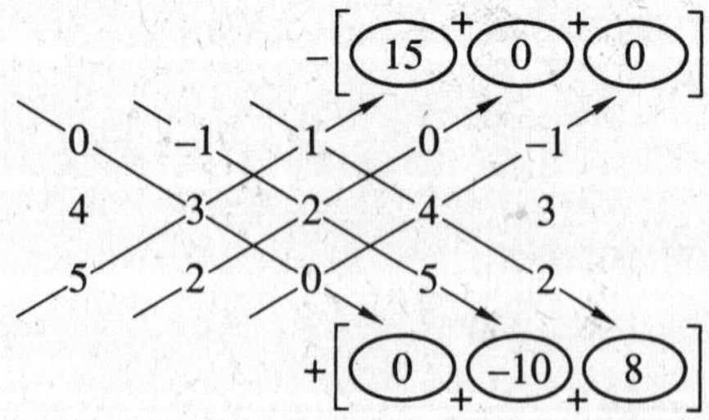

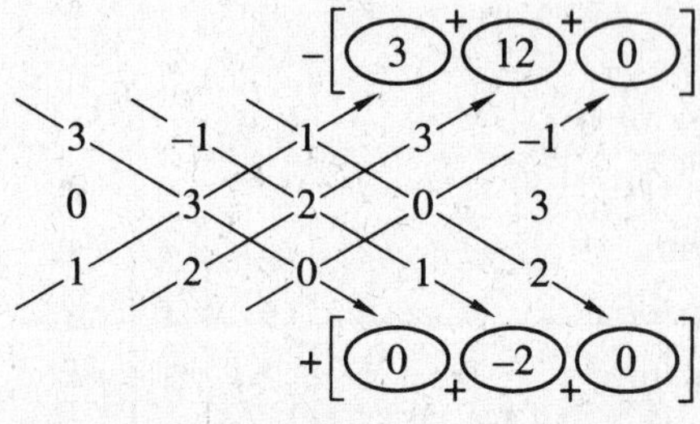

$x = \dfrac{[0 - 10 + 8] - [15 + 0 + 0]}{[0 - 2 + 0] - [3 + 12 + 0]} = \dfrac{-17}{-17} = \mathbf{1}$

11. (100)

$$\begin{aligned} A + B &= x \\ A - B &= y \\ \hline 2A &= x + y \\ A &= \frac{x + y}{2} \end{aligned} \qquad \begin{aligned} A + B &= x \\ -(A - B &= y) \\ \hline 2B &= x - y \\ B &= \frac{x - y}{2} \end{aligned}$$

$$\begin{aligned} \cos(A + B) &= \cos A \cos B - \sin A \sin B \\ \cos(A - B) &= \cos A \cos B + \sin A \sin B \\ \hline \cos(A + B) + \cos(A - B) &= 2\cos A \cos B \end{aligned}$$

$$\cos x + \cos y = 2\cos\frac{x + y}{2}\cos\frac{x - y}{2}$$

12. (100)

$$\begin{aligned} A + B &= x \\ A - B &= y \\ \hline 2A &= x + y \\ A &= \frac{x + y}{2} \end{aligned} \qquad \begin{aligned} A + B &= x \\ -(A - B &= y) \\ \hline 2B &= x - y \\ B &= \frac{x - y}{2} \end{aligned}$$

$$\begin{aligned} -\cos(A - B) &= -\cos A \cos B - \sin A \sin B \\ \cos(A + B) &= \cos A \cos B - \sin A \sin B \\ \hline \cos(A + B) - \cos(A - B) &= -2\sin A \sin B \end{aligned}$$

$$\cos x - \cos y = -2\sin\frac{x + y}{2}\sin\frac{x - y}{2}$$

13. (99) $\begin{cases} \text{(a)}\ -\sqrt{xy} = -12 \\ \text{(b)}\ \dfrac{x + y}{2} = 20 \end{cases}$

(b) $\dfrac{x + y}{2} = 20$

$x + y = 40$

$y = 40 - x$

(a)
$$\begin{aligned} -\sqrt{xy} &= -12 \\ xy &= 144 \\ x(40 - x) &= 144 \\ 40x - x^2 &= 144 \\ x^2 - 40x + 144 &= 0 \\ (x - 4)(x - 36) &= 0 \\ x &= 4, 36 \end{aligned}$$

If $x = 4$,
$y = 40 - x = 40 - (4) = 36.$

If $x = 36$,
$y = 40 - x = 40 - (36) = 4.$

4 and 36

14. (98) (a) $\log_7 40 = x$

$$7^x = 40$$
$$x \log 7 = \log 40$$
$$x = \frac{\log 40}{\log 7}$$
$$x = \mathbf{1.90}$$

(b) $\log_3 40 = x$

$$3^x = 40$$
$$x \log 3 = \log 40$$
$$x = \frac{\log 40}{\log 3}$$
$$x = \mathbf{3.36}$$

15. (98)

$$\sqrt{\log_{81} x} = 2 \log_{81} x$$
$$\left(\sqrt{\log_{81} x}\right)^2 = (2 \log_{81} x)^2$$
$$\log_{81} x = 4(\log_{81} x)^2$$
$$4(\log_{81} x)^2 - \log_{81} x = 0$$
$$\log_{81} x(4 \log_{81} x - 1) = 0$$

$$\log_{81} x = 0 \qquad 4 \log_{81} x = 1$$
$$x = 81^0 \qquad \log_{81} x = \frac{1}{4}$$
$$x = 1 \qquad x = 81^{\frac{1}{4}}$$
$$x = 3$$

$$x = \mathbf{1, 3}$$

16. (98) $\log_2 (x + 1) - \log_2 (x - 1) = 1$

$$\log_2 \left(\frac{x + 1}{x - 1}\right) = 1$$
$$\frac{x + 1}{x - 1} = 2^1$$
$$x + 1 = 2x - 2$$
$$x = \mathbf{3}$$

17. (98) $\ln x + \ln (x - 1) = \ln 20$

$$\ln [x(x - 1)] = \ln 20$$
$$x^2 - x = 20$$
$$x^2 - x - 20 = 0$$
$$(x - 5)(x + 4) = 0$$
$$x = 5, -4 \quad (x \neq -4)$$
$$x = \mathbf{5}$$

18. (97)

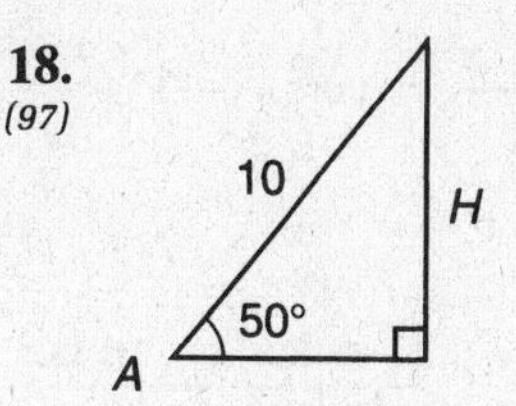

$H = 10 \sin 50° = 7.66$

$H < 9 < 10$, therefore two triangles are possible.

$$\frac{9}{\sin 50°} = \frac{10}{\sin B}$$
$$\sin B = \frac{10 \sin 50°}{9}$$
$$B = 58.34° \text{ or } 121.66°$$

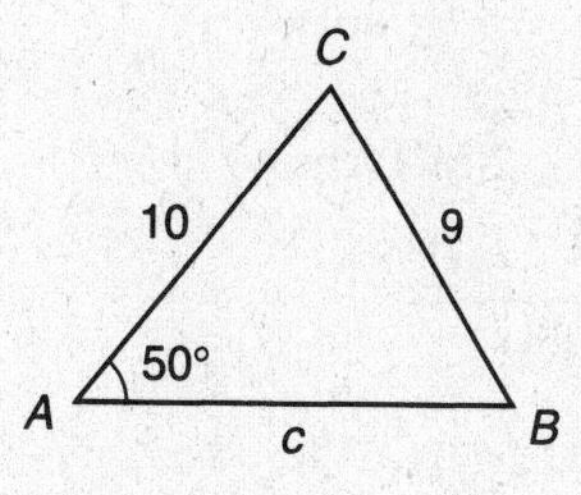

If $B = 58.34°$,

$C = 180° - 50° - 58.34° = 71.66°$

$$\frac{9}{\sin 50°} = \frac{c}{\sin 71.66°}$$
$$c = \frac{9 \sin 71.66°}{\sin 50°}$$
$$c = \mathbf{11.15}$$

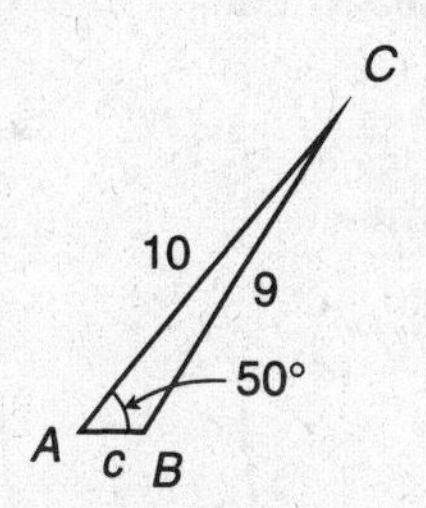

If $B = 121.66°$,

$C = 180° - 50° - 121.66° = 8.34°$

$$\frac{9}{\sin 50°} = \frac{c}{\sin 8.34°}$$
$$c = \frac{9 \sin 8.34°}{\sin 50°}$$
$$c = \mathbf{1.70}$$

19. (84,93) $\dfrac{\sin^2 x}{1 - \cos x} - \dfrac{1}{\sec x}$

$$= \frac{1 - \cos^2 x}{1 - \cos x} - \frac{1}{\sec x}$$
$$= \frac{(1 - \cos x)(1 + \cos x)}{1 - \cos x} - \cos x$$
$$= 1 + \cos x - \cos x = 1$$

20. (96) $\dfrac{\cos 2x}{\sin^2 x} = \dfrac{\cos^2 x - \sin^2 x}{\sin^2 x} = \cot^2 x - 1$

21. (93,96) $\dfrac{1 + \sin x}{\cos x} + \dfrac{\cos x}{1 + \sin x} = \dfrac{(1 + \sin x)^2 + \cos^2 x}{\cos x(1 + \sin x)}$

$= \dfrac{1 + 2\sin x + \sin^2 x + \cos^2 x}{\cos x(1 + \sin x)}$

$= \dfrac{2 + 2\sin x}{\cos x(1 + \sin x)} = \dfrac{2(1 + \sin x)}{\cos x(1 + \sin x)}$

$= \dfrac{2}{\cos x} \cdot \dfrac{2\sin x}{2\sin x} = \dfrac{4\sin x}{2\sin x\cos x} = \dfrac{4\sin x}{\sin 2x}$

22. (81,96) $7^2 = 5^2 + 4^2 - 2(5)(4)\cos A$

$\cos A = \dfrac{7^2 - 5^2 - 4^2}{-2(5)(4)}$

$\mathbf{A = 101.54°}$

$\text{Area} = \dfrac{1}{2}bc\sin A$

$= \dfrac{1}{2}(4)(5)\sin 101.54°$

$= 9.7979 \text{ in.}^2\left(\dfrac{2.54 \text{ cm}}{1 \text{ in.}}\right)\left(\dfrac{2.54 \text{ cm}}{1 \text{ in.}}\right)$

$\textbf{Area} = \mathbf{63.21 \text{ cm}^2}$

23. (94) $y = 2\tan x$

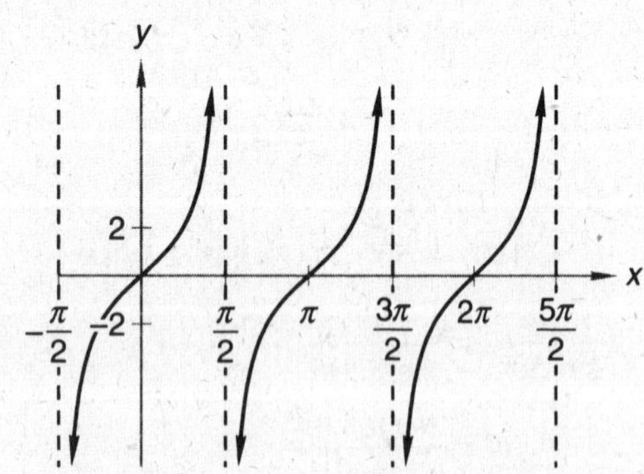

24. (94) (a) $\mathbf{y = 6 + 8\csc x}$

(b) $\mathbf{y = \cot\theta}$

25. (52) $\tan 5\theta = 0$

$5\theta = 0, \pi, 2\pi, 3\pi, 4\pi, 5\pi, 6\pi, 7\pi, 8\pi, 9\pi$

$\theta = \mathbf{0, \frac{\pi}{5}, \frac{2\pi}{5}, \frac{3\pi}{5}, \frac{4\pi}{5}, \pi, \frac{6\pi}{5}, \frac{7\pi}{5}, \frac{8\pi}{5}, \frac{9\pi}{5}}$

26. (85) $3\tan^2 x - 5\sec x + 1 = 0$

$3(\sec^2 x - 1) - 5\sec x + 1 = 0$

$3\sec^2 x - 5\sec x - 2 = 0$

$(3\sec x + 1)(\sec x - 2) = 0$

$3\sec x = -1 \qquad \sec x = 2$

$\sec x = -\dfrac{1}{3} \qquad x = \mathbf{\dfrac{\pi}{3}, \dfrac{5\pi}{3}}$

no answer

27. (95) $-2 + 8i = 8.25 \text{ cis } 104.04°$

$(8.25 \text{ cis } 104.04°)^{\frac{1}{5}}$

$= 8.25^{\frac{1}{5}} \text{ cis}\left(\dfrac{104.04°}{5} + n\dfrac{360°}{5}\right)$

= **1.52 cis 20.81°, 1.52 cis 92.81°, 1.52 cis 164.81°, 1.52 cis 236.81°, 1.52 cis 308.81°**

28. (73) $\text{Side} = \dfrac{30 \text{ ft}}{6} = 5 \text{ ft}$

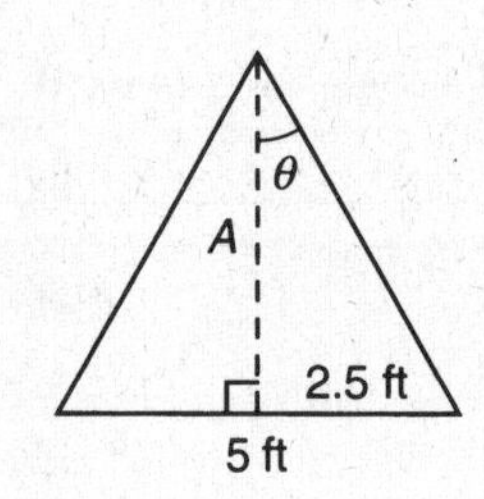

$\theta = \dfrac{1}{2}\left(\dfrac{360°}{6}\right) = 30°$

$A = \dfrac{2.5}{\tan 30°} = 4.33 \text{ ft}$

$\text{Area}_\Delta = \dfrac{1}{2}(2.5)(4.33) = 5.4127 \text{ ft}^2$

$\text{Area} = (12)(5.4127) = \mathbf{64.95 \text{ ft}^2}$

29. (46) $x^3 + 5x^2 + 10x = 0$

$x(x^2 + 5x + 10) = 0$

$x = \dfrac{-5 \pm \sqrt{25 - (4)(10)}}{2} = \dfrac{-5 \pm \sqrt{15}i}{2}$

$\mathbf{\left(x + \frac{5}{2} - \frac{\sqrt{15}}{2}i\right)\left(x + \frac{5}{2} + \frac{\sqrt{15}}{2}i\right)x}$

30. (24) $(f - g)(x) = x^3 - 2x^2 + 1$

$(f - g)(2) = (2)^3 - 2(2)^2 + 1$

$= 8 - 8 + 1 = \mathbf{1}$

Problem Set 103

1. (91) $a_4 = 128\left(\frac{1}{4}\right)^4 = \frac{128}{256} = \mathbf{\frac{1}{2}\ ft}$

2. (83) $\frac{1}{2} \cdot \frac{1}{2} \cdot \frac{1}{2} \cdot \frac{1}{2} = \left(\frac{1}{2}\right)^4 = \mathbf{\frac{1}{16}}$

3. (88) $A_0 = 42$ cubits, $A_{30} = 39$ cubits

$$A_t = A_0 e^{kt}$$
$$39 = 42e^{30k}$$
$$\frac{13}{14} = e^{30k}$$
$$\ln\frac{13}{14} = 30k$$
$$k = -0.0024703$$

$$12.5 = 42e^{-0.0024703t}$$
$$\frac{12.5}{42} = e^{-0.0024703t}$$
$$\ln\frac{12.5}{42} = -0.0024703t$$
$$t = \mathbf{490.61\ min}$$

4. (18) $P = k\frac{1}{V}$

$$4 = k\frac{1}{10}$$
$$k = 40$$
$$P = k\frac{1}{V} = (40)\frac{1}{2} = \mathbf{20\ atm}$$

5. (28) $\text{Rate} = \frac{k}{p}\frac{\text{mi}}{\text{hr}}$

$$\text{New rate} = \left(\frac{k}{p} + s\right)\frac{\text{mi}}{\text{hr}} = \frac{k + sp}{p}\frac{\text{mi}}{\text{hr}}$$

$$\text{New time} = \frac{\text{distance}}{\text{new rate}}$$
$$= \frac{12\text{ mi}}{\frac{k + sp}{p}\frac{\text{mi}}{\text{hr}}} = \mathbf{\frac{12p}{k + sp}\ hr}$$

6. (18) $\text{Acid}_1 + \text{acid added} = \text{acid final}$

$$180(0.35) + A = (180 + A)(0.5)$$
$$63 + A = 90 + 0.5A$$
$$0.5A = 27$$
$$A = \mathbf{54\ ml}$$

7. (103) $\frac{3}{5}\log_3 x + \frac{1}{4}\log_3 y - 3\log_3 z$

$$= \log_3 x^{\frac{3}{5}} + \log_3 y^{\frac{1}{4}} - \log_3 z^3$$
$$= \mathbf{\log_3 \frac{x^{\frac{3}{5}}y^{\frac{1}{4}}}{z^3}}$$

8. (103) $\frac{\sqrt[6]{525{,}000}}{(6300)^{1.5}} = \mathbf{1.80 \times 10^{-5}}$

9. (103) $H^+ = 5.3 \times 10^{-5}$

$$\text{pH} = -\log H^+ = -\log\left(5.3 \times 10^{-5}\right) = \mathbf{4.28}$$

10. (103) $\text{pH} = 6.5$

$$H^+ = 10^{-\text{pH}} = 10^{-6.5} = \mathbf{3.16 \times 10^{-7}\ \frac{mole}{liter}}$$

11. (102) $(x - y)^8$

$F = x,\ S = -y$

There will be $(n + 1)$ or 9 terms. We want the 5th term.

Term	①	②	③	④	⑤	⑥ . . . ⑨
Exp. of F	8	7	6	5	4	3 . . . 0
Exp. of S	0	1	2	3	4	5 . . . 8
Coefficient	1	8	28	56	70	56 . . . 1

$$70F^4S^4 = 70(x)^4(-y)^4 = \mathbf{70x^4y^4}$$

12. (102) $\left(3a^2 - b^3\right)^4$

$F = 3a^2,\ S = -b^3$

Term	①	②	③	④	⑤
Exp. of F	4	3	2	1	0
Exp. of S	0	1	2	3	4
Coefficient	1	4	6	4	1

$$(F + S)^4$$
$$= F^4 + 4F^3S + 6F^2S^2 + 4FS^3 + S^4$$
$$= \left(3a^2\right)^4 + 4\left(3a^2\right)^3\left(-b^3\right) + 6\left(3a^2\right)^2\left(-b^3\right)^2 + 4\left(3a^2\right)\left(-b^3\right)^3 + \left(-b^3\right)^4$$
$$= \mathbf{81a^8 - 108a^6b^3 + 54a^4b^6 - 12a^2b^9 + b^{12}}$$

13. (101)

$$x = \frac{\begin{vmatrix} -3 & 1 & 3 \\ 5 & 3 & 4 \\ 3 & 2 & 0 \end{vmatrix}}{\begin{vmatrix} 2 & 1 & 3 \\ 0 & 3 & 4 \\ 1 & 2 & 0 \end{vmatrix}}$$

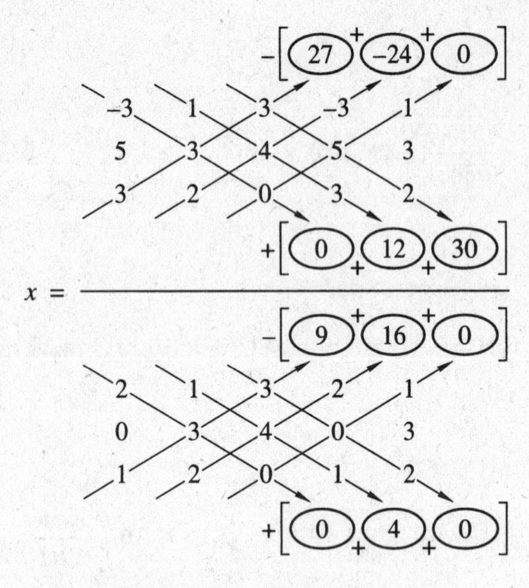

$$x = \frac{(0 + 12 + 30) - (27 - 24 + 0)}{(0 + 4 + 0) - (9 + 16 + 0)}$$

$$x = \frac{42 - 3}{4 - 25} = \frac{39}{-21} = -\mathbf{\frac{13}{7}}$$

14. (100)

$$\begin{aligned} A + B &= x \\ A - B &= y \\ \hline 2A \quad &= x + y \end{aligned} \qquad \begin{aligned} A + B &= x \\ -(A - B &= y) \\ \hline 2B &= x - y \end{aligned}$$

$$A = \frac{x + y}{2} \qquad B = \frac{x - y}{2}$$

$$\begin{aligned} \cos(A + B) &= \cos A \cos B - \sin A \sin B \\ -\cos(A - B) &= -\cos A \cos B - \sin A \sin B \\ \hline \cos(A + B) - \cos(A - B) &= -2 \sin A \sin B \end{aligned}$$

$$\cos x - \cos y = -2 \sin \frac{x + y}{2} \sin \frac{x - y}{2}$$

15. (100)

$$\begin{aligned} \cos(A - B) &= \cos A \cos B + \sin A \sin B \\ -\cos(A + B) &= -\cos A \cos B + \sin A \sin B \\ \hline \cos(A - B) - \cos(A + B) &= 2 \sin A \sin B \end{aligned}$$

$$\sin A \sin B = \frac{\cos(A - B) - \cos(A + B)}{2}$$

$$\sin A \sin B = \frac{1}{2}[\cos(A - B) - \cos(A + B)]$$

16. (98) (a)

$$\begin{aligned} y &= \log_5 22 \\ 5^y &= 22 \\ y \log 5 &= \log 22 \\ y &= \frac{\log 22}{\log 5} \\ y &= \mathbf{1.92} \end{aligned}$$

(b)

$$\begin{aligned} y &= \log_{12} 22 \\ 12^y &= 22 \\ y \log 12 &= \log 22 \\ y &= \frac{\log 22}{\log 12} \\ y &= \mathbf{1.24} \end{aligned}$$

17. (59,98)

$$\begin{aligned} \log_5 \sqrt{x^2 + 16} &= 1 \\ \sqrt{x^2 + 16} &= 5^1 \\ x^2 + 16 &= 25 \\ x^2 &= 9 \\ x &= \mathbf{\pm 3} \end{aligned}$$

18. (59)

$$\begin{aligned} \frac{2}{3} \log_4 8 - \log_4(3x - 2) &= 2 \\ \log_4 8^{\frac{2}{3}} - \log_4(3x - 2) &= 2 \\ \log_4\left(\frac{4}{3x - 2}\right) &= 2 \\ 4^2 &= \frac{4}{3x - 2} \\ 16(3x - 2) &= 4 \\ 48x - 32 &= 4 \\ 48x &= 36 \\ x &= \frac{36}{48} \\ x &= \mathbf{\frac{3}{4}} \end{aligned}$$

19. (59,98)

$$\begin{aligned} 4 \log_3 \sqrt[4]{x} &= \log_3(3x - 1) \\ \log_3\left(\sqrt[4]{x}\right)^4 &= \log_3(3x - 1) \\ \log_3 x &= \log_3(3x - 1) \\ x &= 3x - 1 \\ -2x &= -1 \\ x &= \mathbf{\frac{1}{2}} \end{aligned}$$

20. $\log_2(\log_2 x) = 1$
(98)

$$\log_2 x = 2^1$$
$$x = 2^2$$
$$x = \mathbf{4}$$

21.
(97)

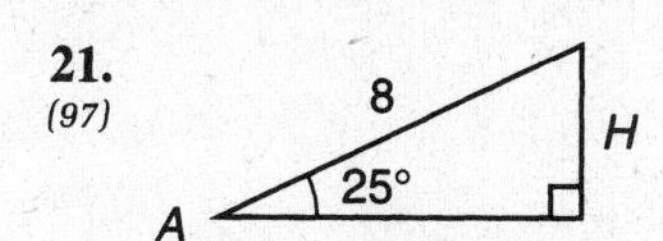

$H = 8 \sin 25° = 3.38$

$H < 5 < 8$, therefore two triangles are possible.

$$\frac{5}{\sin 25°} = \frac{8}{\sin B}$$
$$\sin B = \frac{8 \sin 25°}{5}$$
$$B = 42.55° \text{ or } 137.45°$$

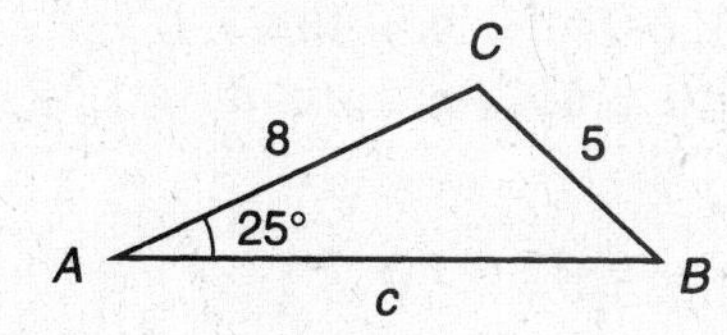

If $B = 42.55°$,

$C = 180° - 25° - 42.55° = 112.45°$

$$\frac{5}{\sin 25°} = \frac{c}{\sin 112.45°}$$
$$c = \frac{5 \sin 112.45°}{\sin 25°}$$
$$c = \mathbf{10.93}$$

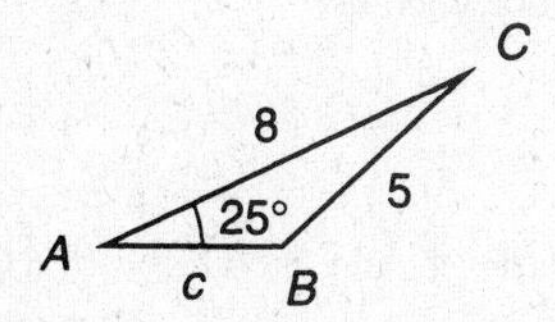

If $B = 137.45°$,

$C = 180° - 25° - 137.45° = 17.55°$

$$\frac{5}{\sin 25°} = \frac{c}{\sin 17.55°}$$
$$c = \frac{5 \sin 17.55°}{\sin 25°}$$
$$c = \mathbf{3.57}$$

22. $\dfrac{\cos 2x + 1}{2} = \dfrac{2\cos^2 x - 1 + 1}{2} = \cos^2 x$
(96)

23. $\dfrac{2 \cot x}{\tan 2x} = \dfrac{2 \cot x}{\dfrac{2 \tan x}{1 - \tan^2 x}}$
(96)

$$= 2 \cot x \cdot \frac{1 - \tan^2 x}{2 \tan x}$$
$$= 2 \cot x \cdot \frac{\frac{(1 - \tan^2 x)}{2}}{\cot x}$$
$$= \cot^2 x(1 - \tan^2 x) = \cot^2 x - 1$$
$$= \csc^2 x - 1 - 1 = \csc^2 x - 2$$

24.
(81,96)

0.05 m, 120°, a, 0.08 m

$$\text{Area} = \frac{1}{2}bc \sin A$$
$$\text{Area} = \frac{1}{2}(0.05)(0.08) \sin 120°$$
$$\textbf{Area} = \mathbf{0.0017\ m^2}$$
$$a^2 = 5^2 + 8^2 - 2(5)(8) \cos 120°$$
$$a = \mathbf{11.36\ cm}$$

25. $y = \dfrac{1}{2}\cot x$
(94)

26. $\cos(345°) = \cos(120° + 225°)$
(87)

$$= \cos(120°)\cos(225°) - \sin(120°)\sin(225°)$$
$$= \left(-\frac{1}{2}\right)\left(-\frac{\sqrt{2}}{2}\right) - \left(\frac{\sqrt{3}}{2}\right)\left(-\frac{\sqrt{2}}{2}\right)$$
$$= \frac{\sqrt{2}}{4} + \frac{\sqrt{6}}{4} = \mathbf{\frac{\sqrt{6} + \sqrt{2}}{4}}$$

27. $1 + \tan 3\theta = 0$
(52)

$$\tan 3\theta = -1$$
$$3\theta = 135°, 315°, 495°, 675°, 855°, 1035°$$
$$\theta = \mathbf{45°, 105°, 165°, 225°, 285°, 345°}$$

28. (60)

$$\sqrt{2}\sec x = \sqrt{\sec x + 1}$$

$$2\sec^2 x = \sec x + 1$$

$$2\sec^2 x - \sec x - 1 = 0$$

$$(2\sec x + 1)(\sec x - 1) = 0$$

$2\sec x = -1$ $\sec x = 1$

$\sec x = -\frac{1}{2}$ $x = \mathbf{0°}$

no answer

29. (95) $-5 + 12i = 13 \text{ cis } 112.62°$

$$(13 \text{ cis } 112.62°)^{\frac{1}{3}} = 13^{\frac{1}{3}} \text{ cis}\left(\frac{112.62°}{3} + n\frac{360°}{3}\right)$$

$= \mathbf{2.35 \text{ cis } 37.54°, 2.35 \text{ cis } 157.54°, 2.35 \text{ cis } 277.54°}$

30. (59) $(x + 1)4^{2\log_4 x - \log_4(x^2 + x)}$

$$= (x + 1)4^{\log_4 x^2 - \log_4(x^2+x)}$$

$$= (x + 1)4^{\log_4\left(\frac{x^2}{x^2+x}\right)}$$

$$= (x + 1)\left(\frac{x^2}{x^2 + x}\right) = (x + 1)\frac{x^2}{(x + 1)(x)} = \mathbf{x}$$

Problem Set 104

1. (92) $P(8 \cup B) = P(8) + P(B) - P(8 \cap B)$

$$= \frac{4}{52} + \frac{26}{52} - \frac{2}{52}$$

$$= \frac{28}{52} = \mathbf{\frac{7}{13}}$$

2. (44) Rate $= 140 + f + s$

Rate $\times N$ = price

$$N = \frac{\text{price}}{\text{rate}} = \mathbf{\frac{4200}{140 + s + f} \textbf{ gal}}$$

3. (18)

$$\begin{cases} \text{(a)} & 0.60N_B = N_W \\ \text{(b)} & 5N_R = 2N_W + 10 \\ \text{(c)} & N_W + N_R + N_B = 140 \end{cases}$$

(a) $0.60N_B = N_W$

$$N_B = \frac{5}{3}N_W$$

(b) $5N_R = 2N_W + 10$

$$N_R = \frac{2}{5}N_W + 2$$

(c) $N_W + N_R + N_B = 140$

$$N_W + \left(\frac{2}{5}N_W + 2\right) + \left(\frac{5}{3}N_W\right) = 140$$

$$\frac{46}{15}N_W = 138$$

$$\mathbf{N_W = 45}$$

(a) $0.60N_B = N_W$

$$0.60N_B = (45)$$

$$\mathbf{N_B = 75}$$

(c) $N_W + N_R + N_B = 140$

$$(45) + N_R + (75) = 140$$

$$\mathbf{N_R = 20}$$

4. (18)

$$\begin{cases} \text{(a)} & T + U = 9 \\ \text{(b)} & 10U + T = 9 + 10T + U \end{cases}$$

(b) $10U + T = 9 + 10T + U$

$$9U - 9T = 9$$

$$U - T = 1$$

$$U = 1 + T$$

(a) $T + U = 9$

$$T + (1 + T) = 9$$

$$2T = 8$$

$$T = 4$$

$U = 1 + T$

$U = 1 + (4)$

$U = 5$

The number was **45**.

5. (91,104) (a) $S_n = a_1 + (a_1 + d) + \ldots + [a_1 + (n - 1)d]$

$$S_n = \frac{n}{2}\{a_1 + [a_1 + (n - 1)d]\}$$

$$\mathbf{S_n = \frac{n}{2}(a_1 + a_n)}$$

(b) $2 + 4 + 6 + \ldots + 98 + 100$

From the arithmetic series we find that $a_1 = 2$, $a_n = 100$ and $d = 2$.

First, determine n using:

$$a_1 + (n - 1)d = a_n$$

$$(2) + (n - 1)(2) = 100$$

$$n - 1 = 49$$

$$n = 50$$

Then, find the sum of the arithmetic series using the equation developed in part (a).

$$S_n = \frac{n}{2}(a_1 + a_n)$$

$$S_{50} = \frac{50}{2}(2 + 100) = \mathbf{2550}$$

6. (104) (a)

$$S_n = a_1 + a_1r + \ldots + a_1r^{n-2} + a_1r^{n-1}$$
$$-rS_n = -a_1r - a_1r^2 - \ldots - a_1r^{n-1} - a_1r^n$$

$$S_n - rS_n = a_1 - a_1r^n$$

$$S_n(1 - r) = a_1(1 - r^n)$$

$$\mathbf{S_n = \frac{a_1(1 - r^n)}{1 - r}}$$

(b) $a_1 = 2,\ r = -2,\ n = 5$

$$S_n = \frac{a_1(1 - r^n)}{1 - r}$$

$$S_5 = \frac{2[1 - (-2)^5]}{1 - (-2)} = \frac{2(1 + 32)}{3} = \frac{66}{3} = \mathbf{22}$$

7. (103) $\log_4 \dfrac{\sqrt[3]{(s - 1)^2(t + 2)}}{\sqrt{s^3}}$

$$= \log_4 \frac{\left[(s - 1)^2(t + 2)\right]^{\frac{1}{3}}}{(s^3)^{\frac{1}{2}}}$$

$$= \log_4\left[(s - 1)^2(t + 2)\right]^{\frac{1}{3}} - \log_4 s^{\frac{3}{2}}$$

$$= \frac{1}{3}\left[\log_4(s - 1)^2(t + 2)\right] - \frac{3}{2}\log_4 s$$

$$= \frac{1}{3}\left[\log_4(s - 1)^2 + \log_4(t + 2)\right] - \frac{3}{2}\log_4 s$$

$$= \frac{1}{3}\left[2\log_4(s - 1) + \log_4(t + 2)\right] - \frac{3}{2}\log_4 s$$

$$= \mathbf{\frac{2}{3}\log_4(s - 1) + \frac{1}{3}\log_4(t + 2) - \frac{3}{2}\log_4 s}$$

8. (103) $\dfrac{\sqrt[3]{53{,}000}}{(3200)^{2.5}} + \sqrt[5]{(75{,}000)^2} = \mathbf{89.13}$

9. (103) $\text{pH} = -\log H^+$

$\text{pH} = -\log(6.2 \times 10^{-4}) = 3.2076 = \mathbf{3.2}$

10. (103) $H^+ = 10^{-\text{pH}} = 10^{-8.5} = \mathbf{3.16 \times 10^{-9}\ \dfrac{mole}{liter}}$

11. (102) $(x + 2y)^6$

$F = x,\ S = 2y$

Term	①	②	③	④	⑤...⑦
Exp. of F	6	5	4	3	2... 0
Exp. of S	0	1	2	3	4... 6
Coefficient	1	6	15	20	15... 1

$20F^3S^3 = 20(x)^3(2y)^3 = \mathbf{160x^3y^3}$

12. (102) $(x^2 - 2y)^3$

$F = x^2,\ S = -2y$

Term	①	②	③	④
Exp. of F	3	2	1	0
Exp. of S	0	1	2	3
Coefficient	1	3	3	1

$$(F + S)^3$$
$$= F^3 + 3F^2S + 3FS^2 + S^3$$
$$= (x^2)^3 + 3(x^2)^2(-2y) + 3(x^2)(-2y)^2 + (-2y)^3$$
$$= \mathbf{x^6 - 6x^4y + 12x^2y^2 - 8y^3}$$

13. (101) $x = \dfrac{\begin{vmatrix} 0 & 3 & 1 \\ 4 & 0 & 2 \\ -4 & 1 & -3 \end{vmatrix}}{\begin{vmatrix} 2 & 3 & 1 \\ 1 & 0 & 2 \\ 0 & 1 & -3 \end{vmatrix}}$

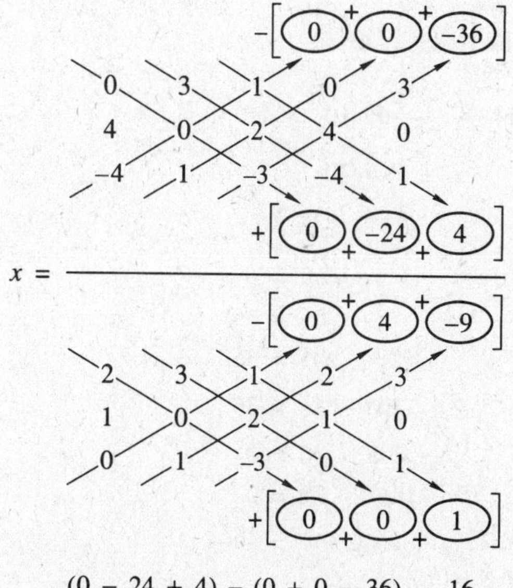

$$x = \frac{(0 - 24 + 4) - (0 + 0 - 36)}{(0 + 0 + 1) - (0 + 4 - 9)} = \frac{16}{6} = \mathbf{\frac{8}{3}}$$

14. (101)
$$\begin{vmatrix} 7 & -2 & 3 \\ 0 & 4 & 4 \\ 0 & 1 & 1 \end{vmatrix}$$

$-[(0)\ (28)\ (0)]$

```
 7   -2   3   7   -2
 0    4   4   0    4
 0    1   1   0    1
```

$+[(28)\ (0)\ (0)]$

$= (28 + 0 + 0) - (0 + 28 + 0)$

$= 28 - 28 = \mathbf{0}$

15. (100)

$$\begin{aligned} A + B &= x \\ A - B &= y \\ \hline 2A &= x + y \end{aligned} \qquad \begin{aligned} A + B &= x \\ -(A - B &= y) \\ \hline 2B &= x - y \end{aligned}$$

$$A = \frac{x + y}{2} \qquad B = \frac{x - y}{2}$$

$$\begin{aligned} \cos(A + B) &= \cos A \cos B - \sin A \sin B \\ \cos(A - B) &= \cos A \cos B + \sin A \sin B \\ \hline \cos(A + B) + \cos(A - B) &= 2 \cos A \cos B \end{aligned}$$

$$\cos x + \cos y = 2 \cos \frac{x + y}{2} \cos \frac{x - y}{2}$$

16. (100)

$$\begin{aligned} \cos(x + y) &= \cos x \cos y - \sin x \sin y \\ \cos(x - y) &= \cos x \cos y + \sin x \sin y \\ \hline \cos(x + y) + \cos(x - y) &= 2 \cos x \cos y \end{aligned}$$

$$\cos x \cos y = \mathbf{\frac{1}{2}[\cos(x + y) + \cos(x - y)]}$$

17. (99)
$$\begin{cases} \text{(a)} & \dfrac{L + S}{2} = 10 \\ \text{(b)} & \sqrt{LS} = 6 \end{cases}$$

(a)
$$\begin{aligned} \frac{L + S}{2} &= 10 \\ L + S &= 20 \\ L &= 20 - S \end{aligned}$$

(b)
$$\begin{aligned} \sqrt{LS} &= 6 \\ LS &= 36 \\ (20 - S)S &= 36 \\ 20S - S^2 &= 36 \\ S^2 - 20S + 36 &= 0 \\ (S - 18)(S - 2) &= 0 \\ S &= 18, 2 \end{aligned}$$

If $S = 18$,
$L = 20 - S = 20 - (18) = 2.$

If $S = 2$,
$L = 20 - S = 20 - (2) = 18.$

2 and 18

18. (98) (a)
$$\begin{aligned} \log_6 45 &= y \\ 6^y &= 45 \\ y \log 6 &= \log 45 \\ y &= \frac{\log 45}{\log 6} \\ y &= \mathbf{2.12} \end{aligned}$$

(b)
$$\begin{aligned} \log_{36} 45 &= x \\ 36^x &= 45 \\ x \log 36 &= \log 45 \\ x &= \frac{\log 45}{\log 36} \\ x &= \mathbf{1.06} \end{aligned}$$

19. (98)
$$\begin{aligned} \sqrt{\log_2 x} &= \frac{1}{2} \log_2 x \\ \log_2 x &= \left(\frac{1}{2} \log_2 x\right)^2 \\ \log_2 x &= \frac{1}{4}(\log_2 x)^2 \end{aligned}$$

$$\frac{1}{4}(\log_2 x)^2 - \log_2 x = 0$$

$$\log_2 x\left(\frac{1}{4} \log_2 x - 1\right) = 0$$

$$\begin{aligned} \log_2 x &= 0 \\ x &= 2^0 \\ x &= 1 \end{aligned} \qquad \begin{aligned} \frac{1}{4} \log_2 x &= 1 \\ \log_2 x &= 4 \\ x &= 2^4 \\ x &= 16 \end{aligned}$$

$x = \mathbf{1, 16}$

20. (98)
$$\begin{aligned} \ln x + \ln x - \ln(2x - 2) &= \ln 2 \\ \ln\left(x^2\right) - \ln(2x - 2) &= \ln 2 \\ \ln\left(\frac{x^2}{2x - 2}\right) &= \ln 2 \\ \frac{x^2}{2x - 2} &= 2 \\ x^2 &= 4x - 4 \\ x^2 - 4x + 4 &= 0 \\ (x - 2)^2 &= 0 \\ x &= \mathbf{2} \end{aligned}$$

21. (98)
$$\log_2(\log_3 x) = 1$$
$$\log_3 x = 2^1$$
$$x = 3^2$$
$$x = \mathbf{9}$$

22. (96)
$$\frac{1 - \cos 2x}{2} = \frac{1 - (1 - 2\sin^2 x)}{2}$$
$$= \frac{2\sin^2 x}{2} = \sin^2 x$$

23. (84,93)
$$\frac{1 - 3\cos x - 4\cos^2 x}{\sin^2 x}$$
$$= \frac{(1 - 4\cos x)(1 + \cos x)}{1 - \cos^2 x}$$
$$= \frac{(1 - 4\cos x)(1 + \cos x)}{(1 - \cos x)(1 + \cos x)}$$
$$= \frac{1 - 4\cos x}{1 - \cos x}$$

24. (84)
$$y = -1 + \sin\frac{1}{2}(x - 90°)$$
$$\text{Period} = \frac{360°}{\frac{1}{2}} = 720°$$

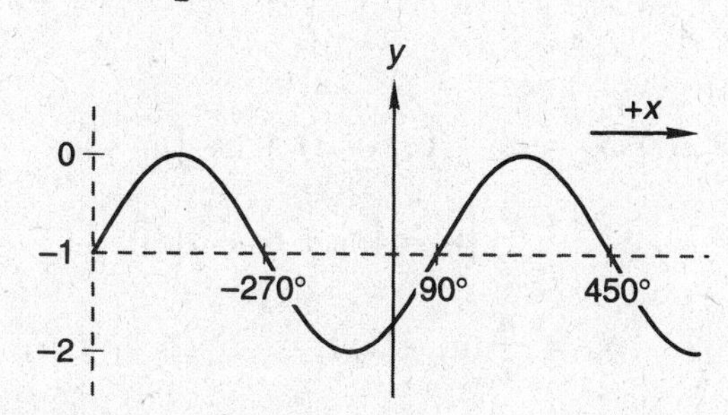

25. (94)
$$y = 3\sec x$$
$$\text{Period} = \frac{2\pi}{1} = 2\pi$$

26. (95)
$$0 + i = 1 \text{ cis } 90°$$
$$(1 \text{ cis } 90°)^{\frac{1}{3}} = 1^{\frac{1}{3}} \text{ cis}\left(\frac{90°}{3} + n\frac{360°}{3}\right)$$
$$= 1 \text{ cis } 30°, 1 \text{ cis } 150°, 1 \text{ cis } 270°$$
$$= \mathbf{\frac{\sqrt{3}}{2} + \frac{1}{2}i, -\frac{\sqrt{3}}{2} + \frac{1}{2}i, -i}$$

27. (52)
$$-1 + \sqrt{2}\sin 3\theta = 0$$
$$\sin 3\theta = \frac{1}{\sqrt{2}}$$
$$3\theta = \frac{\pi}{4}, \frac{3\pi}{4}, \frac{9\pi}{4}, \frac{11\pi}{4}, \frac{17\pi}{4}, \frac{19\pi}{4}$$
$$\theta = \mathbf{\frac{\pi}{12}, \frac{\pi}{4}, \frac{3\pi}{4}, \frac{11\pi}{12}, \frac{17\pi}{12}, \frac{19\pi}{12}}$$

28. (85)
$$3\tan^2 x - 2\sec x + 2 = 0$$
$$3(\sec^2 x - 1) - 2\sec x + 2 = 0$$
$$3\sec^2 x - 2\sec x - 1 = 0$$
$$(3\sec x + 1)(\sec x - 1) = 0$$

$3\sec x = -1$ $\qquad$ $\sec x = 1$

$\sec x = -\frac{1}{3}$ $\qquad$ $x = \mathbf{0}$

no answer

29. (59)
$$te^{-\ln t} = te^{\ln(t^{-1})} = t \cdot t^{-1} = \mathbf{1}$$

30. (46)
$$x^2 + x + 3 = 0$$
$$x = \frac{-1 \pm \sqrt{(1)^2 - 4(1)(3)}}{2(1)}$$
$$x = \frac{-1 \pm \sqrt{-11}}{2}$$
$$x = \frac{-1 \pm \sqrt{11}i}{2}$$
$$\mathbf{\left(x + \frac{1}{2} - \frac{\sqrt{11}}{2}i\right)\left(x + \frac{1}{2} + \frac{\sqrt{11}}{2}i\right)}$$

Problem Set 105

1. (91) $256\left(\frac{1}{4}\right)^4 = \mathbf{1\ ft}$

2. (88) $A_0 = 4200,\ A_{20} = 4100$

$$A_t = A_0e^{kt}$$
$$4100 = 4200e^{20k}$$
$$\frac{41}{42} = e^{20k}$$
$$\ln\frac{41}{42} = 20k$$
$$k = -0.0012049$$

$$2100 = 4200e^{-0.0012049t}$$
$$0.5 = e^{-0.0012049t}$$
$$\ln 0.5 = -0.0012049t$$
$$t = \mathbf{575.28\ min}$$

3. (85) $R_B = 1,\ R_L = \frac{1}{12},\ T_B = T_L$

$$\begin{cases} R_BT_B = S + 5 \\ R_LT_L = S \end{cases}$$

$$R_BT_B = S + 5$$
$$(1)T = (R_LT_L) + 5$$
$$T = \frac{1}{12}T + 5$$
$$\frac{11}{12}T = 5$$
$$T = \mathbf{5\frac{5}{11}\ min}$$

4. (44) $RWT = \text{jobs}$

$$R = \frac{40}{hK}\ \frac{\text{jobs}}{\text{worker-hr}}$$

$$RWT = \text{jobs}$$

$$W = \frac{\text{jobs}}{RT} = \frac{m}{\left(\frac{40}{hK}\right)14} = \mathbf{\frac{mhK}{560}\ workers}$$

5. (75) ${}_{10}C_6 = \frac{10!}{4!6!} = \mathbf{210\ teams}$

6. (62) $N_S = mN_T + b$

$$\begin{aligned} &\text{(a)} \begin{cases} 2 = m40 + b \\ 12 = m80 + b \end{cases} \\ &\text{(b)} \end{aligned}$$

$$\begin{aligned} -\text{(a)}\quad -2 &= -m40 - b \\ \text{(b)}\quad 12 &= m80 + b \\ \hline 10 &= m40 \end{aligned}$$

$$m = \frac{1}{4}$$

$$\text{(a)}\ 2 = m40 + b$$
$$2 = \left(\frac{1}{4}\right)40 + b$$
$$b = -8$$

$$N_S = \frac{1}{4}N_T - 8 = \frac{1}{4}(60) - 8 = \mathbf{7}$$

7. (105) We expand using cofactors of the first row.

$$\begin{vmatrix} -2 & 0 & 1 \\ 3 & 1 & 2 \\ 1 & 1 & 0 \end{vmatrix} = -2\begin{vmatrix} 1 & 2 \\ 1 & 0 \end{vmatrix} - 0\begin{vmatrix} 3 & 2 \\ 1 & 0 \end{vmatrix} + 1\begin{vmatrix} 3 & 1 \\ 1 & 1 \end{vmatrix}$$
$$= -2(0 - 2) - 0 + (3 - 1) = \mathbf{6}$$

8. (105) We expand using cofactors of the first row.

$$\begin{vmatrix} -2 & 0 & 3 \\ 0 & 1 & 2 \\ 3 & 1 & -1 \end{vmatrix} = -2\begin{vmatrix} 1 & 2 \\ 1 & -1 \end{vmatrix} - 0\begin{vmatrix} 0 & 2 \\ 3 & -1 \end{vmatrix} + 3\begin{vmatrix} 0 & 1 \\ 3 & 1 \end{vmatrix}$$
$$= -2(-1 - 2) - 0 + 3(0 - 3) = \mathbf{-3}$$

9. (91,104) (a) $S_n = a_1 + (a_1 + d) + \ldots + [a_1 + (n - 1)d]$

$$S_n = \frac{n}{2}\{a_1 + [a_1 + (n - 1)d]\}$$
$$\mathbf{S_n = \frac{n}{2}(a_1 + a_n)}$$

(b) $-2 + 4 + 10 + \ldots + 46$

From the arithmetic series, we find that $a_1 = -2$, $a_n = 46$, and $d = 6$.

First determine n using:

$$a_1 + (n - 1)d = a_n$$
$$(-2) + (n - 1)6 = 46$$
$$n - 1 = 8$$
$$n = 9$$

Then, find the sum of the arithmetic series using the equation developed in part (a).

$$S_n = \frac{n}{2}(a_1 + a_n)$$
$$S_9 = \frac{9}{2}(-2 + 46) = \mathbf{198}$$

10. (104) (a)

$$S_n = a + ar + ar^2 + \ldots + ar^{n-1}$$
$$-rS_n = -ar - ar^2 - ar^3 - \ldots - a_1r^n$$
$$S_n - rS_n = a - ar^n$$
$$S_n(1 - r) = a(1 - r^n)$$
$$\mathbf{S_n = \frac{a(1 - r^n)}{1 - r}}$$

(b) $a_1 = 1,\ r = -\frac{1}{2},\ n = 6$

$$S_n = \frac{a(1 - r^n)}{1 - r}$$

$$S_6 = \frac{1\left[1 - \left(-\frac{1}{2}\right)^6\right]}{1 - \left(-\frac{1}{2}\right)} = \frac{\frac{63}{64}}{\frac{3}{2}} = \mathbf{\frac{21}{32}}$$

11. (103) $\left(\frac{3300 \times 10^7}{2200 \times 10^3}\right)\sqrt[4]{4400 \times 10^{11}} = \mathbf{68{,}699{,}634.76}$

12. (103) $H^+ = 4.4 \times 10^{-4}$

$pH = -\log H^+ = -\log\left(4.4 \times 10^{-4}\right) = \mathbf{3.36}$

13. (102) $(x - 2y)^6$

$F = x,\ S = -2y$

Term	①	②	③	④	⑤	⑥	⑦
Exp. of F	6	5	4	3	2	1	0
Exp. of S	0	1	2	3	4	5	6
Coefficient	1	6	15	20	15	6	1

$15F^2S^4 = 15x^2(-2y)^4 = \mathbf{240x^2y^4}$

14. (102) $(3x - 2y)^3$

$F = 3x,\ S = -2y$

Term	①	②	③	④
Exp. of F	3	2	1	0
Exp. of S	0	1	2	3
Coefficient	1	3	3	1

$(F + S)^3 = F^3 + 3F^2S + 3FS + S^3$

$= (3x)^3 + 3(3x)^2(-2y) + 3(3x)(-2y)^2 + (-2y)^3$

$= \mathbf{27x^3 - 54x^2y + 36xy^2 - 8y^3}$

15. (101)

$$y = \frac{\begin{vmatrix} 2 & 0 & 1 \\ 0 & 5 & 3 \\ 1 & 1 & 0 \end{vmatrix}}{\begin{vmatrix} 2 & 1 & 1 \\ 0 & 2 & 3 \\ 1 & 3 & 0 \end{vmatrix}}$$

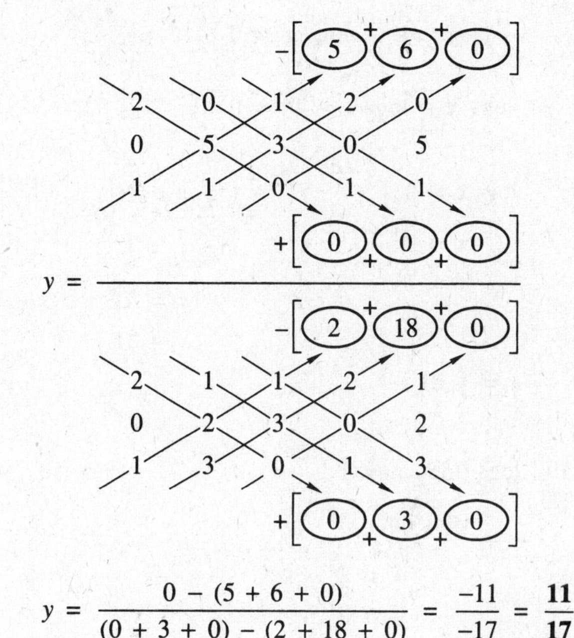

$$y = \frac{0 - (5 + 6 + 0)}{(0 + 3 + 0) - (2 + 18 + 0)} = \frac{-11}{-17} = \mathbf{\frac{11}{17}}$$

16. (100)

$A + B = x$
$A - B = y$
$2A = x + y$
$A = \frac{x + y}{2}$

$A + B = x$
$-(A - B = y)$
$2B = x - y$
$B = \frac{x - y}{2}$

$\sin(A + B) = \sin A\cos B + \cos A\sin B$
$-\sin(A - B) = -\sin A\cos B + \cos A\sin B$
$\sin(A + B) - \sin(A - B) = 2\cos A\sin B$

$$\sin x - \sin y = 2\cos\frac{x + y}{2}\sin\frac{x - y}{2}$$

17. (100) $\cos 75° \sin 15°$

$= \frac{1}{2}[\sin(75° + 15°) - \sin(75° - 15°)]$

$= \frac{1}{2}(\sin 90° - \sin 60°)$

$= \frac{1}{2}\left(1 - \frac{\sqrt{3}}{2}\right) = \mathbf{\frac{2 - \sqrt{3}}{4}}$

18. (98)

$$\sqrt{\log_3 x} = \frac{1}{2}\log_3 x$$
$$\left(\sqrt{\log_3 x}\right)^2 = \left(\frac{1}{2}\log_3 x\right)^2$$
$$\log_3 x = \frac{1}{4}(\log_3 x)^2$$
$$\frac{1}{4}(\log_3 x)^2 - \log_3 x = 0$$
$$\log_3 x\left(\frac{1}{4}\log_3 x - 1\right) = 0$$

$\log_3 x = 0$, $x = 3^0$, $x = 1$

$\frac{1}{4}\log_3 x = 1$, $\log_3 x = 4$, $x = 3^4$, $x = 81$

$x =$ **1, 81**

19. (98) $\log_3(\log_2 x) = 1$
$$\log_2 x = 3^1$$
$$x = 2^3$$
$$x = \mathbf{8}$$

20. (59,98) $\ln(2x + 6) - \ln(x - 3) = \ln x$
$$\ln\frac{2x + 6}{x - 3} = \ln x$$
$$\frac{2x + 6}{x - 3} = x$$
$$2x + 6 = x^2 - 3x$$
$$x^2 - 5x - 6 = 0$$
$$(x - 6)(x + 1) = 0$$
$$x = 6, -1 \quad (x \neq -1)$$
$$x = \mathbf{6}$$

21. (96) $\frac{1}{2}\sin 2x \sec x = \frac{1}{2}2\sin x\cos x\frac{1}{\cos x} = \sin x$

22. (96) $\frac{1 + \cos 2x}{\sin 2x} = \frac{1 + 2\cos^2 x - 1}{2\sin x\cos x}$
$$= \frac{2\cos^2 x}{2\sin x\cos x} = \frac{\cos x}{\sin x} = \cot x$$

23. (96)

6 cm, 8 cm, A, 75°, C

$$\frac{8}{\sin 75°} = \frac{6}{\sin C}$$
$$\sin C = \frac{6\sin 75°}{8}$$
$$C = 46.42°$$
$$A = 180° - 75° - 46.42° = 58.58°$$
$$\text{Area} = \frac{1}{2}bc\sin A$$
$$= \frac{1}{2}(8)(6)\sin 58.58°$$
$$= \mathbf{20.48\ cm^2}$$

24. (94) $y = 2\csc x$

25. (94) (a) $y = \tan\theta$ (bold)

(b) $y = -7 + 4\sec x$ (bold)

26. (52) $1 - \sqrt{2}\cos 3x = 0$
$$\cos 3x = \frac{1}{\sqrt{2}}$$
$$3x = 45°, 315°, 405°, 675°, 765°, 1035°$$
$$x = \mathbf{15°, 105°, 135°, 225°, 255°, 345°}$$

27. (85)
$$\sec^2\theta = 2\tan\theta$$
$$(\tan^2\theta + 1) - 2\tan\theta = 0$$
$$\tan^2\theta - 2\tan\theta + 1 = 0$$
$$(\tan\theta - 1)^2 = 0$$
$$\tan\theta - 1 = 0$$
$$\tan\theta = 1$$
$$\theta = \mathbf{45°, 225°}$$

28. $-1 = 1 \text{ cis } 180°$
(95)

$$(1 \text{ cis } 180°)^{\frac{1}{4}} = 1^{\frac{1}{4}} \text{ cis}\left(\frac{180°}{4} + n\frac{360°}{4}\right)$$

$= 1 \text{ cis } 45°, 1 \text{ cis } 135°, 1 \text{ cis } 225°, 1 \text{ cis } 315°$

$$= \frac{\sqrt{2}}{2} + \frac{\sqrt{2}}{2}i, -\frac{\sqrt{2}}{2} + \frac{\sqrt{2}}{2}i, -\frac{\sqrt{2}}{2} - \frac{\sqrt{2}}{2}i,$$
$$\frac{\sqrt{2}}{2} - \frac{\sqrt{2}}{2}i$$

29. $8x^2 - 50y^2 = -200$
(78)

$$\frac{y^2}{4} - \frac{x^2}{25} = 1$$

Let $x = 0$ Let $y = 0$

$y = \pm 2$ $x = \pm 5i$

Vertices = (0, 2), (0, –2)

Asymptotes: $y = \frac{2}{5}x$; $y = -\frac{2}{5}x$

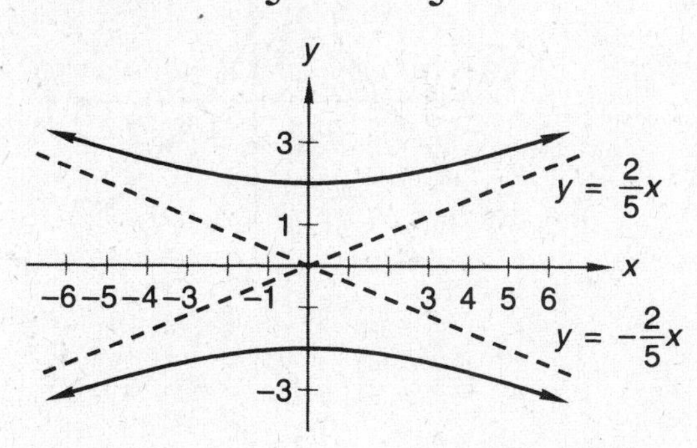

30. $x - 3y + 5 = 0$
(58)

$$y = \frac{1}{3}x + \frac{5}{3}$$

Equation of the perpendicular line:

$y = -3x + b$

$2 = -3(1) + b$

$b = 5$

$y = -3x + 5$

Point of intersection:

$$\frac{1}{3}x + \frac{5}{3} = -3x + 5$$

$$\frac{10}{3}x = \frac{10}{3}$$

$x = 1$

$y = -3x + 5 = -3(1) + 5 = 2$

(1, 2) and (1, 2)

The point is on the line. Therefore, $D =$ **0.**

Problem Set 106

1. $A_0 = 400$, $A_{20} = 380$
(88)

$A_t = A_0 e^{kt}$

$380 = 400e^{20k}$

$0.95 = e^{20k}$

$\ln 0.95 = 20k$

$k = -0.0025647$

$200 = 400e^{-0.0025647t}$

$0.5 = e^{-0.0025647t}$

$\ln 0.5 = -0.0025647t$

$t =$ **270.27 min**

2. $(9 - 1)! = 8! =$ **40,320**
(55)

3. $v = r\omega$
(53)

$$= (1 \text{ m})\left(300 \frac{\text{rad}}{\text{min}}\right)\left(\frac{100 \text{ cm}}{1 \text{ m}}\right)\left(\frac{1 \text{ in.}}{2.54 \text{ cm}}\right)$$
$$\times \left(\frac{1 \text{ ft}}{12 \text{ in.}}\right)\left(\frac{1 \text{ mi}}{5280 \text{ ft}}\right)$$

$$= \frac{(300)(100)}{(2.54)(12)(5280)} \frac{\text{mi}}{\text{min}} = \mathbf{0.19 \frac{mi}{min}}$$

4. $\frac{3}{7}P = 12{,}600$
(25)

$P = 29{,}400$

$\frac{4}{21}(29{,}400) =$ **$5600**

5. $\begin{cases} R_O T_O = 72 \\ R_B T_B = 72 \end{cases}$
(25)

$R_O = 3R_B$

$T_O = 24 - T_B$

$R_O T_O = 72$

$3R_B(24 - T_B) = 72$

$72R_B - 3R_B T_B = 72$

$72R_B - 3(72) = 72$

$72R_B = 288$

$R_B = 4$

$R_O = 3R_B = 3(4) =$ **12 mph**

6. (106)

$$x^2 + 16y^2 - 10x + 32y + 25 = 0$$
$$(x^2 - 10x \quad) + 16(y^2 + 2y \quad) = -25$$
$$(x^2 - 10x + 25) + 16(y^2 + 2y + 1) = 16$$
$$(x - 5)^2 + 16(y + 1)^2 = 16$$
$$\mathbf{\frac{(x-5)^2}{16} + \frac{(y+1)^2}{1} = 1}$$

Center = (5, –1)

$a = 4,\ b = 1$

Length of major axis = 8

Length of minor axis = 2

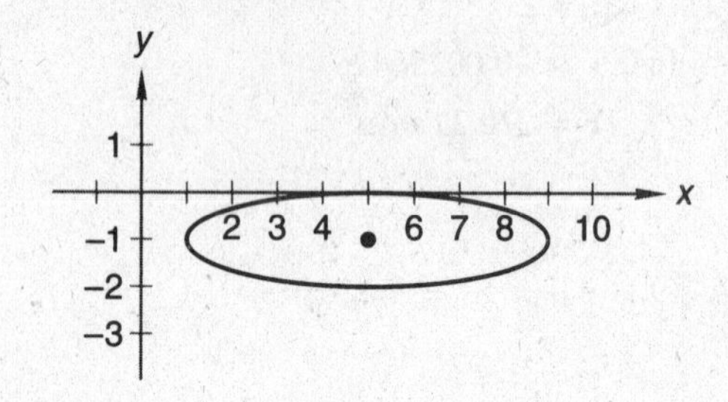

7. (106)

$$9x^2 + 4y^2 + 54x - 8y + 49 = 0$$
$$9(x^2 + 6x \quad) + 4(y^2 - 2y \quad) = -49$$
$$9(x^2 + 6x + 9) + 4(y^2 - 2y + 1) = 36$$
$$9(x + 3)^2 + 4(y - 1)^2 = 36$$
$$\mathbf{\frac{(x+3)^2}{4} + \frac{(y-1)^2}{9} = 1}$$

Center = (–3, 1)

$a = 3,\ b = 2$

Length of major axis = 6

Length of minor axis = 4

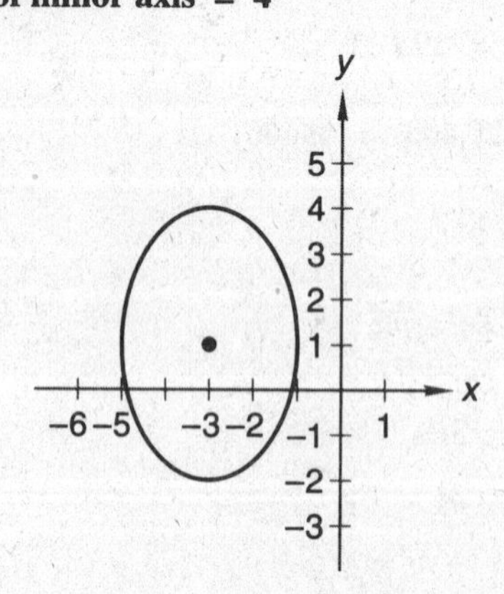

8. (106)

$$-x^2 + y^2 - 2x - 4y + 4 = 0$$
$$-(x^2 + 2x \quad) + (y^2 - 4y \quad) = -4$$
$$-(x^2 + 2x + 1) + (y^2 - 4y + 4) = -1$$
$$-(x + 1)^2 + (y - 2)^2 = -1$$
$$\mathbf{\frac{(x+1)^2}{1} - \frac{(y-2)^2}{1} = 1}$$

From the equation we obtain:

$h = -1,\ k = 2,\ a = 1,\ b = 1$

Center: $(h, k) = \mathbf{(-1, 2)}$

Vertices: $(h + a, k), (h - a, k) = \mathbf{(-2, 2), (0, 2)}$

$$m = \pm\frac{b}{a} = \pm\frac{1}{1} = \pm 1$$

$y = mx + b$	$y = mx + b$
$y = 1x + b$	$y = -1x + b$
$2 = 1(-1) + b$	$2 = -1(-1) + b$
$b = 3$	$b = 1$
$y = x + 3$	$y = -x + 1$

Asymptotes: $\mathbf{y = x + 3;\ y = -x + 1}$

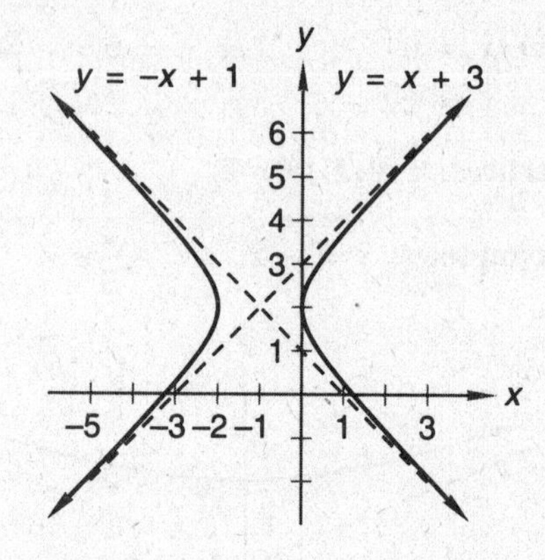

9. (106)

$$4x^2 - y^2 + 8x - 4y - 4 = 0$$
$$4(x^2 + 2x \quad) - (y^2 + 4y \quad) = 4$$
$$4(x^2 + 2x + 1) - (y^2 + 4y + 4) = 4$$
$$4(x + 1)^2 - (y - 2)^2 = 4$$
$$\mathbf{\frac{(x+1)^2}{1} - \frac{(y+2)^2}{4} = 1}$$

From the equation we obtain:

$h = -1,\ k = -2,\ a = 1,\ b = 2$

Center: $(h, k) = \mathbf{(-1, -2)}$

Vertices: $(h + a, k), (h - a, k) = \mathbf{(0, -2), (-2, -2)}$

Asymptotes:

$$m = \pm\frac{b}{a} = \pm\frac{2}{1} = \pm 2$$

$y = mx + b$	$y = mx + b$
$y = 2x + b$	$y = -2x + b$
$-2 = 2(-1) + b$	$-2 = -2(-1) + b$
$b = 0$	$b = -4$
$y = 2x$	$y = -2x - 4$

Asymptotes: $\mathbf{y = 2x;\ y = -2x - 4}$

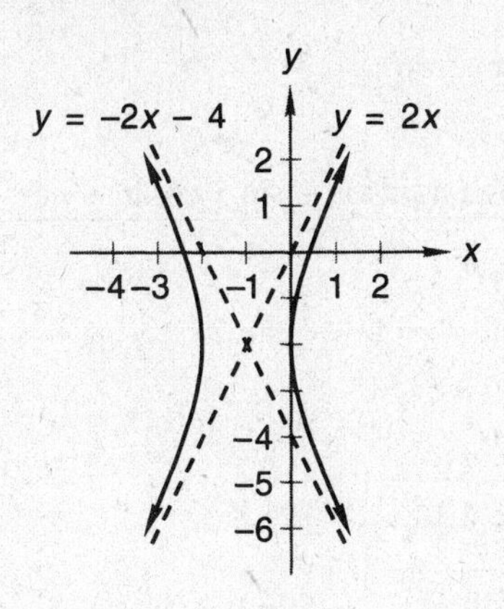

10. (63,106) Center = (−2, 3)

Radius = 3

Standard form: $\mathbf{(x + 2)^2 + (y - 3)^2 = 9}$

General form:

$$x^2 + 4x + 4 + y^2 - 6y + 9 - 9 = 0$$

$$\mathbf{x^2 + y^2 + 4x - 6y + 4 = 0}$$

11. (105) We expand using cofactors of the first row.

$$\begin{vmatrix} 1 & 2 & 1 \\ 0 & 1 & 0 \\ 1 & 2 & 2 \end{vmatrix} = 1\begin{vmatrix} 1 & 0 \\ 2 & 2 \end{vmatrix} - 2\begin{vmatrix} 0 & 0 \\ 1 & 2 \end{vmatrix} + 1\begin{vmatrix} 0 & 1 \\ 1 & 2 \end{vmatrix}$$

$$= 1(2 - 0) - 2(0 - 0) + 1(0 - 1)$$

$$= 2 - 0 - 1 = \mathbf{1}$$

12. (91,104) (a) $S_n = a_1 + (a_1 + d) + \ldots + [a_1 + (n - 1)d]$

$$S_n = \frac{n}{2}\{a_1 + [a_1 + (n - 1)d]\}$$

$$\mathbf{S_n = \frac{n}{2}(a_1 + a_n)}$$

(b) $-8 + (-4) + 0 + \ldots + 24$

$a_1 = -8,\ a_n = 24$, and $d = 4$

$$a_1 + (n - 1)d = a_n$$

$$-8 + (n - 1)4 = 24$$

$$n - 1 = 24$$

$$n = 9$$

$$S_n = \frac{n}{2}(a_1 + a_n)$$

$$S_9 = \frac{9}{2}(-8 + 24)$$

$$S_9 = \mathbf{72}$$

13. (104) (a)

$$S_n = a_1 + a_1 r + a_1 r^2 + \ldots + a_1 r^{n-1}$$

$$-rS_n = -a_1 r - a_1 r^2 - a_1 r^3 - \ldots - a_1 r^n$$

$$S_n - rS_n = a_1 - a_1 r^n$$

$$S_n(1 - r) = a_1(1 - r^n)$$

$$\mathbf{S_n = \frac{a_1(1 - r^n)}{1 - r}}$$

(b) 4, −8, 16, ...

$a_1 = 4,\ r = -2,\ n = 7$

$$S_n = \frac{a_1(1 - r^n)}{1 - r}$$

$$S_7 = \frac{4[1 - (-2)^7]}{1 - (-2)} = \frac{4(129)}{3} = \mathbf{172}$$

14. (103) $\frac{3}{2}\log x^6 - \frac{2}{3}\log y^3 + \frac{3}{4}\log z^2 - \log xy$

$$= \log(x^6)^{\frac{3}{2}} - \log(y^3)^{\frac{2}{3}} + \log(z^2)^{\frac{3}{4}} - \log xy$$

$$= \log x^9 - \log y^2 + \log z^{\frac{3}{2}} - \log xy$$

$$= \log \frac{x^9 z^{\frac{3}{2}}}{y^2 xy} = \mathbf{\log\left(\frac{x^8 z^{\frac{3}{2}}}{y^3}\right)}$$

15. (103) $H^+ = 10^{-pH} = 10^{-3.5} = \mathbf{3.16 \times 10^{-4}\ \frac{mole}{liter}}$

16. (102) $(3x - 2y)^6$

$F = 3x,\ S = -2y$

Term	①	②	③	④	⑤	⑥	⑦
Exp. of F	6	5	4	3	2	1	0
Exp. of S	0	1	2	3	4	5	6
Coefficient	1	6	15	20	15	6	1

$15F^4S^2 = 15(3x)^4(-2y)^2 = \mathbf{4860x^4y^2}$

17. (101)

$$z = \frac{\begin{vmatrix} 2 & 1 & 2 \\ 0 & 2 & 5 \\ 1 & 3 & 2 \end{vmatrix}}{\begin{vmatrix} 2 & 1 & 2 \\ 0 & 2 & 1 \\ 1 & 3 & 0 \end{vmatrix}} = \frac{2\begin{vmatrix} 2 & 5 \\ 3 & 2 \end{vmatrix} - 1\begin{vmatrix} 0 & 5 \\ 1 & 2 \end{vmatrix} + 2\begin{vmatrix} 0 & 2 \\ 1 & 3 \end{vmatrix}}{2\begin{vmatrix} 2 & 1 \\ 3 & 0 \end{vmatrix} - 1\begin{vmatrix} 0 & 1 \\ 1 & 0 \end{vmatrix} + 2\begin{vmatrix} 0 & 2 \\ 1 & 3 \end{vmatrix}}$$

$$= \frac{2(4 - 15) - 1(0 - 5) + 2(0 - 2)}{2(0 - 3) - 1(0 - 1) + 2(0 - 2)} = \frac{-21}{-9} = \mathbf{\frac{7}{3}}$$

18. (100)

$$\sin(x+y) = \sin x\cos y + \cos x\sin y$$
$$\sin(x-y) = \sin x\cos y - \cos x\sin y$$

$$\sin(x+y) + \sin(x-y) = 2\sin x\cos y$$

$$\sin x\cos y = \frac{1}{2}[\sin(x+y) + \sin(x-y)]$$

19. (100) $\cos 195° \cos 105°$

$$= \frac{1}{2}[\cos(195° + 105°) + \cos(195° - 105°)]$$

$$= \frac{1}{2}(\cos 300° + \cos 90°) = \frac{1}{2}\left(\frac{1}{2} + 0\right) = \mathbf{\frac{1}{4}}$$

20. (98) $\frac{1}{2}\log_3 25 - \log_3(2x - 5) = 2$

$$\log_3 \frac{5}{2x-5} = 2$$
$$3^2 = \frac{5}{2x-5}$$
$$18x - 45 = 5$$
$$18x = 50$$
$$x = \mathbf{\frac{25}{9}}$$

21. (98) $\log_2 3 = 2\log_2 x$

$$\log_2 3 = \log_2 x^2$$
$$3 = x^2$$
$$x = \pm\sqrt{3} \qquad (x \neq -\sqrt{3})$$
$$x = \mathbf{\sqrt{3}}$$

22. (98)

$$\sqrt{\ln x} = 3\ln\sqrt{x}$$
$$\sqrt{\ln x} = 3\ln x^{\frac{1}{2}}$$
$$\sqrt{\ln x} = \frac{3}{2}\ln x$$
$$\ln x = \frac{9}{4}(\ln x)^2$$

$$\frac{9}{4}\ln x^2 - \ln x = 0$$

$$\ln x\left(\frac{9}{4}\ln x - 1\right) = 0$$

$\ln x = 0$, $x = e^0$, $x = 1$

$\frac{9}{4}\ln x = 1$, $\ln x = \frac{4}{9}$, $x = e^{\frac{4}{9}}$

$$x = \mathbf{1, e^{\frac{4}{9}}}$$

23. (96) $\dfrac{\sin^3 x + \cos^3 x}{\sin x + \cos x}$

$$= \frac{(\sin x + \cos x)(\sin^2 x - \sin x\cos x + \cos^2 x)}{\sin x + \cos x}$$

$$= 1 - \sin x\cos x = 1 - \frac{1}{2}\sin 2x$$

24. (93)

$$\frac{\cot x + 1}{\cot x - 1} = \frac{\frac{\cos x}{\sin x} + 1}{\frac{\cos x}{\sin x} - 1} \cdot \frac{\frac{1}{\cos x}}{\frac{1}{\cos x}}$$

$$= \frac{\frac{1}{\sin x} + \frac{1}{\cos x}}{\frac{1}{\sin x} - \frac{1}{\cos x}} = \frac{\csc x + \sec x}{\csc x - \sec x}$$

25. (94) $y = \tan\frac{1}{2}x$

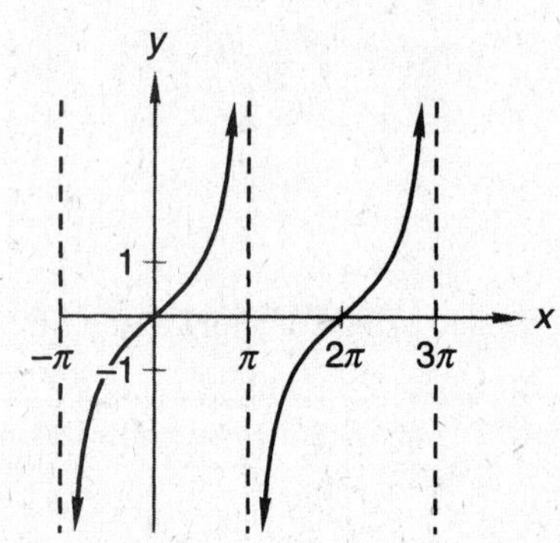

26. (84) $y = -7 + 4\sin\frac{3}{8}\left(x - \frac{2\pi}{3}\right)$

$$\text{Period} = \frac{2\pi}{\frac{3}{8}} = \frac{16\pi}{3}$$

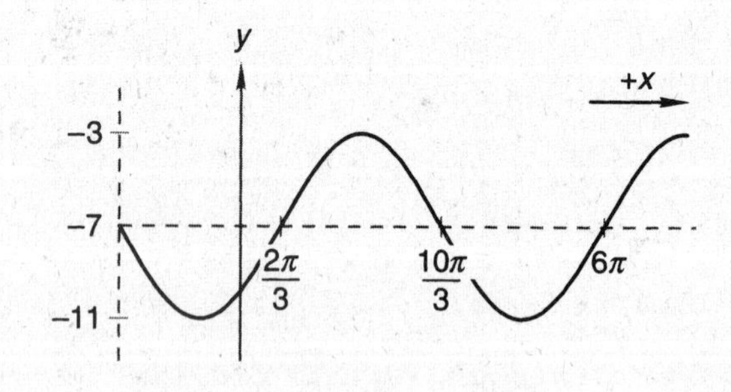

27. (60) $2\sin^2 x + \sqrt{3}\sin x = 0$

$$\sin x(2\sin x + \sqrt{3}) = 0$$

$\sin x = 0$, $x = 0°, 180°$

$2\sin x = -\sqrt{3}$, $\sin x = -\frac{\sqrt{3}}{2}$, $x = 240°, 300°$

$$x = \mathbf{0°, 180°, 240°, 300°}$$

28. (95) $-6 + 4i = 7.21 \text{ cis } 146.31°$

$$(7.21 \text{ cis } 146.31°)^{\frac{1}{4}}$$

$$= 7.21^{\frac{1}{4}} \text{ cis}\left(\frac{146.31°}{4} + n\frac{360°}{4}\right)$$

$$= \mathbf{1.64 \text{ cis } 36.58°, 1.64 \text{ cis } 126.58°,}$$
$$\mathbf{1.64 \text{ cis } 216.58°, 1.64 \text{ cis } 306.58°}$$

29. (73) $\text{Side} = \dfrac{25 \text{ cm}}{5} = 5 \text{ cm}$

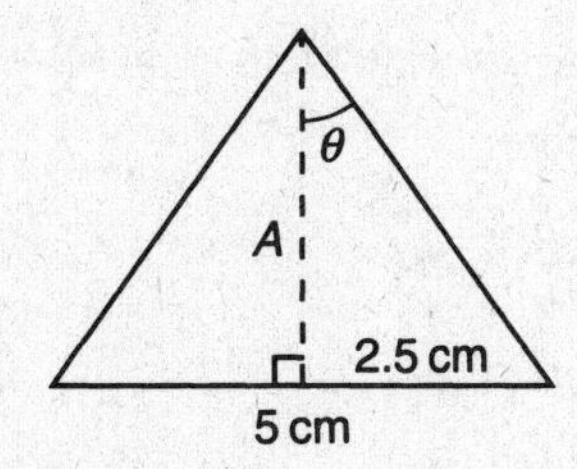

$$\theta = \frac{1}{2}\left(\frac{360°}{5}\right) = 36°$$

$$A = \frac{2.5}{\tan 36°} = 3.441$$

$$\text{Area}_\Delta = \frac{1}{2}(2.5)(3.441) = 4.3013 \text{ cm}^2$$

$$\text{Area} = (10)(4.3013) = \mathbf{43.01 \text{ cm}^2}$$

30. (59) $(t^2 + t)10^{-2\log(t+1)} = (t^2 + t)10^{\log(t+1)^{-2}}$

$$= (t^2 + t)(t + 1)^{-2} = \frac{t(t + 1)}{(t + 1)^2} = \mathbf{\frac{t}{t + 1}}$$

Problem Set 107

1. (85) $R_B = 1,\ R_L = \dfrac{1}{12},\ T_B = T_L$

$$\begin{cases} R_L T_L = S \\ R_B T_B = S + 30 \end{cases}$$

$$R_B T_B = S + 30$$

$$(1)T = (R_L T_L) + 30$$

$$T = \frac{1}{12}T + 30$$

$$\frac{11}{12}T = 30$$

$$T = 32\frac{8}{11} \text{ min}$$

$$\mathbf{6{:}32\frac{8}{11} \text{ p.m.}}$$

2. (44) $R_L T_L + R_E T_E = \text{jobs}$

$$\left(\frac{1}{x}\right)(2) + R_E(2) = 1$$

$$2R_E = 1 - \frac{2}{x}$$

$$2R_E = \frac{x - 2}{x}$$

$$R_E = \frac{x - 2}{2x}$$

$$R_E T_E = 1$$

$$\left(\frac{x - 2}{2x}\right)T_E = 1$$

$$T_E = \mathbf{\frac{2x}{x - 2} \text{ days}}$$

3. (38) $\dfrac{m + x}{2} = A$

$$m + x = 2A$$

$$x = \mathbf{2A - m}$$

4. (38) $\dfrac{1.0 + 0.8 + 0.2 + x}{4} = 0.6$

$$1.0 + 0.8 + 0.2 + x = 2.4$$

$$x = \mathbf{0.4}$$

5. (18)
$$\begin{cases} \text{(a)} & 0.10N_G + 0.20N_B = 16 \\ \text{(b)} & \dfrac{N_G}{N_B} = \dfrac{2}{3} \end{cases}$$

$$\text{(b)} \quad \frac{N_G}{N_B} = \frac{2}{3}$$

$$N_G = \frac{2}{3}N_B$$

$$10\text{(a)} \qquad N_G + 2N_B = 160$$

$$\left(\frac{2}{3}N_B\right) + 2N_B = 160$$

$$\frac{8}{3}N_B = 160$$

$$N_B = \mathbf{60 \text{ boys}}$$

6. (107) $a_1 + a_1r + a_1r^2 + a_1r^3 + \ldots$

$$4 + \frac{4}{3} + \frac{4}{9} + \frac{4}{27} + \ldots$$

$$a_1 = 4$$

$$a_1r = \frac{4}{3}$$

$$r = \frac{4}{3a_1} = \frac{4}{(3)(4)} = \frac{1}{3}$$

Since $|r| < 1$, we can find the sum of the infinite series.

$$S = \frac{a_1}{1 - r} = \frac{4}{1 - \frac{1}{3}} = \frac{4}{\frac{2}{3}} = \mathbf{6}$$

7. (107) $a_1 + a_1r + a_1r^2 + \ldots$

$$4 + 2 + 1 + \ldots$$

$$a_1 = 4$$

$$a_1r = 2$$

$$r = \frac{2}{a_1} = \frac{2}{(4)} = \frac{1}{2}$$

Since $|r| < 1$, we can find the sum of the infinite series.

$$S = \frac{a_1}{1 - r} = \frac{4}{1 - \frac{1}{2}} = \frac{4}{\frac{1}{2}} = \mathbf{8\text{ mi}}$$

8. (106)

$$9x^2 + 4y^2 + 54x - 8y + 49 = 0$$

$$9(x^2 + 6x \quad) + 4(y^2 - 2y \quad) = -49$$

$$9(x^2 + 6x + 9) + 4(y^2 - 2y + 1) = 36$$

$$9(x + 3)^2 + 4(y - 1)^2 = 36$$

$$\mathbf{\frac{(x + 3)^2}{4} + \frac{(y - 1)^2}{9} = 1}$$

Center = (–3, 1)

$a = 3,\ b = 2$

Length of major axis = 6

Length of minor axis = 4

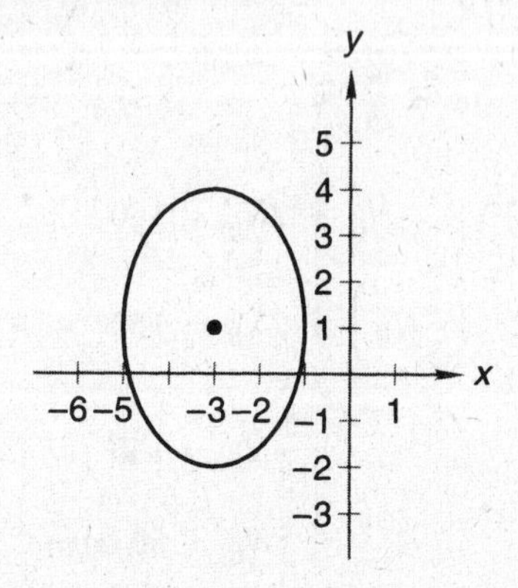

9. (106)

$$x^2 - y^2 - 14x - 8y + 29 = 0$$

$$(x^2 - 14x \quad) - (y^2 + 8y \quad) = -29$$

$$(x^2 - 14x + 49) - (y^2 + 8y - 16) = 4$$

$$(x - 7)^2 - (y + 4)^2 = 4$$

$$\mathbf{\frac{(x - 7)^2}{4} - \frac{(y + 4)^2}{4} = 1}$$

From the equation we obtain:

$h = 7,\ k = -4,\ a = 2,\ b = 2$

Center: $(h, k) =$ **(7, –4)**

Vertices: $(h + a, k), (h - a, k) =$ **(9, –4), (5, –4)**

Asymptotes:

$$m = \pm\frac{b}{a} = \pm 1$$

$y = mx + b$	$y = mx + b$
$y = 1x + b$	$y = -1x + b$
$-4 = 1(7) + b$	$-4 = -1(7) + b$
$b = -11$	$b = 3$
$y = x - 11$	$y = -x + 3$

Asymptotes: $\mathbf{y = x - 11;\ y = -x + 3}$

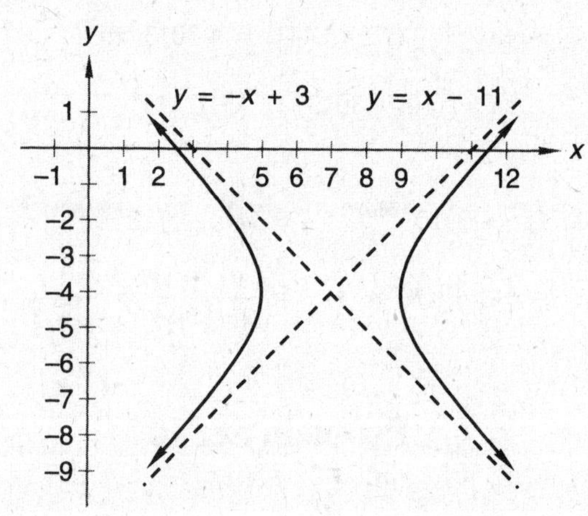

10. (106) Center = (1, –3)

Radius = 1

Standard form: $\mathbf{(x - 1)^2 + (y + 3)^2 = 1}$

$$x^2 - 2x + 1 + y^2 + 6y + 9 = 1$$

General form: $\mathbf{x^2 + y^2 - 2x + 6y + 9 = 0}$

11. (105)

$$\begin{vmatrix} 2 & 3 & -1 \\ 0 & 4 & 1 \\ 1 & 3 & 3 \end{vmatrix}$$

$$= 2\begin{vmatrix} 4 & 1 \\ 3 & 3 \end{vmatrix} - 0\begin{vmatrix} 3 & -1 \\ 3 & 3 \end{vmatrix} + 1\begin{vmatrix} 3 & -1 \\ 4 & 1 \end{vmatrix}$$

$$= 2(12 - 3) - 0 + 1(3 + 4)$$

$$= 2(9) + 7 = 18 + 7 = \mathbf{25}$$

12. (101,105)

$$x = \frac{\begin{vmatrix} 7 & 1 & -1 \\ 1 & 2 & 3 \\ -8 & -3 & 0 \end{vmatrix}}{\begin{vmatrix} -2 & 1 & -1 \\ 0 & 2 & 3 \\ 1 & -3 & 0 \end{vmatrix}}$$

$$x = \frac{7\begin{vmatrix} 2 & 3 \\ -3 & 0 \end{vmatrix} - 1\begin{vmatrix} 1 & 3 \\ -8 & 0 \end{vmatrix} - 1\begin{vmatrix} 1 & 2 \\ -8 & -3 \end{vmatrix}}{-2\begin{vmatrix} 2 & 3 \\ -3 & 0 \end{vmatrix} - 1\begin{vmatrix} 0 & 3 \\ 1 & 0 \end{vmatrix} - 1\begin{vmatrix} 0 & 2 \\ 1 & -3 \end{vmatrix}}$$

$$x = \frac{7(0 + 9) - 1(0 + 24) - 1(-3 + 16)}{-2(0 + 9) - 1(0 - 3) - 1(0 - 2)}$$

$$x = \frac{26}{-13}$$

$\mathbf{x = -2}$

$$y = \frac{\begin{vmatrix} -2 & 7 & -1 \\ 0 & 1 & 3 \\ 1 & -8 & 0 \end{vmatrix}}{-13}$$

$$y = \frac{-2\begin{vmatrix} 1 & 3 \\ -8 & 0 \end{vmatrix} - 7\begin{vmatrix} 0 & 3 \\ 1 & 0 \end{vmatrix} - 1\begin{vmatrix} 0 & 1 \\ 1 & -8 \end{vmatrix}}{-13}$$

$$y = \frac{-2(0 + 24) - 7(0 - 3) - 1(0 - 1)}{-13}$$

$$y = \frac{-26}{-13}$$

$\mathbf{y = 2}$

$$z = \frac{\begin{vmatrix} -2 & 1 & 7 \\ 0 & 2 & 1 \\ 1 & -3 & -8 \end{vmatrix}}{-13}$$

$$z = \frac{-2\begin{vmatrix} 2 & 1 \\ -3 & -8 \end{vmatrix} - 1\begin{vmatrix} 0 & 1 \\ 1 & -8 \end{vmatrix} + 7\begin{vmatrix} 0 & 2 \\ 1 & -3 \end{vmatrix}}{-13}$$

$$z = \frac{-2(-16 + 3) - 1(0 - 1) + 7(0 - 2)}{-13}$$

$$z = \frac{13}{-13}$$

$\mathbf{z = -1}$

13. (99,104) (a) $\mathbf{a_n = a_1 + (n - 1)d}$

(b) $S_n = \frac{n}{2}(a_1 + a_n)$

$= \frac{n}{2}[a_1 + a_1 + (n - 1)d]$

$= \mathbf{\frac{n}{2}[2a_1 + (n - 1)d]}$

14. (103) $H^+ = 10^{-pH} = 10^{-8.5} = \mathbf{3.16 \times 10^{-9} \frac{mole}{liter}}$

15. (103) $\sqrt[4]{42{,}000}\,\sqrt[3]{2300} - \sqrt[5]{540} = \mathbf{185.45}$

16. (102) $(2x - 1)^7$

$F = 2x,\ S = -1$

Term	①	②	③	④	⑤	⑥...⑧
Exp. of F	7	6	5	4	3	2... 0
Exp. of S	0	1	2	3	4	5... 7
Coefficient	1	7	21	35	35	21... 1

$35F^3S^4 = 35(2x)^3(-1)^4 = \mathbf{280x^3}$

17. (100) (a)

$$\begin{aligned} \sin(A + B) &= \sin A \cos B + \cos A \sin B \\ -\sin(A - B) &= -\sin A \cos B + \cos A \sin B \\ \hline \sin(A + B) - \sin(A - B) &= 2\cos A \sin B \end{aligned}$$

$$\cos A \sin B = \frac{1}{2}[\sin(A + B) - \sin(A - B)]$$

(b) $\cos 75° \sin 15°$

$= \frac{1}{2}[\sin(75° + 15°) - \sin(75° - 15°)]$

$= \frac{1}{2}(\sin 90° - \sin 60°)$

$= \frac{1}{2}\left(1 - \frac{\sqrt{3}}{2}\right)$

$= \frac{1}{2} - \frac{\sqrt{3}}{4} = \mathbf{\frac{2 - \sqrt{3}}{4}}$

18. (100) (a)

$$\begin{aligned} A + B &= x \\ A - B &= y \\ \hline 2A &= x + y \end{aligned} \qquad \begin{aligned} A + B &= x \\ -(A - B &= y) \\ \hline 2B &= x - y \end{aligned}$$

$$A = \frac{x + y}{2} \qquad B = \frac{x - y}{2}$$

$$\begin{aligned} \sin(A + B) &= \sin A \cos B + \cos A \sin B \\ \sin(A - B) &= \sin A \cos B - \cos A \sin B \\ \hline \sin(A + B) + \sin(A - B) &= 2\sin A \cos B \end{aligned}$$

$$\sin x + \sin y = 2\sin\frac{x + y}{2}\cos\frac{x - y}{2}$$

(b) $\sin 15° + \sin 75°$

$= 2\sin\frac{15° + 75°}{2}\cos\frac{15° - 75°}{2}$

$= 2\sin 45° \cos(-30°)$

$= 2\left(\frac{\sqrt{2}}{2}\right)\left(\frac{\sqrt{3}}{2}\right) = \mathbf{\frac{\sqrt{6}}{2}}$

19. (98)

$$\ln 22 = x$$
$$e^x = 22$$
$$\log e^x = \log 22$$
$$x \log e = \log 22$$
$$x = \mathbf{\frac{\log 22}{\log e}}$$

20. (82)

$$2^{3x-2} = 4$$
$$2^{3x-2} = 2^2$$
$$3x - 2 = 2$$
$$3x = 4$$
$$x = \mathbf{\frac{4}{3}}$$

21. (98)

$$\log_3 (\log_2 x) = 2$$
$$\log_2 x = 3^2$$
$$\log_2 x = 9$$
$$x = \mathbf{2^9}$$

22. (59)

$$\log_6 (x - 1) + \log_6 (x - 2) = 2 \log_6 \sqrt{6}$$
$$\log_6 (x - 1) + \log_6 (x - 2) = \log_6 6$$
$$\log_6 [(x - 1)(x - 2)] = \log_6 6$$
$$(x - 1)(x - 2) = 6$$
$$x^2 - 3x - 4 = 0$$
$$(x - 4)(x + 1) = 0$$
$$x = 4, -1 \quad (x \neq -1)$$
$$x = \mathbf{4}$$

23. (85) $\sec x - \sin x \tan x$

$$= \frac{1}{\cos x} - \left(\frac{\sin x}{1}\right)\left(\frac{\sin x}{\cos x}\right)$$
$$= \frac{1}{\cos x} - \frac{\sin^2 x}{\cos x}$$
$$= \frac{1 - \sin^2 x}{\cos x} = \frac{\cos^2 x}{\cos x} = \mathbf{\cos x}$$

24. (96)

$$\frac{\sec^2 x}{2 - \sec^2 x} = \frac{\frac{1}{\cos^2 x}}{2 - \frac{1}{\cos^2 x}} = \frac{\frac{1}{\cos^2 x}}{\frac{2\cos^2 x - 1}{\cos^2 x}}$$
$$= \frac{1}{\cos^2 x} \cdot \frac{\cos^2 x}{2\cos^2 x - 1} = \frac{1}{2\cos^2 x - 1}$$
$$= \frac{1}{\cos 2x} = \mathbf{\sec 2x}$$

25. (72,96)

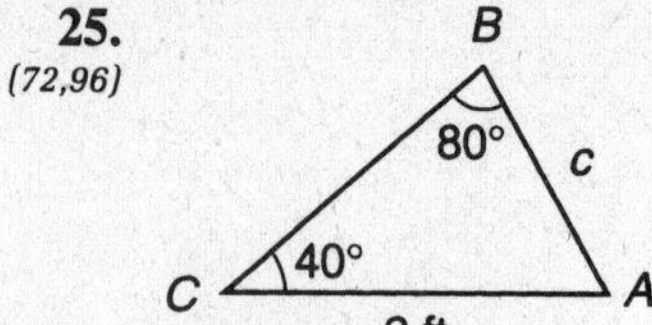

$$\frac{2}{\sin 80°} = \frac{c}{\sin 40°}$$
$$c = \frac{2 \sin 40°}{\sin 80°}$$
$$\mathbf{c = 1.31\ ft}$$

$$A = 180° - 80° - 40° = 60°$$

$$\text{Area} = \frac{1}{2}bc \sin A = \frac{1}{2}(2)(1.31) \sin 60° = 1.1345 \text{ ft}^2$$

$$1.1345 \text{ ft}^2 \left(\frac{12 \text{ in.}}{1 \text{ ft}}\right)^2 \left(\frac{2.54 \text{ cm}}{1 \text{ in.}}\right)^2 = 1053.98 \text{ cm}^2$$

$$\mathbf{Area = 1053.98\ cm^2}$$

26. (94) $y = \cot 2x$

$$\text{Period} = \frac{\pi}{2}$$

27. (84) $y = 2 + \sin \frac{3}{2}(x + 40°)$

$$\text{Period} = \frac{360°}{\frac{3}{2}} = 240°$$

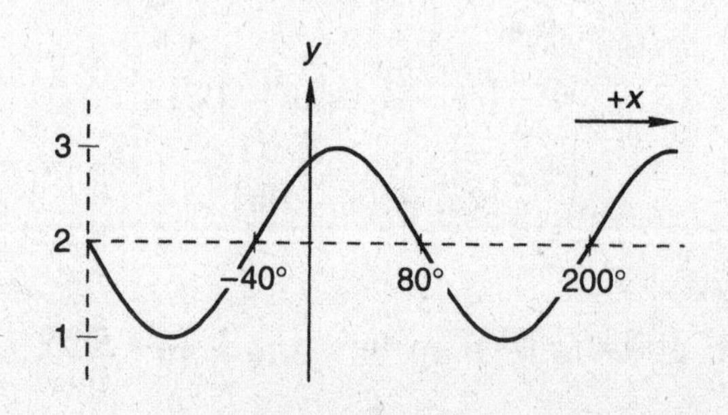

28. (87) $\sin 105° = \sin(60° + 45°)$
$$= \sin 60° \cos 45° + \cos 60° \sin 45°$$
$$= \left(\frac{\sqrt{3}}{2}\right)\left(\frac{\sqrt{2}}{2}\right) + \left(\frac{1}{2}\right)\left(\frac{\sqrt{2}}{2}\right)$$
$$= \mathbf{\frac{\sqrt{6} + \sqrt{2}}{4}}$$

29. (52) $\sqrt{3} \tan \frac{3\theta}{4} - 1 = 0$
$$\tan \frac{3\theta}{4} = \frac{1}{\sqrt{3}}$$
$$\frac{3\theta}{4} = 30°, 210°$$
$$\theta = \mathbf{40°, 280°}$$

30. (95) $0 - i = 1 \text{ cis } 270°$
$$(1 \text{ cis } 270°)^{\frac{1}{3}} = 1^{\frac{1}{3}} \text{ cis}\left(\frac{270°}{3} + n\frac{360°}{3}\right)$$
$$= 1 \text{ cis } 90°, 1 \text{ cis } 210°, 1 \text{ cis } 330°$$
$$= \mathbf{i, -\frac{\sqrt{3}}{2} - \frac{1}{2}i, \frac{\sqrt{3}}{2} - \frac{1}{2}i}$$

Problem Set 108

1. (92) $P(D \cup 6) = P(D) + P(6) - P(D \cap 6)$
$$= \frac{1}{4} + \frac{1}{6} - \frac{1}{4} \cdot \frac{1}{6}$$
$$= \frac{6}{24} + \frac{4}{24} - \frac{1}{24} = \frac{9}{24} = \mathbf{\frac{3}{8}}$$

2. (88) $A_0 = 10{,}000,\ A_5 = 15{,}000$
$$A_t = A_0 e^{\left(\frac{r}{100}\right)t}$$
$$15{,}000 = 10{,}000\, e^{\left(\frac{r}{100}\right)5}$$
$$\frac{3}{2} = e^{\frac{r}{20}}$$
$$\ln \frac{3}{2} = \frac{r}{20}$$
$$r = 20 \ln \frac{3}{2} = 8.11$$
$$r\% = \mathbf{8.11\%}$$

3. (44) $RWT = \text{jobs}$
$$R = \frac{\text{jobs}}{WT} = \frac{1}{3W} \frac{\text{pool}}{\text{worker-day}}$$
$$T = \frac{\text{jobs}}{RW} = \frac{1}{\left(\frac{1}{3W}\right)(W + P)} = \mathbf{\frac{3W}{W + p} \text{ days}}$$

4. (25)
$$\begin{cases} \text{(a)} & (J_N - 3) = 4(K_N - 3) \\ \text{(b)} & (J_N + 5) = 2(K_N + 5) \end{cases}$$

(a) $(J_N - 3) = 4(K_N - 3)$
$$J_N = 4K_N - 9$$

(b) $(J_N + 5) = 2(K_N + 5)$
$$(4K_N - 9) + 5 = 2K_N + 10$$
$$2K_N = 14$$
$$K_N = 7$$

$$J_N = 4K_N - 9$$
$$J_N = 4(7) - 9$$
$$J_N = 19$$

In 10 years:
$K = 17$ yr
$J = 29$ yr

5. (108) $A = \begin{bmatrix} 3 & 7 \\ 1 & 4 \end{bmatrix},\ B = \begin{bmatrix} 1 & 0 \\ -2 & 3 \end{bmatrix}$
$$A + B = \begin{bmatrix} 3+1 & 7+0 \\ 1-2 & 4+3 \end{bmatrix} = \mathbf{\begin{bmatrix} 4 & 7 \\ -1 & 7 \end{bmatrix}}$$
$$A - B = \begin{bmatrix} 3-1 & 7-0 \\ 1+2 & 4-3 \end{bmatrix} = \mathbf{\begin{bmatrix} 2 & 7 \\ 3 & 1 \end{bmatrix}}$$
$$2A = \begin{bmatrix} 2 \cdot 3 & 2 \cdot 7 \\ 2 \cdot 1 & 2 \cdot 4 \end{bmatrix} = \mathbf{\begin{bmatrix} 6 & 14 \\ 2 & 8 \end{bmatrix}}$$

6. (108) $A = \begin{bmatrix} 1 & 0 & 1 \\ 0 & 2 & 3 \\ 1 & 1 & 0 \end{bmatrix},\ B = \begin{bmatrix} -2 & 1 & 0 \\ 3 & 2 & 0 \\ 4 & 4 & 0 \end{bmatrix}$
$$A + B = \begin{bmatrix} 1-2 & 0+1 & 1+0 \\ 0+3 & 2+2 & 3+0 \\ 1+4 & 1+4 & 0+0 \end{bmatrix} = \mathbf{\begin{bmatrix} -1 & 1 & 1 \\ 3 & 4 & 3 \\ 5 & 5 & 0 \end{bmatrix}}$$
$$A - B = \begin{bmatrix} 1+2 & 0-1 & 1-0 \\ 0-3 & 2-2 & 3-0 \\ 1-4 & 1-4 & 0-0 \end{bmatrix} = \mathbf{\begin{bmatrix} 3 & -1 & 1 \\ -3 & 0 & 3 \\ -3 & -3 & 0 \end{bmatrix}}$$
$$2A = \begin{bmatrix} 2 \cdot 1 & 2 \cdot 0 & 2 \cdot 1 \\ 2 \cdot 0 & 2 \cdot 2 & 2 \cdot 3 \\ 2 \cdot 1 & 2 \cdot 1 & 2 \cdot 0 \end{bmatrix} = \mathbf{\begin{bmatrix} 2 & 0 & 2 \\ 0 & 4 & 6 \\ 2 & 2 & 0 \end{bmatrix}}$$

7. (108) $A = \begin{bmatrix} 1 & 1 \\ 2 & 3 \end{bmatrix}, B = \begin{bmatrix} 3 & 2 \\ -1 & 1 \end{bmatrix}$

$$A \cdot B = \begin{bmatrix} 1 & 1 \\ 2 & 3 \end{bmatrix} \cdot \begin{bmatrix} 3 & 2 \\ -1 & 1 \end{bmatrix}$$

$$= \begin{bmatrix} 1 \cdot 3 + 1 \cdot (-1) & 1 \cdot 2 + 1 \cdot 1 \\ 2 \cdot 3 + 3 \cdot (-1) & 2 \cdot 2 + 3 \cdot 1 \end{bmatrix} = \mathbf{\begin{bmatrix} 2 & 3 \\ 3 & 7 \end{bmatrix}}$$

$$B \cdot A = \begin{bmatrix} 3 & 2 \\ -1 & 1 \end{bmatrix} \cdot \begin{bmatrix} 1 & 1 \\ 2 & 3 \end{bmatrix}$$

$$= \begin{bmatrix} 3 \cdot 1 + 2 \cdot 2 & 3 \cdot 1 + 2 \cdot 3 \\ -1 \cdot 1 + 1 \cdot 2 & -1 \cdot 1 + 1 \cdot 3 \end{bmatrix} = \mathbf{\begin{bmatrix} 7 & 9 \\ 1 & 2 \end{bmatrix}}$$

8. (108) $A = \begin{bmatrix} 1 \\ 1 \\ 1 \end{bmatrix}, B = \begin{bmatrix} 0 & 2 & 0 \\ 3 & 1 & 4 \end{bmatrix}$

$\boldsymbol{A \cdot B}$ **does not exist.**

$$B \cdot A = \begin{bmatrix} 0 \cdot 1 + 2 \cdot 1 + 0 \cdot 1 \\ 3 \cdot 1 + 1 \cdot 1 + 4 \cdot 1 \end{bmatrix} = \mathbf{\begin{bmatrix} 2 \\ 8 \end{bmatrix}}$$

9. (107) $a_1 + a_1 r + a_1 r^2 + a_1 r^3 + \ldots$

$$3 - \frac{3}{2} + \frac{3}{4} - \frac{3}{8} + \ldots$$

$$a_1 = 3$$

$$a_1 r = -\frac{3}{2}$$

$$r = -\frac{3}{2a_1} = -\frac{3}{2(3)} = -\frac{1}{2}$$

Since $|r| < 1$, we can find the sum of the infinite series.

$$S_n = \frac{a_1}{1 - r} = \frac{3}{1 - \left(-\frac{1}{2}\right)} = \mathbf{2}$$

10. (106)

$$5x^2 + 3y^2 + 20x - 18y + 32 = 0$$

$$5(x^2 + 4x \quad) + 3(y^2 - 6y \quad) = -32$$

$$5(x^2 + 4x + 4) + 3(y^2 - 6y + 9) = 15$$

$$5(x + 2)^2 + 3(y - 3)^2 = 15$$

$$\mathbf{\frac{(x + 2)^2}{3} + \frac{(y - 3)^2}{5} = 1}$$

Center = (–2, 3)

$a = \sqrt{5},\ b = \sqrt{3}$

Length of major axis = $\mathbf{2\sqrt{5}}$

Length of minor axis = $\mathbf{2\sqrt{3}}$

Vertical

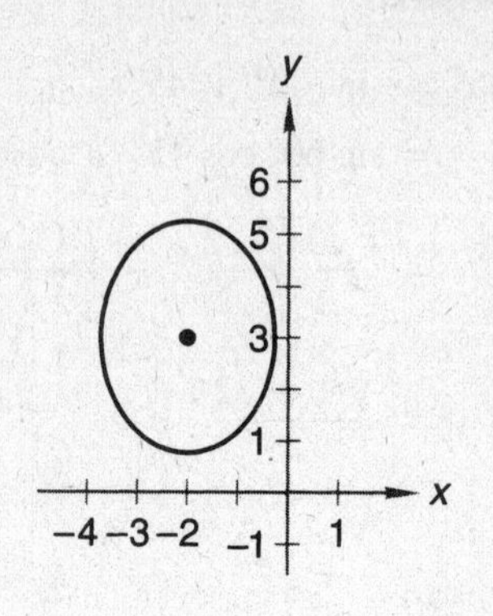

11. (106)

$$9x^2 - 4y^2 - 18x - 8y - 31 = 0$$

$$9(x^2 - 2x \quad) - 4(y^2 + 2y \quad) = 31$$

$$9(x^2 - 2x + 1) - 4(y^2 + 2y + 1) = 36$$

$$9(x - 1)^2 - 4(y + 1)^2 = 36$$

$$\mathbf{\frac{(x - 1)^2}{4} - \frac{(y + 1)^2}{9} = 1}$$

From the equation we obtain:

$h = 1,\ k = -1,\ a = 2,\ b = 3$

Center: $(h, k) = \mathbf{(1, -1)}$

Vertices: $(h + a, k), (h - a, k) = \mathbf{(-1, -1), (3, -1)}$

Asymptotes:

$$m = \pm\frac{b}{a} = \pm\frac{3}{2}$$

$y = mx + b$	$y = mx + b$
$y = \frac{3}{2}x + b$	$y = -\frac{3}{2}x + b$
$-1 = \frac{3}{2}(1) + b$	$-1 = -\frac{3}{2}(1) + b$
$b = -\frac{5}{2}$	$b = \frac{1}{2}$
$y = \frac{3}{2}x - \frac{5}{2}$	$y = -\frac{3}{2}x + \frac{1}{2}$

Asymptotes: $\mathbf{y = \frac{3}{2}x - \frac{5}{2};\ y = -\frac{3}{2}x + \frac{1}{2}}$

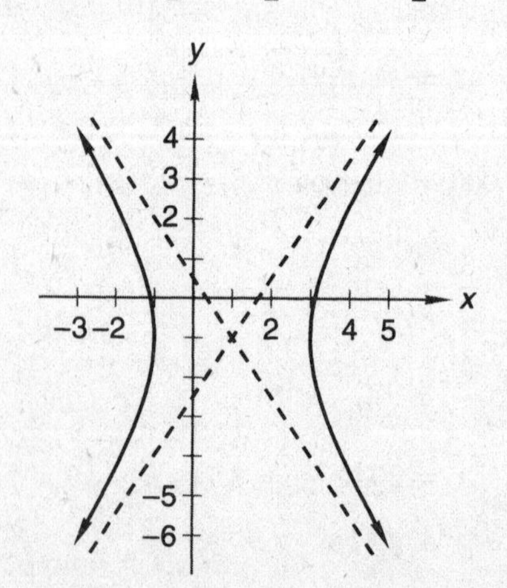

12. (105)
$$\begin{vmatrix} 6 & 8 & 5 \\ 0 & 0 & 4 \\ -3 & 2 & 4 \end{vmatrix}$$
$$= 6\begin{vmatrix} 0 & 4 \\ 2 & 4 \end{vmatrix} - 8\begin{vmatrix} 0 & 4 \\ -3 & 4 \end{vmatrix} + 5\begin{vmatrix} 0 & 0 \\ -3 & 2 \end{vmatrix}$$
$$= 6(0 - 8) - 8(0 + 12) + 5(0 + 0) = \mathbf{-144}$$

13. (101)
$$z = \frac{\begin{vmatrix} 2 & -1 & 5 \\ 1 & 2 & 10 \\ 3 & -3 & 8 \end{vmatrix}}{\begin{vmatrix} 2 & -1 & 3 \\ 1 & 2 & -1 \\ 3 & -3 & 1 \end{vmatrix}}$$
$$z = \frac{2\begin{vmatrix} 2 & 10 \\ -3 & 8 \end{vmatrix} + 1\begin{vmatrix} 1 & 10 \\ 3 & 8 \end{vmatrix} + 5\begin{vmatrix} 1 & 2 \\ 3 & -3 \end{vmatrix}}{2\begin{vmatrix} 2 & -1 \\ -3 & 1 \end{vmatrix} + 1\begin{vmatrix} 1 & -1 \\ 3 & 1 \end{vmatrix} + 3\begin{vmatrix} 1 & 2 \\ 3 & -3 \end{vmatrix}}$$
$$z = \frac{2(16 + 30) + 1(8 - 30) + 5(-3 - 6)}{2(2 - 3) + 1(1 + 3) + 3(-3 - 6)}$$
$$z = -\frac{25}{25} = \mathbf{-1}$$

14. (99) $a_1 = 5,\ a_4 = 12$
$$a_4 = a_1 + 3d$$
$$12 = 5 + 3d$$
$$d = \frac{7}{3}$$
$$a_n = a_1 + (n - 1)d$$
$$\mathbf{a_n = 5 + (n - 1)\frac{7}{3}}$$

15. (103) $\text{pH} = -\log \text{H}^+ = -\log\left(8.3 \times 10^{-9}\right) = \mathbf{8.08}$

16. (103)
$$\log_6\left(\frac{36\sqrt[3]{x^5y^4}}{z^{-3}}\right)$$
$$= \log_6\left(6^2 x^{\frac{5}{3}} y^{\frac{4}{3}} z^3\right)$$
$$= \log_6 6^2 + \log_6 x^{\frac{5}{3}} + \log_6 y^{\frac{4}{3}} + \log_6 z^3$$
$$= \mathbf{2 + \frac{5}{3}\log_6 x + \frac{4}{3}\log_6 y + 3\log_6 z}$$

17. (102) $(3x^2 + y)^9$

$F = 3x^2,\ S = y$

Term	①	②	③...⑧	⑨	⑩
Exp. of F	9	8	7... 2	1	0
Exp. of S	0	1	2... 7	8	9
Coefficient	1	9	36... 36	9	1

$S^9 = \mathbf{y^9}$

18. (100) (a)
$$\cos(A + B) = \cos A \cos B - \sin A \sin B$$
$$\cos(A - B) = \cos A \cos B + \sin A \sin B$$
$$\cos(A + B) + \cos(A - B) = 2\cos A \cos B$$
$$\cos A \cos B = \frac{1}{2}[\cos(A + B) + \cos(A - B)]$$

(b) $\cos 105° \cos 75°$
$$= \frac{1}{2}[\cos(105° + 75°) + \cos(105° - 75°)]$$
$$= \frac{1}{2}(\cos 180° + \cos 30°)$$
$$= \frac{1}{2}\left((-1) + \frac{\sqrt{3}}{2}\right) = \mathbf{\frac{-2 + \sqrt{3}}{4}}$$

19. (100) (a)
$$\begin{aligned} A + B &= x \\ A - B &= y \\ \hline 2A &= x + y \end{aligned} \qquad \begin{aligned} A + B &= x \\ -(A - B &= y) \\ \hline 2B &= x - y \end{aligned}$$
$$A = \frac{x + y}{2} \qquad B = \frac{x - y}{2}$$
$$\cos(A + B) = \cos A \cos B - \sin A \sin B$$
$$-\cos(A - B) = -\cos A \cos B - \sin A \sin B$$
$$\cos(A + B) - \cos(A - B) = -2\sin A \sin B$$
$$\cos x - \cos y = -2\sin\frac{x + y}{2}\sin\frac{x - y}{2}$$

(b) $\cos 165° - \cos 75°$
$$= -2\sin\frac{165° + 75°}{2}\sin\frac{165° - 75°}{2}$$
$$= -2\sin 120° \sin 45°$$
$$= -2 \cdot \frac{\sqrt{3}}{2} \cdot \frac{\sqrt{2}}{2} = \mathbf{-\frac{\sqrt{6}}{2}}$$

20. (98) $\log_7 50 = x$
$$7^x = 50$$
$$x\ln 7 = \ln 50$$
$$x = \mathbf{\frac{\ln 50}{\ln 7}}$$

21. (98) $\log(\ln x)^2 = 2$

$$(\ln x)^2 = 10^2$$
$$\ln x = 10$$
$$x = \mathbf{e^{10}}$$

22. (82) $e^{4x-1} = 3$

$$4x - 1 = \ln 3$$
$$x = \frac{\ln 3 + 1}{4}$$
$$x = \mathbf{0.52}$$

23. (59) $2\log_3(x + 2) + \log_3(x - 1) = 3\log_3 x$

$$\log_3\left[(x + 2)^2(x - 1)\right] = \log_3 x^3$$
$$\left(x^2 + 4x + 4\right)(x - 1) = x^3$$
$$x^3 + 4x^2 + 4x - x^2 - 4x - 4 - x^3 = 0$$
$$3x^2 - 4 = 0$$
$$\left(\sqrt{3}x + 2\right)\left(\sqrt{3}x - 2\right) = 0$$
$$x = \pm\frac{2}{\sqrt{3}}$$
$$x = \mathbf{\frac{2\sqrt{3}}{3}}$$

24. (76,80) $\sec\theta - \cos\theta = \dfrac{1}{\cos\theta} - \cos\theta = \dfrac{1 - \cos^2\theta}{\cos\theta}$

$$= \frac{\sin^2\theta}{\cos\theta} = \sin\theta \cdot \frac{\sin\theta}{\cos\theta} = \sin\theta\tan\theta$$

25. (94) $y = \csc\dfrac{1}{2}x$

$$\text{Period} = \frac{2\pi}{\frac{1}{2}} = 4\pi$$

26. (84) $y = -7 - 2\cos 2\left(x + \dfrac{\pi}{3}\right)$

$$\text{Period} = \frac{2\pi}{2} = \pi$$

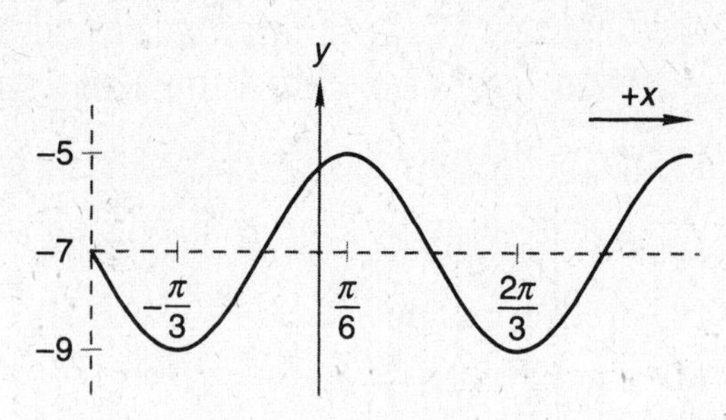

27. (94) $y = \sec 2x$

$$\text{Period} = \frac{360°}{2} = 180°$$

28. (85)

$$\cos^2\theta - \sin^2\theta = 1$$
$$\left(1 - \sin^2\theta\right) - \sin^2\theta = 1$$
$$2\sin^2\theta = 0$$
$$\sin\theta = 0$$
$$\theta = \mathbf{0°, 180°}$$

29. (95) $9 - 9i = \sqrt{162}\text{ cis }315° = 12.73\text{ cis }315°$

$$(12.73\text{ cis }315°)^{\frac{1}{5}} = 12.73^{\frac{1}{5}}\text{ cis}\left(\frac{315°}{5} + n\frac{360°}{5}\right)$$

= **1.66 cis 63°, 1.66 cis 135°, 1.66 cis 207°, 1.66 cis 279°, 1.66 cis 351°**

30. (56)

$$\begin{cases} x^2 + 4x + y^2 + 6y + 13 \geq 16 \\ 6y - 4x + 24 > 0 \\ 4x > 6y - 60 \end{cases}$$

$$x^2 + 4x + y^2 + 6y + 13 \geq 16$$
$$(x^2 + 4x + 4) + (y^2 + 6y + 9) \geq 3 + 4 + 9$$
$$(x + 2)^2 + (y + 3)^2 \geq 4^2 \quad \text{(circle)}$$

$$6y - 4x + 24 > 0$$
$$y > \frac{2}{3}x - 4 \quad \text{(line)}$$
$$4x > 6y - 60$$
$$-6y > -4x - 60$$
$$y < \frac{2}{3}x + 10 \quad \text{(line)}$$

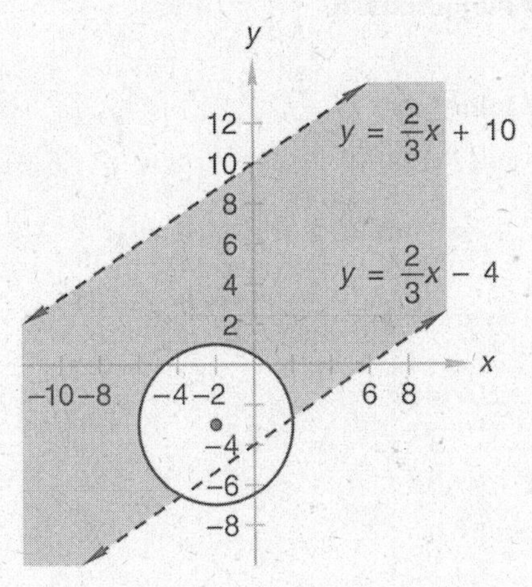

Problem Set 109

1. (83)
$$\frac{3}{7} \cdot \frac{2}{6} = \mathbf{\frac{1}{7}}$$

2. (88) $A_{20} = 1400$, $A_{40} = 1300$
$$A_{40} = A_{20}e^{k(40-20)}$$
$$1300 = 1400e^{k(20)}$$
$$\frac{13}{14} = e^{20k}$$
$$\ln \frac{13}{14} = 20k$$
$$k = -0.0037054$$

$$A_{200} = A_{20}e^{-0.0037054(200-20)}$$
$$A_{200} = 1400e^{-0.0037054(200-20)}$$
$$= 1400e^{-0.0037054(180)} = \mathbf{718.56}$$

3. (44) $\text{Rate} = \frac{d}{n}$

$$\text{New rate} = \frac{d}{n} + 3 = \frac{d + 3n}{n}$$

$$\text{Rate} \times N = \text{price}$$
$$N = \frac{\text{price}}{\text{rate}} = \frac{400}{\frac{d + 3n}{n}} = \mathbf{\frac{400n}{d + 3n}}$$

4. (44) $R_M T_M + R_R T_R = \text{jobs}$
$$\left(\frac{5}{H}\right)3 + (R)3 = 10$$
$$3R = \frac{10H - 15}{H}$$
$$R = \mathbf{\frac{10H - 15}{3H}\ \frac{jobs}{hr}}$$

5. (25)
$$\begin{cases} \text{(a)}\ M_N - 10 = 2(J_N - 10) + 6 \\ \text{(b)}\ 3(M_N + 10) = 4(J_N + 10) + 30 \end{cases}$$

(a)
$$M_N - 10 = 2(J_N - 10) + 6$$
$$M_N = 2J_N - 20 + 6 + 10$$
$$M_N = 2J_N - 4$$

(b)
$$3(M_N + 10) = 4(J_N + 10) + 30$$
$$3M_N + 30 = 4J_N + 40 + 30$$
$$3(2J_N - 4) + 30 = 4J_N + 70$$
$$6J_N - 12 + 30 = 4J_N + 70$$
$$2J_N = 52$$
$$J_N = 26$$

$$M_N = 2J_N - 4$$
$$M_N = 2(26) - 4$$
$$M_N = 48$$

In 15 years:

$\mathbf{J = 41}$ **yr**

$\mathbf{M = 63}$ **yr**

6. (109) $0.000\overline{31}$

$$= 31 \times 10^{-5} + 31 \times 10^{-7} + 31 \times 10^{-9} + \ldots$$
$$= (31 \times 10^{-5}) + (31 \times 10^{-5})(10^{-2})$$
$$+ (31 \times 10^{-5})(10^{-2})^2 + \ldots$$

If $a_1 = 31 \times 10^{-5}$, $r = 10^{-2}$

$$S = \frac{a_1}{1 - r}$$
$$= \frac{31 \times 10^{-5}}{1 - 10^{-2}} = \frac{31 \times 10^{-5}}{0.99} \cdot \frac{10^5}{10^5} = \mathbf{\frac{31}{99{,}000}}$$

7. (109) $6.0\overline{17}$

$$= 6 + 17 \times 10^{-3} + 17 \times 10^{-5} + \ldots$$

$$= 6 + (17 \times 10^{-3}) + (17 \times 10^{-3})(10^{-2}) + (17 \times 10^{-3})(10^{-2})^2 + \ldots$$

If $a_1 = 17 \times 10^{-3}$, $r = 10^{-2}$

$$6 + \frac{a_1}{1 - r} = 6 + \frac{17 \times 10^{-3}}{1 - 10^{-2}}$$

$$= 6 + \frac{17 \times 10^{-3}}{0.99} \cdot \frac{10^3}{10^3} = 6 + \frac{17}{990}$$

$$= \frac{5940}{990} + \frac{17}{990} = \mathbf{\frac{5957}{990}}$$

8. (107) Sum of distance fallen:

$$S = \frac{a_1}{1 - r} = \frac{132}{1 - \frac{1}{4}} = 176$$

Sum of distances rebounded:

$(176 - 132) = 44$

Sum of ups and downs:

$176 + 44 =$ **220 ft**

9. (108) $A = \begin{bmatrix} 2 & 3 & 1 \\ 4 & 0 & 2 \\ 1 & 1 & 1 \end{bmatrix}$, $B = \begin{bmatrix} 0 & -1 & 2 \\ 3 & 5 & 7 \\ -2 & 0 & 6 \end{bmatrix}$

$$A + B = \begin{bmatrix} 2 + 0 & 3 + (-1) & 1 + 2 \\ 4 + 3 & 0 + 5 & 2 + 7 \\ 1 + (-2) & 1 + 0 & 1 + 6 \end{bmatrix}$$

$$= \begin{bmatrix} \mathbf{2} & \mathbf{2} & \mathbf{3} \\ \mathbf{7} & \mathbf{5} & \mathbf{9} \\ \mathbf{-1} & \mathbf{1} & \mathbf{7} \end{bmatrix}$$

$$A \cdot B = \begin{bmatrix} 2 & 3 & 1 \\ 4 & 0 & 2 \\ 1 & 1 & 1 \end{bmatrix} \begin{bmatrix} 0 & -1 & 2 \\ 3 & 5 & 7 \\ -2 & 0 & 6 \end{bmatrix}$$

$$= \begin{bmatrix} 0 + 9 - 2 & -2 + 15 + 0 & 4 + 21 + 6 \\ 0 + 0 - 4 & -4 + 0 + 0 & 8 + 0 + 12 \\ 0 + 3 - 2 & -1 + 5 + 0 & 2 + 7 + 6 \end{bmatrix}$$

$$= \begin{bmatrix} \mathbf{7} & \mathbf{13} & \mathbf{31} \\ \mathbf{-4} & \mathbf{-4} & \mathbf{20} \\ \mathbf{1} & \mathbf{4} & \mathbf{15} \end{bmatrix}$$

10. (106)

$$9x^2 + 25y^2 - 36x + 150y + 260 = 0$$

$$9(x^2 - 4x \quad) + 25(y^2 + 6y \quad) = -260$$

$$9(x^2 - 4x + 4) + 25(y^2 + 6y + 9) = 1$$

$$9(x - 2)^2 + 25(y + 3)^2 = 1$$

$$\mathbf{\frac{(x - 2)^2}{\frac{1}{9}} + \frac{(y + 3)^2}{\frac{1}{25}} = 1}$$

Center = (2, –3)

$a = \frac{1}{3}$, $b = \frac{1}{5}$

Length of major axis = $\mathbf{\frac{2}{3}}$

Length of minor axis = $\mathbf{\frac{2}{5}}$

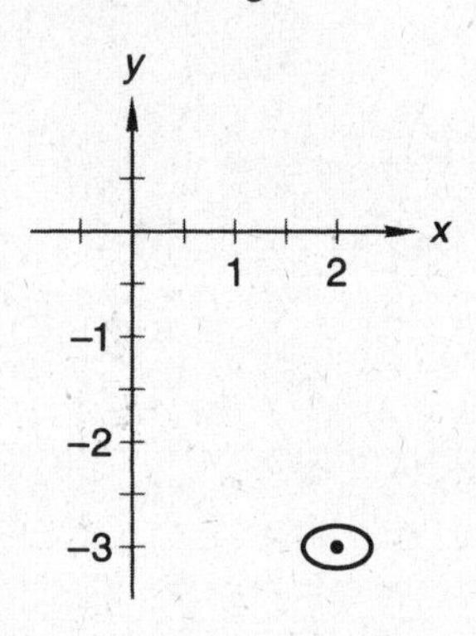

11. (106)

$$4x^2 - y^2 + 24x + 4y + 28 = 0$$

$$4(x^2 + 6x \quad) - (y^2 - 4y \quad) = -28$$

$$4(x^2 + 6x + 9) - (y^2 - 4y + 4) = 4$$

$$4(x + 3)^2 - (y - 2)^2 = 4$$

$$\mathbf{\frac{(x + 3)^2}{1} - \frac{(y - 2)^2}{4} = 1}$$

From the equation we obtain:

$h = -3$, $k = 2$, $a = 1$, $b = 2$

Center: $(h, k) =$ **(–3, 2)**

Vertices: $(h + a, k), (h - a, k) =$ **(–2, 2), (–4, 2)**

Asymptotes:

$$m = \pm\frac{b}{a} = \pm 2$$

$y = mx + b$	$y = mx + b$
$y = 2x + b$	$y = -2x + b$
$2 = 2(-3) + b$	$2 = -2(-3) + b$
$b = 8$	$b = -4$
$y = 2x + 8$	$y = -2x - 4$

Asymptotes: $\mathbf{y = 2x + 8; \; y = -2x - 4}$

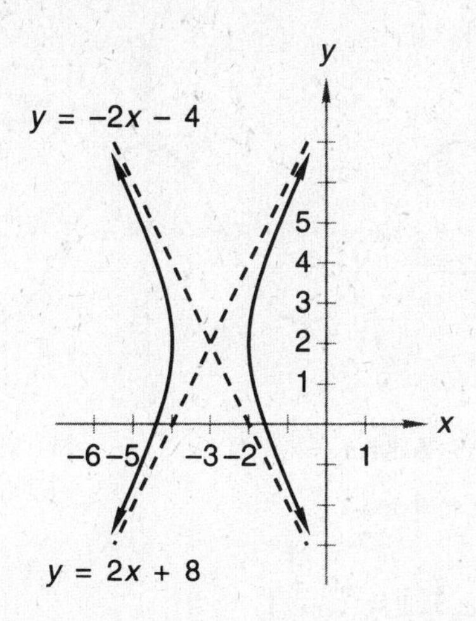

12. (105) $\begin{vmatrix} 9 & 3 & 0 \\ 1 & 5 & 2 \\ 0 & -3 & 7 \end{vmatrix} = 9\begin{vmatrix} 5 & 2 \\ -3 & 7 \end{vmatrix} - 3\begin{vmatrix} 1 & 2 \\ 0 & 7 \end{vmatrix} + 0\begin{vmatrix} 1 & 5 \\ 0 & -3 \end{vmatrix}$

$= 9(35 + 6) - 3(7 - 0) + 0 = \mathbf{348}$

13. (101) $y = \dfrac{\begin{vmatrix} 1 & 1 & -2 \\ 2 & -1 & 3 \\ 5 & 1 & -8 \end{vmatrix}}{\begin{vmatrix} 1 & 1 & -2 \\ 2 & -3 & 3 \\ 5 & 2 & -8 \end{vmatrix}}$

$y = \dfrac{1\begin{vmatrix} -1 & 3 \\ 1 & -8 \end{vmatrix} - 1\begin{vmatrix} 2 & 3 \\ 5 & -8 \end{vmatrix} - 2\begin{vmatrix} 2 & -1 \\ 5 & 1 \end{vmatrix}}{1\begin{vmatrix} -3 & 3 \\ 2 & -8 \end{vmatrix} - 1\begin{vmatrix} 2 & 3 \\ 5 & -8 \end{vmatrix} - 2\begin{vmatrix} 2 & -3 \\ 5 & 2 \end{vmatrix}}$

$y = \dfrac{1(8 - 3) - 1(-16 - 15) - 2(2 + 5)}{1(24 - 6) - 1(-16 - 15) - 2(4 + 15)}$

$y = \dfrac{22}{11} = \mathbf{2}$

14. (103) $\log_5 (x + 1) - 2 \log_5 x - \log_5 (x^2 - 1)$

$+ \log_5 (x^2 + x)$

$= \log_5 \dfrac{(x + 1)(x^2 + x)}{x^2(x^2 - 1)}$

$= \log_5 \dfrac{(x + 1)(x)(x + 1)}{x^2(x + 1)(x - 1)}$

$= \mathbf{\log_5 \left(\dfrac{x + 1}{x^2 - x}\right)}$

15. (104) $a_1 = 2,\ r = -2,\ n = 12$

(a) $a_n = ar^{(n - 1)}$

$a_{12} = ar^{(12-1)} = \mathbf{2(-2)^{11}}$

(b) $S_n = \dfrac{2[1 - r^n]}{1 - r}$

$S_{12} = \dfrac{2[1 - (-2)^{12}]}{1 - (-2)}$

$= \dfrac{2(1 - 4096)}{3} = \dfrac{-8190}{3} = \mathbf{-2730}$

16. (103) $H^+ = 10^{-pH} = 10^{-6.5} = \mathbf{3.16 \times 10^{-7}\ \dfrac{mole}{liter}}$

17. (102) $(3a + bc)^7$

$F = 3a,\ S = bc$

Term	①	②	③	④	⑤	⑥
Exp. of F	5	4	3	2	1	0
Exp. of S	0	1	2	3	4	5
Coefficient	1	5	10	10	5	1

$10F^2S^3 = 10(3a)^2(bc)^3 = \mathbf{90a^2b^3c^3}$

18. (100) (a)

$A + B = x$
$A - B = y$
$2A = x + y^2$
$A = \dfrac{x + y}{2}$

$A + B = x$
$-(A - B = y)$
$B = x - y$
$B = \dfrac{x - y}{2}$

$\sin (A + B) = \sin A \cos B + \cos A \sin B$
$-\sin (A - B) = -\sin A \cos B + \cos A \sin B$
$\sin (A + B) - \sin (A - B) = 2 \cos A \sin B$

$\sin x - \sin y = 2 \cos \dfrac{x + y}{2} \sin \dfrac{x - y}{2}$

(b) $\sin 255° - \sin 15°$

$= 2 \cos \dfrac{255° + 15°}{2} \sin \dfrac{255° - 15°}{2}$

$= 2 \cos 135° \sin 120°$

$= 2\left(-\dfrac{\sqrt{2}}{2}\right)\left(\dfrac{\sqrt{3}}{2}\right) = \mathbf{-\dfrac{\sqrt{6}}{2}}$

19. (98) $x = \ln 42$

$e^x = 42$

$x \log e = \log 42$

$x = \mathbf{\dfrac{\log 42}{\log e}}$

20. (98)
$$x^{\sqrt{\log x}} = 10^{27}$$
$$\sqrt{\log x}\,\log x = 27$$
$$(\log x)^{\frac{3}{2}} = 27$$
$$\log x = 27^{\frac{2}{3}}$$
$$\log x = 9$$
$$x = \mathbf{10^9}$$

21. (59) $\frac{3}{4}\log_5 16 - \log_5 (3x - 2) = -\log_5 (2x + 1)$
$$\log_5 \frac{16^{\frac{3}{4}}}{3x - 2} = \log_5 \frac{1}{2x + 1}$$
$$\frac{8}{3x - 2} = \frac{1}{2x + 1}$$
$$16x + 8 = 3x - 2$$
$$13x = -10$$
$$x = -\frac{10}{13}$$

$-\frac{10}{13}$ is invalid. **No solution.**

22. (93,96) $(\sin x - \cos x)^2$
$$= \sin^2 x - 2\sin x\cos x + \cos^2 x$$
$$= 1 - 2\sin x\cos x = 1 - \sin 2x$$

23. (84) $\frac{\tan^2 x}{\sec x + 1} = \frac{\sec^2 x - 1}{\sec x + 1}$
$$= \frac{(\sec x + 1)(\sec x - 1)}{\sec x + 1} = \sec x - 1$$

24. (81,96)
$$7^2 = 3^2 + 8^2 - 2(3)(8)\cos A$$
$$\cos A = \frac{3^2 + 8^2 - 7^2}{2(3)(8)}$$
$$\cos A = 0.5$$
$$\mathbf{A = 60°}$$

$$\text{Area} = \frac{1}{2}bc\sin A$$
$$= \frac{1}{2}(8)(3)\sin 60°$$
Area = 10.39 ft^2

25. (84) $y = -3 + 5\cos(x + 2\pi)$

Period $= 2\pi$

Phase angle $= -2\pi$

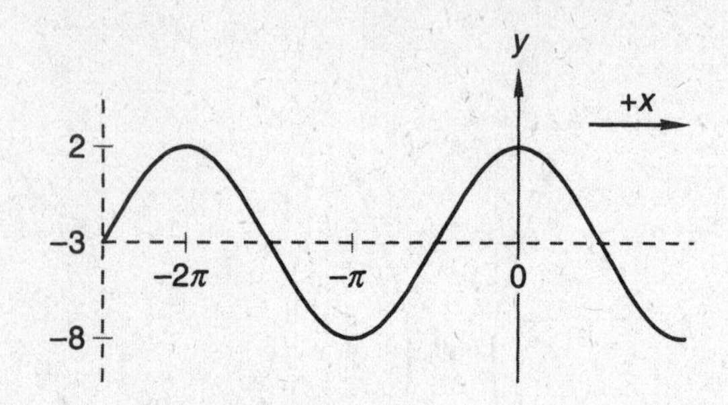

26. (94) (a) $\mathbf{y = 10 + 6\csc x}$

(b) $\mathbf{y = -7 + 4\sec x}$

27. (85) $(\sqrt{2} - 2\cos 3x)(\sin x + 1) = 0$

$$\cos 3x = \frac{\sqrt{2}}{2} \qquad \sin x = -1$$
$$3x = \frac{\pi}{4}, \frac{7\pi}{4}, \frac{9\pi}{4}, \frac{15\pi}{4}, \frac{17\pi}{4}, \frac{23\pi}{4} \qquad x = \frac{3\pi}{2}$$

$$x = \mathbf{\frac{\pi}{12}, \frac{7\pi}{12}, \frac{3\pi}{4}, \frac{5\pi}{4}, \frac{17\pi}{12}, \frac{3\pi}{2}, \frac{23\pi}{12}}$$

28. (95) $1 - i = \sqrt{2}\operatorname{cis} 315° = 1.41\operatorname{cis} 315°$
$$(1.41\operatorname{cis} 315°)^{\frac{1}{3}} = 1.41^{\frac{1}{3}}\operatorname{cis}\left(\frac{315°}{3} + n\frac{360°}{3}\right)$$
$$= \mathbf{1.12\ cis\ 105°, 1.12\ cis\ 225°, 1.12\ cis\ 345°}$$

29. (58) $5x + y + 13 = 0$
$$y = -5x - 13$$

Equation of the perpendicular line:
$$y = \frac{1}{5}x + b$$
$$0 = \frac{1}{5}(0) + b$$
$$b = 0$$
$$y = \frac{1}{5}x$$

Point of intersection:
$$\frac{1}{5}x = -5x - 13$$
$$\frac{26}{5}x = -13$$
$$x = -\frac{5}{2}$$
$$y = \frac{1}{5}x = \frac{1}{5}\left(-\frac{5}{2}\right) = -\frac{1}{2}$$

$\left(-\frac{5}{2}, -\frac{1}{2}\right)$ and (0, 0)

$$D = \sqrt{\left(-\frac{5}{2}\right)^2 + \left(-\frac{1}{2}\right)^2} = \sqrt{\frac{25}{4} + \frac{1}{4}} = \mathbf{\frac{\sqrt{26}}{2}}$$

30. (24)
$$(f \circ g)(x) = 2 - (x + 1)^3$$
$$= 2 - (x^3 + 3x^2 + 3x + 1)$$
$$= \mathbf{-x^3 - 3x^2 - 3x + 1}$$

Problem Set 110

1. (92)
$$P(Q \cup B) = P(Q) + P(B) - P(Q \cap B)$$
$$= \frac{4}{52} + \frac{26}{52} - \frac{2}{52} = \frac{28}{52} = \mathbf{\frac{7}{13}}$$

2. (88) $A_5 = 120$, $A_{10} = 400$

$$A_{10} = A_5 e^{k(10-5)}$$
$$400 = 120e^{k(5)}$$
$$\frac{10}{3} = e^{k(5)}$$
$$\ln \frac{10}{3} = 5k$$
$$k = 0.24079$$
$$A_{30} = A_{10}e^{0.24079(30-10)}$$
$$= 400e^{0.24079(20)} = \mathbf{49{,}382.72}$$

3. (85) $R_B = 1$, $R_L = \frac{1}{12}$, $T_B = T_L$

$$\begin{cases} R_BT_B = S + 30 \\ R_LT_L = S \end{cases}$$

$$R_BT_B = S + 30$$
$$(1)T = (R_LT_L) + 30$$
$$T = \frac{1}{12}T + 30$$
$$\frac{11}{12}T = 30$$
$$T = \frac{360}{11} = \mathbf{32\frac{8}{11}\ min}$$

4. (36)
$$\begin{cases} \text{(a)} & (B + W)T_D = D_D \\ \text{(b)} & (B - W)T_U = D_U \end{cases}$$
$B = 3W$, $T_D = T_U - 1$

(b)
$$(B - W)T_U = D_U$$
$$(2W)T_U = 32$$
$$WT_U = 16$$

(a)
$$(B + W)T_D = D_D$$
$$4W(T_U - 1) = 48$$
$$4WT_U - 4W = 48$$
$$4(16) - 4W = 48$$
$$4W = 16$$
$$\mathbf{W = 4\ mph}$$

$$B = 3W$$
$$B = 3(4)$$
$$\mathbf{B = 12\ mph}$$

5. (25)
$$R_1T_1 + R_2T_2 = \text{tank}$$
$$\frac{1}{10}(T + 2) + \frac{1}{8}T = 1$$
$$8(T + 2) + 10T = 80$$
$$8T + 16 + 10T = 80$$
$$18T = 64$$
$$T = \frac{64}{18} = \mathbf{\frac{32}{9}\ hr}$$

6. (110) (a)

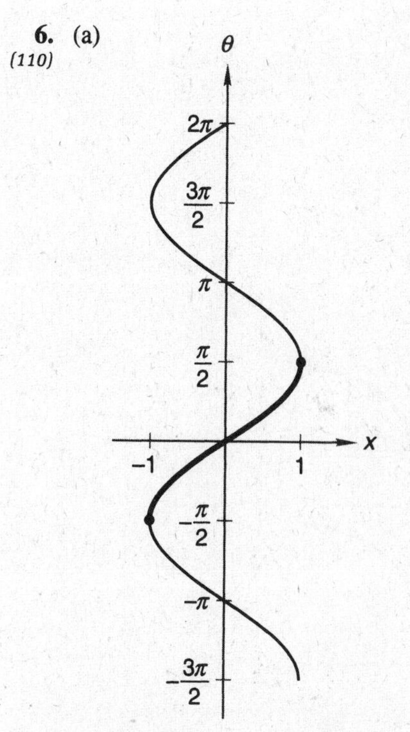

Domain $= \{x \in \mathbb{R} \mid \mathbf{-1 \le x \le 1}\}$

Range $= \left\{\theta \in \mathbb{R} \mid \mathbf{-\frac{\pi}{2} \le \theta \le \frac{\pi}{2}}\right\}$

Arcsin $\mathbf{\frac{\sqrt{3}}{2} = \frac{\pi}{3}}$

(b)

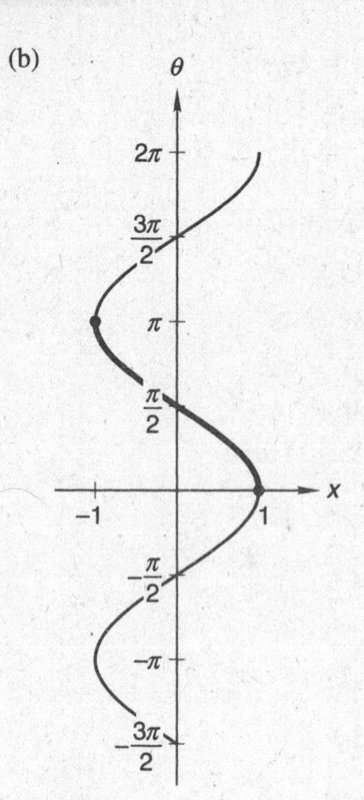

Domain $= \{x \in \mathbb{R} \mid -1 \le x \le 1\}$

Range $= \{\theta \in \mathbb{R} \mid 0 \le \theta \le \pi\}$

Arccos $\left(-\frac{1}{2}\right) = \frac{2\pi}{3}$

(c)

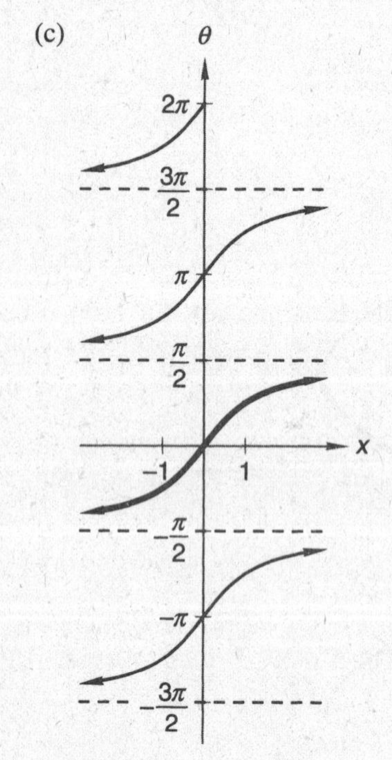

Domain $= \{x \in \mathbb{R}\}$

Range $= \left\{\theta \in \mathbb{R} \mid -\frac{\pi}{2} < \theta < \frac{\pi}{2}\right\}$

Arctan $\left(-\frac{\sqrt{3}}{3}\right) = -\frac{\pi}{6}$

7. (109) $0.00\overline{241}$

$$= 241 \times 10^{-5} + 241 \times 10^{-8} + 241 \times 10^{-11} + \ldots$$

$$= (241 \times 10^{-5}) + (241 \times 10^{-5})(10^{-3}) + (241 \times 10^{-5})(10^{-3})^2 + \ldots$$

If $a_1 = 241 \times 10^{-5}$, $r = 10^{-3}$

$$S = \frac{a}{1 - r}$$

$$= \frac{241 \times 10^{-5}}{1 - 10^{-3}} = \frac{241 \times 10^{-5}}{0.999} \cdot \frac{10^5}{10^5} = \mathbf{\frac{241}{99{,}900}}$$

8. (107) $\left(\frac{1}{4}\right) + \left(-\frac{3}{16}\right) + \left(\frac{9}{64}\right) + \left(-\frac{24}{256}\right) + \ldots$

$$a_1 = \frac{1}{4}$$

$$a_1 r = -\frac{3}{16}$$

$$r = -\frac{3}{16a_1} = -\frac{3}{4}$$

$$S = \frac{a_1}{1 - r} = \frac{\frac{1}{4}}{1 + \frac{3}{4}} = \frac{\frac{1}{4}}{\frac{7}{4}} = \mathbf{\frac{1}{7}}$$

9. (108) $A = \begin{bmatrix} 1 & 0 & 1 \\ -2 & 4 & 3 \\ 6 & 2 & 0 \end{bmatrix}$, $B = \begin{bmatrix} 2 & 2 & 4 \\ 1 & 1 & 0 \\ -2 & 3 & -1 \end{bmatrix}$

$$A + B = \begin{bmatrix} 1+2 & 0+2 & 1+4 \\ -2+1 & 4+1 & 3+0 \\ 6-2 & 2+3 & 0-1 \end{bmatrix} = \begin{bmatrix} \mathbf{3} & \mathbf{2} & \mathbf{5} \\ \mathbf{-1} & \mathbf{5} & \mathbf{3} \\ \mathbf{4} & \mathbf{5} & \mathbf{-1} \end{bmatrix}$$

$$B \cdot A = \begin{bmatrix} 2 & 2 & 4 \\ 1 & 1 & 0 \\ -2 & 3 & -1 \end{bmatrix} \cdot \begin{bmatrix} 1 & 0 & 1 \\ -2 & 4 & 3 \\ 6 & 2 & 0 \end{bmatrix}$$

$$= \begin{bmatrix} 2 - 4 + 24 & 0 + 8 + 8 & 2 + 6 + 0 \\ 1 - 2 + 0 & 0 + 4 + 0 & 1 + 3 + 0 \\ -2 - 6 - 6 & 0 + 12 - 2 & -2 + 9 + 0 \end{bmatrix}$$

$$= \begin{bmatrix} \mathbf{22} & \mathbf{16} & \mathbf{8} \\ \mathbf{-1} & \mathbf{4} & \mathbf{4} \\ \mathbf{-14} & \mathbf{10} & \mathbf{7} \end{bmatrix}$$

10. (106)

$$25x^2 + 9y^2 - 200x + 18y + 184 = 0$$

$$25(x^2 - 8x \quad) + 9(y^2 + 2y \quad) = -184$$

$$25(x^2 - 8x + 16) + 9(y^2 + 2y + 1) = 225$$

$$\mathbf{\frac{(x - 4)^2}{9} + \frac{(y + 1)^2}{25} = 1}$$

Center = (4, –1)

$a = 5,\ b = 3$

Length of major axis = 10

Length of minor axis = 6

Vertical

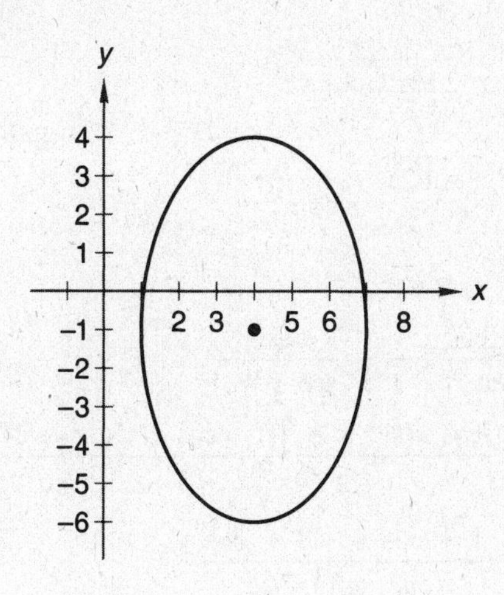

11. (106)

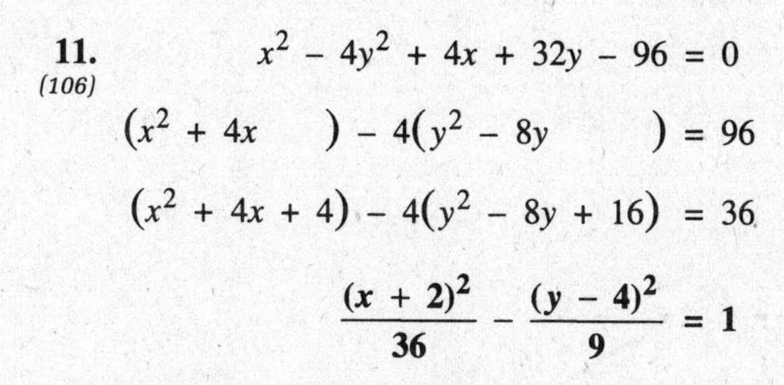

$$x^2 - 4y^2 + 4x + 32y - 96 = 0$$

$$(x^2 + 4x \qquad) - 4(y^2 - 8y \qquad) = 96$$

$$(x^2 + 4x + 4) - 4(y^2 - 8y + 16) = 36$$

$$\mathbf{\frac{(x + 2)^2}{36} - \frac{(y - 4)^2}{9} = 1}$$

From the equation we obtain:
$h = -2,\ k = 4,\ a = 6,\ b = 3$

Center: $(h, k) = \mathbf{(-2, 4)}$

Vertices: $(h + a, k), (h - a, k) = \mathbf{(4, 4), (-8, 4)}$

Asymptotes:

$$m = \pm\frac{b}{a} = \pm\frac{1}{2}$$

$y = mx + b$	$y = mx + b$
$y = \frac{1}{2}x + b$	$y = -\frac{1}{2}x + b$
$4 = \frac{1}{2}(-2) + b$	$4 = -\frac{1}{2}(-2) + b$
$b = 5$	$b = 3$
$y = \frac{1}{2}x + 5$	$y = -\frac{1}{2}x + 3$

Asymptotes: $\mathbf{y = \frac{1}{2}x + 5;\ y = -\frac{1}{2}x + 3}$

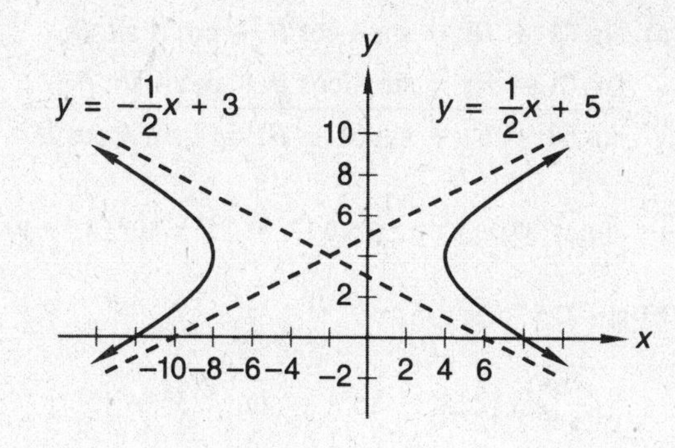

12. (101)

$$x = \frac{\begin{vmatrix} 0 & 2 & 0 \\ 5 & 0 & -3 \\ -2 & 1 & 1 \end{vmatrix}}{\begin{vmatrix} 3 & 2 & 0 \\ 1 & 0 & -3 \\ 1 & 1 & 1 \end{vmatrix}}$$

$$= \frac{0\begin{vmatrix} 0 & -3 \\ 1 & 1 \end{vmatrix} - 2\begin{vmatrix} 5 & -3 \\ -2 & 1 \end{vmatrix} + 0\begin{vmatrix} 5 & 0 \\ -2 & 1 \end{vmatrix}}{3\begin{vmatrix} 0 & -3 \\ 1 & 1 \end{vmatrix} - 2\begin{vmatrix} 1 & -3 \\ 1 & 1 \end{vmatrix} + 0\begin{vmatrix} 1 & 0 \\ 1 & 1 \end{vmatrix}}$$

$$= \frac{0 - 2(5 - 6) + 0}{3(0 + 3) - 2(1 + 3) + 0} = \frac{2}{9 - 8} = \mathbf{2}$$

13. (99) $a_1 = 4,\ r = 2,\ n = 15$

$$a_n = a_1 r^{(n-1)}$$

$$a_{15} = 4(2)^{14} = \mathbf{65{,}536}$$

14. (99) Arithmetic mean $= \frac{x + y}{2} = \frac{7 + 12}{2} = \mathbf{\frac{19}{2}}$

Geometric mean $= \pm\sqrt{xy} = \pm\sqrt{(7)(12)} = \mathbf{\pm 2\sqrt{21}}$

15. (103) $10^{\sqrt{\left(\frac{100}{9}\right)\left(\frac{16}{36}\right)}} = 10^{\left(\frac{10}{3}\right)\left(\frac{4}{6}\right)} = 10^{\frac{20}{9}} = \mathbf{166.81}$

16. (102) $(x^2 - 2y)^3$

$F = x^2,\ S = -2y$

Term	①	②	③	④
Exp. of F	3	2	1	0
Exp. of S	0	1	2	3
Coefficient	1	3	3	1

$$(F + S)^3 = F^3 + 3F^2S + 3FS^2 + S^3$$

$$= (x^2)^3 + 3(x^2)^2(-2y) + 3(x^2)(-2y)^2 + (-2y)^3$$

$$= \mathbf{x^6 - 6x^4y + 12x^2y^2 - 8y^3}$$

17. (100) (a)

$$\begin{aligned} \sin(A+B) &= \sin A\cos B + \cos A\sin B \\ \sin(A-B) &= \sin A\cos B - \cos A\sin B \end{aligned}$$

$$\sin(A+B) + \sin(A-B) = 2\sin A\cos B$$

$$\sin x \cos y = \frac{1}{2}[\sin(x+y) + \sin(x-y)]$$

(b) $\sin 225° \cos 15°$

$$= \frac{1}{2}[\sin(225° + 15°) + \sin(225° - 15°)]$$

$$= \frac{1}{2}(\sin 240° + \sin 210°)$$

$$= \frac{1}{2}\left[-\frac{\sqrt{3}}{2} + \left(-\frac{1}{2}\right)\right] = \mathbf{-\frac{\sqrt{3}}{4} - \frac{1}{4}}$$

18. (98)

$$\log_4 8 = y$$

$$4^y = 8$$

$$y\log 4 = \log 8$$

$$y = \mathbf{\frac{\log 8}{\log 4}}$$

19. (98)

$$\sqrt{\log_2 x} = \log_2 x - 2$$

$$\log_2 x = (\log_2 x)^2 - 4\log_2 x + 4$$

$$(\log_2 x)^2 - 5\log_2 x + 4 = 0$$

$$(\log_2 x - 4)(\log_2 x - 1) = 0$$

$\log_2 x = 4$, $x = 2^4$, $x = \mathbf{16}$

$\log_2 x = 1$, $x = 2$, invalid

20. (98)

$$\log\sqrt[3]{x^2} + \log\sqrt[3]{x^4} = \log 2^{-4}$$

$$\log x^{\frac{2}{3}} + \log x^{\frac{4}{3}} = \log 2^{-4}$$

$$\log\left(x^{\frac{2}{3}} \cdot x^{\frac{4}{3}}\right) = \log 2^{-4}$$

$$\log x^2 = \log 2^{-4}$$

$$x^2 = \frac{1}{16}$$

$$x = \mathbf{\pm\frac{1}{4}}$$

21. (96) $\frac{1}{2}\sec x \csc(-x)$

$$= -\frac{1}{2}\sec x \csc x$$

$$= -\frac{1}{2}\frac{1}{\cos x}\frac{1}{\sin x}$$

$$= -\frac{1}{2\sin x\cos x}$$

$$= -\frac{1}{\sin 2x}$$

22. (93) $\frac{\cos x}{\sec x - 1} - \frac{\cos x}{\sec x + 1}$

$$= \frac{\cos x(\sec x + 1) - \cos x(\sec x - 1)}{\sec^2 x - 1}$$

$$= \frac{1 + \cos x - 1 + \cos x}{\tan^2 x}$$

$$= \frac{2\cos x}{\frac{\sin^2 x}{\cos^2 x}} = \frac{2\cos^3 x}{\sin^2 x}$$

$$= 2\cos^3 x \csc^2 x$$

23. (81,96)

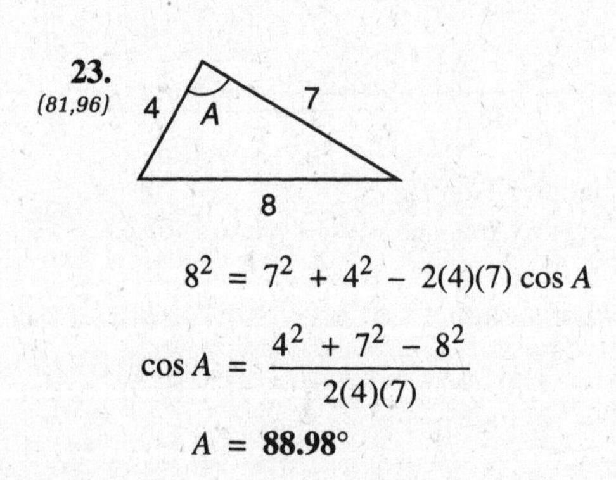

$$8^2 = 7^2 + 4^2 - 2(4)(7)\cos A$$

$$\cos A = \frac{4^2 + 7^2 - 8^2}{2(4)(7)}$$

$$A = \mathbf{88.98°}$$

24. (94) $y = \tan\left(x + \frac{\pi}{2}\right)$

y
1
−1
x
$-\pi$ $-\frac{\pi}{2}$ $\frac{\pi}{2}$ π $\frac{3\pi}{2}$ 2π

25. $y = -4 + 3\sin\frac{1}{3}(x - 2\pi)$
(84)

$$\text{Period} = \frac{2\pi}{\frac{1}{3}} = 6\pi$$

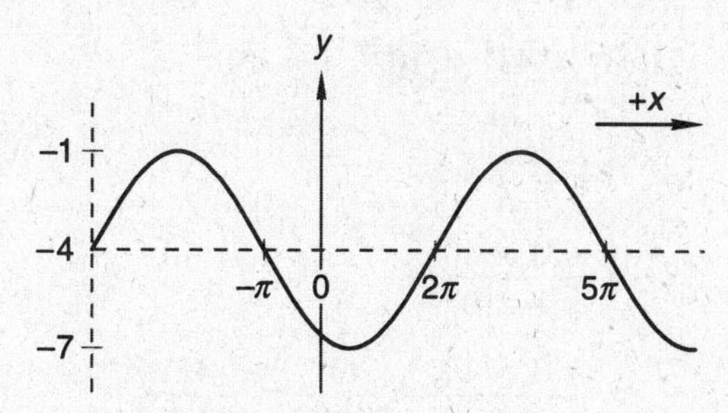

26. $y = \cot\left(x - \frac{\pi}{2}\right)$
(94)

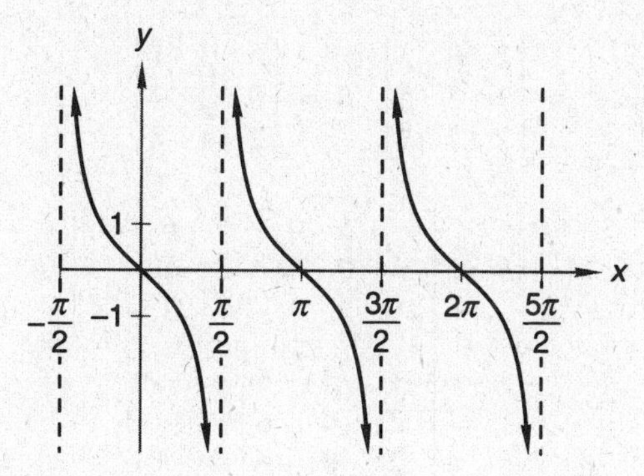

27. $-2\sin 3x = 1$
(52)

$$\sin 3x = -\frac{1}{2}$$

$$3x = 210°, 330°, 570°, 690°, 930°, 1050°$$

$$x = \mathbf{70°, 110°, 190°, 230°, 310°, 350°}$$

28. $1 = 1 \text{ cis } 0°$
(79)

$$(1 \text{ cis } 0°)^{\frac{1}{3}} = 1^{\frac{1}{3}} \text{ cis}\left(\frac{0}{3} + n\frac{360°}{3}\right)$$

$$= 1 \text{ cis } 0°, 1 \text{ cis } 120°, 1 \text{ cis } 240°$$

$$= \mathbf{1, -\frac{1}{2} + \frac{\sqrt{3}}{2}i, -\frac{1}{2} - \frac{\sqrt{3}}{2}i}$$

29.
(73)

$$\theta = \frac{1}{2}\left(\frac{360°}{10}\right) = 18°$$

$$x = 1\sin 18° = 0.3090 \text{ in.}$$

$$\text{Side} = 2x = 0.6180 \text{ in.}$$

$$\text{Perimeter} = 10(0.6180) = \mathbf{6.18 \text{ in.}}$$

30. $x^3 - 1 = (x - 1)\left(x^2 + x + 1\right)$
(46)

$$x = \frac{-1 \pm \sqrt{1 - 4(1)(1)}}{2(1)} = \frac{-1 \pm \sqrt{3}i}{2}$$

$$\mathbf{(x - 1)\left(x + \frac{1}{2} + \frac{\sqrt{3}}{2}i\right)\left(x + \frac{1}{2} - \frac{\sqrt{3}}{2}i\right)}$$

Problem Set 111

1. $D = 256\left(\frac{3}{4}\right)^4 = \mathbf{81 \text{ ft}}$
(91)

2. $\frac{1}{2} \cdot \frac{1}{2} = \mathbf{\frac{1}{4}}$
(83)

3. $A_0 = 10,\ A_2 = 6$
(88)

$$A_t = A_0 e^{kt}$$

$$6 = 10e^{k(2)}$$

$$\frac{3}{5} = e^{2k}$$

$$\ln\frac{3}{5} = 2k$$

$$k = -0.25541$$

$$A_5 = 10e^{-0.25541(5)} = \mathbf{2.79 \text{ grams}}$$

4. Rate $= f$, time $= t$, distance $= ft$
(28)
New time $= t - 2$

$$\text{New rate} = \frac{\text{distance}}{\text{new time}} = \mathbf{\frac{ft}{t-2}\ \frac{ft}{hr}}$$

5. $2(180 - A) = 5(90 - A) + 30$
(1)

$$360 - 2A = 450 - 5A + 30$$

$$3A = 120$$

$$A = \mathbf{40°}$$

6. ${}_8P_6 = \frac{8!}{2!} = \mathbf{20{,}160}$
(92)

$${}_8C_6 = \frac{8!}{2!6!} = \mathbf{28}$$

7. $\log_3(x - 2) < 3$
(111)

$$x - 2 < 3^3 \quad \text{and} \quad x - 2 > 0$$

$$x < 29 \qquad\qquad x > 2$$

$$\mathbf{2 < x < 29}$$

8. $\log_{\frac{1}{2}}(x - 3) > 3$
(111)

$$x - 3 < \left(\frac{1}{2}\right)^3 \quad \text{and} \quad x - 3 > 0$$
$$x > 3$$
$$x < 3\frac{1}{8}$$

$$\mathbf{3 < x < 3\frac{1}{8}}$$

9. (a) $f(x) = \text{Arcsec } x$
(110)

Domain $= \{x \in \mathbb{R} \mid x \leq -1 \text{ or } 1 \leq x\}$

Range

$$= \left\{y \in \mathbb{R} \mid 0 \leq y < \frac{\pi}{2} \text{ or } \frac{\pi}{2} < y \leq \pi\right\}$$

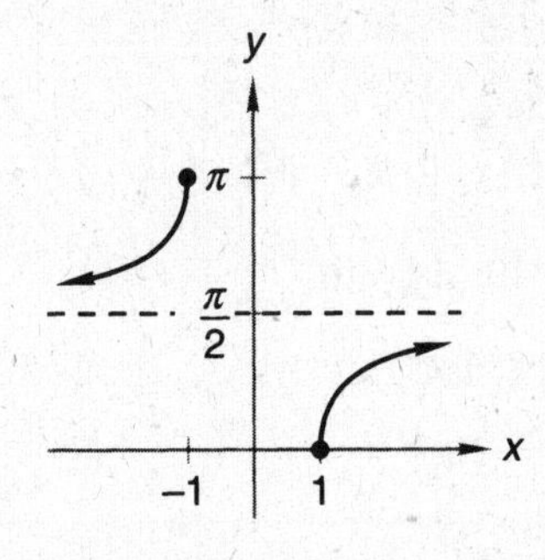

(b) $g(x) = \text{Arccsc } x$

Domain $= \{x \in \mathbb{R} \mid x \leq -1 \text{ or } 1 \leq x\}$

Range

$$= \left\{y \in \mathbb{R} \mid -\frac{\pi}{2} \leq y < 0 \text{ or } 0 < y \leq \frac{\pi}{2}\right\}$$

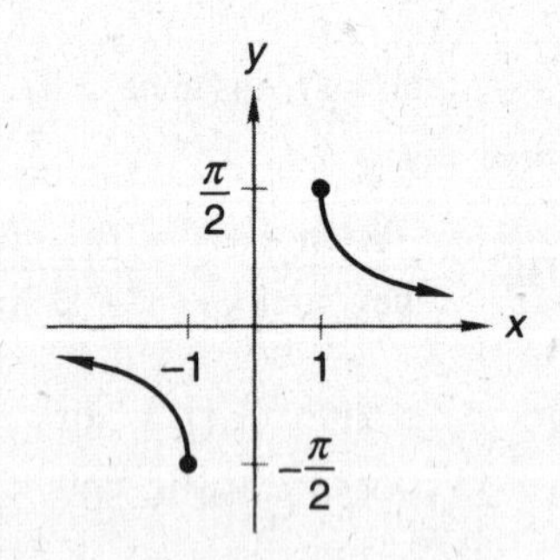

(c) $h(x) = \text{Arccot } x$

Domain $= \{x \in \mathbb{R}\}$

Range $= \{y \in \mathbb{R} \mid 0 < y < \pi\}$

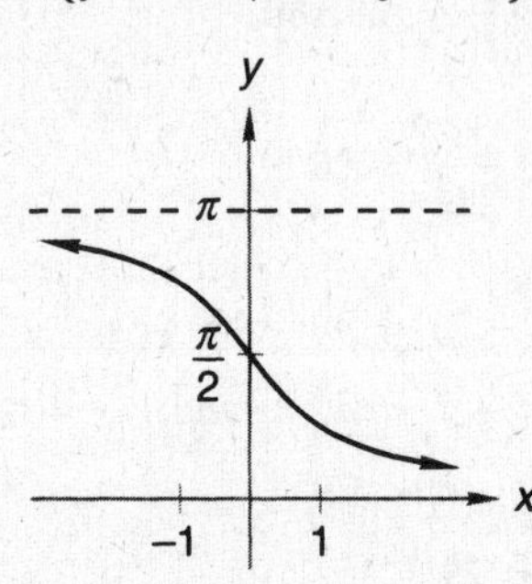

10. $0.00\overline{431}$
(109)

$$= 431 \times 10^{-5} + 431 \times 10^{-8} + 431 \times 10^{-11} + \ldots$$
$$= (431 \times 10^{-5}) + (431 \times 10^{-5})(10^{-3})$$
$$+ (431 \times 10^{-5})(10^{-3})^2 + \ldots$$

If $a_1 = 431 \times 10^{-5}$, $r = 10^{-3}$

$$s = \frac{a_1}{1 - r}$$
$$= \frac{431 \times 10^{-5}}{1 - 10^{-3}}$$
$$= \frac{431 \times 10^{-5}}{0.999} \cdot \frac{10^5}{10^5} = \mathbf{\frac{431}{99{,}900}}$$

11. $A = \begin{bmatrix} 2 & 0 & 3 \\ 1 & 3 & 5 \\ 2 & 2 & 1 \end{bmatrix}$, $B = \begin{bmatrix} 5 & 3 & 1 \\ 0 & 2 & 4 \\ 2 & 4 & 8 \end{bmatrix}$
(108)

$$2A - B = \begin{bmatrix} 4 - 5 & 0 - 3 & 6 - 1 \\ 2 - 0 & 6 - 2 & 10 - 4 \\ 4 - 2 & 4 - 4 & 2 - 8 \end{bmatrix}$$
$$= \begin{bmatrix} \mathbf{-1} & \mathbf{-3} & \mathbf{5} \\ \mathbf{2} & \mathbf{4} & \mathbf{6} \\ \mathbf{2} & \mathbf{0} & \mathbf{-6} \end{bmatrix}$$

$$B \cdot A = \begin{bmatrix} 5 & 3 & 1 \\ 0 & 2 & 4 \\ 2 & 4 & 8 \end{bmatrix}\begin{bmatrix} 2 & 0 & 3 \\ 1 & 3 & 5 \\ 2 & 2 & 1 \end{bmatrix}$$
$$= \begin{bmatrix} 10 + 3 + 2 & 0 + 9 + 2 & 15 + 15 + 1 \\ 0 + 2 + 8 & 0 + 6 + 8 & 0 + 10 + 4 \\ 4 + 4 + 16 & 0 + 12 + 16 & 6 + 20 + 8 \end{bmatrix}$$
$$= \begin{bmatrix} \mathbf{15} & \mathbf{11} & \mathbf{31} \\ \mathbf{10} & \mathbf{14} & \mathbf{14} \\ \mathbf{24} & \mathbf{28} & \mathbf{34} \end{bmatrix}$$

12. $25x^2 + 9y^2 + 50x - 36y - 164 = 0$
(106)

$$25(x^2 + 2x + 1) + 9(y^2 - 4y + 4) = 225$$
$$25(x + 1)^2 + 9(y - 2)^2 = 225$$
$$\mathbf{\frac{(x + 1)^2}{9} + \frac{(y - 2)^2}{25} = 1}$$

Center = (–1, 2)

$a = 5$, $b = 3$

Length of major axis = 10

Length of minor axis = 6

Vertical

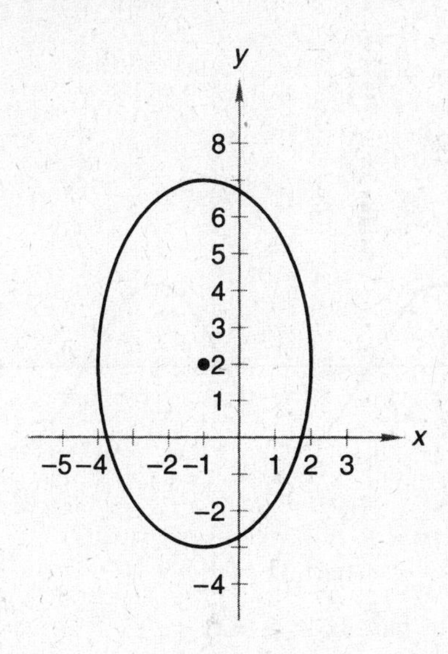

13. (106)

$$4x^2 - 36y^2 - 40x + 216y - 368 = 0$$

$$4(x^2 - 10x + 25) - 36(y^2 - 6y + 9) = 144$$

$$4(x-5)^2 - 36(y-3)^2 = 144$$

$$\mathbf{\frac{(x-5)^2}{36} - \frac{(y-3)^2}{4} = 1}$$

From the equation we obtain:

$h = 5,\ k = 3,\ a = 6,\ b = 2$

Center: $(h, k) = \mathbf{(5, 3)}$

Vertices: $(h \pm a, k) = \mathbf{(11, 3), (-1, 3)}$

Asymptotes:

$$m = \pm\frac{b}{a} = \pm\frac{2}{6} = \pm\frac{1}{3}$$

$y = mx + b$	$y = mx + b$
$y = \frac{1}{3}x + b$	$y = -\frac{1}{3}x + b$
$3 = \frac{1}{3}(5) + b$	$3 = -\frac{1}{3}(5) + b$
$b = \frac{4}{3}$	$b = \frac{14}{3}$
$y = \frac{1}{3}x + \frac{4}{3}$	$y = -\frac{1}{3}x + \frac{14}{3}$

Asymptotes: $\mathbf{y = \frac{1}{3}x + \frac{4}{3};\ y = -\frac{1}{3}x + \frac{14}{3}}$

14. (101)

$$x = \frac{\begin{vmatrix} 10 & 4 & 1 \\ 0 & -1 & 1 \\ 10 & 2 & 3 \end{vmatrix}}{\begin{vmatrix} 3 & 4 & 1 \\ 1 & -1 & 1 \\ 1 & 2 & 3 \end{vmatrix}}$$

$$= \frac{10\begin{vmatrix} -1 & 1 \\ 2 & 3 \end{vmatrix} - 4\begin{vmatrix} 0 & 1 \\ 10 & 3 \end{vmatrix} + 1\begin{vmatrix} 0 & -1 \\ 10 & 2 \end{vmatrix}}{3\begin{vmatrix} -1 & 1 \\ 2 & 3 \end{vmatrix} - 4\begin{vmatrix} 1 & 1 \\ 1 & 3 \end{vmatrix} + 1\begin{vmatrix} 1 & -1 \\ 1 & 2 \end{vmatrix}}$$

$$= \frac{10(-3-2) - 4(0-10) + 1(0+10)}{3(-3-2) - 4(3-1) + 1(2+1)}$$

$$= \frac{-50 + 40 + 10}{-15 - 8 + 3} = \frac{0}{-20} = \mathbf{0}$$

15. (103)

$$2\log(x-1) - 3\log x + \log(x^2 + x) - \log(x^2 - 1)$$

$$= \log(x-1)^2 - \log x^3 + \log[x(x+1)] - \log[(x-1)(x+1)]$$

$$= \log\left[\frac{(x-1)^2 x(x+1)}{x^3(x-1)(x+1)}\right] = \mathbf{\log\frac{x-1}{x^2}}$$

16. (103)

$$\frac{\left(\sqrt[3]{203 \times 10^4}\right)^2}{\sqrt{804 \times 10^{16}}} = \mathbf{5.65 \times 10^{-6}}$$

17. (102) $(2x + y)^4$

$F = 2x,\ S = y$

Term number	①	②	③	④	⑤
Exp. of F	4	3	2	1	0
Exp. of S	0	1	2	3	4
Coefficients	1	4	6	4	1

$$(F + S)^4$$

$$= F^4 + 4F^3S + 6F^2S^2 + 4FS^3 + S^4$$

$$= (2x)^4 + 4(2x)^3y + 6(2x)^2y^2 + 4(2x)y^3 + y^4$$

$$= \mathbf{16x^4 + 32x^3y + 24x^2y^2 + 8xy^3 + y^4}$$

18. (101)

$$\begin{vmatrix} x & 0 & 1 \\ 0 & 2 & 4 \\ 4 & 0 & x \end{vmatrix} = 0$$

$$x\begin{vmatrix} 2 & 4 \\ 0 & x \end{vmatrix} - 0\begin{vmatrix} 0 & 4 \\ 4 & x \end{vmatrix} + 1\begin{vmatrix} 0 & 2 \\ 4 & 0 \end{vmatrix} = 0$$

$$x(2x - 0) - 0 + 1(0 - 8) = 0$$

$$2x^2 - 8 = 0$$

$$2(x^2 - 4) = 0$$

$$(x+2)(x-2) = 0$$

$$x = \mathbf{\pm 2}$$

19. (100) $\sin 255° + \sin 15°$

$$= 2\sin\left(\frac{255° + 15°}{2}\right)\cos\left(\frac{255° - 15°}{2}\right)$$

$$= 2\sin 135° \cos 120° = 2\left(\frac{\sqrt{2}}{2}\right)\left(-\frac{1}{2}\right) = -\frac{\sqrt{2}}{2}$$

20. (98) $\ln(\ln x) = 1$

$$\ln x = e^1$$

$$x = e^e$$

21. (82)

$$4^{x+2} = 2^{3x} \cdot 8$$

$$2^{2(x+2)} = 2^{3x} \cdot 2^3$$

$$2^{2(x+2)} = 2^{3x+3}$$

$$2x + 4 = 3x + 3$$

$$x = \mathbf{1}$$

22. (96) $\sec x \csc x = \frac{1}{\cos x} \cdot \frac{1}{\sin x} = \frac{1}{\cos x \sin x} \cdot \frac{2}{2}$

$$= \frac{2}{2\cos x \sin x} = \frac{2}{\sin 2x} = 2\csc 2x$$

23. (96)

$$\frac{2\tan\theta}{1 + \tan^2\theta} = \frac{\frac{2\sin\theta}{\cos\theta}}{\sec^2\theta} = \frac{2\frac{\sin\theta}{\cos\theta}}{\frac{1}{\cos^2\theta}}$$

$$= \frac{2\sin\theta}{\cos\theta} \cdot \frac{\cos^2\theta}{1} = 2\sin\theta\cos\theta = \sin 2\theta$$

24. (81,96) $a^2 = 8^2 + 10^2 - 2(8)(10)\cos 140°$

$$a^2 = 286.57$$

$$a = \mathbf{16.93\ cm}$$

25. (84) $y = 2 + 4\sin\frac{1}{3}(x - 20°)$

$$\text{Period} = \frac{360°}{\frac{1}{3}} = 1080°$$

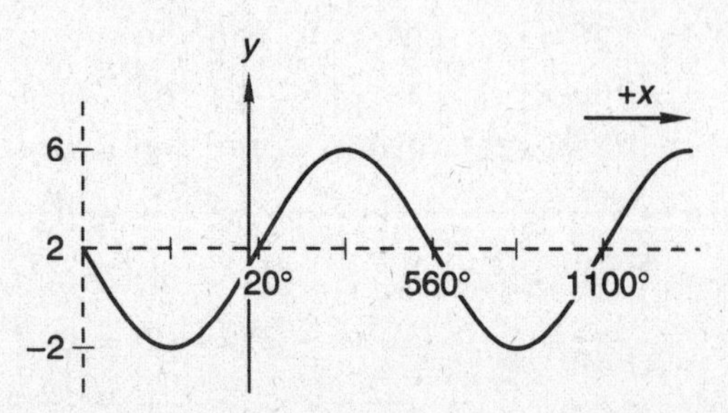

26. (84) $y = \cos\left(x - \frac{\pi}{2}\right)$

Period $= 2\pi$

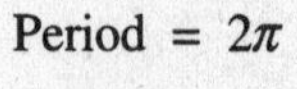

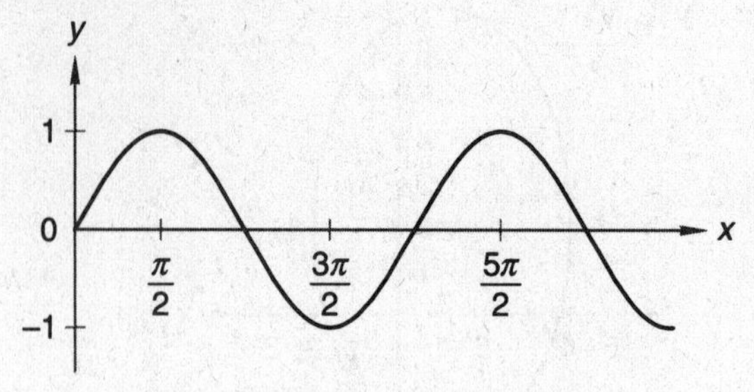

27. (52,60)

$$\tan^2 3x = 1$$

$$\tan^2 3x - 1 = 0$$

$$(\tan 3x - 1)(\tan 3x + 1) = 0$$

$\tan 3x = 1$

$$3x = \frac{\pi}{4}, \frac{5\pi}{4}, \frac{9\pi}{4}, \frac{13\pi}{4}, \frac{17\pi}{4}, \frac{21\pi}{4}$$

$\tan 3x = -1$

$$3x = \frac{3\pi}{4}, \frac{7\pi}{4}, \frac{11\pi}{4}, \frac{15\pi}{4}, \frac{19\pi}{4}, \frac{23\pi}{4}$$

$$x = \frac{\pi}{12}, \frac{\pi}{4}, \frac{5\pi}{12}, \frac{7\pi}{12}, \frac{3\pi}{4}, \frac{11\pi}{12}, \frac{13\pi}{12}, \frac{5\pi}{4}, \frac{17\pi}{12}, \frac{19\pi}{12}, \frac{7\pi}{4}, \frac{23\pi}{12}$$

28. (52,60)

$$\sec^2 2x - 1 = 0$$

$$(\sec 2x - 1)(\sec 2x + 1) = 0$$

$\sec 2x = 1$

$$2x = 0, 2\pi$$

$\sec 2x = -1$

$$2x = \pi, 3\pi$$

$$x = \mathbf{0}, \frac{\pi}{2}, \pi, \frac{3\pi}{2}$$

29. (79) $16 = 16 \text{ cis } 0°$

$$(16 \text{ cis } 0°)^{\frac{1}{4}} = 16^{\frac{1}{4}} \text{ cis}\left(\frac{0°}{4} + n\frac{360°}{4}\right)$$

$$= 2 \text{ cis } 0°, 2 \text{ cis } 90°, 2 \text{ cis } 180°, 2 \text{ cis } 270°$$

$$= \mathbf{2, 2i, -2, -2i}$$

30. (46) $x^3 + 27 = (x + 3)(x^2 - 3x + 9)$

$$x = \frac{3 \pm \sqrt{9 - 4(1)(9)}}{2(1)} = \frac{3}{2} \pm \frac{3\sqrt{3}i}{2}$$

$$(x + 3)\left(x - \frac{3}{2} + \frac{3\sqrt{3}}{2}i\right)\left(x - \frac{3}{2} - \frac{3\sqrt{3}}{2}i\right)$$

Problem Set 112

1. Sum of distances fallen.
(107)

$a_1 = 128$

$r = \frac{1}{3}$

$S = \frac{a_1}{1 - r} = \frac{128}{1 - \frac{1}{3}} = 192$

Sum of distances rebounded:

$S - a_1 = 192 - 128 = 64$

Sum of ups and downs:

$192 + 64 =$ **256 ft**

2. $P(B \cup R) = P(B) + P(R) - P(B \cap R)$
(92)

$= 0.4 + 0.6 - 0.2 = \mathbf{0.8}$

3. ${}_7C_3 = \frac{7!}{4!3!} =$ **35 groups**
(75)

4. $2.20 + (p - 3)(0.90) = 2.20 + 0.9p - 2.70$
(44)

$= \mathbf{\$(0.9p - 0.5)}$

5. $C + 0.8C = 900$
(R)

$1.8C = 900$

$C = 500$

New price $= 500 + (0.5)500 =$ **\$750**

6. $\left[\left(\frac{8{,}000{,}000}{100}\right)0.60\right](0.04) = \mathbf{\$1920}$
(R)

7. $(2x^2 - y)^{10}$
(112)

$F = 2x^2,\ S = -y,\ n = 10,\ k = 4$

$\frac{n!}{(n - k + 1)!(k - 1)!}F^{n-k+1}S^{k-1}$

$= \frac{10!}{(10 - 4 + 1)!(4 - 1)!}F^{10-4+1}S^{4-1}$

$= \frac{10!}{7!3!}F^7S^3 = \frac{10!}{7!3!}(2x^2)^7(-y)^3$

$= 120(128x^{14})(-y^3) = \mathbf{-15{,}360x^{14}y^3}$

8. $(a^2 - 2b)^8$
(112)

$F = a^2,\ S = -2b,\ n = 8,\ k = 5$

$\frac{8!}{(8 - 5 + 1)!(5 - 1)!}F^{8-5+1}S^{5-1}$

$= \frac{8!}{4!4!}F^4S^4$

$= \frac{8!}{4!4!}(a^2)^4(-2b)^4$

$= 70a^8(16)b^4 = \mathbf{1120a^8b^4}$

9. $\log_5 (x - 2) < 3$
(111)

$x - 2 < 5^3$ and $x - 2 > 0$

$x < 127$ $\quad x > 2$

$\mathbf{2 < x < 127}$

10. $\log_{\frac{1}{3}} (x + 1) < 2$
(111)

$x + 1 > \left(\frac{1}{3}\right)^2$ and $x + 1 > 0$

$x > -\frac{8}{9}$ $\quad x > -1$

but $-1 < -\frac{8}{9}$, so $\mathbf{x > -\frac{8}{9}}$

11. $f(x) = \text{Arcsin}\, x$
(110)

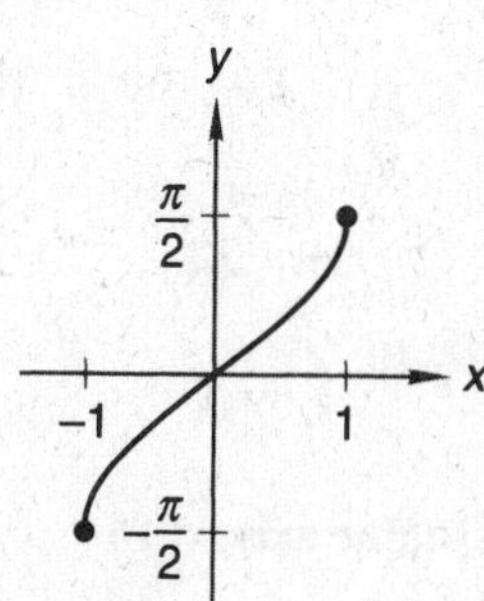

Domain $= \{x \in \mathbb{R} \mid -1 \le x \le 1\}$

Range $= \left\{y \in \mathbb{R} \mid -\frac{\pi}{2} \le y \le \frac{\pi}{2}\right\}$

Arcsin $\left(-\frac{1}{2}\right) = -\frac{\pi}{6}$

Arcsin $\frac{1}{2} = \frac{\pi}{6}$

Arcsin $(-1) = -\frac{\pi}{2}$

12. (109) $0.00\overline{7}$

$$= 7 \times 10^{-3} + 7 \times 10^{-4} + 7 \times 10^{-5} + \ldots$$

$$= (7 \times 10^{-3}) + (7 \times 10^{-3})(10^{-1}) + (7 \times 10^{-3})(10^{-1})^2 + \ldots$$

If $a_1 = 7 \times 10^{-3}$, $r = 10^{-1}$

$$s = \frac{a_1}{1 - r} = \frac{7 \times 10^{-3}}{1 - 10^{-1}} = \frac{7 \times 10^{-3}}{0.9} \cdot \frac{10^3}{10^3} = \mathbf{\frac{7}{900}}$$

13. (108) $A = \begin{bmatrix} 2 & 0 \\ 1 & 4 \end{bmatrix}$, $B = \begin{bmatrix} 5 & 2 \\ 3 & 4 \end{bmatrix}$

$$2A + 3B = \begin{bmatrix} 4 + 15 & 0 + 6 \\ 2 + 9 & 8 + 12 \end{bmatrix} = \mathbf{\begin{bmatrix} 19 & 6 \\ 11 & 20 \end{bmatrix}}$$

$$A \cdot 2B = \begin{bmatrix} 2 & 0 \\ 1 & 4 \end{bmatrix}\begin{bmatrix} 10 & 4 \\ 6 & 8 \end{bmatrix} = \begin{bmatrix} 20 + 0 & 8 + 0 \\ 10 + 24 & 4 + 32 \end{bmatrix} = \mathbf{\begin{bmatrix} 20 & 8 \\ 34 & 36 \end{bmatrix}}$$

14. (106)

$$16y^2 + x^2 - 128y + 20x + 292 = 0$$

$$(x^2 + 20x + 100) + 16(y^2 - 8y + 16) = 64$$

$$(x + 10)^2 + 16(y - 4)^2 = 64$$

$$\mathbf{\frac{(x + 10)^2}{64} + \frac{(y - 4)^2}{4} = 1}$$

Center = (–10, 4)

$a = 8$, $b = 2$

Length of major axis = 16

Length of minor axis = 4

Horizontal

15. (106)

$$x^2 - y^2 - 6x - 10y = 0$$

$$(x^2 - 6x + 9) - (y^2 + 10y + 25) = -16$$

$$(x - 3)^2 - (y + 5)^2 = -16$$

$$\mathbf{\frac{(y + 5)^2}{16} - \frac{(x - 3)^2}{16} = 1}$$

From the equation we obtain:

$h = 3$, $k = -5$, $a = 4$, $b = 4$

Center: $(h, k) =$ **(3, –5)**

Vertices: $(h, k \pm b) =$ **(3, –1), (3, –9)**

Asymptotes:

$$m = \pm\frac{b}{a} = \pm\frac{1}{1} = \pm 1$$

$y = mx + b$	$y = mx + b$
$y = x + b$	$y = -x + b$
$-5 = (3) + b$	$-5 = -(3) + b$
$b = -8$	$b = -2$
$y = x - 8$	$y = -x - 2$

Asymptotes: $\mathbf{y = x - 8;\ y = -x - 2}$

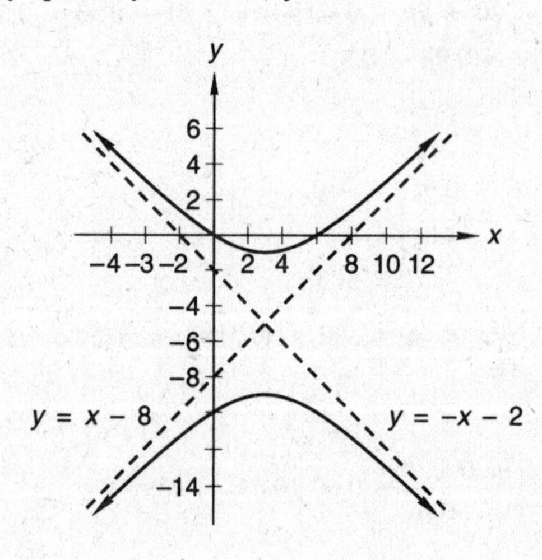

16. (103) $\log\left(\dfrac{\sqrt[4]{x^3yz^5}}{x^2\sqrt{yz}}\right)$

$$= \log \frac{x^{\frac{3}{4}}y^{\frac{1}{4}}z^{\frac{5}{4}}}{x^2y^{\frac{1}{2}}z^{\frac{1}{2}}} = \log \frac{z^{\frac{3}{4}}}{x^{\frac{5}{4}}y^{\frac{1}{4}}}$$

$$= \mathbf{\frac{3}{4}\log z - \frac{5}{4}\log x - \frac{1}{4}\log y}$$

17. (101) $y = \dfrac{\begin{vmatrix} 3 & 1 & 1 \\ 1 & 2 & -4 \\ 2 & 3 & 0 \end{vmatrix}}{\begin{vmatrix} 3 & 2 & 1 \\ 1 & -5 & -4 \\ 2 & 1 & 0 \end{vmatrix}}$

$$y = \frac{3\begin{vmatrix} 2 & -4 \\ 3 & 0 \end{vmatrix} - 1\begin{vmatrix} 1 & -4 \\ 2 & 0 \end{vmatrix} + 1\begin{vmatrix} 1 & 2 \\ 2 & 3 \end{vmatrix}}{3\begin{vmatrix} -5 & -4 \\ 1 & 0 \end{vmatrix} - 2\begin{vmatrix} 1 & -4 \\ 2 & 0 \end{vmatrix} + 1\begin{vmatrix} 1 & -5 \\ 2 & 1 \end{vmatrix}}$$

$$y = \frac{3(0 + 12) - 1(0 + 8) + 1(3 - 4)}{3(0 + 4) - 2(0 + 8) + 1(1 + 10)}$$

$$y = \mathbf{\frac{27}{7}}$$

18. (100) $\cos 75° \sin 15°$

$$= \frac{1}{2}[\sin(75° + 15°) - \sin(75° - 15°)]$$

$$= \frac{1}{2}(\sin 90° - \sin 60°)$$

$$= \frac{1}{2}\left[1 - \frac{\sqrt{3}}{2}\right] = \mathbf{\frac{2 - \sqrt{3}}{4}}$$

19. (98) $x^{\sqrt{\ln x}} = e$

$$\sqrt{\ln x}\ln x = 1$$

$$(\ln x)^{\frac{3}{2}} = 1$$

$$\left[(\ln x)^{\frac{3}{2}}\right]^{\frac{2}{3}} = 1^{\frac{2}{3}}$$

$$\ln x = 1$$

$$x = \mathbf{e}$$

20. (98) $\log(\ln x) = 2$

$$\ln x = 10^2$$

$$x = \mathbf{e^{100}}$$

21. (96) $\tan 2x = \dfrac{\sin 2x}{\cos 2x}$

$$= \frac{2\sin x\cos x}{\cos^2 x - \sin^2 x} \cdot \frac{\frac{1}{\cos^2 x}}{\frac{1}{\cos^2 x}}$$

$$= \frac{2\frac{\sin x}{\cos x}}{1 - \frac{\sin^2 x}{\cos^2 x}} = \frac{2\tan x}{1 - \tan^2 x}$$

22. (93) $\dfrac{1 + \cos\theta}{\sin\theta} + \dfrac{\sin\theta}{\cos\theta}$

$$= \frac{\cos\theta(1 + \cos\theta) + \sin^2\theta}{\sin\theta\cos\theta}$$

$$= \frac{\cos\theta + \cos^2\theta + \sin^2\theta}{\sin\theta\cos\theta}$$

$$= \frac{\cos\theta + 1}{\sin\theta\cos\theta}$$

23. (81,96) $a^2 = 8^2 + 12^2 - 2(8)(12)\cos 150°$

$$a^2 = 374.28$$

$$a = \mathbf{19.35\ m}$$

24. (84) $y = 1 - 2\cos\dfrac{1}{2}\left(x + \dfrac{\pi}{3}\right)$

$$\text{Period} = \frac{2\pi}{\frac{1}{2}} = 4\pi$$

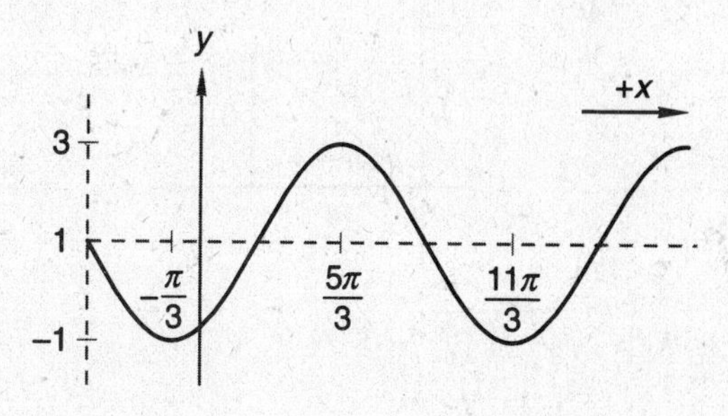

25. (94) $y = \csc(x - \pi)$

$$\text{Period} = 2\pi$$

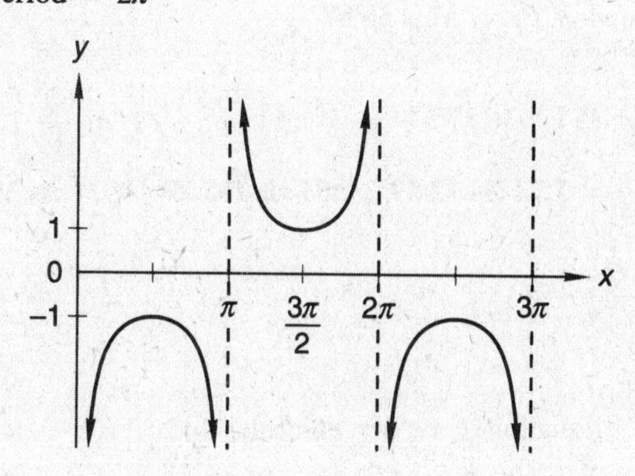

26. (60) $2\sin^2 x + 3\sin x = -1$

$$2\sin^2 x + 3\sin x + 1 = 0$$

$$2\left(\sin x + \frac{1}{2}\right)(\sin x + 1) = 0$$

$$\sin x = -\frac{1}{2} \qquad \sin x = -1$$

$$x = 210°, 330° \qquad x = 270°$$

$$x = \mathbf{210°, 270°, 330°}$$

27. (85)

$$\cos^2 x - \sin^2 x = \frac{1}{2}$$

$$\cos^2 x - (1 - \cos^2 x) - \frac{1}{2} = 0$$

$$2\cos^2 x - \frac{3}{2} = 0$$

$$\cos^2 x - \frac{3}{4} = 0$$

$$\left(\cos x + \frac{\sqrt{3}}{2}\right)\left(\cos x - \frac{\sqrt{3}}{2}\right) = 0$$

$$\cos x = \pm\frac{\sqrt{3}}{2}$$

$$x = \mathbf{30°, 150°, 210°, 330°}$$

28. $4 + 3i$ (95)

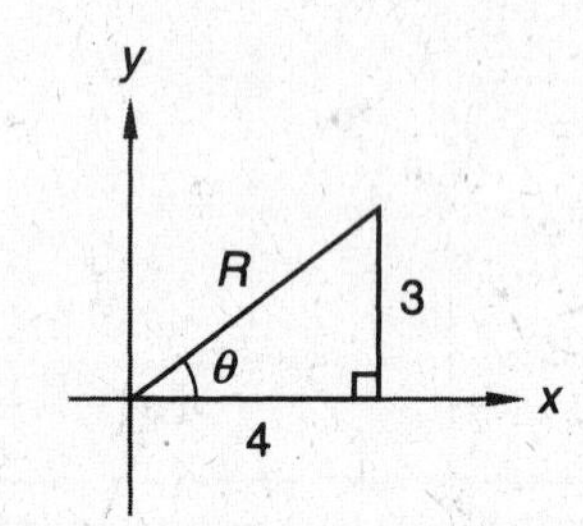

$$R = \sqrt{4^2 + 3^2} = 5$$

$$\tan\theta = \frac{3}{4}$$

$$\theta = 36.87°$$

$$4 + 3i = 5 \text{ cis } 36.87°$$

$$(5 \text{ cis } 36.87°)^{\frac{1}{3}} = 5^{\frac{1}{3}} \text{ cis}\left(\frac{36.87°}{3} + n\frac{360°}{3}\right)$$

$$= \mathbf{1.71 \text{ cis } 12.29°,\ 1.71 \text{ cis } 132.29°,\ 1.71 \text{ cis } 252.29°}$$

29. (58) $y - x = 2$

$y = x + 2$

Equation of the perpendicular line:

$y = -x + b$

$0 = -(1) + b$

$b = 1$

$y = -x + 1$

Point of intersection:

$x + 2 = -x + 1$

$2x = -1$

$x = -\frac{1}{2}$

$y = x + 2 = -\frac{1}{2} + 2 = \frac{3}{2}$

$\left(-\frac{1}{2}, \frac{3}{2}\right)$ and $(1, 0)$

$$D = \sqrt{\left[1 - \left(-\frac{1}{2}\right)\right]^2 + \left(0 - \frac{3}{2}\right)^2}$$

$$= \sqrt{\frac{18}{4}} = \mathbf{\frac{3\sqrt{2}}{2}}$$

30. (46)

$$x^4 - 16 = (x^2)^2 - 4^2$$

$$= (x^2 - 4)(x^2 + 4)$$

$x^2 = 4$ $\qquad$ $x^2 = -4$

$x = \pm 2$ $\qquad$ $x = \pm 2i$

$$\mathbf{(x - 2)(x + 2)(x - 2i)(x + 2i)}$$

Problem Set 113

1. (83) $P(R, \text{then } W) = \frac{7}{11} \cdot \frac{4}{10} = \mathbf{\frac{14}{55}}$

2. (88) $A_0 = 10{,}000$

$A_t = A_0 e^{kt}$

$5000 = 10000e^{14k}$

$\frac{1}{2} = e^{14k}$

$\ln\frac{1}{2} = 14k$

$k = -0.049511$

$4000 = 10000e^{-0.049511t}$

$0.4 = e^{-0.049511t}$

$\ln 0.4 = -0.049511t$

$t = \mathbf{18.51 \text{ yr}}$

3. (28) Total distance = 400

Distance traveled = hk

Distance to go = $\mathbf{(400 - hk) \text{ mi}}$

4. (53) $v = r\omega$

$$v = (k \text{ in.})\left(r\frac{\text{rad}}{\text{sec}}\right)\left(\frac{2.54 \text{ cm}}{1 \text{ in.}}\right)\left(\frac{1 \text{ m}}{100 \text{ cm}}\right)$$

$$\times\left(\frac{60 \text{ sec}}{1 \text{ min}}\right)\left(\frac{60 \text{ min}}{1 \text{ hr}}\right)$$

$$= \mathbf{\frac{(k)(r)(2.54)(60)(60)}{100} \frac{m}{hr}}$$

5. (38)
$$\begin{cases} R_O T_O = 600 \\ R_B T_B = 600 \end{cases}$$

$R_O = 3R_B$, $T_O + T_B = 40$

$$T_O + T_B = 40$$
$$T_O = 40 - T_B$$
$$R_O T_O = 600$$
$$3R_B(40 - T_B) = 600$$
$$120R_B - 3R_B T_B = 600$$
$$120R_B - 3(600) = 600$$
$$120R_B = 2400$$
$$\mathbf{R_B = 20\text{ mph}}$$

$$R_O = 3R_B$$
$$R_O = 3(20)$$
$$\mathbf{R_O = 60\text{ mph}}$$

6. (113) $(x^3 - 5x^2 + 12) \div (x - 2)$

2	1	−5	0	12
		2	−6	−12
	1	−3	−6	0

$$\frac{x^3 - 5x^2 + 12}{x - 2} = \mathbf{x^2 - 3x - 6}$$

7. (113) $(4x^4 - 4x^3 + x^2 - 3x + 2) \div (x + 2)$

−2	4	−4	1	−3	2
		−8	24	−50	106
	4	−12	25	−53	108

$$\frac{4x^4 - 4x^3 + x^2 - 3x + 2}{x + 2}$$
$$= \mathbf{4x^3 - 12x^2 + 25x - 53 + \frac{108}{x + 2}}$$

8. (113) $x^3 + 2x^2 - 3x - 4$

−1	1	2	−3	−4
		−1	−1	4
	1	1	−4	0

Yes, the remainder is 0, so −1 is a zero of the polynomial.

9. (113) $2x^3 - x^2 + 2x - 4 = 0$

−2	2	−1	2	−4
		−4	10	−24
	2	−5	12	−28

No, the remainder is not 0, so −2 is not a root of the polynomial.

10. (113) $x^4 - 10x^2 - 40x - 175 = 0$

5	1	0	−10	−40	−175
		5	25	75	175
	1	5	15	35	0

Yes, the remainder is 0, so 5 is a root of the polynomial.

11. (112) $(2x^2 - y^3)^{15}$

$F = 2x^2$, $S = -y^3$, $n = 15$, $k = 5$

$$\frac{n!}{(n - k + 1)!(k - 1)!}F^{n-k+1}S^{k-1}$$
$$= \frac{15!}{(15 - 5 + 1)!(5 - 1)!}F^{15-5+1}S^{5-1}$$
$$= \frac{15!}{11!4!}F^{11}S^4 = \frac{15!}{11!4!}(2x^2)^{11}(-y^3)^4$$
$$= 1365(2048x^{22})(y^{12}) = \mathbf{2{,}795{,}520x^{22}y^{12}}$$

12. (112) $(3x^3 - y^2)^8$

$F = 3x^3$, $S = -y^2$, $n = 8$, $k = 4$

$$\frac{n!}{(n - k + 1)!(k - 1)!}F^{n-k+1}S^{k-1}$$
$$= \frac{8!}{(8 - 4 + 1)!\,(4 - 1)!}F^{8-4+1}S^{4-1}$$
$$= \frac{8!}{5!3!}F^5S^3 = \frac{8!}{5!3!}(3x^3)^5(-y^2)^3$$
$$= 56(243x^{15})(-y^6) = \mathbf{-13{,}608x^{15}y^6}$$

13. (111) $\log_4(2x - 1) < 2$

$$2x - 1 < 4^2 \quad \text{and} \quad 2x - 1 > 0$$
$$2x < 17 \qquad\qquad x > \frac{1}{2}$$
$$x < \frac{17}{2}$$

$$\mathbf{\frac{1}{2} < x < \frac{17}{2}}$$

14. (111) $\log_{\frac{1}{3}}(2x + 1) < 2$

$$2x + 1 > \left(\frac{1}{3}\right)^2 \quad \text{and} \quad 2x + 1 > 0$$
$$2x > -\frac{8}{9} \qquad\qquad x > -\frac{1}{2}$$
$$x > -\frac{4}{9}$$

but $-\frac{1}{2} < -\frac{4}{9}$, so $\mathbf{x > -\frac{4}{9}}$

15. $f(x) = \text{Arccos } x$
(110)

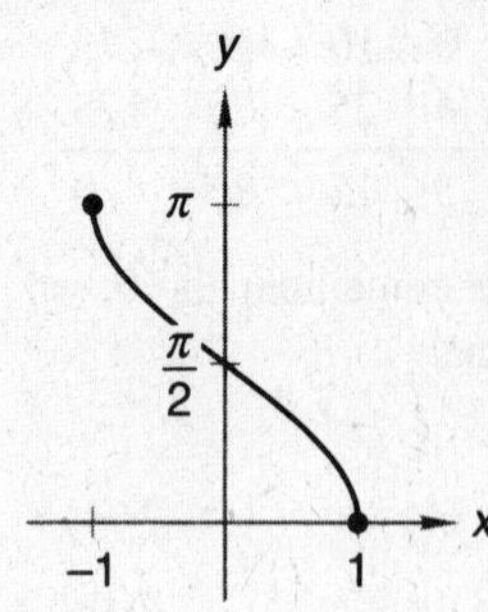

Domain $= \{x \in \mathbb{R} \mid -1 \le x \le 1\}$

Range $= \{y \in \mathbb{R} \mid 0 \le y \le \pi\}$

Arccos $\left(-\frac{1}{2}\right) = \frac{2\pi}{3}$

Arccos $\frac{1}{2} = \frac{\pi}{3}$

Arccos $(-1) = \pi$

16. $0.0\overline{13}$
(109)

$= 13 \times 10^{-3} + 13 \times 10^{-5} + 13 \times 10^{-7} + \ldots$

$= (13 \times 10^{-3}) + (13 \times 10^{-3})(10^{-2})$

$+ (13 \times 10^{-3})(10^{-2})^2 + \ldots$

If $a_1 = 0.013,\ r = 10^{-2}$

$$s = \frac{a_1}{1 - r}$$

$$= \frac{0.013}{1 - 10^{-2}} = \frac{0.013}{0.99} \cdot \frac{10^3}{10^3} = \mathbf{\frac{13}{990}}$$

17. $36x^2 + 9y^2 = 216x$
(106)

$36(x^2 - 6x \quad) + 9y^2 = 0$

$36(x^2 - 6x + 9) + 9y^2 = 324$

$36(x - 3)^2 + 9y^2 = 324$

$$\mathbf{\frac{(x - 3)^2}{9} + \frac{(y - 0)^2}{36} = 1}$$

Center = (3, 0)

$a = 6,\ b = 3$

Length of major axis = 12

Length of minor axis = 6

Vertical

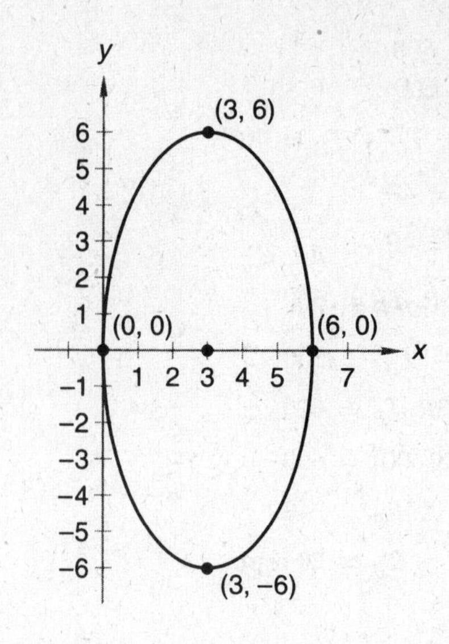

18. (101)

$$z = \frac{\begin{vmatrix} 1 & 0 & 1 \\ 2 & 1 & 5 \\ 3 & 3 & 7 \end{vmatrix}}{\begin{vmatrix} 1 & 0 & -1 \\ 2 & 1 & 2 \\ 3 & 3 & 4 \end{vmatrix}}$$

$$z = \frac{1(7 - 15) - 0 + 1(6 - 3)}{1(4 - 6) - 0 - 1(6 - 3)}$$

$$z = \frac{-8 + 3}{-2 - 3} = \frac{-5}{-5} = \mathbf{1}$$

19. $\ln(xy - x) - \ln(x^2 z) + 2 \ln z - \ln(y^2 - 1)$
(103)

$= \ln(xy - x) - \ln(x^2 z) + \ln z^2 - \ln(y^2 - 1)$

$$= \ln \frac{(xy - x)z^2}{(x^2 z)(y^2 - 1)}$$

$$= \ln \frac{x(y - 1)z^2}{x^2 z(y - 1)(y + 1)}$$

$$= \mathbf{\ln \frac{z}{x(y + 1)}}$$

20. $\sin A \sin B = \frac{1}{2}[\cos(A - B) - \cos(A + B)]$
(100)

$\sin 285° \sin 15°$

$$= \frac{1}{2}[\cos(285° - 15°) - \cos(285° + 15°)]$$

$$= \frac{1}{2}(\cos 270° - \cos 300°)$$

$$= \frac{1}{2}\left(0 - \frac{1}{2}\right) = \mathbf{-\frac{1}{4}}$$

21. (98)

$$\sqrt{\ln x} = \ln\sqrt{x}$$

$$\sqrt{\ln x} = \frac{1}{2}\ln x$$

$$\ln x = \frac{1}{4}(\ln x)^2$$

$$\frac{1}{4}(\ln x)^2 - \ln x = 0$$

$$(\ln x)\left(\frac{1}{4}\ln x - 1\right) = 0$$

$$\ln x = 0 \qquad \frac{1}{4}\ln x = 1$$

$$x = e^0 \qquad \ln x = 4$$

$$x = 1 \qquad x = e^4$$

$$\mathbf{x = 1, e^4}$$

22. (93)

$$\frac{\tan x}{\tan x - 1} - \frac{\cot x}{\cot x + 1}$$

$$= \frac{\tan x \cot x + \tan x - \cot x \tan x + \cot x}{\tan x \cot x + \tan x - \cot x - 1}$$

$$= \frac{\tan x + \cot x}{1 - \cot x + \tan x - 1}$$

$$= \frac{\tan x + \cot x}{\tan x - \cot x}$$

23. (72)

$$\frac{3}{\sin 60°} = \frac{a}{\sin 50°}$$

$$a = \frac{3\sin 50°}{\sin 60°}$$

$$a = \mathbf{2.65\ cm}$$

24. (94) $y = \sec\left(x + \frac{\pi}{2}\right)$

Period $= 2\pi$

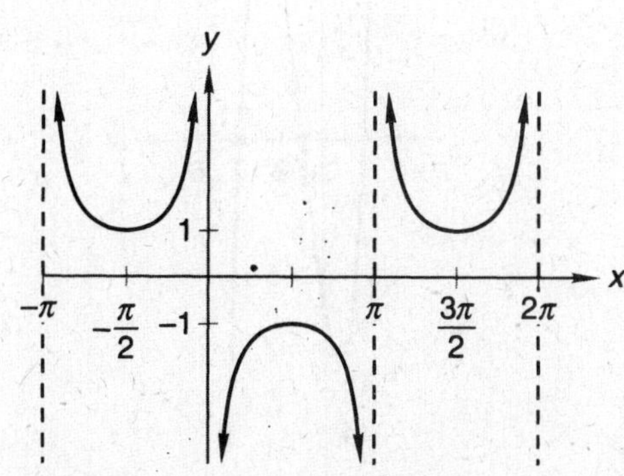

25. (60)

$$2\sin^2 x - \left(2\sqrt{2} + \sqrt{3}\right)\sin x + \sqrt{6} = 0$$

$$\left(2\sin x - \sqrt{3}\right)\left(\sin x - \sqrt{2}\right) = 0$$

$$2\sin x = \sqrt{3} \qquad \sin x = \sqrt{2}$$

$$\sin x = \frac{\sqrt{3}}{2} \qquad \text{no answer}$$

$$x = \mathbf{\frac{\pi}{3}, \frac{2\pi}{3}}$$

26. (95)

$$1 = 1 \text{ cis } 0°$$

$$(1 \text{ cis } 0°)^{\frac{1}{6}} = 1 \text{ cis}\left(\frac{0°}{6} + n\frac{360°}{6}\right)$$

$$= 1 \text{ cis } 0°,\ 1 \text{ cis } 60°,\ 1 \text{ cis } 120°,\ 1 \text{ cis } 180°,\ 1 \text{ cis } 240°,\ 1 \text{ cis } 300°$$

$$= \mathbf{1, \frac{1}{2} + \frac{\sqrt{3}}{2}i, -\frac{1}{2} + \frac{\sqrt{3}}{2}i, -1, -\frac{1}{2} - \frac{\sqrt{3}}{2}i, \frac{1}{2} - \frac{\sqrt{3}}{2}i}$$

27. (59)

$$(a + 3)5^{\log_5(a^2 - 5a + 6) - \log_5(a^2 - 9)}$$

$$= (a + 3)5^{\log_5\left(\frac{a^2 - 5a + 6}{a^2 - 9}\right)}$$

$$= (a + 3)5^{\log_5\left[\frac{(a-3)(a-2)}{(a-3)(a+3)}\right]}$$

$$= (a + 3)5^{\log_5\left(\frac{a-2}{a+3}\right)}$$

$$= (a + 3)\left(\frac{a - 2}{a + 3}\right)$$

$$= \mathbf{a - 2}$$

28. (56)

$$\begin{cases} x^2 + y^2 \le 9 & \text{circle} \\ y^2 - x^2 \ge 1 & \text{hyperbola} \end{cases}$$

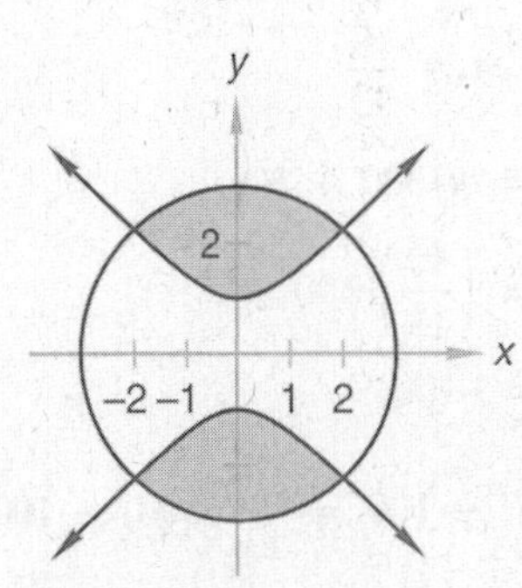

29. $x + y = 5$
(58)

$y = -x + 5$

Equation of the perpendicular line:

$y = x + b$

$-1 = 4 + b$

$b = -4 - 1 = -5$

$y = x - 5$

Point of intersection:

$x - 5 = -x + 5$

$2x = 10$

$x = 5$

$y = x - 5 = (5) - 5 = 0$

(5, 0) and (4, –1)

$D = \sqrt{(5 - 4)^2 + (0 - (-1))^2}$

$= \sqrt{1^2 + 1^2} = \sqrt{2}$

30. $x^4 + 2x^2 + 1 = (x^2 + 1)(x^2 + 1)$
(46)

$x^2 + 1 = 0$

$x^2 = -1$

$x = \pm i$

$(x + i)(x + i)(x - i)(x - i)$

Problem Set 114

1. $\frac{1}{6} \cdot \frac{1}{6} \cdot \frac{1}{6} = \mathbf{\frac{1}{216}}$
(83)

2. $\frac{198}{2} = \mathbf{99}$
(45)

3. $S = mP + b$
(62)

$$\begin{cases} \text{(a)} & 900 = m1300 + b \\ \text{(b)} & 1900 = m2600 + b \end{cases}$$

$$\begin{array}{rrcl} \text{(b)} & 1900 & = & m2600 + b \\ -\text{(a)} & -900 & = & -m1300 - b \\ \hline & 1000 & = & m1300 \end{array}$$

$m = \frac{10}{13}$

(a) $900 = m1300 + b$

$900 = \left(\frac{10}{13}\right)1300 + b$

$b = -100$

$S = \frac{10}{13}P - 100 = \frac{10}{13}(6500) - 100 = \mathbf{4900}$

4. $R_1T_1 + R_2T_2 = \text{pages}$
(25)

$\left(\frac{30}{2}\right)4 + \left(\frac{54}{3}\right)4 = \mathbf{132\ pages}$

5. $y = x(x - 3)(x + 3) = x^3 - 9x$
(114)

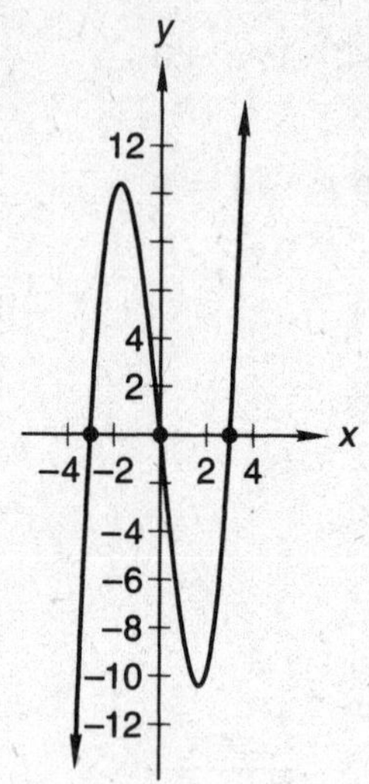

6. $y = x^2(x + 3) = x^3 + 3x^2$
(114)

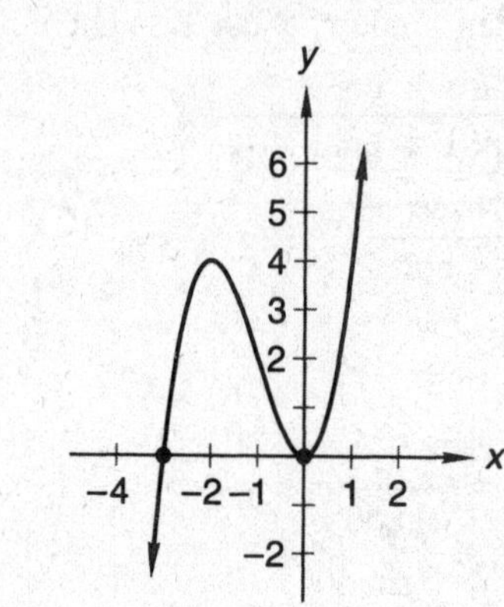

7. $y = (x - 1)(4 - x)(x + 2) = -x^3 + 3x^2 + 6x - 8$
(114)

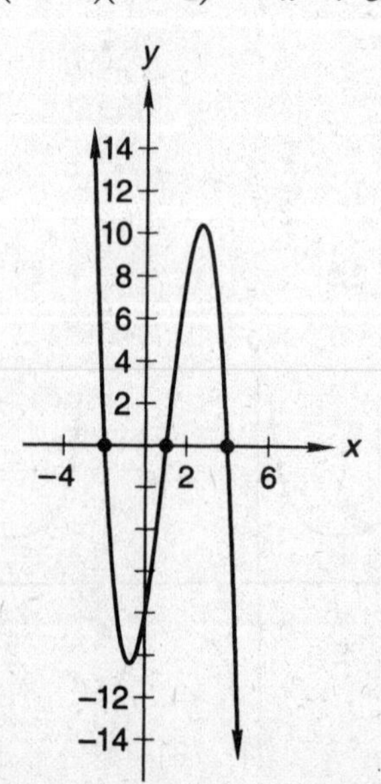

8. (114) The graph intersects the x-axis at 0, 1, –2. The dominant term is x^3. Therefore the answer is **A.**

9. (114) The graph intersects the x-axis at 1, 2, 3 and $h(x)$ is positive when $x = 0$. The dominant term is $-x^3$. Therefore the answer is **D.**

10. (113) $(3x^5 - 2x^3 - 20x - 40) \div (x - 2)$

2⌋	3	0	–2	0	–20	–40
		6	12	20	40	40
	3	6	10	20	20	0

$$\frac{3x^5 - 2x^3 - 20x - 40}{x - 2}$$
$$= \mathbf{3x^4 + 6x^3 + 10x^2 + 20x + 20}$$

11. (113) $(2x^3 - x) \div (x + 3)$

–3⌋	2	0	–1	0
		–6	18	–51
	2	–6	17	–51

$$\frac{2x^3 - x}{x + 3} = \mathbf{2x^2 - 6x + 17 - \frac{51}{x + 3}}$$

12. (113) $x^4 + x^3 - x^2 + x - 172 = 0$

–4⌋	1	1	–1	1	–172
		–4	12	–44	172
	1	–3	11	–43	0

Yes, the remainder is zero, so –4 is a root of the polynomial.

13. (102) $(2x - y^3)^4$

$F = 2x, \; S = -y^3$

Term	①	②	③	④	⑤
Exp. of F	4	3	2	1	0
Exp. of S	0	1	2	3	4
Coefficients	1	4	6	4	1

$(F + S)^4$

$= F^4 + 4F^3S + 6F^2S^2 + 4FS^3 + S^4$

$= (2x)^4 + 4(2x)^3(-y^3) + 6(2x)^2(-y^3)^2$

$\quad + 4(2x)(-y^3)^3 + (-y^3)^4$

$= \mathbf{16x^4 - 32x^3y^3 + 24x^2y^6 - 8xy^9 + y^{12}}$

14. (111) $\log_{\frac{1}{10}} (x - 1) < 2$

$$x - 1 > \left(\frac{1}{10}\right)^2 \quad \text{and} \quad x - 1 > 0$$
$$x > \frac{101}{100} \qquad\qquad x > 1$$

but $1 < \frac{101}{100}$, so $\mathbf{x > \frac{101}{100}}$

15. (110) $f(x) = \text{Arctan}\, x$

Domain $= \{\mathbf{x \in \mathbb{R}}\}$

Range $= \left\{\mathbf{y \in \mathbb{R} \mid -\frac{\pi}{2} < y < \frac{\pi}{2}}\right\}$

Arctan $(-\sqrt{3}) = \mathbf{-\frac{\pi}{3}}$

Arctan $(\sqrt{3}) = \mathbf{\frac{\pi}{3}}$

Arctan (1) $= \mathbf{\frac{\pi}{4}}$

16. (106)
$$60x^2 + 15y^2 - 120x - 60y - 120 = 0$$
$$60(x^2 - 2x + 1) + 15(y^2 - 4y + 4) = 240$$
$$60(x - 1)^2 + 15(y - 2)^2 = 240$$
$$\mathbf{\frac{(x - 1)^2}{4} + \frac{(y - 2)^2}{16} = 1}$$

Center = (1, 2)

$a = 4, \; b = 2$

Length of major axis = 8

Length of minor axis = 4

Vertical

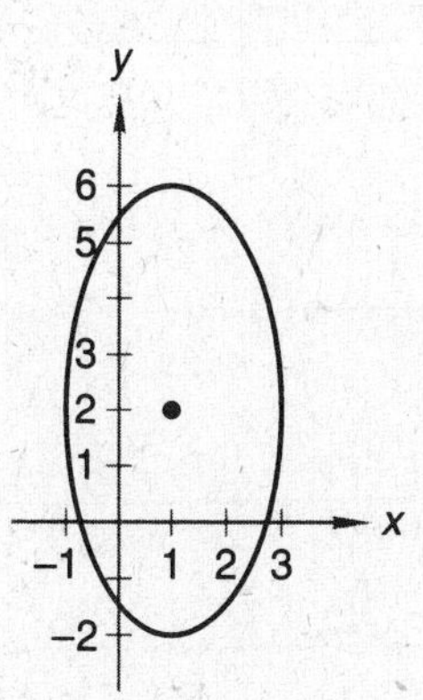

17. (106)
$$3x^2 - 6x - 3y^2 - 6y + 48 = 0$$
$$3(x^2 - 2x + 1) - 3(y^2 + 2y + 1) = -48$$
$$\frac{(y + 1)^2}{16} - \frac{(x - 1)^2}{16} = 1$$

From the equation we obtain:

$h = 1,\ k = -1,\ a = 4,\ b = 4$

Center: $(h, k) = \mathbf{(1, -1)}$

Vertices: $(h, k \pm b) = \mathbf{(1, -5), (1, 3)}$

Asymptotes:

$m = \pm\frac{b}{a} = \pm\frac{1}{1} = \pm 1$

$y = mx + b$	$y = mx + b$
$y = x + b$	$y = -x + b$
$-1 = 1 + b$	$-1 = -(1) + b$
$b = -2$	$b = 0$
$y = x - 2$	$y = -x$

Asymptotes: $\mathbf{y = -x;\ y = x - 2}$

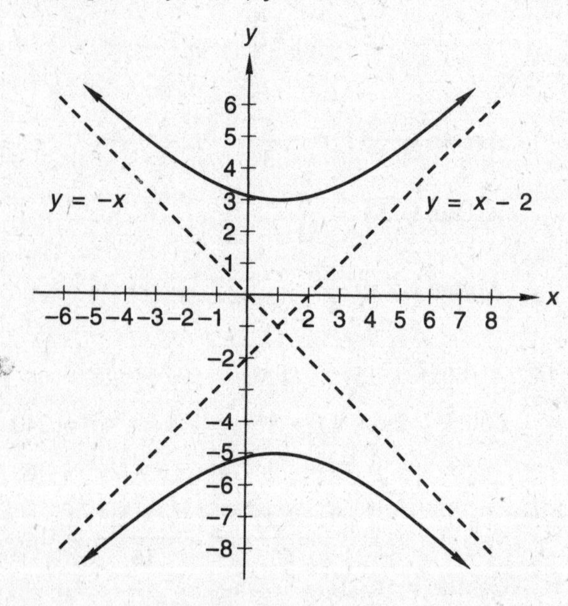

18. (101)
$$x = \frac{\begin{vmatrix} 4 & 2 & 1 \\ 9 & -3 & -2 \\ 3 & 1 & 1 \end{vmatrix}}{\begin{vmatrix} 3 & 2 & 1 \\ 4 & -3 & -2 \\ 1 & 1 & 1 \end{vmatrix}}$$
$$x = \frac{4\begin{vmatrix} -3 & -2 \\ 1 & 1 \end{vmatrix} - 2\begin{vmatrix} 9 & -2 \\ 3 & 1 \end{vmatrix} + 1\begin{vmatrix} 9 & -3 \\ 3 & 1 \end{vmatrix}}{3\begin{vmatrix} -3 & -2 \\ 1 & 1 \end{vmatrix} - 2\begin{vmatrix} 4 & -2 \\ 1 & 1 \end{vmatrix} + 1\begin{vmatrix} 4 & -3 \\ 1 & 1 \end{vmatrix}}$$
$$x = \frac{4(-3 + 2) - 2(9 + 6) + 1(9 + 9)}{3(-3 + 2) - 2(4 + 2) + 1(4 + 3)}$$
$$x = \frac{-4 - 30 + 18}{-3 - 12 + 7} = \frac{-16}{-8} = \mathbf{2}$$

19. (103) $\log(a + b) + 2\log c - \log\left(a^2c - b^2c\right)$
$$= \log\frac{(a + b)c^2}{c\left(a^2 - b^2\right)} = \log\frac{(a + b)c^2}{c(a + b)(a - b)}$$
$$= \mathbf{\log\frac{c}{a - b}}$$

20. (100) $\cos x + \cos y = 2\cos\frac{x + y}{2}\cos\frac{x - y}{2}$

$\cos 105° + \cos 15°$
$$= 2\cos\frac{105° + 15°}{2}\cos\frac{105° - 15°}{2}$$
$$= 2\cos 60°\cos 45° = 2\left(\frac{1}{2}\right)\left(\frac{\sqrt{2}}{2}\right) = \mathbf{\frac{\sqrt{2}}{2}}$$

21. (59) $\ln(x + 8) - \ln(x - 1) = \ln x$
$$\ln\frac{x + 8}{x - 1} = \ln x$$
$$x + 8 = x^2 - x$$
$$x^2 - 2x - 8 = 0$$
$$(x - 4)(x + 2) = 0$$
$$x = \mathbf{4}$$

22. (93) $\tan^4 x + 2\sec^2 x - 1$
$$= \left(\tan^2 x\right)^2 + 2\sec^2 x - 1$$
$$= \left(\sec^2 x - 1\right)^2 + 2\sec^2 x - 1$$
$$= \sec^4 x - 2\sec^2 x + 1 + 2\sec^2 x - 1$$
$$= \mathbf{\sec^4 x}$$

23. (84) $y = -1 + 2\cos 3(x + \pi)$

$$\text{Period} = \frac{2\pi}{3}$$

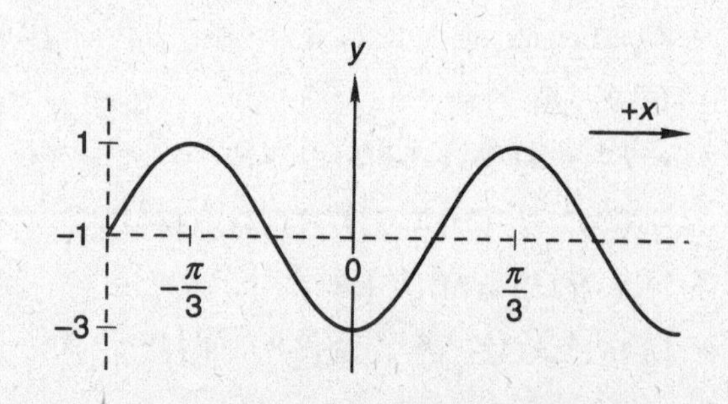

24. (94) $y = \tan(x - \pi)$

Period $= \pi$

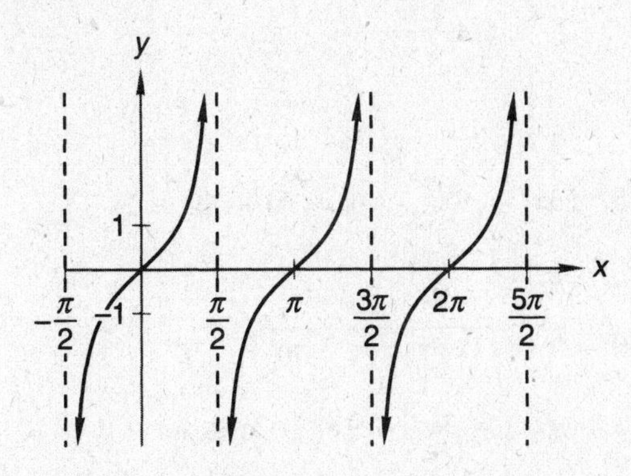

25. (90)

$$\cos 2x - \sin x = 1$$
$$(1 - 2\sin^2 x) - \sin x - 1 = 0$$
$$2\sin^2 x + \sin x = 0$$
$$\sin x(2\sin x + 1) = 0$$

$\sin x = 0$ $2\sin x = -1$

$x = 0°, 180°$ $\sin x = -\frac{1}{2}$

$x = 210°, 330°$

$x =$ **0°, 180°, 210°, 330°**

26. (103) (a) $H^+ = 10^{-pH}$

$H^+ = 10^{-7.4} = \mathbf{3.98 \times 10^{-8} \frac{mole}{liter}}$

(b) $pH = -\log H^+$

$pH = -\log(7.7 \times 10^{-7}) = \mathbf{6.11}$

27. (99) $a_3 = 7,\ a_{12} = 106$

$$\begin{cases} \text{(a)}\ a_1 + 2d = 7 \\ \text{(b)}\ a_1 + 11d = 106 \end{cases}$$

(b) $a_1 + 11d = 106$

$-$(a) $-a_1 - 2d = -7$

$9d = 99$

$d = 11$

(a) $a_1 + 2d = 7$

$a_1 + 2(11) = 7$

$a_1 = -15$

$a_n = a_1 + (n - 1)d$

$a_{100} = a_1 + 99d = -15 + 99(11) = \mathbf{1074}$

28. (99)

$$\begin{cases} \text{(a)}\ \sqrt{xy} = 4\sqrt{6} \\ \text{(b)}\ \dfrac{x + y}{2} = 10 \end{cases}$$

(b) $\frac{x + y}{2} = 10$

$x + y = 20$

$y = 20 - x$

(a) $\sqrt{xy} = 4\sqrt{6}$

$xy = 96$

$x(20 - x) = 96$

$20x - x^2 = 96$

$x^2 - 20x + 96 = 0$

$(x - 8)(x - 12) = 0$

$x = 8, 12$

If $x = 8$, $y = 20 - x = 20 - (8) = 12$

If $x = 12$, $y = 20 - (12) = 8$

8 and 12

29. (59) $3^{\log_3 12 + 2\log_3 2 - \log_3 6} = 3^{\log_3 12 + \log_3 2^2 - \log_3 6}$

$= 3^{\log_3 \frac{(12)(2)^2}{6}} = \frac{12 \cdot 4}{6} = \mathbf{8}$

30. (46) $x^4 - 125x = x(x^3 - 125)$

$= x(x - 5)(x^2 + 5x + 25)$

$$x = \frac{-5 \pm \sqrt{5^2 - 4(1)(25)}}{2(1)} = \frac{-5 \pm 5\sqrt{3}i}{2}$$

$$\mathbf{x(x - 5)\left(x + \frac{5}{2} + \frac{5\sqrt{3}}{2}i\right)\left(x + \frac{5}{2} - \frac{5\sqrt{3}}{2}i\right)}$$

Problem Set 115

1. (92) $P(2 \cup S) = P(2) + P(S) - P(2 \cap S)$

$= \frac{4}{52} + \frac{13}{52} - \frac{1}{52}$

$= \frac{16}{52} = \mathbf{\frac{4}{13}}$

2. (25) Distance $= m$, rate $= k$, time $= \frac{m}{k}$

New time $= \frac{m}{k} - p = \frac{m - pk}{k}$

New rate $= \frac{\text{distance}}{\text{new time}} = \frac{m}{\frac{m - pk}{k}} = \mathbf{\frac{mk}{m - pk}}$ **mph**

3. (53)
$$\omega = \frac{v}{r} = \frac{\frac{60 \text{ ft}}{\text{s}}}{35 \text{ cm}} = \frac{60}{35} \frac{\text{ft}}{\text{s-cm}}$$
$$= \left(\frac{60}{35} \frac{\text{ft}}{\text{s-cm}}\right)\left(\frac{60 \text{ s}}{1 \text{ min}}\right)\left(\frac{12 \text{ in.}}{1 \text{ ft}}\right)\left(\frac{2.54 \text{ cm}}{1 \text{ in.}}\right)$$
$$= \mathbf{\frac{(60)(60)(12)(2.54)}{35} \frac{rad}{min}}$$

4. (18)
$$\begin{cases} \text{(a)} & N_R = 4N_W + 2 \\ \text{(b)} & 7N_W = 4N_P - 14 \\ \text{(c)} & 4(N_W + N_P) = 3N_R + 6 \end{cases}$$

(b)
$$7N_W = 4N_P - 14$$
$$-4N_P = -7N_W - 14$$
$$N_P = \frac{7}{4}N_W + \frac{7}{2}$$

(c)
$$4(N_W + N_P) = 3N_R + 6$$
$$4N_W + 4N_P = 3N_R + 6$$
$$4N_W + 4\left(\frac{7}{4}N_W + \frac{7}{2}\right) = 3(4N_W + 2) + 6$$
$$4N_W + 7N_W + 14 = 12N_W + 6 + 6$$
$$\mathbf{N_W = 2}$$

(a)
$$N_R = 4N_W + 2$$
$$N_R = 4(2) + 2$$
$$\mathbf{N_R = 10}$$

(b)
$$7N_W = 4N_P - 14$$
$$7(2) = 4N_P - 14$$
$$4N_P = 28$$
$$\mathbf{N_P = 7}$$

5. (18)
$$\begin{cases} \text{(a)} & R + T = 200 \\ \text{(b)} & R - \frac{1}{2}R + T - \frac{1}{10}T = 124 \end{cases}$$

(a)
$$R + T = 200$$
$$T = 200 - R$$

(b)
$$R - \frac{1}{2}R + T - \frac{1}{10}T = 124$$
$$\frac{1}{2}R + \frac{9}{10}T = 124$$
$$\frac{1}{2}R + \frac{9}{10}(200 - R) = 124$$
$$\frac{1}{2}R + 180 - \frac{9}{10}R = 124$$
$$-\frac{4}{10}R = -56$$
$$\mathbf{R = \$140}$$

$$T = 200 - R$$
$$T = 200 - (140)$$
$$\mathbf{T = \$60}$$

6. (115) $(3x^3 - 9x^2 - 3x + 4) \div (x + 2)$

```
-2| 3  -9  -3    4
       -6  30  -54
   ---------------
    3 -15  27  -50
```

$$f(x) = 3x^3 - 9x^2 - 3x + 4$$
$$= (3x^2 - 15x + 27)(x + 2) - 50$$
$$f(-2) = 3(-2)3 - 9(-2)2 - 3(-2) + 4$$
$$= (3(-2)^2 - 15(-2) + 27)((-2) + 2) - 50$$
$$f(-2) = -24 - 36 + 6 + 4$$
$$= (12 + 30 + 27)(0) - 50$$
$$f(-2) = -50$$
$$= -50$$
$$\mathbf{f(-2) = -50}$$

7. (115) $x^5 - 3x^4 + 3x - 4$

$x = 2$

```
2| 1  -3   0   0   3   -4
       2  -2  -4  -8  -10
  -----------------------
   1  -1  -2  -4  -5  -14
```

$p(2) = \mathbf{-14}$

8. (113) $(x^2 - 4x + 6) \div (x - 2)$

```
2| 1  -4   6
       2  -4
  ----------
   1  -2   2
```

No, the remainder is not 0, so 2 is not a zero of the polynomial.

9. (113) $(x^6 - 5x^5 + 4x^3 - 2x + 1) \div (x - 1)$

```
1| 1  -5   0   4   0  -2   1
       1  -4  -4   0   0  -2
  --------------------------
   1  -4  -4   0   0  -2  -1
```

$$\frac{x^6 - x^5 + 4x^3 - 2x + 1}{x - 1}$$
$$= \mathbf{x^5 - 4x^4 - 4x^3 - 2 - \frac{1}{x - 1}}$$

10. $y = x(x - 1)(x - 2) = x^3 - 3x^2 + 2x$
(114)

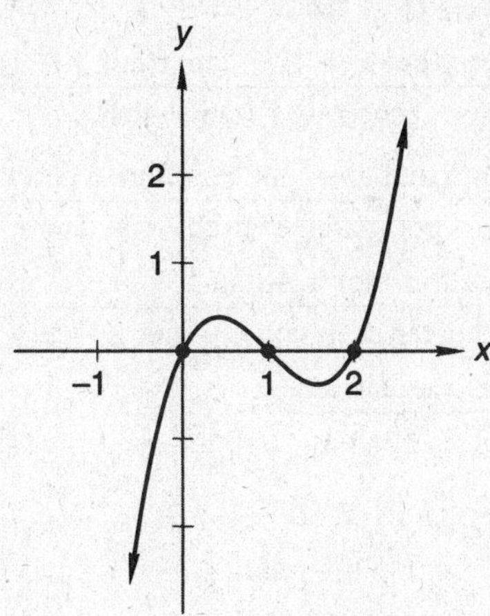

11. $y = (x + 1)^2(x - 1) = x^3 + x^2 - x - 1$
(114)

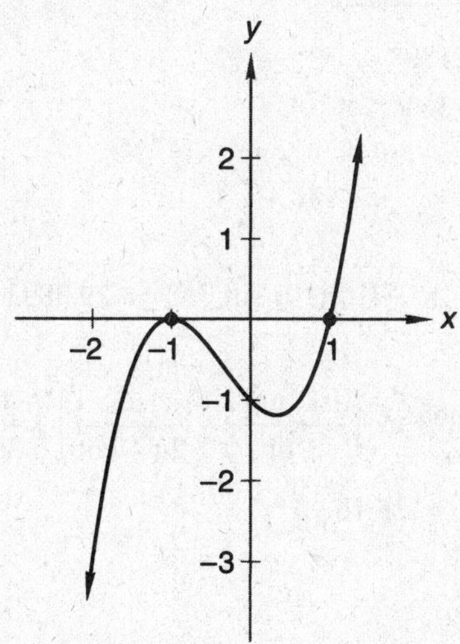

12. The graph intersects the x axis at $x = 0$ and bumps the x axis $x = 2$. The dominant term is x^3. Therefore the answer is **A.**
(114)

13. $(x^4 - y^3)^{10}$
(112)

$F = x^4,\ S = -y^3,\ n = 10,\ k = 7$

$$\frac{n!}{(n - k + 1)!(k - 1)!}F^{n-k+1}S^{k-1}$$

$$= \frac{10!}{(10 - 7 + 1)!(7 - 1)!}F^{10-7+1}S^{7-1}$$

$$= \frac{10!}{4!6!}F^4S^6 = 210(x^4)^4(-y^3)^6 = \mathbf{210x^{16}y^{18}}$$

14. $\log_{\frac{1}{3}}(3x - 2) > 1$
(111)

$3x - 2 < \left(\frac{1}{3}\right)^1$ and $3x - 2 > 0$

$3x < \frac{7}{3}$ $\qquad 3x > 2$

$x < \frac{7}{9}$ $\qquad x > \frac{2}{3}$

$$\mathbf{\frac{2}{3} < x < \frac{7}{9}}$$

15. $f(x) = \text{Arcsin } x$
(110)

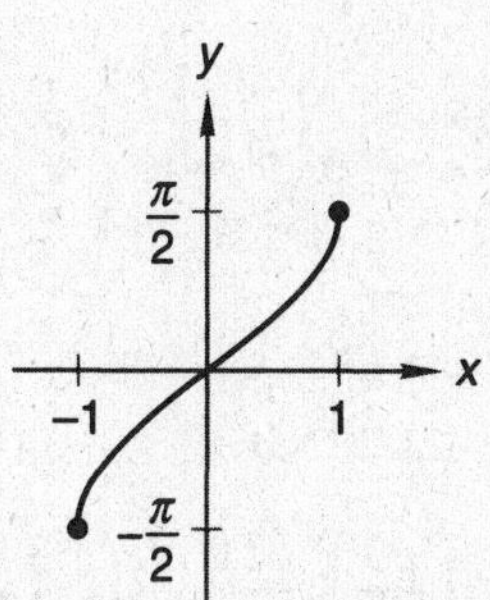

Domain $= \{x \in \mathbb{R} \mid -1 \leq x \leq 1\}$

Range $= \left\{y \in \mathbb{R} \mid -\frac{\pi}{2} \leq y \leq \frac{\pi}{2}\right\}$

Arcsin $\left(\frac{\sqrt{3}}{2}\right) = \frac{\pi}{3}$

Arcsin 1 $= \frac{\pi}{2}$

Arcsin $\left(-\frac{\sqrt{3}}{2}\right) = -\frac{\pi}{3}$

16. $a_1 + a_1r + a_1r^2 + \ldots$
(107)

$\frac{1}{4} + \frac{2}{12} + \frac{4}{36} + \ldots$

$a_1 = \frac{1}{4}$

$a_1r = \frac{2}{12}$

$r = \frac{2}{12a_1} = \frac{2}{3}$

Since $|r| < 1$, we can find the sum of the infinite series.

$$S = \frac{a_1}{1 - r} = \frac{\frac{1}{4}}{1 - \frac{2}{3}} = \frac{\frac{1}{4}}{\frac{1}{3}} = \mathbf{\frac{3}{4}}$$

17. (99) Geometric mean $= \pm\sqrt{6 \cdot 24} = \mathbf{\pm 12}$

Arithmetic mean $= \dfrac{6 + 24}{2} = \mathbf{15}$

18. (103) $\ln\left(\dfrac{\sqrt[4]{(x+1)^2(y-2)^3}}{x(y+2)}\right)$

$= \ln\left(\dfrac{(x+1)^{\frac{1}{2}}(y-2)^{\frac{3}{4}}}{x(y+2)}\right)$

$= \mathbf{\frac{1}{2}\ln(x+1) + \frac{3}{4}\ln(y-2) - \ln x - \ln(y+2)}$

19. (103) $\dfrac{\sqrt{90 \times 10^5}}{\sqrt[3]{40 \times 10^4}} - 3\sqrt[4]{9 \times 10^7} = \mathbf{-251.48}$

20. (87) $\cos 75° = \cos(30° + 45°)$

$= \cos 30° \cos 45° - \sin 30° \sin 45°$

$= \left(\dfrac{\sqrt{3}}{2}\right)\left(\dfrac{\sqrt{2}}{2}\right) - \left(\dfrac{1}{2}\right)\left(\dfrac{\sqrt{2}}{2}\right)$

$= \mathbf{\dfrac{\sqrt{6} - \sqrt{2}}{4}}$

21. (98) $\log_2(\log_3 x) = 2$

$\log_3 x = 2^2$

$x = 3^4$

$x = \mathbf{81}$

22. (98) $\log_2 \sqrt{x} = \sqrt{\log_2 x}$

$\dfrac{1}{2}\log_2 x = \sqrt{\log_2 x}$

$\dfrac{1}{4}(\log_2 x)^2 = \log_2 x$

$\dfrac{1}{4}(\log_2 x)^2 - \log_2 x = 0$

$\log_2 x\left(\dfrac{1}{4}\log_2 x - 1\right) = 0$

$\log_2 x = 0$

$x = 2^0$

$x = 1$

$\dfrac{1}{4}\log_2 x - 1 = 0$

$\dfrac{1}{4}\log_2 x = 1$

$\log_2 x = 4$

$x = 2^4$

$x = 16$

$x = \mathbf{1, 16}$

23. (93) $\dfrac{\cot x}{\cot x - 1} - \dfrac{\tan x}{\tan x + 1}$

$= \dfrac{\cot x(\tan x + 1) - \tan x(\cot x - 1)}{(\cot x - 1)(\tan x + 1)}$

$= \dfrac{\cot x \tan x + \cot x - \tan x \cot x + \tan x}{\cot x \tan x + \cot x - \tan x - 1}$

$= \dfrac{\cot x + \tan x}{\cot x \tan x + \cot x - \tan x - 1}$

$= \mathbf{\dfrac{\cot x + \tan x}{\cot x - \tan x}}$

24. (96)

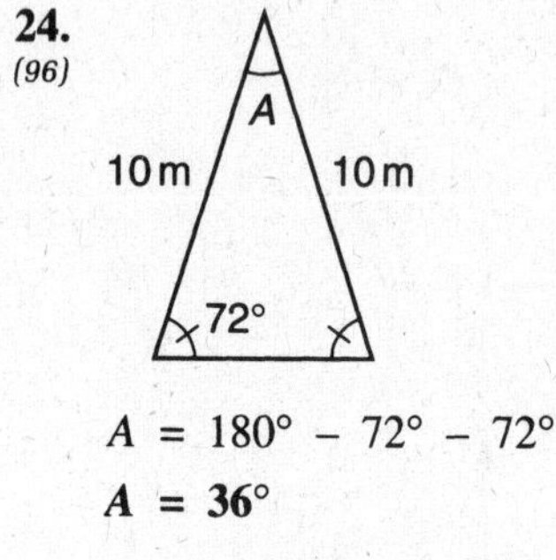

$A = 180° - 72° - 72°$

$\mathbf{A = 36°}$

Area $= \dfrac{1}{2}bc \sin A$

$= \dfrac{1}{2}(10)(10)\sin 36° = 29.3893 \text{ m}^2$

$29.3893 \text{ m}^2\left(\dfrac{100 \text{ cm}}{1 \text{ m}}\right)^2\left(\dfrac{1 \text{ in.}}{2.54 \text{ cm}}\right)^2\left(\dfrac{1 \text{ yd}}{36 \text{ in.}}\right)^2$

Area = 35.15 yd²

25. (94) $y = \cot(x + \pi)$

Phase angle $= -\pi$

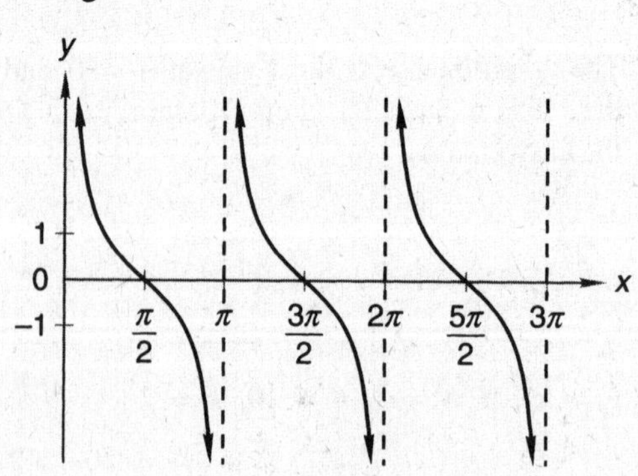

26. (84) $y = 4 - \cos 4\left(x - \dfrac{\pi}{4}\right)$

Period $= \dfrac{2\pi}{4} = \dfrac{\pi}{2}$

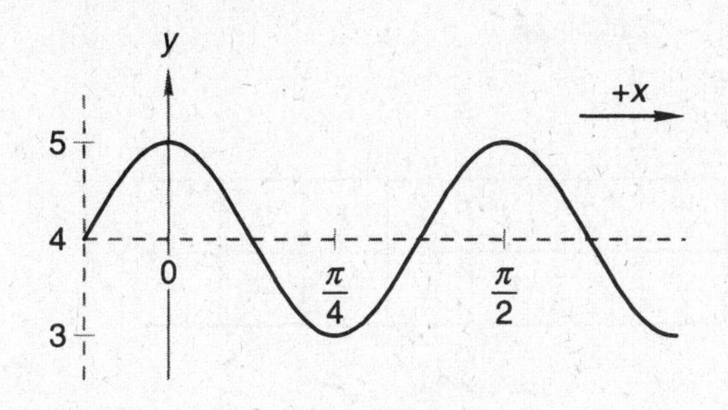

27. (52) $\sqrt{2} - 2\sin\frac{3\theta}{2} = 0$

$$\sin\frac{3\theta}{2} = \frac{\sqrt{2}}{2}$$
$$\frac{3\theta}{2} = 45°, 135°, 405°, 495°$$
$$\boldsymbol{\theta = 30°, 90°, 270°, 330°}$$

28. (90) $\sin 2\theta - \cos 2\theta = 0$

$$\sin 2\theta = \cos 2\theta$$
$$\frac{\sin 2\theta}{\cos 2\theta} = 1$$
$$\tan 2\theta = 1$$
$$2\theta = 45°, 225°, 405°, 585°$$
$$\boldsymbol{\theta = 22.5°, 112.5°, 202.5°, 292.5°}$$

29. (95) $-1 - i = \sqrt{2}\text{ cis } 225°$

$$\left(\sqrt{2}\text{ cis } 225°\right)^{\frac{1}{5}} = \left(\sqrt{2}\right)^{\frac{1}{5}}\text{cis}\left(\frac{225°}{5} + n\frac{360°}{5}\right)$$

$= \mathbf{1.07\ cis\ 45°,\ 1.07\ cis\ 117°,\ 1.07\ cis\ 189°,}$
$\mathbf{1.07\ cis\ 261°,\ 1.07\ cis\ 333°}$

30. (101)
$$x = \frac{\begin{vmatrix} 0 & 1 & -1 \\ 7 & 2 & 3 \\ 7 & -3 & 0 \end{vmatrix}}{\begin{vmatrix} 2 & 1 & -1 \\ 0 & 2 & 3 \\ 2 & -3 & 0 \end{vmatrix}}$$

$$x = \frac{0\begin{vmatrix} 2 & 3 \\ -3 & 0 \end{vmatrix} - 1\begin{vmatrix} 7 & 3 \\ 7 & 0 \end{vmatrix} + (-1)\begin{vmatrix} 7 & 2 \\ 7 & -3 \end{vmatrix}}{2\begin{vmatrix} 2 & 3 \\ -3 & 0 \end{vmatrix} - 1\begin{vmatrix} 0 & 3 \\ 2 & 0 \end{vmatrix} + (-1)\begin{vmatrix} 0 & 2 \\ 2 & -3 \end{vmatrix}}$$

$$x = \frac{0(0 + 9) - 1(0 - 21) - 1(-21 - 14)}{2(0 + 9) - 1(0 - 6) - 1(0 - 4)}$$

$$x = \frac{0 + 21 + 35}{18 + 6 + 4}$$

$$x = \frac{56}{28}$$

$$\boldsymbol{x = 2}$$

$$y = \frac{\begin{vmatrix} 2 & 0 & -1 \\ 0 & 7 & 3 \\ 2 & 7 & 0 \end{vmatrix}}{28}$$

$$y = \frac{2\begin{vmatrix} 7 & 3 \\ 7 & 0 \end{vmatrix} - 0\begin{vmatrix} 0 & 3 \\ 2 & 0 \end{vmatrix} + (-1)\begin{vmatrix} 0 & 7 \\ 2 & 7 \end{vmatrix}}{28}$$

$$y = \frac{2(0 - 21) - 0 - 1(0 - 14)}{28}$$

$$y = -\frac{28}{28}$$

$$\boldsymbol{y = -1}$$

Problem Set 116

1. (83) $P(R, \text{then } G) = \frac{6}{9} \cdot \frac{3}{8} = \boldsymbol{\frac{1}{4}}$

2. (88) $A_7 = 1400,\ A_{24} = 1300$

$$A_t = A_0 e^{kt}$$
$$1300 = 1400e^{k(24-7)}$$
$$\frac{13}{14} = e^{17k}$$
$$\ln\frac{13}{14} = 17k$$
$$k = -0.0043593$$

$$A_{60} = 1400e^{-0.0043593(60-7)}$$
$$A_{60} = \mathbf{1111.19}$$

3. (85) $R_B = 1,\ R_L = \frac{1}{12},\ T_B = T_L$

$$\begin{cases} R_L T_L = S \\ R_B T_B = S + 20 \end{cases}$$

$$R_B T_B = S + 20$$
$$(1)T = R_L T_L + 20$$
$$T = \frac{T}{12} + 20$$
$$\frac{11}{12}T = 20$$
$$T = \frac{240}{11}$$
$$T = \mathbf{21\frac{9}{11}\ min}$$

4. (44)

$$RWT = \text{jobs}$$

$$R(5 + w)(7) = 2$$

$$R = \frac{2}{7(5 + w)} \frac{\text{jobs}}{\text{workers-hr}}$$

$$RWT = J$$

$$T = \frac{J}{RW}$$

$$T = \frac{c}{\frac{2}{7(5 + w)}(m + 4)}$$

$$T = \mathbf{\frac{7c(5 + w)}{2(m + 4)}\ hr}$$

5. (62) $S = mT + b$

$$\begin{cases} \text{(a) } 15 = m20 + b \\ \text{(b) } 25 = m40 + b \end{cases}$$

$$\begin{array}{rrcl} \text{(b)} & 25 & = & m40 + b \\ -2\text{(a)} & -30 & = & -m40 - 2b \\ \hline & -5 & = & -b \\ & b & = & 5 \end{array}$$

(a) $15 = m20 + b$

$15 = m20 + 5$

$10 = m20$

$m = \frac{1}{2}$

$S = \frac{1}{2}T + 5 = \frac{1}{2}(100) + 5 = \mathbf{55}$ **succeeded**

6. (116) $y = x^3 - 5x + 1$

$r = |-5| + 1$

$\boldsymbol{r = 6}$

x	−2	−1	0	1	2
y	3	5	1	−3	−1

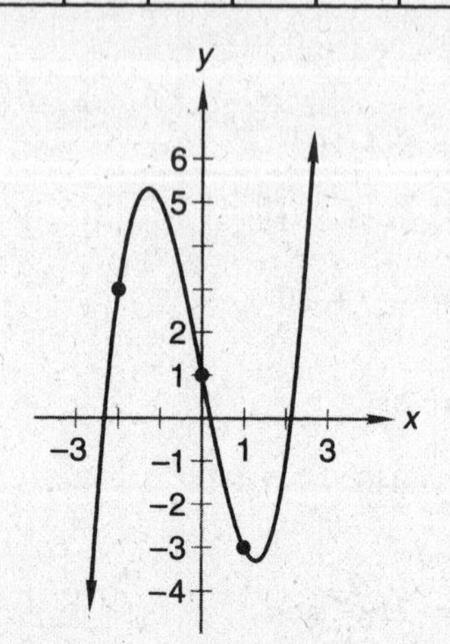

7. (116) $y = x^3 - 2x^2 - 5x + 6$

$r = |6| + 1$

$\boldsymbol{r = 7}$

x	−2	−1	0	1	2	3
y	0	8	6	0	−4	0

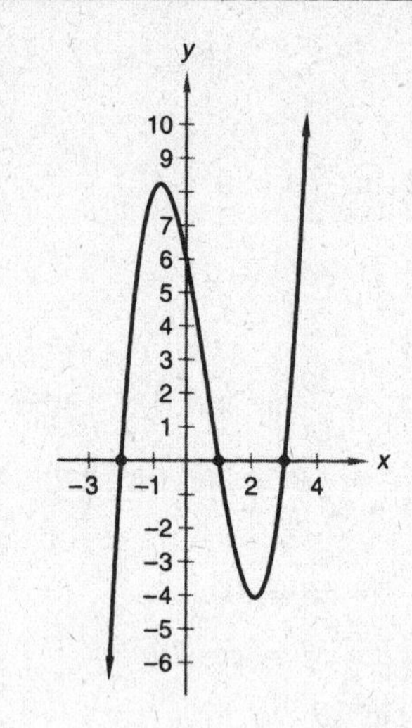

8. (114) $y = x(x - 3)(x - 4) = x^3 - 7x^2 + 12x$

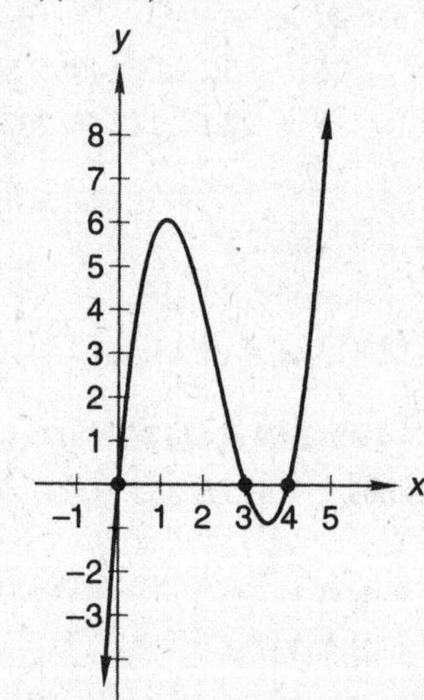

9. (114) $y = (x - 1)(x + 2)^2$

The graph "bumps" the x-axis at $x = -2$.

If $x = 0$, $y = (0)^3 + 3(0)^2 - 4 = -4$.

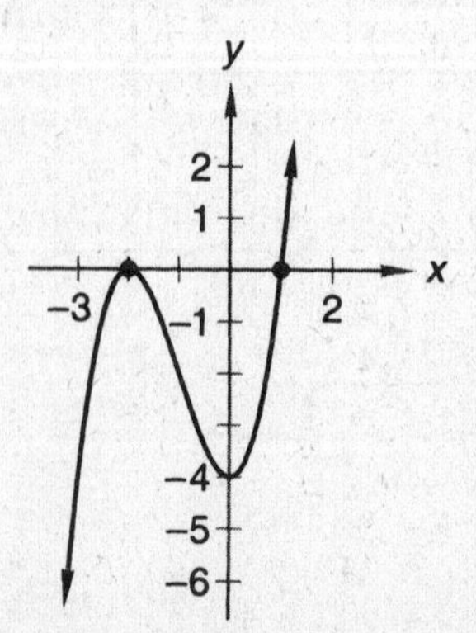

10. (114) $f(x) = (x - 2)^2(x + 1)$

The graph crosses the x-axis at $x = -1$.
The graph "bumps" the x-axis at $x = 2$.

$$f(x) = (x - 2)^2(x + 1) = \left(x^2 - 4x + 4\right)(x + 1) = x^3 - 3x^2 + 4$$

The dominant term is x^3 which is positive for large positive values of x and negative for large negative values of x. Therefore the answer is **B.**

11. (113) $\left(x^4 - 3x^3 + 2x^2 - 3x + 4\right) \div (x - 3)$

```
3| 1  -3   2  -3   4
       3   0   6   9
   ---------------------
   1   0   2   3  13
```

$$\frac{x^4 - 3x^3 + 2x^2 - 3x + 4}{x - 3} = \mathbf{x^3 + 2x + 3 + \frac{13}{x - 3}}$$

12. (113) $\left(x^5 - 1\right) \div (x + 1)$

```
-1| 1   0   0   0   0  -1
       -1   1  -1   1  -1
   --------------------------
    1  -1   1  -1   1  -2
```

$$\frac{x^5 - 1}{x + 1} = \mathbf{x^4 - x^3 + x^2 - x + 1 - \frac{2}{x + 1}}$$

13. (115) $x^4 - 3x^3 + 2x - 1$

$x = 3$

```
3| 1  -3   0   2  -1
       3   0   0   6
   ---------------------
   1   0   0   2   5
```

$p(3) = \mathbf{5}$

14. (112) $\left(x^3 - 2y^3\right)^{12}$

$F = x^3,\ S = -2y^3,\ n = 12,\ k = 6$

$$\frac{n!}{(n - k + 1)!(k - 1)!}F^{n-k+1}S^{k-1}$$

$$= \frac{12!}{(12 - 6 + 1)!\,(6 - 1)!}F^{12-6+1}S^{6-1}$$

$$= \frac{12!}{7!5!}F^7S^5 = 792\left(x^3\right)^7\left(-2y^3\right)^5$$

$$= 792x^{21}\left(-32y^{15}\right) = \mathbf{-25{,}344x^{21}y^{15}}$$

15. (111) $\log_3(2x + 1) < 4$

$2x + 1 < 3^4$ and $2x + 1 > 0$

$2x < 80$ $\qquad x > -\frac{1}{2}$

$x < 40$

$$\mathbf{-\frac{1}{2} < x < 40}$$

16. (111) $2\log(x + 1) > 4$

$\log(x + 1) > 2$

$x + 1 > 10^2$ and $x + 1 > 0$

$x > 99$ $\qquad x > -1$

but $-1 < 99$ so, $\mathbf{x > 99}$

17. (110) $f(x) = \text{Arccos } x$

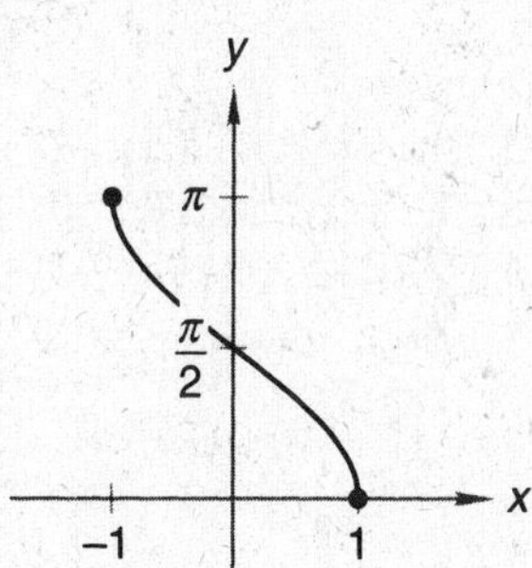

Domain $= \{\mathbf{x \in \mathbb{R} \mid -1 \le x \le 1}\}$

Range $= \{\mathbf{y \in \mathbb{R} \mid 0 \le y \le \pi}\}$

$$\textbf{Arccos } \frac{\sqrt{3}}{2} = \frac{\pi}{6}$$

$$\textbf{Arccos}\left(-\frac{\sqrt{3}}{2}\right) = \frac{5\pi}{6}$$

$$\textbf{Arccos 0} = \frac{\pi}{2}$$

18. (109) $2.\overline{15}$

$$= 2 + \left[0.15 + 0.15\left(10^{-2}\right) + 0.15\left(10^{-4}\right) + \ldots\right]$$

$$= 2 + 0.15 + 0.15\left(10^{-2}\right) + 0.15\left(10^{-2}\right)^2 + \ldots$$

If $a_1 = 0.15,\ r = 10^{-2}$

$$2.\overline{15} = 2 + \frac{a_1}{1 - r} = 2 + \frac{0.15}{1 - 10^{-2}} = 2 + \frac{0.15}{0.99}\cdot\frac{10^2}{10^2} = 2 + \frac{15}{99} = \frac{66}{33} + \frac{5}{33} = \mathbf{\frac{71}{33}}$$

19. (107) $a_1, a_1r, a_1r^2, \ldots$

$$-\frac{1}{3}, \frac{2}{9}, -\frac{4}{27}, \ldots$$

$$a_1 = -\frac{1}{3}$$

$$a_1r = \frac{2}{9}$$

$$r = \frac{2}{9a_1} = -\frac{2}{3}$$

$$S = \frac{a_1}{1-r}$$

$$= \frac{-\frac{1}{3}}{1-\left(-\frac{2}{3}\right)} = \frac{-\frac{1}{3}}{\frac{5}{3}} = -\frac{1}{3}\cdot\frac{3}{5} = \mathbf{-\frac{1}{5}}$$

20. (108) $A = \begin{bmatrix} 2 & 0 & 0 \\ 3 & 1 & 1 \\ 4 & 0 & 2 \end{bmatrix}, B = \begin{bmatrix} 5 & -1 & -1 \\ 0 & 0 & 0 \\ 2 & -3 & 2 \end{bmatrix}$

$$A - B = \begin{bmatrix} (2-5) & (0+1) & (0+1) \\ (3-0) & (1-0) & (1-0) \\ (4-2) & (0+3) & (2-2) \end{bmatrix}$$

$$= \begin{bmatrix} -3 & 1 & 1 \\ 3 & 1 & 1 \\ 2 & 3 & 0 \end{bmatrix}$$

$$B \cdot (A - B) = \begin{bmatrix} 5 & -1 & -1 \\ 0 & 0 & 0 \\ 2 & -3 & 2 \end{bmatrix} \cdot \begin{bmatrix} -3 & 1 & 1 \\ 3 & 1 & 1 \\ 2 & 3 & 0 \end{bmatrix}$$

$$= \begin{bmatrix} -15-3-2 & 5-1-3 & 5-1+0 \\ 0+0+0 & 0+0+0 & 0+0+0 \\ -6-9+4 & 2-3+6 & 2-3+0 \end{bmatrix}$$

$$= \begin{bmatrix} \mathbf{-20} & \mathbf{1} & \mathbf{4} \\ \mathbf{0} & \mathbf{0} & \mathbf{0} \\ \mathbf{-11} & \mathbf{5} & \mathbf{-1} \end{bmatrix}$$

21. (106)

$$4x^2 - 9y^2 - 24x - 36y = 36$$

$$4(x^2 - 6x + 9) - 9(y^2 + 4y + 4) = 36$$

$$4(x-3)^2 - 9(y+2)^2 = 36$$

$$\mathbf{\frac{(x-3)^2}{9} - \frac{(y+2)^2}{4} = 1}$$

From the equation we obtain:

$h = 3,\ k = -2,\ a = 3,\ b = 2$

Center: $(h, k) = \mathbf{(3, -2)}$

Vertices: $(h \pm a, k) = \mathbf{(6, -2), (0, -2)}$

Asymptotes:

$$m = \pm\frac{b}{a} = \pm\frac{2}{3}$$

$y = mx + b$ $\qquad$ $y = mx + b$

$y = \frac{2}{3}x + b$ $\qquad$ $y = -\frac{2}{3}x + b$

$-2 = \frac{2}{3}(3) + b$ $\qquad$ $-2 = -\frac{2}{3}(3) + b$

$b = -4$ $\qquad$ $b = 0$

$y = \frac{2}{3}x - 4$ $\qquad$ $y = -\frac{2}{3}x$

Asymptotes: $\mathbf{y = \frac{2}{3}x - 4;\ y = -\frac{2}{3}x}$

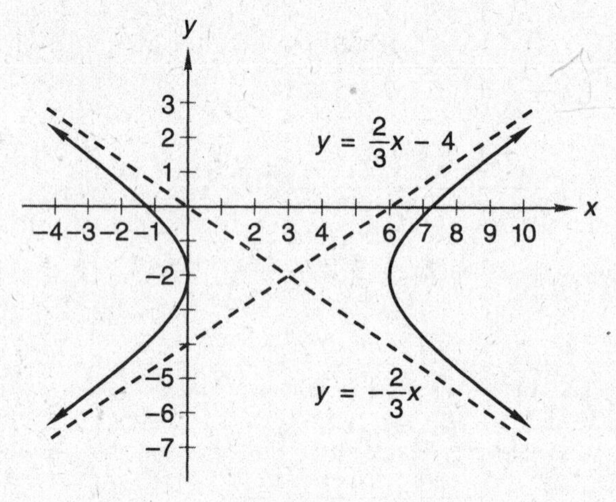

22. (101)

$$x = \frac{\begin{vmatrix} 9 & 2 & 1 \\ 0 & 1 & -2 \\ -2 & 3 & 0 \end{vmatrix}}{\begin{vmatrix} -1 & 2 & 1 \\ 0 & 1 & -2 \\ 2 & 3 & 0 \end{vmatrix}}$$

$$x = \frac{9(0+6) - 2(0-4) + 1(0+2)}{-1(0+6) - 2(0+4) + 1(0-2)}$$

$$x = \frac{54 + 8 + 2}{-6 - 8 - 2}$$

$$x = -\frac{64}{16}$$

$\mathbf{x = -4}$

$$z = \frac{\begin{vmatrix} -1 & 2 & 9 \\ 0 & 1 & 0 \\ 2 & 3 & -2 \end{vmatrix}}{-16}$$

$$z = \frac{-1(-2-0) - 2(0-0) + 9(0-2)}{-16}$$

$$z = \frac{-16}{-16}$$

$\mathbf{z = 1}$

23. (103)

(a) $H^+ = 10^{-pH} = 10^{-9.2} = \mathbf{6.3 \times 10^{-10}\ \dfrac{mole}{liter}}$

(b) $pH = -\log H^+$

$$= -\log\left(1.3 \times 10^{-8}\right) = 7.8861 = \mathbf{7.89}$$

24. (59)

$$2\log_3 x - \log_3\left(\frac{2}{3}x - \frac{1}{3}\right) = 1$$

$$\log_3 x^2 - \log_3\left(\frac{2x-1}{3}\right) = 1$$

$$\log_3\left(\frac{x^2}{\frac{2x-1}{3}}\right) = 1$$

$$\log_3\left(\frac{3x^2}{2x-1}\right) = 1$$

$$\frac{3x^2}{2x-1} = 3^1$$

$$3x^2 = 6x - 3$$

$$3\left(x^2 - 2x + 1\right) = 0$$

$$(x-1)^2 = 0$$

$$x = \mathbf{1}$$

25. (98)

$$x^{\sqrt{\log x}} = x$$

$$\sqrt{\log x}\log x = \log x$$

$$\sqrt{\log x}\log x - \log x = 0$$

$$\log x\left(\sqrt{\log x} - 1\right) = 0$$

$$\log x = 0 \qquad \sqrt{\log x} = 1$$

$$x = 10^0 \qquad \log x = 1$$

$$x = 1 \qquad x = 10$$

$\mathbf{x = 1, 10}$

26. (80,84)

$$\frac{\cos^2 x}{1 - \sin x} = \frac{1 - \sin^2 x}{1 - \sin x}$$

$$= \frac{(1 + \sin x)(1 - \sin x)}{1 - \sin x} = 1 + \sin x$$

$$= 1 + \frac{1}{\csc x} = \frac{1 + \csc x}{\csc x}$$

27. (87)

$$\tan(A + B) = \frac{\tan A + \tan B}{1 - \tan A \tan B}$$

$$\tan 75° = \tan(30° + 45°)$$

$$\tan(30° + 45°)$$

$$= \frac{\tan 30° + \tan 45°}{1 - \tan 30° \tan 45°}$$

$$= \frac{\frac{1}{\sqrt{3}} + 1}{1 - \left(\frac{1}{\sqrt{3}}\right)(1)} = \frac{\frac{3 + \sqrt{3}}{3}}{\frac{3 - \sqrt{3}}{3}}$$

$$= \frac{3 + \sqrt{3}}{3 - \sqrt{3}} \cdot \frac{3 + \sqrt{3}}{3 + \sqrt{3}} = \frac{9 + 6\sqrt{3} + 3}{9 - 3}$$

$$= \frac{12 + 6\sqrt{3}}{6} = \mathbf{2 + \sqrt{3}}$$

28. (52)

$$-\sqrt{2} - 2\sin 3\theta = 0$$

$$-2\sin 3\theta = \sqrt{2}$$

$$\sin 3\theta = -\frac{\sqrt{2}}{2}$$

$$3\theta = \frac{5\pi}{4}, \frac{7\pi}{4}, \frac{13\pi}{4}, \frac{15\pi}{4}, \frac{21\pi}{4}, \frac{23\pi}{4}$$

$$\theta = \mathbf{\frac{5\pi}{12}, \frac{7\pi}{12}, \frac{13\pi}{12}, \frac{5\pi}{4}, \frac{7\pi}{4}, \frac{23\pi}{12}}$$

29. (94) $y = -1 - \csc x$

Period $= 2\pi$

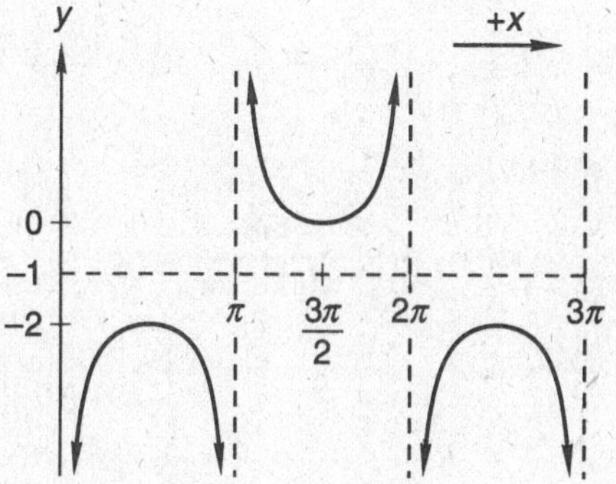

30. (58) $y = 3$

$(1, 5)$

Equation of the perpendicular line: $x = 1$

Point of intersection: $(1, 3)$

$(1, 3)$ and $(1, 5)$

$$D = \sqrt{(1-1)^2 + (5-3)^2} = \sqrt{2^2} = \mathbf{2}$$

Problem Set 117

1. (92) $P(R \cup S) = P(R) + P(S) - P(R \cap S)$
$= 0.4 + 0.7 - 0.25 = \mathbf{0.85}$

2. (92)

	1	2	3	4	5	6
1	2	3	4	5	6	(7)
2	3	4	5	6	(7)	8
3	4	5	6	(7)	8	9
4	5	6	(7)	8	9	10
5	6	(7)	8	9	10	(11)
6	(7)	8	9	10	(11)	12

$P(\text{sum} = 7 \text{ or sum} = 11) = \frac{8}{36} = \mathbf{\frac{2}{9}}$

3. (53) $\omega = \frac{v}{r} = \frac{m \frac{\text{mi}}{\text{hr}}}{c \text{ cm}} = \frac{m}{c} \frac{\text{mi}}{\text{cm-hr}}$

$\omega = \frac{m}{c} \frac{\text{mi}}{\text{cm-hr}} \left(\frac{5280 \text{ ft}}{1 \text{ mi}}\right)\left(\frac{12 \text{ in.}}{1 \text{ ft}}\right)\left(\frac{2.54 \text{ cm}}{1 \text{ in.}}\right) \times \left(\frac{1 \text{ hr}}{60 \text{ min}}\right)\left(\frac{1 \text{ min}}{60 \text{ s}}\right)$

$\omega = \mathbf{\frac{m(5280)(12)(2.54)}{c(60)(60)} \frac{rad}{s}}$

4. (44) $RWT = J$

$R(4)(m) = k$

$R = \frac{k}{4m} \frac{\text{job}}{\text{worker-hr}}$

$RWT = J$

$T = \frac{J}{RW}$

$T = \frac{35}{\left(\frac{k}{4m}\right)(4 + 3)} = \mathbf{\frac{20m}{k} \text{ hr}}$

5. (38) First leg $= \frac{m}{h} \frac{\text{mi}}{\text{hr}}\left(\frac{1 \text{ hr}}{60 \text{ min}}\right) = \frac{m}{60h} \frac{\text{mi}}{\text{min}}$

Second leg $= \frac{k}{m} \frac{\text{mi}}{\text{min}}$

Overall average rate $= \frac{m + k}{60h + m} \frac{\text{mi}}{\text{min}}$

$\text{Time} = \frac{\text{distance}}{\text{rate}}$

$= \frac{30 \text{ mi}}{\frac{m + k}{60h + m} \frac{\text{mi}}{\text{min}}} = \mathbf{\frac{1800h + 30m}{m + k} \text{ min}}$

6. (117) $136 = 2 \cdot 2 \cdot 2 \cdot 17$

$81 = 3 \cdot 3 \cdot 3 \cdot 3$

Yes, the two are relatively prime.

7. (117) **Yes.** Two consecutive integers have no common factors other than one, and therefore, are always relatively prime.

8. (117) $2x^2 + 3x + 5 = 0$

Integral Factors		Possible Quotients
$\frac{\{-1, 1, -5, 5\}}{\{-1, 1, -2, 2\}}$	→	$\mathbf{\pm 1, \pm 5, \pm\frac{1}{2}, \pm\frac{5}{2}}$

9. (117) $4x^{15} - 2x^9 + 6x + 3 = 0$

Integral Factors		Possible Quotients
$\frac{\{1, -1, 3, -3\}}{\{1, -1, 2, -2, 4, -4\}}$	→	$\mathbf{\pm 1, \pm 3, \pm\frac{1}{2}, \pm\frac{3}{2}, \pm\frac{1}{4}, \pm\frac{3}{4}}$

10. (116) $y = x^3 - 3x^2 - x + 3$

$r = |-3| + 1$

$\mathbf{r = 4}$

x	−2	−1	0	1	2	3
y	−15	0	3	0	−3	0

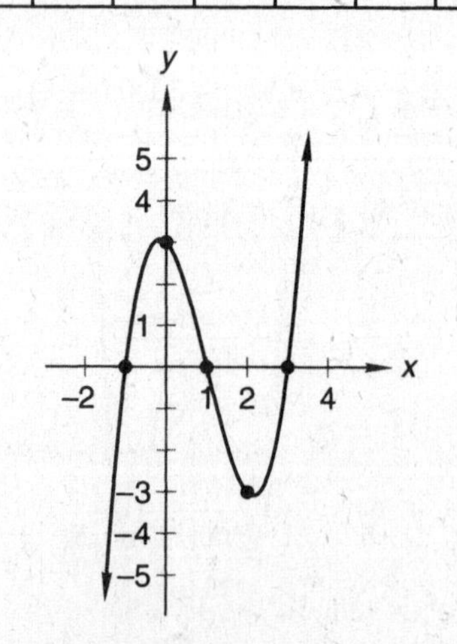

11. (116) $y = -x^3 + 2x - 2$

$r = |2| + 1$

$\boldsymbol{r = 3}$

x	−2	−1	0	1	2
y	2	−3	−2	−1	−6

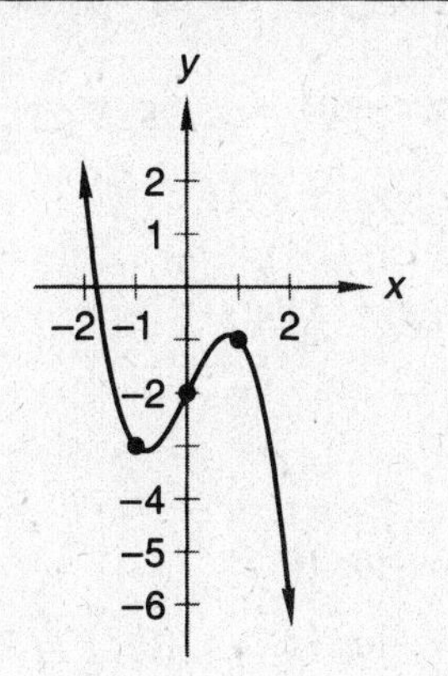

12. (114,116) $y = -x^2(x - 1)(x + 1) = -x^4 + x^2$

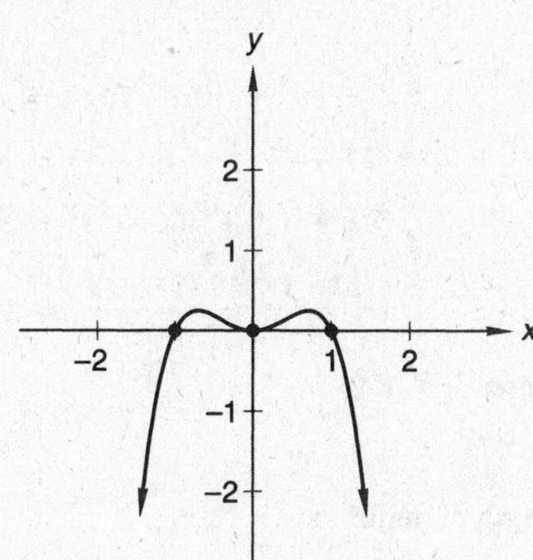

13. (114,116) $y = x^4 - 5x^2 + 4$

$y = (x^2 - 4)(x^2 - 1)$

$\boldsymbol{y = (x - 2)(x + 2)(x - 1)(x + 1)}$

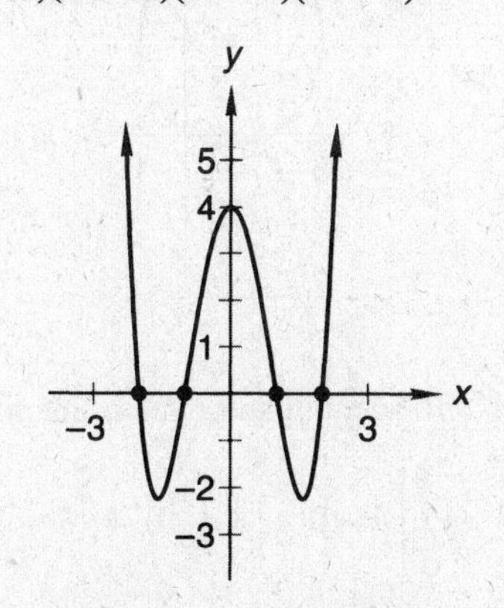

14. (114) It is a quadratic equation, thus we know it may have two zeros and one turning point. Its zeros are at 0 and 2, and it opens upward, since the dominant term is x^2. Therefore the answer is **B.**

15. (113) $(2x^4 - 3x^2 - 4x + 1) \div (x - 2)$

$$\begin{array}{r|rrrrr} 2 & 2 & 0 & -3 & -4 & 1 \\ & & 4 & 8 & 10 & 12 \\ \hline & 2 & 4 & 5 & 6 & 13 \end{array}$$

$$\frac{2x^4 - 3x^2 - 4x + 1}{x - 2}$$

$$= \mathbf{2x^3 + 4x^2 + 5x + 6 + \frac{13}{x - 2}}$$

16. (115) $4x^3 + 2x^2 + 3x + 1$

$x = -1$

$$\begin{array}{r|rrrr} -1 & 4 & 2 & 3 & 1 \\ & & -4 & 2 & -5 \\ \hline & 4 & -2 & 5 & -4 \end{array}$$

$p(-1) = \mathbf{-4}$

17. (77,102) $(x^2 + 2y)^5$

$F = x^2,\ S = 2y$

Term	①	②	③	④	⑤	⑥
Exp. of F	5	4	3	2	1	0
Exp. of S	0	1	2	3	4	5
Coefficients	1	5	10	10	5	1

$(F + S)^5$

$= F^5 + 5F^4S + 10F^3S^2 + 10F^2S^3 + 5FS^4 + S^5$

$= (x^2)^5 + 5(x^2)^4(2y) + 10(x^2)^3(2y)^2 + 10(x^2)^2(2y)^3 + 5(x^2)(2y)^4 + (2y)^5$

$= \mathbf{x^{10} + 10x^8y + 40x^6y^2 + 80x^4y^3 + 80x^2y^4 + 32y^5}$

18. (111) $\log_2 (3x - 1) > 4$

$3x - 1 > 2^4$ and $3x - 1 > 0$

$3x - 1 > 16$ $\qquad 3x > 1$

$3x > 17$ $\qquad x > \frac{1}{3}$

$x > \frac{17}{3}$

but $\frac{1}{3} < \frac{17}{3}$, so $\boldsymbol{x > \frac{17}{3}}$

19. (110) $f(x) = \text{Arctan } x$

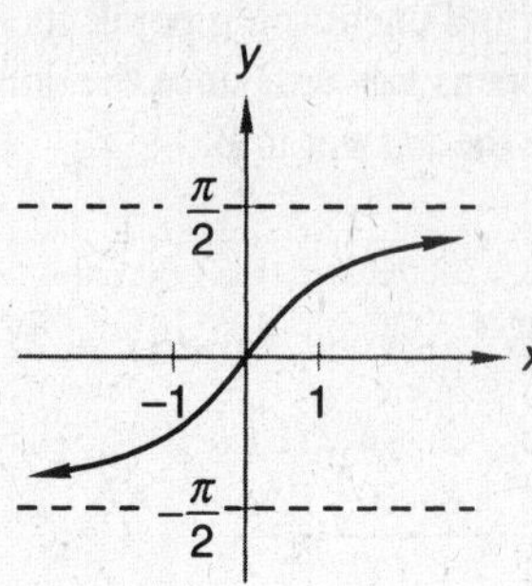

Domain $= \{x \in \mathbb{R}\}$

Range $= \left\{y \in \mathbb{R} \mid -\frac{\pi}{2} < y < \frac{\pi}{2}\right\}$

Arctan $\frac{\sqrt{3}}{3} = \frac{\pi}{6}$

Arctan $\left(-\frac{\sqrt{3}}{3}\right) = -\frac{\pi}{6}$

Arctan $(-1) = -\frac{\pi}{4}$

20. (94) $y = 1 + 2\sec x$

Period $= 2\pi$

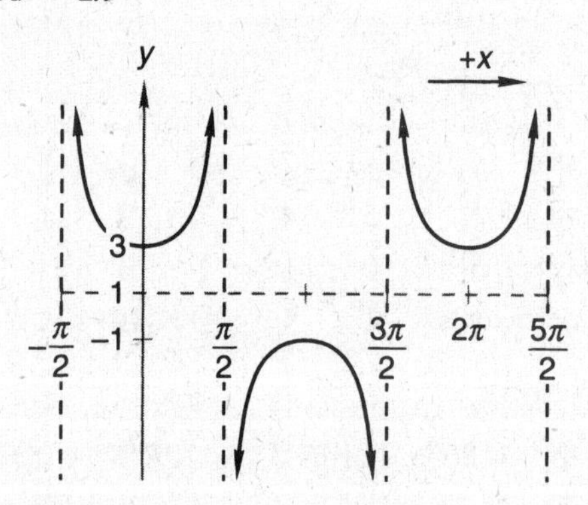

21. (108) $A = \begin{bmatrix} 2 & 3 \\ 1 & 0 \\ 2 & 1 \end{bmatrix}$, $B = \begin{bmatrix} -5 & 2 & 1 \\ 3 & 7 & 0 \end{bmatrix}$

$$A \cdot B = \begin{bmatrix} 2 & 3 \\ 1 & 0 \\ 2 & 1 \end{bmatrix}\begin{bmatrix} -5 & 2 & 1 \\ 3 & 7 & 0 \end{bmatrix}$$

$$= \begin{bmatrix} -10 + 9 & 4 + 21 & 2 + 0 \\ -5 + 0 & 2 + 0 & 1 + 0 \\ -10 + 3 & 4 + 7 & 2 + 0 \end{bmatrix}$$

$$= \begin{bmatrix} \mathbf{-1} & \mathbf{25} & \mathbf{2} \\ \mathbf{-5} & \mathbf{2} & \mathbf{1} \\ \mathbf{-7} & \mathbf{11} & \mathbf{2} \end{bmatrix}$$

22. (103) $\log_2\left(\frac{(x+1)^5\sqrt[3]{y}}{\sqrt{y+2}\,(x+1)^2}\right)$

$$= \log_2 (x+1)^5\sqrt[3]{y} - \log_2 \sqrt{y+2}\,(x+1)^2$$

$$= \log_2 (x+1)^5 + \log_2 \sqrt[3]{y} - \left(\log_2 \sqrt{y+2} + \log_2 (x+1)^2\right)$$

$$= 5\log_2 (x+1) + \frac{1}{3}\log_2 y - \frac{1}{2}\log_2 (y+2) - 2\log_2 (x+1)$$

$$= \mathbf{\frac{1}{3}\log_2 y + 3\log_2 (x+1) - \frac{1}{2}\log_2 (y+2)}$$

23. (103) (a) $\text{pH} = -\log \text{H}^+$

$$= -\log\left(5.2 \times 10^{-8}\right) = 7.2840 = \mathbf{7.28}$$

(b) $\text{H}^+ = 10^{-\text{pH}} = 10^{-10.4} = \mathbf{3.98 \times 10^{-11}\ \frac{mole}{liter}}$

24. (106)

$$5x^2 + 4y^2 + 10x - 16y = -1$$

$$5(x^2 + 2x + 1) + 4(y^2 - 4y + 4) = 20$$

$$5(x+1)^2 + 4(y-2)^2 = 20$$

$$\mathbf{\frac{(x+1)^2}{4} + \frac{(y-2)^2}{5} = 1}$$

Center $= \mathbf{(-1, 2)}$

$a = \sqrt{5}$, $b = 2$

Length of major axis $= \mathbf{2\sqrt{5}}$

Length of minor axis $= \mathbf{4}$

Vertical

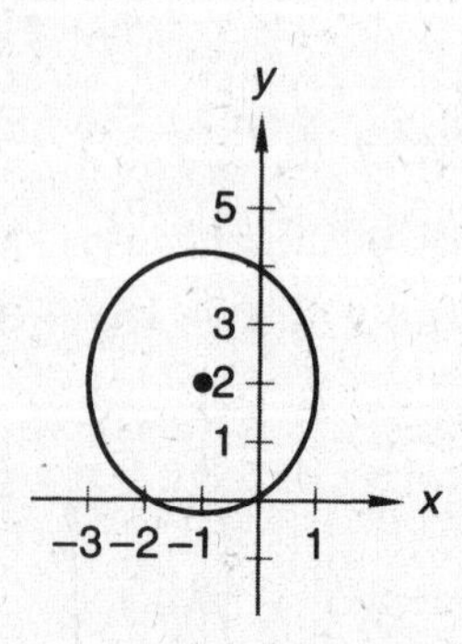

25. (101) First solve the denominator determinant.

$$\begin{vmatrix} 2 & -1 & 1 \\ 0 & 2 & -1 \\ 3 & -2 & 0 \end{vmatrix} = (0 + 3 + 0) - (6 + 4 + 0) = -7$$

$$y = \frac{\begin{vmatrix} 2 & 4 & 1 \\ 0 & -3 & -1 \\ 3 & 5 & 0 \end{vmatrix}}{-7}$$

$$y = \frac{[0 + (-12) + 0] - (-9 - 10 + 0)}{-7}$$

$$y = \frac{-12 - (-19)}{-7}$$

$$\mathbf{y = -1}$$

$$z = \frac{\begin{vmatrix} 2 & -1 & 4 \\ 0 & 2 & -3 \\ 3 & -2 & 5 \end{vmatrix}}{-7}$$

$$z = \frac{(20 + 9 + 0) - (24 + 12 + 0)}{-7}$$

$$z = \frac{29 - 36}{-7}$$

$$\mathbf{z = 1}$$

26. (99)

$$\begin{cases} \text{(a)} & \sqrt{ab} = 15 \\ \text{(b)} & \dfrac{a + b}{2} = 39 \end{cases}$$

$$\text{(b)} \quad \frac{a + b}{2} = 39$$

$$a + b = 78$$

$$a = 78 - b$$

$$\text{(a)} \quad \sqrt{ab} = 15$$

$$\sqrt{(78 - b)(b)} = 15$$

$$(78 - b)(b) = 225$$

$$78b - b^2 = 225$$

$$b^2 - 78b + 225 = 0$$

$$(b - 75)(b - 3) = 0$$

$$b = 75, 3$$

If $b = 75$, $a = 78 - 75 = 3$

If $b = 3$, $a = 78 - 3 = 75$

3 and 75

27. (59)

$$3 \log_2 x + 2 = 2 \log_2 x - 3$$

$$3 \log_2 x - 2 \log_2 x = -3 - 2$$

$$\log_2 x^3 - \log_2 x^2 = -5$$

$$\log_2 \frac{x^3}{x^2} = -5$$

$$\log_2 x = -5$$

$$x = 2^{-5}$$

$$x = \mathbf{\frac{1}{32}}$$

28. (76,80)

$$\frac{\csc x + \cot x}{\tan x + \sin x} = \frac{\dfrac{1}{\sin x} + \dfrac{\cos x}{\sin x}}{\dfrac{\sin x}{\cos x} + \sin x} \cdot \frac{\sin x}{\sin x}$$

$$= \frac{1 + \cos x}{\dfrac{\sin^2 x + \sin^2 x \cos x}{\cos x}} = \frac{1 + \cos x}{\dfrac{\sin^2 x(1 + \cos x)}{\cos x}}$$

$$= \frac{\cos x}{\sin^2 x} = \frac{\cos x}{\sin x} \cdot \frac{1}{\sin x} = \cot x \csc x$$

29. (85)

$$3 \sec^2 x - 2\sqrt{3} \tan x - 6 = 0$$

$$3(\tan^2 x + 1) - 2\sqrt{3} \tan x - 6 = 0$$

$$3 \tan^2 x - 2\sqrt{3} \tan x - 3 = 0$$

$$(\sqrt{3} \tan x + 1)(\sqrt{3} \tan x - 3) = 0$$

$$\tan x = -\frac{1}{\sqrt{3}} \qquad \tan x = \frac{3}{\sqrt{3}}$$

$$x = 150°, 330° \qquad x = \sqrt{3}$$

$$x = 60°, 240°$$

$$x = \mathbf{60°, 150°, 240°, 330°}$$

30. (46) $x^2 + x + 2 = 0$

$$x = \frac{-1 \pm \sqrt{1 - 4(1)(2)}}{2(1)} = \frac{-1 \pm \sqrt{7}i}{2}$$

$$\left(\mathbf{x + \frac{1}{2} + \frac{\sqrt{7}}{2}i}\right)\left(\mathbf{x + \frac{1}{2} - \frac{\sqrt{7}}{2}i}\right)$$

Problem Set 118

1. (83) $\frac{2}{6} \cdot \frac{3}{6} = \mathbf{\frac{1}{6}}$

2. (88) $A_{14} = 400$, $A_{40} = 2800$

$$A_{40} = A_{14}e^{k(40-14)}$$

$$2800 = 400e^{k26}$$

$$7 = e^{26k}$$

$$\ln 7 = 26k$$

$$k = 0.074843$$

$$A_t = A_{14}e^{0.074843(t-14)}$$

$$5000 = 400e^{0.074843(t-14)}$$

$$12.5 = e^{0.074843(t-14)}$$

$$\ln 12.5 = 0.074843(t - 14)$$

$$t - 14 = 33.74716$$

$$t = \mathbf{47.75 \text{ yr}}$$

3. (25)

$$\begin{cases} \text{(a)} & R_W T_W = 400 \\ \text{(b)} & R_B T_B = 1120 \end{cases}$$

$$R_W = R_B - 20,\ T_W = \frac{1}{2}T_B$$

(a)

$$R_W T_W = 400$$
$$(R_B - 20)\left(\frac{1}{2}T_B\right) = 400$$
$$\frac{1}{2}R_B T_B - 10T_B = 400$$
$$\frac{1}{2}(1120) - 10T_B = 400$$
$$10T_B = 160$$
$$\mathbf{T_B = 16\ hr}$$

(b)

$$R_B T_B = 1120$$
$$R_B(16) = 1120\text{ mi}$$
$$\mathbf{R_B = 70\ mph}$$

$$T_W = \frac{1}{2}T_B$$
$$T_W = \frac{1}{2}(16)$$
$$\mathbf{T_W = 8\ hr}$$

$$R_W = R_B - 20$$
$$R_W = (70) - 20$$
$$\mathbf{R_W = 50\ mph}$$

4. (25) $R_1 = \frac{2}{3}\frac{\text{tanks}}{\text{hr}},\ R_2 = \frac{1}{4}\frac{\text{tank}}{\text{hr}}$

$$R_1T_1 + R_2T_2 = \text{tanks}$$
$$\frac{2}{3}T + \frac{1}{4}T = 1$$
$$\frac{11}{12}T = 1$$
$$T = \mathbf{\frac{12}{11}\ hr}$$

5. (18)

$$\begin{cases} \text{(a)} & 10N_F = 2N_S - 180 \\ \text{(b)} & \frac{1}{2}N_S = 3N_F + 40 \end{cases}$$

(b)

$$\frac{1}{2}N_S = 3N_F + 40$$
$$N_S = 6N_F + 80$$

(a)

$$10N_F = 2N_S - 180$$
$$10N_F = 2(6N_F + 80) - 180$$
$$10N_F = 12N_F + 160 - 180$$
$$2N_F = 20$$
$$\mathbf{N_F = 10}$$

$$N_S = 6N_F + 80$$
$$N_S = 6(10) + 80$$
$$\mathbf{N_S = 140}$$

6. (113) $y = x^3 - 2x^2 - 9x + 18$

$x = 3$

$$\begin{array}{r|rrrr} 3 & 1 & -2 & -9 & 18 \\ & & 3 & 3 & -18 \\ \hline & 1 & 1 & -6 & 0 \end{array}$$

$$(x - 3)(x^2 + x - 6) = 0$$
$$(x - 3)(x + 3)(x - 2) = 0$$

The other zeros are: **−3, 2**

7. (118) $4x^3 - 13x + 6 = 0$

$x = -2$

$$\begin{array}{r|rrrr} -2 & 4 & 0 & -13 & 6 \\ & & -8 & 16 & -6 \\ \hline & 4 & -8 & 3 & 0 \end{array}$$

$$(x + 2)(4x^2 - 8x + 3) = 0$$
$$(x + 2)(2x - 3)(2x - 1) = 0$$

The other roots are: $\mathbf{\frac{1}{2}, \frac{3}{2}}$

8. (118) $4x^3 - 4x^2 - 5x + 3 = 0$

Possible rational roots: $\pm 1, \pm 3, \pm\frac{1}{2}, \pm\frac{3}{2}, \pm\frac{1}{4}, \pm\frac{3}{4}$

$$\begin{array}{r|rrrr} -1 & 4 & -4 & -5 & 3 \\ & & -4 & 8 & -3 \\ \hline & 4 & -8 & 3 & 0 \end{array}$$

$$(x + 1)(4x^2 - 8x + 3) = 0$$
$$(x + 1)(2x - 3)(2x - 1) = 0$$

Roots: $\mathbf{-1, \frac{1}{2}, \frac{3}{2}}$

9. (117)

$$96 = 8 \cdot 12$$
$$60 = 5 \cdot 12$$

No, they have the factor of 12 in common.

10. $4x^7 - 8x^3 + 2x - 6 = 0$
(117)

Integral Factors

$$\frac{\{\pm1, \pm2, \pm3, \pm6\}}{\{\pm1, \pm2, \pm4\}}$$

Possible Quotients

$\rightarrow$ $\mathbf{\pm1, \pm2, \pm3, \pm6, \pm\frac{1}{2}, \pm\frac{3}{2}, \pm\frac{1}{4}, \pm\frac{3}{4}}$

11. $y = 4x^3 - 4x^2 - 5x + 3$
(114,116)

$$r = \left|-\frac{5}{4}\right| + 1$$

$$\boldsymbol{r = \frac{9}{4}}$$

x	−2	−1	0	1	2
y	−35	0	3	−2	9

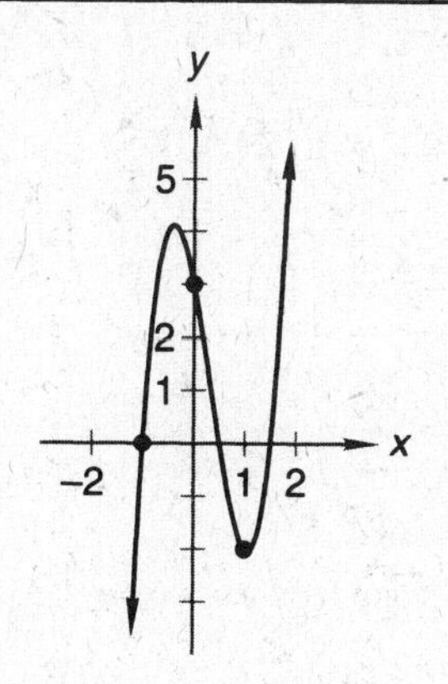

12. $y = x^4 - x^2 + 1$
(114,116)

$r = |1| + 1$

$\boldsymbol{r = 2}$

x	−2	−1	0	1	2
y	13	1	1	1	13

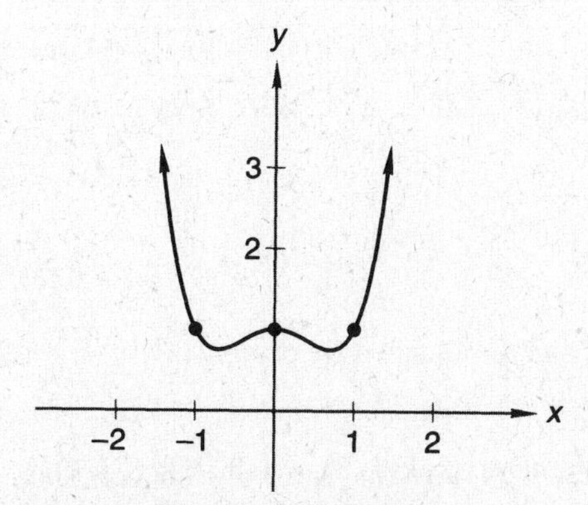

13. $y = (1 - x)(1 + x)(x - 3) = -x^3 + 3x^2 + x - 3$
(114)

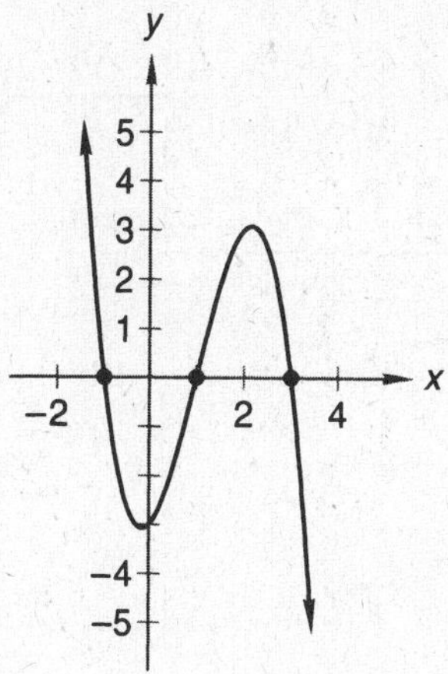

14. $y = x^3 - 2x^2 - 9x + 18$
(114)

$= x^2(x - 2) - 9(x - 2)$

$= (x - 2)(x^2 - 9)$

$= \mathbf{(x - 2)(x - 3)(x + 3)}$

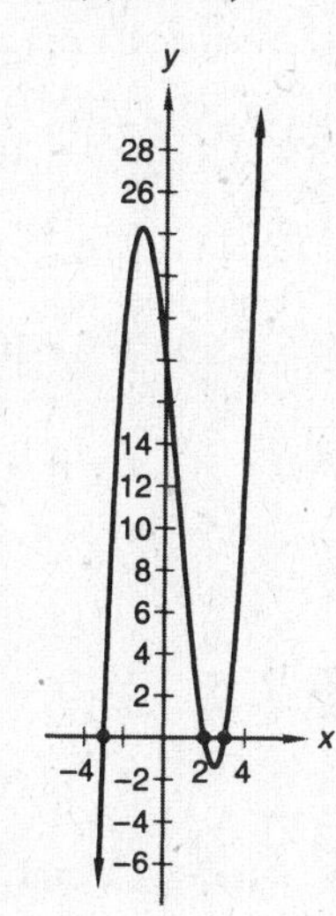

15. $f(x) = x(x - 2)(x + 2)$. Roots are 0, 2, −2.
(114)

The dominant term is x^3 which is positive when x is positive and negative when x is negative. Therefore the answer is **D.**

16. $x^6 - 3x^4 + 2x^2 - 1$
(115)

$x = -1$

$$\begin{array}{r|rrrrrrr} -1 & 1 & 0 & -3 & 0 & 2 & 0 & -1 \\ & & -1 & 1 & 2 & -2 & 0 & 0 \\ \hline & 1 & -1 & -2 & 2 & 0 & 0 & -1 \end{array}$$

$p(-1) = \mathbf{-1}$

17. (113) $(3x^5 - 7x^4 + 9x^3 - 12x + 2) \div (x + 1)$

$$\begin{array}{r|rrrrrr} -1 & 3 & -7 & 9 & 0 & -12 & 2 \\ & & -3 & 10 & -19 & 19 & -7 \\ \hline & 3 & -10 & 19 & -19 & 7 & -5 \end{array}$$

$$\frac{3x^5 - 7x^4 + 9x^3 - 12x + 2}{x + 1}$$

$$= \mathbf{3x^4 - 10x^3 + 19x^2 - 19x + 7 - \frac{5}{x + 1}}$$

18. (112) $(x^3 - y^2)^5$

$F = x^3,\ S = -y^2$

Term	①	②	③	④	⑤	⑥
Exp. of F	5	4	3	2	1	0
Exp. of S	0	1	2	3	4	5
Coefficients	1	5	10	10	5	1

$(F + S)^5$

$= F^5 + 5F^4S + 10F^3S^2 + 10F^2S^3 + 5FS^4 + S^5$

$= (x^3)^5 + 5(x^3)^4(-y^2) + 10(x^3)^3(-y^2)^2$

$\quad + 10(x^3)^2(-y^2)^3 + 5(x^3)(-y^2)^4(-y^2)^5$

$= \mathbf{x^{15} - 5x^{12}y^2 + 10x^9y^4 - 10x^6y^6 + 5x^3y^8 - y^{10}}$

19. (111) $3 \log_{\frac{1}{2}} (x - 2) < 9$

$\log_{\frac{1}{2}} (x - 2)^3 < 9$

$(x - 2)^3 > \left(\frac{1}{2}\right)^9$ and $x - 2 > 0$

$x > 2$

$x - 2 > \left(\frac{1}{2}\right)^3$

$x - 2 > \frac{1}{8}$

$x > \frac{17}{8}$

but $2 < \frac{17}{8}$, so $\mathbf{x > \frac{17}{8}}$

20. (109) $0.42\overline{3} = 0.42 + [0.003 + 0.003(10^{-1})$

$+ 0.003(10^{-2}) + \ldots]$

If $a_1 = 0.003,\ r = 10^{-1}$

$$0.42\overline{3} = 0.42 + \frac{a_1}{1 - r}$$

$$= \frac{42}{100} + \frac{0.003}{1 - 10^{-1}}$$

$$= \frac{42}{100} + \frac{0.003}{0.9} \cdot \frac{10^3}{10^3}$$

$$= \frac{42}{100} + \frac{3}{900} = \frac{126}{300} + \frac{1}{300} = \mathbf{\frac{127}{300}}$$

21. (110) $y = \text{Arcsin}\, x$

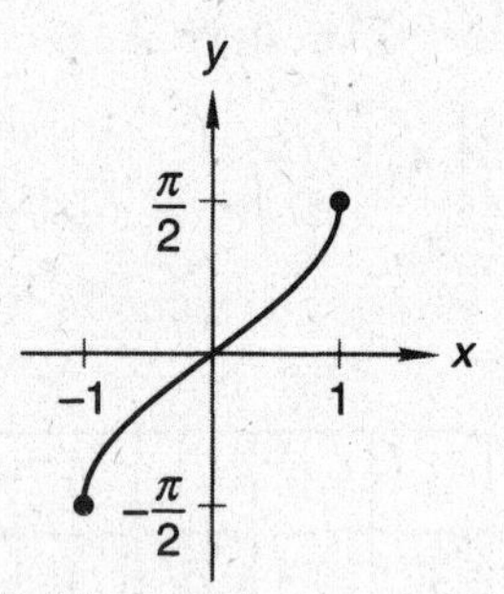

Domain $= \{x \in \mathbb{R} \mid -1 \le x \le 1\}$

Range $= \left\{y \in \mathbb{R} \mid -\frac{\pi}{2} \le y \le \frac{\pi}{2}\right\}$

Arcsin $\frac{\sqrt{2}}{2} = \frac{\pi}{4}$

Arcsin $\left(-\frac{\sqrt{2}}{2}\right) = -\frac{\pi}{4}$

Arcsin 0 = 0

22. (108) $A = \begin{bmatrix} 2 & 1 & 3 \\ 1 & 0 & 1 \\ 2 & 0 & 2 \end{bmatrix},\ B = \begin{bmatrix} 1 & 0 \\ 1 & 0 \\ 1 & 1 \end{bmatrix}$

$$2A \cdot B = \begin{bmatrix} 4 & 2 & 6 \\ 2 & 0 & 2 \\ 4 & 0 & 4 \end{bmatrix} \cdot \begin{bmatrix} 1 & 0 \\ 1 & 0 \\ 1 & 1 \end{bmatrix}$$

$$= \begin{bmatrix} 4 + 2 + 6 & 0 + 0 + 6 \\ 2 + 0 + 2 & 0 + 0 + 2 \\ 4 + 0 + 4 & 0 + 0 + 4 \end{bmatrix} = \mathbf{\begin{bmatrix} 12 & 6 \\ 4 & 2 \\ 8 & 4 \end{bmatrix}}$$

23. (106)

$$25y^2 + 100y - 9x^2 - 54x = 206$$

$$25(y^2 + 4y + 4) - 9(x^2 + 6x + 9) = 225$$

$$25(y + 2)^2 - 9(x + 3)^2 = 225$$

$$\mathbf{\frac{(y + 2)^2}{9} - \frac{(x + 3)^2}{25} = 1}$$

From the equation we obtain:

$h = -3,\ k = -2,\ a = 5,\ b = 3$

Vertices: $(h, k \pm b) = \mathbf{(-3, -5), (-3, 1)}$

$$m = \pm\frac{b}{a} = \pm\frac{3}{5}$$

$$(y - k) = m(x - h) \qquad (y - k) = m(x - h)$$

$$y + 2 = \frac{3}{5}(x + 3) \qquad y + 2 = -\frac{3}{5}(x + 3)$$

$$y = \frac{3}{5}x - \frac{1}{5} \qquad y = -\frac{3}{5}x - \frac{19}{5}$$

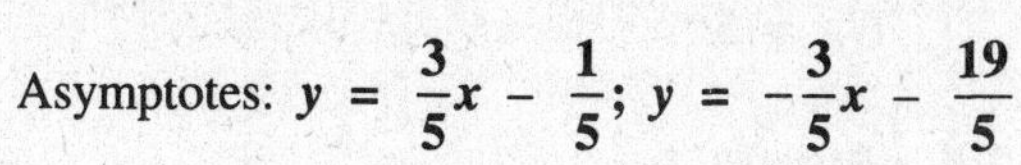

Asymptotes: $\mathbf{y = \frac{3}{5}x - \frac{1}{5}};\ \mathbf{y = -\frac{3}{5}x - \frac{19}{5}}$

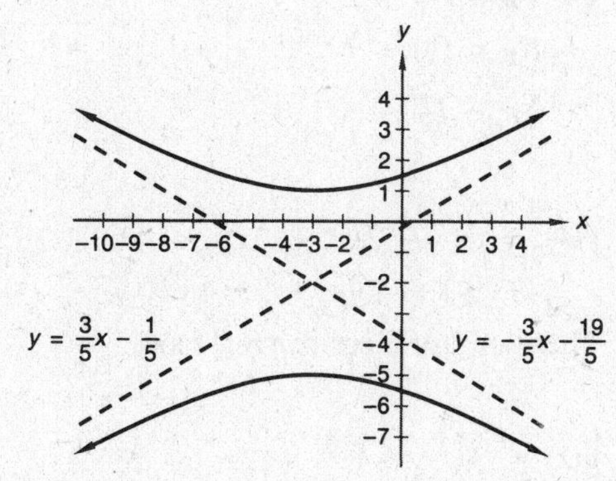

24. (112) $\left(2x^2 - \frac{1}{x}\right)^7$

$$F = 2x^2,\ S = -\frac{1}{x},\ n = 7,\ k = 3$$

$$\frac{n!}{(n - k + 1)!(k - 1)!}F^{n-k+1}S^{k-1}$$

$$= \frac{7!}{(7 - 3 + 1)!(3 - 1)!}F^{7-3+1}S^{3-1}$$

$$= \frac{7!}{5!2!}F^5S^2 = \frac{7!}{5!2!}(2x^2)^5\left(-\frac{1}{x}\right)^2$$

$$= 21(32x^{10})\left(\frac{1}{x^2}\right) = \mathbf{672x^8}$$

25. (103) $\log_3\left(\frac{(x + 1)^2\sqrt[3]{x + 2}}{x - 3}\right)$

$$= \log_3 (x + 1)^2\sqrt[3]{x + 2} - \log_3 (x - 3)$$

$$= \log_3 (x + 1)^2 + \log_3 \sqrt[3]{x + 2} - \log_3 (x - 3)$$

$$= \mathbf{2\log_3 (x + 1) + \frac{1}{3}\log_3 (x + 2) - \log_3 (x - 3)}$$

26. (98) $\log_2(\log_2 x) = 4$

$$\log_2 x = 2^4$$

$$\log_2 x = 16$$

$$x = \mathbf{2^{16}}$$

27. (82) $4^{x+2} = 2^{3x-1}$

$$(2^2)^{x+2} = 2^{3x-1}$$

$$2^{2x+4} = 2^{3x-1}$$

$$2x + 4 = 3x - 1$$

$$x = \mathbf{5}$$

28. (90,96) $\cos 2x + \sin 2x + 2\sin^2 x$

$$= \cos^2 x - \sin^2 x + 2\sin x\cos x + 2\sin^2 x$$

$$= \cos^2 x + 2\sin x\cos x + \sin^2 x$$

$$= (\sin x + \cos x)^2$$

29. (94) $y = -3 + 5\csc x$

Period $= 2\pi$

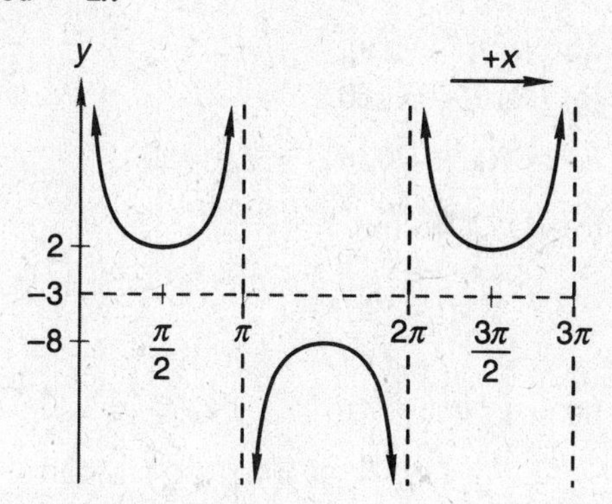

30. (85) $2\cot^2 x + 3\csc x = 0$

$$2(\csc^2 x - 1) + 3\csc x = 0$$

$$2\csc^2 x + 3\csc x - 2 = 0$$

$$(2\csc x - 1)(\csc x + 2) = 0$$

$$\csc x = \frac{1}{2} \qquad \csc x = -2$$

no answer $\qquad x = \mathbf{\frac{7\pi}{6}, \frac{11\pi}{6}}$

Problem Set 119

1. (107) Distance fallen:

$$100 + 100\left(\frac{1}{5}\right) + 100\left(\frac{1}{5}\right)^2 + \ldots$$

$$S = \frac{a_1}{1 - r} = \frac{100}{1 - \frac{1}{5}} = \frac{100}{\frac{4}{5}} = 125 \text{ ft}$$

Distance traveled

= Distance fallen + Distance rebounded

= 125 ft + (125 ft − 100 ft)

= **150 ft**

2. (92)
$$P(A \cup R) = P(A) + P(R) - P(A \cap R)$$
$$= \frac{4}{52} + \frac{26}{52} - \frac{2}{52} = \frac{28}{52} = \mathbf{\frac{7}{13}}$$

3. (88) $A_5 = 800,\ A_7 = 500$

$$A_7 = A_5e^{(7-5)k}$$
$$500 = 800e^{2k}$$
$$\frac{500}{800} = e^{2k}$$
$$\ln\frac{5}{8} = 2k$$
$$k = -0.23500$$
$$A_{10} = A_5e^{-0.23500(10-5)} = 800e^{-0.23500(5)} = \mathbf{247.05}$$

4. (25)
$$\begin{cases} \text{(a)}\ R_CT_C = 240 \\ \text{(b)}\ R_WT_W = 160 \end{cases}$$

$R_C = R_W + 20,\ T_C = T_W - 2$

(b) $R_WT_W = 160$
$$R_W = \frac{160}{T_W}$$

(a)
$$R_CT_C = 240$$
$$(R_W + 20)(T_W - 2) = 240$$
$$R_WT_W - 2R_W + 20T_W - 40 = 240$$
$$160 - 2\left(\frac{160}{T_W}\right) + 20T_W = 280$$
$$-\frac{320}{T_W} + 20T_W = 120$$
$$-320 + 20(T_W)^2 = 120T_W$$
$$20(T_W)^2 - 120T_W - 320 = 0$$
$$(T_W)^2 - 6T_W - 16 = 0$$
$$(T_W - 8)(T_W + 2) = 0$$
$$\mathbf{T_W = 8\ hr}$$

(b) $R_WT_W = 160$
$$R_W(8) = 160$$
$$\mathbf{R_W = 20\ mph}$$

$T_C = T_W - 2$

$T_C = 8 - 2$

$\mathbf{T_C = 6\ hr}$

(a) $R_CT_C = 240$
$$R_C(6) = 240$$
$$\mathbf{R_C = 40\ mph}$$

5. (18)
$$\begin{cases} \text{(a)}\ 10N_R = 2N_B + 26 \\ \text{(b)}\ N_B = N_R + 7 \end{cases}$$

(a) $10N_R = 2N_B + 26$
$$10N_R = 2(N_R + 7) + 26$$
$$10N_R = 2N_R + 14 + 26$$
$$8N_R = 40$$
$$\mathbf{N_R = 5}$$

(b) $N_B = N_R + 7$
$$N_B = (5) + 7$$
$$\mathbf{N_B = 12}$$

6. (119) $p(x) = x^4 - 2x^3 + 3x^2 + 4x - 7$

+ Ⓒ − Ⓒ + + Ⓒ −

There are **1 or 3** positive real roots.

7. (119) $p(x) = 3x^4 - 4x^3 + 2x^2 + 5x - 2$

$p(-x) = 3x^4 + 4x^3 + 2x^2 - 5x - 2$

+ + + Ⓒ − −

There is **1** negative root.

8. (119) $p(x) = 6x^5 - 17x^4 - 3x^3 + 22x^2 - 7x + 20$

+ Ⓒ − − Ⓒ + Ⓒ − Ⓒ +

There are **0, 2, or 4** positive real roots.

9. (119) $p(x) = 2x^5 - 3x^4 - 5x^3 + 11x^2 - 4x + 5$

$p(-x) = -2x^5 - 3x^4 + 5x^3 + 11x^2 + 4x + 5$

− − Ⓒ + + + +

There is **1** negative real root.

10. (119) $x^3 - 3x + 19 = 0$

1⌋	1	0	−3	19
		1	1	−2
	1	1	−2	17

2⌋	1	0	−3	19
		2	4	2
	1	2	1	21

Upper bound = 2

−1⌋	1	0	−3	19
		−1	1	2
	1	−1	−2	21

−2⌋	1	0	−3	19
		−2	4	−2
	1	−2	1	17

−3⌋	1	0	−3	19
		−3	9	−18
	1	−3	6	1

−4⌋	1	0	−3	19
		−4	16	−52
	1	−4	13	−33

Lower bound = −4

11. $x^4 - 2x^3 - 7x + 5 = 0$
(119)

$$\begin{array}{r|rrrrr} 1 & 1 & -2 & 0 & -7 & 5 \\ & & 1 & -1 & -1 & -8 \\ \hline & 1 & -1 & -1 & -8 & -3 \end{array} \qquad \begin{array}{r|rrrrr} 2 & 1 & -2 & 0 & -7 & 5 \\ & & 2 & 0 & 0 & -14 \\ \hline & 1 & 0 & 0 & -7 & -9 \end{array}$$

$$\begin{array}{r|rrrrr} 3 & 1 & -2 & 0 & -7 & 5 \\ & & 3 & 3 & 9 & 6 \\ \hline & 1 & 1 & 3 & 2 & 11 \end{array}$$

Upper bound = 3

$$\begin{array}{r|rrrrr} -1 & 1 & -2 & 0 & -7 & 5 \\ & & -1 & 3 & -3 & 10 \\ \hline & 1 & -3 & 3 & -10 & 15 \end{array}$$

Lower bound = −1

12. $x^3 + 2x^2 - x - 2 = 0$
(118)

Possible rational roots: $\pm 1, \pm 2$

$$\begin{array}{r|rrrr} 1 & 1 & 2 & -1 & -2 \\ & & 1 & 3 & 2 \\ \hline & 1 & 3 & 2 & 0 \end{array}$$

$(x - 1)(x^2 + 3x + 2) = 0$

$(x - 1)(x + 1)(x + 2) = 0$

Roots: **−2, −1, 1**

13. $x^3 - 3x^2 + 4x - 12 = 0$
(118)

Possible rational roots: $\pm 1, \pm 2, \pm 3, \pm 4, \pm 6, \pm 12$

$$\begin{array}{r|rrrr} 3 & 1 & -3 & 4 & -12 \\ & & 3 & 0 & 12 \\ \hline & 1 & 0 & 4 & 0 \end{array}$$

$(x - 3)(x^2 + 4) = 0$

For $x^2 + 4$ use the quadratic formula:

$$x = \frac{0 \pm \sqrt{0 - 16}}{2} = \frac{\pm\sqrt{-16}}{2} = \frac{\pm 4i}{2} = \pm 2i$$

Roots: **$3, 2i, -2i$**

14. $y = x^3 - 6x^2 + 13x - 10$
(113)

$x = 2$

$$\begin{array}{r|rrrr} 2 & 1 & -6 & 13 & -10 \\ & & 2 & -8 & 10 \\ \hline & 1 & -4 & 5 & 0 \end{array}$$

$(x - 2)(x^2 - 4x + 5) = 0$

For $x^2 - 4x + 5$ use the quadratic formula:

$$x = \frac{4 \pm \sqrt{16 - 20}}{2} = \frac{4 \pm \sqrt{-4}}{2} = 2 \pm i$$

The other zeros are: **$2 + i, 2 - i$**

15. $y = 3x^3 - x^2 + 2x - 4$
(113)

$x = 1$

$$\begin{array}{r|rrrr} 1 & 3 & -1 & 2 & -4 \\ & & 3 & 2 & 4 \\ \hline & 3 & 2 & 4 & 0 \end{array}$$

$(x - 1)(3x^2 + 2x + 4) = 0$

For $3x^2 + 2x + 4$ use the quadratic formula:

$$x = \frac{-2 \pm \sqrt{4 - 4(3)(4)}}{2(3)} = \frac{-2 \pm \sqrt{-44}}{6}$$

$$= \frac{-2 \pm 2\sqrt{11}i}{6} = -\frac{1}{3} \pm \frac{\sqrt{11}}{3}i$$

The other zeros are: $\mathbf{-\frac{1}{3} + \frac{\sqrt{11}}{3}i, -\frac{1}{3} - \frac{\sqrt{11}}{3}i}$

16. $8x^7 - 5x^3 + 3 = 0$
(117)

INTEGRAL FACTORS

$$\frac{\{1, -1, 3, -3\}}{\{1, -1, 2, -2, 4, -4, 8, -8\}}$$

POSSIBLE QUOTIENTS

→ $\mathbf{\pm 1, \pm 3, \pm\frac{1}{2}, \pm\frac{3}{2}, \pm\frac{1}{4}, \pm\frac{3}{4}, \pm\frac{1}{8}, \pm\frac{3}{8}}$

17. $102 = 2 \cdot 3 \cdot 17$
(117)

$35 = 5 \cdot 7$

Yes, they are relatively prime.

18. $y = x^3 + 2x^2 - x - 2$
(114,116)

$$r = \left|\frac{2}{1}\right| + 1$$

$\mathbf{r = 3}$

x	−2	−1	0	1
y	0	0	−2	0

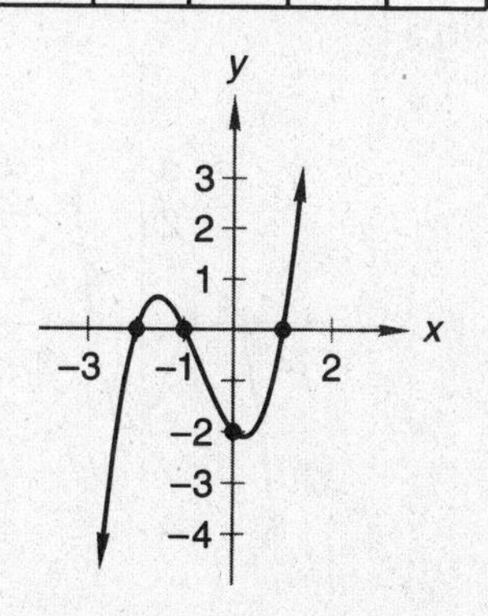

19. $y = -x^4 + x^2 + 2$
(114,116)

$r = \left|\frac{2}{1}\right| + 1$

$\boldsymbol{r = 3}$

x	-1	0	1
y	2	2	2

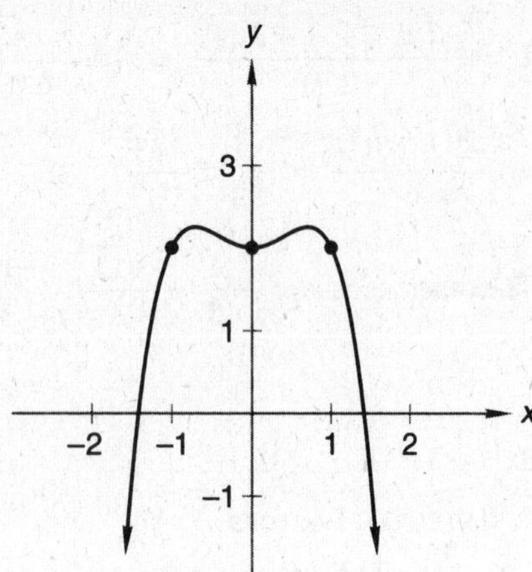

20. $y = (x - 2)(x + 1)^2 = x^3 - 3x - 2$
(114)

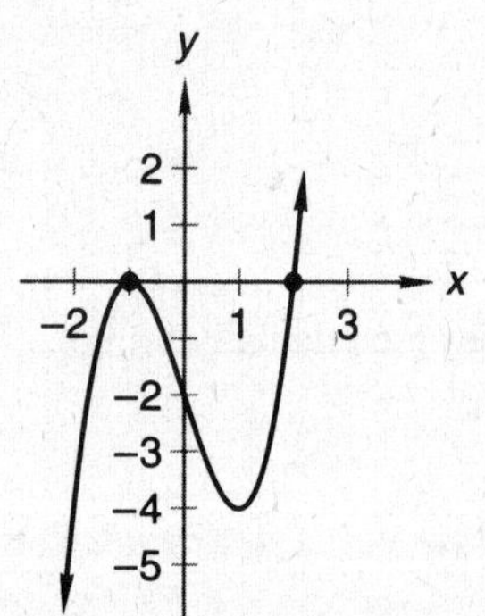

21. $y = x^3 + 2x^2 - x - 2$
(114)

$y = (x + 2)(x^2 - 1)$

$\boldsymbol{y = (x + 2)(x + 1)(x - 1)}$

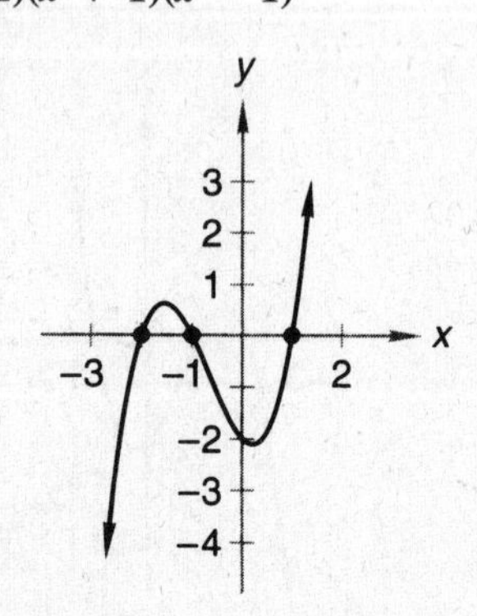

22. $x^5 - 3x$
(115)

$x = -2$

$$\begin{array}{r|rrrrrr} -2 & 1 & 0 & 0 & 0 & -3 & 0 \\ & & -2 & 4 & -8 & 16 & -26 \\ \hline & 1 & -2 & 4 & -8 & 13 & -26 \end{array}$$

$p(-2) = \boldsymbol{-26}$

23. $\left(x + \frac{1}{x}\right)^3$
(77,102)

$F = x,\ S = \frac{1}{x}$

Term	①	②	③	④
Exp. of F	3	2	1	0
Exp. of S	0	1	2	3
Coefficients	1	3	3	1

$(F + S)^3$

$= F^3 + 3F^2S + 3FS^2 + S^3$

$= (x^3) + 3(x)^2\left(\frac{1}{x}\right) + 3(x)\left(\frac{1}{x}\right)^2 + \left(\frac{1}{x}\right)^3$

$\boldsymbol{= x^3 + 3x + \frac{3}{x} + \frac{1}{x^3}}$

24.
(59,98)

$$x^{2\ln x} = x$$
$$\ln x^{2\ln x} = \ln x$$
$$2\ln x \ln x = \ln x$$
$$2\ln x \ln x - \ln x = 0$$
$$\ln x(2\ln x - 1) = 0$$

$\ln x = 0$ $\qquad$ $\ln x = \frac{1}{2}$

$x = e^0$

$x = 1$ $\qquad$ $x = e^{\frac{1}{2}}$

$\boldsymbol{x = 1,\ e^{\frac{1}{2}}}$

25. $y = 1 + \frac{1}{2}\tan x$
(94)

Period $= \pi$

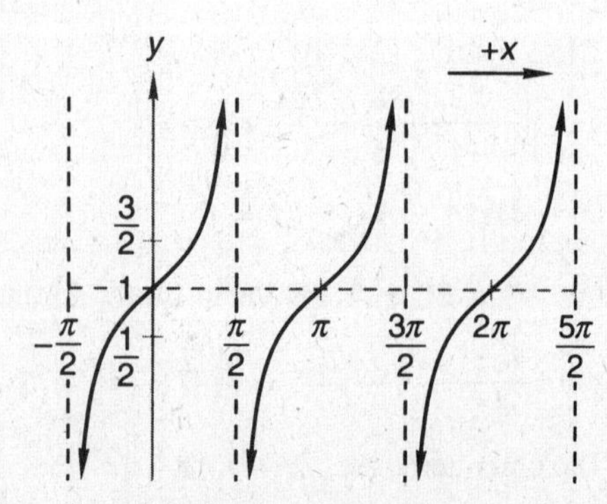

26. (93,96) $\cos^2\theta\left(\frac{1-\cos 2\theta}{1+\cos 2\theta}\right) = \cos^2\theta\,\frac{(1 - 2\cos^2\theta + 1)}{(1 + 2\cos^2\theta - 1)}$

$= \cos^2\theta\,\frac{(2 - 2\cos^2\theta)}{2\cos^2\theta} = \frac{2(1 - \cos^2\theta)}{2}$

$= 1 - \cos^2\theta = \sin^2\theta$

27. (85)

$$\sec^2 x + \tan^2 x = 3$$
$$(1 + \tan^2 x) + \tan^2 x = 3$$
$$1 + 2\tan^2 x = 3$$
$$2\tan^2 x = 2$$
$$\tan^2 x = 1$$
$$\tan x = \pm 1$$
$$x = \mathbf{45°,\ 135°,\ 225°,\ 315°}$$

28. (111) $\log(x - 2) > 2$

$x - 2 > 10^2$ and $x - 2 > 0$

$x > 102$ $\quad$ $x > 2$

but $2 < 102$, so $\boldsymbol{x > 102}$

29. (95) $-i = 1 \text{ cis } 270°$

$(1 \text{ cis } 270°)^{\frac{1}{4}} = 1^{\frac{1}{4}} \text{cis}\left(\frac{270°}{4} + n\frac{360°}{4}\right)$

$=$ **cis 67.5°, cis 157.5°, cis 247.5°, cis 337.5°**

30. (98)

$$\log_6 72 = y$$
$$6^y = 72$$
$$y \log 6 = \log 72$$
$$y = \mathbf{\frac{\log 72}{\log 6}}$$

Problem Set 120

1. (83) $P(\text{not } D) = \frac{39}{52} \cdot \frac{38}{51} = \mathbf{\frac{19}{34}}$

2. (88) $A_0 = 10,\ A_2 = 5$

$$A_t = A_0 e^{kt}$$
$$5 = 10e^{2k}$$
$$\frac{1}{2} = e^{2k}$$
$$\ln\frac{1}{2} = 2k$$
$$k = -0.34657$$

$$A_t = A_0 e^{kt}$$
$$2 = 10e^{-0.34657t}$$
$$0.2 = e^{-0.34657t}$$
$$\ln 0.2 = -0.34657t$$
$$t = \mathbf{4.64\ days}$$

3. (45)

W	W	W	$= 10 \cdot 9 \cdot 8 = 720$
W	M	W	$= 10 \cdot 8 \cdot 9 = 720$
W	W	M	$= 10 \cdot 9 \cdot 8 = 720$
W	M	M	$= 10 \cdot 8 \cdot 7 = 560$
			2720

4. (44) $R_H = \frac{1}{H_1}\frac{\text{ditches}}{\text{hr}},\ R_A = \frac{1}{H_2}\frac{\text{ditches}}{\text{hr}}$

$$R_H T_H + R_A T_A = 1 \text{ ditch}$$
$$\frac{1}{H_1}(1 + T) + \frac{1}{H_2}(T) = 1$$
$$\frac{1}{H_1} + \frac{1}{H_1}T + \frac{1}{H_2}T = 1$$
$$T\left(\frac{1}{H_1} + \frac{1}{H_2}\right) = 1 - \frac{1}{H_1}$$
$$T\left(\frac{H_1 + H_2}{H_1 H_2}\right) = \frac{H_1 - 1}{H_1}$$
$$T = \left(\frac{H_1 - 1}{H_1}\right)\left(\frac{H_1 H_2}{H_1 + H_2}\right)$$
$$= \frac{H_1 H_2 (H_1 - 1)}{H_1(H_1 + H_2)}$$
$$= \mathbf{\frac{H_1 H_2 - H_2}{H_1 + H_2}\ hr}$$

5. (120)

Original Matrix	Inverse Matrix
$\begin{bmatrix} a & b \\ c & d \end{bmatrix}$	$\begin{bmatrix} \frac{d}{ad - cb} & \frac{-b}{ad - cb} \\ \frac{-c}{ad - cb} & \frac{a}{ad - cb} \end{bmatrix}$
$\begin{bmatrix} 1 & 2 \\ 3 & 4 \end{bmatrix}$	$\begin{bmatrix} \frac{4}{(1)(4) - (3)(2)} & \frac{-2}{(1)(4) - (3)(2)} \\ \frac{-3}{(1)(4) - (3)(2)} & \frac{1}{(1)(4) - (3)(2)} \end{bmatrix}$

$$= \begin{bmatrix} \mathbf{-2} & \mathbf{1} \\ \mathbf{\frac{3}{2}} & \mathbf{-\frac{1}{2}} \end{bmatrix}$$

6. (120) ORIGINAL MATRIX

$$\begin{bmatrix} a & b \\ c & d \end{bmatrix}$$

$$\begin{bmatrix} -3 & 2 \\ -2 & -4 \end{bmatrix}$$

INVERSE MATRIX

$$\begin{bmatrix} \frac{d}{ad - cb} & \frac{-b}{ad - cb} \\ \frac{-c}{ad - cb} & \frac{a}{ad - cb} \end{bmatrix}$$

$$\begin{bmatrix} \frac{-4}{(-3)(-4) - (-2)(2)} & \frac{-2}{(-3)(-4) - (-2)(2)} \\ \frac{-(-2)}{(-3)(-4) - (-2)(2)} & \frac{-3}{(-3)(-4) - (-2)(2)} \end{bmatrix}$$

$$= \begin{bmatrix} \mathbf{-\frac{1}{4}} & \mathbf{-\frac{1}{8}} \\ \mathbf{\frac{1}{8}} & \mathbf{-\frac{3}{16}} \end{bmatrix}$$

7. (120) $\begin{bmatrix} 1 & 2 \\ 3 & 4 \end{bmatrix} \cdot \begin{bmatrix} x \\ y \end{bmatrix} = \begin{bmatrix} 4 \\ 6 \end{bmatrix}$

ORIGINAL MATRIX

$$\begin{bmatrix} 1 & 2 \\ 3 & 4 \end{bmatrix}$$

INVERSE MATRIX

$$\frac{1}{(1)(4) - (2)(3)} \begin{bmatrix} 4 & -2 \\ -3 & 1 \end{bmatrix}$$

$$= -\frac{1}{2} \begin{bmatrix} 4 & -2 \\ -3 & 1 \end{bmatrix} = \begin{bmatrix} -2 & 1 \\ 1.5 & -0.5 \end{bmatrix}$$

$$\begin{bmatrix} -2 & 1 \\ 1.5 & -0.5 \end{bmatrix} \cdot \begin{bmatrix} 1 & 2 \\ 3 & 4 \end{bmatrix} \cdot \begin{bmatrix} x \\ y \end{bmatrix} = \begin{bmatrix} -2 & 1 \\ 1.5 & -0.5 \end{bmatrix} \cdot \begin{bmatrix} 4 \\ 6 \end{bmatrix}$$

$$\begin{bmatrix} 1 & 0 \\ 0 & 1 \end{bmatrix} \cdot \begin{bmatrix} x \\ y \end{bmatrix} = \begin{bmatrix} -2 & 1 \\ 1.5 & -0.5 \end{bmatrix} \cdot \begin{bmatrix} 4 \\ 6 \end{bmatrix}$$

$$\begin{bmatrix} x \\ y \end{bmatrix} = \begin{bmatrix} -8 + 6 \\ 6 - 3 \end{bmatrix}$$

$$\begin{bmatrix} x \\ y \end{bmatrix} = \begin{bmatrix} -2 \\ 3 \end{bmatrix}$$

$\mathbf{x = -2;\ y = 3}$

8. (120) $\begin{bmatrix} -3 & 2 \\ -2 & -4 \end{bmatrix} \cdot \begin{bmatrix} x \\ y \end{bmatrix} = \begin{bmatrix} -11 \\ 14 \end{bmatrix}$

ORIGINAL MATRIX

$$\begin{bmatrix} -3 & 2 \\ -2 & -4 \end{bmatrix}$$

INVERSE MATRIX

$$\frac{1}{(-3)(-4) - (2)(-2)} \begin{bmatrix} -4 & -2 \\ 2 & -3 \end{bmatrix}$$

$$= \frac{1}{16} \begin{bmatrix} -4 & -2 \\ 2 & -3 \end{bmatrix} = \begin{bmatrix} -0.25 & -0.125 \\ 0.125 & -0.1875 \end{bmatrix}$$

$$\begin{bmatrix} -0.25 & -0.125 \\ 0.125 & -0.1875 \end{bmatrix} \cdot \begin{bmatrix} -3 & 2 \\ -2 & -4 \end{bmatrix} \cdot \begin{bmatrix} x \\ y \end{bmatrix} = \begin{bmatrix} -0.25 & -0.125 \\ 0.125 & -0.1875 \end{bmatrix} \cdot \begin{bmatrix} -11 \\ 14 \end{bmatrix}$$

$$\begin{bmatrix} 1 & 0 \\ 0 & 1 \end{bmatrix} \cdot \begin{bmatrix} x \\ y \end{bmatrix} = \begin{bmatrix} -0.25 & -0.125 \\ 0.125 & -0.1875 \end{bmatrix} \cdot \begin{bmatrix} -11 \\ 14 \end{bmatrix}$$

$$\begin{bmatrix} x \\ y \end{bmatrix} = \begin{bmatrix} 2.75 - 1.75 \\ -1.375 - 2.625 \end{bmatrix}$$

$$\begin{bmatrix} x \\ y \end{bmatrix} = \begin{bmatrix} 1 \\ -4 \end{bmatrix}$$

$\mathbf{x = 1;\ y = -4}$

9. (119) (a) $p(x) = x^4 + x^3 - 3x^2 - x + 2$

+ + Ⓒ − − Ⓒ +

There are **0 or 2** positive real roots.

(b) $p(x) = x^4 + x^3 - 3x^2 - x + 2$

$p(-x) = x^4 - x^3 - 3x^2 + x + 2$

+ Ⓒ − − Ⓒ + +

There are **0 or 2** negative real roots.

10. (119) (a) $p(x) = 3x^3 - 4x^2 - 2x - 3$

+ Ⓒ − − −

There is **1** positive real root.

(b) $p(x) = 3x^3 - 4x^2 - 2x - 3$

$p(-x) = -3x^3 - 4x^2 + 2x - 3$

− − Ⓒ + Ⓒ −

There are **0 or 2** negative real roots.

11. (119) $x^4 + x^3 - 3x^2 - x + 2 = 0$

```
1| 1   1  -3  -1   2        2| 1   1  -3  -1   2
       1   2  -1  -2               2   6   6  10
   ------------------          ------------------
   1   2  -1  -2   0           1   3   3   5  12
```

Upper bound = 2

```
-1| 1   1  -3  -1   2       -2| 1   1  -3  -1   2
       -1   0   3  -2             -2   2   2  -2
    ------------------          ------------------
    1   0  -3   2   0           1  -1  -1   1   0

-3| 1   1  -3  -1   2
       -3   6  -9  30
    ------------------
    1  -2   3 -10  32
```

Lower bound = −3

12. $x^4 + x^3 - 3x^2 - x + 2 = 0$
(118)

Possible rational roots: ±1, ±2

We use synthetic division to see that 1 is a root.

$$\begin{array}{r|rrrrr} 1 & 1 & 1 & -3 & -1 & 2 \\ & & 1 & 2 & -1 & -2 \\ \hline & 1 & 2 & -1 & -2 & 0 \end{array}$$

$(x - 1)(x^3 + 2x^2 - x - 2) = 0$

$$\begin{array}{r|rrrr} -1 & 1 & 2 & -1 & -2 \\ & & -1 & -1 & 2 \\ \hline & 1 & 1 & -2 & 0 \end{array}$$

$(x - 1)(x + 1)(x^2 + x - 2) = 0$

$(x - 1)(x + 1)(x + 2)(x - 1) = 0$

Roots: **−2, −1, 1, 1**

13. $y = x^3 - x^2 - 4x + 4$
(113)

$$\begin{array}{r|rrrr} 2 & 1 & -1 & -4 & 4 \\ & & 2 & 2 & -4 \\ \hline & 1 & 1 & -2 & 0 \end{array}$$

$(x - 2)(x^2 + x - 2) = 0$

$(x - 2)(x + 2)(x - 1) = 0$

The other zeros are: **−2 and 1**

14. $y = -x^4 - x^3 + 3x^2 + x - 2$
(114,116)

$r = |3| + 1$

$\boldsymbol{r = 4}$

x	−2	−1	0	1	2
y	0	0	−2	0	−12

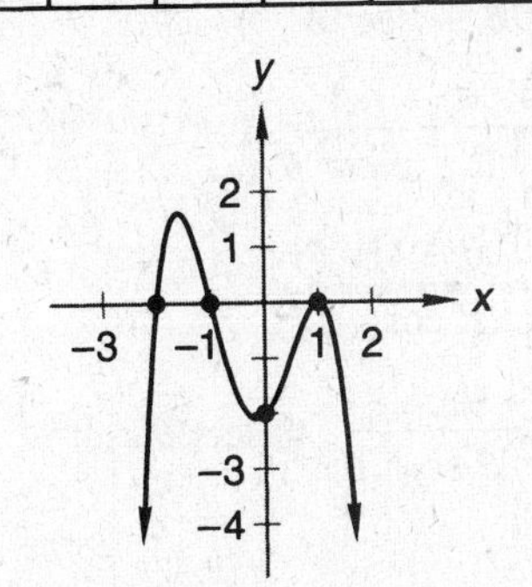

15. $y = x^2(x - 1)(x + 2)$
(114)

$= x^2(x^2 + x - 2)$

$= x^4 + x^3 - 2x^2$

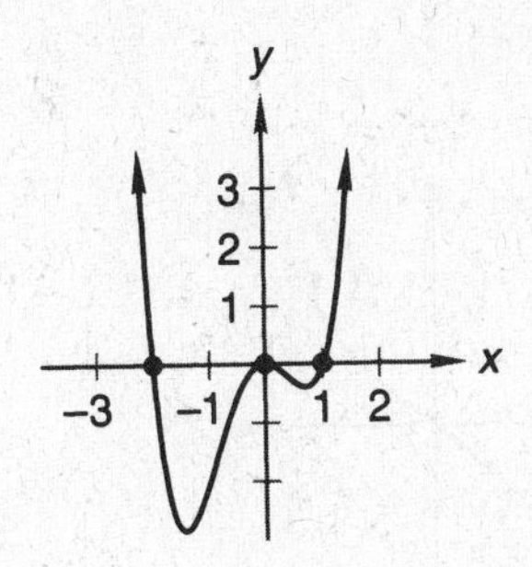

16. $(3xy - 2a^4)^4$
(112)

$F = 3xy,\ S = -2a^4,\ n = 4,\ k = 3$

$$\frac{n!}{(n - k + 1)!(k - 1)!}F^{n-k+1}S^{k-1}$$

$$\frac{4!}{(4 - 3 + 1)!\,(3 - 1)!}F^{4-3+1}S^{3-1} = \frac{4!}{2!2!}F^2S^2$$

$$= 6(3xy)^2(-2a^4)^2 = 6(9x^2y^2)(4a^8) = \mathbf{216x^2y^2a^8}$$

17. $2\log_3 x > \log_3 x$
(111)

$\log_3 x^2 > \log_3 x$

$x^2 > x$

$\boldsymbol{x > 1}$

18. (a) $f(x) = \text{Arccos}\,x$
(110)

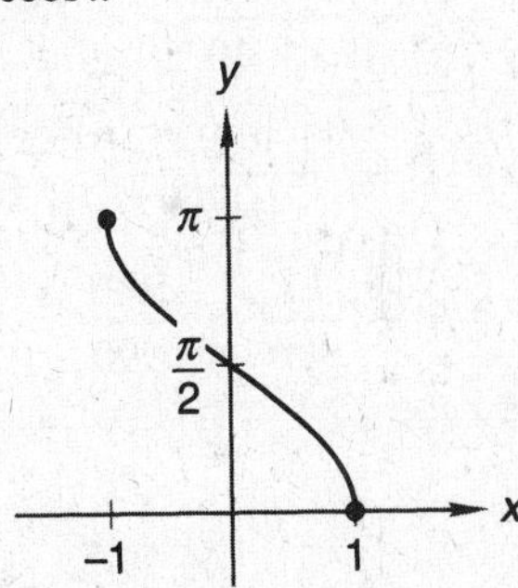

Domain = $\{x \in \mathbb{R} \mid -1 \le x \le 1\}$

Range = $\{y \in \mathbb{R} \mid 0 \le y \le \pi\}$

(b) $g(x) = \text{Arctan}\,x$

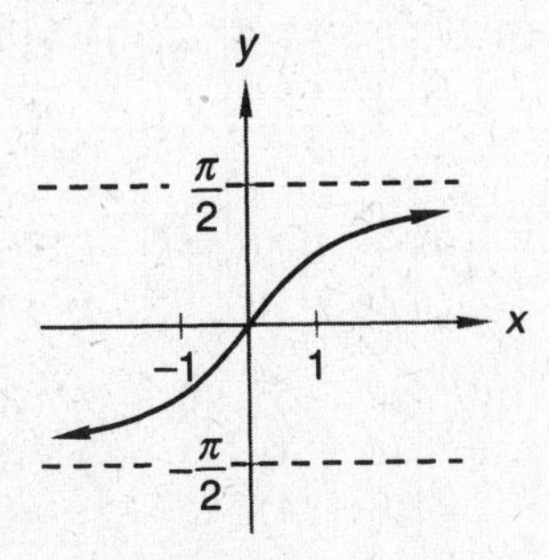

Domain = $\{x \in \mathbb{R}\}$

Range = $\left\{y \in \mathbb{R} \mid -\frac{\pi}{2} < y < \frac{\pi}{2}\right\}$

(c) $h(x) = \text{Arcsin } x$

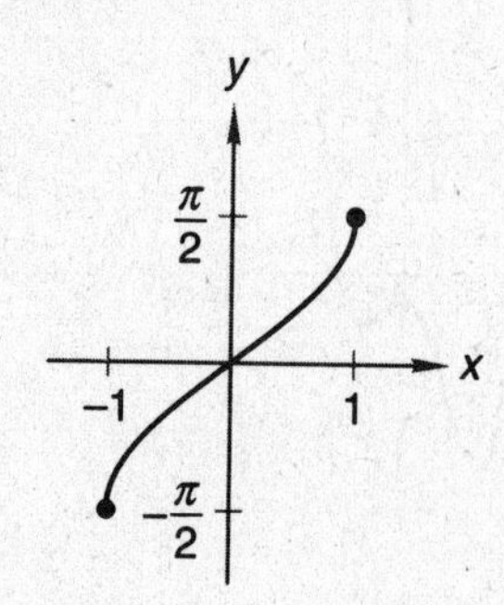

Domain $= \{x \in \mathbb{R} \mid -1 \le x \le 1\}$

Range $= \left\{y \in \mathbb{R} \mid -\frac{\pi}{2} \le y \le \frac{\pi}{2}\right\}$

19. (109) $2.0\overline{11} =$

$$2 + \left[11 \times 10^{-3} + (11 \times 10^{-3})(10^{-2}) + (11 \times 10^{-3})(10^{-2})^2 + \ldots\right]$$

If $a_1 = 11 \times 10^{-3}$, $r = 10^{-2}$

$$2.0\overline{11} = 2 + \frac{a_1}{1 - r}$$
$$= 2 + \frac{11 \times 10^{-3}}{1 - 10^{-2}}$$
$$= 2 + \frac{11 \times 10^{-3}}{0.99} \cdot \frac{10^3}{10^3}$$
$$= 2 + \frac{11}{990} = \frac{1991}{990} = \mathbf{\frac{181}{90}}$$

20. (108) $A = \begin{bmatrix} 2 & 0 & 1 \\ 0 & 1 & 1 \\ 3 & 5 & 0 \end{bmatrix}$, $B = \begin{bmatrix} 5 & 6 & 2 \\ 1 & 7 & -3 \\ 4 & 1 & 2 \end{bmatrix}$

$$A - B = \begin{bmatrix} 2 & 0 & 1 \\ 0 & 1 & 1 \\ 3 & 5 & 0 \end{bmatrix} - \begin{bmatrix} 5 & 6 & 2 \\ 1 & 7 & -3 \\ 4 & 1 & 2 \end{bmatrix}$$
$$= \begin{bmatrix} 2 & 0 & 1 \\ 0 & 1 & 1 \\ 3 & 5 & 0 \end{bmatrix} + \begin{bmatrix} -5 & -6 & -2 \\ -1 & -7 & 3 \\ -4 & -1 & -2 \end{bmatrix}$$
$$= \begin{bmatrix} (2-5) & (0-6) & (1-2) \\ (0-1) & (1-7) & (1+3) \\ (3-4) & (5-1) & (0-2) \end{bmatrix}$$
$$= \begin{bmatrix} \mathbf{-3} & \mathbf{-6} & \mathbf{-1} \\ \mathbf{-1} & \mathbf{-6} & \mathbf{4} \\ \mathbf{-1} & \mathbf{4} & \mathbf{-2} \end{bmatrix}$$

$$A \cdot B = \begin{bmatrix} 2 & 0 & 1 \\ 0 & 1 & 1 \\ 3 & 5 & 0 \end{bmatrix} \cdot \begin{bmatrix} 5 & 6 & 2 \\ 1 & 7 & -3 \\ 4 & 1 & 2 \end{bmatrix}$$
$$= \begin{bmatrix} 10+0+4 & 12+0+1 & 4+0+2 \\ 0+1+4 & 0+7+1 & 0-3+2 \\ 15+5+0 & 18+35+0 & 6-15+0 \end{bmatrix}$$
$$= \begin{bmatrix} \mathbf{14} & \mathbf{13} & \mathbf{6} \\ \mathbf{5} & \mathbf{8} & \mathbf{-1} \\ \mathbf{20} & \mathbf{53} & \mathbf{-9} \end{bmatrix}$$

21. (59)
$$3 \log_2 x - 2 = 2 \log_2 x + 3$$
$$\log_2 x^3 - 2 = \log_2 x^2 + 3$$
$$\log_2 x^3 - \log_2 x^2 = 5$$
$$\log_2 x = 5$$
$$x = 2^5$$
$$x = \mathbf{32}$$

22. (98)
$$\ln\left(x^{\ln x}\right) = 1$$
$$x^{\ln x} = e^1$$
$$\ln x^{\ln x} = \ln e$$
$$(\ln x) \ln x = 1$$
$$(\ln x)^2 = 1$$
$$\ln x = \pm 1$$
$$x = \mathbf{e, e^{-1}}$$

23. (100) $\sec^2\left(\frac{\pi}{2} - \theta\right) + \csc^2\left(\frac{\pi}{2} - \theta\right)$

$$= \csc^2 \theta + \sec^2 \theta$$
$$= \frac{1}{\sin^2 \theta} + \frac{1}{\cos^2 \theta}$$
$$= \frac{\cos^2 \theta + \sin \theta}{\sin^2 \theta \cos^2 \theta}$$
$$= \frac{1}{\sin^2 \theta \cos^2 \theta} = \sec^2 \theta \csc^2 \theta$$

24. (72)
$$\frac{5}{\sin 30°} = \frac{c}{\sin 100°}$$
$$c = \frac{5 \sin 100°}{\sin 30°}$$
$$c = \mathbf{9.85\ ft}$$

25. $y = -3 + \cot(x - \pi)$
(94)

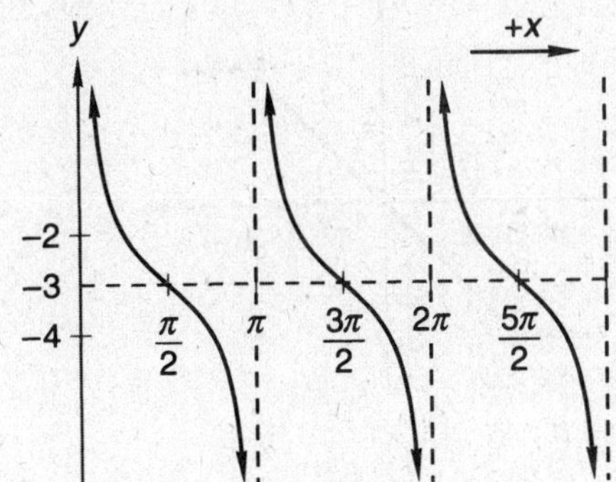

26. $y = -10 + 8\cos 4\left(x + \dfrac{\pi}{3}\right)$
(84)

$$\text{Period} = \frac{2\pi}{4} = \frac{\pi}{2}$$

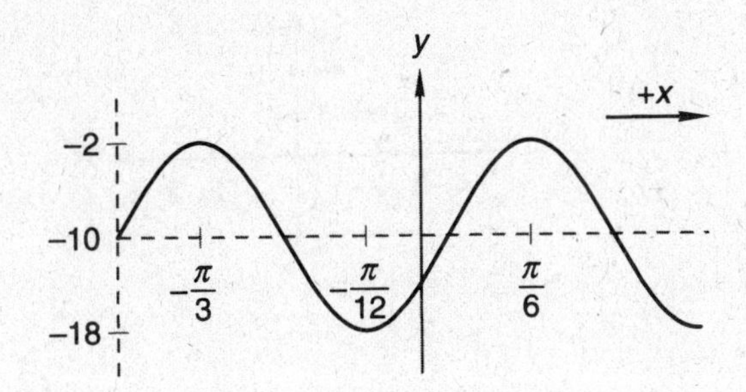

27.
(52,60)

$$2\sin^2\frac{\theta}{4} = 1$$

$$\sin^2\frac{\theta}{4} = \frac{1}{2}$$

$$\sin^2\frac{\theta}{4} - \frac{1}{2} = 0$$

$$\left(\sin\frac{\theta}{4} + \frac{1}{\sqrt{2}}\right)\left(\sin\frac{\theta}{4} - \frac{1}{\sqrt{2}}\right) = 0$$

$$\sin\frac{\theta}{4} = -\frac{1}{\sqrt{2}} \qquad \sin\frac{\theta}{4} = \frac{1}{\sqrt{2}}$$

$$\frac{\theta}{4} = \frac{5\pi}{4} \qquad \frac{\theta}{4} = \frac{\pi}{4}$$

$$\theta = 5\pi \qquad \theta = \pi$$

invalid

28. $\log_5 10 = x$
(98)

$$5^x = 10$$

$$\ln 5^x = \ln 10$$

$$x \ln 5 = \ln 10$$

$$x = \mathbf{\frac{\ln 10}{\ln 5}}$$

29. $-64i = 64 \text{ cis } 270°$
(95)

$$(64 \text{ cis } 270°)^{\frac{1}{6}}$$

$$= 64^{\frac{1}{6}} \text{ cis}\left(\frac{270°}{6} + n\frac{360°}{6}\right)$$

$$= \mathbf{2 \text{ cis } 45°, 2 \text{ cis } 105°, 2 \text{ cis } 165°, 2 \text{ cis } 225°,}$$
$$\mathbf{2 \text{ cis } 285°, 2 \text{ cis } 345°}$$

30. $\text{Side} = \dfrac{18 \text{ in.}}{6} = 3 \text{ in.}$
(73)

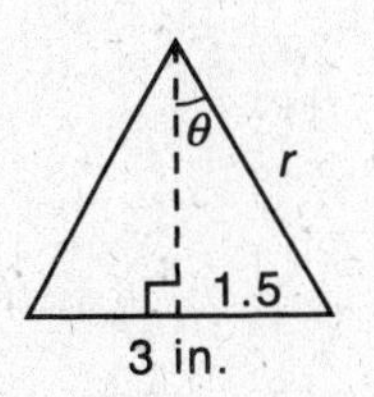

$$\theta = \frac{1}{2}\left(\frac{360°}{6}\right) = 30°$$

$$r = \frac{1.5}{\sin 30°} = \mathbf{3 \text{ in.}}$$

Problem Set 121

1. $P(5 \cup 9) = P(5) + P(9) - P(5 \cap 9)$
(92)

$$= \frac{4}{36} + \frac{4}{36} - 0$$

$$= \frac{8}{36} = \mathbf{\frac{2}{9}}$$

2. $A_0 = 600, A_{60} = 280$
(88)

$$A_t = 600e^{kt}$$

$$280 = 600e^{60k}$$

$$\frac{7}{15} = e^{60k}$$

$$\ln\frac{7}{15} = 60k$$

$$k = -0.012702$$

$$A_t = 600e^{-0.012702t}$$

$$300 = 600e^{-0.012702t}$$

$$0.5 = e^{-0.012702t}$$

$$\ln 0.5 = -0.012702t$$

$$t = \mathbf{54.57 \text{ hr}}$$

3. $v = r\omega$
(53)

$$\omega = 191\frac{\text{rev}}{\text{min}}\left(\frac{2\pi \text{ rad}}{1 \text{ rev}}\right)\left(\frac{1 \text{ min}}{60 \text{ sec}}\right) = \frac{(191)(2\pi)}{60}\frac{\text{rad}}{\text{s}}$$

$$r = \frac{v}{\omega} = \frac{85\,\frac{\text{m}}{\text{s}}}{\frac{(191)(2\pi)}{(60)}\frac{\text{rad}}{\text{s}}} = \frac{(85)(60)}{(191)(2\pi)}\text{ m}$$

$$\text{Diameter} = 2r = \mathbf{\frac{(85)(60)(2)}{(191)(2\pi)}\ m = 8.50\ m}$$

4. When $T = 200°$ C
(18)

$$P = k\frac{1}{V}$$

$$8 = k\frac{1}{60}$$

$$k = 8(60) = 480$$

$$P = (480)\frac{1}{V} = (480)\frac{1}{112} = \mathbf{\frac{30}{7}\ atm}$$

5. (a)
(121)

$$\begin{cases} y = -1 & \text{if } x \le 0 \\ y = x & \text{if } 0 < x \le 1 \\ y = 1 & \text{if } x \ge 1 \end{cases}$$

(b)

$$\begin{cases} y = -2 & \text{if } x \le 1 \\ y = x - 1 & \text{if } 1 < x \le 3 \\ y = 2 & \text{if } x \ge 3 \end{cases}$$

6.
(121)

$$\begin{cases} y = -\frac{1}{2}x & \text{if } x \le 4 \\ y = x - 7 & \text{if } 4 < x < 10 \end{cases}$$

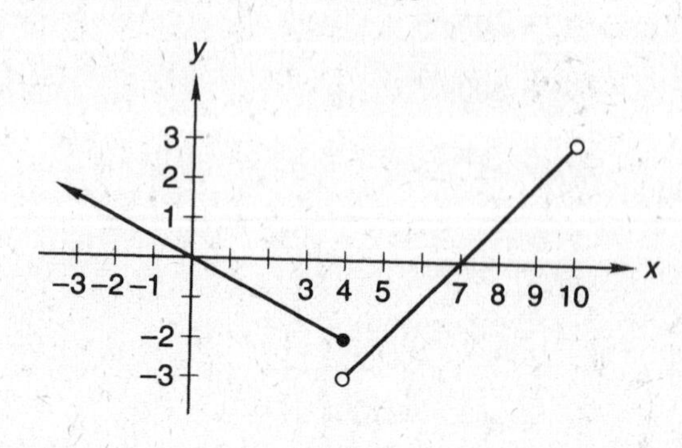

7.
(121)

$$\begin{cases} y = -3 & \text{if } -\infty < x < -3 \\ y = x & \text{if } -3 < x < 3 \\ y = 3 & \text{if } 3 < x < \infty \end{cases}$$

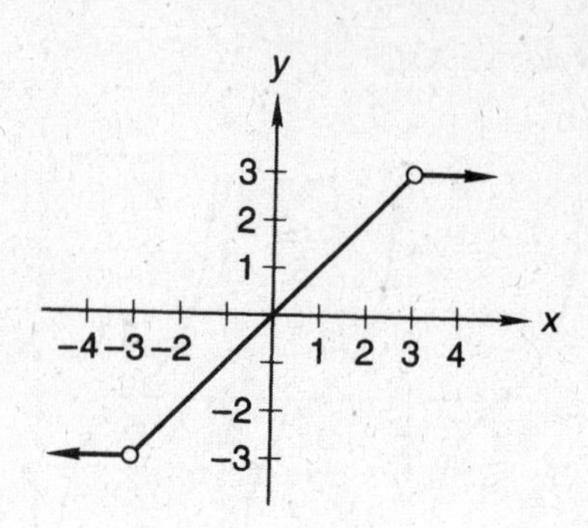

8. $y = [x] + 3$
(121)

This is the graph of $y = [x]$ translated 3 units up.

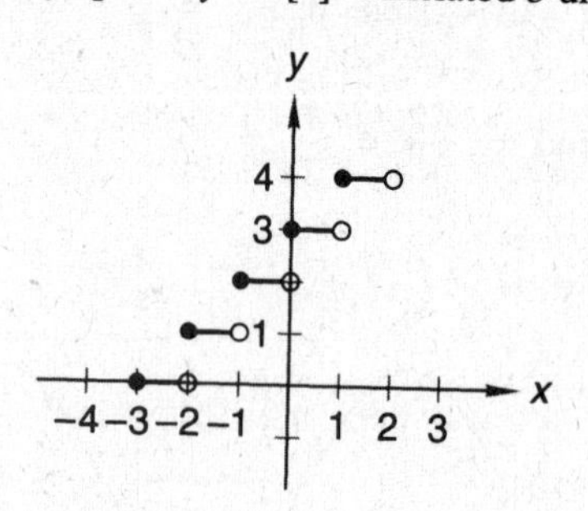

9. $\begin{bmatrix} 4 & 2 \\ 3 & -1 \end{bmatrix} \cdot \begin{bmatrix} x \\ y \end{bmatrix} = \begin{bmatrix} 7 \\ 2 \end{bmatrix}$
(120)

ORIGINAL MATRIX	INVERSE MATRIX
$\begin{bmatrix} 4 & 2 \\ 3 & -1 \end{bmatrix}$	$\frac{1}{(4)(-1) - (2)(3)}\begin{bmatrix} -1 & -2 \\ -3 & 4 \end{bmatrix}$

$$= -\frac{1}{10}\begin{bmatrix} -1 & -2 \\ -3 & 4 \end{bmatrix}$$

$$= \begin{bmatrix} \mathbf{\frac{1}{10}} & \mathbf{\frac{1}{5}} \\ \mathbf{\frac{3}{10}} & \mathbf{-\frac{2}{5}} \end{bmatrix}$$

$$\begin{bmatrix} \frac{1}{10} & \frac{2}{10} \\ \frac{3}{10} & -\frac{4}{10} \end{bmatrix} \cdot \begin{bmatrix} 4 & 2 \\ 3 & -1 \end{bmatrix} \cdot \begin{bmatrix} x \\ y \end{bmatrix} = \begin{bmatrix} \frac{1}{10} & \frac{2}{10} \\ \frac{3}{10} & -\frac{4}{10} \end{bmatrix} \cdot \begin{bmatrix} 7 \\ 2 \end{bmatrix}$$

$$\begin{bmatrix} 1 & 0 \\ 0 & 1 \end{bmatrix} \cdot \begin{bmatrix} x \\ y \end{bmatrix} = \begin{bmatrix} \frac{7}{10} + \frac{4}{10} \\ \frac{21}{10} - \frac{8}{10} \end{bmatrix}$$

$$\begin{bmatrix} x \\ y \end{bmatrix} = \begin{bmatrix} \frac{11}{10} \\ \frac{13}{10} \end{bmatrix}$$

$$\mathbf{x = \frac{11}{10};\ y = \frac{13}{10}}$$

10. (119) (a) $p(x) = 4x^3 + 8x^2 - x - 2$

$+ \quad + \quad ⓒ \quad - \quad -$

There is **1** positive real root.

(b) $p(x) = 4x^3 + 8x^2 - x - 2$

$p(-x) = -4x^3 + 8x^2 + x - 2$

$- ⓒ + \quad + ⓒ -$

There are **0 or 2** negative real roots.

11. (119) $4x^3 + 8x^2 - x - 2 = 0$

$$\begin{array}{r|rrrr} 1 & 4 & 8 & -1 & -2 \\ & & 4 & 12 & 11 \\ \hline & 4 & 12 & 11 & 9 \end{array}$$

Upper Bound = 1

$$\begin{array}{r|rrrr} -1 & 4 & 8 & -1 & -2 \\ & & -4 & -4 & 5 \\ \hline & 4 & 4 & -5 & 3 \end{array} \qquad \begin{array}{r|rrrr} -2 & 4 & 8 & -1 & -2 \\ & & -8 & 0 & 2 \\ \hline & 4 & 0 & -1 & 0 \end{array}$$

$$\begin{array}{r|rrrr} -3 & 4 & 8 & -1 & -2 \\ & & -12 & 12 & -33 \\ \hline & 4 & -4 & 11 & -35 \end{array}$$

Lower Bound = –3

12. (117) $4x^3 + 8x^2 - x - 2 = 0$

INTEGRAL FACTORS → POSSIBLE QUOTIENTS

$$\frac{\{1, -1, 2, -2\}}{\{1, -1, 2, -2, 4, -4\}} \rightarrow \pm 1, \pm 2, \pm\frac{1}{2}, \pm\frac{1}{4}$$

13. (118) $4x^3 + 8x^2 - x - 2 = 0$

Possible rational roots: $\pm 1, \pm 2, \pm\frac{1}{2}, \pm\frac{1}{4}$

From #11 we know that –2 is a root.

$$\begin{array}{r|rrrr} -2 & 4 & 8 & -1 & -2 \\ & & -8 & 0 & 2 \\ \hline & 4 & 0 & -1 & 0 \end{array}$$

$(x + 2)(4x^2 - 1) = 0$

For $4x^2 - 1$ use the quadratic formula:

$$x = \frac{0 \pm \sqrt{0 - 4(4)(-1)}}{2(4)} = \frac{\pm 4}{8} = \pm\frac{1}{2}$$

Roots: $\mathbf{-2, -\frac{1}{2}, \frac{1}{2}}$

14. (117) $128 = 2 \cdot 2 \cdot 2 \cdot 2 \cdot 2 \cdot 2 \cdot 2$

$87 = 3 \cdot 29$

Yes, 128 and 87 are relatively prime.

15. (115) $x^6 - 2x^4 + x^3 - 3x^2 - 30x + 2$

$x = 2$

$$\begin{array}{r|rrrrrrr} 2 & 1 & 0 & -2 & 1 & -3 & -30 & 2 \\ & & 2 & 4 & 4 & 10 & 14 & -32 \\ \hline & 1 & 2 & 2 & 5 & 7 & -16 & -30 \end{array}$$

$p(2) = \mathbf{-30}$

16. (114,116) $y = 4x^3 + 8x^2 - x - 2$

$r = \left|\frac{8}{4}\right| + 1$

$\mathbf{r = 3}$

x	–2	–1	0	1
y	0	3	–2	9

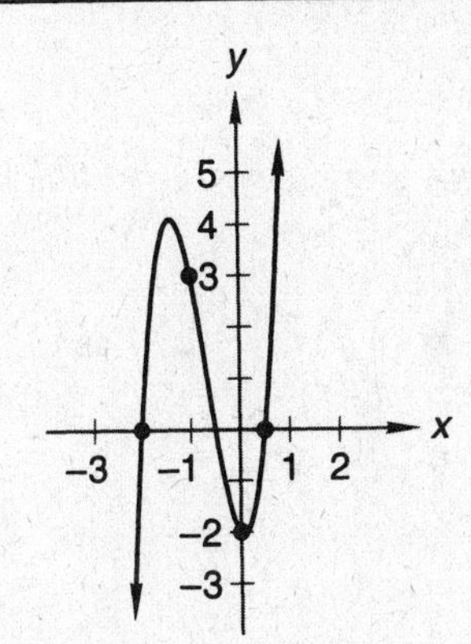

17. (114) $y = (x - 3)^2(x + 1)(1 - x)$

$= -x^4 + 6x^3 - 8x^2 - 6x + 9$

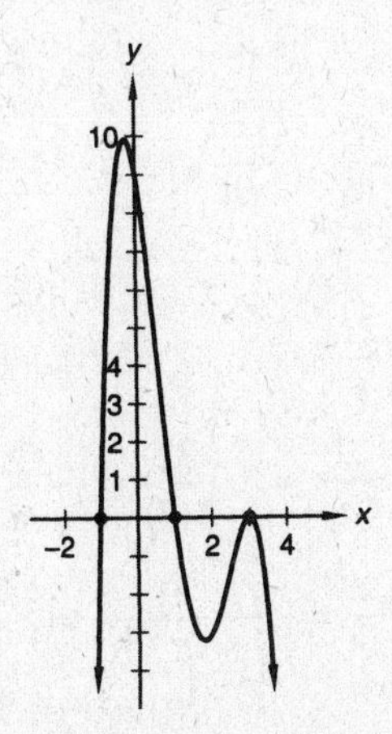

18. $(x^5 - 2x^3 - 4x + 1) \div (x - 3)$
(113)

3	1	0	−2	0	−4	1
		3	9	21	63	177
	1	3	7	21	59	178

$$\frac{x^5 - 2x^3 - 4x + 1}{x - 3}$$

$$= \mathbf{x^4 + 3x^3 + 7x^2 + 21x + 59 + \frac{178}{x - 3}}$$

19. $\left(1 - \frac{2}{p}\right)^{12}$
(112)

$F = 1,\ S = -\frac{2}{p},\ n = 12,\ k = 4$

$$\frac{n!}{(n - k + 1)!(k - 1)!}F^{n-k+1}S^{k-1}$$

$$= \frac{12!}{(12 - 4 + 1)!\,(4 - 1)!}F^{12-4+1}S^{4-1} = \frac{12!}{9!3!}F^9S^3$$

$$= 220(1)^9\left(-\frac{2}{p}\right)^3 = 220\left(-\frac{8}{p^3}\right) = \mathbf{-\frac{1760}{p^3}}$$

20. $\log_{\frac{1}{2}}(3x + 2) > 2$
(111)

$$3x + 2 < \left(\frac{1}{2}\right)^2 \quad \text{and} \quad 3x + 2 > 0$$

$$3x + 2 < \frac{1}{4} \qquad\qquad x > -\frac{2}{3}$$

$$3x < -\frac{7}{4}$$

$$x < -\frac{7}{12}$$

$$\mathbf{-\frac{2}{3} < x < -\frac{7}{12}}$$

21. $7, -\frac{21}{4}, \frac{63}{16}, -\frac{189}{64}, \ldots$
(107)

$a_1 = 7$

$$a_1r = -\frac{21}{4}$$

$$r = -\frac{21}{4a_1} = -\frac{21}{4(7)} = -\frac{3}{4}$$

$$S = \frac{a_1}{1 - r} = \frac{7}{1 - \left(-\frac{3}{4}\right)} = \frac{7}{\frac{7}{4}} = \mathbf{4}$$

22. $\sqrt{\log_6 x} = \log_6 \sqrt[3]{x}$
(98)

$$(\log_6 x)^{\frac{1}{2}} = \log_6 x^{\frac{1}{3}}$$

$$(\log_6 x)^{\frac{1}{2}} = \frac{1}{3}\log_6 x$$

$$\log_6 x = \frac{1}{9}(\log_6 x)^2$$

$$\frac{1}{9}(\log_6 x)^2 - \log_6 x = 0$$

$$(\log_6 x)^2 - 9\log_6 x = 0$$

$$(\log_6 x)(\log_6 x - 9) = 0$$

$$\log_6 x = 0 \qquad\qquad \log_6 x - 9 = 0$$

$$x = 6^0 \qquad\qquad \log_6 x = 9$$

$$x = 1 \qquad\qquad x = 6^9$$

$$\mathbf{x = 1, 6^9}$$

23. $(\sec\theta + \tan\theta)^2$
(87,93)

$$= \sec^2\theta + 2\sec\theta\tan\theta + \tan^2\theta$$

$$= \frac{1}{\cos^2\theta} + 2\left(\frac{1}{\cos\theta}\cdot\frac{\sin}{\cos\theta}\right) + \frac{\sin^2\theta}{\cos^2\theta}$$

$$= \frac{1}{\cos^2\theta} + \frac{2\sin\theta}{\cos^2\theta} + \frac{\sin^2\theta}{\cos^2\theta}$$

$$= \frac{1 + 2\sin\theta + \sin^2\theta}{1 - \sin^2\theta}$$

$$= \frac{(1 + \sin\theta)(1 + \sin\theta)}{(1 + \sin\theta)(1 - \sin\theta)} = \frac{1 + \sin\theta}{1 - \sin\theta}$$

24. $\frac{1 - \tan^2 x}{1 + \tan^2 x} = \frac{1 - \frac{\sin^2 x}{\cos^2 x}}{\sec^2 x}$
(90,96)

$$= \frac{\frac{\cos^2 x - \sin^2 x}{\cos^2 x}}{\frac{1}{\cos^2 x}}$$

$$= \frac{\cos^2 x - \sin^2 x}{\cos^2 x}\cdot\frac{\cos^2 x}{1} = \cos 2x$$

25. $\theta = \text{Arccos } x$
(110)

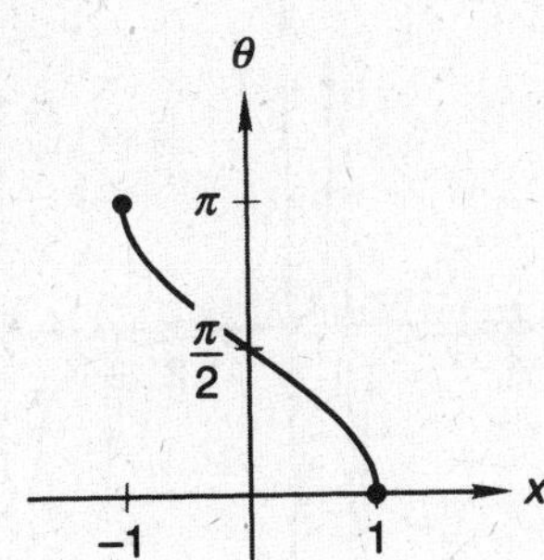

26. $y = \frac{3}{2} - 4\cos\left(\theta - \frac{7\pi}{4}\right)$
(84)

Period $= 2\pi$

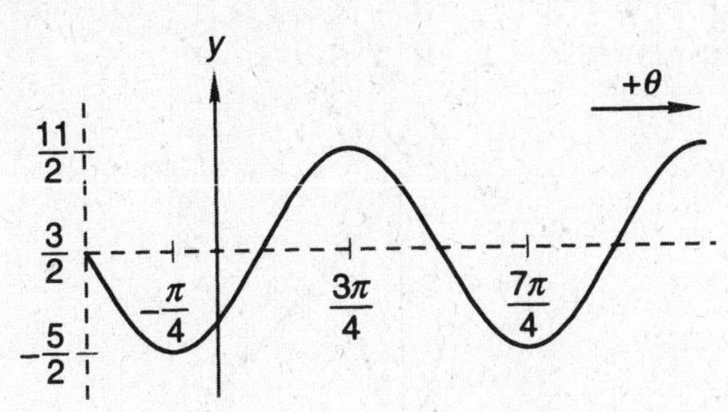

27.
(85)

$$3\csc^2 x - 2\sqrt{3}\cot x - 6 = 0$$
$$3(1 + \cot^2 x) - 2\sqrt{3}\cot x - 6 = 0$$
$$3 + 3\cot^2 x - 2\sqrt{3}\cot x - 6 = 0$$
$$3\cot^2 x - 2\sqrt{3}\cot x - 3 = 0$$
$$(3\cot x + \sqrt{3})(\cot x - \sqrt{3}) = 0$$

$\cot x = -\frac{\sqrt{3}}{3}$ $\cot x = \sqrt{3}$

$x = 120°, 300°$ $x = 30°, 210°$

$\mathbf{x = 30°, 120°, 210°, 300°}$

28.
(85)

$$6\sin^2\theta - 3 = 0$$
$$6\sin^2\theta = 3$$
$$\sin^2\theta = \frac{1}{2}$$
$$\sin^2\theta - \frac{1}{2} = 0$$
$$\left(\sin\theta + \frac{1}{\sqrt{2}}\right)\left(\sin\theta - \frac{1}{\sqrt{2}}\right) = 0$$

$\sin\theta = -\frac{1}{\sqrt{2}}$ $\sin\theta = \frac{1}{\sqrt{2}}$

$\theta = 225°, 315°$ $\theta = 45°, 135°$

$\mathbf{\theta = 45°, 135°, 225°, 315°}$

29. $(64 \text{ cis } 270°)^{\frac{1}{3}} = 64^{\frac{1}{3}} \text{ cis}\left(\frac{270°}{3} + n\frac{360°}{3}\right)$
(79)

$= 4 \text{ cis } 90°, 4 \text{ cis } 210°, 4 \text{ cis } 330°$

$= \mathbf{4i, -2\sqrt{3} - 2i, 2\sqrt{3} - 2i}$

30. $(f \circ g)(x) = \dfrac{2\left(\frac{x}{2x - 1}\right) + 1}{\frac{x}{2x - 1}}$
(40)

$$= \frac{\frac{2x + 2x - 1}{2x - 1}}{\frac{x}{2x - 1}} = \frac{4x - 1}{x} \neq x$$

$$(g \circ f)(x) = \frac{\frac{2x + 1}{x}}{2\left(\frac{2x + 1}{x}\right) - 1} = \frac{\frac{2x + 1}{x}}{\frac{4x + 2 - x}{x}}$$

$$= \frac{2x + 1}{3x + 2} \neq x$$

$(f \circ g)(x) \neq x$

$(g \circ f)(x) \neq x$

No, f and g are not inverse functions.

Problem Set 122

1. $\frac{6}{10} \cdot \frac{5}{9} = \mathbf{\frac{1}{3}}$
(83)

2. $A_1 = 6, A_6 = 192$
(88)

$$A_6 = A_1 e^{k(6-1)}$$
$$192 = 6e^{k5}$$
$$32 = e^{5k}$$
$$\ln 32 = 5k$$
$$k = 0.69315$$

$$A_t = A_0 e^{kt}$$
$$6 = A_0 e^{0.69315(1)}$$
$$A_0 = \frac{6}{e^{0.69315}} = \mathbf{3}$$

3. New distance $= k + 5$
(28)

New time $= p + 2$

New rate $= \mathbf{\frac{k + 5}{p + 2}}$ **mph**

4. $N,\ N + 2,\ N + 4,\ N + 6$
(7)

$$2N(N + 6) + 1 = (N + 2)(N + 4)$$
$$2N^2 + 12N + 1 = N^2 + 6N + 8$$
$$N^2 + 6N - 7 = 0$$
$$(N - 1)(N + 7) = 0$$
$$N = 1$$

1, 3, 5, 7

5. (122)

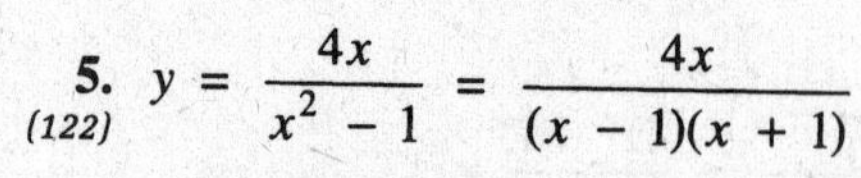

$$y = \frac{4x}{x^2 - 1} = \frac{4x}{(x - 1)(x + 1)}$$

x-intercept: $x = 0$

Asymptotes: $x = -1, 1$

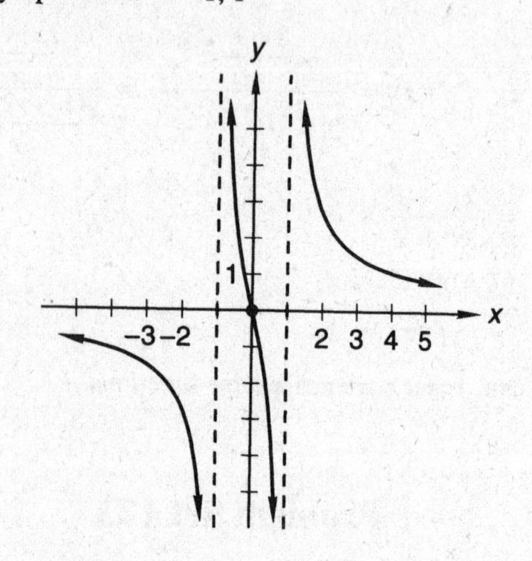

6. $y = \dfrac{(x + 1)(x - 2)}{x(x + 3)(x + 2)(x - 4)}$
(122)

x-intercepts: $x = -1, 2$

Asymptotes: $x = 0, -3, -2, 4$

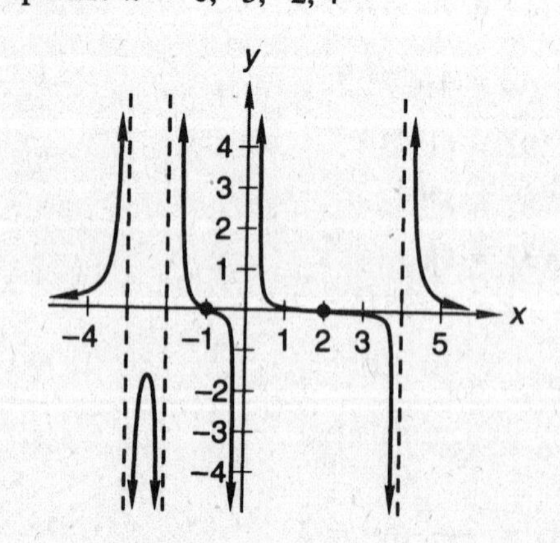

7. $y = \dfrac{x(x - 1)}{(x - 1)(x + 1)(x - 5)}$
(122)

x-intercept: $x = 0$

Asymptotes: $x = -1, 5$

Hole: $x = 1$

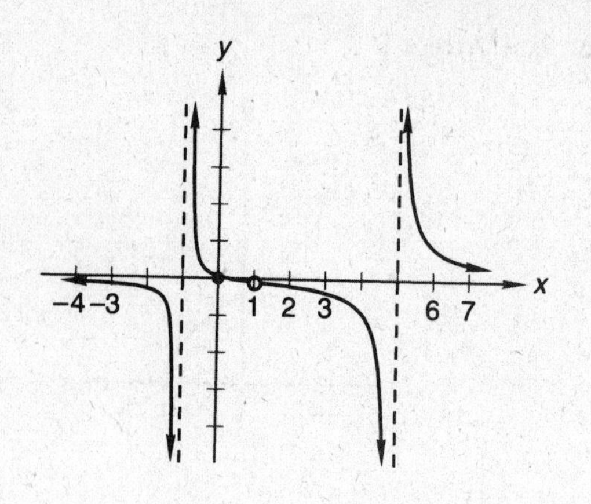

8. $\begin{cases} y = 2x^2 & \text{if } -1 < x < 1 \\ y = 5 - x & \text{if } 1 \le x < \infty \end{cases}$
(121)

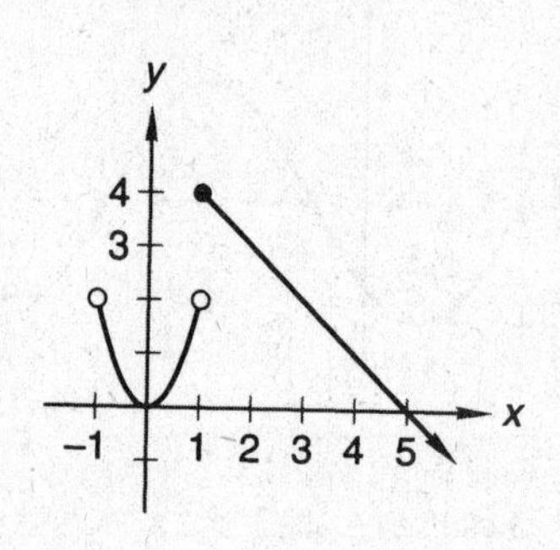

9. (121)

(a) $\begin{cases} y = x + 1 & \text{if } -\infty < x \le -1 \\ y = -2 & \text{if } -1 < x < 1 \\ y = x - 1 & \text{if } 1 \le x < \infty \end{cases}$

(b) $\begin{cases} y = -1 & \text{if } -\infty < x < -2 \\ y = -x - 2 & \text{if } -2 \le x \le 0 \\ y = 2x & \text{if } 0 < x \le 1 \\ y = 2 & \text{if } 1 \le x < \infty \end{cases}$

10. $\begin{bmatrix} 4 & 2 \\ 7 & -5 \end{bmatrix} \cdot \begin{bmatrix} x \\ y \end{bmatrix} = \begin{bmatrix} 10 \\ 9 \end{bmatrix}$
(120)

ORIGINAL MATRIX	INVERSE MATRIX
$\begin{bmatrix} 4 & 2 \\ 7 & -5 \end{bmatrix}$	$\dfrac{1}{4(-5) - 2(7)}\begin{bmatrix} -5 & -2 \\ -7 & 4 \end{bmatrix}$

$$= -\frac{1}{34}\begin{bmatrix} -5 & -2 \\ -7 & 4 \end{bmatrix}$$

$$= \begin{bmatrix} \mathbf{\frac{5}{34}} & \mathbf{\frac{1}{17}} \\ \mathbf{\frac{7}{34}} & \mathbf{-\frac{2}{17}} \end{bmatrix}$$

$$\begin{bmatrix} \frac{5}{34} & \frac{1}{17} \\ \frac{7}{34} & -\frac{2}{17} \end{bmatrix} \cdot \begin{bmatrix} 4 & 2 \\ 7 & -5 \end{bmatrix} \cdot \begin{bmatrix} x \\ y \end{bmatrix} = \begin{bmatrix} \frac{5}{34} & \frac{1}{17} \\ \frac{7}{34} & -\frac{2}{17} \end{bmatrix} \cdot \begin{bmatrix} 10 \\ 9 \end{bmatrix}$$

$$\begin{bmatrix} 1 & 0 \\ 0 & 1 \end{bmatrix} \cdot \begin{bmatrix} x \\ y \end{bmatrix} = \begin{bmatrix} \frac{68}{34} \\ \frac{34}{34} \end{bmatrix}$$

$$\begin{bmatrix} x \\ y \end{bmatrix} = \begin{bmatrix} 2 \\ 1 \end{bmatrix}$$

$\mathbf{x = 2; y = 1}$

11. (119) (a) $p(x) = 5x^3 + 4x - 9$

$p(-x) = -5x^3 - 4x - 9$

There are **0** negative real roots.

(b) $p(x) = 5x^3 + 4x - 9$

$+ \quad + \text{ⓒ} \; -$

There is **1** positive real root.

12. (119) (a) $p(x) = 5x^3 + 3x^2 + 6x + 8$

$p(-x) = -5x^3 + 3x^2 - 6x + 8$

$- \text{ⓒ} + \text{ⓒ} - \text{ⓒ} +$

There are **1 or 3** negative real roots.

(b) $p(x) = 5x^3 + 3x^2 + 6x + 8$

There are **0** positive real roots.

13. (113,119) $x^3 - 3x^2 - 6x + 8 = 0$

$$\begin{array}{r|rrrr} 4 & 1 & -3 & -6 & 8 \\ & & 4 & 4 & -8 \\ \hline & 1 & 1 & -2 & 0 \end{array} \qquad \begin{array}{r|rrrr} 5 & 1 & -3 & -6 & 8 \\ & & 5 & 10 & 20 \\ \hline & 1 & 2 & 4 & 28 \end{array}$$

Upper bound = 5

$$\begin{array}{r|rrrr} -2 & 1 & -3 & -6 & 8 \\ & & -2 & 10 & -8 \\ \hline & 1 & -5 & 4 & 0 \end{array} \qquad \begin{array}{r|rrrr} -3 & 1 & -3 & -6 & 8 \\ & & -3 & 18 & -36 \\ \hline & 1 & -6 & 12 & -28 \end{array}$$

Lower bound = –2

14. (113,119) $3x^3 - 5x^2 + 6x - 10 = 0$

$$\begin{array}{r|rrrr} 1 & 3 & -5 & 6 & -10 \\ & & 3 & -2 & 4 \\ \hline & 3 & -2 & 4 & -6 \end{array} \qquad \begin{array}{r|rrrr} 2 & 3 & -5 & 6 & -10 \\ & & 6 & 2 & 16 \\ \hline & 3 & 1 & 8 & 6 \end{array}$$

Upper bound = 2

$$\begin{array}{r|rrrr} -1 & 3 & -5 & 6 & -10 \\ & & -3 & 8 & -14 \\ \hline & 3 & -8 & 14 & -24 \end{array}$$

Lower bound = –1

15. (118) $5x^3 - 2x^2 - 5x + 2 = 0$

Possible rational roots:

$\pm 1, \pm 2, \pm\frac{1}{5}, \pm\frac{2}{5}$

$$\begin{array}{r|rrrr} 1 & 5 & -2 & -5 & 2 \\ & & 5 & 3 & -2 \\ \hline & 5 & 3 & -2 & 0 \end{array}$$

$(x - 1)(5x^2 + 3x - 2) = 0$

For $5x^2 + 3x - 2$ use the quadratic formula:

$$x = \frac{-3 \pm \sqrt{3^2 - 4(5)(-2)}}{2(5)}$$

$$= \frac{-3 \pm \sqrt{49}}{10} = \frac{-3 \pm 7}{10} = \frac{2}{5}, -1$$

Roots: $\mathbf{-1, \frac{2}{5}, 1}$

16. (118) $5x^3 + 3x^2 + 6x + 8 = 0$

Possible rational roots:

$\pm 1, \pm 2, \pm 4, \pm 8, \pm\frac{1}{5}, \pm\frac{2}{5}, \pm\frac{4}{5}, \pm\frac{8}{5}$

$$\begin{array}{r|rrrr} -1 & 5 & 3 & 6 & 8 \\ & & -5 & 2 & -8 \\ \hline & 5 & -2 & 8 & 0 \end{array}$$

$(x + 1)(5x^2 - 2x + 8) = 0$

For $5x^2 - 2x + 8$ use the quadratic formula:

$$x = \frac{2 \pm \sqrt{(-2)^2 - 4(5)(8)}}{2(5)} = \frac{1}{5} \pm \frac{\sqrt{39}}{5}i$$

Roots: $\mathbf{-1, \frac{1}{5} \pm \frac{\sqrt{39}}{5}i}$

17. (113) $y = x^3 + 4x^2 + 3x - 8$

$$\begin{array}{r|rrrr} 1 & 1 & 4 & 3 & -8 \\ & & 1 & 5 & 8 \\ \hline & 1 & 5 & 8 & 0 \end{array}$$

$(x - 1)(x^2 + 5x + 8) = 0$

For $x^2 + 5x + 8$ use the quadratic formula:

$$x = \frac{-5 \pm \sqrt{5^2 - 4(1)(8)}}{2(1)} = -\frac{5}{2} \pm \frac{\sqrt{7}}{2}i$$

The other zeros are: $\mathbf{-\frac{5}{2} \pm \frac{\sqrt{7}}{2}i}$

18. (113) If $-i$ is a root, then $+i$ is another.

$x = \pm i$

$x^2 = -1$

$(x^2 + 1)$ is a factor

$$\begin{array}{r} x^2 - 4 \\ x^2 + 1 \overline{)x^4 - 0x^3 - 3x^2 - 4} \\ \underline{x^4 \quad\quad + x^2} \\ -4x^2 \\ \underline{-4x^2 - 4} \\ 0 \end{array}$$

$(x^2 + 1)(x^2 - 4) = 0$

$(x^2 + 1)(x + 2)(x - 2) = 0$

The other roots are: $\mathbf{i, -2, 2}$

19. (117) $120 = 8 \cdot 15$

$195 = 13 \cdot 15$

No, 15 is a common factor.

20. (114,116) $y = -x^3 - 2x^2 + 5x + 6$

$r = |-6| + 1$

$\mathbf{r = 7}$

x	-3	-2	-1	0	1	2
y	0	-4	0	6	8	0

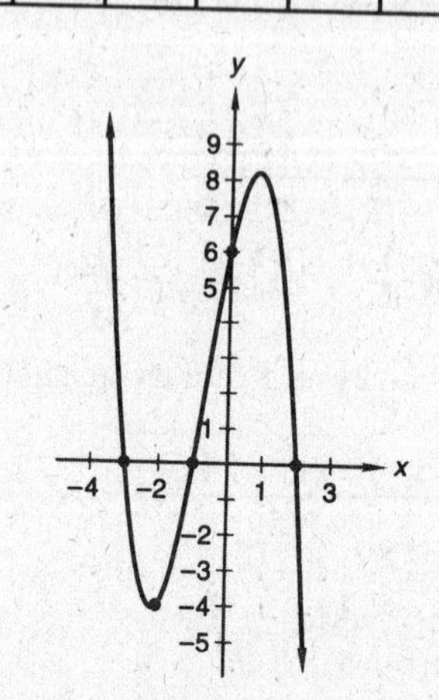

21. (114) $y = x(x - 2)^2(x + 2) = x^4 - 2x^3 - 4x^2 + 8x$

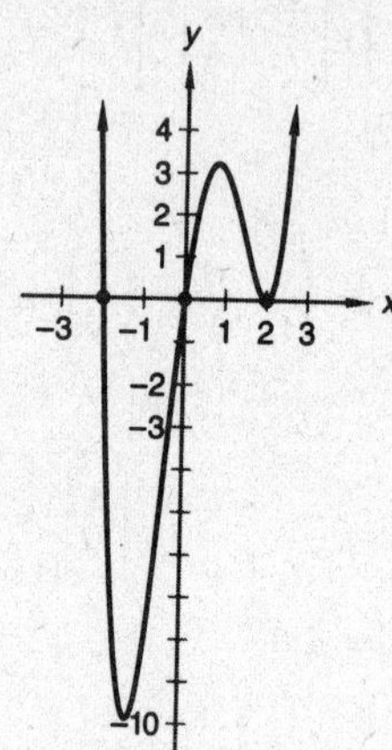

22. (114) $f(x) = x^2(1 - x)^2$

$f(x)$ is always positive. Therefore the answer is **C.**

23. (115) $3x^6 - 5x^4 + x^2$

$x = 1$

$$\begin{array}{r|rrrrrrr} 1 & 3 & 0 & -5 & 0 & 1 & 0 & 0 \\ & & 3 & 3 & -2 & -2 & -1 & -1 \\ \hline & 3 & 3 & -2 & -2 & -1 & -1 & -1 \end{array}$$

$\mathbf{p(1) = -1}$

24. (112) $\left(2x + \frac{3}{x^2}\right)^3$

$F = 2x,\ S = \frac{3}{x^2}$

Term	①	②	③	④
Exponents of F	3	2	1	0
Exponents of S	0	1	2	3
Coefficients	1	3	3	1

$(F + S)^3$

$= F^3 + 3F^2S + 3FS^2 + S^3$

$= (2x)^3 + 3(2x)^2\left(\frac{3}{x^2}\right) + 3(2x)\left(\frac{3}{x^2}\right)^2 + \left(\frac{3}{x^2}\right)^3$

$\mathbf{= 8x^3 + 36 + \frac{54}{x^3} + \frac{27}{x^6}}$

25. (111) $\log_5 (x - 5) < 1$

$x - 5 < 5$ and $x - 5 > 0$

$x < 10$ $\quad x > 5$

$\mathbf{5 < x < 10}$

26. (109)
$$3.\overline{1} = 3 + \left[1 \times 10^{-1} + (1 \times 10^{-1})10^{-1} + (1 + 10^{-1})10^{-2} + \ldots\right]$$

If $a_1 = 1 \times 10^{-1}$, $r = 10^{-1}$

$$3.\overline{1} = 3 + \frac{a_1}{1 - r}$$
$$= 3 + \frac{1 \times 10^{-1}}{1 - 10^{-1}}$$
$$= 3 + \frac{1 \times 10^{-1}}{0.9} \cdot \frac{10^1}{10^1}$$
$$= 3 + \frac{1}{9} = \frac{27}{9} + \frac{1}{9} = \mathbf{\frac{28}{9}}$$

27. (98)
$$\log x^3 = (\log x)^2$$
$$3 \log x = (\log x)^2$$
$$(\log x)^2 - 3 \log x = 0$$
$$\log x(\log x - 3) = 0$$

$\log x = 0$ $\qquad$ $\log x = 3$

$x = 10^0$ $\qquad$ $x = 10^3$

$x = 1$ $\qquad$ $x = 1000$

$\mathbf{x = 1, 1000}$

28. (59)
$$2 \log x - \log (x - 1) = \log 4$$
$$\log x^2 - \log (x - 1) = \log 4$$
$$\log \frac{x^2}{x - 1} = \log 4$$
$$\frac{x^2}{x - 1} = 4$$
$$x^2 = 4x - 4$$
$$x^2 - 4x + 4 = 0$$
$$(x - 2)^2 = 0$$
$$x = \mathbf{2}$$

29. (96)
$$\cos 3x = \cos (2x + x)$$
$$= \cos 2x \cos x - \sin 2x \sin x$$
$$= (2 \cos^2 x - 1)\cos x - (2 \sin x \cos x)\sin x$$
$$= 2 \cos^3 x - \cos x - 2 \cos x \sin^2 x$$
$$= 2 \cos^3 x - \cos x - 2 \cos x(1 - \cos^2 x)$$
$$= 2 \cos^3 x - \cos x - 2 \cos x + 2 \cos^3 x$$
$$= 4 \cos^3 x - 3 \cos x$$

30. (85,90)
$$\cos 2x - \cos x = 0$$
$$(2 \cos^2 x - 1) - \cos x = 0$$
$$2 \cos^2 x - \cos x - 1 = 0$$
$$(2 \cos x + 1)(\cos x - 1) = 0$$

$\cos x = -\frac{1}{2}$ $\qquad$ $\cos x = 1$

$x = \frac{2\pi}{3}, \frac{4\pi}{3}$ $\qquad$ $x = 0$

$$\mathbf{x = 0, \frac{2\pi}{3}, \frac{4\pi}{3}}$$

Problem Set 123

1. (62) $C = mN + b$

(a) $60 = m4 + b$
(b) $30 = m2 + b$

(–b) $-30 = -m2 - b$
(a) $60 = m4 + b$

$30 = m2$

$m = 15$

(b) $30 = m2 + b$

$30 = (15)2 + b$

$b = 0$

$C = 15N = 15(10) =$ **150 units**

2. (18)
(a) $0.1P_N + 0.4D_N = 0.3(300)$
(b) $P_N + D_N = 300$

(b) $P_N + D_N = 300$
$$P_N = 300 - D_N$$

(a)
$$0.1P_N + 0.4D_N = 0.3(300)$$
$$0.1(300 - D_N) + 0.4D_N = 90$$
$$30 - 0.1D_N + 0.4D_N = 90$$
$$0.3D_N = 60$$
$$D_N = \textbf{200 ml of 40\%}$$

$$P_N = 300 - D_N$$
$$P_N = 300 - (200)$$
$$P_N = \textbf{100 ml of 10\%}$$

3. (44)
$$\begin{cases} R_T T_T = D \\ R_S T_S = D \end{cases}$$

$$T_S = T_T - 3$$

$$R_T = 50$$

$$R_S = 2R_T = 2(50) = 100$$

$$R_T T_T = R_S T_S$$
$$(50)T_T = (100)(T_T - 3)$$
$$50T_T = 100T_T - 300$$
$$50T_T = 300$$
$$T_T = 6$$

$$R_T T_T = (50)(6) = \textbf{300 mi}$$

4. (7) $N,\ N + 2,\ N + 4$

$$N(N + 4) = 9(N + 2) - 24$$
$$N^2 + 4N = 9N + 18 - 24$$
$$N^2 - 5N + 6 = 0$$
$$(N - 3)(N - 2) = 0$$
$$N = 2$$

2, 4, 6

5. (25)
$$\begin{cases} \text{(a) } J_N = B_N - 10 \\ \text{(b) } 3(B_N + 10) = 4(J_N + 10) - 16 \end{cases}$$

(b) $3(B_N + 10) = 4(J_N + 10) - 16$
$$3B_N + 30 = 4J_N + 40 - 16$$
$$3B_N = 4J_N - 6$$
$$3B_N = 4(B_N - 10) - 6$$
$$3B_N = 4B_N - 40 - 6$$
$$B_N = \textbf{46 yr}$$

(a) $J_N = B_N - 10$
$$J_N = (46) - 10$$
$$J_N = \textbf{36 yr}$$

6. (123) (a) $xy = 4$

$$y = \frac{4}{x},\ x \neq 0;\ x = \frac{4}{y},\ y \neq 0$$

Asymptotes: $y = 0,\ x = 0$

x	-4	-2	-1	1	2	4
y	-1	-2	-4	4	2	1

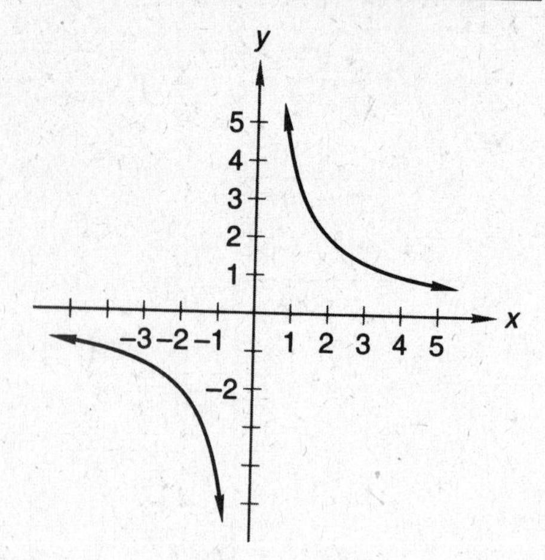

(b) $xy = -4$

$$y = \frac{-4}{x},\ x \neq 0;\ x = \frac{-4}{y},\ y \neq 0$$

Asymptotes: $y = 0,\ x = 0$

x	-4	-2	-1	1	2	4
y	1	2	4	-4	-2	-1

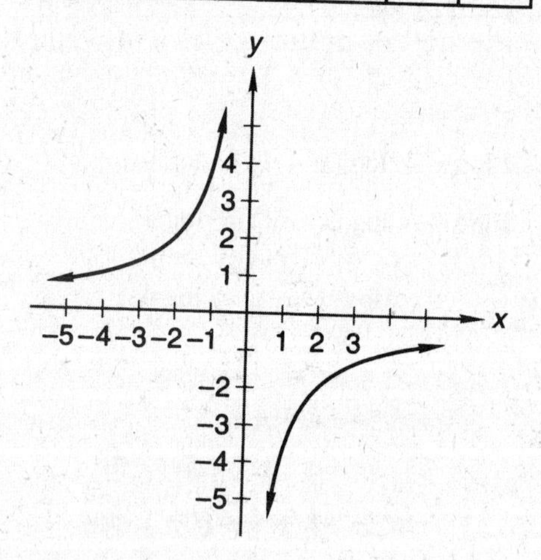

7. (123) $ax^2 + bxy + cy^2 + dx + ey + f = 0$

(a) $x^2 + y^2 - 4x = 0$

This is a **circle** since $b = 0$ and $a = c \neq 0$.

(b) $4x^2 + 9y^2 = 1$

This is an **ellipse** since $b = 0$, a and c have the same sign, and $a \neq c$.

(c) $4x^2 - y^2 = 4$

This is a **hyperbola** since $b = 0$ and a and c have opposite signs.

(d) $x^2 + 2x - 8y - 3 = 0$

This is a **parabola** since $b = c = 0$, $a \neq 0$ and $e \neq 0$.

(e) $x^2 + y^2 + 8x - 6y - 15 = 0$

This is a **circle** since $b = 0$ and $a = c \neq 0$.

8. (106,123)

$$x^2 + y^2 + 8x - 6y - 15 = 0$$
$$(x^2 + 8x \quad) + (y^2 - 6y \quad) = 15$$
$$(x^2 + 8x + 16) + (y^2 - 6y + 9) = 40$$
$$(x + 4)^2 + (y - 3)^2 = 40$$
$$\mathbf{(x + 4)^2 + (y - 3)^2 = (\sqrt{40})^2}$$

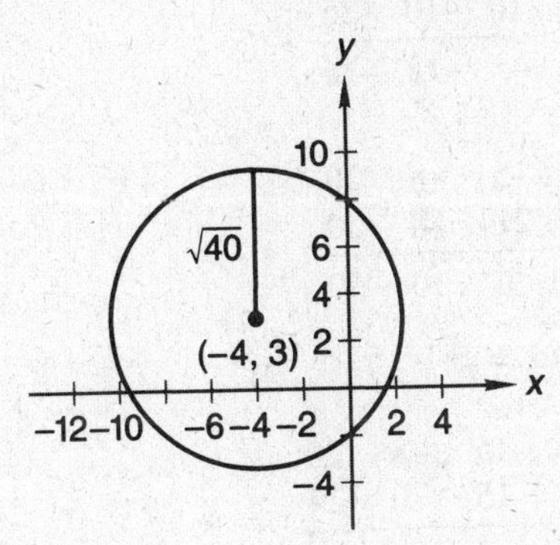

9. (123) $ax^2 + bxy + cy^2 + dx + ey + f = 0$

(a) $4x^2 + 36y^2 + 40x - 288y + 532 = 0$

This is an **ellipse** since $b = 0$, a and c have the same sign and $a \neq c$.

(b) $24x^2 - 16y^2 - 100x - 96y - 444 = 0$

This is a **hyperbola** since $b = 0$ and a and c have opposite signs.

(c) $x^2 + y^2 - 10x - 8y + 16 = 0$

This is a **circle** since $b = 0$ and $a = c \neq 0$.

(d) $y = -2x^2 + 12x - 16$

This is a **parabola** since $b = c = 0$, $a \neq 0$, and $e \neq 0$.

(e) $x^2 + y^2 - 10x + 8y + 5 = 0$

This is a **circle** since $b = 0$ and $a = c \neq 0$.

10. (106,123)

$$y = -2x^2 + 12x - 16$$
$$y = -2(x^2 - 6x \quad) - 16$$
$$y = -2(x^2 - 6x + 9) - 16 + 18$$
$$\mathbf{y = -2(x - 3)^2 + 2}$$

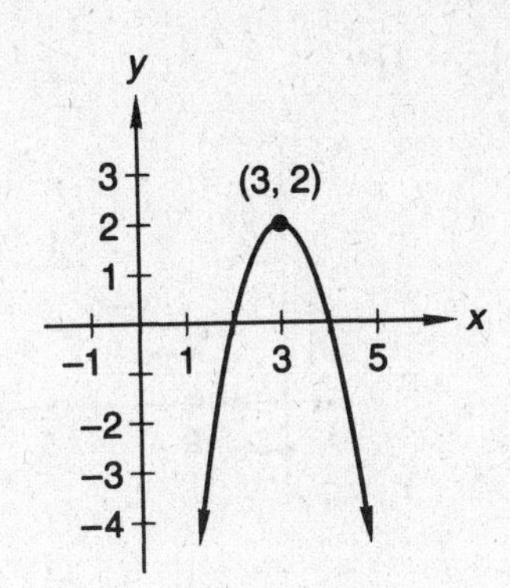

11. (122) $y = \dfrac{(x + 2)(x - 3)}{(x - 1)(x + 3)(x - 5)}$

x-intercepts: $x = -2, 3$

Asymptotes: $x = 1, -3, 5$

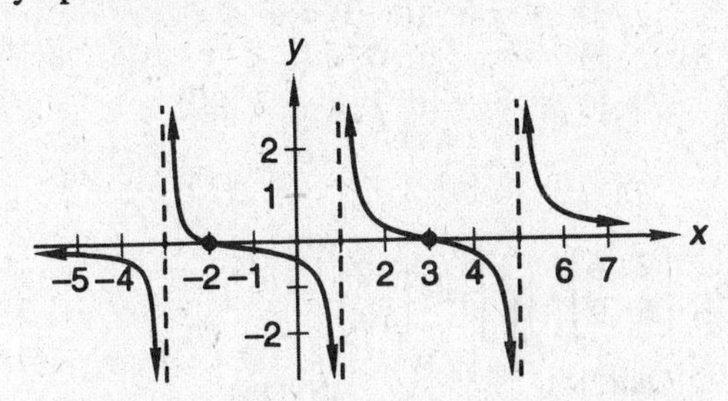

12. (122)

$$y = \frac{x^2 + 2x}{(x - 1)(x + 2)(x + 3)}$$
$$= \frac{x(x + 2)}{(x - 1)(x + 2)(x + 3)}$$

x-intercept: $x = 0$

Asymptotes: $x = 1, -3$

Hole: $x = -2$

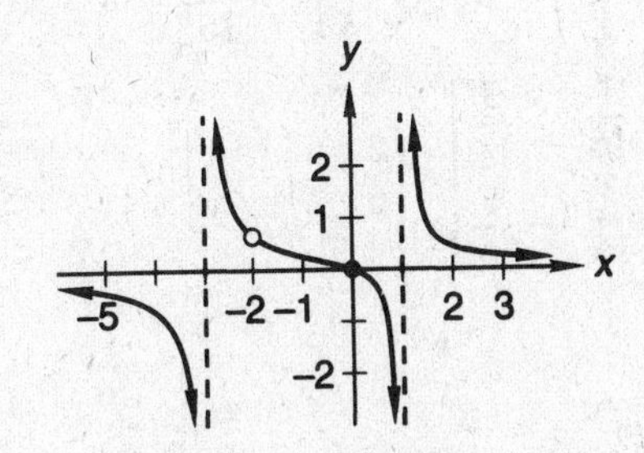

13. (121)

$$\begin{cases} y = 2 & \text{if } -\infty < x < 0 \\ y = 0 & \text{if } x = 0 \\ y = x + 1 & \text{if } 0 < x < \infty \end{cases}$$

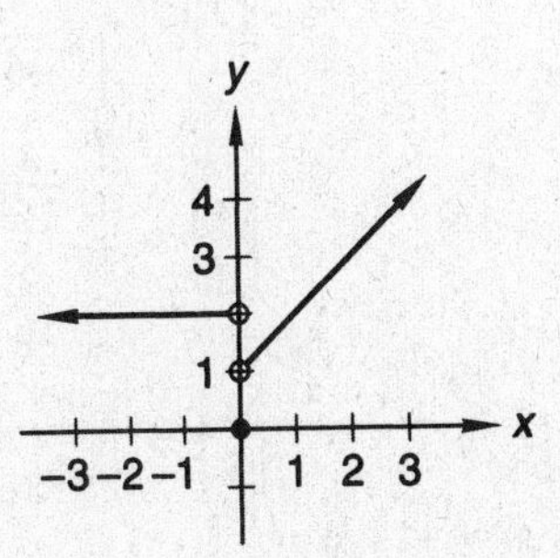

14. (121) $y = [x - 1]$

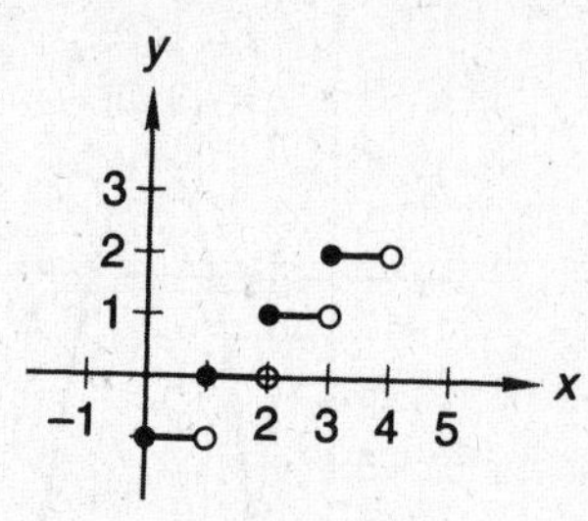

15. (121) (a) $\begin{cases} y = -x - 1 & \text{if } -\infty < x \le -1 \\ y = -2 & \text{if } -1 < x < 1 \\ y = x - 1 & \text{if } 1 \le x < \infty \end{cases}$

(b) $\begin{cases} y = -4 & \text{if } -\infty < x \le -3 \\ y = -2 & \text{if } -3 < x \le 0 \\ y = x & \text{if } 0 < x \le 2 \\ y = 3 & \text{if } 2 < x < \infty \end{cases}$

16. (120) $\begin{bmatrix} 2 & 6 \\ 6 & 2 \end{bmatrix} \cdot \begin{bmatrix} x \\ y \end{bmatrix} = \begin{bmatrix} -2 \\ 10 \end{bmatrix}$

ORIGINAL MATRIX: $\begin{bmatrix} 2 & 6 \\ 6 & 2 \end{bmatrix}$

INVERSE MATRIX:

$$\frac{1}{(2)(2) - (6)(6)} \begin{bmatrix} 2 & -6 \\ -6 & 2 \end{bmatrix}$$

$$= -\frac{1}{32} \begin{bmatrix} 2 & -6 \\ -6 & 2 \end{bmatrix}$$

$$= \begin{bmatrix} -\mathbf{\frac{1}{16}} & \mathbf{\frac{3}{16}} \\ \mathbf{\frac{3}{16}} & -\mathbf{\frac{1}{16}} \end{bmatrix}$$

$$\begin{bmatrix} -\frac{1}{16} & \frac{3}{16} \\ \frac{3}{16} & -\frac{1}{16} \end{bmatrix} \cdot \begin{bmatrix} 2 & 6 \\ 6 & 2 \end{bmatrix} \cdot \begin{bmatrix} x \\ y \end{bmatrix}$$

$$= \begin{bmatrix} -\frac{1}{16} & \frac{3}{16} \\ \frac{3}{16} & -\frac{1}{16} \end{bmatrix} \cdot \begin{bmatrix} -2 \\ 10 \end{bmatrix}$$

$$\begin{bmatrix} 1 & 0 \\ 0 & 1 \end{bmatrix} \cdot \begin{bmatrix} x \\ y \end{bmatrix} = \begin{bmatrix} -\frac{1}{16} & \frac{3}{16} \\ \frac{3}{16} & -\frac{1}{16} \end{bmatrix} \cdot \begin{bmatrix} -2 \\ 10 \end{bmatrix}$$

$$\begin{bmatrix} x \\ y \end{bmatrix} = \begin{bmatrix} \frac{2}{16} + \frac{30}{16} \\ -\frac{6}{16} - \frac{10}{16} \end{bmatrix}$$

$$\begin{bmatrix} x \\ y \end{bmatrix} = \begin{bmatrix} 2 \\ -1 \end{bmatrix}$$

$\mathbf{x = 2; y = -1}$

17. (119) (a) $p(x) = 2x^4 + x^3 - 15x^2 - 8x + 20$

+ + Ⓒ − − Ⓒ +

There are **0 or 2** positive real roots.

(b) $p(x) = 2x^4 + x^3 - 15x^2 - 8x + 20$

$p(-x) = 2x^4 - x^3 - 15x^2 + 8x + 20$

+ Ⓒ − − Ⓒ + +

There are **0 or 2** negative real roots.

18. (119) $2x^4 + x^3 - 15x^2 - 8x + 20 = 0$

```
2| 2   1  -15   -8   20
       4   10  -10  -36
   ----------------------
   2   5   -5  -18  -16

3| 2   1  -15   -8   20
       6   21   18   30
   ----------------------
   2   7    6   10   50
```

Upper Bound = 3

```
-3| 2   1  -15   -8   20
       -6   15    0   24
    ----------------------
    2  -5    0   -8   44
```

Lower Bound = −3

19. (118) $2x^4 + x^3 - 15x^2 - 8x + 20 = 0$

Possible rational roots:

$\pm 1, \pm 2, \pm 4, \pm 5, \pm 10, \pm 20, \pm\frac{1}{2}, \pm\frac{5}{2}$

```
 1| 2   1  -15   -8   20
        2    3  -12  -20
    ----------------------
-2| 2   3  -12  -20    0
       -4    2   20
    -----------------
-2| 2  -1  -10    0
       -4   10
    ------------
    2  -5    0
```

$(x - 1)(x + 2)(x + 2)(2x - 5) = 0$

Roots: $\mathbf{-2, -2, 1, \frac{5}{2}}$

20. (114,116) $y = x^4 - 5x^3 + 6x^2$

$r = |6| + 1$

$\mathbf{r = 7}$

x	0	1	2	3
y	0	2	0	0

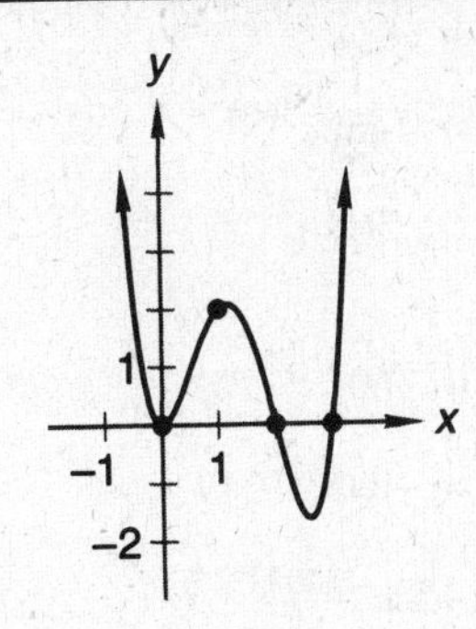

21. (114) $y = x^2(2 - x) = -x^3 + 2x^2$

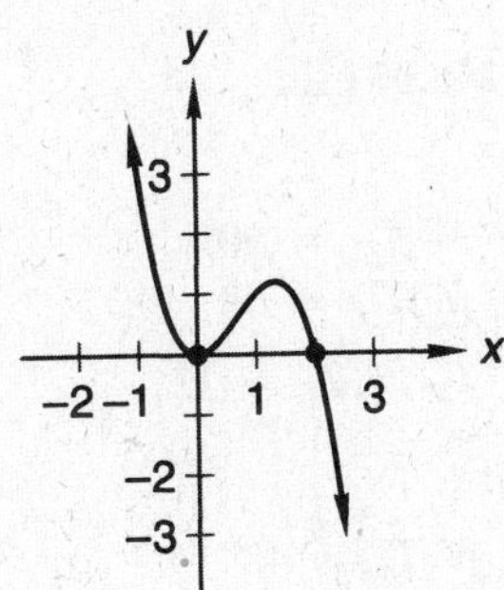

22. (113) $\left(x^4 - 3x^2 - 2x + 1\right) \div (x - 1)$

$$\begin{array}{r|rrrrr} 1 & 1 & 0 & -3 & -2 & 1 \\ & & 1 & 1 & -2 & -4 \\ \hline & 1 & 1 & -2 & -4 & -3 \end{array}$$

$$\frac{x^4 - 3x^2 - 2x + 1}{x - 1}$$

$$= \mathbf{x^3 + x^2 - 2x - 4 - \frac{3}{x - 1}}$$

23. (112) $\left(1 - \frac{1}{m}\right)^{12}$

$$F = 1,\ S = -\frac{1}{m},\ n = 12,\ k = 7$$

$$\frac{n!}{(n - k + 1)!(k - 1)!}F^{n-k+1}S^{k-1}$$

$$\frac{12!}{(12 - 7 + 1)!\,(7 - 1)!}F^{12-7+1}S^{7-1} = \frac{12!}{6!6!}F^6S^6$$

$$= 924(1)^6\left(-\frac{1}{m}\right)^6 = 924(1)\frac{1}{m^6} = \mathbf{\frac{924}{m^6}}$$

24. (111) $\log_{\frac{1}{2}}(x - 3) < 3$

$$x - 3 > \left(\frac{1}{2}\right)^3 \quad \text{and} \quad x - 3 > 0$$

$$x > 3$$

$$x - 3 > \frac{1}{8}$$

$$x > \frac{25}{8}$$

$$\text{but } 3 < \frac{25}{8},\ \text{so } \mathbf{x > \frac{25}{8}}$$

25. (98)

$$x^{\log x} = 10$$

$$\log x^{\log x} = \log 10$$

$$(\log x)(\log x) = 1$$

$$(\log x)^2 = 1$$

$$\log x = \pm 1$$

$$\log x = 1 \quad \text{or} \quad \log x = -1$$

$$x = 10 \qquad\qquad x = \frac{1}{10}$$

$$\mathbf{x = 10, \frac{1}{10}}$$

26. (80) $\dfrac{\cot^2 x + \sec^2 x + 1}{\cot^2 x}$

$$= \frac{\csc^2 x + \sec^2 x}{\cot^2 x}$$

$$= \frac{\dfrac{1}{\sin^2 x} + \dfrac{1}{\cos^2 x}}{\dfrac{\cos^2 x}{\sin^2 x}}$$

$$= \frac{\cos^2 x + \sin^2 x}{\sin^2 x \cos^2 x} \cdot \frac{\sin^2 x}{\cos^2 x}$$

$$= \frac{1}{\cos^4 x} = \sec^4 x$$

27. (81) $a^2 = b^2 + c^2 - 2bc\cos A$

$$a^2 = 6^2 + 8^2 - 2(6)(8)\cos 20^\circ$$

$$a^2 = 36 + 64 - 90.21$$

$$a^2 = 9.79$$

$$\mathbf{a = 3.13\ in.}$$

28. $y = 6 + 2\cos\frac{1}{5}(x - 40°)$
(84)

$$\text{Period} = \frac{360°}{\frac{1}{5}} = 1800°$$

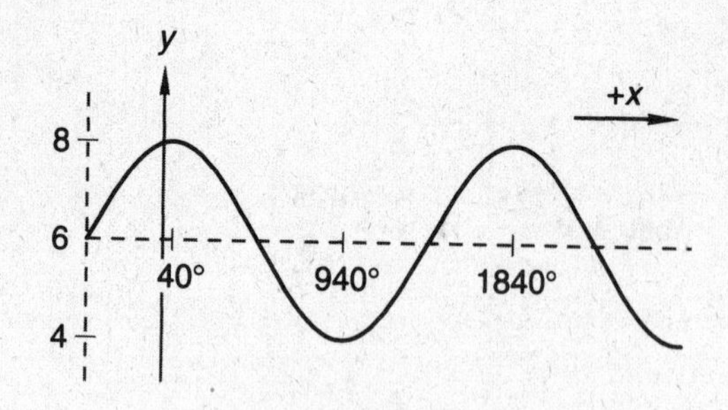

29. $-\sqrt{3} - \tan 3\theta = 0$
(52)

$$\tan 3\theta = -\sqrt{3}$$

$$3\theta = 120°, 300°, 480°, 660°, 840°, 1020°$$

$$\boldsymbol{\theta = 40°, 100°, 160°, 220°, 280°, 340°}$$

30. $x^4 + x^3 + 3x^2 + 2x + 2$
(46)

$$= (x^2 + 2)(x^2 + x + 1)$$

For $x^2 + 2$,

$$x = \frac{0 \pm \sqrt{0 - 4(1)(2)}}{2(1)}$$

$$= \pm\frac{\sqrt{-8}}{2} = \pm\sqrt{2}i$$

For $x^2 + x + 1$,

$$x = \frac{-1 \pm \sqrt{1^2 - 4(1)(1)}}{2(1)}$$

$$= \frac{-1 \pm \sqrt{-3}}{2}$$

$$= \frac{-1 \pm \sqrt{3}i}{2}$$

$$x^4 + x^3 + 3x^2 + 2x + 2$$

$$= \boldsymbol{(x + \sqrt{2}i)(x - \sqrt{2}i)}$$

$$\boldsymbol{\times \left(x + \frac{1}{2} + \frac{\sqrt{3}}{2}i\right)\left(x + \frac{1}{2} - \frac{\sqrt{3}}{2}i\right)}$$

Problem Set 124

1. $A_{40} = 1460,\ A_{400} = 440$
(88)

$$A_{400} = A_{40}e^{k(400-40)}$$

$$440 = 1460e^{360k}$$

$$\frac{440}{1460} = e^{360k}$$

$$\ln\frac{440}{1460} = 360k$$

$$k = -0.0033317$$

$$730 = 1460e^{-0.0033317t}$$

$$\frac{1}{2} = e^{-0.0033317t}$$

$$\ln\frac{1}{2} = -0.0033317t$$

$$t = \textbf{208.05 min}$$

2. (a) (b)
(36)

$$\begin{cases} (B + W)T_D = 90 & \text{(a)} \\ (B - W)T_U = 48 & \text{(b)} \end{cases}$$

$$B = 3W,\ T_D = T_U - 1$$

Upstream:

(b) $(B - W)T_U = 48$

$$(2W)T_U = 48$$

$$WT_U = 24$$

Downstream:

(a) $(B + W)T_D = 90$

$$(4W)(T_U - 1) = 90$$

$$4WT_U - 4W = 90$$

$$4(24) - 4W = 90$$

$$4W = 6$$

$$\boldsymbol{W = \frac{3}{2}}\ \textbf{mph}$$

$$B = 3W$$

$$B = 3\left(\frac{3}{2}\right)$$

$$\boldsymbol{B = \frac{9}{2}}\ \textbf{mph}$$

3. $P = \frac{k}{V}$
(18)

$$5 = \frac{k}{40}$$

$$k = 200$$

$$P = \frac{k}{1} = \frac{200}{1} = \textbf{200 atm}$$

4. (18)

$$\begin{cases} \text{(a)} & \dfrac{N_R}{N_B} = \dfrac{1}{2} \\ \text{(b)} & N_G = N_R + 7 \\ \text{(c)} & N_R + N_B = N_G + 1 \end{cases}$$

(a) $\dfrac{N_R}{N_B} = \dfrac{1}{2}$

$N_B = 2N_R$

(c) $N_R + N_B = N_G + 1$

$N_R + (2N_R) = (N_R + 7) + 1$

$2N_R = 8$

$\mathbf{N_R = 4}$

(b) $N_G = N_R + 7$

$N_G = (4) + 7$

$\mathbf{N_G = 11}$

$N_B = 2N_R$

$N_B = 2(4)$

$\mathbf{N_B = 8}$

5. (1)

$$\begin{cases} 4(90° - A) = 2(180° - B) + 40° \\ 180° - A = 2(90° - B) + 70° \end{cases}$$

(b) $180° - A = 2(90° - B) + 70°$

$180° - A = 180° - 2B + 70°$

$180° - A = -2B + 250°$

$A = 2B - 70°$

(a) $4(90° - A) = 2(180° - B) + 40°$

$360° - 4A = 360° - 2B + 40°$

$360° - 4(2B - 70°) = 400 - 2B$

$360° - 8B + 280° = 400 - 2B$

$6B = 240°$

$\mathbf{B = 40°}$

$A = 2B - 70°$

$A = 2(40°) - 70°$

$\mathbf{A = 10°}$

6. (124) $P_1 = (x_1, y_1),\ P_2 = (x_2, y_2)$

$x_F = x_1 + \Delta x$

$= x_1 + \dfrac{3}{4}(x_2 - x_1)$

$= x_1 + \dfrac{3}{4}x_2 - \dfrac{3}{4}x_1$

$= \dfrac{1}{4}x_1 + \dfrac{3}{4}x_2$

$y_F = y_1 + \Delta y$

$= y_1 + \dfrac{3}{4}(y_2 - y_1)$

$= y_1 + \dfrac{3}{4}y_2 - \dfrac{3}{4}y_1$

$= \dfrac{1}{4}y_1 + \dfrac{3}{4}y_2$

$\mathbf{\left(\dfrac{1}{4}x_1 + \dfrac{3}{4}x_2,\ \dfrac{1}{4}y_1 + \dfrac{3}{4}y_2\right)}$

7. (124) $P_1 = (x_1, y_1),\ P_2 = (x_2, y_2)$

$x_F = x_1 + \Delta x$

$= x_1 + \dfrac{4}{11}(x_2 - x_1)$

$= x_1 + \dfrac{4}{11}x_2 - \dfrac{4}{11}x_1$

$= \dfrac{7}{11}x_1 + \dfrac{4}{11}x_2$

$y_F = y_1 + \Delta y$

$= y_1 + \dfrac{4}{11}(y_2 - y_1)$

$= y_1 + \dfrac{4}{11}y_2 - \dfrac{4}{11}y_1$

$= \dfrac{7}{11}y_1 + \dfrac{4}{11}y_2$

$\mathbf{\left(\dfrac{7}{11}x_1 + \dfrac{4}{11}x_2,\ \dfrac{7}{11}y_1 + \dfrac{4}{11}y_2\right)}$

8. (124) $P_1 = (2, 4),\ P_2 = (6, 6)$

$x_F = x_1 + \Delta x$

$= 2 + \dfrac{1}{4}(6 - 2)$

$= 2 + \dfrac{1}{4}(4) = 3$

$y_F = y_1 + \Delta y$

$= 4 + \dfrac{1}{4}(6 - 4)$

$= 4 + \dfrac{1}{4}(2)$

$= 4\dfrac{1}{2} = \dfrac{9}{2}$

$\mathbf{\left(3, \dfrac{9}{2}\right)}$

9. (124)
$$x_F = x_1 + \Delta x$$
$$= -2 + \frac{3}{5}(-5 + 2)$$
$$= -2 + \frac{3}{5}(-3)$$
$$= -2 - \frac{9}{5} = -\frac{19}{5}$$

$$y_F = y_1 + \Delta y$$
$$= -3 + \frac{3}{5}(4 + 3)$$
$$= -3 + \frac{3}{5}(7)$$
$$= -3 + \frac{21}{5} = \frac{6}{5}$$

$$\left(-\frac{\mathbf{19}}{\mathbf{5}}, \frac{\mathbf{6}}{\mathbf{5}}\right)$$

10. (123) $ax^2 + bxy + ay^2 + dx + ey + f = 0$

(a) $x^2 + 10x + y^2 + 2y = 5$

This is a **circle** since $b = 0$ and $a = c \neq 0$.

(b) $16x^2 - 49y^2 - 200x - 100 = 98y$

This is a **hyperbola** since $b = 0$ and a and c have opposite signs.

(c) $-y^2 + 9x^2 - 90x + 4y + 302 = 0$

This is a **hyperbola** since $b = 0$ and a and c have opposite signs.

(d) $y = x^2 - 4x + 3$

This is a **parabola** since $b = c = 0$, $a \neq 0$ and $e \neq 0$.

(e) $16x^2 + 25y^2 + 500 = 300y$

This is an **ellipse** since $b = 0$, a and c have the same sign and $a \neq c$.

11. (106,123)
$$16x^2 + 25y^2 + 500 = 300y$$
$$16x^2 + 25(y^2 - 12y \quad\quad) = -500$$
$$16(x)^2 + 25(y^2 - 12y + 36) = 400$$
$$16x^2 + 25(y - 6)^2 = 400$$
$$\frac{\mathbf{x^2}}{\mathbf{25}} + \frac{\mathbf{(y - 6)^2}}{\mathbf{16}} = \mathbf{1}$$

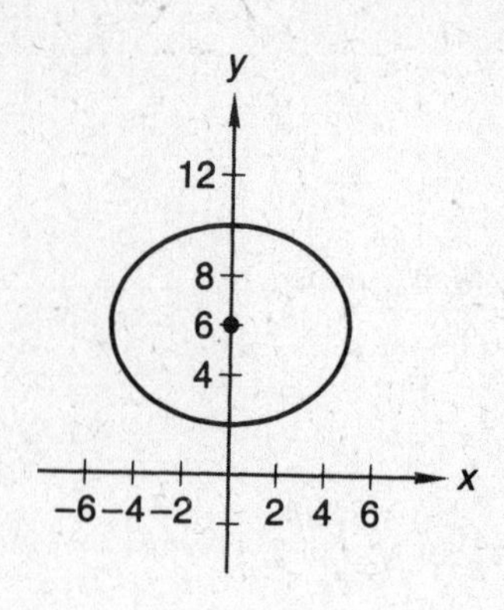

12. (123) $ax^2 + bxy + ay^2 + dx + ey + f = 0$

(a) $x^2 + y^2 - 12x - 4y + 8 = 0$

This is a **circle** since $b = 0$ and $a = c \neq 0$.

(b) $12x^2 - y^2 = 48$

This is a **hyperbola** since $b = 0$ and a and c have opposite signs.

(c) $9x^2 + 81y^2 + 18x - 162y - 100 = 0$

This is an **ellipse** since $b = 0$, a and c have the same sign and $a \neq c$.

(d) $16x^2 - 9y^2 + 144 = 0$

This is a **hyperbola** since $b = 0$ and a and c have opposite signs.

(e) $10y = -x^2$

This is a **parabola** since $b = c = 0$, $a \neq 0$ and $e \neq 0$.

13. (106,123)
$$16x^2 - 9y^2 + 144 = 0$$
$$16x^2 - 9y^2 = -144$$
$$-\frac{x^2}{9} + \frac{y^2}{16} = 1$$
$$\frac{\mathbf{y^2}}{\mathbf{16}} - \frac{\mathbf{x^2}}{\mathbf{9}} = \mathbf{1}$$

Center = (0, 0)

Vertices = (0, 4), (0, –4)

$$m = \frac{4}{3}, -\frac{4}{3}$$

Asymptotes: $y = \frac{4}{3}x$; $y = -\frac{4}{3}x$

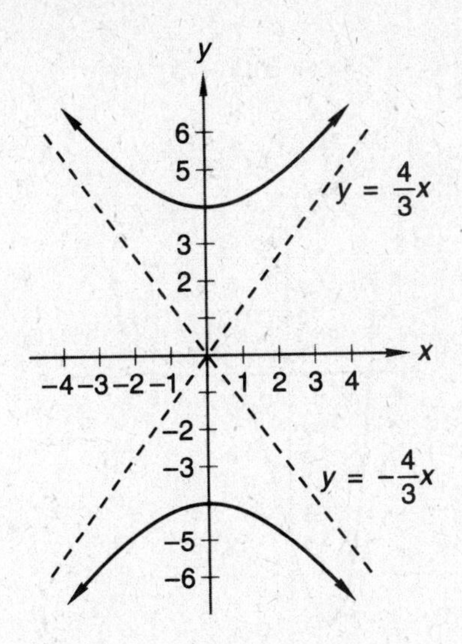

14. (122) $y = \dfrac{3x}{(x - 2)(x + 1)(x + 3)}$

x-intercept: $x = 0$

Asymptotes: $x = -3, -1, 2$

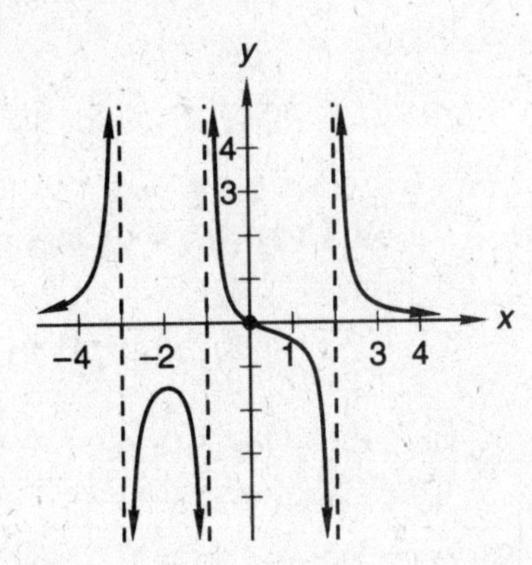

15. (122) $y = \dfrac{x - 2}{x^2 - 3x + 2} = \dfrac{x - 2}{(x - 1)(x - 2)}$

Asymptotes: $x = 1$

Hole: $x = 2$

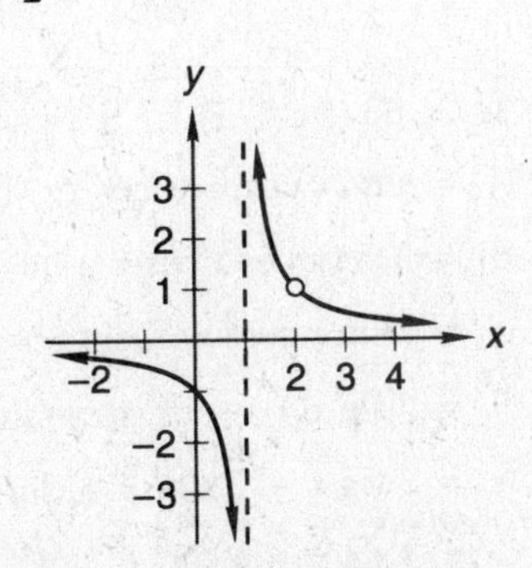

16. (121) $\begin{cases} y = -x & -\infty < x \le -1 \\ y = 1 & -1 < x < 1 \\ y = x & 1 < x < \infty \end{cases}$

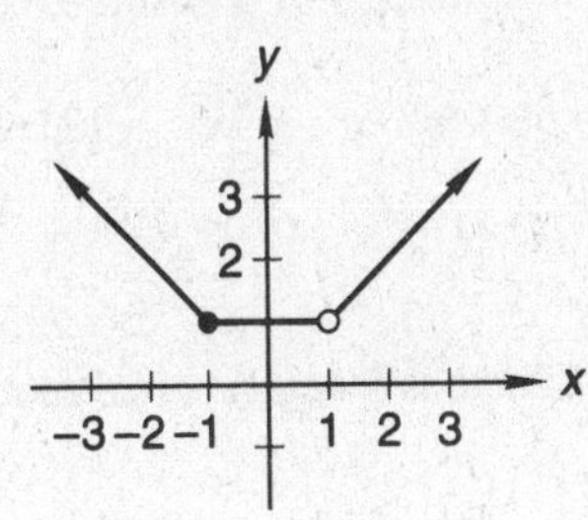

17. (121) (a) $\begin{cases} y = -1 & \text{if } -\infty < x < -3 \\ y = x - 1 & \text{if } -3 \le x \le 1 \\ y = 1 & \text{if } 1 < x \le 3 \\ y = 3 & \text{if } 3 < x < \infty \end{cases}$

(b) $\begin{cases} y = -1 & \text{if } -\infty < x < -1 \\ y = x + 1 & \text{if } -1 < x < 0 \\ y = x & \text{if } 0 < x < 1 \\ y = x - 1 & \text{if } 1 < x < 2 \\ y = -1 & \text{if } 2 < x < \infty \end{cases}$

18. (120) $\begin{bmatrix} 5 & -7 \\ 12 & 3 \end{bmatrix} \cdot \begin{bmatrix} x \\ y \end{bmatrix} = \begin{bmatrix} 3 \\ -51 \end{bmatrix}$

ORIGINAL MATRIX	INVERSE MATRIX
$\begin{bmatrix} 5 & -7 \\ 12 & 3 \end{bmatrix}$	$\dfrac{1}{(5)(3) - (-7)(12)} \begin{bmatrix} 3 & 7 \\ -12 & 5 \end{bmatrix}$

$$= \frac{1}{99}\begin{bmatrix} 3 & 7 \\ -12 & 5 \end{bmatrix}$$

$$= \begin{bmatrix} \mathbf{\frac{1}{33}} & \mathbf{\frac{7}{99}} \\ -\mathbf{\frac{4}{33}} & \mathbf{\frac{5}{99}} \end{bmatrix}$$

$$\begin{bmatrix} \frac{1}{33} & \frac{7}{99} \\ -\frac{4}{33} & \frac{5}{99} \end{bmatrix} \cdot \begin{bmatrix} 5 & -7 \\ 12 & 3 \end{bmatrix} \cdot \begin{bmatrix} x \\ y \end{bmatrix} = \begin{bmatrix} \frac{1}{33} & \frac{7}{99} \\ \frac{-4}{33} & \frac{5}{99} \end{bmatrix} \cdot \begin{bmatrix} 3 \\ -51 \end{bmatrix}$$

$$\begin{bmatrix} 1 & 0 \\ 0 & 1 \end{bmatrix} \cdot \begin{bmatrix} x \\ y \end{bmatrix} = \begin{bmatrix} \frac{3}{33} - \frac{357}{99} \\ -\frac{12}{33} - \frac{255}{99} \end{bmatrix}$$

$$\begin{bmatrix} x \\ y \end{bmatrix} = \begin{bmatrix} -\frac{116}{33} \\ -\frac{97}{33} \end{bmatrix}$$

$\mathbf{x = -\frac{116}{33};\ y = -\frac{97}{33}}$

19. (119) (a) $p(x) = 3x^3 - 7x^2 - 12x + 28$

$+ \quad ⓒ \quad - \quad\quad - \quad ⓒ \quad +$

There are **0 or 2** positive real roots.

(b) $p(x) = 3x^3 - 7x^2 - 12x + 28$

$p(-x) = -3x^3 - 7x^2 + 12x + 28$

$- \quad\quad - \quad ⓒ \quad + \quad\quad +$

There is **1** negative real root.

20. (119) $3x^3 - 7x^2 - 12x + 28$

$$\begin{array}{r|rrrr} 3 & 3 & -7 & -12 & 28 \\ & & 9 & 6 & -18 \\ \hline & 3 & 2 & -6 & 10 \end{array} \qquad \begin{array}{r|rrrr} 4 & 3 & -7 & -12 & 28 \\ & & 12 & 20 & 32 \\ \hline & 3 & 5 & 8 & 60 \end{array}$$

Upper bound = 4

$$\begin{array}{r|rrrr} -1 & 3 & -7 & -12 & 28 \\ & & -3 & 10 & 2 \\ \hline & 3 & -10 & -2 & 30 \end{array} \qquad \begin{array}{r|rrrr} -2 & 3 & -7 & -12 & 28 \\ & & -6 & 26 & -28 \\ \hline & 3 & -13 & 14 & 0 \end{array}$$

Lower bound = −2

21. (118) $3x^3 - 7x^2 - 12x + 28 = 0$

$$\begin{array}{r|rrrr} -2 & 3 & -7 & -12 & 28 \\ & & -6 & 26 & -28 \\ \hline 2 & 3 & -13 & 14 & 0 \\ & & 6 & -14 & \\ \hline & 3 & -7 & 0 & \end{array}$$

$(x + 2)(x - 2)(3x - 7) = 0$

Roots: $\mathbf{-2, 2, \frac{7}{3}}$

22. (114,116) $y = -2x^4 + 8x^2$

$r = |-4| + 1$

$r = 5$

x	−3	−2	−1	0	1	2	3
y	−90	0	6	0	6	0	−90

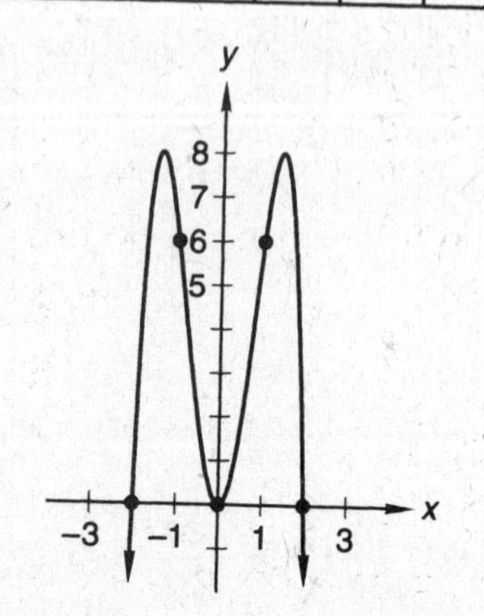

23. (114) $y = \frac{1}{25}(x - 2)(x + 3)(x - 5)^2$

$= \frac{1}{25}(x^4 - 9x^3 + 9x^2 + 85x - 150)$

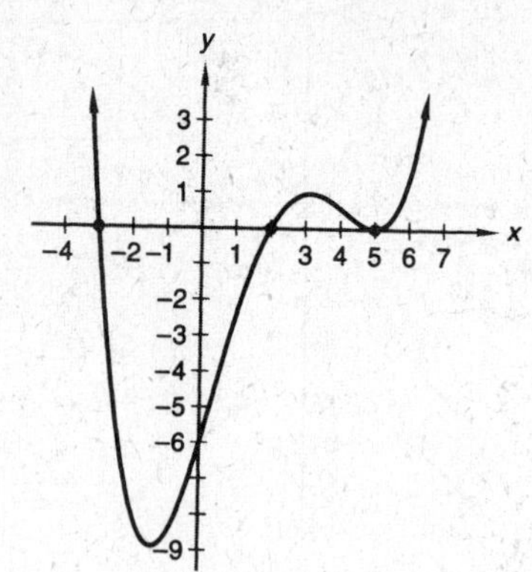

24. (111) $\log_5 (x - 2) > 2$

$x - 2 > 5^2$ and $x - 2 > 0$

$x > 27 \qquad x > 2$

but $2 < 27$, so $\mathbf{x > 27}$

25. (98)

$$\sqrt{\log_3 x} = \log_3 \sqrt[4]{x}$$

$$(\log_3 x)^{\frac{1}{2}} = \frac{1}{4} \log_3 x$$

$$\log_3 x = \frac{1}{16}(\log_3 x)^2$$

$$\frac{1}{16}(\log_3 x)^2 - \log_3 x = 0$$

$$\log_3 x\left(\frac{1}{16} \log_3 x - 1\right) = 0$$

$\log_3 x = 0 \qquad \frac{1}{16}(\log_3 x) - 1 = 0$

$x = 1 \qquad \log_3 x = 16$

$x = 3^{16}$

$\mathbf{x = 1, 3^{16}}$

26. (96,100) $\sin 3x = \sin (2x + x)$

$= \sin 2x \cos x + \cos 2x \sin x$

$= 2 \sin x \cos x \cos x + (1 - 2 \sin^2 x) \sin x$

$= 2 \sin x \cos^2 x + \sin x - 2 \sin^3 x$

$= 2 \sin x(1 - \sin^2 x) + \sin x - 2 \sin^3 x$

$= 2 \sin x - 2 \sin^3 x + \sin x - 2 \sin^3 x$

$= 3 \sin x - 4 \sin^3 x$

27. $\theta = \text{Arctan}\, x$
(110)

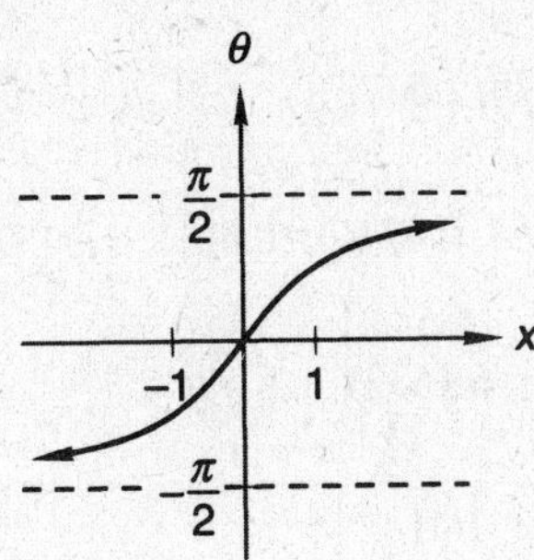

28. $y = \frac{1}{2} + 2\sin(x - 10°)$
(84)

Period $= 2\pi$

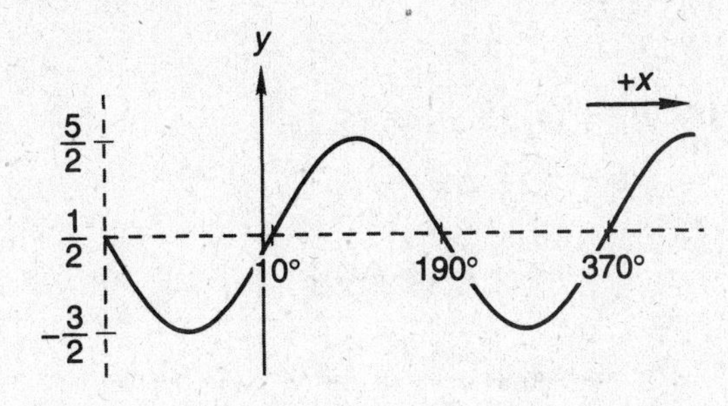

29.
(52,60)

$$2\cos^2 2\theta - 1 = 0$$

$$\cos^2 2\theta - \frac{1}{2} = 0$$

$$\left(\cos 2\theta + \frac{1}{\sqrt{2}}\right)\left(\cos 2\theta - \frac{1}{\sqrt{2}}\right) = 0$$

$$\cos 2\theta = \pm\frac{1}{\sqrt{2}}$$

$$2\theta = \frac{\pi}{4}, \frac{3\pi}{4}, \frac{5\pi}{4}, \frac{7\pi}{4}, \frac{9\pi}{4}, \frac{11\pi}{4}, \frac{13\pi}{4}, \frac{15\pi}{4}$$

$$\theta = \mathbf{\frac{\pi}{8}, \frac{3\pi}{8}, \frac{5\pi}{8}, \frac{7\pi}{8}, \frac{9\pi}{8}, \frac{11\pi}{8}, \frac{13\pi}{8}, \frac{15\pi}{8}}$$

30. $\log 10^{y^2 - 5xy} + 4^{\log_4 5xy}$
(59)

$= y^2 - 5xy + 5xy = \mathbf{y^2}$

Problem Set 125

1. $A_3 = 10,\ A_5 = 6$
(88)

$$A_5 = A_3 e^{k(5-3)}$$

$$6 = 10e^{2k}$$

$$\frac{3}{5} = e^{2k}$$

$$\ln\frac{3}{5} = 2k$$

$$k = -0.25541$$

$$A_3 = A_0 e^{(-0.25541)3}$$

$$A_0 = \frac{10}{e^{-0.766}}$$

$$A_0 = \mathbf{21.52\ g}$$

$$10.76 = 21.52e^{-0.25541t}$$

$$0.5 = e^{-0.25541t}$$

$$\ln 0.5 = -0.25541t$$

$$t = \mathbf{2.71\ days}$$

2. $R_B = 1,\ R_L = \frac{1}{12},\ T_B = T_L$
(85)

$$\begin{cases} R_L T_L = S \\ R_B T_B = S + 30 \end{cases}$$

$$R_B T_B = S + 30$$

$$(1)T = (R_L T_L) + 30$$

$$T = \frac{T}{12} + 30$$

$$\frac{11}{12}T = 30$$

$$T = \mathbf{32\frac{8}{11}\ min}$$

3. ${}_{15}C_3 = \frac{15!}{3!12!} = \mathbf{455}$
(75)

4. Circular permutations
(55)

$(n - 1)! = (4 - 1)! = 3! = \mathbf{6}$

5. $y = 2x^3 + 4x$
(125)

6. (125) $y = x^3 + 2x^2 - x - 4$

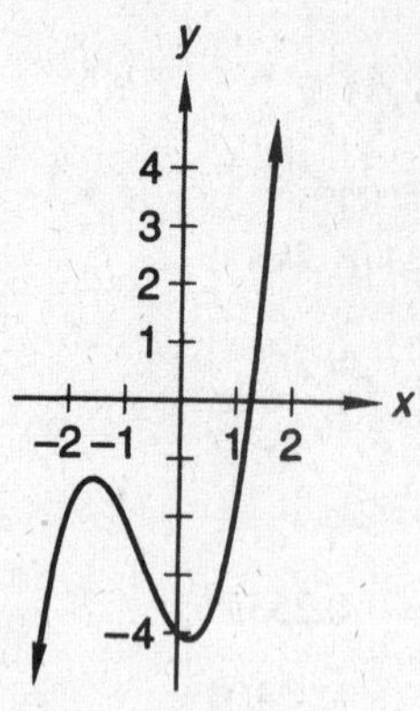

7. (125) $y = 0.04(x - 2)(x + 3)(x - 5)^2$

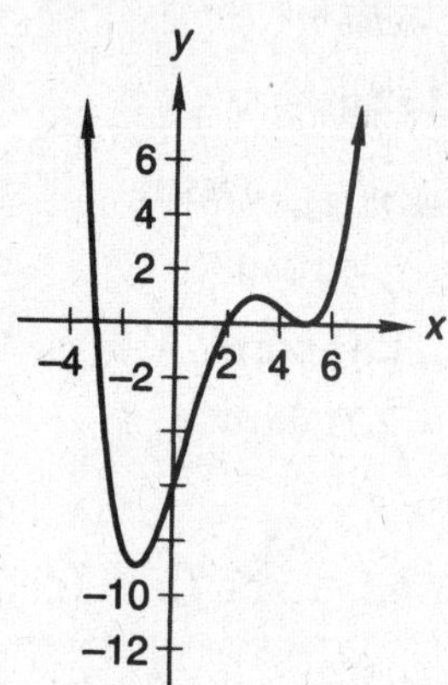

8. (125) $y = \dfrac{3x}{(x - 2)(x + 1)(x + 3)}$

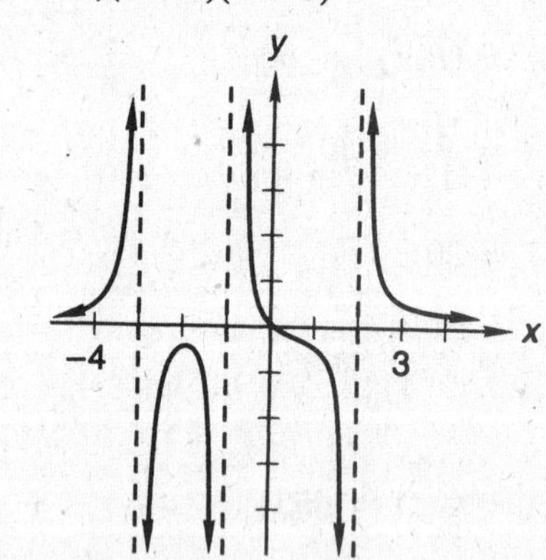

9. (125) $y = \dfrac{x + 2}{(x - 1)(x - 5)}$

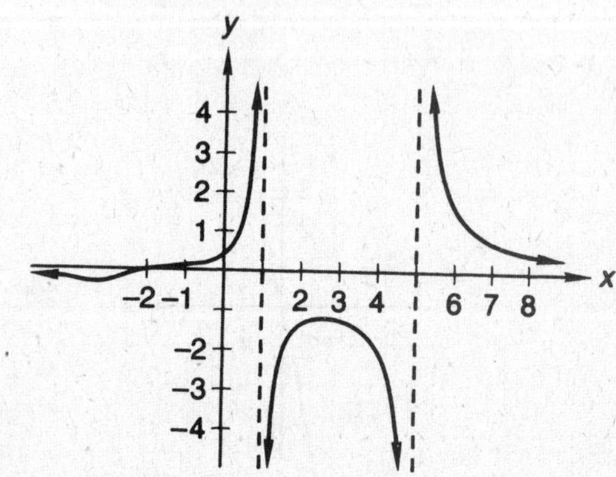

10. (125) (–1.5538, –0.5538), (1.5538, 2.5538)

11. (125) (–1.1478, 1.3173)

12. (125) (–0.7454, 0.4745), (0.4918, 1.6352)

13. (125) (1.3553, 6.1632)

14. (125) –1.4142, 1.4142

15. (125) –0.4142, 1, 2.4142

16. (125) –2.8846, 1.0815

17. (125) –2, 2

18. (124) $P_1 = (4, 7),\ P_2 = (5, 2)$

$$x_F = x_1 + \Delta x$$
$$= -4 + \frac{5}{8}[5-(-4)]$$
$$= -4 + \frac{45}{8} = \frac{13}{8}$$

$$y_F = y_1 + \Delta y$$
$$= 7 + \frac{5}{8}(2 - 7)$$
$$= 7 + \left(\frac{-25}{8}\right) = \frac{31}{8}$$

$$\left(\mathbf{\frac{13}{8}, \frac{31}{8}}\right)$$

19. (123) $ax^2 + bxy + cy^2 + dx + ey + f = 0$

(a) $4y = -2x - 7 + x^2$

This is a **parabola** since $b = c = 0$, $a \neq 0$, and $e \neq 0$.

(b) $y + 4x^2 = 20x - 16$

This is a **parabola** since $b = c = 0$, $a \neq 0$, and $e \neq 0$.

(c) $-y^2 + 20x + 100 - 2y = -25x^2$

This is a **hyperbola** since $b = 0$ and a and c have opposite signs.

(d) $36x^2 + 9y^2 - 288x = -90y - 553$

This is an **ellipse** since $b = 0$, a and c have the same sign and $a \neq c$.

(e) $-14y + y^2 + 48 = -x^2$

This is a **circle** since $b = 0$ and $a = c \neq 0$.

20. (106,123)

$$4y = -2x - 7 + x^2$$
$$= x^2 - 2x - 7$$
$$= (x^2 - 2x \quad) - 7$$
$$= (x^2 - 2x + 1) - 7 - 1$$
$$= (x - 1)^2 - 8$$
$$\mathbf{y = \frac{1}{4}(x - 1)^2 - 2}$$

Vertex $(1, -2)$

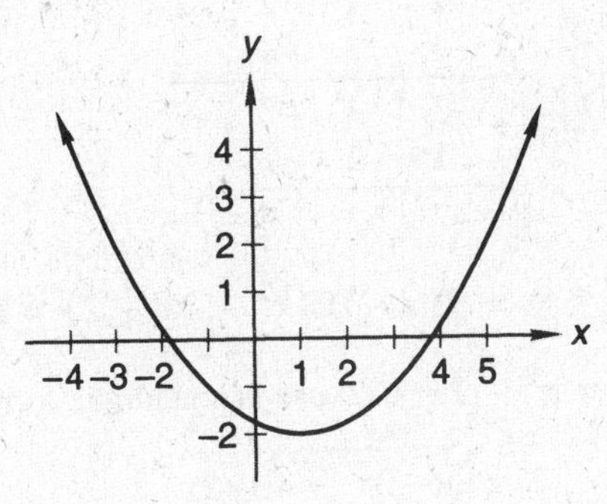

21. (123)

(a) $xy = 8$

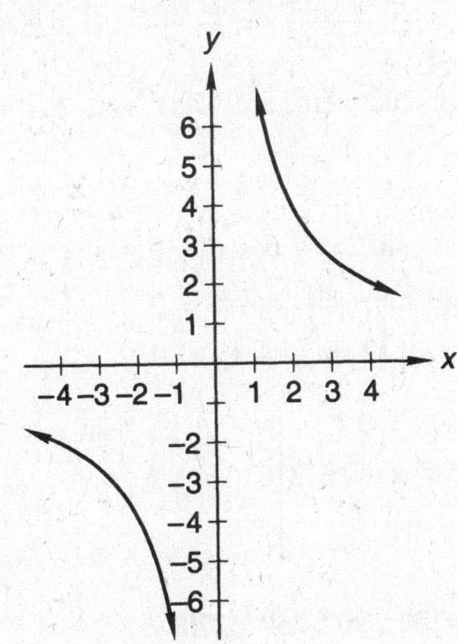

(b) $xy = -8$

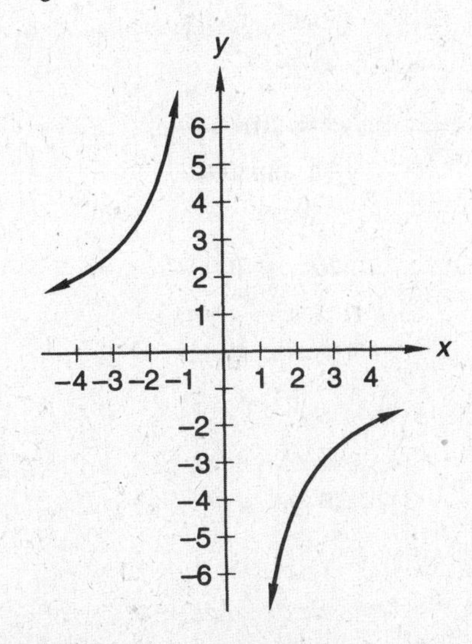

22. (122)

$$y = \frac{x^2 - 3x}{(x + 1)(x - 3)(x - 2)}$$
$$= \frac{x(x - 3)}{(x + 1)(x - 3)(x - 2)}$$

x-intercept: $x = 0$

Asymptotes: $x = -1, 2$

Hole: $x = 3$

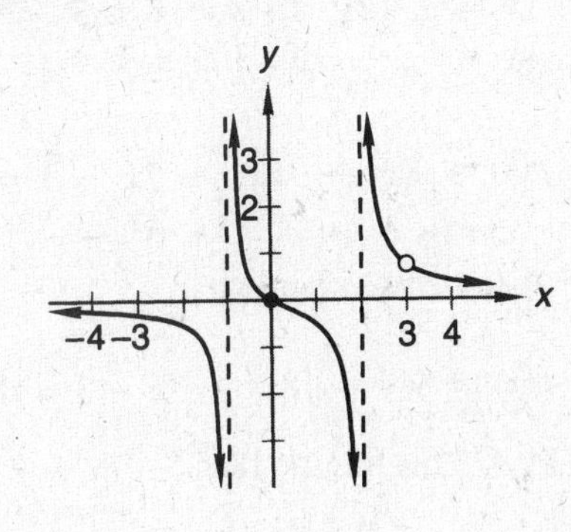

23. (122)

$$y = \frac{x - 2}{x^2 - x - 2}$$
$$= \frac{x - 2}{(x - 2)(x + 1)}$$

Asymptotes: $x = -1$

Hole: $x = 2$

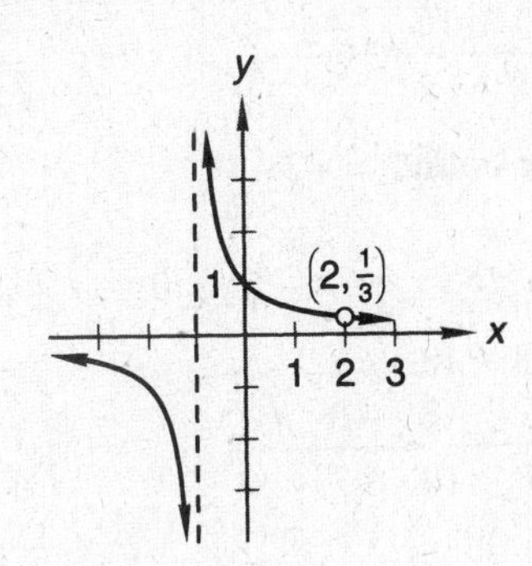

24. (121)

$$\begin{cases} y = 1 & \text{if } -\infty < x \le -1 \\ y = |x| & \text{if } -1 < x < 1 \\ y = 1 & \text{if } 1 \le x < \infty \end{cases}$$

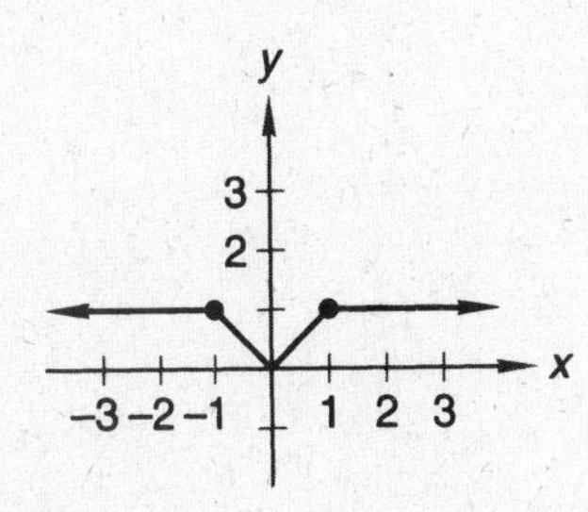

25. (121) (a) $\begin{cases} y = \dfrac{1}{2}x & \text{if } -2 < x \le 0 \\ y = -\dfrac{1}{2}x + 1 & \text{if } 0 < x \le 2 \\ y = 1 & \text{if } 2 < x < \infty \end{cases}$

(b) $\begin{cases} y = -1 & \text{if } -\infty < x < 1 \\ y = x - 1 & \text{if } 1 < x < 2 \\ y = 3 - x & \text{if } 2 < x < 3 \\ y = 1 & \text{if } 3 \le x < \infty \end{cases}$

26. (119) (a) $p(x) = 8x^3 + 18x^2 + 7x - 3 = 0$

$\quad + \quad + \quad + \text{ⓒ} -$

There is **1** positive real root.

(b) $p(x) = 8x^3 + 18x^2 + 7x - 3 = 0$

$p(-x) = -8x^3 + 18x^2 - 7x - 3$

$\quad - \text{ⓒ} + \text{ⓒ} - \quad -$

There are **0 or 2** negative real roots.

27. (119) $8x^3 + 18x^2 + 7x - 3 = 0$

$$\begin{array}{r|rrrr} 1 & 8 & 18 & 7 & -3 \\ & & 8 & 26 & 33 \\ \hline & 8 & 26 & 33 & 30 \end{array}$$

Upper bound = 1

$$\begin{array}{r|rrrr} -2 & 8 & 18 & 7 & -3 \\ & & -16 & -4 & -6 \\ \hline & 8 & 2 & 3 & -9 \end{array} \qquad \begin{array}{r|rrrr} -3 & 8 & 18 & 7 & -3 \\ & & -24 & 18 & -75 \\ \hline & 8 & -6 & 25 & -78 \end{array}$$

Lower bound = −3

28. (118) $8x^3 + 18x^2 + 7x - 3 = 0$

$$\begin{array}{r|rrrr} -1 & 8 & 18 & 7 & -3 \\ & & -8 & -10 & 3 \\ \hline & 8 & 10 & -3 & 0 \end{array}$$

$(x + 1)(8x^2 + 10x - 3) = 0$

$(x + 1)(2x + 3)(4x - 1) = 0$

Roots: $\mathbf{-\frac{3}{2}, -1, \frac{1}{4}}$

29. (118) $y = x^4 - 2x^3 + x^2 + 2x - 2$

$x = -1$

$$\begin{array}{r|rrrrr} -1 & 1 & -2 & 1 & 2 & -2 \\ & & -1 & 3 & -4 & 2 \\ \hline 1 & 1 & -3 & 4 & -2 & 0 \\ & & 1 & -2 & 2 & \\ \hline & 1 & -2 & 2 & 0 & \end{array}$$

$y = (x + 1)(x - 1)(x^2 - 2x + 2) = 0$

For $x^2 - 2x + 2$ use the quadratic formula:

$$x = \frac{2 \pm \sqrt{(-2)^2 - 4(1)(2)}}{2(1)}$$

$$x = \frac{2 \pm \sqrt{-4}}{2} = \frac{2 \pm 2i}{2} = 1 \pm i$$

The other zeros are: $\mathbf{1, 1 + i, 1 - i}$

30. (85,90)

$\sin 2x - \cos x = 0$

$2 \sin x \cos x - \cos x = 0$

$\cos x (2 \sin x - 1) = 0$

$\cos x = 0$

$x = 90°, 270°$

$2 \sin x - 1 = 0$

$\sin x = \dfrac{1}{2}$

$x = 30°, 150°$

$x = \mathbf{30°, 90°, 150°, 270°}$